National
Fire Protection
Association

Engine Company Fireground Operations

FOURTH EDITION

Jones & Bartlett Learning
World Headquarters
25 Mall Road
Burlington, MA 01803
978-443-5000
info@jblearning.com
www.jblearning.com
www.psglearning.com

National Fire Protection Association
1 Batterymarch Park
Quincy, MA 02169-7471
www.NFPA.org

Jones & Bartlett Learning books and products are available through most bookstores and online booksellers. To contact the Jones & Bartlett Learning Public Safety Group directly, call 800-832-0034, fax 978-443-8000, or visit our website, www.psglearning.com.

Substantial discounts on bulk quantities of Jones & Bartlett Learning publications are available to corporations, professional associations, and other qualified organizations. For details and specific discount information, contact the special sales department at Jones & Bartlett Learning via the above contact information or send an email to specialsales@jblearning.com.

Production Credits
Chief Product Officer, Professional Education: Kimberly Brophy
VP, Product Development: Christine Emerton
Executive Editor: Bill Larkin
Senior Development Editor: Janet Morris
Editorial Assistant: Alexander Belloli
Vice President of Sales: Phil Charland
Manager, Project Management: Jessica deMartin
Manager, Project Management: Kristen Rogers
Digital Project Specialist: Rachel Reyes
Director of Marketing Operations: Brian Rooney
Content Services Manager: Colleen Lamy
VP, Manufacturing and Inventory Control: Therese Connell
Composition: S4Carlisle Publishing Services
Project Management: S4Carlisle Publishing Services
Cover Design: Kristin E. Parker
Text Design: Kristin E. Parker
Senior Media Development Editor: Shannon Sheehan
Rights Specialist: John Rusk
Cover Image (Title Page, Chapter Opener):
© Rick McClure, Los Angeles Fire Department, Retired
Printing and Binding: Lakeside Book Company
Cover Printing: Lakeside Book Company

Library of Congress Cataloging-in-Publication Data
Names: Angulo, Raul A, author. | Richman, Harold, 1926-1999. Engine company fireground operations. | National Fire Protection Association. Research Division, author.
Title: Engine company fireground operations / Raul A. Angulo, National Fire Protection Association.
Description: 4th edition. | Burlington, MA : Jones & Bartlett Learning, 2021. | Includes bibliographical references and index. | Summary: "Engine company personnel are a fundamental part of fireground operations. This new edition emphasizes the importance of understanding fire science and behavior based on the latest research, which results in the most effective firefighting while realizing the highest survivability potential" Provided by publisher.
Identifiers: LCCN 2019033878 | ISBN 9781284023855 (paperback)
Subjects: LCSH: Fire extinction--Handbooks, manuals, etc. | Engine companies--Handbooks, manuals, etc.
Classification: LCC TH9310.5 .R5 2021 | DDC 628.9/25--dc23
LC record available at https://lccn.loc.gov/2019033878

6048

Printed in the United States of America

28 27 26 25 24 10 9 8 7 6 5 4 3 2

© Rick McClure, Los Angeles Fire Department, Retired.

Brief Contents

CHAPTER 1 **Introduction and Overview** 1

CHAPTER 2 **Fire Dynamics** 18

CHAPTER 3 **Building Construction** 71

CHAPTER 4 **An Analysis of the UL/NIST Experiments** 112

CHAPTER 5 **Equipment and Initial Hose-Lay Operations** 149

CHAPTER 6 **Apparatus Positioning** 193

CHAPTER 7 **Water Supply** 211

CHAPTER 8 **Initial Size-Up and Developing a Quick Incident Action Plan** 237

CHAPTER 9 **Search and Rescue** 271

CHAPTER 10 **Initial Fire Attack and Ventilation** 309

CHAPTER 11 **Backup Lines** 376

CHAPTER 12 **Exposure Protection** 387

CHAPTER 13 **Master Stream Appliances** 420

CHAPTER 14 **Fire Protection Systems** 442

CHAPTER 15 **High-Rise Firefighting** 465

CHAPTER 16 **Salvage and Overhaul** 545

GLOSSARY 573

INDEX 579

Contents

© Rick McClure, Los Angeles Fire Department, Retired.

Acknowledgments xii
Preface xiv

CHAPTER 1
Introduction and Overview 1

Introduction 2
Emergency Medical Services 2
Advancements in Technology 3
Where Are We Now? 5
UL/NIST 8
National Incident Management System/ Incident Command System (NIMS/ICS) 8
The Incident Priorities 9
RECEO/VS and BRECEO/VS 10
Decision-Making Model 11
Strategy, Tactics, Tasks, and Resources 11
Initial Role of the First-In Engine Officer 11
Value-Time-Size (VTS) 12
Identifying Problems 13
Initial Fireground Assignments 14
Engine Company Operations and Responsibilities 14

CHAPTER 2
Fire Dynamics 18

Introduction 19
The Fire Tetrahedron 19
What Is Fire? 19
Fire Chemistry, Physics, and Matter 20
Sources of Heat Energy 22
Vapor Pressure and Vapor Density 24
Boiling Liquid Expanding Vapor Explosion (BLEVE) 25
Boiling Point 27
Flammable or Explosive Limits of Gases 27
The Burning Process: Characteristics of Fire Behavior 27
Incipient (Ignition) Stage 28
Initial Growth Stage 29
Ventilation-Limited Stage 31
Explosive Growth Stage (New) 31
Fully Developed Stage 32
Decay Stage 33
Modes of Heat Transfer 34
Conduction 34
Convection 36
Path of Least Resistance and Flow Paths 39
Radiation 40
The Physical State of Fuels and Its Effect on Combustion 42
Solids 42
Liquids 44
Gases 45
Smoke: By-Product of Combustion 46
The Theory of Extinguishment 46
Reducing the Temperature 46
Removing the Fuel 48
Eliminating the Available Oxygen 48
Interrupting the Chemical Chain Reaction 48
Fire Behavior Events 49
Thermal Layering 49
Rollover 50
Flashover 50
Backdraft 51
Rapid Fire Growth 53

Behavior of Ventilation-Limited Fires 53
Smoke Explosions 53
Wind-Driven Fire 55
Fire Behavior in Modern Structures 56
Classes of Fire 57
Class A 57
Class B 57
Class C 57
Class D 57
Class K 58
Reading Smoke at Structural Fires 58
The Four Factors 59
The Four Steps of Reading Smoke 60
Smoke Reading through the Door 63
Health and Safety 65

CHAPTER 3
Building Construction 71

Introduction 72
Occupancy 72
Contents and Fire Loads 73
Occupancy Sprinkler Regulations 73
Types of Construction Materials 73
Masonry 74
Concrete 74
Steel 75
Glass 77
Gypsum Board 79
Wood 79
Engineered Wood Products 80
Fire-Retardant Wood 80
Plastics 80
Building Construction Terms and Mechanics 81
Types of Loads 81
Imposition of Loads 82
Forces 83
Floors and Ceilings 83
Fire-Resistive Floors 84
Wood-Supported Floors 84
Ceiling Assemblies 85
Roofs 85
Trusses 86
Wood Trusses 87
Steel Trusses 89
Void Spaces 90
Stairs 90
Parapet Walls and Cornices 91
Signs of Collapse 92
Buildings under Construction 92
Preparing for Collapse 92
Concern for Collapse Determines the Strategy 94
Collapse Times and Operational Periods 94
Strategic Considerations 95
Types of Building Construction 97
Determining the Time (T) in VTS 98
Type I: Fire Resistive 98
Type II: Noncombustible 99
Type III: Ordinary 100
Type IV: Heavy Timber 101
Type V: Wood Frame 101
Tactical Consideration for Balloon-Frame Construction 103
Platform Frame 104
Plank and Beam 104
Lightweight Wood Construction 105
Other Types of Construction 108
Prefires and New Construction Inspections 109

CHAPTER 4
An Analysis of the UL/NIST Experiments 112

Introduction 113
Fire Behavior 116
Changing Fire Environment 118
Survivability Profiles 121
Smoke Toxicity 122
Smoke Temperature 122
Ventilation, Flow Paths, and How Fires Spread 124
Wind-Driven Fires 125
Water Application and Coordinated Fire Attack 130

Search and Rescue Safety Considerations 136
Reading the Windows 137
Basement Fires 139
Tactical Considerations for Basement Fires 144
The Parker Doctrine 144

CHAPTER 5
Equipment and Initial Hose-Lay Operations **149**

Introduction 150
Section 1: Engine Company Equipment 150
Hose Storage 152
Fire Hose 153
Nozzles 157
Master Stream Appliance 160
Pre-Plumbed Master Stream Appliance 160
Soft-Suction Hookups 161
Pump Intake Connections 161
2½-Inch Pump Intake Connections 162
Hydrant Assist Valve 162
Valves and Hydrant Gates 163
Double-Male and Double-Female Couplings 165
Ground Ladders 165
Bolt Cutters 166
Modern Advancements in Nozzle Technology and Equipment 166
TFT Flip-Tip 167
TFT Vortex 167
The Blitzfire Portable Monitor 168
The Chimney Snuffer 169
The HydroVent™ 169
Specialty High-Rise Nozzles 170
Flamefighter and Transformer Piercing Nozzles 173
Thermal-Imaging Camera (TIC) 174
Hose Alert Hose Restraint Safety System 174
Firefighting Foam 176
History, Science, and Development of Foam 177
Difference between Class A and Class B Foams 177
Emulsifiers 178
Other Advantages to Class A Foam 179
Firefighter Safety and Foam 180
Overhaul and Rekindles 182
CAFS 182
Section II: Initial Hose Operations 182
Safety First 182
Single Engine Hose Lays 183
Forward Lay—Wet 183
Forward Lay—Dry 184
Split Lay 186
Direct-to-Engine—No-Line-Laid Approach 186
Reverse Lay Using a Charged Line 189

CHAPTER 6
Apparatus Positioning **193**

Introduction 194
Pre-Incident Planning 194
Building versus Structure 194
Basic Coverage Responsibility 195
Coverage Assignments 195
Positioning: Alpha Side Front 196
Positioning: Charlie Side Rear 199
Positioning: Bravo and Delta Sides 202
General Front, Rear, and Side Operations 203
Problem Buildings 204
Setback Buildings 204
Shopping Malls 204
Standard Shopping Centers 205
Mercantile Areas 206
Garden Apartments 206
Central Corridor Construction 208
High-Rise Buildings 208

CHAPTER 7
Water Supply **211**

Introduction 212
Inherent Risks 212
The Water Source 213
Hydrants and Water Main Systems 214
Reading Pressure Levels 216
Estimating Attack Lines 217
Static Water Sources 217

Apparatus Water Tanks 218
Mobile Water Supply Apparatus (Tankers and Tenders) 218
Water Shuttle Apparatus Corridor 220
Typical Water Supply Evolution 221
Engines and Pumpers 221
Rated Capacity 221
Intakes 223
Discharges 223
Pump Speed and Capacity 224
Residual Pressure 224
The Hose 226
2½-Inch Hose Line (Single Supply Line) 226
Two 2½-Inch Hose Lines (Dual Supply Lines) 227
3-inch and 3½-inch Hose Lines 228
Large Diameter Hose (LDH) 228
Supply Line Procedures 229
Forward and Reverse Lays 229
Pumper Relays (Tandem Pumping) 231
Setting Up the Relay 231
Relay Pumping 232
Increasing Water Flow 233

CHAPTER 8
Initial Size-Up and Developing a Quick Incident Action Plan 237

Introduction 238
The IC 238
Size-Up 238
Risk Management 239
The Risk Management Plan 239
The National Fire Academy Command Sequence Model 242
The Three Phases of Size-Up 242
Phase 1: Pre-Incident Size-Up 242
Phase 2: Initial Size-Up 247
Performing the On-Scene Size-Up 250
Consequences of an Inadequate Size-Up 250
360-Degree Walk-Around Size-Up 250
Setting Up the Fireground 252
Phase 3: Ongoing Size-Up 268

CHAPTER 9
Search and Rescue 271

Introduction 272
Risk Management 272
Survivability Profile 272
Smoke and Room Temperatures 273
Smoke Toxicity 274
Forcible Entry Effects on Search and Rescue 274
The Chronology of Rescue 275
Before the Alarm 275
Receipt of the Alarm 275
At the Fire Scene 275
Defensive Search Tactics 278
Two In/Two Out Rule 279
Rapid Intervention Crew 280
Water Supply and Hose Stream Placement 283
Hose Stream Placement 283
Rescue Profile Considerations 286
Primary Search 287
Secondary Search 287
Ventilation 287
Fire Attack for Rescue 287
Single-Family Dwellings 288
Multiple-Family Residences 289
Hospitals, Schools, Institutions 289
Fire-Resistant Construction 290
Search 292
Oriented Search 292
Required Tools 292
Typical Search 293
Value of the Standard Search Procedure 294
Standard Search Patterns 295
What to Check 297
Indicating That a Room Has Been Searched 298
Vent-Enter-Isolate-Search (VEIS) 299
Search and Rescue Safety Considerations 299
Reading the Windows 300
Commercial Structures and Large-Area Search 303
Public Assemblies 303
Rescue Drags and Carries 303
Emergency Medical Services (EMS) 306

CHAPTER 10 Initial Fire Attack and Ventilation 309

Introduction 310

Three Strategic and Tactical Areas for Initial Fire Attack 310

What You Need to Know Before You Even Pull a Hose Line 312

- Flashover 312
- Rapid Fire Growth 314

Introduction to Initial Attack Lines 315

Locating the Fire 316

- Direct Attack 316
- Indirect Attack 317
- Combination Attack 320
- Transitional Attack 320
- Defensive Attack 323

Choosing Attack Lines 324

Sizes of Attack Lines 324

- Attack Lines and Nozzles 325
- Solid Streams Versus Straight Streams 326
- Combination Spray Nozzles 329

Effective Stream Operations During Initial Attack 329

- Sizes and Number of Hose Lines 331
- Master Streams for Initial Fire Attack 332

Initial Attack Operations 334

- Attack Lines 334
- Advancing the Hose Line 336

Basement Fires 337

- The Bricelyn Street Fire 338
- Identifying the Presence of a Basement 338
- Smoke Behavior When Making Entry 339
- Making Entry Down the Stairs 340
- First-Floor Collapse 342
- Exterior Window Indirect or Transitional Attack 343
- Cellar Pipes and Distributor Nozzles 343
- Introducing a Basement Sprinkler System Using an Indirect Attack 345
- Controlling Basement Fires 346
- Protecting Exposures 350

Advancing Attack Lines to Upper Floors 351

- Standpipe Operations 353
- Making Entry into the Fire Room for Initial Attack 358

Ventilation 359

- Vertical Ventilation 360
- Horizontal Ventilation 364

Ventilation-Limited, Smoldering, and Decay-Stage Fires 367

- Indications of a Ventilation-Limited Smoldering Fire 369
- Backdraft 369
- Ventilation for Preventing a Backdraft 370
- Smoke Explosions 371
- Ventilation of a Smoldering Fire 371
- Initiating the Coordinated Attack 373

CHAPTER 11 Backup Lines 376

Introduction 377

Sizes and Purposes of Backup Lines 377

Positioning of Backup Lines 379

- Hose Order and Hose Layering 379

Sizes of Backup Lines 381

- Backing Up 1¾-Inch Lines 381
- Backing Up 2½-Inch Hose Lines 383
- Master Streams for Backup 384

Use of Backup Lines 384

Personnel 385

CHAPTER 12 Exposure Protection 387

Introduction 388

Definition of Terms in Exposure Protection 388

Exterior Exposures 389

- Initial Response Considerations 390
- Fire Patrols 394
- Radiant Heat Exposures 395
- Exposure Coverage 396
- Hose Lines and Nozzles 397

Interior Exposures 400

Fire in Concealed Spaces 403
Vertical Fire Spread 404
Horizontal Fire Spread 408
Open Interior Spread 410
Basement Fires 411
Controlling Basement Fires 411
Protecting Exposures 416

CHAPTER 13
Master Stream Appliances 420

Introduction 421
Transitional Attack 421
Defensive Operations 421
Types of Master Stream Appliances 423
Portable Master Stream Appliance 424
Fixed Master Stream Appliances 426
Elevated Master Stream Appliances 426
Nozzles for Master Streams 427
Water Supply for Master Stream Appliances 429
Scenario 1 429
Scenario 2 429
Scenario 3 430
Pumper-to-Pumper Operation 430
Adequate Number of Supply Lines 430
Standard Operating Guidelines 431
Use of Master Stream Appliances 432
Solid-Stream Nozzles Versus Spray Nozzles 432
Positioning the Master Stream Appliance 433
Directing a Master Stream 434
Shutdown 435
Elevated Master Streams 437
Structural Collapse 439
Signs of Collapse 439
Collapse Zones 439

CHAPTER 14
Fire Protection Systems 442

Introduction 443
Standpipe Systems 444
Classes of Standpipe Systems 444
Types of Standpipe Systems 445
Dry System Features and Uses 446
Wet System Features and Uses 446
Water Supply to Wet Standpipe Systems 447
Automatic Sprinkler Systems 452
Sprinkler System Theory 453
Sprinkler Heads 453
Fusible Link and Glass Bulb Sprinkler Heads 455
Types of Sprinkler Systems 456
Water Supply for Sprinkler Systems 458
Exposure Protection 461
Out-of-Service Sprinkler Systems 461
Sprinkler System Failure 461
Shutting Down the System 461
Placing the Sprinkler System Back in Service 461
Other Fire Protection Systems 462

CHAPTER 15
High-Rise Firefighting 465

Introduction 466
Findings from the NFPA Research Report and Others 466
Specialized Problems and Hazards 467
Variables in High-Rise Firefighting 468
Successful High-Rise Firefighting 468
General Information on High-Rise Building Construction 469
Center-Core and Side-Core Design 469
Floor Area 470
Heating, Ventilation, and Air Conditioning (HVAC) Systems 470
Automatic Sprinkler Systems 471
NFPA 14 473
Strategic and Tactical Limitations 474
Tactical Clarifications 474
Firefighter Safety Considerations 475
The Utilization and Preservation of Firefighters 476
Conservative Approach for Smaller Fire Departments 476
Firefighter Accountability 476
Command Priorities 477

Initial Size-Up for High-Rise Buildings 478
Lobby Control 479
Verifying the Fire Location 479
Identifying the Firefighting and Evacuation Stairways 481
Evacuation Control 482
Gaining Control of the Elevators 482
Directing Crews and the Initial Action Plan 484
Engine Apparatus/Pumper Positioning 485
First-In Engine 485
Second- and Third-In Engines 487
Stairwell Support Group 489
Use of Elevators 489
Operating in Phase II 491
The Staging Floor 492
Hose, Nozzles, and Equipment 493
Connecting to Standpipe Discharge Outlets 497
Pressure-Reducing Valves (PRVs) 497
Connecting to Below-Grade Standpipe Discharge Outlets 498
Dangers in the Firefighting Stairway 499
Initial Hose Operations in Residential High-Rise Buildings 502
Beginning the Direct Attack 502
Ascending the Stairs 503
Tactical Objectives 503
Rapid Intervention Teams (RITs) 504
Hooking Up below the Fire Floor 505
1½-inch and 2½-inch Gated Wyes 506
1¾-inch Attack in Residential High-Rise Apartments 508
Gateway Apartments and Townhomes Fire in San Francisco, California 509
Single-Hose Method (2½-inch) 511
Hallway Corridor Standpipe Connections 513
Stretching Hose from the Standpipe 514
Making Entry and Attacking the Fire 514
Connect the RIT Line to the Fire Floor 515
Wind-Driven Fires in Residential High-Rises 516
Fire Curtains 517
Residential High-Rise Nozzles 518
Indirect Flanking Attack 519
Commercial High-Rise Fire Attack 519
Exterior Recon 520
Direct Attack 520
Pincer or Flank Attack 521
Defensive Interior Attack 521
Portable Master Streams 521
Relying on Type I Construction 522
No-Attack Strategy 523
Attacking from Below the Fire Floor 523
Attacking from the Floor Above the Fire 525
Protecting the Floor Above the Fire 526
Final Strategy at the First Interstate Fire 527
Checking for Extension Below the Fire Floor 529
Other Uses for Standpipe Systems 529
Fire Attack in Adjoining Buildings 529
Exposure Protection 529
Use of Water from Uninvolved Buildings 530
Search-Evacuate-Rescue (SER) 530
Shelter and Defend in Place 532
SER Areas of Emphasis 534
SER in Open-Space High-Rise Office Buildings 535
Door Control 535
Firefighter Air Management 535
Air Standpipe Systems 536
Ventilation 538
Post-Fire Operations 541

CHAPTER 16
Salvage and Overhaul 545

Introduction 546
Salvage 546
Overhaul 547
Rekindles 547
Rehabilitation 549
Arson and Scene Preservation 550
Pre-Overhaul Safety Inspection 551
Smoke Explosions 551
Health Risks 555

Personnel 556
Control of Overhaul Personnel 557
Disposable Buildings 557
The Overhaul Operation and Work Assignments 558
Securing Utilities 558
Cleanup 559
Removing Items from the Structure 560
Overhaul Operations 560
What to Look for and Where to Look 561
Checking Above the Fire 563
Debris Pile 565
Final Atmospheric Air Monitoring 565
Water Removal 565
Securing the Property 568
Fire Watch 568
Company Drive-By 569
Some Final Thoughts on Salvage 569
Owner Walkthroughs 570

Glossary 573

Index 579

Acknowledgments

© Rick McClure, Los Angeles Fire Department, Retired.

Jones & Bartlett Learning and the National Fire Protection Association would like to thank the editors, authors, contributors, and reviewers of *Engine Company Fireground Operations, Fourth Edition.*

Reviewers and Contributors

Andrew Arndt
Assistant Fire Chie, Operations (Retired)
Great Lakes Naval Station
Great Lakes, Illinois

David Blair
Lieutenant, City of Columbus Division of Fire, Columbus, Ohio
Career and Technologies Education Centers of Licking County (CTEC), Fire I and II Instructor Newark, Ohio

Jason Byington
Training Officer
Holston Army Ammunition Plant Fire Department
Kingsport, Tennessee

Doug deBest
Deputy Chief/Training Officer
Baroda Fire Department
Baroda, Michigan

Brad Dougherty
Captain
Norfolk Naval Station Fire-Rescue
Norfolk, Virginia

Brian Dudley
Jones County Fire Department and Rescue
Jones County, Georgia

Brian Dulay
Deputy Chief
Honesdale Fire Department Training Center
Honesdale, Pennsylvania

Allen Evans
Firefighter
Alvin Volunteer Fire Department
Alvin, Texas

Christopher Farrell
National Fire Protection Association
Quincy, Massachusetts

Eric D. Gartley
Safety Officer
Arnold Volunteer Engine Company #2
Arnold, Pennsylvania

Brian S. Gettemeier
Captain
Cottleville Fire Protection District
St. Charles, Missouri

Brandon Harrill
Support Services Division Chief
Rutherfordton Fire and Rescue
Rutherfordton, North Carolina

Barry Hendren
Assistant Fire Chief
Asheville Fire Department
City of Asheville, North Carolina

Derek Kaucher, BS, MA
Captain/Paramedic–CC
Wacker Fire Department
Charleston, Tennessee

Bob Lloyd
Fire Chief
Fairport Harbor Fire Department
Fairport Harbor, Ohio

J. J. Lyons
Firefighter/Engineer
Williamsport Bureau of Fire
Williamsport, Pennsylvania

Andrew Mihans
Captain
Arlington Fire District
Poughkeepsie, New York

Carl J. Mack
OFE, OFC Fire Science Coordinator
Lorain County Community College
Elyria, Ohio

Carter Pittman
Assistant Fire Chief
Woodlawn Fire Department
Woodlawn, Ohio

Guy Peifer
Lieutenant
Yonkers Fire Department
Yonkers, New York

Christopher Rutola
Orange Fire Department
Orange, Massachusetts

Brian Serowinski
Bucks County Community College
Bucks County, Pennsylvania

Ali Shah, LP, MPH
San Jacinto College
Pasadena, Texas

Chad Smith
Captain
Georgetown Fire Department
Georgetown, Kentucky

Curt Smith, BS, CFO
Assistant Chief
Hastings Fire and Rescue
Hastings, Nebraska

Al Wickline
Adjunct Faculty
CCAC–Boyce Campus
Monroeville, Pennsylvania

David Wiklanski, MA
FF/EMT
New Brunswick Fire Department
New Brunswick, New Jersey

Christopher J. Yoch, MS, FO
Firefighter
Keystone Fire Co. No. 1 Shillington
Shillington, Pennsylvania

Preface

© Rick McClure, Los Angeles Fire Department, Retired.

I am privileged to have been selected to author the 4th Edition of *Engine Company Fireground Operations.* This is a complete and extensive rewrite of the original Harold Richman series. It is an honor to continue his legacy while making my own contribution to the fire service.

This textbook covers the basic strategies, tactics, methods, and responsibilities for engine companies at a structure fire. It also covers the tactics and assignments traditionally assigned to ladder companies, but the primary focus is on the engine company. The 4th Edition incorporates the Underwriters Laboratories/National Institute of Standards and Technology (UL/NIST) recommendations from the experiments and studies on modern fuels and fire behavior, into the strategies and tactics presented within the chapters. It also includes the recommendations from the recent studies and experiments on ventilation conducted by Underwriters Laboratories/Firefighter Safety Research Institute (UL/FSRI).

When we talk about the *basics*, the reader must understand that the basics actually start with modern fire dynamics which incorporate fire science, chemistry, physics, modern fire behavior, building construction and its effects on fire, and the most recent scientific evidence-based practices. Engine company officers should thoroughly understand fireground size-up, risk assessment, firefighter accountability, civilian survivability profiles, and rescue profiles before they pick up a hose and nozzle or lead a crew into a burning structure. Developing size-up skills requires continuous study and practice.

Each chapter is a stand-alone chapter of the subject, hence many critical teaching points, illustrations, photographs, and tactical solutions are covered more than once and repeated within the selected subject matter. A plethora of relevant examples and case studies have been included to give my perspectives their relevance and credibility to drive home these concepts along with the lessons learned. Obviously, many illustrations are not to scale or realistically correct. They are conceptual drawings to illustrate a fireground tactic or situation that can occur.

There are also many variables on the fireground such as: weather, wind, and terrain; whether the structure is a single-family residence, a multi-family residence, or a multi-story structure; and whether it's a commercial structure or a highrise building. Other factors include the size of the responding fire department, the type of fire apparatus, the number of fire apparatus on scene, the number of firefighters on scene to advance hose lines and carry out other assigned tasks, the availability of additional staffing and resources, the firefighting equipment used – the size of nozzles and hoses, the condition of personal protective equipment, whether its used, new, wet, dry, or dirty; the career experience of firefighters and pump operators, hydraulics, the type of fuel burning, the fire load within the burning compartment, and how close a firefighter can get to the seat of the fire in order to apply water effectively from a safe position. All these variables affect time duration measurements, listed temperature measurements, distance measurements and gallonage requirements. Hence the listed gallonage requirements, time, temperature, and distance measurements and numerical figures within this text are not absolutes, but close estimates to what is accurate, effective, and acceptable.

Whatever fireground risk situation you enter, your probability of success and survival in a dangerous, uncontrolled environment is directly related to your frequency of practice and the caliber of realistic training.

Acknowledgements

It starts with my greatest gratitude and appreciation to my friend, Bill Manning, who saw my potential and gave me a chance by opening the doors to Fire Engineering and FDIC, and for giving me a platform on the national stage.

I would like to especially thank Deputy Chief Stewart Rose (SFD retired). He has been my teacher and mentor from the day I entered the Seattle Fire Department and remains so today. This book is as much yours as it is mine and I hope it meets with your approval.

I would like to thank three people who invested in my career and served as mentors and colleagues. They gave me access, knowledge, wisdom, laughs, and friendship. I only wish they were around to see how their efforts paid off: Chief Alan Brunacini, Phoenix (AZ) FD, and Chief Tommy Brennan FDNY and Waterbury (CT) FD. I hope this book would have made them proud. And thanks to Deputy Chief Vincent Dunn (FDNY ret.) who got me started writing in the first place.

I owe a special thanks to the following colleagues and friends who contributed their expertise to this textbook project. They answered numerous phone calls, texts, and emails at all hours of the day and night; conversations that usually started with "Hey chief…?" or "What do you think about this?" I am sure I was a pest, but I could not have finished this textbook without your guidance, opinions, and input. You took my calls every time. Thank you!

Battalion Chief William "*Smokey*" Simpson, Seattle Fire Department
Captain Stephen Bernocco, Engine 38, Seattle Fire Department
Chief Glenn Corbett, PE, Waldwick (NJ) FD, Associate Professor – Fire,
John Jay College of Criminal Justice
Deputy Chief Nick Visconti, FDNY (Retired)
Alan M. Petrillo, Ex-chief, Verdoy (NY) FD, *Fire Apparatus and Emergency Equipment Magazine*
Fire Chief Brent Batla, Horseshoe Bay (TX) Fire Dept.
Fire Marshal Paul T. Dansbach, Rutherford (NJ) Fire Dept.
Dr. Burton Clark, PhD, EFO, National Fire Academy (Retired)
Dr. Denis Onieal, PhD, Deputy Fire Administrator, U.S. Fire Administration
District Chief Matt Stuckey, Houston (TX) Fire Dept. (Retired)
Chief Dan Madrzykowski, PhD. P.E. Senior Director of Research UL/FSRI
Chief Stephen Kerber, PhD. Vice-President of Research, UL/FSRI
Deputy Chief Jaime Moore, Los Angeles Fire Department
Division Chief J. Gordon Routley, Montreal Fire Department, Canada
Battalion Chief Stephen Marsar (FDNY)
Battalion Chief Frank Montagna (FDNY Retired)
Fire Chief Jay Hagen, Bellevue (WA) Fire Dept.
Lt. Mike Ciampo (FDNY)
Battalion Chief Jerry Tracy (FDNY)
Captain John Ceriello, Rescue 1 (FDNY)
Captain Bill Gustin, Metro Dade (FL) Fire and Rescue
Battalion Chief Michael Wielgat, Chicago (IL) Fire Dept. (Ret.)
Lieutenant Kevin O'Donnell, Hoffman Estates (IL) Fire Dept. (Ret.)
Fire Chief Joe Starnes, Sandy Ridge VFD (Ret.) co-founder, Kill the Flashover
Battalion Chief Jon Gillis, Seattle Fire Dept. (Ret.)
Battalion Chief Andy Starnes, Charlotte (NC) Fire Dept.
Deputy Chief Gary English, Seattle Fire Dept. (Ret.)
Captain James Hilliard, Engine 27, Seattle Fire Dept.
Captain Steve Baer, Engine 32, Seattle Fire Dept.
Lieutenant Eric Jensen, Engine 2, Ladder 6, Seattle Fire Dept.
Firefighter Galen Thomaier, Engine 18 Seattle Fire Dept. (Ret.)
Firefighter Dennis Stanley, Engine 33, Ladder 1, Seattle Fire Dept.
Firefighter Chris Quinlan, Engine 33, Seattle Fire Dept.
Firefighter Don Henry Jr. Ladder 6, Seattle Fire Dept.
Firefighter William Cleary, Ladder 6, Seattle Fire Dept.
Assistant Chief of Operations, Chuck Kahler (Ret.), South King Fire and Rescue (WA.)
Robert Tutterow, President, F.I.E.R.O.
Health and Safety Officer, Charlotte (NC) Fire Dept. (Ret.)
David "*Super Dave*" Cress, Fire Garage Master Mechanic, Seattle Fire Dept.

I must thank the Seattle firefighters and crews of Ladder 6, Ladder 1, Ladder 8, Engine 8, Engine 33, Engine 18, Engine 9, Engine 20, and Engine 2; and the crews of Engine 24 of Hoffman Estates (IL) Fire Department who were always willing (and sometimes not so willing) to pose for and stage training photographs. I could not have completed this monumental project without your help.

I want to thank all my editors who have worked with me throughout the years to help make me a better writer:

Kathleen Hjørdis Knudsen, thank you for your many hours of tedious work and willingness to help me succeed.
Diane Rothschild, *Fire Engineering*
Ed Ballam, *Fire Apparatus and Emergency Equipment Magazine*
Lyn Bixby, *Fire Apparatus and Emergency Equipment Magazine*
Peter Prochillo, *Fire Engineering.com*
Gaius and Sue Reynolds, Editors of F.C.F. *The Encourager*

Finally, I wish to dedicate this book to my children, Allison, Andrew, and Emily, and to my loving and long-suffering wife, Jan, for their love and patient support while I put them on hold and neglected them so I could write this important and life-saving book. It's finally finished!

About the Author

Captain Raul A. Angulo is a 40-year veteran of the fire service. He was born in Los Angeles, California, graduated from Cathedral High School, and attended Humboldt State University where he majored in music, becoming an accomplished classical pianist and Latin Jazz percussionist. He served in the U.S. Coast Guard as a rescue swimmer and attained the rank of Boatswain's Mate 3rd Class. His training included the Marine Firefighting Schools at Treasure Island in San Francisco and in Yorktown, Virginia where his passion for firefighting was ignited. He served two short stints as a firefighter with Vernon (CA) FD and Farmington (NM) FD before he landed his career job with the Seattle (WA) Fire Department. As a firefighter, he was assigned to E2, L4, E20, and L1, then trained to be a dispatcher at the Fire Alarm Center (FAC). After being promoted to lieutenant, he remained at the FAC until he was assigned to E24 and E9, including two assignments at Training Division, where he served as a recruit academy drill instructor.

Promoted to captain, he was assigned to E33, E18, and finished his career at Ladder Company 6. Captain Angulo was an emergency medical technician (EMT) throughout his career and served on the SFD's Hazardous Materials Unit, Wildland Firefighting Team, and the FEMA/Puget Sound Urban Search and Rescue Team. After more than 35 years of dedicated and honorable service, he retired as the senior captain of the SFD.

Captain Angulo's leadership positions included the Board of Directors for the Sumner (WA) School District, Board of Directors for the Fellowship of Christian Firefighters International, and a founding board member for the National Fire Academy Alumni Association. He was an advisory board member for the Fire Department Instructors Conference (FDIC) and is currently on the editorial advisory board for *Fire Apparatus and Emergency Equipment Magazine*. He has over 330 articles published in all the major trade magazines including

Fire Engineering, *Fire Engineering.com*, *Fire Apparatus and Emergency Equipment*, *Firehouse Magazine*, *Fire Rescue Magazine*, *International Fire Fighter Magazine*, and *The Encourager*. Captain Angulo has been teaching at FDIC International for over three decades and has taught throughout the United States, Canada, Mexico, and Latin America. He teaches and lectures on fire service leadership, crew development, Calling the Mayday, Drills You're Not Going to Find in the Books, and firefighting strategy and tactics, with an emphasis on firefighter accountability systems.

Captain Angulo now resides with his family in Huntington Beach, California.

© Rick McClure, Los Angeles Fire Department, Retired.

CHAPTER

Introduction and Overview

LEARNING OBJECTIVES

- Identify and explain the responsibilities of the engine company.
- Understand the premise of the UL/NIST studies.
- Recognize that all fireground operations comply and work within NIMS/ICS.
- List the four incident priorities.
- Compare RECEO/VS and BRECEO/VS.
- Describe the differences between strategy, tactics, tasks, and resources.
- Describe the role of the first-in company officer.
- Explain the purpose of a QIAP.
- List the top-tier problems on the fireground.
- List the basic fireground assignments.

Introduction

The engine company is the basic unit of a fire department. According to the U.S. Fire Administration (USFA) website, there are 29,727 fire departments. Of that, 9% are all career, 6% are mostly career, 18% are mostly volunteer, and 67% are all volunteer. It is estimated that there are 1,160,450 firefighters in the United States; 345,600 are career (30%) and 814,850 are volunteers (70%). The career fire departments (the 30%) protect our major metropolitan cities with the highest populations.

Many fire departments consist of only engine companies. The reason is simple: The engine company provides the primary firefighting agent—water—and the firefighters to apply it properly to a burning structure. In most cities, the engine company is also the first responding unit providing rescue and emergency medical services (EMS) for medical emergencies, motor vehicle accidents, industrial accidents, mass casualty incidents (MCIs), active shooter incidents, hazardous materials (hazmat) incidents, and natural disasters, but the primary responsibility of the engine company is to respond to and extinguish structure fires **FIGURE 1-1**.

The engine apparatus is an emergency vehicle that must be relied on to transport firefighters safely to and from an incident and to operate reliably and properly to support the mission of the fire department. A piece of fire apparatus that breaks down at any time during an emergency operation not only compromises the success of the operation but might jeopardize the safety of the firefighters relying on that apparatus to support their role in the operation. An old, worn-out, or poorly maintained fire apparatus that frequently breaks down has no role in providing emergency services to a community **FIGURE 1-2(A)** (see NFPA 1901, D.6). Regularly scheduled preventative maintenance inspections not only protect the investments of expensive and reliable fire apparatus, they also extend the in-service time and longevity of the engines **FIGURE 1-2(B)**.

FIGURE 1-1 The primary responsibility of the engine company is to respond to and extinguish structure fires.

Courtesy of Rick McClure (LAFD ret.).

A

B

FIGURE 1-2 **A**. An old or poorly maintained fire apparatus has no role in providing emergency services to a community. **B**. Regularly scheduled inspections extend the in-service time and longevity of the engine.

Courtesy of Raul Angulo.

Emergency Medical Services

EMS has become our primary emergency response service provided to our cities and communities. According to the USFA, 13,500 (45.4%) fire departments

FIGURE 1-3 EMS has become our primary emergency response service provided to our cities and communities.
Courtesy of John Odegard.

provide basic life support (BLS) services, 4,617 (15.5%) fire departments provide advanced life support (ALS) paramedic services, and 11,610 (39.1%) fire departments do not provide emergency medical services; these are handled by other emergency health care agencies FIGURE 1-3. EMS accounts for about 80% of emergency responses. The numbers may vary between cities, but it is fair to say that, at best, fires account for only 20% of emergency responses. This supermajority percentage differential has led many organizations to change their name from *fire department* to *emergency services*. But this hasn't happened without a fight and resistance from hardliners who want to protect the rich history and traditions called the fire department.

Annual required training and in-service continuing education along with recertification in areas of EMS, hazmat, and other disciplines constantly compete with training in the area that kills and injures firefighters the most: the fireground. This text will help provide that training. It is important to emphasize that most, if not all, fire incidents eventually evolve into EMS incidents, from treating injured civilians and firefighters to simply setting up rehabilitation (rehab) stations for firefighters to take a break and rehydrate after a difficult and physically stressful fire fight. Although we acknowledge that EMS may be 80% of our work, this text will focus on the other 20%.

Advancements in Technology

The market is constantly looking at ways to improve technology in fire apparatus manufacturing, firefighting equipment, self-contained breathing apparatus (SCBA), and personal protective equipment (PPE) and clothing. All these advancements help firefighters advance to the seat of the fire by providing maximum protection from flames, heat, smoke, gases, and physical injury. Over the past decade, there have been technical advances made on equipment such as nozzles and associated appliances; lightweight fire hoses, including large-diameter hose (LDH), which is used for supply lines and for providing water to master stream appliances; and on a wide variety of hand and power tools and equipment.

Since the advent of SCBA, no other electronic device has changed the fire service more than the thermal-imaging camera (TIC). Thermal imaging technology has improved so much that handheld units are becoming more compact without losing image resolution quality. TICs are now being integrated into the SCBA pressure gauge control module units, part of the backpack assembly, and the newest technology has the TIC integrated into the SCBA facepiece FIGURE 1-4. The ability to see in the dark and through smoke has changed the way we do business. The TIC gives us the ability to see an unconscious victim from across the room, thus allowing us to perform primary searches

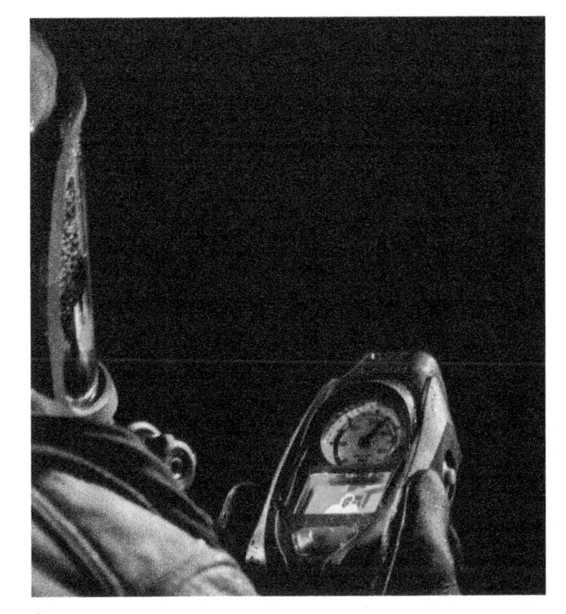

A

B

FIGURE 1-4 A. The MSA *G1* TIC. B. The Scott *Sight Pro* TIC.
A: Photo Courtesy of MSA Fire; B: Photo Courtesy of Adrian Schooler, Scott Fire and Safety.

rapidly and efficiently. It also allows us to locate a downed firefighter quickly during a rapid intervention rescue. With heat and temperature readings, the TIC also helps us locate the fire for a safer, more effective fire attack; check for extension; and uncover hidden fires and hot spots during overhaul. There are many uses for the TIC, including in size-up, which will be covered in this text. The current edition of NFPA 1801 is the *Standard on Thermal Imagers for the Fire Service.*

Nozzle technology has also changed. The nozzles have become considerably lighter and more durable. Structural firefighting nozzles still fall into three basic categories:

- Straight-stream smooth-bore nozzles
- Combination spray and straight-stream nozzles
- Specialty nozzles (i.e., piercing nozzles, distributor nozzles, round nozzles, and foam nozzles)

The most significant changes are the breakaway nozzle, which has a smooth-bore tip within a combination fog nozzle **FIGURE 1-5**, and the dual-attack nozzle like the *HydroVent*™. The latter hangs from an exterior windowsill and sprays multiple straight streams of water into the burning compartment while simultaneously aiming a second attached fog nozzle out the same window. It also performs hydraulic ventilation of the same space at the same time. No other nozzle can do this **FIGURE 1-6**.

More efficient extinguishing agents have also been developed over the last decade. **Universal extinguishing foam (UEF)** works on all classes of fires, and biodegradable surfactants have minimal impact on the environment **FIGURE 1-7**. The newer foams stick better to Class A fuel surfaces, which minimizes the risk of rekindles and can be used on most existing nozzles and compressed air foam systems (CAFS). These improvements are designed to supplement the use of water against fire and to increase its efficiency by reducing water surface tension for deeper penetration and absorption into the burning materials.

Another new chemical multipurpose extinguishing agent is F-500, a micelle-encapsulating fire suppression agent developed by Hazard Control Technologies, Inc. and Energetech, LLC. The fire suppression

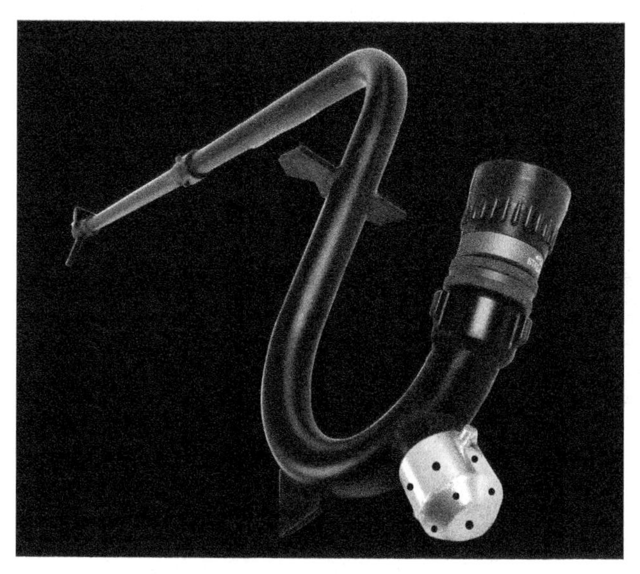

FIGURE 1-6 The *HydroVent*™ is a dual-attack nozzle that hangs from an exterior windowsill and sprays multiple straight streams into the burning compartment while simultaneously aiming a second attached fog nozzle out the same window for hydraulic ventilation.

Photo Courtesy of Kevin O'Donnell.

FIGURE 1-5 The TFT *Flip-Tip* offers the choice of a combination spray nozzle or a smooth-bore nozzle. Both tips are integrated into a single handheld nozzle; firefighters do not have to unscrew anything to exchange tips.

Photo Courtesy of Task Force Tips.

FIGURE 1-7 Universal extinguishing foam (UEF) works on all classes of fires.

Photo courtesy of Troy Carothers and W.S. Darley.

mechanics and technology utilized by F-500 are different than conventional foam because it forms and maintains micelles, "chemical cocoons," around the hydrocarbon fuel molecules, neutralizing the fuel leg and interrupting the chemical chain reaction of the fire tetrahedron rather than forming a blanket to deprive the fire of oxygen. Since F-500 is not a foam, maintaining the integrity of a foam blanket or barrier is not an issue.

F-500 is Underwriters Laboratories (UL) listed for Class A and Class B (polar and nonpolar solvent) fires at 1%, 3%, and 6%. It can also be used on Class D fires by increasing its concentration. The agent is nontoxic, noncorrosive, and fully biodegradable. It can be added to a 2.5-gallon pressurized water extinguisher, used with an eductor, or premixed and delivered through the pump with standard fire hoses and nozzles. Like foam, it also breaks the surface tension of water molecules, so extinguishment is achieved faster than with plain water. In conjunction with the chemical cocoons that suppress flammable vapors, faster extinguishment reduces heat temperatures and stops the production of toxic smoke.

All these technical advancements in firefighting are not a substitute for the primary responsibility of an engine company: to put water on the fire. Fires don't get worse when water is applied in any form, so let's put the fire out!

To supply and use water properly at a fire, firefighters must have a keen sense of situational awareness and be safety conscious. They must have considerable skills and knowledge, especially in the areas of fire behavior and building construction, which will be covered in subsequent chapters. Firefighters must also possess a certain amount of brawn, with the ability to withstand physical, mental, and environmental stress. The engine company must be well trained and proficient in any task required on the fireground. Fire departments do not respond to fires on a daily basis; therefore, firefighters need to hone their skills by continually training. The current National Fire Protection Association (NFPA) 1410, *Standard on Training for Initial Emergency Scene Operations* is a training standard that provides an objective method of measuring performance for initial fire suppression and rescue procedures used by fire department personnel engaged in emergency scene operations, using available personnel and equipment. The standard specifies basic evolutions that can be adapted to local conditions and serves as a standard mechanism for the evaluation of minimum acceptable performance during training for initial fire suppression and rescue activities.

Where Are We Now?

Structures, whether small bungalows, commercial warehouses, or multistory high-rise condominiums, are a series of boxes **FIGURE 1-8**. The boxes stand alone, are stacked next to each other, or are stacked on top of each other. If a fire starts inside a box, it produces flames, smoke, and heat within the box. If left unattended, the fire gets out of control and spreads to other boxes (exposures) until it runs out of air or fuel. Most of us have flown in a jet plane before. As you approach your destination to any city in the world, look down through the window. What do you see? Boxes! Rows and stacks of boxes. In fact, we now call large retail warehouses "big-box stores." Our job as firefighters is to rescue people trapped in the box, put water into the box, get the smoke out of the box, and protect the unburned boxes. That's basically it.

What has changed? Fast-forward to the present. The construction of the boxes and the fuels we put inside them have changed. For example, modern residential wood-frame construction has evolved to include prefabricated engineered materials and supporting brackets that do not resist fire, unlike structures built many years ago, which used dimensional lumber. Prefabricated lumber, which isn't lumber at all, uses glues, adhesives, and resins to laminate layers of thin wood and wood chips together to create the mass that is used in place of what used to be entire lengths of dimensional lumber. Modern construction is more insulated and energy efficient. Single-pane windows are now double- and triple-pane, shatter-resistant thermoplastic polycarbonates, such as Lexan™ **FIGURE 1-9**. These windows can withstand the heat of residential fires for extended periods of time before they fail. Often, the fire runs out of air before these windows fail, creating the condition known as a **ventilation-limited fire**.

Today's household furnishings (fuel load or fire load) are primarily made of petroleum-based plastics and synthetics known as modern fuels. Modern-fuel fires burn faster and hotter, and generate higher heat-release rates that generate black toxic smoke with deadly products of combustion like carbon monoxide, hydrogen cyanide, phosgene, benzene, formaldehyde, and acrolein. The list of the toxic components of smoke is much more detailed, and scientific research now confirms a direct correlation between the exposure to smoke and the increasing cancer rate in firefighters compared to the general public. In addition to lung, prostate, and brain cancer, esophageal cancer is one of the more prominent cancers affecting firefighters.

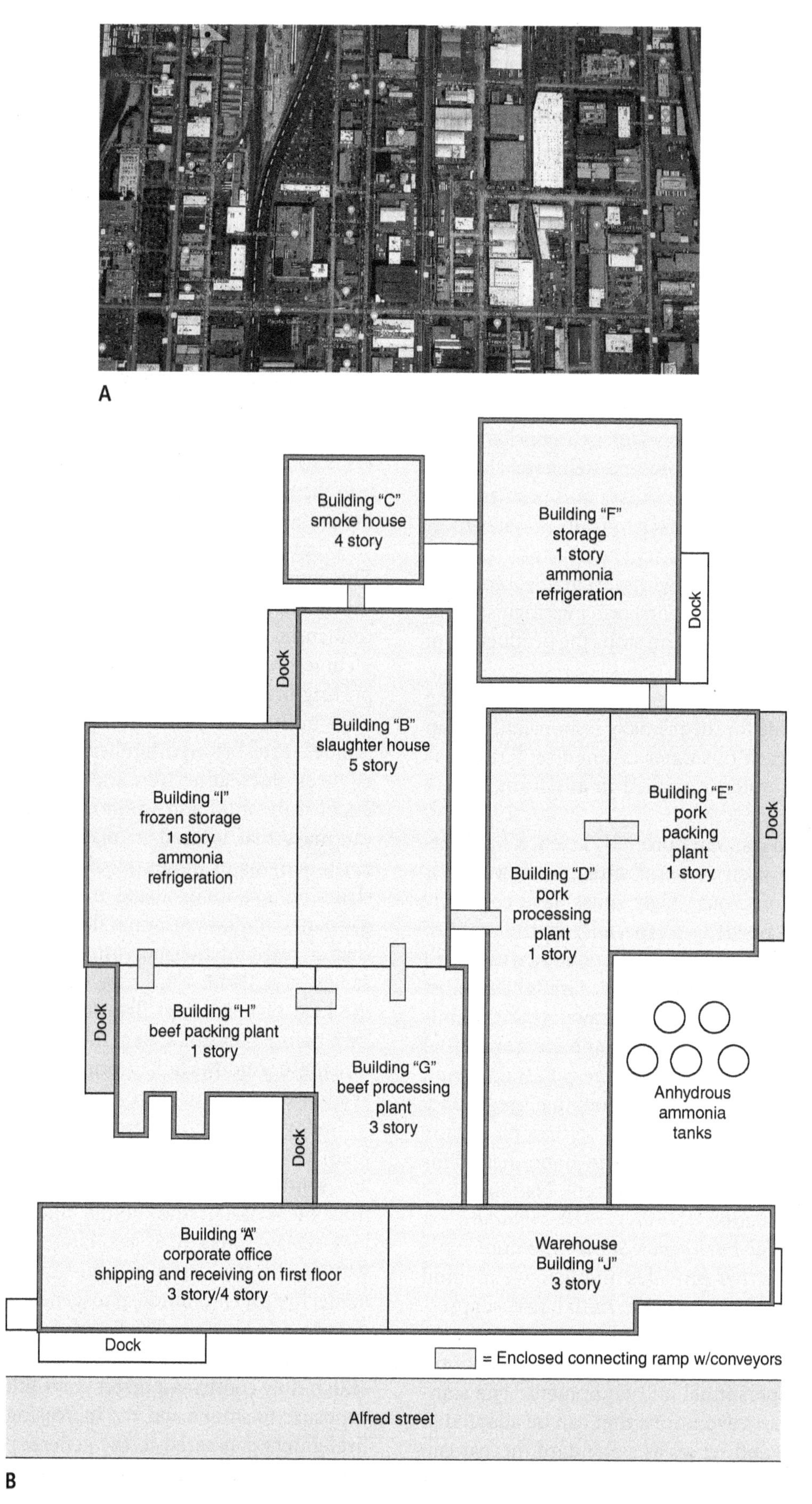

FIGURE 1-8 **A**. Most structures are just a series of boxes—as reflected in **B**. a pre-incident planning map.

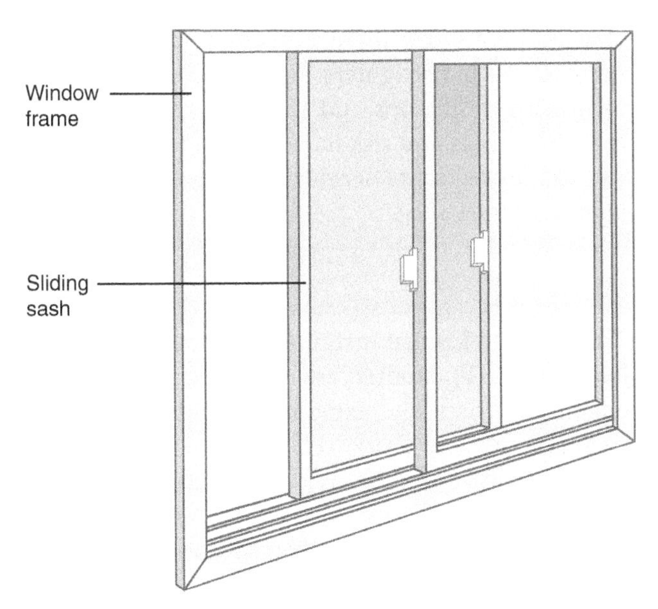

FIGURE 1-9 Modern construction uses stronger, more insulated, energy-efficient windows.

Firefighters work in uncontrolled atmospheres where they are exposed to high levels of contaminants for short periods of time, unlike most occupations where exposure hazards can be controlled through engineering and procedures. We are uniquely exposed to a variety of hazards that are often not well identified or quantified. According to the National Institute for Occupational Safety and Health (NIOSH) and the National Fallen Firefighters Foundation, who include these scientific and medical studies on their websites, digestive cancers; genitourinary (kidney, bladder, testicular, and prostate) cancers; and oral, respiratory, and mesothelioma cancers are affecting firefighters at a higher rate compared to the general public. This is causing great concern for the fire service.

The ability to inhale and tolerate smoke and to wear grimy gear and a smoked-up helmet were all signs of grit, toughness, and bravery. Firefighters sometimes wore them like badges of honor to show *they were there*. Those days are over. Even with the protection of SCBA, these toxins' path of entry into the body is through skin absorption. Our firefighting ensemble gets saturated with sweat, water, and the products of combustion, and every time we put on our pants and coats, our skin is exposed to these carcinogenic chemicals through cross-contamination. Surviving an incident is no longer making it safely to the end of a shift. Smoke exposure is cumulative, and the consequences may not be evident for years, even after retirement. Therefore, a philosophical shift has occurred in safety and risk management about whether it is even worth it to expose firefighters to residual smoke after the fire

FIGURE 1-10 The rapid heat buildup in fires involving modern fuels has produced more flashovers today than in previous decades, often before the arrival of the fire department.

has been extinguished, especially when a property no longer has any value.

Modern fuels generate and put out a tremendous amount of energy, almost double the British thermal unit (BTU) potential of legacy fuels like wood, paper, and natural textile fibers. Compared to legacy fuels, modern fuels can produce higher temperatures in excess of 1,000°F (538°C), which leads to rapid flashover potential. In fact, we are seeing more flashovers today than in previous decades because flashover temperatures have caught up to fire department on-scene arrival times **FIGURE 1-10**. Combine this with energy-efficient windows, which can withstand interior temperatures for an extended period of time, and we now have high-temperature, ventilation-limited fire conditions that are simply waiting for the introduction of a new air source. That new fresh air source is usually introduced by firefighters, and interior conditions can quickly change into an extremely dangerous situation for unsuspecting firefighters. That is why we need to reevaluate our traditional approach to interior search and rescue, interior fire attack, water application, and coordinated ventilation. It is extremely important to adjust our strategy and tactics to the trends and advancements made in modern living.

Although residential and commercial fires are more volatile than ever, fires are still on a downward trend. Contributing factors include improvements and expansion of the fire and building codes. Many new construction occupancies are required to have automatic sprinkler systems and monitored automatic fire alarm systems. In some cities, these requirements have been made retroactive. Fire prevention inspections, public education, and free smoke detector installation programs have all contributed to higher levels of fire safety awareness and early fire detection. Smoke detectors are now as common as televisions, with many doubling as carbon monoxide detectors **FIGURE 1-11**.

Electrical, heating, and cooking appliances are much safer now. Many have automatic safety shutoff features. Gypsum board, drywall, and flame-retardant interior finishes are common in the home construction market. Monitored home security systems are reasonably priced and can protect against home invasion, medical emergencies, and fires. All these factors have led to a decrease in fires and fire fatalities. That's good news, but it also means less real-world experience for company officers and firefighters.

Although residential fires are down, firefighter line-of-duty deaths are disproportionately high and, on average, remain relatively unchanged. Therefore, we need to rely more on live-fire training (which is designed for firefighter safety, so it is not truly a live fire), the study of building construction and fire behavior, prefire planning, drilling, Internet videos, case studies, and computer fire simulators. We also need to teach firefighters and officers how to perform proper size-ups and learn how to apply risk-benefit analysis and risk management to every action we take. Today's fire service is being driven by risk management.

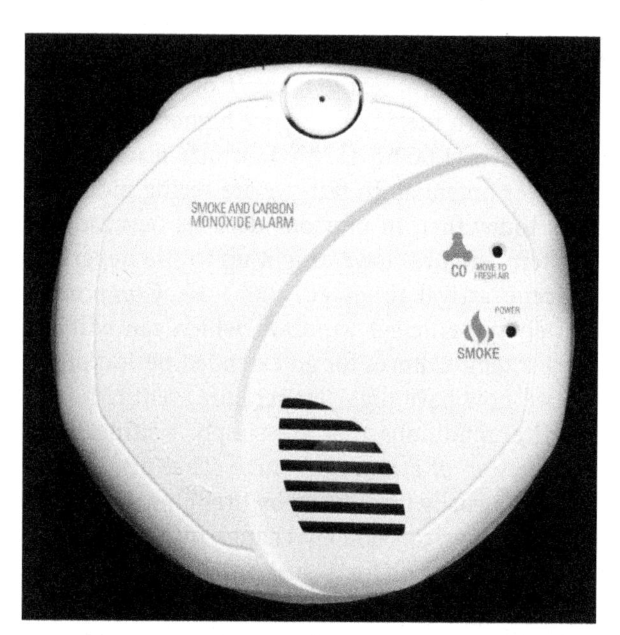

FIGURE 1-11 A combination carbon monoxide and smoke detector is required by law for residential rental properties and multifamily occupancies like apartments, condominiums, and hotels.

UL/NIST

The UL and National Institute of Standards and Technology (NIST) studies are the recently published ongoing research and experiments with the Fire Department of New York (FDNY) and Chicago Fire Departments on fire behavior, water application, fire suppression, ventilation, search and rescue, and survivability profiles as a result of changes in modern building construction and the modern fuels of home interior furnishings. As a result of these changes, fires now burn hotter and faster, releasing tremendous amounts of energy, thus reaching flashover sooner and with higher temperatures **FIGURE 1-12**. The results of these experiments provide the most current, scientifically validated information on fire behavior and clearly show that reevaluation of fireground practices is well overdue. Chapter 4 is dedicated to the UL/NIST studies, the experiments, the recommended fire attack methods for structures and basements, and accompanying ventilation practices. These recommendations will be applied throughout this text.

National Incident Management System/ Incident Command System (NIMS/ICS)

A fire department must operate using a structured incident management system (IMS) on the fireground; it is the law. The US government has mandated the implementation of the National Incident Management System (NIMS) for all fire and emergency service providers that receive federal funding or subsidies. NIMS and the incident command system (ICS) together can be used to organize and manage emergency incidents from a small local scale operation to large multigovernmental, multijurisdictional, and multiagency emergencies. NIMS/ICS is an organized agreed-on system of roles, authority, responsibilities, definitions, communications, and standard operating guidelines used to develop strategies and tactics; manage emergency incidents safely; and manage the scene, dedicated resources, and personnel efficiently. Engine companies must work within the overall strategic plan

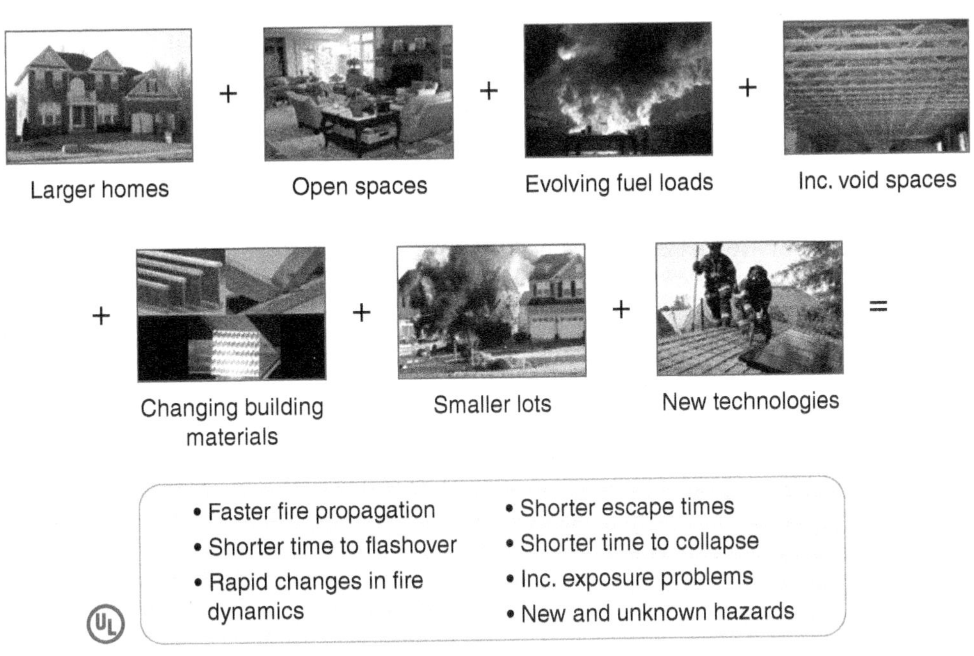

FIGURE 1-12 These factors change the way fires burn in homes furnished with modern fuels.
Courtesy of UL.

developed by the incident commander (IC), and each engine company must work within the ICS and follow the plan by performing tactics and tasks assigned by the IC, thus ensuring that firefighter safety and crew accountability is maintained.

It is understood that, per an IMS, the first-in company officer establishes command and becomes the initial IC. This individual is responsible for all the incident activities, including the initial size-up, risk-benefit analysis, development of strategy and tactics, ordering and assignments made in relation to resources required to carry out such strategy and tactics, and maintenance of a functional accountability system for the overall safety of firefighters working at the incident.

As its title clearly states, this book is about engine company operations, and the text is dedicated to the specific tactics that engine company personnel will perform in any given fire situation. Since this is *not* a command manual, it is assumed that readers have already completed the USFA/National Fire Academy (NFA), as well as the National Fire Service Incident Management Consortium workbooks. All real-world actions and incidents involving any fire department should adhere to the current edition of *Incident Command System Model Procedures Guide for Incidents Involving Structural Firefighting and High-Rise.* Therefore, these IMS topics are no longer covered in this text. This text focuses on the responsibilities of the initial IC, who is typically the first-in engine officer, and covers all the areas necessary for an effective size-up and developing a quick initial action plan (QIAP) for the first-alarm assignment.

The Incident Priorities

The engine company should be equipped and adequately staffed to carry out all of the objectives of a firefighting operation. These objectives, which have evolved over decades of firefighting experience, were developed in the early versions of the ICS. Traditionally, they are:

- Life safety
- Incident stabilization
- Property conservation

In recent publications, *extinguishment* is often substituted as the second incident priority instead of incident stabilization. Clearly, two schools of thought are developing; many share the opinion that either of these two are implied or mean the same thing. They do not. You can have incident stabilization—meaning, things are not getting worse—without extinguishing the fire. For example, the initial actions at a large defensive fire would be to get master streams in place to protect exposures. Once the surrounding exposure buildings are protected, the incident is contained and stabilized before any water is directed onto the main body of the fire. The fire can be attacked defensively or allowed to burn itself out, but it shouldn't get any worse. Incident stabilization is certainly an ICS objective, whereas extinguishment may or may not be. For the engine company on a tactical level, extinguishment is an incident priority, but it may not always be possible. Every fire goes out sooner or later, even if an engine company takes no action. There are no fires still burning today because they weren't extinguished by an engine company.

If anything is "implied," it should be the expectation that engine companies extinguish the fire. However, there is a growing movement to include this change. For the advancement and good of the fire service, there is no reason to resist or reject this shift in strategic philosophy and choose one over the other. It is sometimes better to be inclusive and seek a win-win by changing the number instead of rejecting the implied incident priority. So, the four major incident priorities on the fireground are as follows:

- Life safety
- Incident stabilization
- Extinguishment
- Property conservation

The four incident priorities are listed in the order that they should be carried out. This does not mean that the first priority must be accomplished before another is started, but the priorities are listed in order of importance and form the basis for any fire attack plan. These fireground incident priorities may be occurring simultaneously to ensure the protection and safety of both occupants and firefighters. Thus, firefighters must understand modern fire dynamics which is the nature of fire and the factors that affect its spread, including building construction, type of occupancy, and type of fuel available to the fire. All these subjects are covered in more detail throughout this text.

RECEO/VS and BRECEO/VS

In his 1953 book, *Fire Fighting Tactics*, Chief Lloyd Layman discussed tactics for directing fireground operations that are as relevant today as they were in the past because they are well thought out, and the sequence makes sense. Rescue, exposures, confinement, extinguishment, overhaul/ventilation and salvage (RECEO/VS) follows a logical and systematic approach to accomplishing the four incident priorities of life safety, incident stabilization, extinguishment, and property conservation. In fact, the UL/NIST recommendations for extinguishing fires are actually returning to the methods first talked about by Lloyd Layman in the 1950s.

The recent UL/NIST studies also include numerous experiments on basement fires. New construction plays a big factor here. Many spec-homes around the country leave the garages and basement ceilings unfinished. A fire that originates in the basement can quickly involve the floor system above, leading to early collapse of the first floor. This situation has been deadly for unsuspecting firefighters making entry on floor 1. It has happened enough that we need to say, "Enough!" and reexamine how we attack basement fires. It is paramount that the first-in engine officer immediately determines whether there is a basement fire or not and announces this game-changing information over the radio to incoming units. Basements are the most dangerous part of any structure due to limited and reduced access. It is a good idea to announce on the radio the existence of a basement or below-grade levels, whether they're involved in fire or not **FIGURE 1-13**. For example, "All units, be advised this structure has a basement." or "All units, be advised, this is a basement fire."

There are too many acronyms out there to help us remember the sequence of important information, but this one is worth keeping in mind. The UL/NIST recommendations on basement fires cannot be ignored. The identification of a basement is often overlooked. RECEO/VS is so well-engrained into the fire service that adding the "B" for *basement* or *below-grade* at the beginning of the acronym RECEO/VS is easy to remember and easy to say: **BRECEO/VS** (basement (or below-grade), rescue, exposures, confinement, extinguishment, overhaul/ventilation and salvage); linguistically, it fits. In addition, if there is a basement fire, this little memory tool puts it at the front of the list for

FIGURE 1-13 Announce on the radio the existence of a basement or any below-grade levels, whether they're involved in fire or not.

Courtesy of Raul Angulo.

initial fireground actions. Having a basement fire is a game-changer for initial fire attack actions. Company officers should be thinking BRECEO/VS as they approach the structure to perform a 360-degree size-up. Ruling out or confirming the existence of a basement can save firefighter lives and prevent injuries.

Decision-Making Model

Success or failure on the fireground depends on the first-in engine officer's ability to recognize the situation presented on arrival. The fire service is often compared to the military to illustrate a point. If you think of engine companies as the infantry, the ladder companies could be compared to the combat engineers. An infantry unit commander does not merely deploy soldiers into battle. There needs to be a size-up and reconnaissance of the situation; challenges and problems need to be identified, a plan developed, and then a decision made. There are many decision-making models that are very similar to the eight-step method. Learning to make systematic small decisions makes it easier when you have to make larger decisions. This also applies to the IMS system; if used on small incidents, the system prepares company officers to apply it with confidence on larger incidents. If you get used to applying a decision-making model on small problems, it is easier to apply to large problems, even those of an emergency nature. Before you can solve anything, you have to identify the problem. For fires, start by:

1. Identifying the problem.
2. Thinking of possible solutions.
3. Weighing the pros and cons. Are resources on scene to accomplish the solution?
4. Performing a risk-benefit analysis.
5. Selecting the best option.
6. Implementing.
7. Evaluating the results. If the desired results are accomplished, the problem is solved.
8. If the desired results are not accomplished, implement the next best option and repeat this process until the problem is solved.

Strategy, Tactics, Tasks, and Resources

Strategy is the action plan with broad general goals. There are basically four: offensive, defensive, transitional, or ignore (take no action). The tactics are the actions taken to support and accomplish the strategy. The tasks are the actual tools and procedures to perform the tactics, and the resources are the actual firefighters and apparatus who carry out the tactics.

Initial Role of the First-In Engine Officer

Typically, the first-in unit on an emergency incident is an engine company. Therefore, the engine officer needs to take command to become the initial IC. The initial IC should always keep in mind that the first and primary mission is to prevent additional loss of lives and property. The IC needs to make an initial 360-degree size-up and provide an arrival radio report describing the type of occupancy and visible fire conditions involving the structure. **Size-up** is an analysis of the current fire conditions and the process of identifying problems. Standard operating procedures (SOPs) and the pre-incident data help take away the guesswork for initial decision making and actions. The first-in company officer also must consider other factors such as the time of day, day of the week, and weather conditions; these are all components of size-up. An inadequate, improper size-up is one of the leading causes of firefighter line-of-duty deaths (LODDs). Size-up is an ongoing process, resulting in operational adjustments to match changes in the fire situation. Because size-up is so important to the success of engine company operations, an entire chapter (Chapter 8) of this text is dedicated to it. Engine company operations is more than just pulling hose and spraying water. Size-up, risk-benefit analysis, and risk management form the basis on which all engine company operations are carried out. Therefore, it is implicit in every sentence written in this book.

The full size-up continues with a quick initial action plan (QIAP). The QIAP is a written plan on a whiteboard or tactical worksheet to make first-alarm assignments on the fireground **FIGURE 1-14**. The IC starts the QIAP by performing a risk-benefit analysis and risk assessment, determines rescue and survivability profiles, and considers the rules of engagement before making initial assignments. Tactical worksheets list sequential step-by-step essential fireground considerations so that nothing is forgotten or omitted during the emergency. The use of such worksheets is a must. The size-up does not negate the need for that first-in company to apply water from the outside on any visible fire or to protect exposures.

The IC needs to comply with the two in/two out rule by establishing an initial standby team (IST), a dedicated crew of at least two firefighters in full PPE with SCBA located outside the immediately dangerous to life or health (IDLH) area. They are exclusively to the rescue of injured or trapped firefighters prior to the assignment of a rapid intervention team or crew (RIT, RIC) for immediate rapid deployment. The IC also needs to establish a functional firefighter accountability system from the start with the first company

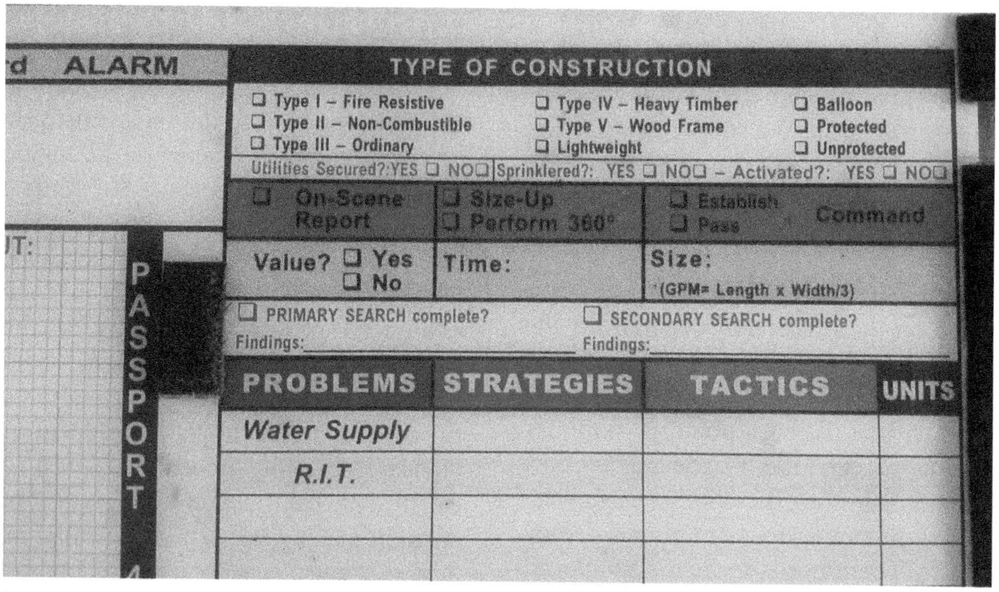

A

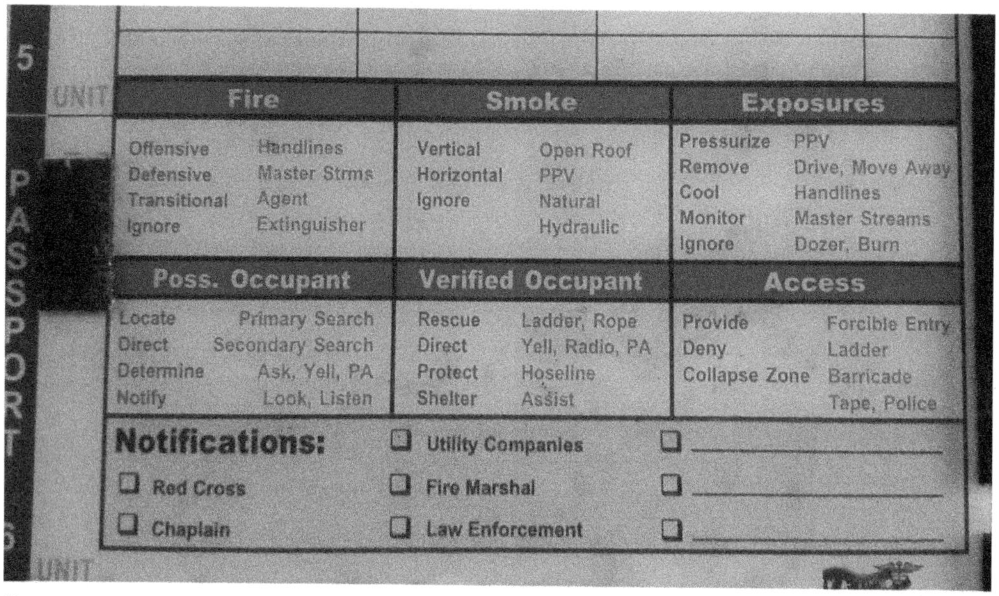

B

FIGURE 1-14 **A**. A QIAP. **B**. Tactical worksheets list sequential step-by-step essential fireground considerations so that nothing is forgotten or omitted during the emergency.

Courtesy of Raul Angulo.

assignment. Tactical worksheet boards can also incorporate a firefighter accountability system. The IC needs to know the name of the company, the names of the crew members, their assignment and location, and the time they entered the structure at any time during the incident in case a Mayday is declared **FIGURE 1-15**.

Value-Time-Size (VTS)

Today's fire service operations are driven by risk management. Value-time-size (VTS) refers to the risk-benefit assessment considerations by which to justify the selection of a strategy.

Value deals with life value and property value. If there is possible or certain life value at stake, the strategy would most likely be an offensive strategy to effect rescue. If there is a savable property value, the IC would be leaning toward an offensive strategy. If there is no life and no property value at stake, a defensive strategy would be the appropriate action to take.

Time refers to how much time is available to fight the fire based on general collapse times of the five building construction types and their fire-rated characteristics. (Note that *collapse times* are a bit of a misnomer. There are no accurate collapse times. These times are based on controlled tests to establish the

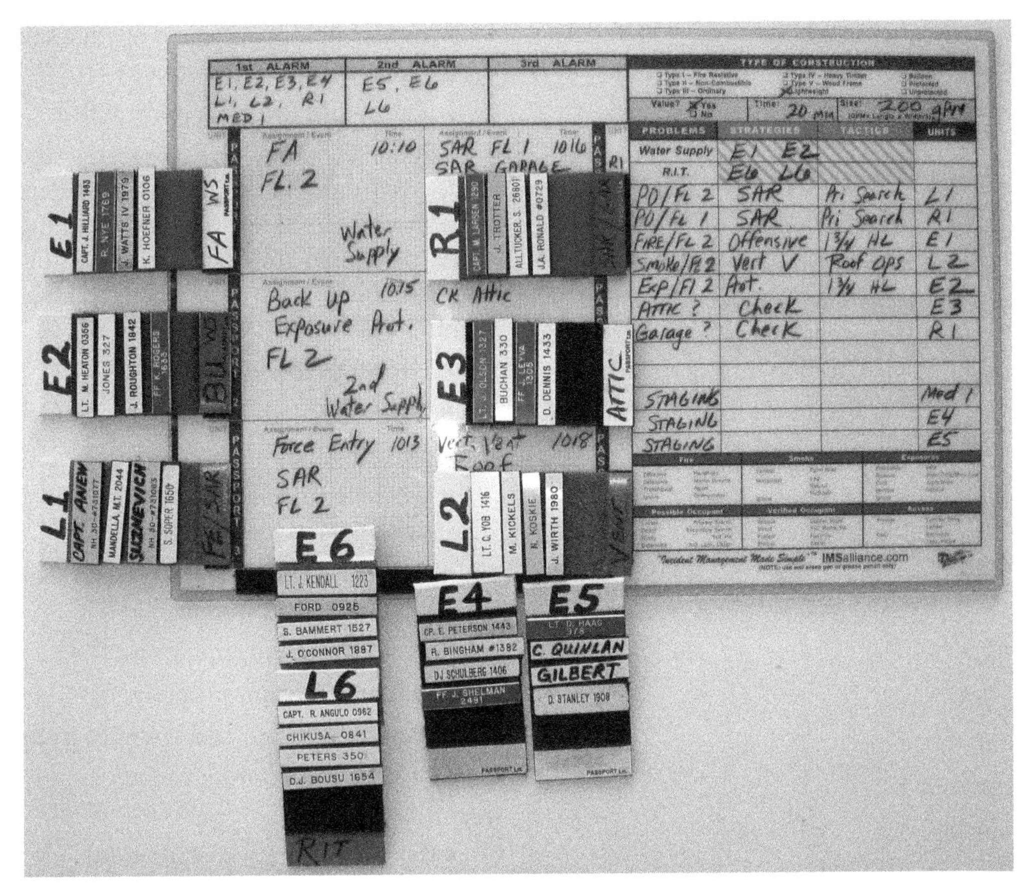

FIGURE 1-15 A tactical worksheet with a firefighter accountability status board.

Courtesy of Raul Angulo.

fire rating of classifications.) The point of discussing *time* in this text is for the company officer to realize that some construction types are more fire resistant than others; therefore, some windows for interior operations are wider than others for taking calculated risks. For example, interior fire attacks can be sustained for longer periods in a Type I fire-resistive high-rise building than they can be in a Type V wood-frame house.

Size refers to the size of the fire and how much water is needed to extinguish it. The required number of gallons is determined by applying the National Fire Academy Fire flow formula (L × W ÷ 3) or the Royer/Nelson formula (L × W × H ÷ 100) to the burning structure. The water delivery rate must exceed the number of BTUs generated for the fire to go out. If you have the apparatus and personnel to meet the required fire flow, you can justify an interior offensive attack pending other no-go criteria. Without sufficient resources, the strategy must remain defensive. After considering VTS, the company officer has the information to declare a strategy: offensive, defensive, or transitional. Chapters 3 and 8 cover VTS in more detail.

Identifying Problems

Before you can take any actions or implement a solution, you need to identify the problems. Fireground problems fall into two categories: top tier and second tier. The **top-tier problems** on the fireground require immediate intervention and typically fall into the following seven categories:

- Possible occupants
- Visible occupants
- Access problems
- Smoke
- Fire
- Exposures
- Possible hazardous materials

Once the problems that exist are identified and listed, the IC can finish formulating the QIAP and start to assign units with strategy and tactics for each problem, as well as determine how many resources are needed to accomplish the tasks.

Second-tier problems are less urgent but still require attention. They include salvage, securing utilities, overhaul, discovering household hazardous materials, water removal, and so on. Our goal should always be to solve the most important (top-tier) problems first.

Initial Fireground Assignments

There are only so many assignments that can be accomplished by a first-alarm unit. Whether the first-alarm unit is a large city fire department or a small rural volunteer fire department, the following initial tactical assignments need to be accomplished:

- Transitional exterior fire attack line (defensive to offensive)
- Water supply
- Forcible entry and gaining access
- Interior fire attack line
- Formation of a RIT
- Backup line
- Exposure line
- Search and rescue
- Ladder placement
- Ventilation

Engine Company Operations and Responsibilities

The primary responsibilities of the engine company are life safety and applying water to the fire. During an offensive strategy and interior operations, an engine company must be able to:

- Execute a transitional attack by applying water from the exterior to any visible flames coming from a door or window to knock down the fire and reduce interior temperatures.
- Place a hose line between the fire and occupants to protect them and provide a safe evacuation route from the building.
- Maintain the integrity of interior stairways for entrance and egress.
- Perform a primary search while advancing a hose line.
- Advance a hose line to the seat of the fire for extinguishment.
- Initiate hydraulic ventilation.
- Open walls and check for extension and hot spots.

During a defensive strategy and exterior operations, an engine company must be able to:

- Set up and operate master stream devices to protect exposures.
- Establish, maintain, and operate safely from outside the collapse zone.
- Extinguish the main body of fire.

Engine company fire apparatus and equipment have been designed to allow firefighters to function quickly and effectively. Through training and experience, personnel must acquire the knowledge, skills, and judgment in performing the following basic firefighting tasks:

- Carry out forcible entry.
- Perform search and rescue operations.
- Perform and function as a RIT.
- Establish an adequate water supply (hydrants, drafting, relay pumping).
- Connect various sizes of fire hose using couplings.
- Use nozzles and appliances.
- Use hose streams properly.
- Deploy, advance, and operate an attack line.
- Deploy and position backup lines and exposure lines.
- Perform hydraulic ventilation.
- Protect exposures.
- Establish a collapse zone.
- Use master stream appliances.
- Use foam properly.
- Operate power tools.
- Use ground extension ladders.
- Follow proper ventilation procedures for vertical, horizontal, positive-pressure ventilation (PPV), and hydraulic.
- Conserve property.
- Perform overhaul operations.
- Wear all required PPE.
- Use and troubleshoot SCBA.
- Follow proper procedures for calling a Mayday.
- Perform self-rescue firefighter survival techniques.
- Provide EMS.

A single engine company is not expected to perform every one of these operations at every fire, nor are the operations necessarily to be carried out in the order given here, but firefighters should be ready to carry out any one of these tasks, with particular attention focused on life safety, incident stabilization, extinguishment, and property conservation. Each strategy and tactic listed above will be covered in detail in the chapters that follow. All personnel should be trained to the applicable NFPA professional qualifications (NFPA 1001 and beyond).

In addition to the basic operations that are directly related to the control of the fire, engine company personnel must also be proficient in property conservation operations. Even though property conservation has no bearing on the spread of the fire, it is very important in terms of total fire loss, including damage by water. It has devastating impacts on families as well as the local economy.

Knowledge of the district from information gathered during building inspections and pre-incident planning is of great help in operating efficiently on the fireground. This information is also useful in developing and improving SOPs. Engine company personnel should have a good working knowledge of their first-alarm district, especially regarding routes of travel and water supply, including the location of hydrants, static water sources, and fire department connections (FDCs). They should know about the occupancies and building construction classifications within their districts, along with associated life hazards. They should be aware of buildings with unique or unusual features, new construction projects, as well as abandoned and dilapidated buildings in their area. They should also be aware of target hazards (arson threats), exposure hazards, or any hazard location particularly dangerous to firefighters during emergencies and firefighting operations. Perhaps it is impossible to know everything about one's company response area; however, unusual situations should be identified and carefully examined with anticipated special procedures developed ahead of time.

In the following chapters, we will cover all the roles and responsibilities of the engine company, the engine officer, and the firefighters. We'll start with the basics: fire science and behavior, building construction, apparatus and equipment, arrival of the first-in engine company, hose lays, apparatus positioning, and water supply. We'll explain the essentials of size-up, risk-benefit analysis, risk management, developing a QIAP, and making the initial fireground assignments with the first-alarm units. We will also demonstrate how to establish an accurate and functional firefighter accountability system. We'll cover the actions taken for residential fires, basement fires, commercial fires, and high-rise fires, including master streams and fire protection systems and how they come into play in attacking the fire. We will cover all the tasks and assignments an engine company may be charged with—from a basic incipient room fire to a large "surround and drown" defensive fire—and discuss bringing these incidents to an end, from fire attack to property conservation.

The fire service is a paramilitary organization, and comparisons to the military are often made, but unlike the military, there are no "acceptable losses." The goals are (1) to have a company officer or acting officer run a basic fire safely with expert knowledge and confidence until command is transferred to a chief officer and (2) to have firefighters function in any capacity required under the responsibilities of an engine company.

After-Action REVIEW

IN SUMMARY

- The engine company is the basic unit of a fire department.
- Many fire departments consist of only engine companies.
- The primary responsibility of the engine company is to respond to and extinguish structure fires.
- EMS has become the primary emergency response service provided to cities and communities.
- EMS accounts for about 80% of emergency responses. Fires account for only 20%.
- Since the advent of SCBA, no other electronic device has changed the fire service more than the thermal-imaging camera.
- Today's household furnishings (fuel load or fire load) are primarily made of petroleum-based plastics and synthetics known as modern fuels.
- Modern-fuel fires burn faster and hotter, and they generate higher heat-release rates that in turn generate black toxic smoke with deadly products of combustion.
- Scientific research confirms a direct correlation between the exposure to smoke and the increasing rates of different types of cancer in firefighters compared to the general public. In addition to lung, prostate, and brain cancer, esophageal cancer is one of the more prominent cancers affecting firefighters.

- Modern fuels generate and put out a tremendous amount of energy, almost double the BTU potential of legacy fuels like wood, paper, and natural textile fibers. Compared to legacy fuels, modern fuels can produce higher temperatures in excess of 1,000°F (538°C), which leads to rapid flashover potential.
- More flashovers occur today than in previous decades because flashover temperatures have caught up to fire department on-scene arrival times.
- The UL/NIST studies include ongoing research and experiments with the FDNY and Chicago Fire Department on fire behavior, water application, fire suppression, ventilation, search and rescue, and survivability profiles as a result of changes in modern building construction and the modern fuels of home interior furnishings.
- The four major incident priorities on the fireground are life safety, incident stabilization, extinguishment, and property conservation.
- Adding the "B" for *basement* or *below-grade* at the beginning of the RECEO/VS acronym to create the BRECEO/VS acronym is an easy way to remember to rule out or confirm the existence of a basement. If there is a basement fire, it puts it at the front of the list for initial fireground actions.
- A QIAP is a written plan on a whiteboard or tactical worksheet to make first-alarm assignments on the fireground.
- Tactical worksheets list sequential, step-by-step essential fireground considerations so that nothing is forgotten or omitted during the emergency, and such worksheets should be used.
- The IC needs to comply with the two in/two out rule by establishing an IST, a dedicated crew of at least two firefighters in full PPE with SCBA located outside the IDLH area exclusively assigned to rescue injured or trapped firefighters prior to the assignment of a RIT for immediate rapid deployment. The IC also needs to establish a functional firefighter accountability system from the start with the first company assignment.
- Today's fire service operations are driven by risk management. VTS refers to risk-benefit assessment considerations made during the size-up to justify the selection of a strategy.
- Fireground problems fall into two categories: top tier and second tier. Top-tier problems on the fireground require immediate intervention. Second-tier problems are less urgent but still require attention.

KEY TERMS

BRECEO/VS A memory aid to help company officers confirm or rule out the existence of a basement or below-grade occupancy; it stands for basement (or below-grade), rescue, exposures, confinement, extinguishment, overhaul/ventilation and salvage. The abbreviation places the letter B before the tactical objectives acronym, RECEO/VS.

rehabilitation An intervention designed to mitigate against the physical, physiological, and emotional stress of firefighting in order to sustain a firefighter's energy, improve performance, and decrease the likelihood of on-scene injury or death. (NFPA 1521)

second-tier problem A problem that is less urgent than a first-tier problem but still requires attention, for example, salvage, securing utilities, overhaul, discovering household hazardous materials, and water removal.

size-up The process of gathering and analyzing information to help fire officers make decisions regarding the deployment of resources and the implementation of tactics. (NFPA 1410)

thermal-imaging camera (TIC) A type of thermographic camera used in firefighting and search and rescue. By rendering infrared radiation as visible light, the camera allows firefighters to see areas of heat through smoke, darkness, or heat-permeable surfaces.

top-tier problem A problem on the fireground that requires immediate intervention. It may fall into one of the following seven categories: possible occupants, visible occupants, access problems, smoke, fire, possible presence of hazardous materials, and exposures.

universal extinguishing foam (UEF) A firefighting extinguishing foam that works on all classes of fires.

value-time-size (VTS) The three risk-benefit assessment considerations made during size-up to justify the selection of a strategy.

ventilation-limited fire A fire in which the heat release rate and fire growth is controlled by the amount of air available to the fire. (NFPA 1403)

REFERENCES

National Fire Protection Association. NFPA 1901, *Standard for Automotive Fire Apparatus*, 2016.

National Fire Protection Association. NFPA 1801, *Standard on Thermal Imagers for the Fire Service*, 2018.

U.S. Fire Department Profile 2017. Data compiled by Ben Evarts and Gary Stein. National Fire Protection Association, March 2019.

CHAPTER 2

Fire Dynamics

LEARNING OBJECTIVES

- Comprehend the fundamentals of fire science and modern fire behavior.
- Explain the difference between organic and inorganic materials.
- Explain the difference between endothermic and exothermic reactions.
- Describe the oxidation and combustion processes.
- Explain how equilibrium plays a role in understanding the basics of fire behavior.
- Describe what is happening during a BLEVE and how it can be prevented.
- Explain the need to consider the new six stages of the burning process.
- List the modes of heat transfer and how they contribute to fire spread.
- Describe how a thermal column or thermal plume develops.
- Compare the two types of flow paths.
- Explain the behavior of ventilation-limited fires.
- Compare legacy fuels versus modern fuels and their effects on fire behavior.
- Analyze how orientation of fuel, fuel configuration, and surface-to-mass ratio affects fire spread.
- Explain the relevance of understanding BTUs.
- Explain how rapid fire growth develops.
- List the classes of fire.
- Explain the four conceptual factors in reading smoke.
- Describe how reading smoke helps determine survivability profile.

Introduction

Fire dynamics is the study of the characteristics of fire and the burning process, which includes how fires start, develop, and spread. Under its umbrella, it also includes the detailed study of fire science, fire chemistry, physics, heat transfer, fuels, materials, hydraulics, building construction, and how all these combined forces interact to influence affect fire behavior within a structure. We have two enemies: the fire, and the building that is on fire. As in military battles, tacticians need to gather and analyze information about the enemy. They have to understand the needs and capabilities of the enemy and be able to predict what the enemy will do. The decision may be to attack the enemy or cut off its supply line. In the same way, fire officers need to learn and understand how fire begins, develops, spreads, and is effectively extinguished.

In this chapter, we'll study the basics of fire science and fire behavior. The recent research from Underwriters Laboratories (UL) and the National Institute of Standards and Technology (NIST) is covered in Chapter 4. To understand the UL/NIST data fully, your commitment to understanding fire behavior has to go far beyond being content with the simple fallback position of "just put the wet stuff on the red stuff." If you can grasp fire behavior science and apply it, there isn't any fire that cannot be extinguished.

We'll confine our focus to what we need to know about fire behavior from the incipient stage; through the growth stage; to the steady free-burning, fully developed stage; and finally to the decay stage within a structure. It is extremely important for your professional development and to the safety and survival of your firefighters that you understand the dynamic behavior of fire inside a building during the offensive interior attacks and, to a lesser degree, defensive attacks. It's not as critical during a defensive fire because you should be operating from a safe position, away from the burning building and outside the collapse zone; however, a defensive fire is still dangerous and without the proper tactics and control measures, buildings could collapse, and building exposures subjected to radiant heat can lead to a conflagration.

The Fire Tetrahedron

Every fire needs three components: fuel, heat, and oxygen (air) in proper proportions. There must be combustible fuel present. Fuels are materials that store energy. There must be oxygen in sufficient quantity, and there must be a heat source for ignition.

Researchers in the field of fire science now recognize there is a fourth essential element. In order to have rapid self-sustained combustion and increasing fire development, there must be a chemical chain reaction to perpetuate the flaming mode of combustion. This spontaneous self-sustained chemical chain reaction keeps the fire burning and growing. The energy released in the form of heat and light has been stored in the fuel before it started to burn. The chemical chain reaction is the fourth element. The illustrated geometric design puts the chemical chain reaction element at the base, forming a pyramid which is now known as the fire tetrahedron **FIGURE 2-1**.

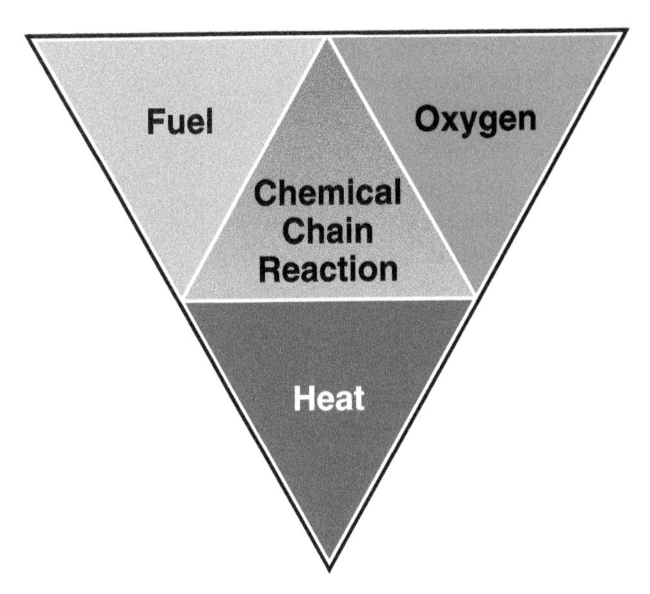

FIGURE 2-1 The fire tetrahedron.

In addition to water application, special extinguishing agents like dry chemicals and foam can also interrupt the chemical chain reaction, causing rapid fire extinguishment. Water is the primary extinguishing agent used by fire departments, and we regularly observe its cooling effect on burning fuels when we see the conversion to steam. The steam absorbs the heat leg of the fire tetrahedron, and the water cools the fuels below their ignition temperature. This makes it easier to understand at least two elements of the fire tetrahedron, but it's a little more complicated than that. Studying the new research and recommendations from UL and NIST will help you understand how to use the oxygen (air) leg of the fire tetrahedron to control and extinguish fires.

What Is Fire?

We all know what fire is when we see it. When something is burning with visible flames, we say, "It's on fire." Fire is also referred to as combustion and as the process of rapid oxidation. At the chemical level, fire and combustion do not mean the same thing: The terms are very closely related and are often used interchangeably within the fire service, but they are not exactly synonymous. In fire, there is a rapid, persistent, self-sustaining, chemical reaction that requires heat, fuel, and oxygen. The fire generates more heat (radiation), emits light (flames), and releases energy as a result of that reaction. Its colors range from yellow at its coolest, to red, orange, blue, and then white at its hottest. Fire is not a solid or a liquid; it is a by-product of the combustible vapors that are released from solid and liquid fuels during this chemical process. Wood is a solid fuel, gasoline is a liquid fuel, and propane is a gaseous fuel. They all burn, but solid and liquid fuels have to vaporize into a gaseous fuel state in order to burn.

In combustion, the heat energy that is released from the reaction is reabsorbed and accumulates within the process pushing the chemical reaction to continue. Increased energy is released, causing heat to be transferred to surrounding solid or liquid fuels to maintain both their vaporization into a gaseous fuel and their fuel temperature, thus ensuring that the chemical chain reaction continues. As more heat, fuel, and oxygen are available, the chemical reaction accelerates, resulting in increased fire growth and intensity. If nothing interferes with this process, the fuel and the oxygen will continue burning, and the fire will continue growing with greater access to fuel, oxygen, and heat. The growth will continue to accelerate and intensify until the supply of fuel or oxygen is exhausted.

It is important in combating fires that we understand the basic concepts of fire behavior and science as it relates to:

- Matter and energy
- The chemical chain reaction of fire
- How the fire progresses
- The products of combustion
- How the fire is extinguished

Understanding fire behavior is the foundation for all firefighting principles, strategies, and tactics.

Fire Chemistry, Physics, and Matter

Matter is every physical object that you see and everything that exists in the universe. Matter is anything that occupies space; has mass, size, weight, or volume; and can be perceived by one or more senses. In science, matter is made up of atoms and molecules that exists in three states: solids, liquids, and gases **FIGURE 2-2**. Matter exists in many forms. It can be configured, changed, and destroyed, but it doesn't disappear. The classic example is a piece of iron that is left exposed to the outdoor elements. Over the years, the iron slowly oxidizes and deteriorates. The oxidation is commonly seen as rust. Even though it may look like it rusted away and disappeared, it merely changed its form.

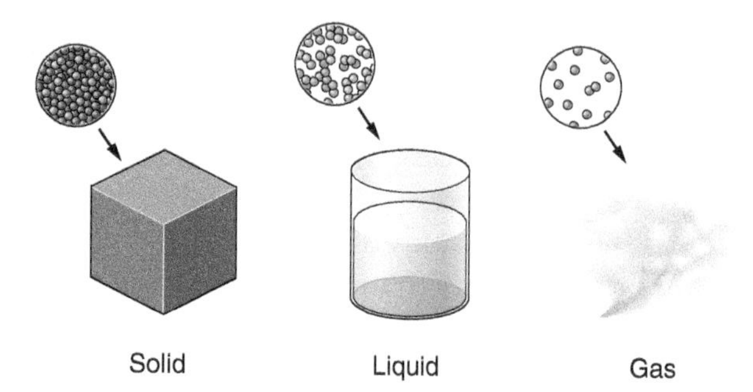

FIGURE 2-2 The state of matter or the physical state of a substance can be classified as a solid, liquid, or gas.

Another popular example is a large wood-frame house that can be dismantled piece by piece. Building components can be separated into piles of roofing materials, plumbing, electrical and heating materials, glass, window and door frames, fixtures, furniture, carpeting, nails, screws, and stacks of lumber. Still the same house, but in a different form. Each component can be broken down even further, all the way to the molecular level. The wood that framed this house is made up of material called cellulose. Because the wood was once a tree—a living organism—it is made up of millions of cells, which are made up of compounds, which are made up of molecules, which in turn are composed of atoms of two or more elements. Because these substances originated from a living organism, they are referred to as **organic** compounds. Organic compounds usually contain elements of carbon, hydrogen, and oxygen.

Hydrocarbons are also organic compounds. Although they have never been living organisms in and of themselves, their chemical makeup comes from organisms that were once alive. Plastics were never alive, but they're made from oil (hydrocarbons). Crude oil fields are subterranean pools of once-living, prehistoric life forms that have been dead and decaying for millions of years. Oil companies drill for this once-living material and through other chemical processes, create a nonliving material. The end results are plastics, petroleum, diesel, and gasoline products.

Matter in the universe that was never alive or part of a living organism is called **inorganic**. Minerals like iron, quartz, and granite are examples of inorganic substances and do not contribute to the combustion process. With few exceptions, the only difference between organic and inorganic materials is that *organic materials will burn.*

The smallest unit of matter is an atom. Atoms are made up of protons, neutrons, and electrons. The nucleus of an atom contains the protons and neutrons. The

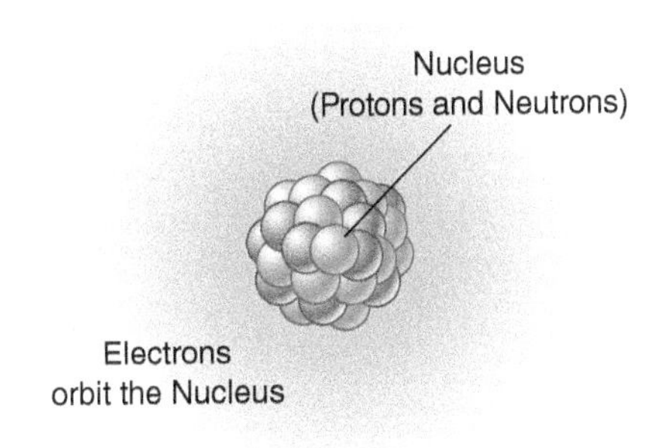

FIGURE 2-3 The smallest unit of matter is an atom, which is made up of protons, neutrons, and electrons.

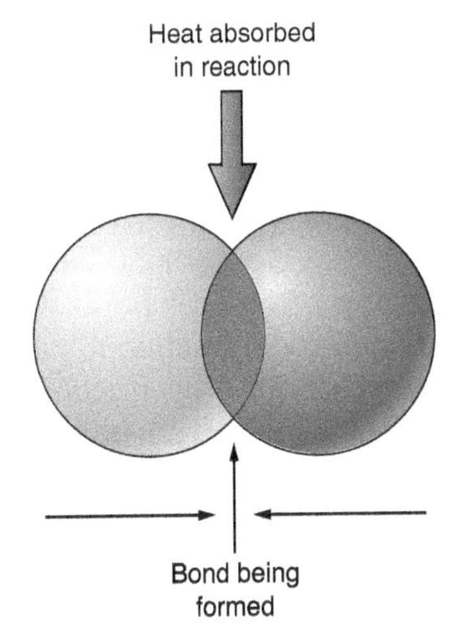

FIGURE 2-4 Endothermic reaction.

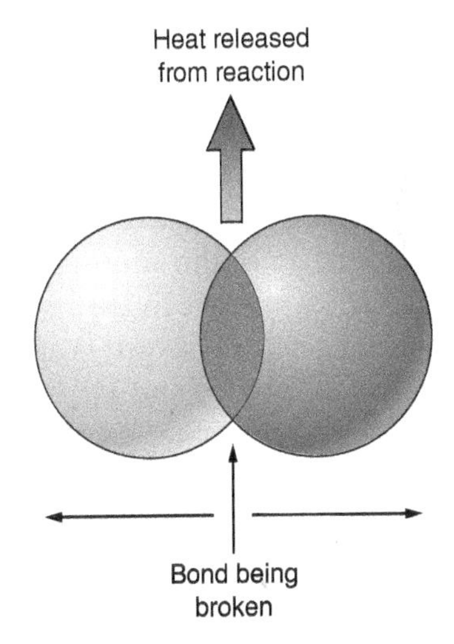

FIGURE 2-5 Exothermic reaction.

electrons orbit the nucleus FIGURE 2-3. Like everything in the universe, atoms want to move to a more stable state where the nucleus is balanced and there is not an excess of energy. If the forces between the neutrons and the protons in the nucleus are unbalanced or the atom is lacking or has an excess of electrons, the atom is unstable. Atoms become chemically stable by gaining, losing, or sharing electrons with other atoms to fill their outermost electron shell. Once the atoms are stabilized, they can retain their formations indefinitely.

Atoms join to form molecules. The most common example is two oxygen atoms joining to form an oxygen molecule with the chemical formula written as O_2. When atoms of one element combine chemically with atoms of another element, they produce a compound that is made up of molecules. For example, when an atom of oxygen (O) combines with two atoms of hydrogen (H_2), the resulting compound is water (H_2O). **Note:** Almost all fuels consist of hydrogen (H) and carbon (C); hence they are called hydrocarbons.

Atoms bond together to form molecules. Chemical compounds are made up of molecules that are joined or separated by a bonding action that uses a form of electricity as energy. The formation or separation of these molecules and atoms provides the essence of what causes combustion and oxidation. The "atomic glue" that holds these molecules together is called a bond. Fire, or combustion, has its origin in the bonds formed between atoms and molecules. When atoms and molecules join together, a certain amount of heat is absorbed into that bond, which creates and holds together the newly formed element. This process is called an endothermic reaction, meaning that heat is absorbed when the bond is created FIGURE 2-4. When the bond is forcefully broken, energy is released in the form of heat and light, called an exothermic reaction FIGURE 2-5. If the release is rapid enough to sustain a continuous reaction, we can see the energy as fire and feel the energy as heat.

As stated previously, atoms and molecules that are stable can retain their form indefinitely, so something needs to be introduced in order to change it. An oxidizer acts as the catalyst to break down what otherwise would remain a stable molecule. The most common oxidizer available is oxygen. An oxidizer possesses a chemical property that can pull apart a molecule and break up the bond that previously existed. The release of energy in the form of light and heat then causes the chemical compound to break apart other compounds, releasing additional light and heat. If this reaction is slow, a gradual deterioration of the material occurs over time almost without detection. Going back to our example of iron undergoing deterioration, the oxygen in the atmosphere combines with the properties of iron and gradually pulls apart the bonds that are keeping the iron atoms together. Although not visible to the naked eye, heat and light are being emitted during this process. If heat-sensitive measuring equipment were attached to the iron, it would most likely be able to measure the heat and light. (A light coat of oil applied on metal prevents rusting because air is unable to come into direct contact with the surface of the metal, so it cannot react and oxidize the metal.) The oxidation process is much more evident during combustion. The presence of oxygen, in combination with heat from a previous bond-break (an exothermic reaction), causes a rapid chain reaction to take place. Because of the speed at which this reaction takes place, the light is visible, and the heat can be perceived by the human senses.

This energy intensifies, which causes additional molecular bonds to break apart, and so on, until a rapid, spontaneous, self-sustaining chain reaction occurs in a process called **combustion**. It is also the description and definition of what happens within the fire tetrahedron. The chemical reaction that occurs during combustion is called **oxidation**. The amount of oxygen present affects the process and speed of oxidation as well as the process and speed of combustion. It can be very slow, as in the case of rusting, or it can occur very fast, as in the case of an explosion. The concentration of oxygen in the atmosphere is important to the combustion process. In Earth's atmosphere, the normal concentration of oxygen is approximately 21%. Nitrogen makes up 78% and 1% is water vapor, neon, carbon dioxide, and other gases. Combustion occurs in oxygen concentrations between 14% and 100%. In atmospheres where the oxygen concentration is below 14%, combustion cannot occur. With diminished concentrations of oxygen, the combustion process is slowed and decelerated. With abundant concentrations of oxygen, the chemical chain reaction is fast and accelerated.

Sources of Heat Energy

Heat is the most common source of energy on Earth. Heat is energy transferred from one body to another when the temperatures of those bodies are different. Heat is the energy source that comes from the sun and powers the entire universe. It cannot be created or destroyed. It's a phenomenon that merely changes from one form to another. Heat forges atoms and molecules together to form compounds (endothermic reaction), and heat is released when those bonds are separated (exothermic reaction). Heat is a form of energy that consists of matter in motion. When a body of matter is heated, the movement and speed of the molecules increase, and the temperature rises.

Temperature is an indicator and measurement (Fahrenheit or Celsius) of that heat. There are five basic sources of heat energy: chemical, electrical, mechanical, solar, and nuclear. The first four are common everyday sources of heat. As we look for ways to change our reliance on fossil fuels, oil, gas, and coal, nuclear power may become a viable alternative for a cleaner and more reliable source of domestic energy in the near future. They're all initiators of fire, so it is important to be aware of them.

Chemical

The most common heat source encountered is the chemical chain reaction of the fire tetrahedron, where the heat release is a by-product of the combustion process. As previously stated, only organic materials that include hydrocarbons and petroleum chemicals burn. As the molecular bonds break apart, heat is released, and the burning material is decomposed by a process call **pyrolysis**. Decomposition, or pyrolysis, is the breaking down of the molecular compounds **FIGURE 2-6**.

Mechanical

A mechanical heat source is caused by friction, which is the breakdown of two materials that are compressed together or rapidly rubbed against each other. Friction can cause heat temperatures sufficient to ignite surrounding combustible materials. The heat buildup from friction is often the cause of fires in machinery. Water or other coolants are used in the mechanical process to transfer or disburse heat buildup to prevent a fire from occurring. Understanding how mechanical fires can occur is important in recognizing and identifying hazards during fire prevention inspections. Knowing how the heat source is being generated and how it can be eliminated is also critical in formulating the tactics for extinguishment.

Electrical

Electricity is a heat source that has the ability to generate high temperatures capable of igniting any combustible material near the heated area. Electric energy

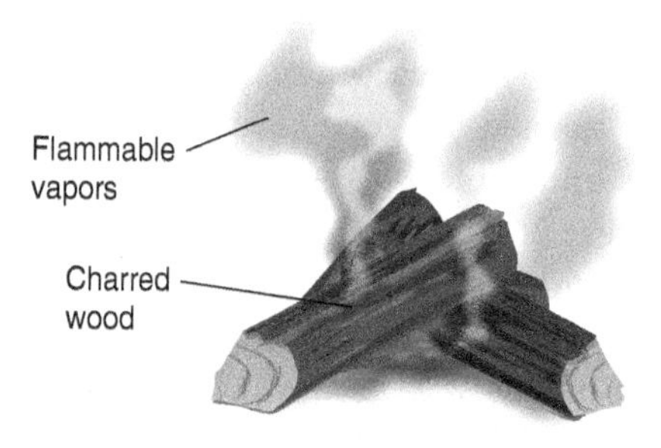

FIGURE 2-6 When wood is sufficiently heated, pyrolysis is evident as the wood decomposes and breaks down into hemicellulose, cellulose, and lignin, and then into vapors and char.

is probably the most familiar to the general public because it is the primary power source for our homes. We use it to light and heat our homes, run our appliances, and cook our food. An electric heat source can occur in several ways: current flowing through a resistance, overcurrent or overload, arching, sparking, static charge, and lightning. An electric current can generate heat. Simply put, electricity is the flow of electrons from a source with an excess of electrons to an area that is lacking in electrons. The source where there is an excess of electrons is said to be negatively charged. Conversely, the area where electrons are lacking is said to be positively charged. As in many examples in physics, it is the nature of electricity to equalize and balance the charges. An electrical conductor allows the flow of electrons from one place to another.

When electrons are traveling through a conductor, the atoms that make up the conductor trade off their electrons from one atom to another. During this process, billions of electrons are in violent motion and collide with each other. When the collision sends the electrons astray, they collide with molecules and cause them to break apart. By breaking apart the bonds that hold the molecules together, the bond releases heat (exothermic reaction).

The amount of heat released depends on the density of the conductor and the insulation surrounding it. As a general rule, the heavier the material, the more efficient it is as a conductor. That's why metal is a good conductor and cotton isn't. An efficient conductor has less heat because more of the electrons are being transferred downline with fewer collisions.

Static Electricity

Static electricity occurs when two different materials are rubbed together, scraped together, or are suddenly joined or separated. This action creates energy in the form of mechanical heat, but on a molecular level; it also creates an unbalanced electrical charge due to the sudden separation of the two surfaces. Electrons from one surface are taken by the other, and when enough of them collect, they attempt to equalize by jumping across the space in the form of a static charge. Though the static charge is small, it can be seen with the naked eye. Nevertheless, it is very powerful and can reach temperatures in excess of 2,000°F (1,093°C). Although it dissipates quickly before it can heat any surrounding combustibles, in an atmosphere of flammable gas, this small charge can ignite the gas with explosive force. Lightning is static electricity on a grand scale that can generate heat energy in excess of 60,000°F (33,300°C).

An interesting note: Static electricity is created when flammable liquids are transferred from one storage container to another and has been the ignition source for many flammable liquid fires. This is why fuel tankers and aircraft have to be properly grounded and bonded by a wire during fueling and fuel-transfer operations. Extra bonding is not necessary when you fill your car at a gas station because gasoline has special additives that make the fuel act as a ground and the metal nozzle in contact with the fill pipe creates an effective electrical bond.

In the case of microwave ovens (dielectric heat), waves of electric energy subject the food to exposure. Fired microwaves alternate in every direction at super-high speed, which results in molecular bombardment, kind of like bombarding the food with many little lightning bolts. Again, the collision of electrons breaks up the molecular bonds, releasing energy that cooks the food.

Solar Energy

This is the most powerful and most common source of heat energy. Electromagnetic radiation is constantly emitted from the sun and is fairly evenly distributed over the Earth to heat it. The sun does not randomly ignite combustible material, but it can burn surfaces like the human skin. It can also preheat natural fuels, as in the case of wildland fires. When solar energy is directed through a lens such as a magnifying glass, the focused beam of light can easily ignite combustible material.

Nuclear Energy

Nuclear energy is the least understood by the general public. The fact that it is used in the atomic bomb frightens people, but it is a safe source of energy if properly controlled. Many US military ships and submarines are powered by nuclear energy. Nuclear energy is created by splitting the nucleus of an atom into two smaller nuclei (nuclear fission) or by combining two small nuclei into a larger nucleus (nuclear fusion). All forms of atomic energy are first converted to heat. Radioactive materials are very unstable and are constantly breaking down, trying to achieve a stable molecular composition. In this process, atomic particles are randomly flying in every direction. When encased in a heavy lead shield called the core, the radioactive material remains safely confined. When energy is needed, the radioactive material, usually in the form of rods, is pulled out from the protective core. A controlled amount of energy is permitted to be transferred into the surrounding water. The heat turns the water into steam, which in turn powers the turbines that generate electricity. When the need for steam is complete, the rods are inserted back into the protective core, and the nuclear bombardment is again contained.

As firefighters, little can be done to safely fight a fire caused by a nuclear heat source. In the Chernobyl (Ukraine) incident of 1986, a meltdown of the core occurred when the water level was too low to keep up with the emitted heat from the rods. Once exposed without a coolant, the rods had no controlling mechanisms and a runaway reaction occurred. This event is above the everyday experience and norm of average firefighters. On-site skilled hazmat technicians must handle this kind of incident. The real danger in a nuclear fuel fire is the exposure to radiation. Without proper specialized protection, firefighters can sustain long-term illness or even a fatal injury due to extreme radiation exposure.

In all the sources of heat energy cited above, heat energy is generally the necessary ignition factor that ignites the surrounding combustibles, whether they are solids, liquids, or gases. Heat causes pyrolysis, the decomposition and chemical breakdown of compounds, causing the vaporization of solid and liquid fuels into ignitable vapors or gases for a fire. Heat causes the continuous production and ignition of fuel vapors so that combustion can continue. With one molecule breaking apart from the bombardment of electrons, heat is released, which causes surrounding molecules to break apart and release more heat. The chemical chain reaction continues until all the fuel is exhausted. Once the heat source is removed, the fire should start to go out. Otherwise the fire is extinguished by removing the oxygen, the fuel, or interrupting the chemical chain reaction.

Vapor Pressure and Vapor Density

Vapor Pressure

The terms *vapor pressure* and *vapor density* are sometimes confused. Pressure is the application of continuous force by one object against another object that it is touching. **Vapor pressure** is the measurable amount of pressure being exerted by a liquid as it converts to a gas and exerts pressure against the sides of a confined closed container. Density is the weight of a material and depends on the number of molecules and atoms that occupy a given volume. The more molecules in a given volume, the denser is the material. In addition, the heavier each individual molecule is, the heavier is the material. The more molecules there are, the greater the frequency of molecular collisions and hence the release of heat. The denser a material, the greater its ability to conduct heat.

A **liquid** is a collection of molecules that occupy a space in a fluid state and tend not to separate. A liquid does not have a fixed shape but conforms to the shape of its container; it retains a nearly constant volume and density independent of pressure. However, molecules are in constant motion, circulating and colliding against the sides of the container. When the molecules circulate beyond the surface of the liquid, they escape into the space within the container. Some molecules are reabsorbed into the liquid, while others remain in a gaseous state within the container. This process is called **diffusion**. If the gases are lighter than air and aren't confined inside the container, all the liquid will eventually evaporate. If the gases are heavier than air, the molecules remain suspended in the space between the surface of the liquid and the top of the container, even in an open container. At the point where there are as many molecules being liberated from the liquid as being reabsorbed by the liquid, it is referred to as being in a state of **equilibrium**.

Temperature and pressure affect the rate of evaporation. As a liquid is heated, the molecules become more active and the speed of evaporation increases **FIGURE 2-7(A)**. If the surrounding atmospheric pressure is greater than normal, it can act as a lid, reducing the liberation of molecules from the surface of the liquid. Reducing the temperature of the liquid also slows molecular activity within the liquid, and reduces the pressure and the number of molecules being liberated into the air space. This is why cooling a container with water is an important tactic to remember. Cooling the container inhibits the liquid's ability to diffuse.

Vapor Density

Vapor density is the weight of a gas compared to normal air. The normal atmospheric pressure of air at sea level is 14.7 psi and is given the designated number of 1 when measured at 32°F (0°C). Gases that weigh less than the same volume of air are lighter than air and tend to rise. Their vapor density is designated as a number less than 1. If the gas is heavier than air, the number designation is greater than 1, and these vapors tend to sink **FIGURE 2-7(B) TABLE 2-1**.

Two flammable gases that we encounter frequently in the fire service are methane (natural gas) and propane. They are the primary heating and cooking fuels in modern society. Although the end use is the same for these two flammable gases, the difference in their

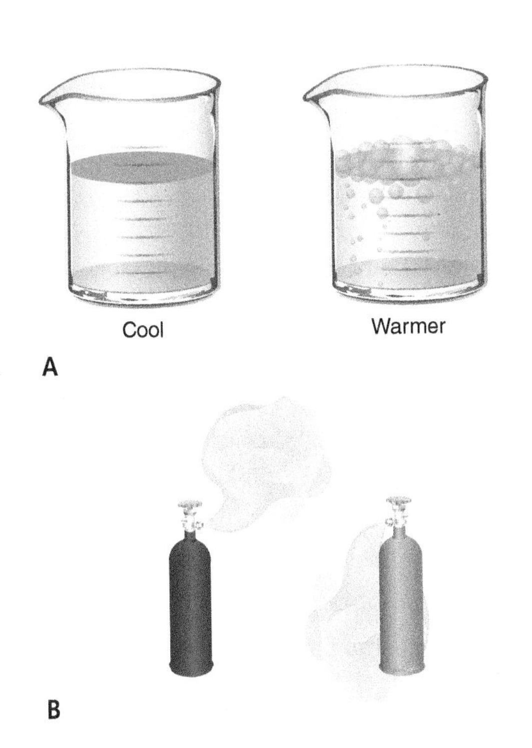

FIGURE 2-7 A. As a liquid is heated, the molecules become more active, and the speed of evaporation increases B. The red cylinder contains a gas with a vapor density less than 1. The yellow cylinder contains a gas with a vapor density greater than 1.

TABLE 2-1 Vapor Density of Common Gases

Gaseous Substance	Vapor Density
Gasoline	>3.0
Ethanol	1.6
Methane	0.55
Propane	1.55

vapor density can create flammable atmospheric hazards if an accidental leak occurs from a container or a product delivery system.

Methane has a vapor density of 0.6, making it lighter than normal air (1), so a product released into the open air will rise, mix, and harmlessly dissipate in the atmospheric air. With an accidental release inside a structure, the gas will rise and accumulate at the highest point, then start to bank down. This is why it is so important to ventilate high and eliminate sources of ignition with natural gas leaks inside a structure.

On the other hand, propane has a vapor density of 1.6, making it heavier than normal air (1). This means if the product is leaking, the vapors sink and collect in low areas such as structural depressions, basements, cellars, sewers, and ravines. This is why it is critical for firefighters to understand the importance of vapor pressure and vapor density. If a firefighter is using a gas detector at a propane leak incident and doesn't realize the readings must first be taken in low areas close to the source, then he may fail to properly identify flammable gas hazards, thus endangering the rest of the crew. A smart firefighter knows that propane vapors tend to collect in low areas and would start taking readings there as well as eliminate low-area ignition sources.

Boiling Liquid Expanding Vapor Explosion (BLEVE)

If a liquid is flammable and stored under pressure, like propane, the bottom half of the container is liquid propane while the top of the container is vaporized propane FIGURE 2-8. Under normal circumstances where atmospheric temperatures fluctuate, a pressure relief valve allows the expanding product vapor to escape into the atmosphere. If this pressurized container is subjected to the heat from a fire, the rapid pressure buildup can exceed the relief valve's ability to vent the flammable vapors safely. The expanding liquid is prevented from evaporating, and if the heating continues, the temperature inside the container reaches a level above the boiling point of the liquid. Eventually the vapor pressure becomes so great that the container will fail, releasing all the propane in a massive explosion. The liquid propane instantly vaporizes into a gigantic fireball, releasing tremendous radiant heat that adds a fire-extension element to the problem. This is called a BLEVE. If the liquid is a hazardous material beyond being flammable, yet another set of incident priorities must be addressed.

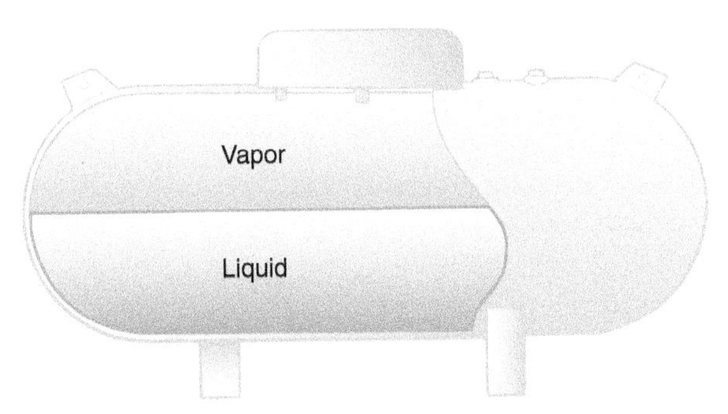

FIGURE 2-8 A propane tank contains both liquid and vapor. Therefore, it has vapor pressure and vapor density.

A **boiling liquid expanding vapor explosion (BLEVE)** occurs when a vessel holding liquid ruptures as a result of pressure being exerted on or against the interior walls when the liquid it is holding begins to boil. Eventually, the increasing pressure exceeds the container's ability to hold it. This usually occurs when intense heat is applied to a closed container or vessel. When the container fails, the ensuing release of vapor and liquid is very violent, resulting in an explosive release of tremendous forces. Even if the liquid in the vessel is not flammable, the container rupture can still be violent, resulting in a force that can send container fragments and shrapnel great distances (½ mile [.8 k]) accompanied by a massive shock wave **FIGURE 2-9**.

FIGURE 2-9 A boiling liquid expanding vapor explosion (BLEVE).

The strategy for preventing a BLEVE is to cool the top of the tank that contains all the pressurized vapor with copious amounts of water, preferably from unmanned monitors or deluge appliances **FIGURE 2-10**. This tactic prevents the fuel from building up unsustainable vapor pressures within the tank and thus causing catastrophic failure of the container. If a reliable water source cannot be maintained or if the risk factor is too great, another tactic would be to evacuate the area and let the product burn until the container fails and a BLEVE occurs or the product is consumed. (Appropriate evacuation distances are found in the current edition of the Department of Transportation's *Emergency Response Guidebook*.)

BLEVEs don't always have to be spectacular fiery events. In fact, every person in the United States has witnessed thousands of BLEVEs since childhood. The most common example of a BLEVE is the popping of popcorn. The liquid inside the hard kernel shell heats up, boils, and exerts increasing pressure against the interior wall of the kernel until the shell fails. The result is a cooked kernel that escaped its container while the pressure inside escaped and equalized.

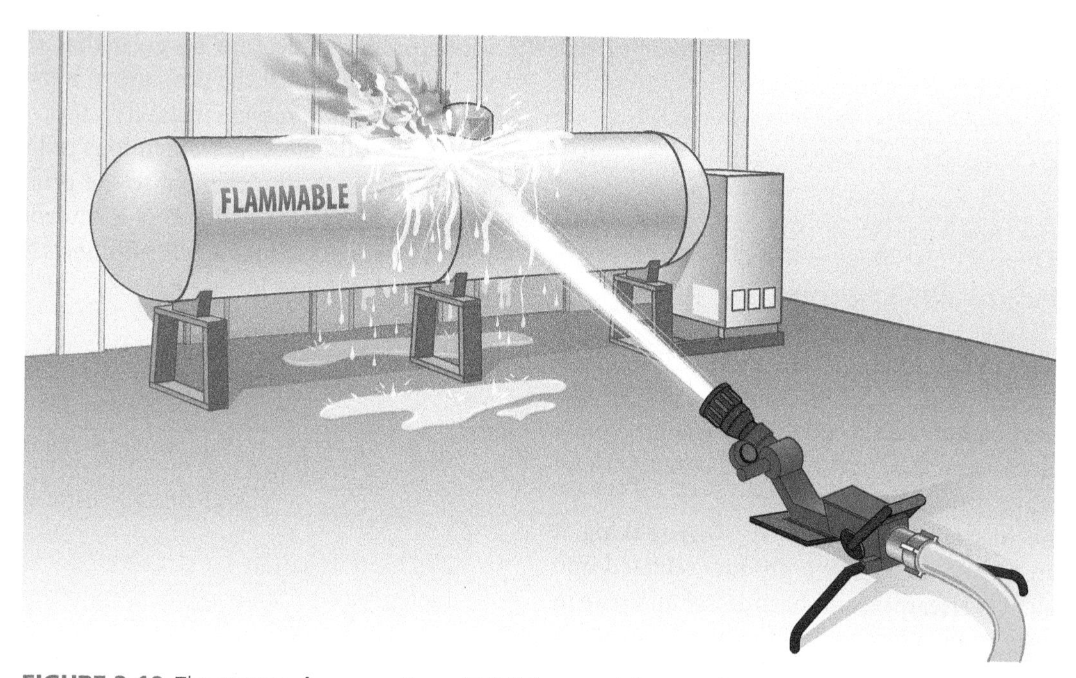

FIGURE 2-10 The strategy for preventing a BLEVE is to cool the top of the tank, which contains all the pressurized vapor, with copious amounts of water, preferably from an unmanned monitor or deluge appliance.

Boiling Point

All materials obey the laws of physics and exist in one of three states: solid, liquid, or gas. Pressure and temperature affects the state of matter. When gases are compressed under great pressure into liquids, they generate heat. When liquids boil, they absorb heat. When a material absorbs heat, it usually does so at its boiling point. The state of conversion from a liquid to a gas under normal atmospheric conditions is called evaporation. Ice is the solid state, water is the liquid state, and steam is the gas state.

Flammable or Explosive Limits of Gases

Because combustion can take place only when a substance is in a gas state, the levels of the gases are described in terms of percentages. Gases can ignite only when certain concentrations of that substance are present in air. If there isn't enough combustible gas present to ignite, the mixture is said to be *too lean* to burn. If there is too much gas present, or the concentration too excessive, it is said to be *too rich* to burn. When a concentration of a gas falls into the range where it can ignite, it is said to be within its flammable or explosive limits. The abbreviations FL (flammable limits) or EL (explosive limits) can be used interchangeably because if the gas can ignite, it can also explode. The lower end of this range is called the lower explosive limit (LEL), and the upper end of the range is called the upper explosive limit (UEL). Everything in between is the sweet spot where the flammable vapors-to-air mixture is perfect for combustion.

Gas detectors are specialized instruments that have been developed to detect and display measurements of flammable gas and explosive atmospheres. These instruments also measure the oxygen content of the immediate area to determine oxygen-rich or oxygen-deficient atmospheres **FIGURE 2-11**. Atmospheres that are below 19% oxygen require the use of a self-contained breathing apparatus (SCBA). The gas detectors are compact handheld units. Many can detect up to five different gases, which helps firefighters better evaluate their surroundings, have a better understanding of the enemy, and be better prepared to engage in firefighting activities safely. For example, carbon monoxide (CO) is a flammable gas sufficiently present at every fire. The flammable range is between 12.5% and 74%. The range for natural gas is between 6.5% and 17%, and for gasoline (92 octane), it is between 1.5% and 7.6%. The range for CO to ignite is much greater than both these flammable products, making it the more volatile fuel. This is important to understand when trying to recognize conditions that may lead to a smoke explosion. It is also important to understand that lower flammable ranges act as ladder fuels for other flammable gases in smoke.

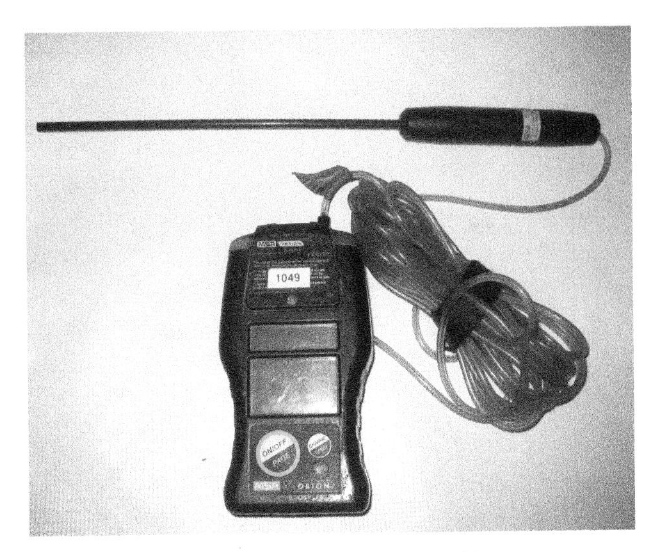

FIGURE 2-11 Gas detectors are specialized instruments that detect and display percentages of flammable gas and explosive atmospheres. These instruments also measure the oxygen content of the immediate area to determine if it is oxygen-rich or oxygen-deficient.

Courtesy of Raul Angulo.

The Burning Process: Characteristics of Fire Behavior

One often hears from firefighters that "every fire is different," and although that statement is true, fires still follow the laws of physics, and for the student of fire, most sequences are predictable. Older references called the stages of fire the incipient stage, free-burning stage, and smoldering stage. Over the years, fire-behavior scientists have redefined the burning stages as follows **FIGURE 2-12**:

- Ignition stage
- Growth stage
- Fully developed stage
- Decay stage

The narrow window of interior operations for fire departments lies somewhere between the ignition stage and the fully developed stage. This is where trapped occupants can be rescued, and the structure is still stable. Unfortunately, this is also the window where extreme fire behavior occurs and can kill firefighters. With the drastic increase in close-call

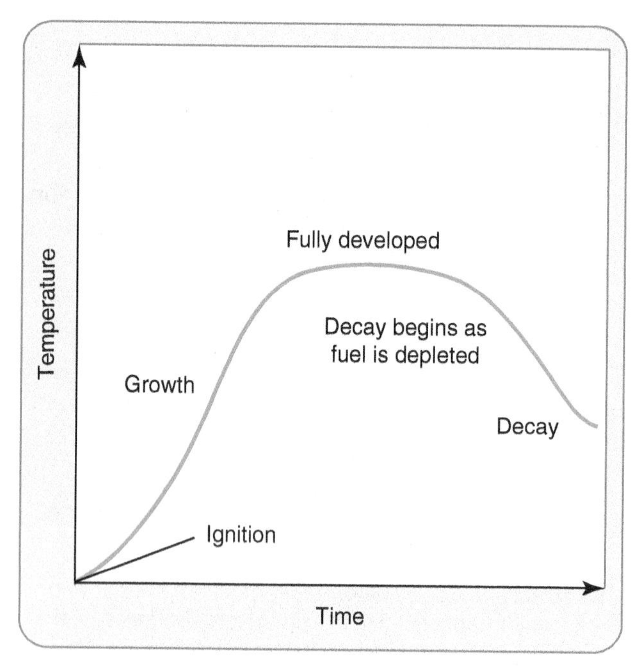

FIGURE 2-12 A graph of the traditional time-temperature fire growth curve with a fuel-limited fire.
Courtesy of NIST.

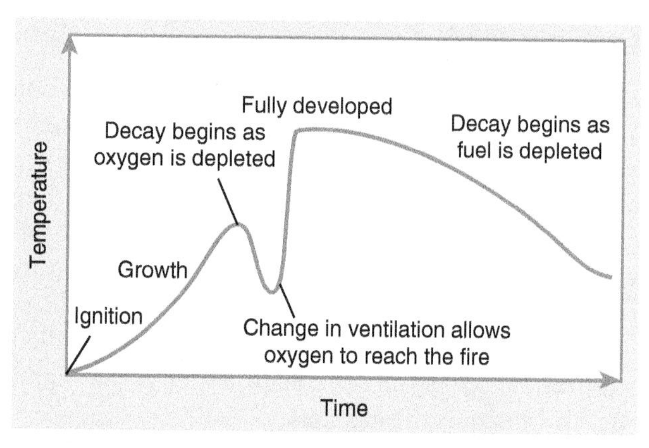

FIGURE 2-13 Graph of a ventilation-limited fire.
© Jones & Bartlett Learning.

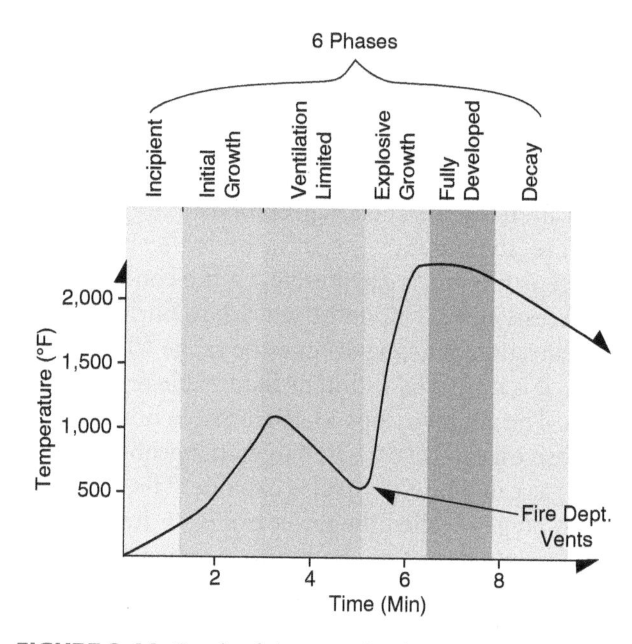

FIGURE 2-14 Graph of the new six-phase ventilation-controlled fire growth model.
© Jones & Bartlett Learning.

incidents and line-of-duty deaths resulting from firefighters being caught in flashovers, along with traditional methods of ventilation and interior fire attacks proving to be ineffective or inadequate for certain fires in relatively small structures, there had to be a way to figure out what was happening. Real scientific research and data on modern fire behavior was needed. The effects of modern synthetic fuels on fire behavior is being better understood thanks to the most recent UL and NIST experiments. The answer lies in the smoke.

The four-stage model is based on a fuel-controlled (available fuel to burn) fire with sufficient oxygen to carry the combustion process through the decay stage. Although many new homes use lightweight construction assemblies, they also use double- and triple-paned glass for insulation. These formidable windows often endure the heat of the fire during the incipient and growth stages without failing, which would provide the fire with additional oxygen. With the pyrolysis of modern fuels, which produce more dense, volatile smoke, a fire in a well-insulated interior compartment retains its heat but often runs out of air before it runs out of fuel. The dense smoke displaces the available oxygen, inhibiting the growth of the fire. This creates the condition of a ventilation-limited or ventilation-controlled fire **FIGURE 2-13**. The atmosphere is oxygen-deficient, so firefighters making entry often unexpectedly provide that source, triggering the rapid acceleration to flashover. Identifying two new stages in fire growth has led to the development of the six-stage **ventilation-controlled fire growth model** **FIGURE 2-14**:

- Ignition (incipient) stage
- Initial growth stage
- Ventilation-limited stage
- Explosive growth stage
- Fully developed stage
- Decay stage

Incipient (Ignition) Stage

Incipient stage and ignition stage are still used interchangeably, but the shift is to use the word *incipient*. When a combustible material heats up, it starts to liberate gases that can burn. The initial heat source usually comes from an external origin, such as a flame or spark from a fire that is already burning. The heat source can also come from radiation, conduction, or convection. At the point when the amount of heat being created exceeds the amount of heat being dispersed, combustion starts and self-generates the necessary heat to sustain the burn.

When the necessary components of a self-sustaining chemical chain reaction are present, ignition occurs. Ignition is that point where the need for a separate external heat source is no longer necessary because the ability for the material to sustain combustion comes from the heat generated by the material itself. Simply put, the ignition stage of a fire is the point at which the four elements of the fire tetrahedron come together. The combustible fuels reach their ignition temperatures, and a fire is started. It's typically small, limited in surface area, and easy to control and extinguish **FIGURE 2-15**.

At the ignition stage of a fire, the flame temperature is between 700°and 1,000°F (371° and 538°C), but the room temperature is relatively unchanged; it may be just above ambient. The fire is emitting light gray smoke, and the oxygen level is still around 20% to 21%. The carbon monoxide (CO) output is still low, so the survivability profile is very good. From the outside of the structure, the gray smoke haze can be seen through the window, and condensation or water droplets can form on the glass. Moisture buildup on the glass is a clue that this is an incipient fire because the temperature is beginning to change. There is very little, if any, smoke coming from the building. Although the terms *ignition stage* and *incipient stage* are used to describe the starting point of a fire, ignition is when the fire actually ignites. *Incipient* can refer to the stage of the fire all the way up to the initial growth stage.

Initial Growth Stage

The word *initial* has been added to *growth stage* because, in certain fires that end up being ventilation-limited, there's going to be two growth stages; hence, we have initial growth stage. From the moment of ignition, a fire begins to grow. Starting with a spark or a small flame, the fire is limited to the initial burning material. Inside a growth-stage room fire, the surface-controlled fire is getting larger. Convected air currents are causing the fire to grow with intensity as the exposed solid materials start to burn **FIGURE 2-16**. Pyrolysis begins and more smoke is being produced, changing its color from light gray to dark gray. As the smoke rises to the ceiling, it spreads horizontally and starts to bank down the walls. The visual density of the smoke increases, and the layers are becoming quite defined. The oxygen content is dropping below 20%. The flames are extending beyond the object of origin, but this fire can still be quickly controlled and extinguished with a hand line. Temperatures at floor level are still tenable, with a survivability profile that is good to moderate; however, that window is quickly closing **FIGURE 2-17**.

FIGURE 2-16 Convected air currents in a room and contents fire during the initial growth stage.

FIGURE 2-15 The incipient stage of a fire.

Courtesy of NIST.

FIGURE 2-17 The initial growth stage of a fire.

Courtesy of NIST.

Other combustibles heat up, liberating flammable gases that also ignite, and rollover begins to occur, which is the momentary ignition of flammable gases that spreads the chemical chain reaction to other adjacent flammables, increasing the size of the fire. Temperatures are quickly rising, and flashover is becoming eminent. The speed of the growth and size of the fire depend on several factors:

- **Fuel supply:** The size, quantity, quality, and class (A, B, C, or D) of the fuel naturally determines the fire's ability to consume it. For example, moisture content is important because wet fuels don't burn as readily as dry fuels. The density and surface area of the fuel also determine growth rate and speed of the fire's spread. For example, a flammable gas is already in the proper state for immediate ignition; conversely, it will be nearly impossible to light a log with a match. If sufficient heat and air is available, a fire is fuel-controlled and proceeds right into the fully developed stage.
- **Container size:** Wildland fires aside, open-air, unconfined areas serve to inhibit fire ignition because heat quickly rises and dissipates into the atmosphere. Short of direct flame contact or superheated ignition sources like a lightning strike, it is difficult for heat to accumulate fast enough to ignite outdoor combustible fuels in an open container. Every other type of fire is in some sort of enclosed container whether it's a residential or commercial structure, vessel, vehicle, or shipping container. In a structure, the container is the four surrounding walls, the floor, ceiling, and any partitions or obstructions. A large container permits the heat to rise and dissipate, slowing the growth of a fire. With a smaller container, heat can rise and radiate back down into the space, preheating and pyrolyzing uninvolved fuel sources.
- **Insulation:** Heat that is radiated back into unburned areas accelerates growth. However, if the container walls and ceiling are insulated, the heat is kept trapped within the container and isn't permitted to escape via radiation, conduction, or convection—at least for the time being. Eventually, all insulation fails if the fire isn't extinguished.
- **Oxygen supply:** The speed at which a fire grows and increases in size depends on the amount of oxygen available. Any limitation of the oxygen or the percentage of oxygen in the air supply curtails the growth and speed of the fire and can even result in extinguishment. With ample air, the fire proceeds to the fully developed stage.

FIGURE 2-18 A significant amount of dark gray or brown laminar smoke coming from the window area of the building is an indication of a growth stage fire.

Courtesy of Raul Angulo.

FIGURE 2-19 The water vapor has dried, leaving stains, and the window appears to be covered in a gray, brown, or black film—another indication of a growth stage fire.

Courtesy of Raul Angulo.

From the outside of the structure, two major clues indicate a growth stage fire because flames may still not be visible if the fire hasn't self-vented yet. The first is the smoke: A significant amount of dark gray or brown smoke should be coming from the building **FIGURE 2-18**. The second is the window to the compartment: flames may be visible through the windows. If it is not, look for random cracks or crazing (a network of fine cracks on the surface) along the glass. Glass cracks when the interior air temperature is between 500° and 600°F (260° and 316°C). The water vapor has dried, leaving stains, and the window appears to be covered in a gray, brown, or black film **FIGURE 2-19**. (The growth stage was previously known as the incipient stage of a fire.)

Ventilation-Limited Stage

Not all fires reach the ventilation-limited stage, but if the conditions listed above create a ventilation-limited fire, the progression of combustion is interrupted, and the fire enters into an early decay stage, even though there is plenty of heat and fuel available. Pyrolysis continues, producing more smoke and vapors, all of which are flammable and thus create an extremely dangerous atmosphere where the only ingredient needed to restart the combustion process is oxygen. That source is often the firefighters making entry into the structure FIGURE 2-20. Because that entry is usually a door, the open space created from top to bottom allows for air to rush in to the structure. The transition back into combustion happens with explosive force.

Explosive Growth Stage (New)

The explosive growth stage is the second growth stage of a fire. In the initial growth stage, the room temperature is low. The flammable vapors of solid fuel are low also because pyrolysis is just starting. Extreme fire events don't occur in the initial growth stage. However, in the explosive growth stage, the fire gains significant headway before running out of oxygen. It happens after the fire has become ventilation-limited. The room temperature could be 500°F (260°C) or higher; surrounding fuels are already preheated, and there is plenty of volatile smoke. When air is introduced, the transition can lead to rapid fire growth with explosive force FIGURE 2-21. The smoke contains many chemicals well within their flammable range, and gases with lower ignition temperatures, for example, acrolein (450°F [232°C]), act as ladder fuels to quickly ignite the rest of the smoke. Firefighters who have entered through the door and are advancing down a smoky hallway have created an oxygen-rich, low-pressure entry point behind them. They cannot reverse and outrun this hostile thermal assault without injury and sometimes death. This is why this stage is called the explosive growth stage. Flashovers tend to occur within the fire room, but modern smoke is fuel, and once ignited, the flames follow the convected currents beyond the room of origin. This is called flameover.

Some may want to dismiss the explosive growth stage as merely a "hot smoldering" backdraft, but this is an oversimplification of the thermal event and demonstrates a lack of understanding of the new scientific data. True backdrafts are high-temperature events and are still rare. The fire has reached the top of the time-temperature curve, the building is pressurized with smoke, and fuels are at or above their ignition temperatures. The tactic for preventing a backdraft is

A

B

FIGURE 2-21 A. The explosive growth stage happens after the fire becomes ventilation-limited and the temperature is high and surrounding fuels are preheated. B. The introduction of air can lead to rapid fire growth with explosive force, often resulting in flashover.

FIGURE 2-20 In a ventilation-limited fire, the only ingredient needed to restart the combustion process is air. That source is often the firefighters making entry into the structure.

to vertically ventilate at the highest point possible to relieve the pressurized smoke.

In the explosive growth stage, the building may not be significantly pressurized, and fuel ignition temperatures are warm. This is not a super-high-temperature event, but the temperature is higher than the initial growth stage and the compartment is pressurized. The explosion, although rapid, is not instantaneous. A low-pressure flow path has to develop to carry the fresh air to the affected area. Firefighters may have time to make it well into the structure before zero-visibility smoke conditions rapidly change into a wave of fast-moving fire they cannot outrun. In addition to becoming engulfed in flames, the heat release rate of modern smoke is so tremendous that a regular 1¾-inch hand line is often ineffective to overwhelm the fire—if it can even be operated at all after the explosive conversion. It is this single event that injures or kills firefighters by thermal assault.

The tactic to prevent the explosive growth stage from overtaking firefighters is to break out a window or open a door and spray a straight stream into the structure from the outside to cool the interior atmosphere and lower the ignition temperature of the smoke. The other method is to enter the structure with the wind at your back and use the hose line to spray a straight stream into the ceiling area to cool as the attack team advances. The first method is obviously safer and recommended because it can be accomplished outside the immediately dangerous to life and health (IDLH) zone **FIGURE 2-22**.

FIGURE 2-22 The tactic for preventing the explosive growth stage from overtaking firefighters is to break out a window or open a door and spray a straight stream into the structure from the outside to cool the interior atmosphere and lower the ignition temperature of the smoke.

Fully Developed Stage

The **fully developed stage** is where flashover occurs. This stage is recognized as the point at which all contents within the perimeter of the fire's boundaries are burning. Inside a structure, this would mean the entire contents of a room. Ceiling temperatures are well over 1,400°F (760°C), and the floor temperature is around 600°F (316°C). The smoke from modern furnishings is black, and the survivability profile is zero. From outside the structure, the glass in the windows fails, and flames are visible from the fire room **FIGURE 2-23**. With window failure, a new source of fresh air is introduced to the fire, causing it to grow and increasing the temperature, affecting exposures from the radiant heat. Increasing interior heat and smoke is evident on other windows in the structure as they change from clear to gray to dark brown, until they fail. Large volumes of smoke are coming from the structure, and it can be dense, dark, and turbulent.

In an outdoor fire, *fully developed* would mean all combustible material within the fire's farthest reach is on fire. There is sufficient fuel, heat, and oxygen for

FIGURE 2-23 A fire in the fully developed stage.

steady and increased combustion. All the components of the fire tetrahedron are complete. The fire is growing, heat and energy are being created, and flammable vapors are being emitted because there is a steady flow of fresh oxygen reaching the seat of the fire.

The fully developed stage is controlled and regulated by one of two methods. In a structure, the amount of air being introduced or supplied to the fire area make it an air-dependent fire. In an outdoor fire, the amount of air is unlimited; therefore, the amount of fuel dictates the size of the fire, making it a fuel-dependent fire. The best example of this is a forest fire. This stage was formerly known as the free-burning stage. It was also referred to as the flame spread phase.

Decay Stage

At the point, where most of the fuel and available oxygen has been consumed, the fire begins to decrease in size. Ultimately it extinguishes itself when the fuel or oxygen supply is exhausted. Obviously, this can take a considerable amount of time. The largest body of flames decrease in intensity and become a series of separate smaller flame fronts that darken down to glowing embers. Eventually those too disappear until all that is left are small pillars of light gray, lazy smoke **FIGURE 2-24**. As the decay stage progresses, a fire that was once oxygen controlled is now a fuel-dependent fire. The most familiar illustration of this entire sequence on a smaller scale is sitting around a campfire. Once the last log is consumed, the fire goes out, even though heat and air are still present. At that point, all the fire needs is additional fuel to continue. Without additional fuel, the temperature continues to drop. After approximately two hours, the temperature drops to ambient.

In an enclosed structure, such as, a well-insulated home, the available oxygen to support combustion may be depleted before the fuel load is consumed. Oxygen percentages are around 15%, and carbon monoxide concentrations can be as high as 35,000 ppm. Oxygen levels below 14% cannot support combustion. The survivability profile is zero. Fire intensity decreases, as does the smoke production, changing its color from dark smoke back toward light gray. In this case, the decay stage fire is *oxygen-deficient*, *oxygen-limited*, or *oxygen-depleted*. The term *ventilation-limited* has also been applied to describe this situation, although it is technically incorrect. Ventilation is merely the mechanism that moves and exchanges the atmosphere to deliver and reintroduce fresh oxygen to a smoldering fire. In a fuel-dependent or oxygen-limited fire, the end result is the same: The fire starts to go out.

The decay stage was previously known as the hot-smoldering stage. There is a misconception that a fire in the decay stage remains hot. Hot to touch, yes, but from a combustion standpoint, when the fire starts to go out from a lack of fuel or oxygen, heat is no longer being generated with the same energy, and the temperature starts to drop back around the 500°–600°F (260°–316°C) range. The decay stage is where smoke explosions typically occur. The decay stage is on the back, downward side of the standard time-temperature curve; the smoke is cooler, but it still contains fire gases, like acrolein, with lower ignition temperatures and well within their flammable range. If fresh air is reintroduced during the decay stage and if the fire still has burning embers throughout,

FIGURE 2-24 A fire in the decay stage. In a well-insulated home, the available oxygen to support combustion may be depleted before the fuel is consumed.

these lower-ignition-temperature gases can reignite, serving as a pre-igniter or ladder fuel to the higher-ignition-temperature gases. Another way a smoke explosion can occur is by uncovering a heat source, like burning embers that were covered up with fallen debris. When the embers are exposed to the air, they will glow, providing sufficient heat and an ignition source for smoke within its flammable range.

Within seconds, all the gases can reignite with explosive force, producing a shock wave that can destabilize or collapse already weakened building components. The difference between a smoke explosion and a flashover is slight, but a flashover can occur on both sides of the standard time-temperature curve. In a smoke explosion, increased flame intensity quickly follows and consumes the pockets of smoke, then self-extinguishes. But the flames may also be sufficient to spread fire to uninvolved preheated areas within the structure.

An excellent example in understanding the decay stage of a fire is a wood stove with air intake vents; when the stove is operating at maximum efficiency with all the components of the fire tetrahedron in place and the air intake vents wide open, inside temperatures can reach as high as 1,000°F. Radiant heat from the stove can be so intense, one cannot stand next to it. Notice the changes when the air vents are closed: The color of the smoke from the chimney turns gray, the radiant heat from the stove decreases, and within two hours the outside temperature of the stove's surface is around 100°F (38°C).

Depending on which stage the fire is in, the removal of a side of the fire tetrahedron will dictate the tactics that should be employed when fighting a fire. In some cases, such as a fire in the hold of a ship, cutting off the oxygen supply is usually the tactic of choice. If a ship's watertight spaces can be sealed, the supply of oxygen can be cut off. Ultimately this causes the fire to self-extinguish. In fighting a wildland fire, the use of fire breaks to cut off access to additional fuel is often key to the attack.

The burning process occurs in clearly defined stages. By interrupting any of the sequential steps, the fire can be controlled or completely extinguished.

Modes of Heat Transfer

Heat is the energy released during combustion, which allows the fire to sustain itself and to extend. When unheated or uninvolved materials are exposed to the heat of burning substances, certain changes occur that can make the new substance a contributing factor in extending a fire. Heat is transferred from one object to another when the temperatures of those objects are different. Heat always moves from warmer objects to cooler objects. The rate at which heat is transferred is related to the temperature differential of the objects. The greater the temperature differential, the greater the heat transfer rate. The term used to describe heat transfer is **heat flux**. Heat flux is the measure of the rate transfer from one surface to another. For example, if a sofa is on fire and the heat is moving to the ceiling, the heat flux would indicate how much heat was being transferred to the ceiling. Temperature is the measurement of heat within a body and is measured in degrees by a thermometer, a thermal coupling, thermal-imaging camera, or other instrument. Temperature is measured on the Fahrenheit scale (F) or the Celsius scale (C).

Higher temperatures reflect greater energy of a material. Therefore, knowing how heat is transmitted from one place to another is one of the first steps in knowing how to control the extension of fire and one of the first steps in extinguishing it.

The three modes by which heat transfers its energy from one substance to another are through conduction, convection, and radiation. The type of exposed material being heated, the distance between the fire and the exposure material being heated, and the ability of the exposure to absorb and retain the heat are factors in the spread of fire.

Conduction

Conduction is the process of transferring heat by direct contact through one solid object to another. Heat always moves from a warmer object to a cooler one, seeking equilibrium **FIGURE 2-25**. All matter conducts

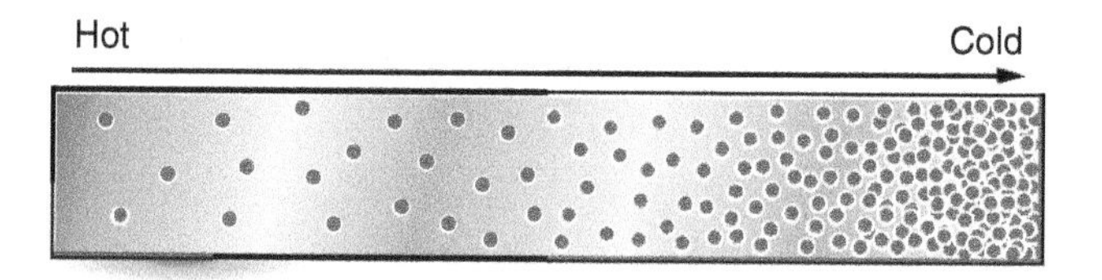

FIGURE 2-25 Conduction is the process of transferring heat by direct contact through one solid object to another.

heat to some degree. The ability of a material to conduct thermal energy depends on its molecular density. Because heat transmission is actually a transfer of energy from one molecule to another (kinetic energy), molecules must be present. The less dense a material is, the fewer molecules it has; hence the more difficult it is for heat to be transferred through it. In the absence of molecules—which is a vacuum—there is no transmission of heat except by radiation. The denser a material is, the more molecules it has in its makeup, and the greater the thermal conductivity potential. If a substance has a great deal of open air space between its molecules, the molecules can become agitated without transferring that agitation energy to other molecules. Materials such as mineral wool and cellulose, which have void spaces in their makeup, contain a great deal of air space. These spaces serve to insulate the heat from being transferred to the unheated side of the material.

When a hot object transfers its heat, conduction is taking place. The transfer could be to a different object or to a portion of the same object. We know combustion occurs on a molecular level, so when an object heats up, the atoms and molecules begin to collide with one another. A chain reaction of molecules and atoms occurs and causes these agitated molecules to pass the heat energy to areas of cooler temperatures, much like a billiard ball transfers energy from one billiard ball to another. As the heat rate increases, so do the energy waves traveling to cooler areas. Unless the heat energy can be dissipated in a sink hole, the internal temperature of the substance increases until it reaches its boiling point if it is a liquid, or its ignition temperature if it is a solid.

The denser the material, the better conductor it is. Because density is a function of weight, heavier substances are generally better conductors. Metals that are among the densest are better conductors of heat; the heavier the metal, the better its ability to transfer heat. For example, in structures, heat is conducted through beams, pipes, walls, and floors in all directions **FIGURE 2-26**. Another example is a concrete wall made up of sand, rocks, and cement. It is dense enough to absorb and retain heat and in time conducts heat with enough energy to ignite combustibles adjacent to it or on the other side of the wall. A steel wall bulkhead in a ship is much denser than concrete. The heat rate (speed) is much greater and can easily ignite combustibles on the opposite side of the bulkhead. Essentially, the ability of a material to transfer heat depends on that material's ability to keep the heat energy accumulating faster than it can be dissipated.

If denser materials are good conductors, then conversely, less dense materials are poor conductors and better insulators. Insulating materials are designed to limit heat transfer. Their diminished ability to conduct heat is due to the air trapped within the insulating material. The air or other gases trapped in small spaces prevents the degree of heat transfer that occurs in more densely constructed materials. Gases also insulate heat from one solid to another because their gas molecules are farther apart than molecules in denser material. This is why thermal-paned windows are so energy efficient.

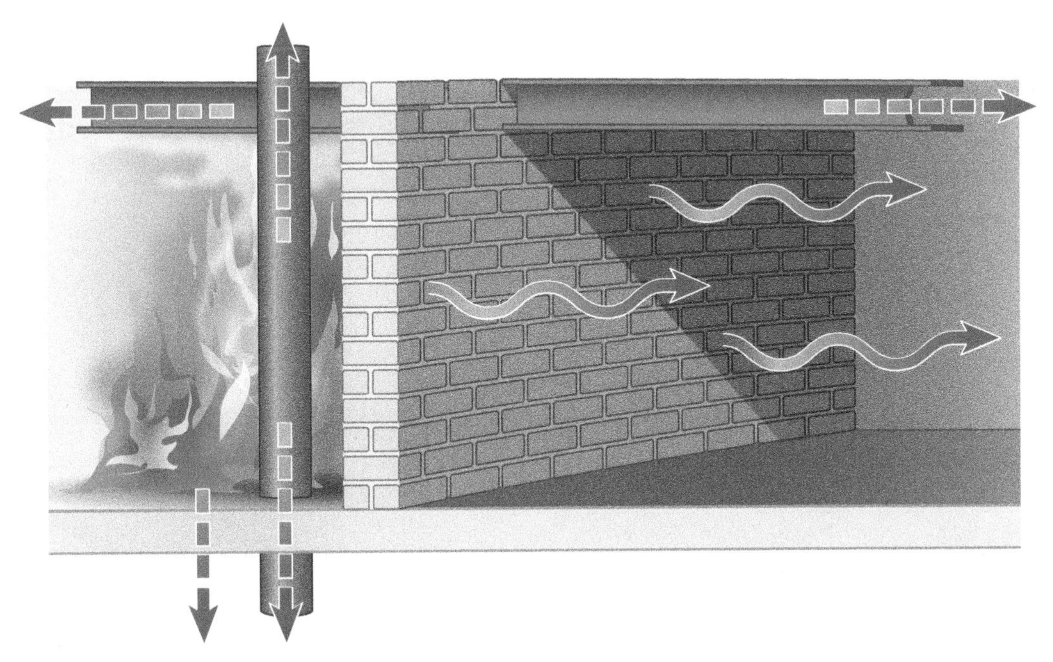

FIGURE 2-26 Metals that are the densest are better conductors of heat.

Lighter materials whose ignition temperatures are very high are also insulators. For example, mineral wool, used in soundproofing, insulation, and fireproofing, is a form of rock spun into web-like material with a lot of air space between the fibers, making it an excellent insulator because heat energy relies on molecular collisions to pass along heat. In the absence of these collisions, the heat energy dissipates. There is no conduction whatsoever in a vacuum because there are no molecules or atoms to contribute to the collision-reliant heating process. The fewer the molecules in a material, the poorer conductor it is. Therefore, the ability of a substance to be a good or poor conductor is a function of time and application of heat. If the application of heat to a substance over a given period of time exceeds its ability to dissipate or shed that heat, the substance begins to heat up.

If a material is a poor conductor because of its light density, it will be able to shed the heat without having its internal temperature rise. If it is a good conductor, like steel, it absorbs the heat and transfers that heat to distant internal locations. A piece of heated steel takes a long time to cool because the heat energy must come to the surface and dissipate in the less dense air. Conduction through exposed steel trusses may cause them to expand, warp, and possibly fail, sometimes taking out the exterior wall with it **FIGURE 2-27**.

Less dense materials like mineral wool, sheet rock, or gypsum board conversely cool almost as quickly as the heat source is removed because of their excellent ability to dissipate heat, owing to their light density. Heat transfer early in the development of a fire is almost entirely due to conduction. As the fire grows, convection and radiation come into play.

Convection

Convection is the transfer of heat by the circulatory movement of air or water. Air and smoke that is hotter than their surroundings rise until they reach an obstruction, then they move horizontally while still seeking a path to continue to rise **FIGURE 2-28**. Air that is cooler than its surroundings sinks. As heated air moves up, cooler air takes its place at the bottom. Even though air molecules flow about freely, they still have weight. When air is heated, the molecules become agitated and begin to collide into each other, necessitating more space to accommodate the increasing motion. When this happens, the air density is reduced so it weighs less, causing the air to rise until it reaches an obstruction or reaches a new level of equilibrium: the level at which the weight is equal to the surrounding atmosphere. As heated air rises due to a fire, it forms a **thermal column**, also known as a **thermal plume**. On the way up, the air around the perimeter of the plume mixes with cooler air. When equilibrium is reached, the agitated molecules calm down and the air begins to cool, returning to its original density. As

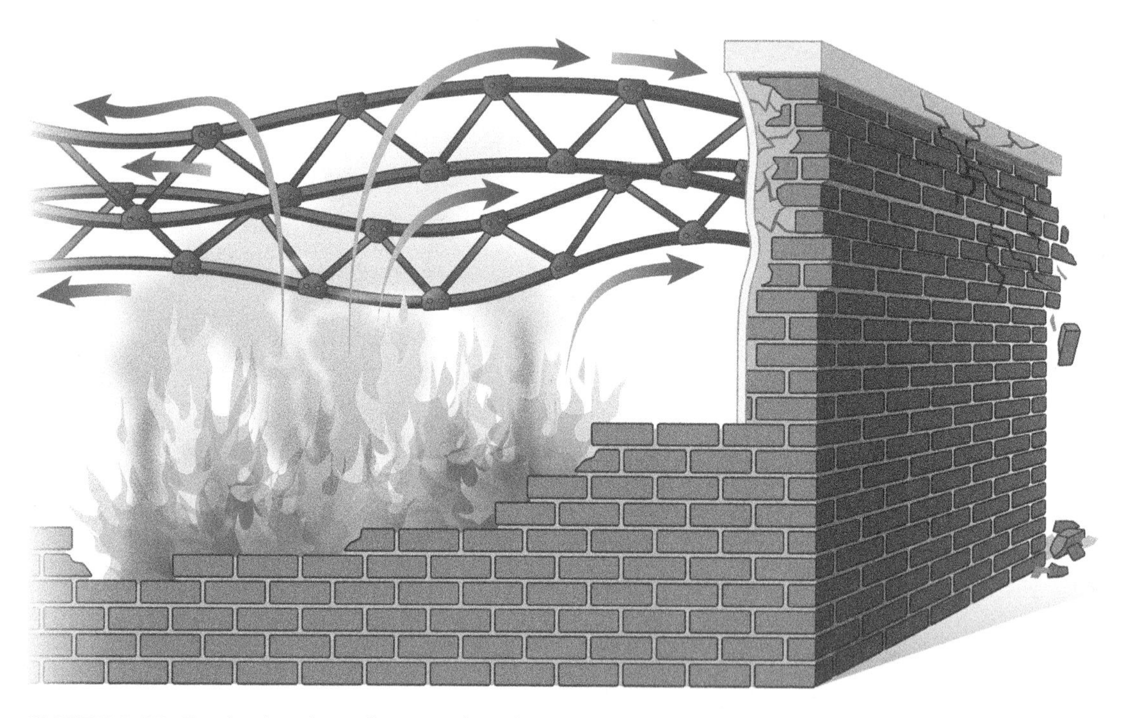

FIGURE 2-27 Conduction through exposed steel trusses may cause them to expand, warp, and possibly fail, sometimes taking out the exterior wall with it.

it loses buoyancy, the cooler air begins to drop back down outside the perimeter of the plume and falls to the bottom. This air is drawn back into the fire and is reheated by the rising temperature of the flames, and the molecular agitation starts all over again. The air is sucked back into the base of the plume. As the air temperature increases, air density decreases, and the air rises once more back to the top of the thermal column **FIGURE 2-29**.

In an enclosed structure fire, pressurized convection currents that are composed primarily of heated air, smoke, and fire gases expand and rise, raising the interior atmospheric pressure. Atmospheric pressure is higher toward the top of the structure and lower near the bottom of the structure. The point where the interior pressure is equal to the pressure outside the structure is called the **neutral plane** and lies near the vertical middle of the structure. The neutral plane can move up or down depending on temperature, ventilation, and wind.

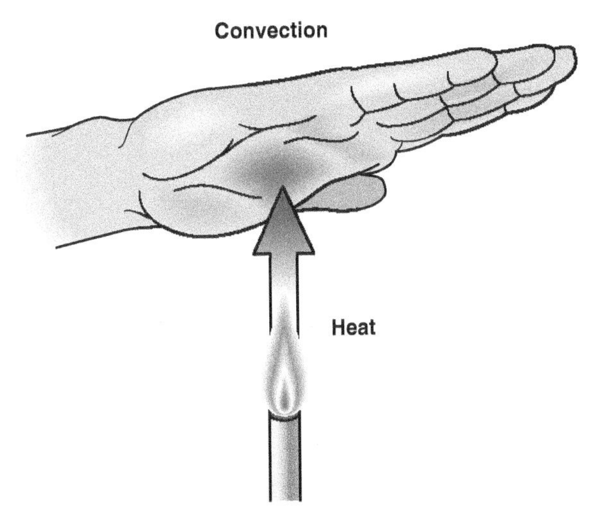

FIGURE 2-28 Convection is the transfer of heat by the circulatory movement of air or water.

Fire spread by convection moves predominantly in an upward direction, but pressurized air currents—natural or mechanical—can carry heat, smoke, and burning embers in any direction. If a thermal plume is confined by a vertical obstruction, like a ceiling, before equilibrium is reached, the pressurized heated air will look for a cooler space to equalize and follow the horizontal path of least resistance.

Convection is the most common method of heat and fire travel within a structure, generally carrying heat from room to room and floor to floor. The vertical and horizontal spread of heat contributes to the spread of fire on upper floors. The spread of fire through corridors; up stairwells, elevator shafts, pipe chases; between walls; and through attic spaces is caused mostly by convection of heat currents. If the convecting heat encounters a ceiling or other vertical barrier that prevents it from rising, it spreads out horizontally along the ceiling **FIGURE 2-30**. When it reaches a vertical wall, it travels down toward the floor, being pushed by more heated pressurized air that is rising behind it. This process is referred to as **mushrooming**. These currents of hot gases eventually heat the combustible room contents enough to ignite them **FIGURE 2-31**.

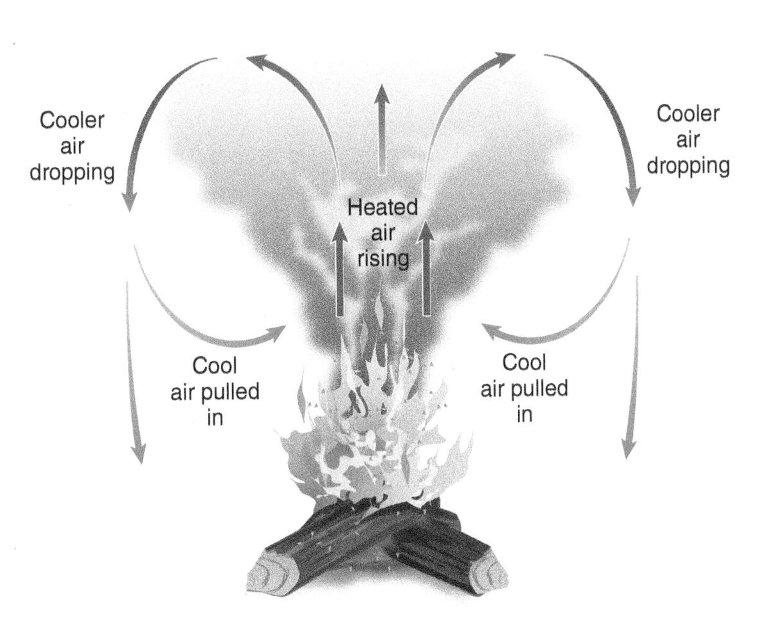

FIGURE 2-29 As heated air rises due to a fire, it forms a thermal column or thermal plume.

FIGURE 2-30 Convection is the most common method of heat and fire travel within a structure, generally carrying heat from room to room and floor to floor.

FIGURE 2-31 Mushrooming.

Path of Least Resistance and Flow Paths

Convection currents are present at every fire. The most important thing to remember about convection currents and predicting where the fire is going to spread is knowing that convection currents are pressurized and *always take the path of least resistance.* Therefore, *the smoke and fire also take the path of least resistance.* As the fire grows in a fire room, it creates increasing volumes of hot gases and smoke. The atmospheric pressure inside the fire room increases and the hot smoky gases push out horizontally toward an opening (either a door or a window) to find equilibrium. Pressurized fire gases always want to equalize. Higher pressure always moves toward lower pressure to achieve equilibrium. Because the pressure at the bottom of the fire room is less, cooler air is drawn in low toward the seat of the fire. The volume of space in which the cooler air being drawn in creates a clear open tunnel to the seat of the fire. Competing for the same space, hotter gases and smoke are escaping under pressure from the top of the same opening. This is called the **flow path**. It is the volume of space in which the movement of cooler air enters through an inlet opening and the space where hot gases and smoke exit through an exhaust outlet. The exhaust outlet can be another opening, like a window on the opposite side of the structure, or it can be the same (single) opening as the intake, like the front door. A flow path is also the boundary for the currents of gases within the building.

Two types of flow paths are important to recognize in engine company and ladder company operations: **bidirectional flow path** and **unidirectional flow path**. A bidirectional flow path is a two-way, or high-low, air current of fresh air and smoke competing for the same opening. The lower air intake is fresh air from the outside being entrained or sucked in toward the seat of the fire. The high point or exhaust path is where the smoke and other heated gases are using the same opening (door or window) to escape **FIGURE 2-32**. These gases are under pressure and seek the path of least resistance in order to equalize (move toward equilibrium). It is very important to remember that wherever heated gases and pressurized smoke are escaping, flames will shortly follow.

As the fire increases in intensity and other openings are created, either intentionally or due to the fire, the ventilation profile changes. When this happens, that same high-low bidirectional flow path can change to a unidirectional flow path. A unidirectional flow path can be a fresh-air intake opening or an exhaust exit port. It simply means that the convected air currents are flowing in the same direction through the entire opening and exit port **FIGURE 2-33**. For example, a vertical ventilation hole at the highest point of the pressurized structure is always a unidirectional exhaust opening. However, in horizontal ventilation,

FIGURE 2-32 A bidirectional flow path.

Courtesy of John Odegard.

FIGURE 2-33 A high-low unidirectional flow path.

Courtesy of Raul Angulo.

you can have a window on the windward side of a building be a unidirectional intake flow path, and on the leeward side of the building, a window on the same floor can be a unidirectional exhaust flow path. A unidirectional flow path can also be created using a positive-pressure fan.

The flow path transfers hot gases and smoke from the higher pressure within the fire area toward the lower-pressure areas—again, following the path of least resistance through doorways, window openings, and roof openings whether designed or created by firefighting tactics. The fire growth progresses along the exhaust side of the flow path. Flow paths are dependent on the location of higher pressure and lower pressure within the building as well as on external factors such as ventilation and wind direction. In addition, convection currents are influenced by the layout of the building. For example, air flows quickly through a structure with an open floor plan versus a structure made of small enclosed rooms, partitions, and cubicles, which interrupt air currents and flow paths. (Additional information about flow paths will be provided in Chapter 4.)

Radiation

Radiation is the combined process of emission, transmission, and absorption of energy traveling by electromagnetic wave propagation (i.e., infrared radiation) between a region of higher temperature and a region of lower temperature. As stated earlier, heat wants to travel from an area of higher temperature to an area of lower temperature to seek equilibrium. When electrons collide with other molecules, breaking apart the molecular bonds, heat energy in the form of heat waves is released in an exothermic reaction. When combustion occurs, light is also produced. Light travels through space in the form of light waves. Light waves range from ultraviolet to infrared. Contained within the light waves are electromagnetic forces, such as radio waves and X-rays, that can even travel through a vacuum without an intervening medium and deposit themselves on remote objects. The energy within electromagnetic waves travels in a straight line at the speed of light. For example, the heat you feel by extending your hands out to a bonfire is heat through radiation. You can sense the strength and concentration of the heat waves by moving closer to the fire or creating distance by moving away from the fire. Another example is the sun and its ability to heat the Earth. Although air is a poor conductor, it's obvious that radiant heat can travel where matter does not exist. Solar-radiated heat waves can reach us even though the sun is not in direct contact with the Earth (that would be conduction) and it is not heating gases that travel through space (that would be convection).

Heat is radiated evenly in all directions. With enough concentration, infrared light waves can permit a fire to jump from the source to a distant object (an exposure); raise its temperature; and, if intense enough, cause it to ignite **FIGURE 2-34**. Several factors must be in place for this to occur. The energy source must be strong enough to sustain the bombardment of the light waves, the exposure must be able to absorb the heat energy, and the exposure must be able to retain the heat buildup without losing it through dissipation.

Radiation, in combination with convection, is the major contributor to flashover. Heat builds up at upper levels of a compartment and simultaneously radiates back down and across the room. Every object in the room is subject to the radiation and begins to absorb the heat energy. Eventually every object in the room pyrolyzes, reaches its respective ignition temperature, and flashover occurs **FIGURE 2-35**. Flashover, heated gases, and radiant heat traveling through a building can also cause ignition of preheated fuels at some distance from the fire **FIGURE 2-36**.

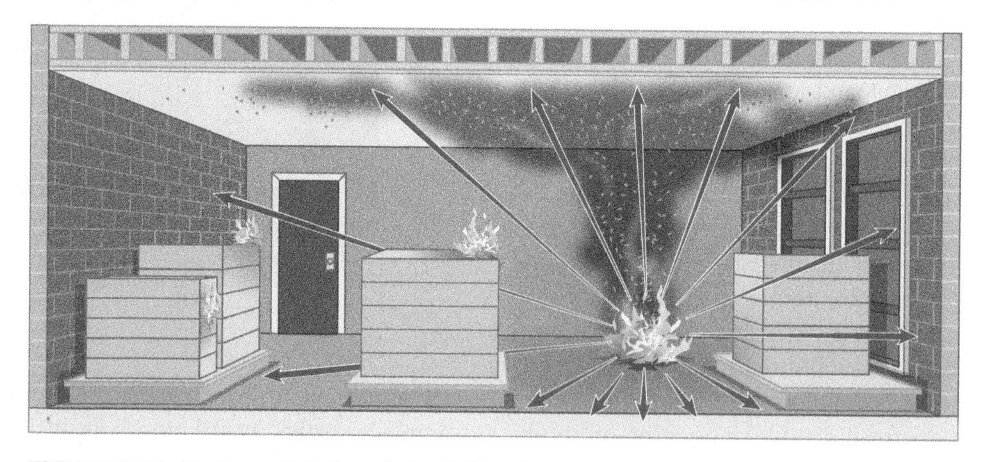

FIGURE 2-34 Heat is radiated evenly in all directions.

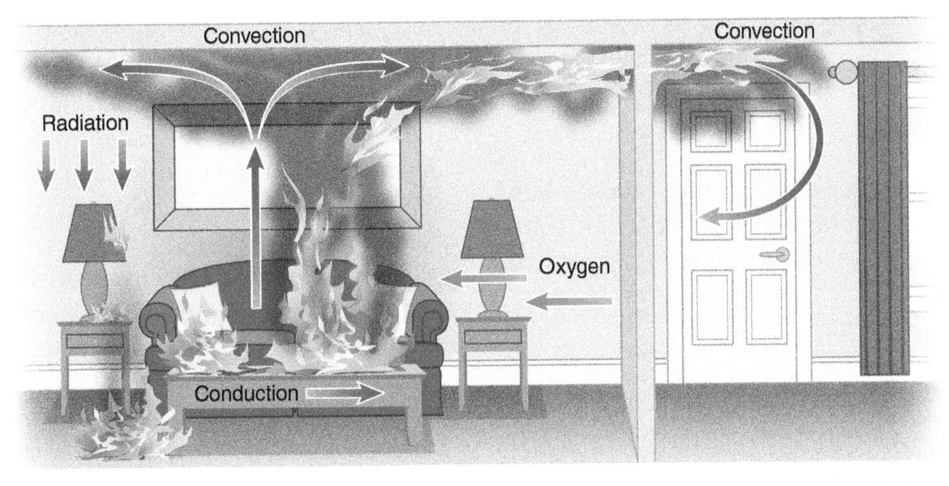

FIGURE 2-35 Radiation, in combination with convection, is the major contributor to flashover.
© Jones & Bartlett Learning.

FIGURE 2-36 Flashover, heated gases, and radiant heat traveling through a building may cause ignition of preheated fuels some distance from the fire.
© Jones & Bartlett Learning.

Radiation is also the cause of most exposure fires. As the fire grows, it radiates more heat and energy. Exposure buildings that absorb more heat faster than they can dissipate it eventually ignite **FIGURE 2-37**. Unlike your hand, which you can move away from a bonfire when the radiated heat becomes too intense, exposure fuels continue to heat up and pyrolyze until they reach their ignition temperature. Only materials that reflect radiant heat interfere with the transmission of this type of energy.

Firefighters must understand the effects of radiated heat because fire can extend through radiation as much as through conduction or convection. Water is used to absorb heat and carry it away from surrounding materials. Water absorbs the accumulating heat from exposed objects, thereby preventing them from reaching their ignition temperatures. By controlling this element of the fire tetrahedron, the fire cannot extend. To a limited extent, water can also be employed to combat conduction extension. In that case, the water carries away the heat, keeping the unburned or unheated portions of the material cool. Things that are kept wet cannot burn. Moisture and fire cannot exist in the same molecular space. Water cools and prevents the transfer of heat to another exposure substance.

FIGURE 2-37 As the fire grows, it radiates more heat and energy.
Courtesy of Raul Angulo.

The Physical State of Fuels and Its Effect on Combustion

We've shown that matter exists in three states: solids, liquids, and gases (Figure 2-2) but in order for combustion to take place, solid and liquid fuels must first convert to a gaseous state. Flammable gases are naturally ready for combustion.

Solids

Starting on a molecular level, a solid consists of atoms and molecular bonds that are packed tightly together. This type of molecular packaging gives the material its physical form; its density; and, to a great extent, its shape.

When heat is applied, the molecules become agitated and begin to collide with one another. The molecular bonds begin to break apart, changing the state of the material. Molecules must be in a free-floating state before oxidation reactions can take place. The molecules that readily accept oxygen atoms become oxidized, and a heat-producing exothermic reaction occurs, causing the free-floating molecules to develop a greater affinity for the oxygen. The newly formed by-products of oxidation release even more heat, and the self-sustained chain reaction called combustion is under way.

Pyrolysis

When heat is applied to a solid material, some of the heat is absorbed into the mass, while the rest of it dissipates to some degree. When the amount of heat applied to the solid can no longer be absorbed and exceeds its ability to dissipate, chemical breakdowns and decomposition begin to occur. A self-sustaining burning reaction results and continues until it is either interrupted or runs out of fuel.

The breakdown of solid fuels to ignitable gases—pyrolysis—is the chemical decomposition of a solid substance through the application of heat. Simply put, as solid fuels are heated, they're slowly being destroyed and converted to liquid, then combustible vapors **FIGURE 2-38**. Pyrolysis occurs shortly before combustion.

Let's examine the pyrolysis of wood. Wood contains varying amounts of moisture content. Water contained in wood or in any solid fuel source serves as a built-in cooling agent. That is why green fire wood recently cut from a live tree is hard to burn. The moisture content is very high—or you can say the wood is well-hydrated—whereas fire wood that has been allowed to dry will ignite much more easily and burn faster with a higher heat release rate. When wood is heated, the first change that occurs is that the water content is vaporized and escapes from the wood. The energy required to vaporize the water delays the beginning of pyrolysis. As the wood is heated to approximately 425°F (218°C), pyrolysis begins. Once the moisture content is vaporized, the wood begins to char. This charring is a natural mechanism in wood in an attempt to protect itself. As the wood chars, it begins to lose mass, which also weakens its load-carrying ability. Various degrees of alligatoring occurs, which is the formation of deep black and shallow cracks along the surface of the wood that resembles the skin of an alligator. Wood burns away approximately ⅛ of an inch every 5 minutes **FIGURE 2-39**. With sustained heat, the process of pyrolysis continues to generate burnable gases that ignite if the other elements of the fire tetrahedron are present.

Characteristics of Solid-Fuel Fires

Fuel Composition. Most of the fuel load in structure fires that firefighters will face consists of solid fuels, including the structure itself. Solid fuels have defined sizes and shapes and include wood, wood-based products, fabrics, paper, carpeting, and

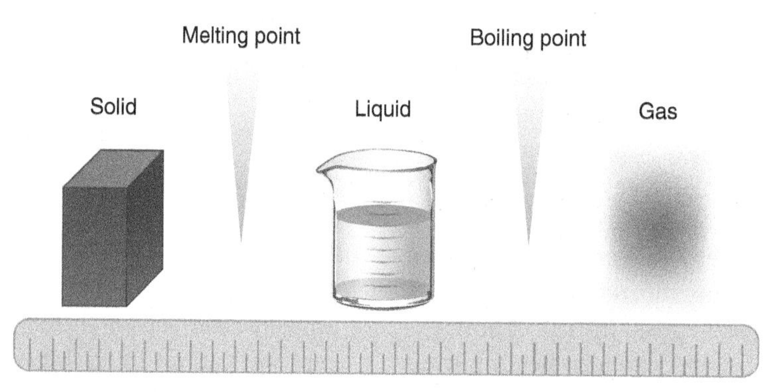

FIGURE 2-38 States of matter are temperature-dependent.

FIGURE 2-39 Alligatoring. As the wood is heated to approximately 425°F (218°C), pyrolysis begins.

Courtesy of Raul Angulo.

FIGURE 2-40 The legacy fuel room has little flame showing 3 minutes and 25 seconds after ignition, while the modern fuel room has flashed over in the same time period.

Courtesy of UL.

many petroleum-based fuels such as plastics and foams. Older fuels, including wooden building construction, are known as legacy fuels. This means wooden building construction utilized solid dimensional lumber, and household wooden furniture was actually made from solid wood. Natural fabrics and textiles like cotton, wool, and paper were used in household furnishings. Today's construction uses engineered lumber and wood products like glue-laminated beams (glulams), oriented strand board (OSB), plywood, and pressboard, all of which have industrial glues holding them together. The majority of household furnishing, appliances, coverings, and decorations are constructed from petroleum-based products like plastics, foams, and synthetics. These are known as **modern fuels**.

All of these building materials and contents influence how a structure fire burns. Plastic and petroleum-based solid fuels pyrolyze faster than wood-based solid fuels and require less heat to decompose and ignite than does wood. Modern fuels contain greater potential heat energy than products made from natural materials. This means that fires fueled by modern petroleum-based products have a higher heat release rate than fires fueled by legacy fuels such as wood and cotton **FIGURE 2-40**.

Fuel Configuration: Surface-to-Mass Ratio

Surface-to-mass ratio is the surface area of the fuel in proportion to the mass. The size and shape of solid fuels determines their ability to ignite, the time they take to be consumed, and the heat release rate of the burning fuel. By dividing up the mass, the surface area is increased. As the surface area increases, more of the material is exposed to the heat and thus generates more burnable gases due to pyrolysis.

Let's go back to our wood example, but now consider a 10-foot (3-m) log. It has mass and weight; however, the surface area is relatively small compared to its mass. If the log were to be cut into 2-foot (60-cm) sections, the mass would remain the same, but the exposed surface area would be greatly increased. If these sections were cut down even further to wooden shingles, again the original mass of the 10-foot (3-m) log still exists, but the surface area is increased even greater than the 2-foot (60-cm) sections. The energy required to ignite the low surface-to-mass ratio, 10-foot (3-m) log is much greater than the energy required to ignite the higher surface-to mass ratio shingles. The higher the surface-to-mass ratio, the less heat energy it takes to light the fuel. For example, holding a match to the 10-foot (3-m) log isn't sufficient energy to ignite it, but the shingle has more surface area exposed to the match, making it easier to light.

Orientation of the Fuel

The fuel's actual position also affects the way it burns. If a solid fuel is in a vertical position, the fire spread is more rapid than if the solid fuel were in a horizontal position **FIGURE 2-41**. Horizontal and vertical sheets of plywood are the same panels. A panel in the vertical position burns more rapidly due to the increased heat transfer and absorption through convection as well as conduction and radiation.

Continuity of the Fuel

Continuity refers to how close the fuel load is arranged. The closer together the different parts of the fuel are, the easier and more quickly they can ignite. Continuity

FIGURE 2-41 Fire burning a solid fuel in a vertical position will spread faster than a fire burning a solid fuel in a horizontal position.
Courtesy of Raul Angulo.

can occur in a horizontal direction along floors, ceilings, or horizontal surfaces of building contents; it can also occur in a vertical direction along walls or the vertical surfaces of building contents. Fuel that is in contact with other fuel ignites faster and reaches a peak heat release rate more quickly.

Clearly, the amount of fuel available to burn is a factor in fire growth and intensity. In a fuel-limited fire, such as a campfire with one burning log horizontally placed, all other factors being equal, produces a much smaller fire than a campfire with six burning logs configured into a standing teepee. Likewise, a fire fueled with 20 wooden pallets burns with greater intensity and has a greater heat release rate than a fire fueled with two wooden pallets. When more fuel is available, there is a higher heat release rate than when less fuel is available. Forest fires happen because there is no shortage of trees to burn, and conflagrations happen because there's no shortage of buildings to burn.

Liquids

In most cases, there is a physical state transformation from a solid to a liquid to a gas before combustion can take place. A liquid's ability to burn depends on it molecules being placed into suspension. A liquid cannot burn unless it is in suspension, which is also referred to as atomization. Fuel gases from liquids are generated by the process called **vaporization**. Vaporization is the transformation of a liquid to its vapor or gaseous state. The transformation from liquid to gas or vapor occurs as molecules of the substance escape from the liquid's surface into the surrounding atmosphere. In order for the molecules to break free from the liquid's surface, there must be some energy input. In most cases, this energy source is provided in the form of heat. Similar to a solid, a liquid acts as a heat sink and dissipates the introduced heat into the cooler areas of the liquid. When the entire pool of liquid is heated, or the ability to dissipate is overcome by the application of heat, a rise in temperature occurs, resulting in the boiling point of the liquid. All liquids eventually reach their boiling points during a fire. As the boiling point is reached, the amount of flammable vapors generated significantly increases. Because most liquids are a mixture of compounds, there isn't one exact boiling point. For example, gasoline contains approximately 100 different compounds, so the flammability of the mixture is determined by the compound with the lowest ignition temperature. The ignition temperature is the temperature at which the fuel spontaneously ignites. The liquid then becomes engaged in the self-sustaining combustion process if it has the affinity for oxygen and permits oxidation. Not all liquids possess that property. Some boil, evaporate, and never engage in combustion. Water is a prime example.

With flammable liquids, the amount of liquid that is vaporized is also related to the volatility of the liquid. The higher the temperature, the more liquid that evaporates. Liquids that have a lower molecular weight tend to evaporate more readily than liquids with a higher molecular weight. As more of the liquid vaporizes, the mixture reaches a point where enough vapor is present in the air to create a flammable vapor-to-air mixture.

Flash point and *flame point* are two other terms used to describe the flammability of a liquid. The **flash point** is the lowest temperature at which a liquid produces a flammable vapor sufficient to "flash" upon the application of a flame or spark. This small flame is momentary and goes out once the flammable vapors burn off. This is clearly seen in rollover, which will be discussed later in this chapter. The **flame point** (also known as the *fire point*) is the temperature to which a flammable liquid must be heated in order to produce sufficient vapors to burn continuously. The flame point is a few degrees above the flash point of a liquid.

Vaporization of liquid fuels generally requires less energy input than does pyrolysis for solid fuels. This is primarily caused by the different densities of substances in solid and liquid states. Solids absorb more of the heat energy because of the mass.

Like the surface-to-mass ratio for solid fuels, the surface-to-volume ratio of liquids is an important factor in their ignitability. A liquid assumes the shape of its container. Thus, when a spill occurs, the liquid assumes the shape of the ground; in other words, it becomes flat. The surface ratio increases significantly, as does the amount of fuel vaporized from the substance. Liquids that easily give off quantities of flammable or combustible vapors can be very dangerous in this spill configuration **FIGURE 2-42**.

Gases

A gas, apart from the other two states, is essentially primed for combustion. Its natural physical property is in a ready-made state that permits the chemical reaction to occur, making gaseous fuels the most dangerous of all the fuel states. No pyrolysis or vaporization is needed to ready the fuel for combustion and less heat energy is required for ignition **FIGURE 2-43**.

For combustion to occur after a fuel has been converted into a gaseous state, it must be mixed with air (the oxidizer) in the proper ratio. The range of concentrations of the fuel vapor and air (oxidizer) is the flammable range or explosive range. These terms are used interchangeably because, if the flammable gas-to-air mixture does not ignite, it also does not explode. The flammable range of a fuel is reported using the percentage by volume of gas or vapor in air for the lower flammable limit (LFL, or LEL) and for the upper flammable limit (UFL, or UEL). The LFL is the minimum concentration of fuel vapor and air that supports combustion. Concentrations below the LFL are said to be *too lean* to burn. The UFL is the concentration above which combustion cannot take place. Concentrations that are above the UFL are said to be *too rich* to burn. As previously mentioned, CO has an LFL of 12.5% and a UFL of 74%. Variations in temperature and pressure can also cause the flammable range to vary considerably. Generally, higher temperatures and pressures

FIGURE 2-42 When a spill occurs, the surface ratio increases significantly, as does the amount of fuel vaporized from the liquid.

FIGURE 2-43 A gas, apart from the other two states, is essentially primed for combustion.

broaden the flammable range, and lower temperatures and pressures narrow the flammable range. Handheld gas detectors that measure the percentages of the most common gas fuels-to-air ratios should be used to determine the atmospheric hazards and explosive environments.

Remember, when a substance changes from one state to another, heat is either given off (exothermic reaction) or absorbed (endothermic reaction). During a fire attack, you can see heat being absorbed when water turns to steam. You can also see heat being absorbed when you discharge a carbon dioxide (CO_2) extinguisher. The CO_2 actually creates snow and collects frost on the cylinder wall when the liquid is released as a gas. On the flip side, when pressure is applied, heat is released. If enough pressure is applied to a gas, it turns into a liquid. For example, CO_2 gas is compressed into liquid CO_2, which is used in extinguishers. When additional pressure is applied to CO_2, the liquid turns into a solid (dry ice) with the release of additional heat. Conversely, when the solid becomes liquid, it absorbs great quantities of heat (ice to water) and again when the liquid becomes a gas (water to steam). The heat-absorbing capabilities of the water to steam conversion are employed when extinguishing fire in ordinary combustibles. Although it varies on atmospheric pressure and temperature, under ideal conditions, the expansion ratio between water in its liquid form and steam is 1:1,700: one part of liquid water expands to 1,700 times its volume as steam when the liquid is boiled. As stated, moisture and fire cannot exist in the same molecular space. The steam robs the fire of its heat and reduces the temperature so that self-sustaining combustion is no longer possible.

Awareness of these properties and their place in the combustion process is another weapon in the arsenal for fighting fire. Keeping a liquid cool enough prevents vaporization, and by excluding oxygen or heat, ignition can be minimized or even prevented. This information is valuable when dealing with an incident such as a tank truck fire or a liquid fuel spill.

Smoke: By-Product of Combustion

The most visible by-product of combustion is smoke, which is hazardous and deadly to firefighters. It is present at every fire. Smoke consists of solids (ashes), vapors, aerosols, and gases. Carcinogenic chemicals contained within smoke are numerous and include carbon monoxide, carbon dioxide, hydrogen cyanide, hydrogen chloride, phosgene, benzene, acrolein, ammonia, formaldehyde, other fire gases, and unburned hydrocarbon materials—and many more chemicals can be added to this list. Seventy percent of smoke is particulate matter (solids). Particulates in smoke are what allow you to see smoke **FIGURE 2-44**. Working with the necessary equipment from a relatively safe position with an adequate number of firefighters makes smoke management through ventilation one of the most challenging strategies on the fireground. The increased cancer rate among firefighters due to smoke exposure is alarming. Minimizing the needless exposure of firefighters to smoke is now a tactical objective in firefighting.

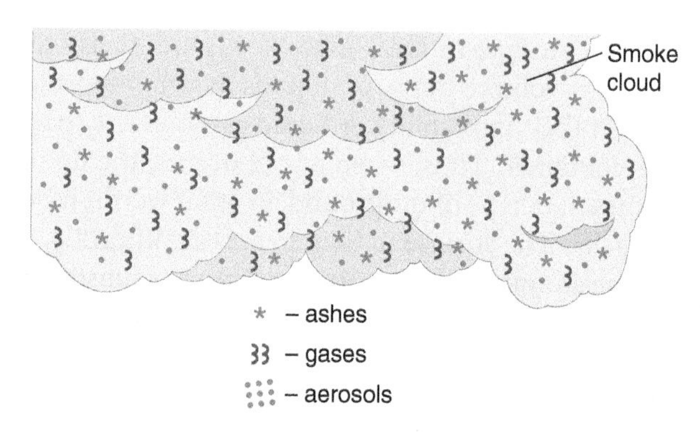

FIGURE 2-44 Smoke consists of particulates (solids and ash), vapor, aerosols, and numerous fire gases.

The Theory of Extinguishment

Every strategy in extinguishing a fire involves limiting, interrupting, or removing one or more of the essential elements in the combustion process. Heat, fuel, oxygen, and a continuing chemical chain reaction are the four elements in the fire tetrahedron. Remove any one of them, and the fire goes out. A fire is extinguished by reducing the temperature, eliminating the available oxygen, removing the fuel, and/or stopping the self-sustained chemical chain reaction **FIGURE 2-45**.

Reducing the Temperature

The primary method of extinguishment in the fire service is cooling with water; it is the most commonly used extinguishing agent. Water reduces the temperature of fuels to a point where they do not release sufficient vapors to burn. Solid and liquid fuels with high flash points can be extinguished by cooling with water. However, water cannot cool or reduce vapor production sufficiently to extinguish fires involving flammable liquids and flammable gases with low flash points.

Smoke and heated gases produced by a fire rise to the ceiling and begin to bank down. When they reach their ignition temperature, they ignite. Flowing

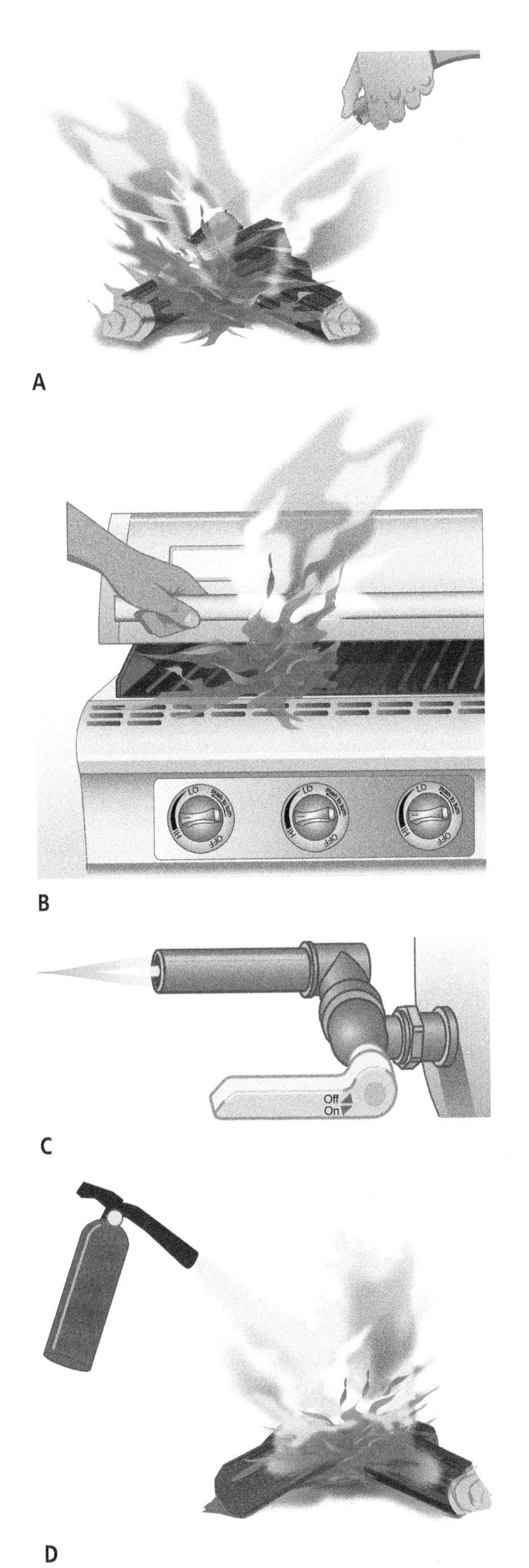

FIGURE 2-45 The four basic methods of fire extinguishment: **A**. Cool the burning material. **B**. Exclude oxygen from the fire. **C**. Remove fuel from the fire. **D**. Interrupt the chemical reaction with a flame inhibitor.

FIGURE 2-46 Flowing water aimed toward the ceiling can cool the smoke and gases, preventing them from igniting.

water aimed toward the ceiling can cool the smoke and gases, preventing them from igniting. This tactic is called *cooling the ceiling* or *cooling the atmosphere* **FIGURE 2-46**.

Applying water is also the most effective method for cooling and extinguishing fires, including smoldering fires. To extinguish a fire by sufficiently reducing the fuel temperature, water must be applied directly onto the burning fuel to absorb the heat being generated by combustion. Fire extinguishers can be the best tool for the job, but they are often underutilized because many firefighters do not realize how effective they can be in extinguishing the incipient fires to which they respond. Also, the logistics and costs associated with recharging, other than 2½-gallon water-pressurized pump cans, make training with CO_2 and dry chemical extinguishers inconvenient.

There's more to flowing water than simply opening the nozzle. Measurements are an important part of firefighting, and there are scientific and mathematic measurements for everything. In this chapter, we have referred a lot to temperature, which is the measurement of heat. In the theory of extinguishment, there is another form of heat measurement that is important to understand: the British thermal unit (BTU). In the fire service, the BTU is used as a quantitative specification for the energy production or energy transfer capability of a fire. For example, you can set a certain temperature for your furnace, but the power of the furnace's heat output is measured in BTU. Therefore, the definition of a BTU is the amount of heat energy required to raise the temperature of one pound of water, one degree Fahrenheit; for example:

- One cubic foot natural gas ± 1,075 BTU
- One pound of dry wood ± 8,030 BTU
- One pound red oak wood ± 8,500 BTU
- One pound of coal ± 12,500 BTU
- One pound of foam and synthetic ± 15,000 BTU
- One pound of propane ± 21,600 BTU
- One gallon of propane ± 91,000 BTU

The delivery of water through a pumper and fire hose is measured in gallons per minute (gpm), or liters per minute. Understanding these two measurements and how they relate to each other is important for applying fire flow water formulas and selecting the proper strategy and tactics for firefighting. Simply put, we fight the BTU output of a fire with gpm. *If the gallons per minute of water cannot overwhelm and exceed the amount of BTUs generated by the fire, the fire will not go out.* This is where the old fire department adage, "*Little fire, little water, big fire, big water*" is soundly rooted in science. Any novice or civilian can understand this theory of extinguishment.

Removing the Fuel

Removing the fuel source effectively extinguishes the fire. A fuel source may be removed by shutting off or capping the flow of a liquid and gaseous fuel, or by physically removing a solid fuel source in the path of the fire. When wildland firefighters cut a gigantic fire line ahead of the fire, that is exactly what they're doing: removing additional fuel from the path of the fire. Another method of fuel removal is to allow the fire to consume the fuel source entirely and burn itself out, or accelerate the combustion process so the fire consumes all the fuel sooner rather than later. This is the theory behind back-burning in wildland firefighting or applying hose streams (from a safe distance) onto water-reactive metals like magnesium. Accelerating the combustion process in a magnesium fire is a strategy that may be viewed as unorthodox, but it should be considered in certain situations because it doesn't prolong the incident and frees up valuable equipment and resources sooner. If this tactic is selected, prepare for a spectacular fire.

Eliminating the Available Oxygen

Reducing the available oxygen for combustion (below 15%) reduces fire growth and can totally extinguish the fire as long as there are no new sources of oxygen. Even smoke with an oxygen content below 15% can act as a buffer to prevent fire spread. The best example, one that is familiar to all of us, is placing a lid over a burning pan of food. Another example is closing the lid to a flaming barbeque unit. Oxygen can also be reduced by flooding the area with an inert gas such as carbon dioxide (CO_2), which displaces the oxygen and interrupts the combustion process. This tactic is frequently used in shipboard firefighting. In fighting flammable liquid fires, oxygen can be separated from the burning fuel by flowing a blanket of firefighting foam on top of the fuel. *Note:* Neither of these methods work with exotic fuels that are self-oxidizing.

Interrupting the Chemical Chain Reaction

Extinguishing agents such as dry chemicals interrupt the combustion process to stop flaming. This method of extinguishment is effective for gas and liquid fuels because they must flame in order to generate heat and continue to burn.

Many flammable liquids have a specific gravity less than 1, so if water is used as the extinguishing agent, the fuel will float on the water while continuing to burn. If the fuel is unconfined, using water could unintentionally spread the fire.

The solubility of a liquid fuel (the ability of a substance to mix with water) is also an important factor in extinguishment. Liquids of similar molecular structure tend to be soluble with each other, while those with different molecular structures and electrical charges tend not to mix. Liquids that readily mix with water are called **polar solvents**. Alcohol and other polar solvents dissolve in water. If large volumes of water are used, alcohol and other polar solvents can be diluted to the point where they do not produce sufficient vapors to burn. As a general rule, hydrocarbon liquid fuels like gasoline are called nonpolar solvents (they are not soluble in water) and they do not dissolve in water and in fact float on top of water. This is why water alone cannot wash oil off your hands. Soap must be added to water to break the surface tension of the oil and dissolve it.

Fire Behavior Events

Several unique events can occur during a structure fire that firefighters must be familiar with, including thermal layering, rollover, flashover, backdraft, rapid fire growth, behavior of ventilation-limited fires, smoke explosions, and wind-driven fires.

Thermal Layering

As a fire continues to grow from the ignition stage toward the growth stage, hot air and gases that are lighter and buoyant continue to rise until they reach equilibrium or balance with the surrounding atmosphere. Thermal layering occurs when the gases produced by the fire stratify into layers based on their temperatures; it is also referred to as heat stratification. The hottest gases rise to the highest point of the room, while the cooler gases, which are heavier, stay closer to the bottom of the room, nearer the floor.

When confined to a structure, the hottest of the gases and air accumulate at the ceiling of the room. As the amount of by-products from the fire accumulate, the interior pressure in the room increases and the gases, heated air, and smoke start to bank down until they find the path of least resistance and escape through any available opening, like a door or a window, while cooler air is being drawn in toward the seat of the fire.

The boundary between the hot gases and the cooler gases is the neutral plane, and it can change throughout the fire. In a bidirectional flow path, the neutral plane is the level in a compartment opening (door or window) at which the pressure of the hot smoke and gases leaving the compartment and the pressure of the cooler gases entering the compartment are equal **FIGURE 2-47**.

Thermal layering is an important safety concept for firefighters to observe and understand. For an offensive interior attack, it can determine if firefighters should enter the *clear tunnel* and operate in a room that is on fire or, if the tunnel is narrow, operate the hose line from a safe distance **FIGURE 2-48**. For a fire that has just been extinguished, the thermal layer and its temperature could provide an indication about whether the smoke can quickly reignite or if it has significantly cooled **FIGURE 2-49**. Modern personal protective equipment (PPE) and SCBA that are NFPA compliant encapsulate and insulate firefighters from sensing or feeling the heat in the upper levels of the thermal layers, giving them a false sense of security while working in a volatile, smoke-filled, high-temperature environment. Thermal-imaging cameras can be used to check the temperatures toward the ceiling in a room, but it is still smarter and safer to stay low to the floor and enter beneath the neutral plane for interior attacks in structural fires because the temperature is lower, and the visibility is better closer to the floor. It is safer still to flow water from the outside through a window to knock down, cool, and take the energy away from the fire. This tactic "resets" the fire.

FIGURE 2-47 In a bidirectional flow path, the neutral plane is the level in a compartment opening (door or window) at which the pressure of the hot smoke and gases leaving the compartment and the pressure of the cooler gases entering the compartment are equal.

Courtesy of Raul Angulo.

FIGURE 2-48 If the thermal layer provides a clear tunnel to enter the structure, an interior attack may be warranted. If the tunnel is narrow, as in this training fire, operate the hose line into the structure from a safe distance to cool the interior atmosphere.

Courtesy of Raul Angulo.

As the fire continues to grow and interior temperatures increase, flammable gases are produced and small dancing flames start to appear in the upper layers near the ceiling. As the flammable vapors burn off,

FIGURE 2-49 For a fire that has just been extinguished, the thermal layer and its temperature could provide an indication about whether the smoke can quickly reignite.

Courtesy of Raul Angulo.

FIGURE 2-50 Rollover (or flameover).

the flames go out. These isolated flames are an indication that the gases are within their flammable range and that this layer is hot enough to cause ignition. Often these flames appear around the edges of the hot plume because this area has sufficient oxygen for combustion. There is not a sufficient amount of vapor being generated from the fuel to sustain the flames. This is the start of rollover. The ceiling atmosphere must be cooled with water to stop rollover.

Rollover

Rollover happens during the growth stage of the fire on the upside of the standard time-temperature curve. In a room fire, the hottest gases rise to the top and accumulate at the ceiling. The superheated gases stratify, and thermal layering occurs. If these superheated gases reach ignition temperature, they spontaneously ignite through radiation coming from the fire or from direct contact with open flames in the uppermost thermal layers, causing the fire to travel across the top of the compartment. This is also known as flameover—where the upper layer of flammable vapors catch fire. These convected waves of flames can flicker across the ceiling and then go out when the flammable vapors are consumed or the convecting currents reach a balance between the heat and air. The process repeats itself in shorter intervals because the fire is gaining headway and producing more energy.

Rollover is a sign that the room temperature is rising, and if it continues to rise, more heated gases are generated. The hottest layer of gases light off again, sending flames that extend along the ceiling level of the container. Remember that under pressure, smoke, fire gases, and vapors always take the path of least resistance, so a rollover can cause the fire to spread horizontally to unaffected low-pressure areas of the compartment, and move towards the source of oxygen, thus expanding the amount of area involved. The superheated gases are a reservoir of fuel. The flames at the upper level radiate energy downward, increasing the pyrolyzation and temperature of lower-level fuels in the compartment. When the fire has brought the entire contents to their ignition temperature, flashover occurs.

Rollover can be very dangerous. It can occur quickly and cause the fire to travel like a canopy over the top of the firefighting team and sometimes get behind them, cutting off their egress from the compartment. Rollover is different from flashover in that only the ceiling vapors are burning in addition to the seat of the fire, not the entire contents of the room **FIGURE 2-50**.

Rollover is your last warning sign before flashover. Flashover is imminent if actions are not taken to change the fire conditions. You must apply water on the fire or cool the gases at the ceiling with a hose line to prevent flashover.

Flashover

A **flashover** is a high-temperature event. As the fire continues to burn in a room, additional heat and smoke are generated. The hottest layers of fire gases, which are at the ceiling, begin to bank down toward the floor. These superheated burning gases raise the temperature of all the surrounding unburned contents within the room. When the contents in the room have absorbed all the heat energy they can withstand from the fire, they reach their ignition temperature, and flashover occurs. Flashover is the rapid transition from the growth stage to the fully developed stage. In a flashover, the entire contents of a room ignite almost simultaneously, releasing a tremendous amount of energy and generating intense heat and flames **FIGURE 2-51**. This rapid change in the compartment occurs in seconds.

FIGURE 2-51 A flashover is a high-temperature event.
Courtesy of Raul Angulo.

The threshold temperature for a flashover to occur is approximately 1,000°F (538°C). Once this temperature is reached, all the fuels in the room are involved in fire, including the floor coverings, so the temperature at floor level may be as high as temperatures at the ceiling before the flashover occurred.

Even with all their structural PPE, which is rated to withstand temperatures up to 500°F (260°C) if it's NFPA 1971 compliant, firefighters cannot survive for more than a few seconds in a flashover. Considering reaction time, a firefighter would have to be between 6 and 10 feet from an exit door for any chance of surviving this dangerous and deadly occurrence.

There are some fire conditions in which flashover is delayed or does not occur. Flashovers are less likely to occur if the fire is in a large room with high ceilings that supply more air during fire development. Larger areas with high ceilings entrain more air in the fire plume, which tends to cool the plume and thus reduces the radiation and heat convection toward other contents in the room. With high ceilings, it takes longer for accumulated heat to radiate back down to other fuels at floor level. This doesn't mean the fire isn't going to continue to grow; it simply means a flashover may not occur.

Flashovers do not occur at every fire. In fires that remain ventilation-limited or oxygen-deficient, a flashover may not occur. A ventilation-limited fire produces a limited amount of heat energy. If the oxygen supply is not increased, the fire starts to enter the decay stage. During the decay stage, heat continues to pyrolyze fuels in the room. The pyrolysis produces additional vaporized fuel in the form of smoke. A fire in the decay stage can quickly flash if oxygen is introduced to the smoldering fire. The fire gases in smoke are already heated to their ignition temperatures (or just below) in the decay stage. This is also where smoke explosions can occur. Keeping doors and windows closed in the fire room helps reduce the amount of oxygen available to the fire and may help prevent flashover. Flashovers can be prevented by cooling the fuel with water; cooling the ceiling atmosphere with water, which reduces the temperature of the gases; removing the fuel; or reducing the amount of oxygen present by controlling ventilation.

Backdraft

A backdraft is another high-temperature event caused by the sudden introduction of oxygen into an enclosure where superheated smoke, gases, and contents are at or above their ignition temperature (+1,200°F/+649°C) but do not have sufficient oxygen to burn, or the gases are too rich to burn. The fire tetrahedron shows that four elements must be present to support combustion: oxygen, fuel, heat, and a continuing chemical reaction. A fire burning in a "closed box," meaning a room or structure that is ventilation-limited, can consume all the available oxygen within the compartment, slowing the combustion process and the intensity of the fire; however, the room is filled with superheated fuel in the form of smoke. At this stage, the fire is "starving" for oxygen. The sudden opening of a door or a window (often by firefighters making the initial entry) introduces a fresh supply of oxygen to a superheated, fuel-rich environment, and the four elements of the fire tetrahedron quickly come together, causing the room to burst violently into flames. Fire and smoke conditions within the structure should be evaluated for signs of backdraft before any type of ventilation opening is made **FIGURE 2-52**.

A backdraft is sometimes confused with a flashover. They are not the same event, even though both occur in very high temperatures on the upward side of the standard time-temperature curve. They usually can occur between the growth stage and fully developed stage of a fire. A backdraft is also sometimes called a smoke explosion, but that is also incorrect; again, they are not the same thing. A backdraft involves superheated temperatures, whereas a smoke explosion typically occurs during the decay stage of a fire with cooler atmospheric temperatures of the smoke. Signs of a potential backdraft include (**FIGURE 2-53**):

- Extreme heat (+1,200°F)
- Any confined fire with a large heat buildup in a tightly sealed building

A

B

FIGURE 2-52 **A**. Fire and smoke conditions within the structure should be evaluated for signs of backdraft before any type of ventilation opening is made. **B**. Improperly venting a fire in a ventilation-limited atmosphere can cause a backdraft.

FIGURE 2-53 No visible flames from the exterior of the building, smoke that has filled the compartment, and heavy black smoke pushing out under pressure are all signs of a backdraft condition.

- No visible flame from the exterior of the building
- Dull red flame in the thick of the smoke
- Puffing of smoke from seams and cracks of windows and doors that is then sucked back into the building
- Heavy black smoke pushing out under pressure
- Smoke-stained windows that are hot to the touch (an indication of a significant fire), or windows may be bowed
- Turbulent smoke
- Thick yellow-brown or yellowish-gray smoke (containing sulfur compounds)
- No smoke showing
- Large open-area structures, for example, supermarkets, bowling alleys, department stores
- Large open-void spaces, for example, giant attic spaces, cocklofts between hanging ceilings

A backdraft is a dangerous situation. This type of explosive combustion can exert enough force to cause structural damage to the building as well as severe injury and/or death to firefighters. A backdraft can occur naturally when a window fails due to excess heat and pressurized smoke, or it can occur by accident when a firefighter opens a closed door or breaks a window for ventilation.

There are basically two strategies to prevent a backdraft from developing; both need to release as much heat and unburned fuel gases as possible. The first is to ventilate vertically at the highest point of the structure. This draws the pressurized hot gases and smoke and allows them to escape into the atmosphere, but only if ventilation is directly above the smoke-charged compartment. Vertical ventilation does not help if the potential backdraft exists on the third floor of a six-story apartment building.

The second strategy, which is much more dangerous, is to ventilate the fire room horizontally from a safe distance with overwhelming water using straight streams from 2½-inch nozzles, a deluge monitor, and/or a deck gun. Keep personnel clear of all building openings. This defensive tactic may intentionally cause the backdraft to occur, but with the immediate deluge of water application, the interior temperatures of the gases and fuel are reduced immediately, lessening the effects of the backdraft. Remember, there is no chance of survival inside a pre-backdraft compartment. Because this is a deliberate tactic, the backdraft won't catch firefighters off guard, but it can still produce a shock wave that can destabilize the building. There is apparently no other way to prevent one, other than standing back and letting events naturally play out.

Firefighters must train themselves to look for the signs of a potential backdraft situation when they are on the scene of fires in closed-up areas or structures. They must resist the sense of urgency to quickly rush in.

Rapid Fire Growth

Some old fire service clichés were rooted in solid fire science, while others probably were not. "Put the wet stuff on the red stuff" still holds true and yields positive results, while "vent early and vent often" can have disastrous results. Perhaps the latter was first said tongue-in-cheek, but many firefighters took it as gospel, as is evident in many videos that still show firefighters indiscriminately breaking every window in their path, thinking this will improve interior smoke condition when, in reality, it creates ample openings for air to enter the structure, allowing the fire to grow rapidly. A fire that could have been easily controlled with one well-placed hand line ends up burning the structure to the ground. These firefighters are not necessarily careless. They are merely acting on the misinformation they have been told. The fire service has not traditionally taught that making entry into a building is part of ventilation, but that now needs to change.

Introducing air into a ventilation-limited (oxygen-deficient) fire can result in explosive fire growth. Research has demonstrated that firefighters making entry through the front door of a structure can introduce enough air into the fire area to produce rapid fire growth and flashover. This change can occur within two minutes and progress so fast that it is impossible for firefighters to escape this deadly environment. We need to carefully consider what constitutes ventilation. Opening any door, window, skylight, or roof introduces oxygen into a burning building. Studies have shown that opening the front door of a closed structure has a significant impact on the growth of the fire. The top-to-bottom open space creates a large opening for air to rush in. Repeated experiments have produced violent flashovers or backdrafts after the front door is opened.

Remember, smoke and fire gases are fuel. Rapid fire growth can be prevented by cooling the superheated ceiling environment, cooling the solid fuels, removing the fuel, or controlling the amount of oxygen entering the structure. During rapid fire growth, a fire can progress to its maximum heat release rate unless immediate actions are taken to limit the flow path of oxygen or to apply water on the fire. Hot smoke flowing swiftly through hallways, corridors, and stairways immediately spreads the fire. Smoke ignition is causing this rapid fire spread. Know the difference between heat-pushed smoke versus volume-pushed smoke. Take notice and be forewarned of smoke moving faster than firefighters can. **TABLE 2-2** is a list of hostile fire events with warning signs and subject notes for quick reference.

Behavior of Ventilation-Limited Fires

Recent experiments conducted by UL, NIST, and the New York City Fire Department (FDNY) have demonstrated that many building fires become ventilation-limited because of a limited supply of oxygen. Newer homes are tightly sealed with added insulation, caulking, double- and triple-paned windows, and storm windows. A fire in a structure during the growth stage can consume all the available oxygen to reduce the concentration below the level needed to sustain a growing fire (an oxygen-deficient atmosphere). When this occurs, the fire is said to be *ventilation-limited* versus a fire that has an unlimited supply of oxygen.

A ventilation-limited fire may appear to be a small incipient fire, but there may be a large amount of energy in the form of smoke and heated gases, which is actually vaporized fuel. Flames may not be visible. If the oxygen supply is reintroduced, a ventilation-limited fire can quickly progress to a flashover in less than two minutes! Research indicates that fires in modern residential structures are likely to enter a ventilation-limited, decay stage prior to the arrival of the first-in engine company. A ventilation-limited fire can be deceiving. From the exterior, what appears to be a "small and relatively insignificant" fire may in fact be a ventilation-limited fire containing enormous amounts of energy that only needs a new supply of fresh air for explosive fire growth. A thorough 360-degree size-up is the only way to identify this situation.

Smoke Explosions

The standard time-temperature (bell-shaped) curve of a fire is a graph firefighters are familiar with. The temperature differential between a flashover and a backdraft are very close together. Both events are high-temperature events on the same side of the curve: the upward incline of the graph where the temperatures are high. When a fire starts, the temperature is just above ambient and the pressure inside the room is low. As the temperature rises, so does the pressure. The upside incline of the graph indicates the pressure inside the structure is building but it is not yet pressurized.

TABLE 2-2 Hostile Fire Events

Event	Warning Signs	Notes
Flashover	■ Turbulent smoke flow that has filled a compartment ■ Ghosting ■ Vent-point ignition (exterior autoignition of the exhaust flow path) ■ Rapid change in smoke volume and velocity (getting worse in seconds)	Flashover is an event triggered by radiant heat reflected within a room or space. All surfaces reach their ignition temperature at virtually the same time due to rapid heat buildup in the space. If air is present, the room becomes fully involved. If air is not present, the flashover is delayed until air is introduced (ventilation-limited phase).
Explosive growth phase	■ Dense smoke appears to have totally filled a building, floor to ceiling ■ Slow but steady smoke flowing from closed doors or windows ■ Smoke that rapidly speeds up when an exterior door is opened ■ The development of an air track below the smoke when an opening is made ■ Exhaust flow paths that intermittently puff or try to suck air (open doors and windows) ■ Smoke flameover upon the breaking of windows or opening of doors (late sign)	Explosive growth is a phase that occurs when air is introduced to a ventilation-limited fire. It can include smoke flameover in flow paths and flashover of individual rooms that are heat saturated.
Backdraft	■ Yellowish-gray smoke emitting from cracks and seams ■ Bowing, black-stained windows ■ Closed pressurized box with signs of extreme heat ■ Sucking and puffing from the cracks and seams of a closed box	Backdraft occurs when oxygen is introduced into a closed, pressurized space where fire products are above their ignition temperature. *Note:* Sucking or puffing of smoke from an *open* door or window is a warning for explosive growth. Sucking and puffing of the cracks and seams of a *closed* pressurized space is a backdraft warning.
Smoke explosion	■ Smoke that is being trapped in a separate space above the fire ■ Signs of a growing fire ■ Signs of smoke starting to pressurize	A smoke explosion occurs when a spark or flame is introduced into trapped smoke that is below its ignition temperature but above its flash point. A proper air mix is necessary. Carbon monoxide (CO) is ignitable (with a spark or flame) at around 300°F (149°C) and has a wide flammable range with the proper air mixture. As trapped CO heats up, its flammable range widens between 12.5% and 74%, making it easier to ignite with a spark or flame as it gets hotter. Smoke explosions typically happen in spaces away from a growing fire.
Flameover	■ Increase in smoke speed ■ Ghosting ■ Laminar flow of smoke that is becoming turbulent ■ Smoke flowing from hallways and stairways faster than a firefighter can move	Flameover occurs when smoke reaches sustaining temperatures that are above the fire point of prevalent gases. The gases can suddenly ignite when touched by an additional spark or flame. Fire spread changes from flame contact across content surfaces to fire spread through the smoke. This marks a significant change in fire spread behavior.

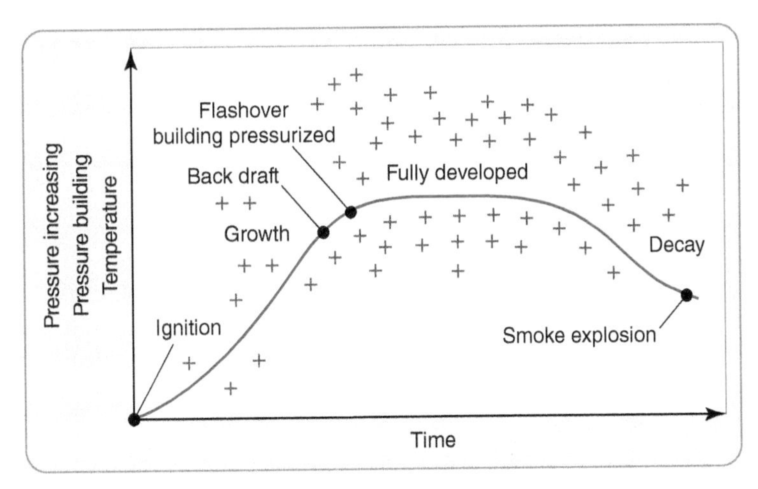

FIGURE 2-54 Smoke explosions typically happen during the decay stage of a fire which is on the back side (decline) of the time-temperature curve. Temperatures are cooling, but the building is now pressurized.

Courtesy of the National Institute of Standards and Technology (NIST).

A **smoke explosion** typically happens during the decay stage of a fire, which is on the back side (decline) of the time-temperature curve **FIGURE 2-54**. The fire has already used up the available oxygen, and incomplete combustion is occurring or has occurred. As the oxygen levels decrease in the structure, the production of carbon monoxide (CO) increases. Temperatures are cooling, but the building is now pressurized. A smoke explosion occurs when a mixture of CO, flammable gases, and oxygen is present, usually in a void space or other pocketed areas separate from the fire room. Because the building is as pressurized as it can be, smoke can travel quite a distance away from the fire. The smoke can be right at its flash point or cooled just below its ignition temperature. Many of the flammable gases are already within their flammable range. Remember, CO has a very wide flammability range (12.5%–74%). When it comes in contact with an ignition source, like a smoldering ember, the flammable mixtures can quickly ignite, often in a violent manner.

The conditions needed to produce a smoke explosion often include the presence of void spaces, combustible building materials, a ventilation-controlled fire that produces unburned fuel (primarily carbon monoxide), relatively cool temperatures, and an ignition source. Although smoke explosions usually occur during the decay stage of a fire, they are not limited to this stage. Smoke explosions result in a violent explosion during final extinguishment or overhaul when smoldering embers may be uncovered by firefighters, thus providing the ignition source. In a smoke explosion, there is no change to the ventilation profile, such as opening a door or window; rather, it occurs from pressurized smoke traveling within the structure to an ignition source.

Wind-Driven Fire

One factor we cannot control that greatly influences fire behavior is the wind. NIST research has shown that wind speeds as low as 10 mph can create wind-driven fire conditions inside a structure. When wind is present, it can be a powerful, pressure-imposed force that can drastically change the direction of a flow path. Whether by course of the fire or by deliberate actions of firefighters making entry or ventilating the building, the strong sudden increase of voluminous wind currents upon a fire can result in rapid fire growth and extreme fire behavior. This is referred to as a wind-driven fire.

Air currents always move from high pressure to low pressure. For example, when the air is relatively still, a room fire may generate an interior atmospheric pressure that is greater than the ambient exterior atmospheric pressure. A strategically placed ventilation hole close to the seat of the fire draws the flow path right out of the structure. The best approach to this fire may be from the uninvolved side toward the involved side of the structure. However, if the wind is blowing 20 to 25 mph (32 to 40 kph) from the opposite side of the building, say, the C side of the building, entering from the A side may be a deadly miscalculation because the wind would push the fire toward the entry team **FIGURE 2-55**. The impact of the strong wind currents would be similar to placing a mobile ventilation unit (MVU) on the C side of the building and pushing huge quantities of air, smoke, and fire toward the crews entering on the A side of the building.

FIGURE 2-55 Interior fire conditions can be affected and intensified by an exterior wind, which can produce rapid fire growth and extreme fire behavior.

The best advantage for firefighters making entry into a structure for fire attack is to keep the wind at their backs. This is a safe approach, but it may require a positive pressure fan behind the crews while they make entry.

The potential effect of the wind may not be evident during the initial size-up when the structure is closed up tightly. However, the wind can have a tremendous effect on fire behavior, so the initial size-up of a structure fire must always include an evaluation of the impact of the wind. This evaluation sometimes changes the initial strategy and tactics for attacking the fire.

Fire Behavior in Modern Structures

Modern building construction practices, along with plastic and synthetic home furnishings, have influenced fire behavior in today's structure fires. Homes built in the last 20 to 30 years are built to be energy and cost efficient. They contain insulation to prevent heat loss in the winter and maintain cool air in the summer. In addition, they are sealed to prevent air exchange through the walls and around the door enclosures and window openings. As a result, when a fire occurs in a modern house, only a limited amount of air can enter the structure when the doors and windows are closed. In many parts of the country, doors and windows are kept closed year-round due to installed heating and cooling systems.

New homes are constructed with lightweight engineered lumber and building materials. These manufactured building components and composites contain many highly flammable products such as glues, resins, and other coatings, most of which are made from petroleum products. Engineered lumber has less mass than houses built with dimensional lumber. Petroleum-based products pyrolyze faster and contain more potential heat energy; thus, they have a higher heat release rate than fires in older homes constructed primarily of solid wood. These products are not only flammable but also burn faster and hotter and are less predictable than legacy fuels. The lightweight frames of newer homes fail much faster than the heavier wood frames of older homes. The lightweight construction has less mass, provides greater surface area of structural members, and provides numerous void spaces between floors.

The widespread use of plastics and petroleum-based products for furniture, home furnishings, coverings, and accessories provide considerably more energy release potential than wood furniture or natural fiber coverings. They ignite easily; reach high temperatures quickly; have a high heat release rate; and release huge quantities of thick, black, acrid smoke when burning. The rapid combustion rate requires large amounts of oxygen to remain burning and larger amounts of water to extinguish. These factors in combination have changed the dynamics of fire behavior and have challenged our traditional understanding of fire behavior inside a structure.

Fires in well-insulated modern structures progress to the growth stage quickly. In most cases, these fires then enter a ventilation-limited stage because there is a limited supply of oxygen coming in from the outside. During this ventilation-limited stage, the heat from the fire continues to pyrolyze available fuels, thereby increasing the supply of carbon monoxide and other vaporized fuel. All that is needed is a new supply of fresh air to change the fire back into a rapid growth stage, often in the form of a flashover or a backdraft, and then on to the fully developed stage.

Often, the fire department arrives on the scene while the fire is in a ventilation-limited decay stage. One of the first actions taken by the first-in engine crew is to open the front door to gain access to the house. Unfortunately, opening the door has the unintentional effect of introducing a fresh supply of oxygen to the fire. When fresh air comes in contact with a smoldering fire and flammable fuel in the form of smoke and gases, the result can be rapid fire growth in the form of a flashover.

To prevent this scenario from happening, firefighters must limit the ventilation and the introduction of fresh air to the fire by keeping doors closed until a hose line can spray water from the outside through a window to knock down the fire. This is called a transitional defensive to offensive attack. These actions should be coordinated with the ladder company. The other tactic being used by some fire departments is the placement of a smoke curtain at the front door before making entry. This tactic takes some practice (and convincing for some firefighters), but it is effective in limiting the air intake as the hose team enters the structure for a direct attack and prevents the creation of the flow path, which would intensify the fire.

Classes of Fire

Fires have been divided into five classifications based on the type of fuel burning: Class A, B, C, D, and K. These classifications allow the correct extinguishing agent to be applied by firefighters (and the public). Engine company fireground operations are more than simply pulling fire hose and applying water, although water is the most common extinguishing agent used in the fire service. Because each type of fuel has different burning characteristics, it is important for safety and effectiveness that firefighters know and understand which method of extinguishment is appropriate. The five categories are discussed here.

Class A

Class A fires are fueled by ordinary combustibles, for example, wood, paper, cloth, cellulose, plastics, and other organic materials FIGURE 2-56.

FIGURE 2-56 A Class A fire involves wood, paper, and other ordinary combustibles.

FIGURE 2-57 Class B fires involve flammable liquids such as gasoline.

Class B

Class B fires are fueled by flammable and combustible liquids and gases such as gasoline, diesel, oil, alcohol, and other petrochemicals FIGURE 2-57.

Class C

Class C fires are basically fueled by electricity or an energized heat source. Although we call them electrical fires, the electricity is actually just the heat source energizing electrical equipment or generating the combustion process on a Class A or Class B fuel. Once the source of electricity is shut off, the fire becomes either a Class A or a Class B fire FIGURE 2-58.

Class D

Class D fires are chemical reaction fires fueled by combustible exotic metals, which are a particular class of heavy metals that can be identified in the periodic table of elements and found primarily in the alkali

FIGURE 2-58 A Class C fire involves energized electrical equipment.

© Shay Levy/PhotoStock-Israel/Alamy Stock Photo.

metal group. Magnesium, titanium, lithium, calcium, zirconium, sodium, and potassium fall into this group. They burn at extremely high temperatures and emit brilliant light. Many of these metals react violently, with spectacular flashes, when water is used as the extinguishing agent **FIGURE 2-59**.

FIGURE 2-59 A Class D fire involves exotic metals such as magnesium, titanium, and sodium. Many automotive manufacturers use magnesium components to reduce the weight while increasing the strength of vehicles.

© Albert Russ/Shutterstock.

Class K

Class K fires are primarily commercial kitchen fires fueled by cooking oils and fats. Restaurants now operate with modern high-efficiency, commercial, deep-fat fryers that use vegetable oils instead of animal fats to fry food. This has resulted in higher cooking temperatures and required the development of a new class of wet-chemical extinguishing agent. A Class K extinguisher is more effective than a Class B extinguisher on commercial cooking fires during the ignition or growth stages **FIGURE 2-60**.

FIGURE 2-60 Class K fires involve combustible cooking oils and fats that produce higher cooking temperatures.

© Kathie Nichols/Shutterstock.

Reading Smoke at Structural Fires

Learning the science, physics, and principles of fire behavior helps firefighters understand how a fire burns. As author Dave Dodson explains in his book *Reading Smoke*, one important practical application of these principles is learning how to *read* the smoke at a fire. Traditionally, firefighters have referred to smoke as "light smoke coming from . . ." or "heavy smoke showing. . . ." These phrases may be fine for an enroute radio report, but they are not descriptive enough during an initial radio report when trying to describe for incoming units what is actually happening with the fire. In the past, if all that was showing was smoke, we often thought it was a positive sign—the absence of flames meant the fire hadn't actually started yet because smoke was defined as incomplete combustion. Today, with the understanding of pyrolysis, we now know that the accumulation of smoke is an accumulation of fuel. Materials that are producing smoke are actually burning, but the rapid oxidation process is not yet the chemical chain reaction we see in fire. The accumulation of dense unburned smoke can be more dangerous for firefighters than the fire. Flames visible is actually a good thing. When the fire self-vents, the combustion process is more complete. Although the fire intensifies with radiant heat affecting fuels, the danger associated with unignited smoke accumulation within the compartment has been minimized.

Being able to read smoke gives clues and potentially enables you to learn which stage the fire is in; what type of fuel is burning; the exact location, size, and speed of the fire; the pathway it's taking; and what direction the fire is heading toward; in other words, it allows you to answer the question, Where is the fire going? Most important, reading smoke helps the company officer recognize the likelihood of explosive and rapid fire events (as shown in Table 2-2).

At a fire, it's easy to focus on the flames. Flames indicate where the fire is now, but they do not tell you how big the fire is or where it is headed; you need to pay attention to the smoke. Fires are dynamic events—what you see at the moment will probably change quickly. Before developing a strategy of attack, the company officer must be able to anticipate where the fire will be in a few minutes. At a structure fire, sometimes there are no visible flames because the seat of the fire is located inside the building, or it has already entered the early decay stage. The ability to read smoke gives the company officer the information and warning signs to:

- Predict fire behavior.
- Determine the location and intensity of the fire.
- Determine the stage of the fire.
- Save the lives of building occupants and firefighters.
- Make better strategic and tactical decisions.
- Mount a more effective attack on the fire.
- Control the effects of ventilation on the fire.
- Warn crews of hostile fire events like flashover, backdraft, smoke explosions, and rapid fire spread.
- Recognize building collapse potential.

The Four Factors

Four conceptual factors need to be remembered when reading smoke, as described in the following subsections.

Factor 1: Smoke Is Fuel

As mentioned earlier, smoke is fuel! It is composed of three major components: particulates, vapors (aerosols), and gases. The particulates are solids that consist of soot (carbon), ash, dust, and fibers that are suspended within the smoke; they make smoke visible. Particulates have a high surface-to-mass ratio, which means they have the ability to absorb heat. Soot, a prevalent particulate, is carbon, which is technically called *carbon black* and can support flaming. Ash, another prevalent particulate, includes the trace metals and minerals (depleted salts) that can no longer support flaming but can still retain a significant amount of heat that can ignite other vapors and gases within the smoke. Anyone who has knocked the ash off a cigar knows there's enough retained heat to burn a hole in a shirt or burn the surface of the skin, although the ash itself no longer burns. Carbon particulates give smoke a spectrum of black color; ash gives a spectrum of white color. Along with dust and fibers, particulates give smoke its *density* and thus the ability to displace air, which is a contributing factor in creating a ventilation-limited fire.

Vapors, or aerosols, within the smoke are primarily water vapor, which is a by-product of combustion; when solid fuels heat up, they release their moisture content. But vapors also come from pyrolysis, the chemical breakdown of solid fuels. Within the moisture are oils and tar (hydrocarbons), along with acids, aldehydes, ketones, and other chemicals. Synthetic fuels (plastics) release the hydrocarbons, which gives smoke its glossy black color. With the right air mixture, often provided by the firefighters making entry, some of these hydrocarbon vapors can self-ignite in temperatures as low as 450°F (232°C).

The *gases* within smoke are numerous and are already in the perfect state for explosive combustion. No more decomposition needs to occur. The five most flammable gases prevalent in the smoke that firefighters encounter on regular structure fires are carbon monoxide (CO), hydrogen cyanide (HCN), benzene (C_6H_6), acrolein (C_3H_4O), and hydrogen sulfide (H_2S). These five gases in particular affect the behavior of fire because ignition temperatures are as low as 450°F (232°C) and have a collective flammable range in air between 1% and 74%. In addition to their flammability, these gases are deadly for occupants and extremely toxic for firefighters. The use of SCBA is necessary for protecting the respiratory tract, but these chemicals can also be absorbed through the skin, which can lead to long-term chronic health issues, including cancer.

Combined, these three components of smoke (particulates, vapors, and gases) represent the fuels that burn when heated to their ignition temperatures in an oxygen-rich environment. Therefore, hot smoke is extremely flammable and ultimately dictates the fire's behavior. Smoke today is not the smoke of yesterday.

Factor 2: Fuels Have Changed

In addition to being lightweight, having less mass, and having more surface area, most of today's building materials and contents are synthetics and plastics, which are made from petroleum products. As a consequence, these materials give off more toxic gases and burn faster at higher temperatures than materials used for construction and building contents in earlier decades.

Factor 3: Fuels Have Trigger Points

The triggers for hostile fire events are the right temperatures and the right mixtures for gases. The gases have to be within their flammable range, with the right percentage of oxygen to ignite. They can't be too rich or too lean. If the gases can ignite, they can explode. If they cannot ignite, they cannot explode. Firefighters making entry into a structure can introduce fresh air to create the right mixture of oxygen to cause these gases to ignite or explode.

Ignition temperature is the lowest temperature at which a fuel produces an ignitable mixture that can self-ignite. Ignition temperature triggers autoignition for flashover, backdraft, and smoke explosions. Fire point is the lowest temperature at which a fuel produces an ignitable mixture that ignites and continues to burn, given an ignition source like a spark or a flame. Fire point triggers rapid fire spread. Flash point is the lowest temperature at which a fuel produces an ignitable mixture that simply flashes but does not sustain a spark or flame. Flash points can trigger a smoke explosion.

Factor 4: Occupant Survivability Profile

Survivability profile is a relatively new term that has emerged in the fire service lexicon. The number one mission of the fire service is to save lives. However, the preservation of the lives of firefighters is paramount. In addition to the incalculable worth of every individual, fire departments invest years of training in each individual to carry out their primary mission. Firefighters *are* our most valuable resource, especially those with years of experience. So let's protect them, and we do that through education in fire behavior, building construction, and effective size-ups; implementing effective strategy and tactics; and through risk-benefit management.

It's easy for an incident commander to write off a building as lost; it's not so easy to write off a life that may be trapped inside a fire. There is a moral dilemma involved in sacrificing the life of a civilian occupant, a decision most incident commanders would prefer to avoid—and that's understandable. However, their solution often is to risk firefighter lives in unjustifiable interior searches without realizing there is a moral dilemma when the survivability profile is zero for both firefighters and civilians. The vast majority of firefighter line-of-duty deaths do not occur during the actual rescue of a trapped occupant. These deaths occur because firefighters are getting caught in a hostile fire event. Understanding fire behavior, learning how to read smoke, and knowing *human life thresholds* can help the first-in company officer make smart decisions based on fire science rather than on emotion.

For purposes of initial size-up and decision making, the human life threshold is knowing that the human trachea or airway can only be exposed to or withstand temperatures around 195°F (91°C) for approximately one minute. The unprotected human skin, the largest organ in the body, can only be exposed to or withstand temperatures around 300°F (149°C) for approximately one minute. After that, survival is dismal or temporary at best.

The Four Steps of Reading Smoke

A smoke column is one of the first signs visible to the first-in unit. One of the most important size-up criteria that the first-in company officer needs to identify when pulling up to the location is whether the smoke is laminar or turbulent. Laminar means the fire is still in its early growth stage. A positive survivability profile exists in a room fire during the ignition and early growth stages of a fire when floor temperatures are still below 195°F (91°C). Turbulent means there's heat behind the smoke, and the fire is close to flashover (temperatures are high). Noting the smoke characteristics determines initial actions.

Reading smoke includes four steps. The best location for the incident commander to observe smoke patterns and behavior is from the outside perimeter of the fire building. Compare the smoke coming out of the openings from different parts of the building. Again, the smoke tells you more about interior fire conditions than simply looking at the flames.

Step 1: Determine the Key Elements of Smoke

The first step in reading smoke is determining the four key elements of smoke: volume, velocity, density, and color. The company officer also needs to factor in the size of the building, the layout of the building, the wind, and any other characteristics of the building that might change the appearance of the smoke's key elements.

Volume. The volume of smoke coming from the fire gives some idea of how much fuel is being heated to the point that it is releasing vapors and gases (off-gassing). As you assess the smoke volume, also consider how much smoke it takes to fill the box. It takes relatively little smoke to fill up a small building, but a substantial amount of smoke is required to fill up a large building. Volume is relative to the box, so assessing the volume of smoke alone does not provide a complete picture, but it sets the stage for a better understanding of the fire. A well-ventilated, free-burning fire in the fully developed stage may produce very little smoke volume, whereas a ventilation-limited fire in an enclosed container can produce a significant volume of smoke. Once the unventilated container fills with smoke, it's pressurized. Pressure (and smoke) seeks the path of least resistance. The pressurized smoke is one of the factors that determines the velocity of the smoke as it attempts to escape the container.

Velocity. The velocity (speed) at which smoke leaves the building suggests how much pressure is accumulating inside. Only volume and heat can cause pressure; therefore, smoke is pushed either by heat or by volume. When smoke is pushed by heat, it rises and then slows down gradually as it cools. When smoke is pushed by volume, it slows down almost immediately once the pressure is equalized. Assess the smoke velocity to determine whether it is being pushed by heat or volume.

Next, consider whether the smoke has a laminar flow or a turbulent flow. **Laminar smoke flow** is a smooth, lazy, streamlined flow produced by pyrolysis **FIGURE 2-61**. It indicates that the box (the building and its contents) is still absorbing heat and that the temperature and pressure in the box are relatively low. This is the stage of the fire where the survivability profile of an occupant is high. An interior search may yield a positive outcome. Identifying laminar smoke correctly can justify initiating an interior search for victims. Water application through a window from the outside is advantageous because it cools the atmosphere and "resets" the fire, making it safer to commence an interior search for victims.

FIGURE 2-61 Laminar smoke.
Courtesy of Bill Strite, Cincinnati, Ohio.

FIGURE 2-62 Turbulent smoke.
Courtesy of Captain Nick Morgan, St. Louis Fire Department.

By contrast, **turbulent smoke flow** is fast, agitated, and boiling **FIGURE 2-62**. Turbulent smoke is generated by high heat temperatures: The faster the smoke, the more heat it has pushing it. The molecular expansion of gases occurs when the box cannot absorb any more heat, thus increasing the pressure. In other words, the box is pressurized. In such circumstances, the radiant heat that the box has absorbed is radiated back into the smoke. Turbulent smoke contains an immense amount of energy. When this energy reaches a point where the smoke is heated to its ignition temperature, flashover occurs. The survivability profile in the box is zero. When you see turbulent smoke, realize it is heat-pushed and pressurized; flashover is imminent.

Next, compare the velocity or speed of the smoke leaving different openings of the building. By comparing the speed and type of flow from similar-sized openings, the officer can get a good idea *where* the fire is located. Compare cracks to cracks, windows to windows, and doors to doors. A similar-sized opening with a higher velocity of smoke is closer to the location of the fire. Look for the fastest smoke coming from the most restrictive opening: That's where the fire is. Studying the velocity of smoke provides valuable information on what's happening right now as firefighters prepare to attack the fire.

Density. Next, evaluate the density or thickness of the smoke. Smoke density suggests how much fuel is contained within the smoke. Dense smoke can produce a powerful flashover when conditions are right. In fact, thick smoke can flash without turbulent flow because the supply of fuel reaches from the seat of the fire to the location where the smoke is escaping from the building. So the thicker the smoke, the more spectacular the flashover will be.

Once you start thinking of smoke as fuel, you quickly realize that firefighters who are in zero visibility, surrounded by dense smoke, are working in a pool of flammable fuel. While no one would send a firefighter into a pool of gasoline, we are doing just that when we send firefighters into thick, dense smoke. In addition, dense smoke contains many poisonous substances. When firefighters are in an environment of dense smoke that has banked down to the floor, they are in an environment that does not support life for a building occupant. Just a few breaths of this toxic brew can result in death. If a flashover occurs, this environment immediately changes to a deadly environment for firefighters, even for those with full turnout gear and SCBA.

Color. The color of smoke may give the company officer some indication of which stage the fire is in and what materials are burning. The majority of residential and commercial fires involve multiple fuel materials, so using smoke color alone to determine what is burning may not always yield a clear answer. Most solid materials such as wood emit a white-colored smoke when they are first heated. This white smoke is primarily a result of pyrolysis and moisture being released from the material. The same effect is apparent when firefighters apply water to a fire; the smoke goes from black to white. Generally, white smoke is cooler; black smoke is hotter. As the material dries out, the color of the smoke emitted changes. For example, smoke from burning wood changes from tan to brown. In an attic fire, if you see brown smoke coming from the gable ends of a house, but no flames, it's an indication that combustion is incomplete because the attic space is oxygen-deficient **FIGURE 2-63**. Opening the roof creates a unidirectional flow path, brings flames into the attic, and involves the trusses, and their structural integrity starts to deteriorate. Firefighters should not be on top of the roof or under the trusses. The best tactic is to extinguish this fire using a piercing nozzle or a fog applicator. Any vertical ventilation or water application with a piercing nozzle should be made from an aerial ladder or tower platform.

FIGURE 2-63 As the moisture content of wooden structural members dries out, the color of the smoke emitted changes from light gray to tan to brown.

© Keith Muratori. Used with permission.

Plastics and painted or stained surfaces emit a gray smoke, which is a combination of black from the hydrocarbons and white from the escaping moisture. When synthetic materials are heated, the white vapor created is unburned fuel. Thick black smoke plumes are still classic signs that a flammable petroleum liquid is involved, and unusual yellow-green colors indicate you may be dealing with an exotic chemical or hazardous materials.

The color of smoke can also help firefighters determine the location of the fire. As pressure pushes the smoke away from the fire, its heat evaporates moisture, and passing through building materials filters the smoke. The change in moisture content can change black smoke to lighter smoke the farther it travels from the fire. In addition, carbon particles suspended in black smoke settle out and are also filtered by the materials they pass through; therefore, carbon-rich black smoke often becomes lighter in color as it travels farther from the fire area.

Although white smoke is generally cooler than black smoke, the concern for white smoke shouldn't be discarded. White smoke that is lazy or slow indicates early heating. White smoke that has its own pressure indicates smoke from a hot fire that has already pressurized the building; the smoke can travel for some distance. This is the classic sign for a decay stage fire with conditions ready for a smoke explosion.

Black fire is a high-volume, high-velocity, turbulent, ultra-dense black smoke. Think of it as a form of fire because, in many ways, it is. Black fire temperatures can reach up to 1,000°F (538°C), creating a deadly environment. Black fire can produce charring and heat damage to steel and concrete. Its presence indicates impending autoignition and flashover. Thus, black fire is often just seconds away from becoming a deadly environment, even for firefighters wearing full

PPE and SCBA. Because black fire is not tenable for unprotected occupants of a building, the survivability profile is zero.

Step 2: Determine What Is Influencing the Key Elements

Consider the size of the box that is on fire when reading the smoke. It does not take a very big fire to fill a small building with smoke. Conversely, in a large building (such as a big-box store or a home improvement center), it takes a large fire to overwhelm the sprinkler system and produce enough smoke to fill and pressurize the whole building. A company officer who pulls up to a large big-box building and sees signs of pressurized smoke venting to the outside should realize the existence of a significant fire, call for additional alarms, and prepare for defensive operations.

Realize that the wind and weather affect smoke elements. Wind can change the direction in which smoke travels or scatter the smoke, making it difficult to determine its origin. A hot-humid forecast gives way to a narrow-defined smoke column; a hot-dry forecast makes for a cone-shaped smoke column. A cold-humid forecast causes the smoke to drop and hang low to the ground; a cold-dry forecast also causes the smoke to drop but is easily dispersed. Below-freezing air temperatures can make black smoke look whitish within a few feet of the outside vent opening because the moisture content in the smoke is condensing. These smoke conditions are well known to firefighters in Alaska and Northern Canada. Frozen moisture in the air and smoke is known as ice fog.

Firefighting efforts, which include fire streams, horizontal and vertical ventilation, and the operation of sprinkler systems, all affect the characteristics of smoke. The normal thermal balance in a room or building can be affected by the heating and air-conditioning (HVAC) systems, especially in a high-rise building, not to mention the stack effect, which is covered in Chapter 15. Ventilation openings are made with the intention of changing the direction of the smoke and the fire, so the characteristics of the smoke naturally are expected to change as the building is ventilated, especially when using positive pressure fans (blowers). An activated sprinkler system cools smoke, causing it to hang close to the bottom of the room and making it difficult to see. Use every thermal-imaging camera available.

Step 3: Determine the Rate of Change

To complete the smoke reading, determine the rate of change for the event. Remember that flames indicate what is happening now, whereas smoke gives a more complete picture of the characteristics of the fire and where it is going. Is the volume, velocity, density, and color of smoke changing? How are the characteristics changing? How rapidly are these changes occurring? What do these changes suggest about the progression of the fire? Are things getting better or worse?

Effective strategies and tactics should affect all four elements of smoke. Volume should increase as water converts to steam, velocity should surge with the initial attack and then slow down, density should immediately decrease, and the color should go to white. If the changes are not occurring, the fire attack is inadequate.

Step 4: Predict the Event

The first three steps should help firefighters determine the location and size of the fire, and predict the potential for the hostile fire events: flashover, backdraft, smoke explosion, and rapid fire spread (Table 2-2). Remember, reading smoke is trying to understand what the smoke is saying about the fire. It's all about comparing smoke characteristics escaping from similar openings or cracks in the structure. Retired battalion chief John Mittendorf (Los Angeles Fire Department [LAFD]) says, "The smoke is the fire talking to you."

One way to become more proficient in smoke reading is to review online videos of fires. Assess the smoke at the beginning of the video and try to identify the location of the fire and the stage of burning. This information about reading smoke was not taught in the training academies in the past, but it has to be taught now because today's smoke is deadly. If firefighters are to survive these events, they need to understand smoke. **TABLE 2-3** provides some shortcuts for reading smoke.

Smoke Reading through the Door

Spraying a quick burst of water onto the surface of the entry door before making entry can give you a clue as to how hot the fire is. If the water on the door turns to steam, it can indicate high heat **FIGURE 2-64**. When indications of a hot fire, such as darkened windows with little visible smoke coming from the edges of closed doors and closed windows, are evident, you're encountering a fire that is in the decay stage due to lack of sufficient oxygen. This is a red flag for danger. As soon as this fire receives a fresh supply of oxygen, it will likely produce a violent backdraft or accelerate quickly to flashover. Fires can be dangerous even when little smoke is showing. The company officer must be vigilant and take measures to control a hostile fire event. It's an excellent practice to have a charged hose line at the ready whenever opening any door to a structure that's involved in fire.

TABLE 2-3 Shortcuts for Reading Smoke

What You See	What It Can Mean
Turbulent smoke that fills a box	Warning sign of impending flashover
Thick, black, fast smoke	Close to the seat of the fire, superhot smoke capable of instant ignition; maybe a ventilation-limited fire that needs air
Thin, black, fast smoke	Flame-pushed smoke; fire nearby that is well ventilated
Dirty white smoke with velocity	Heat-pushed smoke that has traveled a distance or has had the carbon/hydrocarbon filtered (like smoke through a crack)
Same color (white/gray) of smoke and same velocity from multiple openings	Deep-seated fire, possibly located well within a building or in combustible voids and concealed spaces
Low-volume white smoke from more than one location of a large building	Working fire deep within
Brown smoke	Unfinished wood reaching late heating (can support flame)—usually a sign that a contents fire is transitioning into a structure fire; *when coming from structural spaces of lightweight wood structures, a warning sign of collapse!*
Yellowish-gray smoke from cracks or seams	Sulfur compounds: a warning sign of impending backdraft

FIGURE 2-64 Spraying a quick burst of water onto the surface of the entry door before making entry can give you a clue about how hot the fire is.

When you open a door, watch and see what the smoke does. If smoke exits through the top half of the door and a tunnel of clean air enters through the bottom half of the door, then the fire is probably on the same level as the entry point. This phenomenon is a bidirectional flow path **FIGURE 2-65**. If fresh air rushes in with considerable speed, the fire is starving for oxygen and is indicative of a ventilation-limited fire. Expect rapid fire growth to develop.

If smoke rises and the opening clears out, then pulls fresh air through the entire doorway back into the building, then this is an intake unidirectional flow path, which indicates that the fire is probably above the level of the opening **FIGURE 2-66**. Following this flow path will lead you directly to the seat of the fire.

Don't enter the structure blind. Use a thermal-imaging camera or a flashlight to see what the smoke is doing. Establish the direction of the smoke current or flow path. If there is no movement, you are likely in a dead-end corridor. If there is no smoke movement with minimal heat, you are probably below the fire floor. If there is little or no smoke movement but temperatures are hot, the fire floor is probably below you.

If smoke thins briefly (because the backup pressure is being released) when the front door is opened, but then steady smoke flows out through the entire doorway, this is an exhaust point of a unidirectional flow path. This means that the fire is below the level of the opening. This type of unidirectional flow path is basically a chimney indicating that this is a basement fire or below-grade fire: one of the most dangerous fires we face **FIGURE 2-67**. Singular stairway access, small windows, or no windows, and the inability to quickly ventilate smoke make basement fires extremely difficult to escape from. Other than a visual confirmation, this may be your only indication that you have a basement fire. You need to find another way to attack it. Do not attempt to make entry down this basement stairway! Regardless of what you have been told in the past, consider everything that has been discussed in this chapter. Making entry through the exhaust portal down into the flow path is placing you and your hose team in the middle of the chimney. This is a high-low flow path, and you have placed yourself and your crew between the fire and the ventilation exhaust portal–where the fire is, and where the fire wants to go. Natural convection air currents are traveling 10 to 15 mph. You cannot outrun this. If the smoke moves faster than you can, consider yourself warned because if this smoke ignites, any victim or firefighter will most likely not survive this event. Remember, smoke is fuel. There are many line-of-duty death case studies to confirm this deadly decision. The energy of the fire must be knocked down from another vantage point before entry can be made down the stairs. If you have been successful in this firefighting tactic in the past, it's not because it was the right decision—it is because you were lucky. Nothing triggered the ignition of the smoke within the flow path.

FIGURE 2-65 A bidirectional flow path indicates that the fire is on the same level as the entry point.

A

B

FIGURE 2-66 **A**. An intake unidirectional flow path indicates that the fire is probably above the level of the opening or point of entry. **B**. This room has already flashed and is fully involved. Note the absence of smoke at the front door. Air is rushing through the entire door space, fueling the intensity of this fire and illustrating a classic unidirectional flow path.

Health and Safety

One last note on smoke: In addition to the flammable, acrid, toxic makeup of smoke that has already been

FIGURE 2-67 Steady smoke flowing out through the entire doorway is the exhaust portal of a unidirectional flow path indicating a basement fire, or that the fire is below you.

covered, the chemicals in smoke are also known to be carcinogens. The fire service has turned the corner on wearing SCBA during the fire attack. It's a rare event to see a firefighter making entry into a fire without SCBA that isn't compliant to the NFPA 1981 standard. The days of the leather lung smoke-eater are over. But not during overhaul. That is a battle we're still fighting. Many firefighters think that, because the fire is out and the atmosphere has cleared, wearing an SCBA is no longer necessary. Not so. Even after complete extinguishment, burned building contents can still off-gas chemicals for days, long after the fire department has left the scene. Many fire department policies state that, if air-reading samples from gas detectors register CO at 10 ppm or greater, SCBA must still be worn during overhaul. But more progressive departments, like Seattle, are taking an even more conservative approach and moving toward a policy that requires their personnel to wear SCBA inside the structure until the building is released back to the owner.

These carcinogens not only enter the body through the airway; they are also absorbed through the skin. Although wearing SCBA can protect us from breathing these carcinogens, there is no way around having them permeate our skin except to reduce our exposure to smoke, especially heavy smoke. The smoke permeates our bunker gear and on a busy shift, we could be wearing our gear all day long.

After-Action REVIEW

IN SUMMARY

- In order to have rapid self-sustained combustion and increasing fire development, there must be a chemical chain reaction to perpetuate the flaming mode of combustion. This spontaneous, self-sustained, chemical chain reaction keeps the fire burning and growing.
- Substances that originated from a living organism are referred to as organic compounds. Organic compounds usually contain elements of carbon, hydrogen, and oxygen.
- Matter in the universe that was never alive or part of a living organism are called inorganic substances. Minerals like iron, quartz, and granite are examples of inorganic substances and do not contribute to the combustion process.
- The "atomic glue" that holds the molecules together is called a bond. Fire, or combustion, has its origin in the bonding of atoms and molecules.
- An oxidizer acts as the catalyst to break down what otherwise would remain a stable molecule. The most common oxidizer available is oxygen.
- There are five basic sources of heat energy: chemical, electrical, mechanical, solar, and nuclear.
- Vapor pressure is the measurable amount of pressure being exerted by a liquid as it converts to a gas and exerts pressure against the sides of a confined closed container.
- Density is the weight of a material and depends on the number of molecules and atoms that occupy a given volume. The more molecules in a given volume, the denser the material is. In addition, the heavier each individual molecule is, the heavier is the material.

- Vapor density is the weight of a gas compared to normal air. The normal concentration of air at sea level is 14.7 psi and is given the designated number 1 when measured at 32°F (0°C). Gases that weigh less than the same volume of air are lighter than air and tend to rise. Their vapor density is designated as a number less than 1. If the gas is heavier than air, the number designation is greater than 1. Such vapors tend to sink.
- Pressure and temperature affect the state of matter. When gases are compressed under great pressure into liquids, they generate heat. When liquids boil, they absorb heat. When a material absorbs heat, it usually does so at its boiling point. The state of conversion from a liquid to a gas under normal atmospheric conditions is called evaporation.
- Because combustion can take place only when a substance is in a gas state, the levels of the gases are described in terms of percentages. Gases can ignite only when certain concentrations of that substance are present in air.
- If there isn't enough combustible gas present to ignite, the mixture is said to be *too lean* to burn. If there is too much gas present, or the concentration is too excessive, the mixture is said to be *too rich* to burn. When a concentration of a gas falls into the range where it can ignite, it is said to be within its flammable, or explosive, limits.
- The new six-stage ventilation-controlled fire growth model consists of the incipient stage, initial growth stage, ventilation-limited stage, explosive growth stage, fully developed stage, and decay stage.
- In a ventilation-limited fire, the progression of combustion is interrupted, and the fire enters into an early decay stage, even though there is plenty of heat and fuel available. Pyrolysis continues, producing more smoke and vapors. All are flammable, creating an extremely dangerous atmosphere where the only ingredient needed to restart the combustion process is air. That source often comes when firefighters make entry into the structure. The transition back into combustion happens with explosive force.
- Heat is transferred from one object to another when the temperatures of those objects are different. Heat always moves from warmer objects to cooler objects.
- The three modes by which heat transfers its energy from one substance to another are conduction, convection, and radiation.
- Convection currents are present at every fire. The most important thing to remember about convection currents and predicting where the fire is going to spread is knowing that convection currents are pressurized and always take the path of least resistance. Therefore, the smoke and fire also take the path of least resistance.
- Pressurized fire gases always want to equalize. Higher pressure always moves toward lower pressure to achieve equilibrium.
- There are two kinds of flow paths: bidirectional and unidirectional. Unidirectional flow paths can be either intake ventilation portals or exhaust ventilation portals.
- The breakdown of solid fuels to ignitable gases is called pyrolysis—the chemical decomposition of a solid substance through the application of heat. Simply put, as solid fuels are heated, they're slowly being destroyed and converted to liquid and then to combustible vapors. Pyrolysis occurs shortly before combustion.
- Surface-to-mass ratio is the surface area of the fuel in proportion to the mass. The size and shape of solid fuels determines their ability to ignite, the time it takes them to be consumed, and the heat release rate of the burning fuels. By dividing the mass into smaller pieces, the surface area is increased. As the surface area increases, more of the material is exposed to the heat and thus generates more burnable gases due to pyrolysis.
- The fuel's actual position affects the way it burns. If a solid fuel is in a vertical position, the fire spread is more rapid than if the fuel were in a horizontal position.
- For combustion to occur after a fuel has been converted into a gaseous state, it must be mixed with air (the oxidizer) in the proper ratio. The range of concentrations of the fuel vapor and oxygen (oxidizer) is the flammable range or explosive range. These terms are used interchangeably because, if the flammable gas-to-air mixture does not ignite, it also won't explode.
- The flammable range of a fuel is reported using the percentage by volume of gas or vapor in air for the lower flammable limit (LFL) or lower explosive limit (LEL), and for the upper flammable limit (UFL) or the upper explosive limit (UEL). The LFL is the minimum concentration of fuel vapor and air that supports combustion.
- Hostile or extreme fire behavior events include thermal layering, rollover, flashover, backdraft, rapid fire growth, behavior of ventilation-limited fires, smoke explosions, and wind-driven fires.

- An important practical application skill is learning how to read the smoke at a fire. Flames indicate where the fire is now; smoke tells you where the fire is headed.
- The first step in reading smoke is determining the four key elements of smoke: volume, velocity, density, and color. Remember that smoke is fuel.
- The velocity (speed) at which smoke leaves the building suggests how much pressure is accumulating inside. Only volume and heat can cause pressure; therefore, smoke is pushed by either heat or volume.
- Laminar smoke flow is a smooth, lazy, streamlined flow produced by pyrolysis. It indicates that the box (the building and its contents) is still absorbing heat and that the temperature and pressure in the box is relatively low.
- Turbulent smoke is generated by high heat temperatures: the faster the smoke, the more heat it has pushing it. It also means that the box is pressurized. It contains an immense amount of energy. The survivability profile in the box is zero and flashover is imminent.
- Black fire is a high-volume, high-velocity, turbulent, ultra-dense, black smoke. Black fire temperatures can reach up to 1,000°F (538°C), creating a deadly environment. Black fire can produce charring and heat damage to steel and concrete. Its presence indicates impending autoignition and flashover. Survivability profile is zero.
- Reading smoke through the door is observing the development of the flow path. A bidirectional flow path indicates that the fire is on the same level as the entry point; a unidirectional flow path with fresh air rushing in indicates that the fire is most likely above the point of entry; a unidirectional exhaust portal indicates that the fire is below the point of entry, most likely a basement fire or below-grade fire.

KEY TERMS

backdraft A deflagration (explosion) resulting from the sudden introduction of air into a confined space containing oxygen-deficient products of incomplete combustion. (NFPA 1403)

bidirectional flow path A two-way or high-low air current of fresh air and smoke competing for the same opening. The lower air intake is fresh air from the outside being entrained or sucked in toward the seat of the fire. The high point or exhaust path is where the smoke and other heated gases are using the same opening (door or window) to escape, seeking equilibrium.

black fire A hot, high-volume, high-velocity, turbulent, ultra-dense black smoke that indicates an impending flashover or autoignition.

boiling liquid expanding vapor explosion (BLEVE) An explosion that occurs when pressurized liquefied materials (e.g., propane or butane) inside a closed container are exposed to a source of high heat. Eventually, the increasing pressure will exceed the container's ability to hold it.

bond When atoms and molecules have an affinity for one another, they form a chemical bond. The chemical bond is the "atomic glue" that holds these molecules together. Fire, or combustion, has its origin in the bonding of atoms and molecules.

combustion A chemical process of oxidation that occurs at a rate fast enough to produce heat and usually light in the form of either a glow or a flame. (NFPA 1)

conduction Heat transfer to another body or within a body by direct contact. (NFPA 921)

convection Heat transfer by circulation within a medium such as a gas or a liquid. (NFPA 921)

diffusion When molecules circulate beyond the surface of the liquid and escape into the space within the container. Some molecules are reabsorbed back into the liquid, while others remain in a gaseous state within the container. If the gases are lighter than air and aren't confined inside the container, all the liquid eventually evaporates.

endothermic reaction A chemical reaction that absorbs heat or requires heat to be added.

equilibrium The chemical state of balance in various forms of matter. A state in which physical opposing forces or influences are in balance. For example, in a liquid, the point where there are as many molecules being liberated from a liquid as being reabsorbed by the liquid.

evaporation The state of conversion from a liquid to a gas under normal atmospheric conditions.

exothermic reaction A chemical reaction that results in the release of energy in the form of heat.

explosive growth stage The second growth stage of a fire. In the initial growth stage, the room temperature is low. The explosive growth stage happens after the fire becomes ventilation-limited and the temperature is high and surrounding fuels are preheated. The introduction of air can lead to rapid fire growth with explosive force, often resulting in flashover.

fire dynamics The study of the characteristics of fire and the burning process, which includes how fires start, develop, and spread. It also includes the detailed study of fire science, chemistry, physics, heat transfer, fuels, materials, hydraulics, and building construction, and how these forces interact to affect fire behavior within a structure.

fire tetrahedron A geometric shape used to depict the four components required for a fire to occur: fuel, oxygen, heat, and chemical chain reactions.

flameover *See* rollover.

flame point The lowest temperature at which a liquid ignites and achieves sustained burning when exposed to a test flame in accordance with ASTM 92, Standard Test Method for Flash and Fire Points by Cleveland Open Cup Tester. (NFPA 1). Also known as *fire point*.

flashover A transition phase in the development of a compartment fire in which surfaces exposed to thermal radiation reach ignition temperature more or less simultaneously, and fire spreads rapidly throughout the space, resulting in full room involvement or total involvement of the compartment or enclosed space. (NFPA 921)

flash point The minimum temperature at which a liquid or a solid emits vapor sufficient to form an ignitable mixture with air near the surface of the liquid or the solid. (NFPA 115)

flow path The movement of heat and smoke from the higher pressure within the fire area toward the lower pressure areas accessible via doors, window openings, and roof structures. (NFPA 1410)

fully developed stage The stage of fire development where heat release rate has reached its peak within a compartment. (NFPA 1410)

heat flux The measure of the rate of heat transfer to a surface, typically expressed in kilowatts per meter squared (kW/m^2) or BTU/ft^2. (NFPA 268)

ice fog Frozen moisture in the air and in smoke.

ignition temperature Minimum temperature that a substance should attain in order to ignite under specific test conditions. (NFPA 402)

inorganic Being or composed of matter other than hydrocarbons and their derivatives, or matter that is not of plant or animal origin and does not contribute to the combustion process. (ASTM D1079: 2.1)

laminar smoke flow Smooth or streamlined movement of smoke, which indicates that the pressure in the building is not excessively high.

liquid A fluid (such as water) that has no independent shape but takes the shape of its container. It has a definite volume, does not expand indefinitely, and is only slightly compressible.

matter Anything that occupies space; has mass, size, weight, or volume; and can be perceived by one or more senses. It cannot be destroyed and exists in three states: solids, liquids, and gases.

modern fuels The household furnishing, appliances, coverings, and decorations that are constructed from petroleum-based products, like plastics, foams, and synthetics.

mushrooming When convecting heat currents and smoke encounter a ceiling or other vertical barrier that prevents it from rising, it spreads out horizontally along the ceiling until it reaches a vertical wall; then it travels down toward the floor, being pushed by more heated pressurized air that is rising behind it.

neutral plane The interface at a vent, such as a doorway or a window opening, between the hot gas flowing out of a fire compartment and the cool air flowing into the compartment where the pressure difference between the interior and exterior is equal.

organic Being or composed of hydrocarbons or their derivatives, or matter of plant or animal origin. (ASTM D1079: 2.1)

oxidation Reaction with oxygen either in the form of the element or in the form of one of its compounds. (NFPA 53)

oxidizer Any material that readily yields oxygen or other oxidizing gas, or that readily reacts to promote or initiate combustion of combustible materials.

polar solvent A water-soluble flammable liquid such as alcohol, acetone, ester, and ketone.

pyrolysis A process in which material is decomposed, or broken down, into simpler molecular compounds by the effects of heat alone; pyrolysis often precedes combustion. (NFPA 921)

radiation The combined process of emission, transmission, and absorption of energy traveling by electromagnetic wave propagation (e.g., infrared radiation) between a region of higher temperature and a region of lower temperature. (NFPA 550)

rollover The condition in which unburned fuel from the fire has accumulated at the ceiling level to sufficient concentration, at or above the lower flammable limit, that it ignites and burns. This can occur without ignition of, or prior to the ignition of, other fuels separate from the origin. Also known as *flameover*. (NFPA 921)

smoke explosion A violent release of confined energy that occurs when a mixture of flammable gases and oxygen is present, usually in a void or other area separate from the fire compartment, and comes in contact with a source of ignition. In this situation, there is no change to the ventilation profile, such as an open door or window; rather, it occurs from the travel of smoke within the structure to an ignition source.

survivability profile An assessment that weighs the risks likely to be taken versus the benefits of those risks for the viability and survivability of potential fire victims under the current conditions in the structure.

thermal column A cylindrical area above a fire in which heated air and gases rise and travel upward. Also known as *thermal plume.*

thermal layering The stratification (heat layers) that occurs in a room as a result of a fire. Also known as *heat stratification.*

thermal plume *See* thermal column.

turbulent smoke flow Agitated, boiling, angry-looking smoke, which indicates great heat in the burning building. It is a precursor to flashover.

unidirectional flow path A fresh-air intake opening or an exhaust exit port. It simply means the convected air currents are flowing in the same direction through the entire opening and exit port.

vapor density The weight of an airborne concentration (vapor or gas) compared to an equal volume of dry air.

vaporization The transformation of a liquid to its vapor or gaseous state.

vapor pressure The pressure, measured in pounds per square inch absolute (psia), exerted by a liquid, as determined by ASTM D323, Standard Test Method for Vapor Pressure of Petroleum Products (Reid Method).

ventilation-controlled fire growth model A six-stage model that includes the following stages: incipient stage, initial growth stage, ventilation-limited stage, explosive growth stage, fully developed stage, and decay stage.

REFERENCES

Dodson, Dave. *The Art of Reading Smoke,* Fire Engineering/Pennwell. Tulsa, OK. 2007.

Dodson, Dave. *Fire Department Incident Safety Officer,* Third Edition. Burlington, MA: Jones & Bartlett Learning, 2017.

Evidence-Based Practices for Strategic and Tactical Firefighting. Burlington, MA: Jones & Bartlett Learning, 2016.

Fundamentals of Fire Fighter Skills and Hazard Materials Response, Fourth Edition. Burlington, MA: Jones & Bartlett Learning, 2019.

Fundamentals of Fire Fighter Skills, Evidence-Based Practices, Enhanced Third Edition. Burlington, MA: Jones & Bartlett Learning, 2017.

Firefighters Handbook, Essentials of Firefighting and Emergency Response, Second Edition. Delmar/Thompson Learning, 2004.

IFSTA Essentials, Fourth Edition, Stillwater, OK, 1998.

IFSTA *Structural Firefighting Truck Company Skills and Tactics,* Second Edition, Stillwater, OK, 2010.

National Fire Protection Association, NFPA 1981, *Standard on Open-Circuit Self-Contained Breathing Apparatus (SCBA) for Emergency Services,* 2019.

Rose, Stewart E., *Strategy and Tactics MCTO,* InSource Inc. Salem, OR, 2015.

© Rick McClure, Los Angeles Fire Department, Retired.

CHAPTER

Building Construction

LEARNING OBJECTIVES

- Recognize that the anatomy of a building collapse follows a specific fire behavioral sequence.
- List the five building construction classification strengths and weaknesses when buildings are on fire.
- Explain how occupancy classifications indicate what firefighters may expect to find in relation to building contents, furnishings, and fire loads.
- Recognize the collapse hazards associated with each building classification.
- Recognize signs and situations when it is relatively safe to enter a burning structure and when it is time to evacuate the building.
- List the various temperatures and warning signs that lead to collapse in Type II steel structures.
- Describe the differences between legacy construction and modern contemporary construction.
- Describe the types of loads.
- Describe the types of forces.
- Describe the purpose and benefits of trusses as well as the dangers associated with them.
- Describe the difference between an offensive-oriented fire department and a defensive-oriented fire department.
- Explain the difference between and the significance of a structure fire, and a room and contents fire.

Introduction

If fire is the enemy we fight, then the building is the battlefield. Understanding modern fire behavior is understanding the weapons the fire uses against us in the battlefield. Once you study and comprehend modern fire behavior, it is much easier to understand how the fire acts and reacts within the building. When we think of a building collapse, we tend to think of a catastrophic structural failure: A building attacked by fire starts to deteriorate and finally gives out, resulting in a spectacular roof collapse followed by a dramatic wall collapse. But if we look at a building collapse on a molecular level, the anatomy of a building collapse actually follows a specific *fire behavioral sequence*, giving the firefighter a whole new perspective of understanding **FIGURE 3-1**.

The scariest thing firefighters (and their families) think about is getting trapped in a burning building or a structural collapse. Line-of-duty death (LODD) investigations always conclude with a series of preventative recommendations that usually include additional training and education on fire behavior, building construction, and the effects of fire behavior in the building. Many fire departments pride themselves on their reputation of making aggressive interior fire attacks. Unfortunately, some of those same departments have suffered LODDs and preventable firefighter injuries upholding that reputation—often without engaging in the actual process of making a physical rescue.

An exhaustive study of building construction in the fire service is beyond the scope of this book. As mentioned in the first chapter, value-time-size (VTS) includes size-up and risk-benefit considerations that go into selecting the proper strategy for a fire. *Time* deals with how much time we can reasonably operate in an offensive strategy. This is determined by the building construction classification, and although there are no absolutes due to many variables, we must have some reliable assumptions in order to make a decision. This chapter covers the *time* consideration category of VTS by explaining how the building construction classifications relate to fire behavior during the attack. Any window of opportunity and operational time period for a successful fire attack within a reasonable margin of safety are limited by the building's ability to resist the effects of the fire. Some buildings, based on their construction classifications, can resist fire better than others. All buildings in the different classifications have their strengths and weaknesses when they're on fire. But nothing is certain; misapplication of fire resistive coverings, improper assembly and installation, poor maintenance, renovations, and natural deterioration always weaken structures and can render fire-resistive ratings and time factors irrelevant.

FIGURE 3-1 The anatomy of a building collapse on a molecular level follows a specific fire behavioral sequence.

The primary focus of this chapter is examining how building construction comes into play and how it's applied during the first-in engine officer's size-up, risk-benefit analysis, decision-making process, and development of the appropriate strategy and tactics within a predictable time frame of operation. The chapter also looks at the collapse hazards associated with each building classification. Firefighters, especially company officers, must understand what they might fall through or what might be about to fall on top of them. They must rely on their knowledge of building construction to determine when it is relatively safe to enter a burning structure and when it is time to evacuate. It is essential that we learn when to give up on the building before the building gives up on the fire.

This is not a command book. Nevertheless, the first-in engine officer is charged with setting up the fireground as the initial incident commander (IC). There are two versions of a saying in the fire service:

- "As goes the first 5 minutes, so goes the fire."
- "The battle is won or lost in the first 5 minutes of the fire."

The first 5 minutes, which includes the size-up, are so important to include in this text; those first 5 minutes mean more than just pulling hose.

Occupancy

Firefighters begin gathering size-up information for a fire from the time they leave home for the fire station. For example, they already know the weather forecast, the day of the week, time of day, and their district, but when the bell hits, the first thing they want to know after the address is the occupancy. The **occupancy** refers to how the building is being used, with the expected hazards and regulatory requirements, as determined by the occupancy classifications established in the *International Building Code*, the *International Fire Code*, and echoed in the current edition of NFPA 101, *Life Safety Code*. The major occupancy categories cover places of assembly, commercial and industrial businesses, schools, institutional facilities like hospitals, and residential—from single-family dwellings and apartments to high-rise condominiums. The major categories are broken down into more specific subcategories.

The occupancy classification helps firefighters anticipate the number and types of occupants who are likely to be at risk in case of a fire. For example, hospitals and nursing homes have many elderly ambulatory and nonambulatory occupants 24 hours a day, 7 days a week. Office buildings have high occupancy loads on weekdays during regular business hours but substantially less during evening hours and weekends. Schools are at capacity with children during weekdays but probably not in the evenings and weekends, although many schools now double as community and recreation centers after regular school hours. Nightclubs and restaurants are busy in the evening, and single- and multifamily residential occupancies have sleeping residents at night.

Societal norms have changed over the decades. *Normal business hours* is a relative term. Many people work from home, and some kids are homeschooled, so a life hazard is always a possibility in residential occupancies at any time of the day. Stores and businesses that used to close after normal business hours may now be open into the evening hours or even around the clock. Although there may not appear to be any activity going on, some factories may be in full production 24 hours a day. You need to know your district and the work practices of your community.

You must also be aware that older buildings constructed before modern fire codes are probably grandfathered in and may have unique hazards for firefighters, or perhaps the occupancies have changed. For example, many cities are trying to revitalize their economies by turning old vacant buildings into urban villages. They convert many large industrial mill-constructed buildings into business lofts, apartments, and condominiums. The first level may have a multitude of different business spaces. These structures along with their commerce activity need to be identified during fire prevention inspections and pre-incident planning.

Contents and Fire Loads

Occupancy classifications indicate what firefighters may expect to find in relation to building contents, furnishings, and fire loads. For example, we're all familiar with the contents in big-box shopping warehouses and big-box home improvement centers. The flammable and combustible inventories are vast. But the contents are much different than what we would expect to find in an elementary school room, a restaurant, or a business office. The hazards and contents are different in an auto repair garage than the familiar home furnishings and appliances we all possess. Knowing the usual contents associated with the occupancy classification helps us prepare for the expected fire load and the amount of energy (British thermal units [BTUs]) these contents are likely to produce when they're on fire. A library has a tremendous fire load, and because much of its contents are considered valuable and irreplaceable, they're vulnerable to water damage. For this reason, many libraries do not have sprinklers.

Occupancy Sprinkler Regulations

Knowing the basic regulations for occupancy classifications also helps firefighters consider the proper strategy and tactics. For example, most single-family residences do not have automatic sprinkler systems, but modern mid-rise and high-rise apartments, condominiums, and hotels do. Standpipe and sprinkler Siamese connections, outside screw and yoke (OS&Y) valves, post-indicator valves, and wall post-indicator valves on the exterior walls of occupancies are clear indications that there are fire suppression systems in place to support engine company operations.

Types of Construction Materials

Buildings are primarily constructed to shelter humans, their possessions, and their livelihoods from the elements. They're designed to give shade and protection from the sun and the elements. Most important, they must resist the forces of gravity. Under normal circumstances, buildings can resist the elements for decades, but from fire, they can be destroyed in minutes. Understanding building construction in relation to fire starts with examining the building materials used in construction. These materials determine the basic fire characteristics of the buildings themselves.

The most common building materials are wood, engineered or prefabricated (prefab) wood, masonry, concrete, steel, aluminum, glass, gypsum board, and plastics. The four concerns that firefighters have with these materials are their ability to burn, whether they conduct heat, and whether they lose their strength and/or expand at high temperatures. These concerns lead to fire spread and /or building collapse.

Masonry

Masonry includes bricks, blocks, stone, and concrete, and all are inherently fire resistant. They do not burn or deteriorate in temperatures normally encountered in structure fires, and they are poor conductors of heat; therefore, masonry materials are used in the construction of fire walls. A **fire wall** is a significant wall that extends the entire depth of the building—usually from the A side to the C side—and extends vertically beyond the roofline **FIGURE 3-2**. It is a barrier designed to prevent the spread of fire from one side of the wall to the other. Although masonry walls do not burn, they can collapse under heavy fire conditions. Masonry walls have very little lateral stability, and in many cases, the roof or floor assemblies holds the wall in place. Steel beams or joists expand during a fire, creating lateral loads that the walls were not designed to withstand. The collapse of a roof or floor assembly can exert lateral forces against masonry walls, which can cause them to collapse and kill firefighters. Many building codes require that wood and heavy timber floor joists set in masonry walls have a fire cut. The end of the joist is cut at an angle so that a fire-damaged floor that is sagging slides out of the pocket by gravity to preserve the wall instead of pushing it out **FIGURE 3-3**. Hose streams hitting a weakened wall can create an impact load sufficient to collapse the wall. Straight streams can wash out the deteriorated mortar from a brick wall, causing failure or collapse **FIGURE 3-4**.

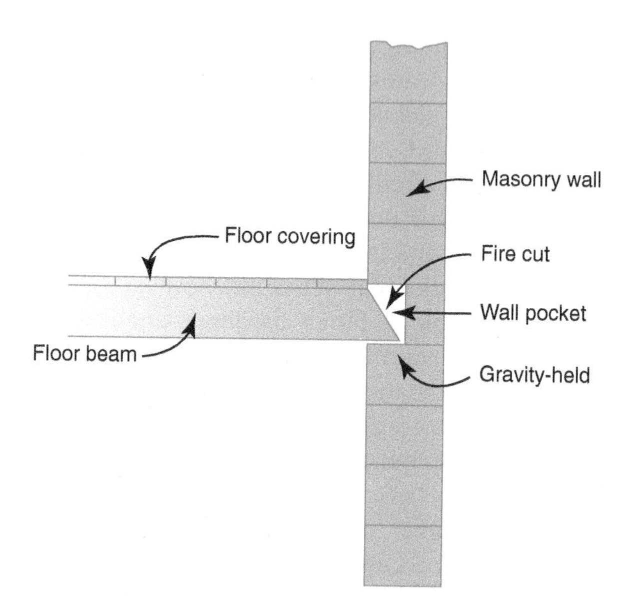

FIGURE 3-3 A fire cut allows a fire-damaged floor to fall into the building via gravity instead of laterally pushing the wall out.

FIGURE 3-4 Hose streams hitting a weakened wall can create an impact load and wash out deteriorated mortar sufficient to collapse the wall.

FIGURE 3-2 A fire wall extends the entire depth of the building and vertically beyond the roofline to prevent the spread of fire from one side of the wall to the other.

Concrete

Concrete is made from a mixture of Portland cement and aggregates like sand and gravel. It's mixed with water and, when cured, it is extremely strong. Under compression, it can support a tremendous amount of weight and is used for foundations, columns, floors, walls, roofs, and pavement. Under tension, however, it is very weak. If it is to be used in construction under tension, it must be reinforced by embedding long, cold-drawn steel rods called rebar FIGURE 3-5. Concrete is also a fire-resistive material and a poor conductor of heat. It doesn't lose its strength when exposed to high temperatures; therefore, as the steel strengthens the concrete, the concrete insulates the steel rods from heat.

Because concrete is mixed with water, moisture can get trapped during the curing process. In extreme fire temperatures, this moisture can convert to steam and expand within the concrete, increasing the internal pressure. This pressure can fatigue the bond, causing chunks of concrete to break off, a process called spalling. Spalling is a concern to firefighters because it can weaken the concrete and possibly expose the interior steel rods to heat. Although a rare event, this can be a prelude warning to structural collapse FIGURE 3-6.

FIGURE 3-5 If concrete is used in construction under tension, it must be reinforced by embedding long, cold-drawn steel rods called rebar.

FIGURE 3-6 Spalling of concrete.

Steel

Steel is an alloy of iron and carbon. It has both tension and compression strength, making it the strongest material used in modern building construction. It is used in the structural framework of buildings to support floor and roof assemblies using heavy beams and columns FIGURE 3-7(A). It is resistant to aging and does not rot; however, most steel rusts if it is not protected from exposure to moisture and air.

Steel is an excellent conductor of heat, so it starts to expand and starts to lose its strength at temperatures in the range of 800°F to 1,000°F and above (427°C to 538°C and above), depending on the thickness of the steel. For this reason, masonry, concrete, or layers of gypsum board are often used to protect steel from the thermal effects of fire. Sprayed-on coatings of mineral or cement-like insulation, often referred to as fireproofing coating, insulate the steel members from the heat of a fire FIGURE 3-7(B). This fireproof covering is

A

B

FIGURE 3-7 A. Steel is an alloy of iron and carbon and has both tension and compression strength, making it the strongest material used in modern building construction. B. Sprayed-on coatings of mineral or cement-like insulation protects the steel members from the heat of a fire in Type I construction.

FIGURE 3-8 The mass in heavier steel I-beams resists the effects of fire; lighter steel-bar joist trusses absorb heat more quickly.

easily seen on the steel framework of Type I high-rise buildings. The amount of heat absorbed by the steel depends on its mass and the amount of fire protection surrounding it. Smaller lighter pieces of steel, like those used in steel bar joist trusses, absorb heat more quickly than do larger, heavier I-beams **FIGURE 3-8**.

Steel expands and loses strength as it is heated. Consequently, an unprotected steel roof beam directly exposed to fire may elongate sufficiently to push out a supporting wall, causing it to collapse. Heated steel beams often sag and twist, whereas steel columns tend to buckle as they lose strength. The bending and distortion are caused by the uneven heating that occurs during an actual fire **FIGURE 3-9**.

Light-gauge steel is now being used for metal studs instead of wood in the walls of many commercial and residential occupancies. Although this reduces fire spread, the concern for firefighters is that wall

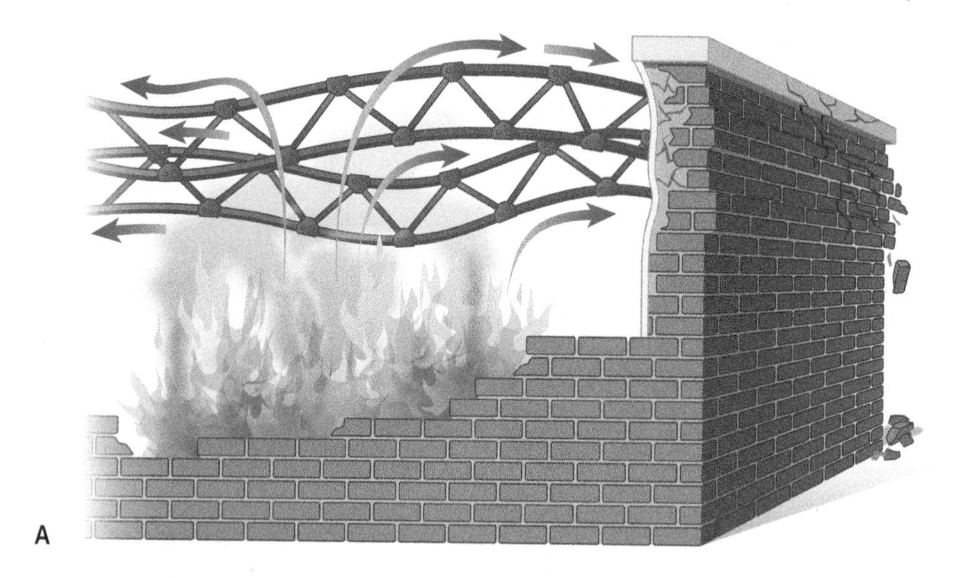

A

B

FIGURE 3-9 **A**. The effects of high heat and conduction through exposed supports may cause the steel to expand, warp, and twist. **B**. Unprotected steel used in Type II noncombustible construction loses strength at temperatures above 1,000°F (538°C).

breaches just became harder. Whether light-gauge steel is breached for rescues, fire attack, or escape, firefighters encounter a greater challenge than they do when wooden studs are used **FIGURE 3-10**.

The strength of steel remains essentially unchanged until 600°F. It retains approximately 50% of its strength at 1,100°F. However, the failure of a steel structural member depends on three factors: the mass of the steel components, the load it is supporting, and the connectors in the assembly. Unfortunately, there are no accurate indicators to warn firefighters or to predict when a steel beam is ready to fail. Any sign or sound of bending, sagging, or elongating of steel structural members should be considered a warning of an imminent collapse. These instances are also indications that temperatures at the ceiling are well over 800°F and probably closer to 1,000°F, so firefighters should immediately evacuate the building and transition to a defensive attack.

A note on aluminum: Aluminum is occasionally used in structural members in building construction; however, it is more expensive and not as strong as steel, so it is generally limited to light-duty applications. When heated, aluminum expands more than steel does and loses its strength quickly when exposed to fire (annealing process). Aluminum has a lower melting point than steel, so it often melts and drips in a fire. Aluminum begins to anneal at the same temperature at which cookies are baked in an oven: at 350°F! This is an easy way to remember this critical temperature.

FIGURE 3-10 Light-gauge steel studs.

Glass

Practically every building utilizes glass in windows, doors, and skylights. Ordinary plate glass breaks easily with any accidental or deliberate impact. Plate glass also breaks with a loud pop from the increase of interior pressures and thermal effects of a fire. Large, razor-sharp shards are the greatest hazard to firefighters encountering broken glass **FIGURE 3-11**.

A variety of specialized manufactured glasses resist breakage from impacts and high temperatures in a fire. Tempered glass, laminated glass, wired glass, and glass blocks are commonly used in building construction. Tempered glass is stronger than ordinary glass but can still be broken with a spring-loaded center punch. The glass shatters into many small pieces but does not have sharp edges like ordinary glass **FIGURE 3-12**.

Laminated glass is made with a thin sheet of plastic between two sheets of glass. It is difficult to break with hand tools and tends to warp with impacts **FIGURE 3-13**. When exposed to fire, laminated glass cracks but most likely remains in place unless it is subjected to extreme heat; in which case, it melts.

Glass blocks are thick pieces of glass the size of bricks or tiles. They are often used within masonry

FIGURE 3-11 Ordinary plate glass breaks with a loud pop from the increase of interior pressures and thermal effects of a fire.

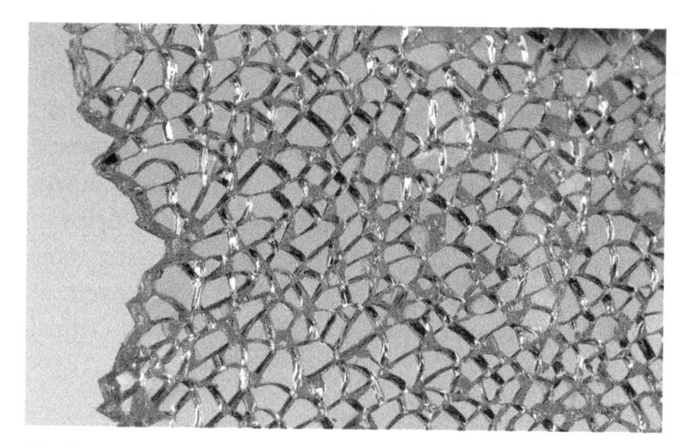

FIGURE 3-12 Tempered glass is stronger than ordinary glass, but it can still be broken with a spring-loaded center punch, or the pick of an ax.

FIGURE 3-14 Glass blocks should not be selected as an access for rescue or as a ventilation exhaust portal except as a last resort.

FIGURE 3-13 Laminated glass is made with a thin sheet of plastic between two sheets of glass.

Courtesy of David Sweet.

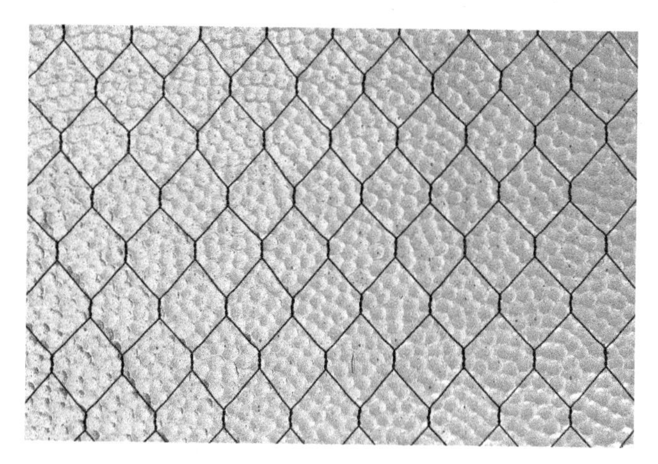

FIGURE 3-15 Wired glass is often used in fire doors and other windows designed to prevent fire spread.

walls to allow light into the structure. Although they are strong, their strength is limited, and they are not designed to carry a load within the wall; however, they do well in resisting fire **FIGURE 3-14**. Glass blocks should not be selected as an access for rescue or as a ventilation exhaust portal except as a last resort. They are extremely difficult to break out with hand tools, and the effort is time consuming. The use of power tools creates fine glass dust that can be inhaled by any firefighter not covered with self-contained breathing apparatus (SCBA) and embeds glass particulates into personal protective equipment (PPE).

Wired glass is manufactured using a molding process that embeds a wire mesh within two sheets of tempered glass. The wire holds the glass together. It is often used in fire doors and other windows designed to prevent the spread of fire **FIGURE 3-15**. There is no reason to break out a wired glass panel within a fire door, but if a wired glass window needs to be taken out, it is best to attack the entire window frame assembly instead of the wired glass panel.

In modern residential construction, energy-efficient homes use double- and triple-pane glass windows to reduce noise and ultraviolet (UV) rays. Triple-pane layered glass units are approximately 1 inch thick. The space between the glass layers is filled with argon or krypton gas, which are great insulators and counteract heat and cold conduction **FIGURE 3-16**. That makes for a very durable window that is hard to break, and these windows often do not completely fail during a house fire. The main concern for firefighters is their resistance to breakage. Some types of glass are nearly impossible to break with traditional hand tools. This can pose an unexpected problem for firefighters who need to

FIGURE 3-16 In modern residential construction, energy-efficient homes use double- and triple-pane glass windows.
Courtesy of Raul Angulo.

ventilate or escape through a window. The end result is a well-sealed home that is able to endure the four burning stages of a room fire, consuming all the available oxygen. The fire enters the decay stage and becomes a ventilation-limited fire. With temperatures sufficiently high, all that is needed is a fresh source of oxygen, which is often provided by the first-in entry team. This can result in rapid fire development or a violent backdraft if the attack is not coordinated with ventilation.

Gypsum Board

Gypsum is a naturally occurring mineral composed of calcium sulfate and water molecules. It is used to make plaster of paris. Gypsum is an excellent insulator and is also noncombustible. It will not burn, even in atmospheres of pure oxygen. Gypsum board, also called drywall, sheetrock, or plasterboard, is commonly used to cover the interior wooden walls and ceilings of residential living spaces and in certain parts of commercial occupancies. To protect wood-frame structural members from fire, drywall sheets are nailed to the wood, providing a fire-resistive protective barrier. Once the gypsum board is properly installed, wood-frame buildings that typically have many rooms are compartmentalized and can withstand and contain room and content fires within their rating. However, drywall isn't really dry. The sheets of gypsum board have a moisture content from infused water of crystallization, a chemical hydration process used during manufacturing. This is why sheetrock feels moist and is cool to the touch. A horizontal ⅝-inch sheet of gypsum board has a 30-minute fire rating. While the sheetrock retains its chemical moisture, the wood-frame assembly is protected. As more and more heat is absorbed, the water of crystallization evaporates, leaving only gypsum dust and filler, which is usually paper. The sheets begin to bow on the walls, or sag from the ceiling. They become very brittle (from a process called calcination) and eventually fail, exposing the wood to fire.

Tactical Considerations

Gypsum board is required in residential structures with attached garages on the walls that share the bearing wall to the house: The gypsum board creates a fire-rated barrier. A residential fire-rated door is also required between the garage and the entrance to the house, which is usually a utility room or the kitchen. It is important for firefighters to know this because many fires that occur inside an attached garage are often attacked by fire crews starting from the inside of the residence, going from the uninvolved toward the involved. Getting a line between the fire and the exposure is a sound tactic except in this scenario. When firefighters enter the garage from the interior fire door of the house, they allow smoke and heat to enter the uninvolved residence, causing needless smoke damage and fire spread. The gypsum board walls and fire door separations are there for a reason: Use them to your advantage. Attached garage fires should be attacked by opening the main garage door and using a straight stream direct attack or transitional attack, defensive to offensive, starting from the involved area. An exposure line should be stretched into the house, but the interior garage door should remain closed until the fire in the garage is extinguished and ventilated.

Wood

Wood is probably the most common material used in building construction. It is a relatively inexpensive product and easy to cut for use in a range of structural configurations like structural framing, roof decking, trusses, columns, beams, walls, floor joists, and siding. Wood can be milled for heavy structural supports to thin strips of decorative molding. Soft woods like pine and harder woods like oak are used in building construction. Although wood structures obviously burn easier than concrete or masonry structures, they tend to fare better in earthquakes because wood is more flexible than concrete and masonry.

Older buildings, built before the 1960s, were built with lumber that measured true to its stated dimensions and is referred to as legacy construction. A 2-inch × 4-inch stud actually measured 2 inches by 4 inches (5 cm × 10 cm). Modern construction practices now use presurfaced wood where the sides are planed for smoothness. S2S, S3S, and S4S lumber refers to how many sides have been presurfaced for smoothness. Approximately ½ inch is taken off each side of the board in the process. An S4S (surface 4 sides) 2-inch × 4-inch stud actually measures

1½ inches × 3⅜ inches (4 cm × 8 cm) or less. Remember, in a fire, wood burns away ⅛ inch every 5 minutes when directly exposed to the flames, so in 20 minutes, 1 inch of a 2-inch × 4-inch stud has burned away. However, if fire is burning on both sides of the board, it burns away 2 inches—the entire thickness—in the same 20 minutes. This is where wood-frame construction gets its 20-minute rating. A true dimensional 2-inch × 4-inch stud resists fire longer than the modern 2-inch × 4-inch studs.

Engineered Wood Products

Like the legacy and modern fuels discussed in Chapter 1, building construction has legacy construction and contemporary construction. Engineered wood products are also known as manufactured wood, manmade wood, laminated wood, or composite wood. They include plywood, fiberboard, oriented strand board (OSB), and particle board. These products are manufactured from small pieces of wood that are held together with glues or adhesives. The adhesives include urea-formaldehyde resins, phenol-formaldehyde resins, melamine-formaldehyde resins, and polyurethane resins—all of which are flammable and toxic.

Engineered wood products are preferred over natural wood for a variety of reasons, but the primary reason is the cost; they are usually less expensive than larger solid planks of natural wood. They can also be fabricated into longer lengths for specific applications that are not possible with natural wood. Entire homes are constructed using engineered lumber. Although there are many benefits from a contractor's perspective, there are many problems with engineered lumber from a firefighter's standpoint.

In parts of the country where there is high humidity and rain, many of these products can warp and delaminate when they become wet. These wood products also burn quickly, with higher heat release energy than natural wood. The high heat release rate preheats the surrounding wood surfaces, causing the fire to spread rapidly. Because the materials are less dense, more surface area is exposed and preheated, which leads to easier ignition. They fail quickly under fire conditions in as little as 2 to 5 minutes when exposed to direct flame contact, and often without warning. Their collapses produce large void spaces. Decomposition with exposure to high temperatures generated by the fire can also reduce the strength of the wood through the process of pyrolysis. A burning structure with engineered wood construction becomes weaker by the minute; firefighters must operate using a wide margin of safety when fighting these fires.

Fire-Retardant Wood

Fire-retardant wood is permeated with mineral salts, which makes the wood more difficult to ignite. Although treating the wood does not make it completely noncombustible, it does slow down the burn rate, significantly reducing the hazards of wood construction. However, the fire-retardant treatment can also reduce the strength of the wood roof rafters and decking. At an actual fire, it is impossible to tell if the wood is treated or not. This is something that must be identified during new-construction inspections. But firefighters should know that adding extra weight by standing on roof decking and trusses that have been constructed with fire-retardant treated wood makes them weaker than roof assemblies constructed with nontreated wood.

Plastics

The use of plastics and synthetics (known as modern fuels) in home interior furnishings was discussed in Chapter 1. Plastics are now being incorporated into structural support members. Fiber-reinforced products (FiRPs) are becoming more common in building construction. FiRPs are plastic fibers mixed with wood to give the wood increased tensile strength. Engineered lumber is often stronger than natural wood. As with most plastics, however, exposure to fire causes the plastic to melt, which leads to quick failure of FiRP assemblies. Plastic is also used for pipes and fittings for water, electric, and natural gas utilities. It's used in vinyl siding, window frames, skylights, insulation, and fixtures. In addition to all the household furniture, appliances, and coverings that may be plastics and synthetics, these plastic building components add to the overall fire load of the structure, causing a tremendous release of heat energy.

The combustibility of plastics varies. Some plastics ignite easily and burn very quickly, while others with large mass soften and melt without igniting. Plastics produce heavy, dense black smoke and release high concentrations of toxic gases as they burn. The smoke resembles smoke from a petroleum flammable liquid fire because plastics are made from petroleum.

Thermoplastic materials melt and drip when exposed to high temperatures and rapidly spread the fire, even in temperatures as low as 500°F (260°C). Thermoset materials are softened and fused together by the heat of the fire. They do not melt the same way that thermoplastic materials do; however, their strength decreases dramatically. These materials soften when heated and harden when cooled.

There is nothing "fireproof" in building construction. Eventually, everything burns or is destroyed by the fire.

Building Construction Terms and Mechanics

Firefighters need to understand certain terms and concepts associated with building construction. The sole purpose of a building is to provide a protected space to shelter occupants, their belongings, contents, and materials from the elements. A building must be built to resist wind, rain, and snow. Most important, it must resist the forces of gravity. The intended purpose of a building can also add a tremendous amount of weight to the structure, placing additional stress on its ability to resist gravity. In construction terms, these elements are known as building loads. Loads are imposed on building materials. This imposition causes stress on the materials, called force. Forces must be transmitted to structural members, then be delivered to the building foundation, and then to the Earth in order for the building to be structurally sound. The primary purpose of a building foundation is to support the entire weight of the building and ensure that the base of the building is securely planted in a fixed location, which helps keep all the building components firmly connected.

Types of Loads

A variety of loads (and terms) are associated with building construction. The two major types of loads are dead loads and live loads. The weight of the building itself is called the **dead load**, and the weight of the building contents is called the **live load**. Dead loads include the weight of all the materials and equipment that was designed into the building and is permanently attached to the structure. A building marquee or an exterior fire escape are examples of dead loads **FIGURE 3-17**. Live loads include the movement of people, equipment, and materials that come into, come onto, and go out of a building. Live loads are not attached to the building. A helicopter landing on the roof helipad of a building is an example of a live load.

Live and dead loads can be broken down into more specific groups, as described here.

Concentrated Load

A **concentrated load** is applied on to a small area. A heating, ventilation, and air-conditioning (HVAC) unit on a roof or a large commercial safe on a floor are examples of concentrated loads. The reason concentrated loads are a concern to firefighters is that they do not want to be standing underneath them when roof or floor supports are being subjected to fire. Concentrated loads should be identified during pre-incident inspections, but they can also be readily identified during a fire. Firefighters encountering concentrated loads involved in or threatened by fire should immediately report the situation over the radio so all crews on the fireground can hear the warning and are made aware of this hazard.

FIGURE 3-17 This fire escape is an example of a dead load.

Distributed Load

A **distributed load** is a load applied equally over a broad area. For example, the weight of tiles on a tiled roof are evenly distributed over the roof decking **FIGURE 3-18**. Snow falling and accumulating on a roof can also be considered a distributed load. Conversely, rainwater accumulating on flat roof can be considered a distributed load but because it is more fluid than snow, the water can run and flow in any direction based on the downward slope of the structure; therefore, it is more accurately considered a live load. This should be a concern for firefighters because they need to recognize conditions that can create an additional distributed load or that create a sudden undesigned load. For example, 1 gallon of water weighs 8.34 pounds (3.78 kg). One foot of standing water on a 20-foot × 20-foot roof adds an extra 25,000 pounds

FIGURE 3-18 A tiled roof and wood shakes or shingles are examples of a distributed load.

(11, 340 kg) of weight to the roof. A ladder pipe master stream flowing 1,000 gallons per minute (gpm) into a building adds 8,340 pounds (3,782 kg) of weight every minute into that building. In an effort to extinguish a massive defensive fire, firefighters could unsuspectingly be creating a live load so extreme that it could actually collapse the building!

Impact Load

An impact load is in motion when its applied. A wrecking ball swung against a wall is an impact load. Wind gusts against a building create an impact load. Large crowds of people running or stomping while cheering inside a stadium at a sporting event is another example of an impact load. Firefighters can impose impact loads on a building by jumping onto a roof from an aerial ladder, directing straight streams against the wall of a building, or hitting structural members inside the building. The impact load from a solid straight-stream pattern from a master stream appliance can easily knock over a weakened masonry wall.

Designed and Undesigned Loads

A designed load is an engineered structural design that has been planned for the intended use of the building. An undesigned load refers to the weight that a building supports for which the building was not designed or an added weight that was not anticipated. Buildings that are altered or refurbished for an occupancy other than the original intent can create an undesigned load. For example, an older mill construction/heavy timber factory that is converted to residential lofts is not a big deal, but a wood-frame residential house converted to a commercial security safe showroom or a music school with numerous grand pianos can add tremendous weight that the building was not designed to safely support. Under normal circumstances, such an occupancy may go years without incident, but if the same structure is on fire, conditions could be deadly.

Fire Load

Fire load is not a term used within the building construction industry, but it is used extensively in the fire service and fire engineering lexicon. Fire load describes the fuel (building contents) and the number of BTUs that will be generated when that fuel is on fire. Fire load is referred to as fuel load.

Imposition of Loads

A load must be transmitted to structural components. This is called *imposition of loads.* Three terms are associated with imposition loads: axial loads, eccentric loads, and torsional loads. An axial load is transmitted through the center of the structural member and runs parallel to ground through that member. An

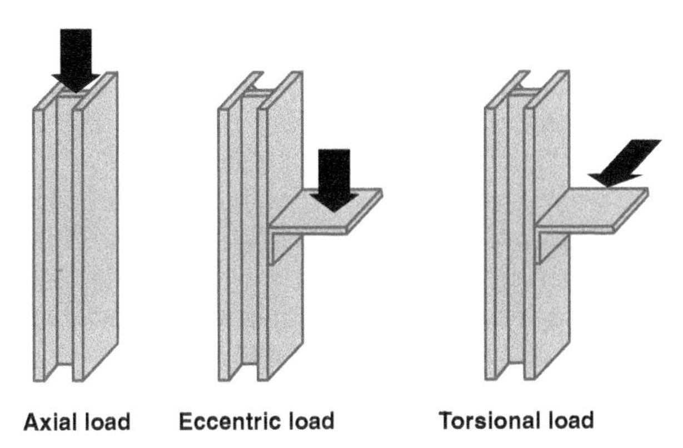

FIGURE 3-19 Loads can be imposed three ways: as axial, eccentric, and torsional loads.

eccentric load is applied perpendicular to a structural member and subsequently does not pass through the center of that member. A torsional load is applied offset to the structural member, causing a twisting stress **FIGURE 3-19**.

Forces

Loads imposed on structural members create stress and strain on the materials used to make the connections. Stress and strain are defined as forces applied through those materials. These forces are defined as compression, tension, and shear. In compression force, forces push materials together. Tension force occurs when forces are pulling the materials and connectors apart. Shear force occurs when the force tends to tear a material apart and the molecules of the material are sliding past each other **FIGURE 3-20**.

All loads, and the forces they create, must eventually pass through the structure, down to the foundation, and then be delivered to the Earth. Under normal conditions, structures resist failure, but under fire conditions, the material and connectors used to resist forces start breaking down through pyrolysis and lose their strength. This causes a shift in the forces and the way the design loads are applied. It is important for firefighters to understand the difference between these forces because it gives them an indication of the type of collapse to anticipate. When the load supports fail, the forces give way in their respective directions. Eventually, gravity takes over and pulls the building to the ground.

The time it takes for gravity to overcome the structure during a fire is not predictable. A number of variables determine the amount of time that a material can resist gravity and fire degradation, including:

- Material mass
- Surface-to-mass ratio
- Overall load being imposed

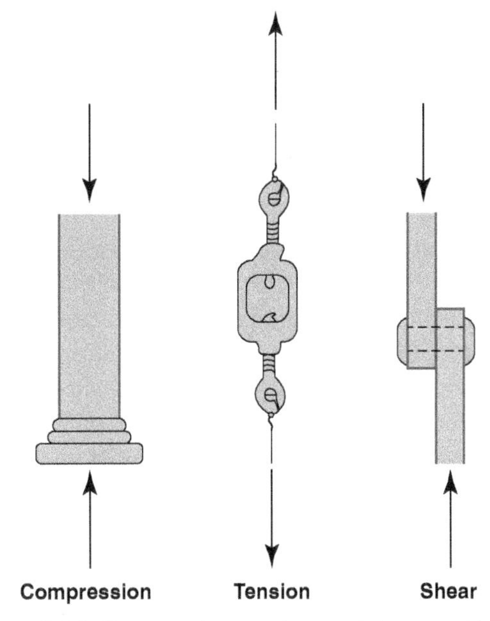

FIGURE 3-20 Compression, tension, and shear are the three types of forces created when a material receives an imposed load.

- BTU potential (fire load)
- Type of construction (assembly method)
- Building modifications and remodels
- Undesigned loads
- Age, deterioration, care, and maintenance of the structure
- Condition of fire-resistive barriers (intact or breached)
- Firefighting impact loads (by personnel, tools, and/or water streams)

Floors and Ceilings

Floor and ceiling construction is very important for firefighters because that's what firefighters are standing on, what they are going to fall through, or what will fall on top of them. Firefighters working inside the building must rely on the integrity of the floor to support their weight. A floor that fails could drop firefighters into a burning basement or into a fire on the floor below. In a multistory building, firefighters may be working below the fire floor or below an attic space or a roof, which if it fails, collapses right on top of them. Failure of a roof or floor assembly can also push out walls, causing a collapse. Floor construction influences whether a fire spreads vertically, from floor to floor within a building, or whether it remains on the floor of origin. Some floor systems are designed to resist fire spread, whereas others have no capability to resist fire spread at all.

Fire-Resistive Floors

In multistory buildings, floors and ceilings are generally considered a combined structural system; that is, the structure that supports a floor also supports the ceiling of the story below. In a fire-resistive building, this system is designed to prevent a fire from spreading vertically and to prevent a collapse when a fire occurs in the space below a floor-ceiling assembly. Fire-resistive floor-ceiling systems are rated in terms of hours of fire resistance, based on a standard test fire.

Concrete floors can be either self-supporting or supported by a system of steel beams or trusses and columns. In the latter case, the steel can be protected by sprayed-on insulating materials, or covered with concrete or gypsum board. If the ceiling is part of the fire-resistive rating, it provides a thermal barrier to protect the steel members from a fire in the area below the ceiling.

The ceiling below a fire-resistive floor can be constructed from plaster or gypsum board, or it can be a system of tiles suspended from the floor structure. In many cases, a void space between the ceiling and the floor above it, called the plenum, contains building systems and equipment, such as electrical, telecommunication, and computer wiring; HVAC ducts; and plumbing and fire sprinkler system pipes **FIGURE 3-21**. If the plenum space above the ceiling is not subdivided by fire-resistant partitions or protected by automatic sprinklers, a fire can quickly extend horizontally across the large space.

Wood-Supported Floors

Wood-floor structures are common in non-fire-resistive construction. Such wooden-floor systems range from heavy timber mill construction to modern lightweight construction. Heavy timber construction uses posts and beams that are atleast 8 inches (20.3 cm) on the smallest side and often as large as 14 inches (35.5 cm) in depth. The floor decking is often assembled from solid wood boards, 2 to 3 inches thick (5.1 to 7.6 cm), which are covered by an additional 1-inch of finished wood flooring. The depth of the wood in this system often contains a fire for more than an hour before the floor fails or burns through **FIGURE 3-22**.

Conventional wood flooring, which was the standard for many decades, is much lighter than heavy timber but uses solid lumber as beams, floor joists, decking, and finished flooring. It burns readily when exposed to a fire. It generally takes approximately 20 minutes to burn through or reach structural failure. This time estimate is only a general unscientific guideline. The actual burn rate depends on many other factors.

Modern lightweight construction uses structural elements that are much less substantial than conventional lumber. For example, lightweight wooden trusses or engineered wooden I-beams are often used as supporting structures **FIGURE 3-23**. Thin sheets of plywood are used as decking, and the top layer often consists of a thin layer of concrete or wood covered by carpet. This floor construction provides little resistance to fire. The lightweight structural elements can fail, or the fire can burn through the decking quickly. In National Institute for Occupational Safety and Health (NIOSH) Publication No. 2009-114, NIOSH reported that a recent test demonstrated that it took 19 minutes for a traditional unprotected residential floor assembly to fail. Under the same test conditions, an unprotected lightweight engineered I-joist failed in only 6 minutes.

Unfortunately, it is impossible to tell how a floor is constructed by looking at it from above. The important information about a floor can be observed only from below, if it is visible at all. The building's

FIGURE 3-21 The plenum space between the ceiling and the floor above can contain building systems and equipment, such as electrical wiring and HVAC ducts.

FIGURE 3-22 Type IV heavy timber flooring and beams, supported by steel columns.

age (era) and local construction methods can provide significant clues to a floor's stability, but you should be aware that many older buildings have been renovated using modern lightweight systems and materials. Firefighters should use pre-incident surveys to gather essential structural information about the buildings. In some renovations for office or computer rooms, the floors have been replaced with removable floor panels to hide the different types of cables under the floor.

Ceiling Assemblies

From a construction standpoint, the ceiling is considered part of the floor assembly, and some ceilings are fire-rated as part of that assembly. However, ceilings are primarily there for aesthetic reasons. Their function is to support light fixtures in place and diffuse light. Ceilings also conceal electrical wiring, HVAC, and automatic sprinkler systems. Ceilings can be part of the fire-resistive features. For example, they can be covered with plaster or gypsum board, or they can consist of dropped ceiling mineral tiles. Hollow spaces between the floor and the ceiling can contribute to the horizontal spread of fire, and penetrations in the floor can also contribute to the vertical spread of fire. A variety of fiberglass insulation materials are used to insulate the spaces between floor joists, cocklofts, and attic spaces. Although they don't readily burn, they do smolder and generate smoke. Older homes, on the other hand, often used ground-up newspaper for insulation, which does burn and leads to fire spread.

FIGURE 3-23 When exposed to direct flame contact, lightweight wooden trusses can fail quickly, in as soon as 2 to 5 minutes.

Courtesy of APA - The Engineered Wood Association.

Roofs

Roofs come in three primary design features: pitched, curved, and flat **FIGURE 3-24**. Several methods and materials are used for roof construction. The major components of a roof assembly are the supporting structure, the roof decking, and the roof covering **FIGURE 3-25**.

Safe interior firefighting operations depend on a stable roof. If the roof collapses, the firefighters inside the building may become trapped, injured, or killed. Interior fires generate hot gases that accumulate under the roof. Thus, firefighters who are performing ventilation activities on the roof to release the heated gases must also depend on the structural integrity of the roof.

The primary purpose of a roof is to protect the inside of a building from the elements and the weather. Roofs are not designed to be as strong as floors, especially in warm climates that do not experience heavy snowfall. If the cockloft or attic space below the roof is used for storage, or if extra HVAC equipment is mounted on top of the roof, the load can exceed the designed strength. Adding the weight of several firefighters and the impact loads associated with rooftop operations to an already overloaded roof can be risky.

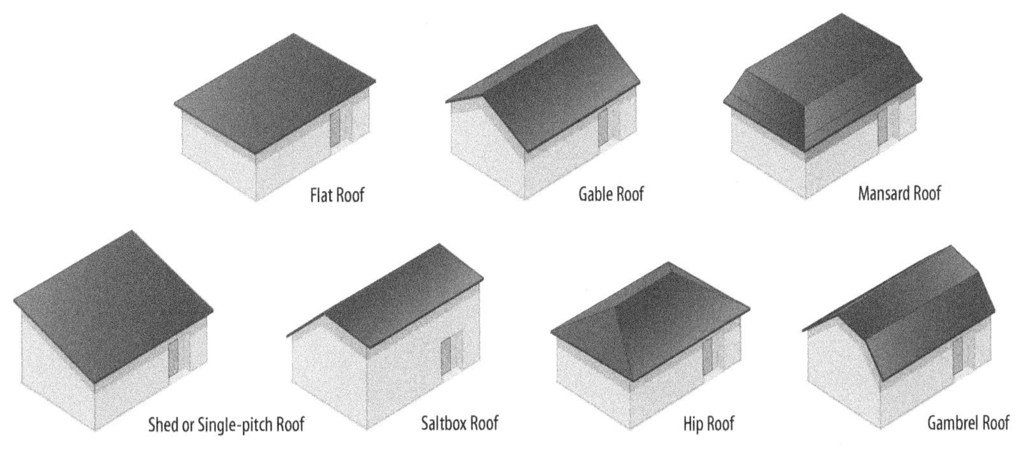

FIGURE 3-24 Different roof designs.

Trusses

Roof and floor truss systems subjected to fire are probably the deadliest building components for firefighters. Roof truss collapses have killed many firefighters, so it is important to cover this subject. Trusses are everywhere so it is essential that you learn how to recognize them and know how they are designed. In some buildings, the trusses are exposed and readily visible from inside the building. In other buildings, the floor or roof trusses are not visible because they are enclosed in attic or floor spaces.

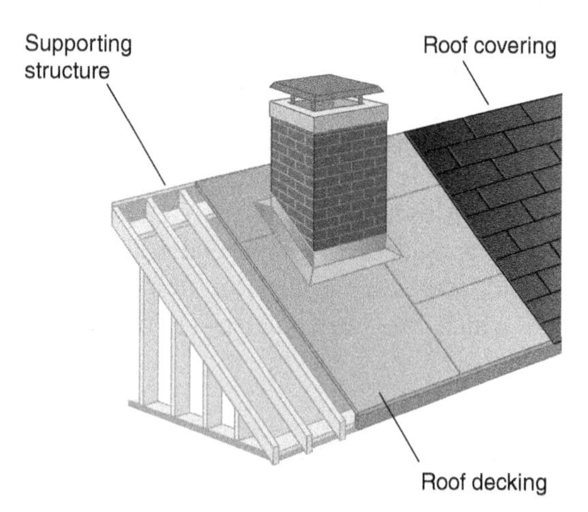

FIGURE 3-25 The major components of a roof assembly are the supporting structure, the roof decking, and the roof covering.

A truss is actually a fake beam. A truss is a structural component that is composed of smaller pieces of boards, timbers, beams, or steel bars that are joined together to form triangle configurations or a series of triangles, as well as rectangles, to create a rigid framework that supports a load **FIGURE 3-26**. Trusses are used extensively in support systems for both roofs and floors. They are common in modern construction for several reasons. The triangular geometry creates a strong rigid structure that can support a load much greater than its own mass. For example, both a solid beam and a simple truss with the same overall dimensions can support the same load, but the truss is considerably stronger. The truss has several advantages over a solid beam: It requires much less material

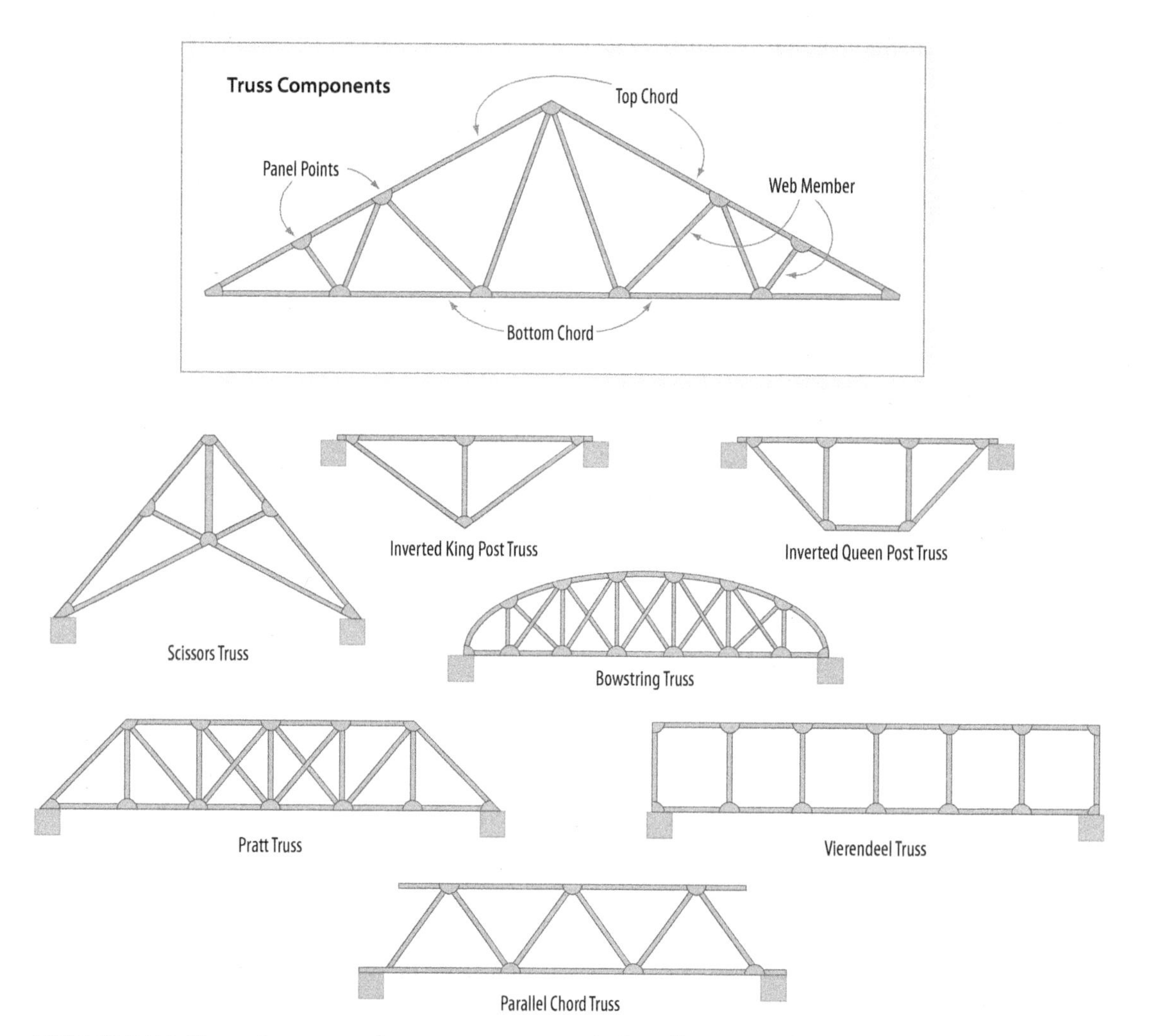

FIGURE 3-26 The various types of trusses are composed of smaller pieces of boards, timbers, beams, or steel bars that are joined to form triangle configurations or a series of triangles.

than the solid beam, it is much lighter, and it can span longer distances without supports. Trusses are usually prefabricated and assembled, then delivered directly to the building site for quick installation.

Trusses are frequently used for floor and roof assemblies of new residential construction, apartment buildings, small office buildings, commercial buildings, warehouses, fast-food restaurants, airplane hangars, churches, and even firehouses. They may be clearly visible, or they may be concealed within the construction. Trusses are often used to replace heavier solid beams and joists in renovated or modified buildings.

The strength of a truss depends on both its members and the connections between them. A properly assembled and installed truss is very strong under normal load conditions. However, if any member or connector fails, the strength of the entire truss is compromised.

The truss roof and floor systems are either heavy timber, lightweight wood, lightweight steel, or a combination of wood and steel. Three of the most commonly used truss configurations are the parallel chord truss, the pitched chord truss, and the bowstring chord truss.

A **parallel chord truss** has two parallel horizontal members connected by a system of diagonal and sometimes vertical members. The top and bottom members are called the *chords* and the connecting pieces are called the *web* members. Parallel chord trusses are typically used to support flat roofs and floors **FIGURE 3-27**. In lightweight construction, an engineered **wood truss** is often assembled with wood chords and either wood or lightweight steel web members. A steel bar joist is another example of a parallel chord truss.

A **pitched chord truss** is typically used to support a sloping roof. Most modern residential construction uses a series of prefabricated wood-pitched chord trusses to support the roof. The roof deck is supported by the top chords, and the ceilings of the occupied room are attached to the bottom chords. In this way, the trusses define the shape of the attic and the roof **FIGURE 3-28**.

The roof of a building with a **bowstring truss** has a distinctive curved shape. The bowstring truss has the same shape as an archery bow, where the top chord represents the curved bow and the bottom chord represents the straight bowstring, hence its name **FIGURE 3-29**. The top chord resists compression forces, and the bottom chord resists tension forces. Bowstring

FIGURE 3-27 A parallel chord truss has two parallel horizontal members connected by a system of diagonal and sometimes vertical web members.

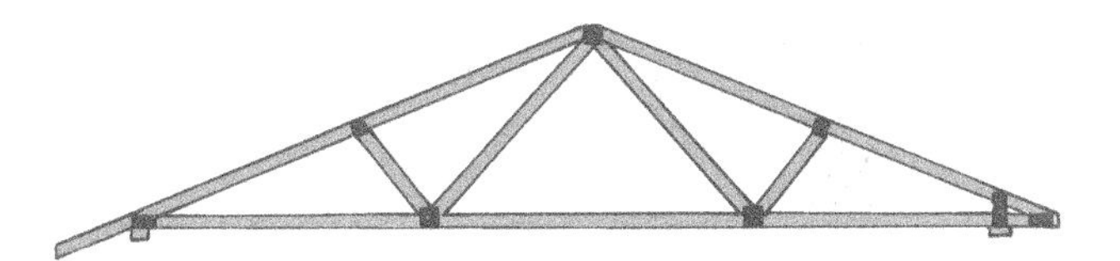

FIGURE 3-28 A pitched chord truss is typically used to support a sloping roof. The truss defines the shape of the attic and the roof.

FIGURE 3-29 A bowstring truss has a distinctive curved shape. Bowstring trusses are prevalent in warehouses, supermarkets, bowling alleys, and similar buildings with large open floor spaces.

Courtesy of Captain David Jackson, Saginaw Township Fire Department.

trusses are usually quite large and widely spaced apart in the structure. They are prevalent in warehouses, supermarkets, bowling alleys, and similar buildings with large open floor spaces.

All trusses rely on each other, including every part of the web members, to carry their portion of the imposed load. Like a beam, the *top chord of the truss is under compression while the bottom chord is under tension*. The web members between transfer the two forces, creating balanced stress and strain and thus giving stability to the truss. Failure of one part of the truss likely causes the entire truss to fail. When one truss fails, that specific load is forcefully transferred to the other trusses, which may not have the capacity to support the sudden additional weight, thereby leading to a domino effect collapse of additional trusses.

Wood Trusses

Wood trusses are an assembly of many pieces of wood. Some may even be press-glued particles. These pieces are connected using gusset plates. A **gusset plate** is a simple and very thin galvanized steel plate with teeth (gang nail perforations) along its surface. The gusset plates are hammered, or machine pressed into the wood holding all the members of the truss together **FIGURE 3-30**. The teeth or gang nail perforations only penetrate the wood about ⅜ inch, in other words, not much. During a fire, the steel gusset plates heat up and transfer heat into the wood fibers. If heating is slow

FIGURE 3-30 Engineered wood trusses are connected using gusset plates.

© Photodisc.

FIGURE 3-32 If heating is fast, steel gusset plates expand, pull away, and pop out.

Courtesy of Raul Angulo.

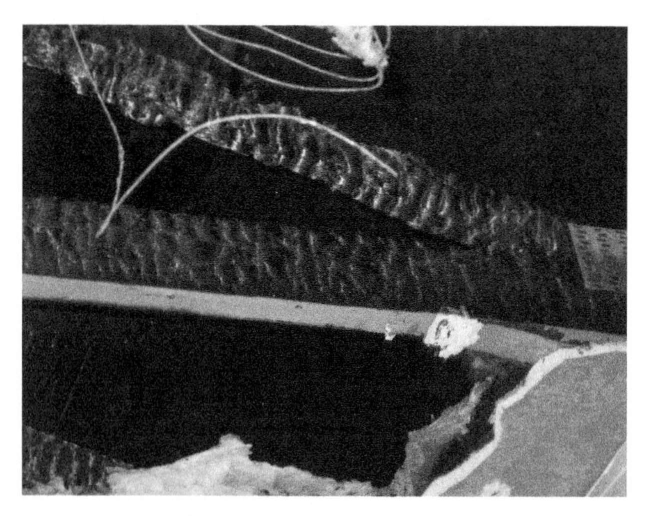

FIGURE 3-31 If heating is slow (like in a smoke-filled attic), the wood pyrolyzes, creating alligatoring.

Courtesy of Raul Angulo.

FIGURE 3-33 The truss may stay together until a sudden impact load, like a crew sounding the roof, causes the truss system to give way.

Courtesy of Raul Angulo.

(like in a smoke-filled attic), the wood pyrolyzes, creating alligatoring, and the gusset plates simply fall out **FIGURE 3-31**. If the heating is fast, like sudden exposure to flames, the steel gusset plates expand, pull away, and pop out. Either way, the fastener holding the truss together is lost and failure is imminent **FIGURE 3-32**. Sometimes the truss gingerly stays together due to the weight of the roofing or flooring materials, but a sudden impact load like a firefighter walking around wearing size 14 boots or a crew sounding the roof can cause the truss system to give way and suddenly collapse **FIGURE 3-33**.

Because wood trusses are mass produced at a factory, quality control may not be the best. Truss gusset plates can be damaged or weakened from loading and unloading and vibration during transport to a lumber yard, then to a jobsite. Contractors may use shortcuts to lift trusses into position, furthering the damage to the ⅜-inch gusset plates. These and other factors can lead to early collapse during a fire. Wood trusses provide a large surface-to-mass ratio, fuel load, and void spaces—three of the worst conditions a firefighter can encounter during a structure fire.

Firefighters have been injured and killed when fire-damaged floor and roof truss systems have collapsed in as little as 2 to 5 minutes of fire involvement, sometimes without warning. The NIOSH Alert, *Preventing Injuries and Deaths of Firefighters Due to Truss System Failure* (Publication 2005-132) is a must-read document for all firefighters, but especially for company officers. According to the alert, more than 60% of all the roof systems in the United States are built with wood truss systems. Unprotected steel trusses are also prone to failure under direct fire conditions and may fail in less time than a wooden truss under the same conditions.

According to Chief Vincent Dunn (Fire Department of the City of New York [FDNY], Retired), three scenarios can occur with trusses that can injure or kill firefighters:

- While firefighters are operating above a burning roof or floor truss, they may fall into a fire as the sheathing or the truss system collapses below them.
- Trusses may collapse onto firefighters while they are operating underneath the roof or floor inside a building with a burning truss floor or roof assemblies.
- While firefighters are operating outside a building with burning trusses, the floor or roof trusses may collapse and cause a secondary wall collapse.

The NIOSH Alert lists the following steps that firefighters should take to minimize the risk of injury and death during structural firefighting operations involving roof and floor truss systems:

- Know how to identify roof and floor truss construction.
- Use the thermal-imaging camera during size-up to help locate fires in concealed spaces.
- Immediately report the presence of truss construction and fire involvement to the IC.
- Use extreme caution and follow standard operating procedures when operating on or under a truss system.
- Utilize the reach of your hose stream to operate from a safe area.
- Immediately open ceilings and other concealed spaces whenever you suspect that a fire is in a truss system.
- Use extreme caution and be prepared when opening concealed spaces, which can result in a backdraft, flashover, or smoke explosion.
- Always have a charged hose line ready.
- Position yourself between the nearest exit and the concealed space to be opened.
- Be aware of the location of other firefighters working in the area.
- Understand that fire ratings may not be truly representative of real-time fire conditions and that the performance of a truss system may be affected by the severity of the fire.
- Fire officers performing operations under or above trusses must quickly withdraw their crews and move to a defensive attack as soon as it is determined that the trusses are exposed to fire.
- Use defensive operating procedures after extinguishing fire in a building with fire-damaged truss construction.
- Use outside master streams to soak the smoldering trusses to prevent rekindles.

Steel Trusses

Steel trusses are just as susceptible to collapse as wood trusses. Like wood, steel trusses are an assembly of pieces, typically of angle iron for the chords and cold-drawn round stock for the web. The pieces are tack-welded together to form the truss. Sometimes the construction industry may refer to these units as open-web steel joist or open-web bar joist—but they're still trusses **FIGURE 3-34**. These trusses expose a large surface area to heat during a fire. Given the lack of mass, the truss heats quickly and softens and expands. The expansion can exert lateral pressure and wall movement, causing a collapse. If the wall does not move, the steel truss twists and buckles to allow for expansion. Cold-drawn steel used for cables, bolts, rebar, and lightweight fasteners loses about 55% of its strength at 800°F (427°C). Unprotected, open-web steel joists lose 50% of their strength at temperatures of 1,000°F (538°C). Unprotected structural steel used for beams and columns loses about 50% of its strength at 1,100°F (593°C). As the temperature rises, structural steel expands and elongates; at 1,100°F, a 100-foot (30 m) beam can elongate 10 inches (25 cm) and knock down

FIGURE 3-34 Type II noncombustible open-web steel joist, or open-web bar joist.

FIGURE 3-35 Applying water directly into the Type II trusses cools the steel, preventing elongation and expansion.
Courtesy of Keith Cullom

FIGURE 3-36 Wood trusses provide a large surface-to-mass ratio, fuel load, and void spaces.

FIGURE 3-37 The open-web spaces between the chords allows fire to spread horizontally throughout the trusses.
Courtesy of Raul Angulo.

an entire wall. Keeping steel trusses cool with hose streams is an important tactical consideration because it prevents elongation and buys time by extending the strength of the steel during a fire. Use the reach of the hose stream and do not stand underneath the steel truss **FIGURE 3-35**.

Void Spaces

Trusses create large void spaces **FIGURE 3-36**. The open-web spaces between the chords allow fire to spread horizontally **FIGURE 3-37**. Although most codes require fire stopping in floor joist spaces, they allow for wide-open attic spaces. Electrical fires can start in void spaces because that is where contractors run the electrical wires. Chimney fires can also spread into attic void spaces for a variety of reasons. In Type III ordinary construction row houses and taxpayers, the voids are numerous. Some voids pass through masonry walls, causing horizontal fire spread from one occupancy to the next. The obvious collapse danger with void spaces is that the fire may go undetected, simultaneously destroying structural components of more than one building.

Stairs

First-arriving firefighting crews rely on internal stairways to help them gain access for rescue and fire attack. For the rescue of occupants, using the stairs is the safest, fastest, and most reliable means of access and egress compared to ground ladders. For years, firefighters have found stairways to be durable and stronger than other interior components. This is a dangerous assumption with newer wood-frame buildings. Like prefabricated trusses, stair stringers are now being built off-site and then delivered to the jobsite, where they are simply hung in place using light metal strapping, brackets, or hangers **FIGURE 3-38**. Stairs today are often made using lightweight, engineered wood products that fail quickly when exposed to fire. Remember, press-glued wood-chip products can fail from the heat in smoke: No flame is required, but direct flame contact does speed up the process.

It's important to address the subject of stairs because some firefighters believe the best access for attacking a basement fire is down the basement stairs from the floor above. This is *not* the way to attack a basement or below-grade fire. When you open the door at the top of the stairs of a basement fire or below-grade fire, you create a high-low flow path to the floor above. As you descend the stairs, you are actually going down the chimney of the fire. You and your crew are in the

FIGURE 3-38 New prefabricated I-stairs use connections of galvanized steel. These steps (tread) and risers are made of OSB and are particularly vulnerable underneath the assembly, especially if left unfinished.

Courtesy of Green Maltese, Retrieved from http://greenmaltese.com/

flow path where the fire wants to go. The fire behavior triggered by this tactic is the exact circumstances that killed Lt. Vincent A. Perez and Firefighter/Paramedic Anthony M. Valerio of the San Francisco, California, Fire Department on June 2, 2011. We must not let the lessons learned from their supreme sacrifice be in vain by repeating the same mistake. We must change our tactical approach to basement and below-grade fires, which has to include a 360-degree size-up in order to identify and confirm their existence.

Flow path issues aside, prefabricated stairs are no longer reliable. Sounding the stairs with a tool does not give any reliable indication of weakened stairs because the firefighter usually sounds the run of the steps. The weak links in the system are the connectors and hangers, not the wooden steps. Once the connectors give way to the heat, the stair stringers collapse, sending the attack crew into the basement without any other means of escape.

A more acceptable practice is sending a crew with an exposure hose line to protect the basement doorv and the first floor from the fire in the basement. Many firefighters take the exposure line right to the basement door and stand guard. Again, when entering a residential structure with lightweight prefabricated building components, we need to be aware of what we can fall through or what can fall on top of us. Prefab floor joists and stairwells can quickly become compromised by the time fire crews are making entry into the occupancy. Exposures and doors can still be protected from a safe and stable area by utilizing the reach of a hose stream without standing directly above the fire area below. The initial fire attack should commence from the outside. (Attacking basement fires will be covered more thoroughly in Chapters 4 and 10.)

Parapet Walls and Cornices

A parapet wall is an extension of a wall past the top of the roofline **FIGURE 3-39**. Parapets are used to help conceal roof equipment like HVAC systems, duct work, elevator housing, and so on, giving the building a finished aesthetic look. They typically consist of masonry and are freestanding, with little stability. Collapse may be caused by the failure of the roof structure. Business signs, theater marquees, utility connections, and other eccentric loads on the parapet wall are secured with steel cables, brackets, and bolts. Rooftop electrical service entrances and weather heads are also eccentric loads that can contribute to an early failure of a parapet wall. During a fire, the steel cables and bolts holding these accessories weaken and subsequently pull down the parapet.

A cornice is a horizontal decorative molding that crowns the rooftop edge of a flat roof building. It can be simple or ornate **FIGURE 3-40**. The function of a cornice is to direct rainwater away from the building walls. Cornices can be massive, extremely heavy masonry or wood sections. They usually impose a cantilevered load onto a building. With a collapse of the roof structure, these giant sections could come crashing down before a wall collapse or along with a wall collapse. Either way, falling cornices are deadly, especially when they land on your head.

FIGURE 3-39 A parapet wall is an extension of a wall past the top of the roofline.

Courtesy of Raul Angulo.

FIGURE 3-40 A cornice is a horizontal decorative molding that crowns the rooftop edge of a flat roof building.

Signs of Collapse

Firefighters must rely on building material knowledge, building construction principles, and an understanding of how fire affects buildings in order to predict or anticipate collapse. Waiting for a visual sign that a building will collapse is dangerous, especially in newer buildings. However, some factors and observations can be used to help anticipate collapse, including:

- Overall age and condition of the building
- Size of the building
- Deterioration of mortar joints and masonry
- Cracks in any of the material
- Signs of building repair, including reinforcing cables, tie-rods, and stars
- Large open spans
- Bulges and bowing of walls
- Sagging floors
- Loud noises, vibration, shaking, and rumblings
- Sudden movement of floors and walls
- Abandoned buildings with major structural elements missing
- Large volume of fire
- Long firefighting operations
- Smoke coming from cracks in the walls
- Dark smoke coming from truss roof or floor spaces. (Brown smoke indicates that wood is being heated significantly; black smoke means combustibles have ignited or are near ignition.)
- Multiple fires burning in the same building
- Damage from previous fires

Buildings under Construction

Buildings during new construction, remodeling, and historical restoration are loaded with temporary hazards and are especially unsafe for firefighters when they're involved in fire. The word *unsafe* not only applies to fire operations but also to rescues, searches, odor investigations, and on-site inspections. Buildings need to meet the fire and life safety codes only when they are completed. During construction, structural components may not be fully connected and may only be held in place with temporary bracing and scaffolding. Installation of many of the fire-protective features may be in progress, or fire-resistive components may be incomplete. Stacked lumber and construction material may overload other structural components with a concentrated load. This is not to say contractors are using unsafe practices; it's simply the reasonable nature of the construction process. Exposed structural elements, incomplete assemblies, and stacked materials contribute to a rapid collapse if a fire were to develop or the construction site was subjected to high winds.

Some fires that occur at new construction sites are deliberately set. When responding fire companies arrive at a fire at a new construction site, the officer must consider the possibility of arson. If there are no life hazards to mitigate, a defensive strategy should be implemented. The same considerations apply to buildings that are being dismantled or demolished—they no longer have any value. Buildings under construction are usually insured and can be replaced—a firefighter's life cannot, so go defensive.

Preparing for Collapse

There are no specific time limits for firefighting operations within a building. Tests have shown that the age-old 20-minute rule used by previous fire officers is no longer accurate. Roofs and walls can collapse within minutes of fire involvement given certain conditions. An overloaded truss (due to improper storage between the web members or other factors) can collapse quickly when exposed to high heat and flames. This situation cannot be determined from exterior roof operations.

Every firefighter must understand two rules about structural collapse during fire operations: recognize the potential for structural failure at every fire and establish a collapse zone. Do not set artificial time limits based on experience. Once a building has been searched and cleared of occupants and victims, the only life hazard and risks that exist are the lives of the firefighters themselves. Therefore, that risk should be reduced—after all, it is now a property and contents fire. If there are indications that the building is in danger of collapsing, act on it. Many firefighters have been injured and killed fighting interior fires only to have the building torn down after the investigation.

The second rule to keep in mind about structural collapse during fire operations is to establish a collapse zone **FIGURE 3-41**, which is the area around and away from the building where debris will land and scatter if the building falls to the ground. The absolute minimum is 1½ times the height of the building. The walls may crumble into a pile or they may tip out the full height of the building. Provide

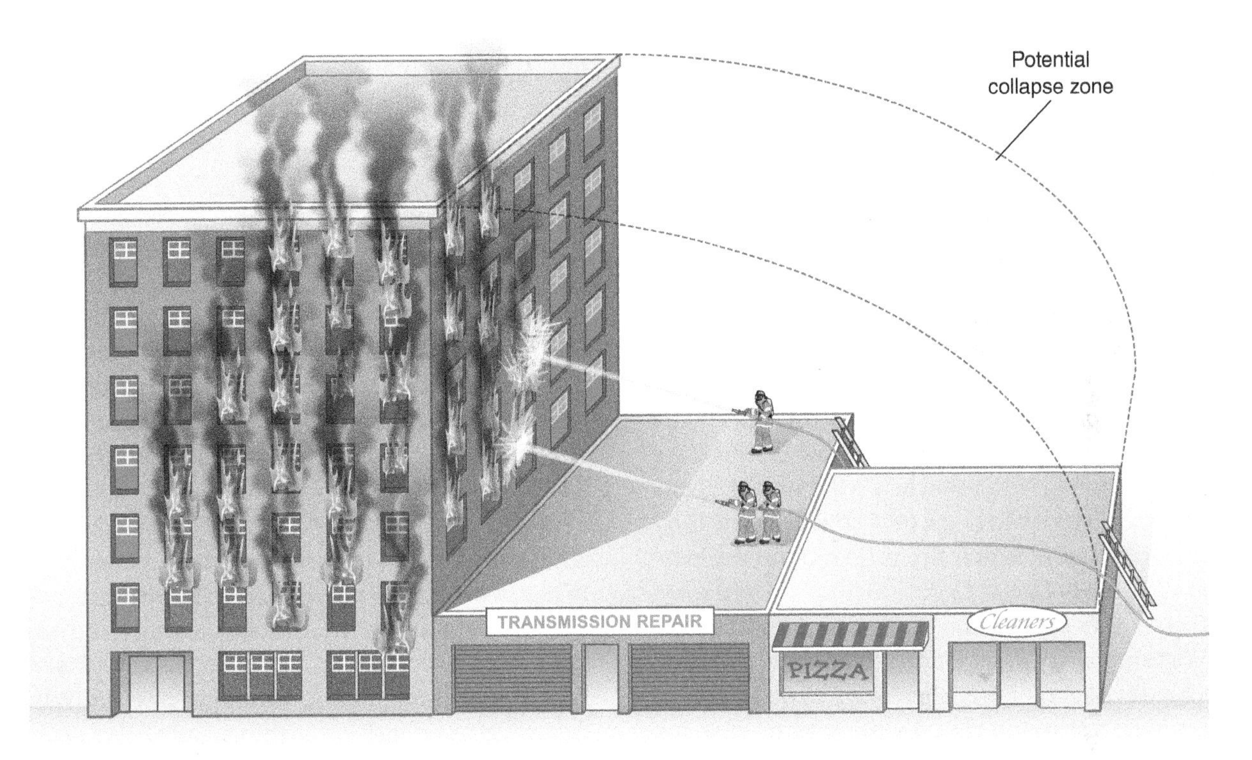

A

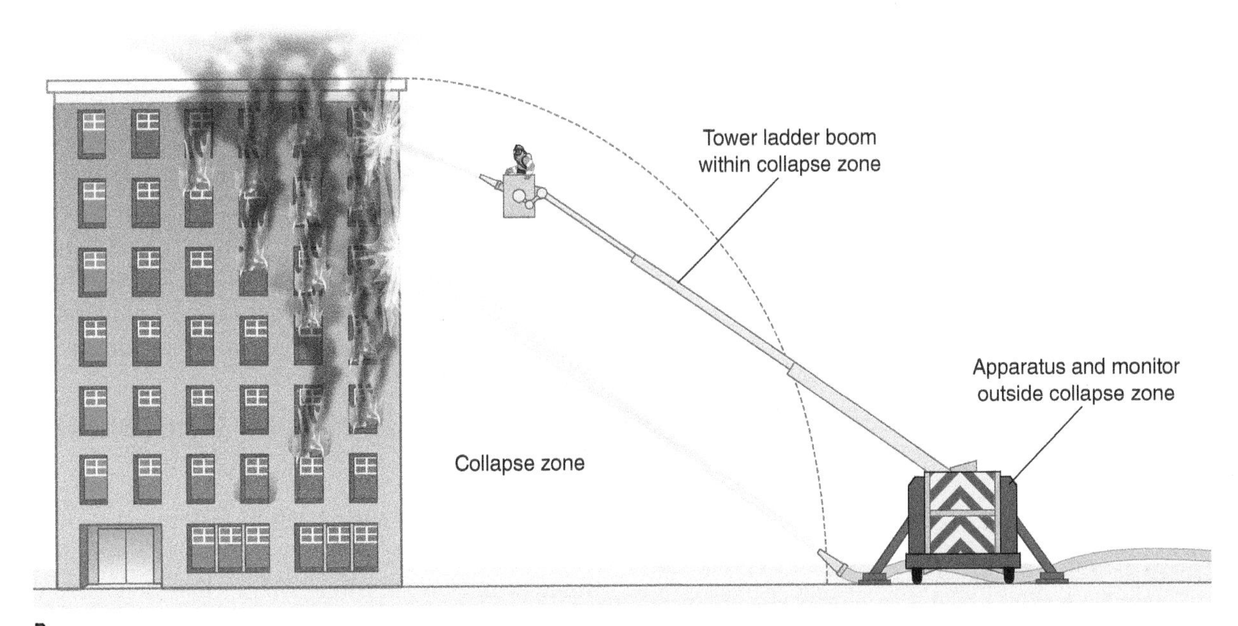

B

FIGURE 3-41 A. At a minimum, a collapse zone should be 1½ times the height of the involved building. **B**. Care must be taken to ensure that the tip of an aerial ladder or the boom of a tower ladder does not enter into the collapse zone, although the apparatus may be clear of the building.

FIGURE 3-42 Mark off a collapse zone by laying uncharged fire hose around the perimeter.
Courtesy of Raul Angulo.

extra room for cascading and scattering debris. Outside defensive firefighting operations can be equally dangerous if firefighters wander in the collapse zone. Use fireground scene tape or rope to cordon off the collapse zone. In windy conditions or when there are no stationary objects to use when tying off fire scene tape, an effective way to mark off a collapse zone is by laying uncharged 2½-inch fire hose around the perimeter **FIGURE 3-42**.

When I was the captain of Engine 33 in Seattle, Washington, my crew responded to an early morning McDonald's restaurant fire. There were no occupants in the fast-food restaurant; the only life hazards were my firefighters. We were first in, and with flames visible, we made an aggressive interior attack. This incident was handled with the first-alarm assignment. We were very proud of our efforts. We made an excellent stop and, from a fire department point of view, this was a great save. By the end of the month, the McDonald's Corporation came in and knocked the entire building down and cleared the lot for the construction of a brand-new, state-of-the-art McDonald's. We were lucky; no injuries or fatalities happened at this fire. Around the same time, a very similar fire occurred at a Houston, Texas, McDonald's with disastrous results. On February 14, 2000, Firefighter Lewis Mayo III, and Firefighter Kimberly Ann Smith were killed when the lightweight roof trusses collapsed on top of them. In the end, both McDonald's structures were torn down and rebuilt.

Concern for Collapse Determines the Strategy

If buildings never collapsed during a fire, there would be no need to have any other strategy except offensive interior attacks. Maybe firefighters would need a defensive-to-offensive transitional attack to knock down a building, but the operational period would continue until the fire was completely extinguished. Collapses happen, however, and this concern about collapse led to the three basic fire attack strategies.

Not all fire departments have the resources of the largest metropolitan cities. Small volunteer or rural departments may have only one or two stations. Smaller fire departments with limited resources must decide if they are going to be an offensive-oriented or a defensive-oriented fire department. Offensive-oriented fire departments rely on firefighters, apparatus, equipment, and training. In addition to training, defensive-oriented fire departments need to rely more on strong fire prevention programs and code enforcement inspections; strict construction regulations; and automatic fire alarm systems, including smoke detection, heat detection, and automatic sprinkler systems. After the life hazard has been addressed, defensive-oriented fire departments may have to use a surround-and-drown strategy by establishing a safe perimeter around a structure fire and let it consume itself while protecting exposures. Offensive-oriented fire departments implement risk-benefit analysis to determine how aggressive a fire attack will be.

There are basically three attack strategies: defensive, transitional (defensive to offensive or offensive to defensive), and offensive. A defensive attack includes surrounding the fire and letting it burn itself out while protecting exposures because there is no life hazard and the building no longer has any value. A defensive-to-offensive transitional attack is a defensive exterior attack using master streams to knock down a large fire back to a hand-line-size fire, after which an interior attack can continue. An offensive-to-defensive transitional attack is riskier because firefighters launch an interior attack to effect a confirmed rescue, then withdraw to a defensive surround-and-drown strategy. An offensive attack—which is the most dangerous for firefighters—is an interior attack using fire crews and hose lines. The information gathered during the initial size-up; the available resources, including the water supply; and performing a risk-benefit analysis determine the decision to go offensive or stay defensive.

Collapse Times and Operational Periods

Today's fire service is driven by risk management and the concern of legal liability. The safest action for firefighters is to do nothing other than defensive operations. This strategy may not be in the best interests of the public, the local economy, or people trapped in a burning building. The opposite extreme is for firefighters to charge into any structure that is on fire because it must be extinguished at any cost. Realistic and responsible firefighting efforts lie somewhere between these two options.

Due to the factual evidence of early collapse with lightweight construction and the fact that National Fire Protection Association (NFPA) fire resistance times were established under laboratory conditions, which cannot be applied in the field due to the numerous variables in existing construction, it is understandable why recent fire service texts and publications shy away from giving any credibility and reliability to collapse times or operational periods. However, it's obvious that certain types of construction endure the effects of fire better than others. Type I and Type IV collapse times range from 2 to 4 hours. All the other construction types put collapse times anywhere from 2 to 20 minutes. The first-in fire officer needs to make a quick strategic decision by properly processing and understanding the information obtained during the initial 360-degree size-up and risk-benefit analysis.

Strategic Considerations

In relation to collapse times or the operational time available before building components are compromised, it is important for the first-in company officer to determine if the fire he or she is looking at is a **structure fire** or simply a **room and contents fire**. There is a difference between the two. A contents fire can appear quite spectacular without involving the actual structure. The only items burning are the interior fuel load. For example, a large waste paper basket or a small sofa burning next to a window can look like a well-involved room fire when in fact the fire is still in the incipient stage, or early growth stage, and all the fire-resistive characteristics of the building, including the gypsum board–covered walls, are still intact.

Although the room is absorbing the radiant heat, the integrity of all the fire-resistive characteristics of the building material remain unaffected and the smoke is still light gray. The time clock toward structural collapse has not started because the structure is not yet involved **FIGURE 3-43(A)**. Therein lies the window of opportunity for a successful interior attack with a high degree of safety and a favorable survivability profile. It isn't until the wood structural members start to char and burn that it becomes a structure fire and the collapse time clock starts on Types III, IV, and V construction. Once the fire is in the growth stage, the room will flash, and the fire will enter the fully developed stage, especially if there is an ample amount of air.

FIGURE 3-43(B) shows the room fire is in the fully developed stage for a lightweight construction Type V residence. The room has flashed over, and all the contents of the room are on fire. The fire has self-vented and the survivability profile in this room is zero. Structural members are receiving direct flame contact. Now you have a structure fire. The best tactic is to aim a straight stream into the window from the outside to knock down and reset the fire, but there is still time to wage a reasonably safe interior attack. Taking a line through the front door, up the stairs, and going with a direct attack from the uninvolved to the involved would also work. The danger to fully protected firefighters at this time is minimal. Note that the front door to the house is still closed.

FIGURE 3-43A A room and contents fire.
Courtesy of Raul Angulo.

FIGURE 3-43B The room fire is in the fully developed stage. It has flashed over, and all the contents of the room are on fire. This is now a structure fire.
Courtesy of Raul Angulo.

FIGURE 3-43C The firefighters have decided on a direct attack up the stairs from the uninvolved toward the involved.
Courtesy of Raul Angulo.

In **FIGURE 3-43(C)**, the firefighters have decided on a direct attack up the stairs from the uninvolved toward the involved. By opening the front door, they have created a high-low uni-directional flow path fueling the intensity of the fire. The air intake portal is low, and the exhaust portal is high. This is also an indication that the bedroom door is open. The radiant heat has ignited the roof of the porch, and the siding and the soffits are exposed to intense direct flame contact. The smoke has changed to dark gray and black. This is a structure fire, and the time clock to collapse has started. Because there are survivable spaces on Floor 1 and Floor 2, the survivability profile for possible occupants is high and an offensive interior attack is warranted. The structure also still has value. The best tactic is taking away the energy from the fire by using a transitional attack, defensive to offensive: by flowing a straight stream through the window from the outside to knock down and reset the fire. But since this fire has self-vented through the window and the flow path is working in favor of the attack team with the wind at their backs, an interior attack from the uninvolved to the involved will work. The lightweight trusses are most vulnerable to the heat and the flames, if they become involved in fire. Collapse can occur within 2 to 5 minutes.

In the acquired structure shown in **FIGURE 3-44(A)**, the fire is on Floor 1, A-B corner. It is in the growth stage. The flames have rolled across the ceiling and have broken out the windows: The fire is self-venting. The smoke is light; the color is light gray. This is still a room and contents fire, and all the fire-resistive separations are intact. The house is Type V wood construction, but due to its age, it likely does not have lightweight structural components. So far, the structural members are not involved, and the time clock to collapse has not started. With dimensional lumber, the collapse time would be approximately 20 minutes, resisting fire a little longer than the lightweight Type V structure. Note the front door to the house is closed.

FIGURE 3-44A This is still a room and contents fire, and all the fire-resistive separations are intact.
Courtesy of Raul Angulo.

FIGURE 3-44B This is now a structure fire in the fully developed stage.
Courtesy of Raul Angulo.

In **FIGURE 3-44(B)**, the fire has spread to Floor 2 through any number of vertical pathways. The fire could have traveled up through the walls, but it is definitely involving the A side exterior siding. Firefighting teams again, decided on an interior direct attack. By opening the front door, they created another high-low flow path directly affecting fire behavior. Note the flame intensity on Floor 2. The bedroom door on floor 2 is open creating a uni-directional flow path. The intense flames are involving exterior walls, the soffit, and the roof. This is now a structure fire in the fully developed stage. The time clock to collapse has started (**FIGURE 3-44(C)**).

It is true: There are no scientific and absolute collapse times or operational periods for construction types. Unlike laboratory tests, actual fires are uncontrolled emergencies, and the condition of the building is anybody's guess. It can be a deadly mistake to rely on such times, but listed times should be used as a guideline, a reference, or an indication to illustrate how long each of

FIGURE 3-44C Once load-bearing structural members are involved in fire, the time clock to collapse has started and may occur quickly.

Courtesy of Raul Angulo.

the five construction types can sustain fire involvement before structural failure can be expected. This information should help the initial incident commander decide whether to launch an interior offensive attack on the fire within a reasonable operating period with some margin of safety or to hold back and set up a defensive perimeter strategy. Quickly applying water from the outside can knock down the fire and reduce the energy emitted from the fire. This resets the fire, slowing down the time to collapse and buying time for firefighters and civilians. We must use some rational criteria to weigh the risks when civilian lives are at stake.

Types of Building Construction

Building and structures are classified according to their type of construction using the following type numbers: Type I, Type II, Type III, Type IV, and Type V. The Building codes and the current edition of NFPA 220, *Standard on Types of Building Construction,* define and establish the fire-resistive ratings of structural elements based on their combustibility. NFPA 220 promotes the protection from fire and the associated hazards on each type of construction and is an extract of Section 7.2 of NFPA 5000, *Building Construction and Safety Code.* The Standard and codes also defines occupancy classifications and means of egress based on these five classification types. These basic categories give firefighters an understanding of the strength of structural elements and the materials used to construct the building. Unfortunately, these broad classifications are not absolute because the construction industry keeps evolving. Remodels and renovations can alter the fire-resistive features of a classification, and new construction projects—like Type III strip malls with commercial occupancies on the ground floor and residential occupancies on upper floors—often combine the classifications. This may lead to deadly assumptions about the makeup of a building because, from the outside, the building may appear to be more fire-resistive than it actually is.

It is important to note that these standards and codes are designed to give occupants time to escape during a fire. They are not designed to establish safe operational timelines for firefighting. Ensuring firefighters can safely operate inside a structurally sound building during fire attack is not part of these classifications.

Table 4.1.1 from NFPA 220, *Standard on Types of Building Construction*, outlines the number of hours that a structural element needs to be protected for each classification. Fire-resistive ratings are established under laboratory conditions, so in the real world, fire resistance rating could "underperform" due to any number of factors. For example, a structural element with a 2-hour fire rating may fail in 30 minutes if it is not assembled correctly or if it was not inspected properly. **TABLE 3-1** is from NFPA 5000, *Building Construction and Safety Code*, and mirrors NFPA 220. Engine company officers should be familiar with both documents.

The US Fire Administration (USFA) and UL, along with other agencies, conducted a series of tests to gain more information about establishing general collapse times for various floor assemblies. The results were presented at the *Changing Severity of Home Fires Workshop* in December 2012. The information is listed in **TABLE 3-2** and should help with the initial IC's size-up, risk-benefit analysis, and decision-making process when considering the possibility of structural collapse.

Fire resistance for structural members can be achieved by various methods, including using drywall (gypsum board), spray-on coatings, and concrete. Concrete is fire-resistive, meaning that it has the capability to withstand the effects of fire. Other materials like steel and wood need fire-resistive assistance to give occupants a chance to escape. Aging, alterations, and wear can damage fire-resistive methods to the point that structural elements have no fire resistance protection.

Noncombustible material is material that, in the form of its anticipated use, will not ignite, burn, support combustion, or release flammable vapors when subjected to fire or heat. **TABLE 3-3** shows the performance of common building materials under stress and fire conditions.

The following section outlines the basic definition of each building type, its general configurations, its typical fire spread problems, and the hour ratings that floors should withstand before collapsing. Only the floors are emphasized because that is what firefighters will fall through or what will fall on top of them.

TABLE 3-1 Fire Resistance Ratings for Types I–V Construction

	Type I		Type II			Type III		Type IV	Type V	
	442	**332**	**222**	**111**	**000**	**211**	**200**	**2HH**	**111**	**000**
Exterior Bearing Walls[a]										
Supporting more than one floor, columns, or other bearing walls	4	3	2	1	0[b]	2	2	2	1	0[b]
Supporting one floor only	4	3	2	1	0[b]	2	2	2	1	0[b]
Supporting a roof only	4	3	1	1	0[b]	2	2	2	1	0[b]
Interior Bearing Walls										
Supporting more than one floor, columns, or other bearing walls	4	3	2	1	0	1	0	2	1	0
Supporting one floor only	3	2	2	1	0	1	0	1	1	0
Supporting a roof only	3	2	1	1	0	1	0	1	1	0

TABLE 3-2 Collapse Times for Various Floor Assemblies

Structural Elements	Ceiling (Protecting the Underside of the Elements)	Firefighter Breach (Collapse) Time (Minutes:Seconds)
2- × 10-inch solid wood joist	None	18:35
2- × 10-inch solid wood joist	Lath and plaster	> 79 minutes
2- × 10-inch solid wood joist	Gypsum wallboard	44:40
Engineered wooden I-beam	None	6:00
Engineered wooden I-beam	Gypsum wallboard	26:43
Wood and metal hybrid truss	None	5:30

USFA/Underwriters Laboratory as presented at the Changing Severity of Home Fires Workshop, Dec. 2012.

Remember, these are only guidelines for rapid decision making on the fireground during the size-up process in considering the "time" category of VTS.

Determining the Time (T) in VTS

Type I: Fire Resistive

Type I fire-resistive construction is a type in which structural elements are of an approved noncombustible or limited combustible material with sufficient fire-resistive rating to withstand the effects of fire and prevent its spread from story to story. Concrete-encased steel, monolithic-poured cement, and steel with sprayed-on fire protection coatings are typical Type I. Generally, the fire-resistive rating must be 3 to 4 hours depending on the specific structural element. Fire-resistive construction is used for high-rise buildings (seven stories and above) and structures with high-volume occupancies like stadiums and convention centers. Most Type I buildings are typically large, multistory structures with multiple elevators, stairways, and exit points.

Fires are difficult to fight due to the height of these buildings, the large square footage of each floor, the open floor space, and subsequent high fire load. Type I

TABLE 3-3 Performance of Common Building Material under Stress and Fire Conditions

Material	Compression	Tension	Shear	Fire Exposure
Brick	Good	Poor	Poor	Fractures, spalls, crumbles
Masonry block	Good	Poor	Poor	Fractures, spalls
Concrete	Good	Poor	Poor	Spalls
Reinforced concrete	Good	Fair	Fair	Spalls
Stone	Good	Poor	Fair	Fractures, spalls
Wood	Good across grain Marginal with grain	Marginal	Poor	Burns, loss of material
Engineered wood	Good	Fair	Fair	Burns, rapid loss of materials, heat can degrade glue, causing rapid failure
Structural steel	Good	Good	Good	Softens, elongates, loses strength
Cast iron*	Good	Fair	Fair	Easily fractures when heated

*Some cast iron may be ornamental in nature and not part of the structure or load bearing.

buildings rely on automatic fire protection systems to rapidly detect and extinguish fires. These systems include heat detectors, smoke detectors, automatic sprinklers, standpipes, fire pumps, and HVAC systems. If these systems do not contain the fire, a difficult firefight will be required. Fire can spread from floor to floor as windows break and the windows on the next floor fail, allowing for the fire to jump. Fire can also make vertical runs through utility and elevator shafts. Regardless, firefighters are relying on the fire-resistive features to protect the structure from collapsing.

Although collapse in fire-resistive structures is rare, it can be massive, as we are reminded from the September 11th attack on the World Trade Center (WTC) in New York City. This was an unusual event, but we can no longer rule out jetliners, terrorist attacks, explosives, and bombings as the cause of a high-rise fire. Nevertheless, even in this extreme example, WTC Tower 2 stood for 56 minutes after it was violently struck, and WTC Tower 1 stood for 102 minutes after it was hit. The point is that Type I buildings are the strongest buildings constructed and give us the longest operating time for fire operations. Although other sections are rated at 3 and 4 hours, the floor system is rated at 2 hours. Since this is what firefighters are standing on or standing underneath, it's best to be conservative and use the lesser of the three ratings. Therefore, the operational period or collapse time for Type I buildings is approximately 2 hours.

Type II: Noncombustible

Type II noncombustible construction consists of structural elements that do not qualify as Type I construction but are of an approved noncombustible or limited combustible material with sufficient fire-resistive rating to withstand the effects of fire and prevent its spread from story to story. The majority of Type II buildings are steel. Modern and big-box warehouses, small arenas, and newer churches and schools are built as noncombustible. Because the steel is not required to have fire-resistive coatings, and many codes do not require automatic sprinkler systems, Type II buildings are susceptible to steel deformation and fatigue from high fire temperatures, resulting in early collapse **FIGURE 3-45**. Fire spread in Type II buildings is influenced by the contents. While the structure itself will not burn, rapid interior collapse is possible from the fire load BTU release, which expands, stresses, and weakens the steel.

Suburban strip malls with concrete block, load-bearing walls and steel roof structures can be classified as Type II. Fires can spread from store to store through wall openings and unsprinklered common (shared) ceiling and attic spaces. The roof structure often consists of lightweight steel bar trusses that fail rapidly when exposed to flames.

More often than not, Type II buildings are one story, and the fire-resistive technique used to protect

FIGURE 3-45 This beam and supporting column are unprotected steel, which elongates, fatigues, and quickly collapses once ceiling temperatures rise above 1,000°F (538°C).

the roof structure is a dropped ceiling. Missing ceiling tiles, damaged drywall, and utility penetrations can render the steel web members vulnerable to fire if they are not protected by an automatic sprinkler system. For example, many big-box warehouse stores have automatic sprinkler systems and automatic ventilation roof hatches that cool the steel web trusses and dissipate the heat, but many Type II noncombustible occupancies do not require an automatic sprinkler system. Therefore, there is no fire-resistive time rating given to the roof structure. As soon as the ceiling temperature reaches approximately 1,000°F (538°C), steel begins to expand and lose strength. When the ceiling temperature reaches approximately 1,200°F (649°C), steel can collapse under its own weight. Both situations can lead to structural collapse. The phase of the fire and how long it has been burning is critical; a building fire in the decay stage can last a long time, whereas the same building fire in the fully developed stage with ceiling temperatures close to 1,500°F is extremely dangerous and should not be entered.

A defensive position is the course of action. The stability of Type II construction is unpredictable under fire conditions; therefore, the floor and roof system collapse time and operational period is zero.

Type III: Ordinary

The term *Type III ordinary construction* is often misapplied to wood-frame construction buildings. By definition, ordinary construction includes buildings where the load-bearing walls are noncombustible, and the roof and floor assemblies are dimensional wood. Most commonly, this is load-bearing brick or concrete block with wood roofs and floors **FIGURE 3-46**. Ordinary construction is prevalent in most downtown areas of established towns and villages. Firefighters have referred to ordinary construction as taxpayers. This slang term is derived from buildings with businesses on the first floor and apartments above to maximize income and thus help landlords pay property taxes. Newer Type III buildings include strip malls with veneer masonry block walls and wood truss roofs.

Ordinary construction presents many challenges for firefighters. In older buildings, numerous remodels, restorations, and repairs have created suspect wall stability and other hidden dangers. These modifications use lightweight building materials; therefore, the building is no longer true ordinary construction but a hybrid using Type V wood-frame components **FIGURE 3-47**. Ordinary construction has many void spaces where fire can spread undetected. Common hallways, utilities, and attic spaces can spread fire rapidly. Masonry walls hold heat inside, making for difficult firefighting.

FIGURE 3-46 Older Type III buildings have load-bearing brick or concrete block walls with wood roofs and floors providing structural mass; therefore, they burn for a longer period of time than modern Type III construction, which incorporates Type V lightweight components. Note the visible decorative stars, which are tensioned steel tie-rods used to "pull in" and support the walls and floors.

FIGURE 3-47 A brick veneer wall in a modern hybrid Type III construction.

Wood floors and roof beams are often gravity-fit within the masonry walls. These can release quickly and cause a general collapse, leaving an unsupported masonry wall. Older Type III buildings have structural mass; therefore, they burn for a longer period of time.

Remember, dimensional lumber burns ⅛ of an inch every 5 minutes when it is subject to direct flame contact. Thus, a true 2-inch × 4-inch burning from both sides loses half its entire thickness to fire in 20 minutes. This fact is a consideration that goes into the 20-minute fire resistance rating for ceiling and floor assemblies in Type III construction. Taking into consideration possible hybrid renovations, Type III estimated operational period and collapse time is 2–20 minutes.

Type IV: Heavy Timber

Type IV heavy timber construction can be defined as those buildings that have block or brick exterior load-bearing walls and interior structural members, roofs, floors, and arches of solid or laminated wood without concealed spaces. The minimum dimensions for structural wood must meet certain criteria. Heavy timber buildings, as the name suggests, are quite stout and are used for warehouses, manufacturing buildings, and some older churches. In many ways, a Type IV building is like a Type III with larger dimensional lumber instead of common wood beams and trusses FIGURE 3-48.

Type IV heavy timber construction is sometimes called mill construction. However, mill construction is a much stouter, more collapse-resistive building that may or may not have block walls. This type of construction was very common in the 1920s and 1930s, and it is still in use today. A new Type IV building is hard to find. The cost of large-dimensional lumber and laminated wood beams makes this type of construction cost-prohibitive, but many churches are Type IV using big glue-laminated beams with 2-inch decking, and they often incorporate large steel columns to support the beams FIGURE 3-49.

Fire in a heavy timber building can spread quickly due to wide-open areas and content exposure. The exposed timbers contribute BTUs to the fire. Because of the mass and the large quantity of exposed structural wood, fires burn for a long time. If the building housed machinery at one time, oil-soaked floors add more heat to the fire and accelerate the time to collapse. Once floors and roofs start to sag, heavy timber beams may release from the walls, which happens by making a fire-cut on the beam. The beam is gravity-fit into a pocket within the exterior load-bearing masonry wall. As the floor sags, it loses its contact point with the wall and simply slides out of its pocket in order to preserve the wall without damage. In other words, you can have interior localized collapses within the structure, but the actual building can remain standing until the fire is extinguished or the fuel burns itself out. It is important to remember that a freestanding masonry wall has little lateral support and requires compressive weight from floors and roofs to make it sound.

Type IV buildings, heavy timber construction, have large wood structural elements with great mass. The mass of these structural members requires a long burn time for failure. The columns supporting floor loads are 8-inch × 8-inch minimum in any dimension. Columns supporting roof loads are a minimum 6 to 8 inches. Tongued and grooved planks for floor decking is 3-inch minimum thickness with additional 1-inch boards at right angles. Tongue and groove planks for roof decking is 2-inch minimum thickness FIGURE 3-50. The connections, usually steel, are the weak links in this type of construction. The fire resistance rating for Type IV heavy timber is 1 hour. The estimated operational period and collapse time is 1 hour.

Type V: Wood Frame

Type V wood-frame construction is perhaps the most common construction type. Homes, newer

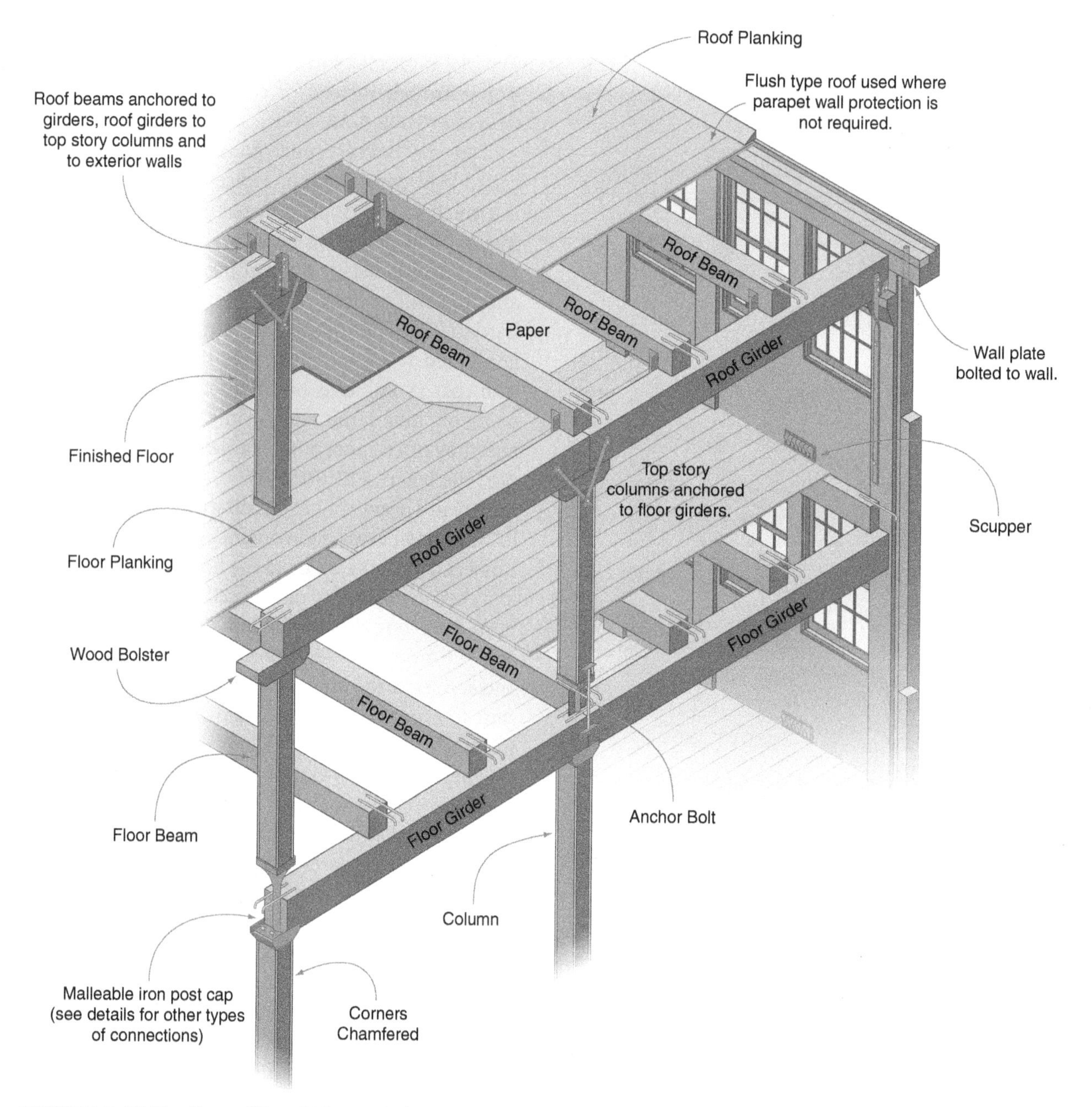

FIGURE 3-48 The floor of Type IV heavy timber mill construction.

FIGURE 3-49 New Type IV construction is combined with Type II, often incorporating large steel columns to support the massive wood beams.

small businesses, and even chain hotels are built primarily with wood. Starting in 1833, older wood-frame, multistory buildings were built as **balloon frame**; that is, the wood studs ran from the foundation, two or more stories high, to the eave line **FIGURE 3-51**. At the floor line, a horizontal board, called a stringer or a ribbon board, was nailed to the wall studs. The floor joists were hung from the ribbon board (rarely from the studs). The channels between the wall studs were usually open from the basement up to the attic, and the channels between the floor joist were also open; fire could travel vertically and horizontally through all the interconnected channel spaces. As you can imagine, fire spread was rapid because it could enter the wall space and run

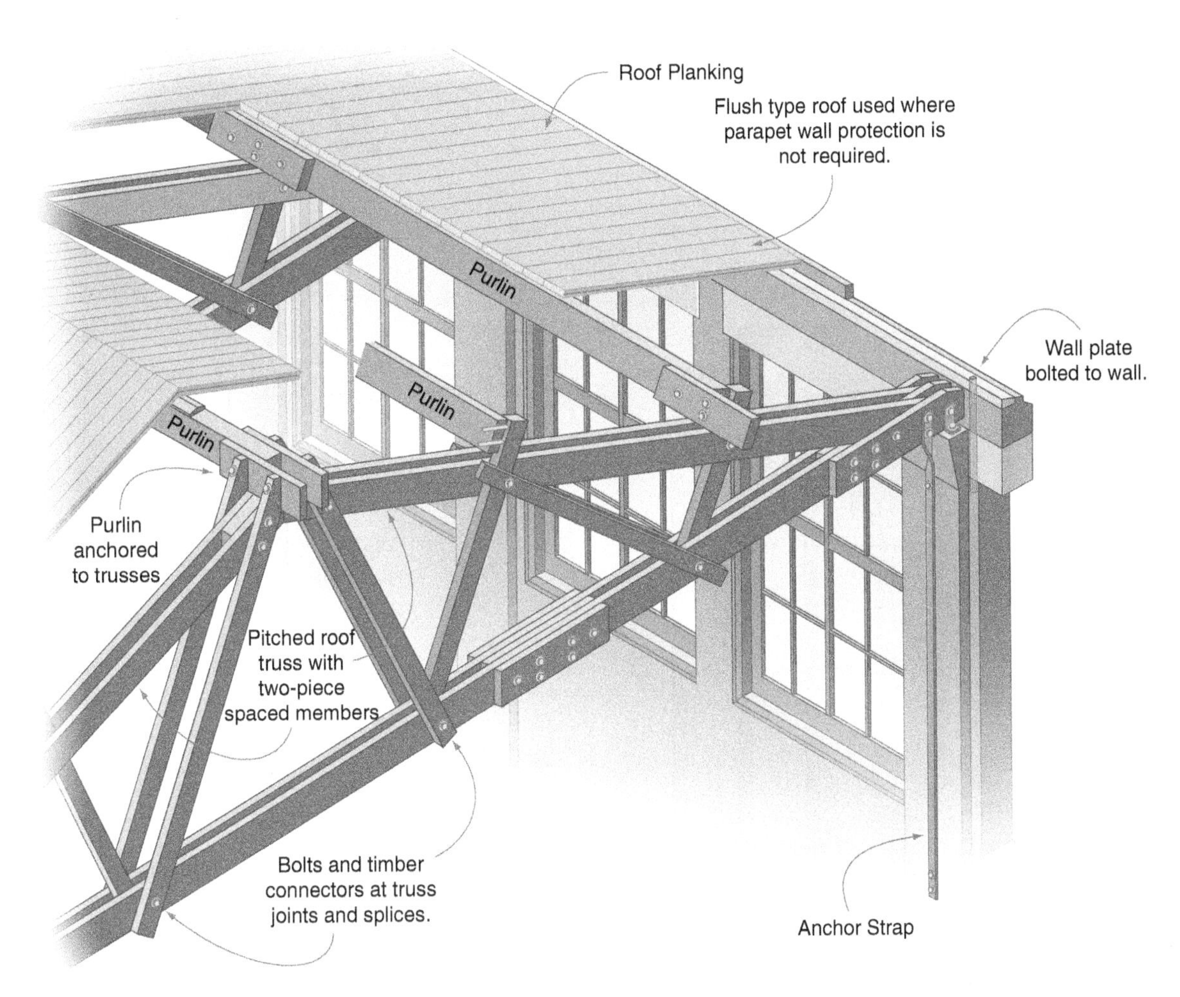

FIGURE 3-50 A Type IV heavy timber mill construction roof assembly.

straight up to the attic and across the floor and ceiling joists **FIGURE 3-52**.

Tactical Consideration for Balloon-Frame Construction

When a fire enters the interior walls of a balloon-frame structure, the fire can spread to every part of the building in any direction due to the lack of fire stops and all the interconnected void spaces between the studs and the joists. The problem with attacking this wall fire from the inside is that the fire moves quickly up through numerous channels simultaneously. As soon as the wall is opened, smoke enters the room, obscuring visibility. Channels may be missed, and the fire can get past firefighters and extend to the next floor and across the ceiling. Fire moving up the channels between the studs can best be stopped by removing the exterior siding and exposing the wall from the outside. Remove the siding at the base of the foundation, at the floor lines and at the eaves. Because firefighters are working from the outside where visibility is clear and there is essentially no heat, they can quickly expose all the wall stud channels by using long pike poles and rubbish hooks. A charged hose line should apply water as they go.

Don't hesitate or wait to check the attic space. Use the thermal-imaging camera to check for heat buried under insulation. It is best to check the attic from an inside scuttle. Opening the roof draws the fire into the attic space. The attic may be charged with heat and smoke, but it is still oxygen-deficient to support combustion. This scenario is when fog applicators, piercing nozzles, or round nozzles should be used to cool the attic space and keep the fire gases below their ignition temperature until the main body of fire is extinguished.

Positive pressure ventilation should not be used on balloon-frame construction during fire attack because the pressure can blow the fire unchecked into every vertical and horizontal channel. Even after the fire has been extinguished, care should be taken until all the void spaces that were in the path of fire spread have been exposed and wetted down. It is best to use smoke ejectors or hydraulic ventilation.

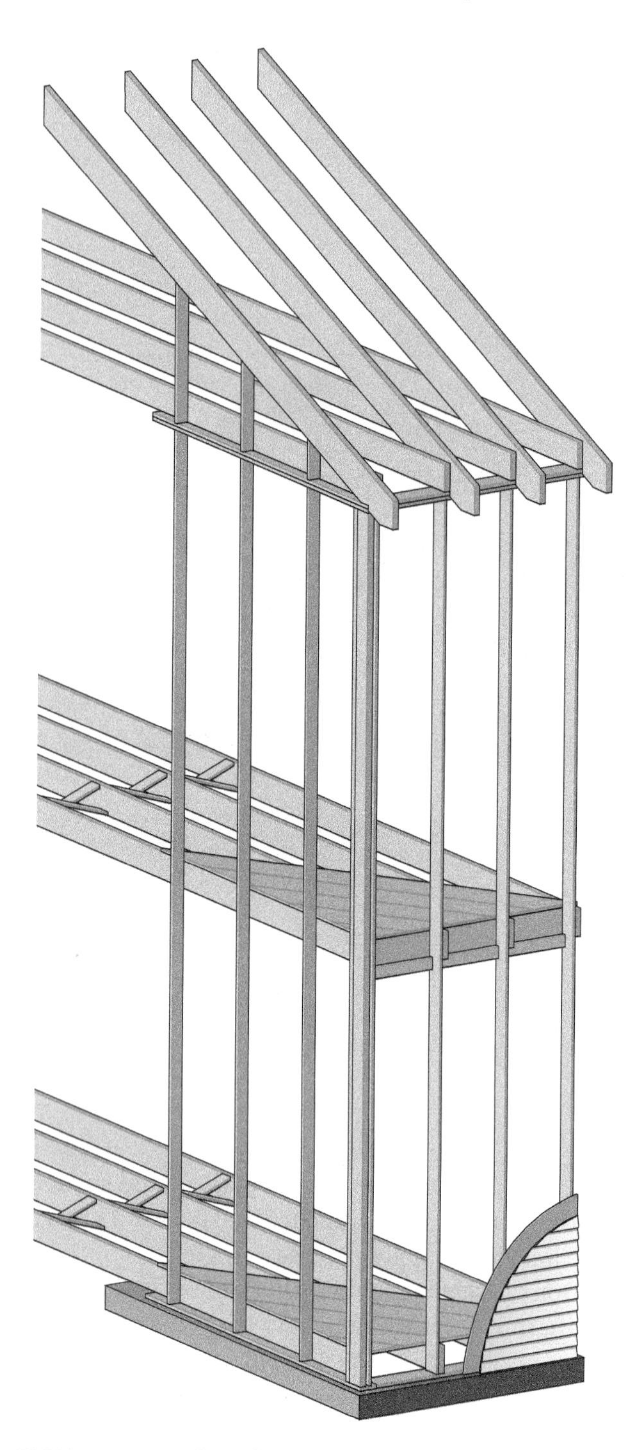

FIGURE 3-51 Balloon frame. Note that the open channels between the wooden studs are continuous from the basement to the attic.

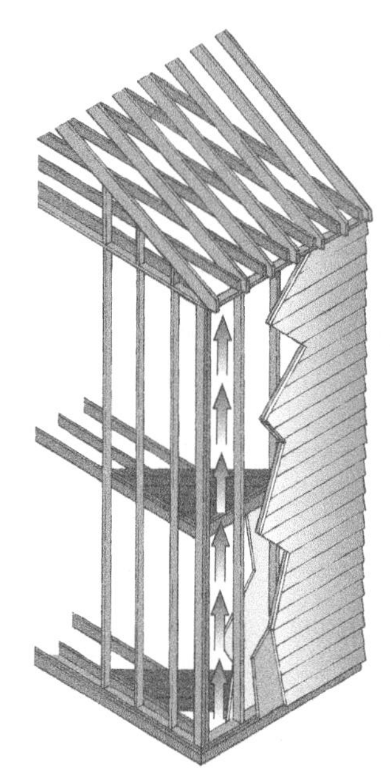

FIGURE 3-52 The open channels between the wood studs, and thus the lack of fire stops, in balloon-frame construction allows a basement fire to extend unchecked into the attic space.

Platform Frame

Balloon frame was the dominant construction method used until the early 1950s, when builders started using **platform frame**, an arrangement where one floor was built as a platform for the next floor **FIGURE 3-53**. This created **fire stopping** to help minimize fire spread. If constructed correctly, there was a good chance of confining a fire to the room of origin, even if it got into the walls **FIGURE 3-54**.

Plank and Beam

Plank-and-beam construction is beefier than platform-frame construction. Dimensional lumber and heavier beams, rafters, and joists are used so they can be spaced farther apart, creating a more open floor plan. Instead of thin plywood decking, finished tongue-and-groove planks are used so they can span a greater distance without flexing. The finished underside is the ceiling. The benefit of plank and beam compared to wood frame is that there are fewer concealed spaces for fire to travel in, and the dimensional mass of the wood withstands fire longer than lightweight platform construction, although it still has only a Type V, 20-minute rating. The disadvantages to plank and beam are the finishes used on the exposed wood. Although the oils enhance the grain and beauty of the wood, they also produce high-flame surface spread, along with thick smoke production. These structures often utilize an open space concept, so the kitchen, living areas, family room, and sometimes sleeping areas are all contained within the same open space. The lack of compartmentation usually provided by non-bearing walls and doors allows the fire and smoke to spread throughout the entire open space. Depending on the fuel load, this can make for a large residential fire and a serious life hazard for possible occupants **FIGURE 3-55**.

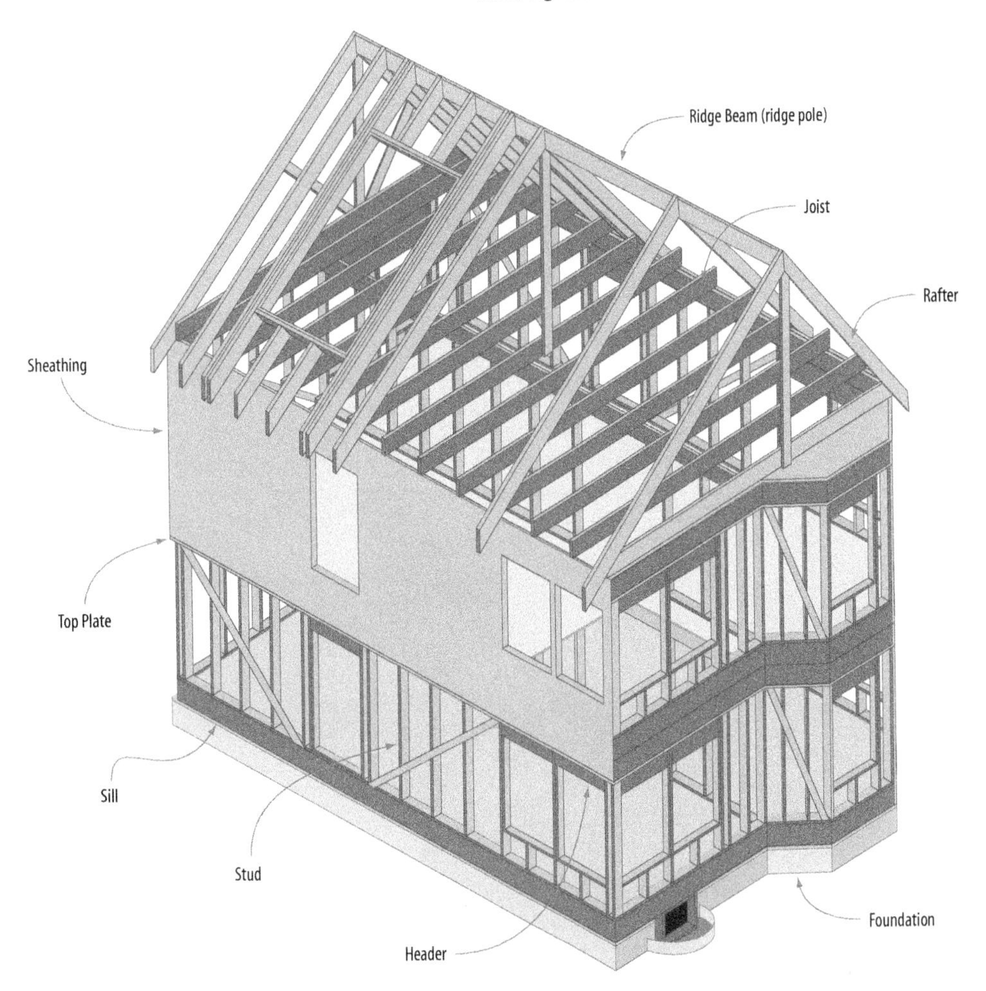

FIGURE 3-53 Platform-frame construction.

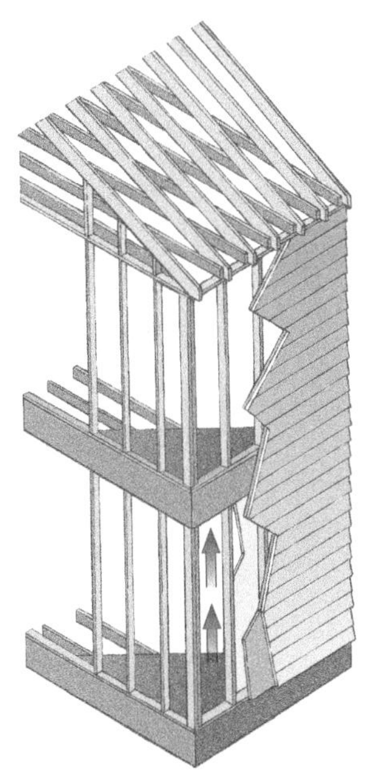

FIGURE 3-54 The fire stops used in platform-frame construction create a horizontal barrier in the wall space between the wood studs, preventing vertical fire spread from one floor to another.

Lightweight Wood Construction

Lightweight wood construction does not have its own separate construction type; it is a method of construction that uses lightweight materials and connectors and is included in the Type V wood-frame category. The old and the new methods are combined, yet older wood-frame structures using dimensional lumber can endure fire for a longer period of time **FIGURE 3-56**. Some wood-frame structures may appear more like a Type III ordinary building because of a brick-wall appearance. Don't be fooled; the brickwork is simply veneer, added for aesthetics.

Newer wood-frame buildings utilize lightweight wood trusses for roofs and wooden I-beams for floors **FIGURE 3-57**. This in essence creates horizontal balloon-frame conditions that can allow quick horizontal fire spread. Coupled with high surface-to-mass wood exposure, collapse becomes a real possibility. Some codes require floor joist spaces to have fire stopping every 500 square feet in single-family residential construction, and every 1,000 square feet in other buildings. Even with the fire stopping, it remains

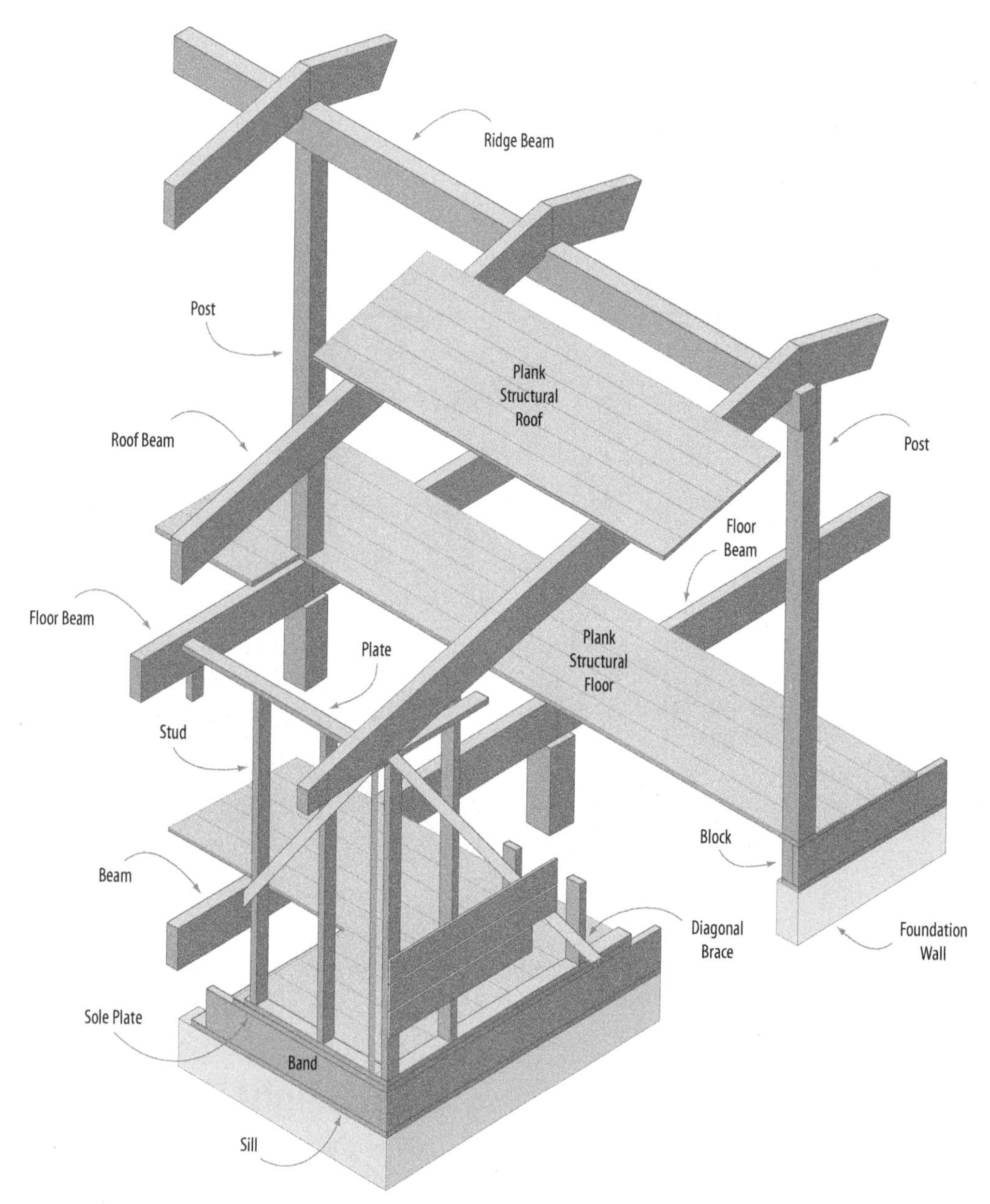

FIGURE 3-55 This illustration shows the larger-dimensional lumber and the wider spacing of rafters and floor joists used in plank-and-beam construction.

A

B

FIGURE 3-56 **A**. Conventional wood-frame construction. Note the ridge beams and the lack of gusset plates. **B**. Lightweight truss construction. Note the numerous gusset plates used to hold the trusses together.

FIGURE 3-57 Newer wood-frame buildings use lightweight wood trusses for roofs and wooden I-beams for floors.

© Jones & Bartlett Learning.

FIGURE 3-58 As long as the gypsum board retains its chemical moisture, the wood-frame assembly is protected.

Courtesy of Raul Angulo.

dangerous to step onto a 500-square foot platform where the fire is burning directly underneath it.

Gypsum board is used to protect wooden structural members from fire. Once the gypsum board is properly installed, it can withstand and contain room and content fires within its rating. As stated in this chapter, drywall isn't really dry. As long as the gypsum board retains its chemical moisture, the wood-frame assembly is protected. As more heat is absorbed, calcination occurs; the sheets eventually become brittle and fail, exposing the wood to fire **FIGURE 3-58**. Many factors affect the fire rating in real life, including the installation process and whether the gypsum walls were properly taped (sealed), or whether the sheetrock was doubled up, adding to the thickness. Company officers need to keep in mind that modern fuels release more heat energy, so the *calcination* process is sped up. UL tests show that drywall delays collapse of an engineered wooden I-beam floor assembly by only 20 minutes. Subtract alarm notification, dispatch sequence, response times, and fireground setup, and you can see that only a small

A

B

FIGURE 3-59 A. Fires that penetrate walls, floors, or attic spaces become significant collapse threats in newer structures. B. This attic fire was extinguished without opening the roof, depriving the fire of oxygen. A hole was punched into the roof decking and a round nozzle was inserted into the attic space, creating a sprinkler system.

A: © Keith Muratori. Used with permission; B: Courtesy of Raul Angulo.

FIGURE 3-60 Cutting a vertical ventilation hole from the safety of an aerial ladder.

window of opportunity is available for rescue and interior fire attack with any margin of safety.

Fires that penetrate walls, floors, or attic spaces become significant collapse threats in newer structures. Often, the only warning that fire has penetrated these void spaces is the issuance of smoke from crawl space vents, roof or gable end vents, and the eaves. If smoke is showing, these spaces need to be checked before continuing an interior fire attack that will put firefighters on top of or beneath these feeble burning assemblies **FIGURE 3-59**. The NFPA fire resistance rating for Type V wood frame is 20 minutes. However, the combination of grouping traditional wood frame with lightweight prefabricated wood frame means that collapse can happen in 10, 5, and even as early as 2 minutes when subjected to high heat and flame contact. Basements and attics, once involved in fire, can collapse in *3 to 4 minutes*. Note that common lightweight roof systems of Type II, Type III, and Type V construction can fail after only a few minutes of direct flame contact. Even when these types of roof systems are not directly involved in fire, the decision of putting fire crews to work on these types of roofs for vertical ventilation should be carefully weighed through risk-benefit analysis. Cutting a vertical ventilation hole draws the fire into the space beneath firefighters. It is always safer working from an aerial ladder or platform when vertically ventilating lightweight roof assemblies **FIGURE 3-60**. The other option is to use horizontal positive pressure ventilation.

Lightweight gang-nail trusses can fail in 3 to 7 minutes, and plywood engineered truss joist I beams (TJI wooden I-beams) can fail in 2 to 3 minutes of fire exposure. The estimated operational period and collapse time for true Type V wood frame with dimensional lumber supporting a 1-inch roof and 1-inch floors is 2 to 20 minutes.

Other Types of Construction

As mentioned earlier, the five broad building types can actually lead to dangerous assumptions. Newer construction and alternative building methods may not fit cleanly into one of the five types of construction. Some buildings combine the construction type classifications.

New lightweight steel homes resemble wood-frame homes. These buildings are actually a post-and-beam steel building with lightweight steel studs to help make the house stiffer and increase wind-load strength. Another interesting construction type uses foam blocks to make a form for a lightweight concrete-mud mixture. The concrete is not continuous: There are many void spaces, utility runs, and foam block spacers (made from plastic or galvanized steel). These structures are called insulated concrete formed (ICF). However, it is important not to be fooled by any claims that these buildings are concrete or less combustible. In reality, these composite buildings are assembled with plastics, polystyrene, lightweight steel, and lightweight concrete. When finished, these buildings may resemble wood frame or even ordinary construction. Extended windows and door jambs are clues that indicate that the wall is thicker than that of typical wood or masonry-built buildings.

The fire service has very little research information on the stability of these new types of buildings during fires. One thing is certain: Firefighters should expect rapid collapse due to the low mass and high surface-to-mass exposure of structural elements. Hybrid-new technology buildings are driven by competitive market costs, not by any desire to increase their resistance to fire. Structures built with new synthetic materials but built without sprinkler systems are built to burn.

Manufactured buildings can be defined as those structures that are built at a factory and then trucked to the jobsite for assembly. These buildings are quite light with little mass. Where a stick-built home uses 2-inch × 4-inch and 2-inch × 6-inch lumber, the manufactured home uses 1-inch × 2-inch and 2-inch × 2-inch lumber. These buildings use galvanized strapping to give required strength. In any case, these buildings burn quickly and collapse equally fast. Lightweight hybrids and manufactured buildings should be considered disposable buildings. Unless a fire is caught in the incipient stage, it progresses quickly, and structural compromise and failure will come quickly after that. These buildings are cheaper and easier to tear down and replace with a new structure rather undergo repairs when damaged by fire.

Prefires and New Construction Inspections

One way to be certain of the building classification or any combination of the fire classes is to check the certificate of occupancy during the inspection. It will be clearly stated on the certificate which areas of the building fall into which categories. Visiting new construction sites during the various construction phases allows firefighters to examine and view the internal membranes and assemblies of floors, walls, stairs, and roofs before they're sealed off and finished. Take advantage of these opportunities. You can't ask for a better classroom for studying the battlefield.

After-Action REVIEW

IN SUMMARY

- Although masonry walls do not burn, they can collapse under heavy fire conditions. Masonry walls have very little lateral stability. Roof and floor assemblies hold the walls in place. When a roof or floor collapses, it often can topple the wall in a secondary collapse.
- Concrete under compression can support a tremendous amount of weight. Under tension, it is very weak. If concrete is to be used in construction under tension, it must be reinforced by embedding long cold-drawn steel rods called rebar.
- A variety of specialized manufactured glass resists breakage from impacts and resists the effects of high temperatures in a fire. The end result is a well-sealed home that can endure the four burning stages of a room fire because all the available oxygen is consumed.
- Many loads are associated with building construction. The two major types of loads are dead loads and live loads.
- A load must be transmitted to structural components. This is called imposition of loads. There are three imposition loads: axial loads, eccentric loads, and torsional loads.
- Loads imposed on structural members create stress and strain on the materials used to make the connections. Stress and strain are defined as forces applied through those materials. These forces are defined as compression, tension, and shear.
- Roofs have three primary designs: pitched, curved, and flat. The major components of a roof assembly are the supporting structure, the roof decking, and the roof covering.
- A truss is a geometric structural component that is composed of smaller pieces of boards, timbers, beams, or steel bars, joined together to form a triangle or a series of triangles and thus create a rigid framework that supports a load greater than its own mass.
- Trusses are used in almost all new buildings. Trusses have high surface-to-mass characteristics that rapidly absorb heat and subsequently fail quickly when involved in fire. Failure of one truss can cause the failure of other trusses.
- Trusses are often used to replace heavier solid beams and joists in renovated or modified buildings. A truss is actually a fake beam.
- New prefabricated I-stairs use connections of galvanized steel, similar to the gusset plates used in wooden trusses. The steps and risers are made of OSB and are particularly vulnerable to fire underneath an unprotected assembly, such as in a basement. Press-glued wood chip products can fail from the heat in smoke alone; no flame is required.
- Steel has both tension and compression strength, making it the strongest material used in modern building construction. Steel beams or joists expand during a fire, creating lateral loads that the walls were not designed to withstand. The collapse of a roof or floor assembly can exert lateral forces against masonry walls that can cause them to collapse and kill firefighters.
- Steel trusses are just as susceptible to collapse as wood trusses. Unprotected structural steel used for beams and columns loses about 50% of its strength at 1,100°F (593°C). As the temperature rises, structural steel expands and elongates; at 1,100°F, a 100-foot (30 m) beam can elongate 10 inches (25 cm) and knock down an entire wall. When the temperature reaches approximately 1,200°F (649°C), steel can collapse under its own weight.
- Keeping steel trusses cool with hose streams is an important tactical consideration because it prevents elongation and buys time by extending the strength of steel during a fire.

- There is a difference between a room and contents fire and a structure fire. In relation to collapse times and the operational time available before building components are compromised, it is important for the first-in company officer to determine if the fire is a structure fire or a room and contents fire.
- For Type I fire-resistive construction, the estimated operational period or collapse time is approximately 2 hours.
- For Type II noncombustible construction, there is no fire-resistive time rating given to an unprotected steel roof structure. As soon as the ceiling temperature reaches approximately 1,000°F (538°C), steel begins to expand and lose strength. When the temperature reaches approximately 1,200°F (649°C), steel can collapse under its own weight. Both can lead to structural collapse.
- Type III ordinary construction has an estimated operational period and collapse time of 2–20 minutes.
- The fire resistance rating for Type IV heavy timber is 1 hour. The estimated operational period and collapse time is 1 hour.
- Lightweight wood construction does not have its own separate construction type; it is a method of construction that uses lightweight materials and connectors and is included in the Type V wood-frame category. The old and the new methods are included in the same category, yet older wood-frame structures using dimensional lumber can endure fire for a longer period of time.
- The NFPA fire resistance rating for Type V wood frame is 20 minutes. However, the combination of traditional wood frame with lightweight prefabricated wood frame can collapse in 10, 5, and even as soon as 2 minutes when subjected to high heat and flame contact.
- Lightweight gang-nail trusses can fail in 3–7 minutes, and plywood engineered TJI floor joist (wooden I-beams) can fail in 2–3 minutes of fire exposure. Basements and attics that are constructed with lightweight trusses and I-beams, once involved in fire, can collapse in 3–4 minutes. Therefore, the estimated operational period and collapse time for true Type V wood frame with dimensional lumber supporting a 1-inch roof and 1-inch floors is 2–20 minutes.
- Risk management now drives the modern fire service, and all decision making must include a risk-benefit analysis.

KEY TERMS

alligatoring When wood pyrolyzes during a fire, creating black and both deep and shallow cracks along the surface of the wood that resembles the skin of an alligator.

axial load A compression load imposed through the center of an object.

balloon frame An older type of wood-frame construction in which the wall studs extend vertically from the basement of a structure to the roof without any fire stops.

bowstring truss A truss with a distinctive curved shape.

collapse zone The area around and away from the building where an exterior wall will land and scatter during a collapse. The minimum clearance from the building should be 1 1/2 times the height of the building.

compression force A force that causes material to be crushed or flattened axially through the material.

concentrated load A load applied on a small area.

contemporary construction Buildings constructed since about 1970 that incorporate lightweight construction techniques and engineered wood components. These buildings exhibit less resistance to fire than older buildings.

dead load A load that consists of the weight of all materials of construction incorporated into the building, including, but not limited to, walls, floors, roofs, ceilings, stairways, built-in partitions, finishes, cladding, and other similarly incorporated architectural and structural items, and fixed service equipment, including the weight of cranes. (NFPA 5000)

designed load A load in engineered structural designs that has been planned for the intended use of the building.

distributed load A load applied equally over a broad area.

eccentric load A compression load imposed off center to another object.

fiber-reinforced products (FiRPs) Plastic fibers mixed with wood to give the wood increased tensile strength.

fire load The total energy content of combustible materials in a building, space, or area including furnishing and contents and combustible building elements. (NFPA 557)

fire stopping The placement of horizontal wood members between the spaces of wall studs and flooring to prevent the vertical spread of fire from floor to floor.

fire wall A wall separating buildings or subdividing a building to prevent the spread of fire and having a fire-resistance rating and structural stability. (NFPA 5000)

forces Loads imposed on structural members that create stress and strain on the materials used to make the connections. Stress and strain are defined as forces applied through those materials.

gusset plate Connecting plate made of a thin sheet of steel used to connect the components of a truss.

gypsum board The generic name for a family of sheet products consisting of a noncombustible core, primarily gypsum with paper surfacing. (NFPA 5000)

impact load A load that is in motion when it is applied.

legacy construction An older type of construction that used sawn lumber and was built before about 1970.

live load The load produced by the use and occupancy of the building or other structure. It does not include construction or environmental loads such as wind load, snow load, rain load, earthquake load, flood load, or dead load. Live loads on a roof are those produced (1) during maintenance by workers, equipment, and materials and (2) during the life of the structure by movable objects such as planters and by people. (NFPA 5000)

occupancy The purpose for which a building or other structure, or part thereof, is used or intended to be used. (NFPA 5000)

parallel chord truss A truss in which the top and bottom chords are parallel.

pitched chord truss A type of truss typically used to support a sloping roof.

platform frame Construction technique for building the frame of the structure one floor at a time. Each floor has a top and bottom plate that acts as a firestop.

pyrolysis A process in which material is decomposed, or broken down, into simpler molecular compounds by the effects of heat alone; pyrolysis often precedes combustion. (NFPA 921)

rebar Cold-drawn steel rods embedded to reinforce the tensile strength of a concrete slab when it will be used under tension.

room and contents fire A fire in the incipient stage and early growth stage. The only fuels burning are the furnishings in the room. Although the entire room is absorbing the heat of the fire, structural members and assemblies are still intact.

shear force The force that causes a material to be torn in opposite directions, perpendicular or diagonal to the material.

spalling Chipping or pitting of concrete or masonry surfaces. (NFPA 921)

structure fire Any fire inside, on, under, or touching a structure. (NFPA 901)

tension force A force that causes a material to be stretched or pulled apart in line with the material.

thermoplastic materials Plastic material capable of being repeatedly softened by heating and hardened by cooling and that, in the softened state, can be repeatedly shaped by molding or forming. (NFPA 5000)

torsional load A load imposed in a manner that causes another object to twist.

undesigned load The weight that a building supports for which the building was not designed or an added weight that was not anticipated. Buildings that are altered or refurbished for an occupancy other than the original intent can create an undesigned load.

ventilation-limited fire A fire in which the heat release rate and fire growth is controlled by the amount of air available to the fire. (NFPA 1403)

wood truss An assembly of small pieces of wood or wood and metal. A truss is a fake beam.

REFERENCES

Corbett, Glenn P., and Francis L. Brannigan, *Brannigan's Building Construction for the Fire Service*, 6th ed., Burlington, MA: Jones & Bartlett Learning, 2019.

Dodson, David W., *Fire Department Incident Safety Officer*, 3rd ed., Burlington, MA: Jones & Bartlett Learning, 2017.

Fundamentals of Fire Fighting Skills and Hazardous Materials Response, 4th ed., Burlington, MA: Jones & Bartlett Learning, 2019.

Kriss, Garcia, Reinhard Kauffmann, and Ray Schelble, *Positive Pressure Attack for Ventilation and Firefighting*, Fire Engineering and Pennwell Corp. Tulsa, OK: 2006.

National Fire Protection Association. NFPA 101, *Life Safety Code*. 2018.

National Fire Protection Association. NFPA 220, *Standard on Types of Building Construction*. 2018.

National Fire Protection Association. NFPA 5000, *Building and Safety Code*. 2018.

Stewart E. Rose, *Strategy and Tactics MCTO*, InSource, Inc. Salem, OR: 2015.

CHAPTER

An Analysis of the UL/NIST Experiments

LEARNING OBJECTIVES

- Understand why the UL/NIST research recommendations should be applied and implemented in your fire department.
- Identify areas within your fire department SOPs that need to be examined and modified to include the UL/NIST recommendations.
- Explain the reason for the increase in the number of flashovers in recent years.
- List the components of today's modern fire environment and how they affect fire behavior.
- Gain a better understanding about operating safely in today's modern fire environment.
- Explain the difference between legacy construction and modern construction and how they affect fire behavior.
- Differentiate between legacy room furnishings and modern room furnishings and how they affect fire behavior.
- Define the two types of flow paths and describe how they spread fire.
- Explain the difference between a fuel-limited time-temperature graph and a ventilation-limited time-temperature graph.
- Interpret the various time-temperature graphs and explain what happens during the different stages of fire development.
- Recognize the importance of applying water quickly on a fire and the immediate results throughout the structure.
- Explain the importance of a coordinated attack.
- Describe the rescue and safety considerations produced by the experiments in relation to VES and VEIS.
- Describe the attack and safety considerations produced by the experiments in relation to fighting basement fires.
- Articulate the Parker Doctrine and how it comes into play within the fire service culture.

Introduction

This chapter is an overview of the recent evidence obtained from the test fires conducted by Underwriters Laboratories (UL) and the National Institute of Standards and Technology (NIST). The chapter offers additional commentary as it sets the foundation for the rest of the chapters in this text.

Three significant trajectories have finally intersected, and they should cause grave concern to firefighters and push fire departments to reevaluate how they fight residential and commercial structure fires using offensive interior attacks. The first trajectory starts with a study that introduced the term *survivability profile* to the US fire service lexicon. In 2009, Captain Stephen Marsar (now a battalion chief with the Fire Department of the City of New York [FDNY]), wrote an award-winning thesis for his National Fire Academy Executive Fire Officer certification. In his work, he discovered that, between 1990 and 2009, there were 32 FDNY line-of-duty firefighter deaths where zero civilians were killed in those same fires. These 32 firefighters were actively engaged in fire attack, primary searches, overhaul, and other fire suppression activities, and became caught or trapped in these structure fires. The zero civilian fatality rate for these fires is not credited to successful rescues but rather to the fact that there was—in the overwhelming majority of cases—no civilian life hazard present at all. The FDNY started tracking actual saves in 2010 as a result of Marsar's study. In a later article, Marsar cited a 2010 National Firefighter statistic from the National Fallen Firefighters Foundation (NFFF), which stated less than half of 1% of civilian fatalities occur in the same structural fires where firefighters are killed.

In 2005, the *Boston Globe* examined the federal investigations of 52 fires associated with 80 firefighter line-of-duty deaths (LODDs) between 1997 and 2004. Fourteen of the 52 fires had possible occupants trapped, and six of the 52 fires had confirmed trapped occupants. Not one of the 52 fires resulted in a civilian death. It is unclear if the six surviving victims were rescued by fire department personnel or escaped on their own because actual rescues are not tracked; it is implied they were not rescued by fire department personnel. Nevertheless, the number of firefighter deaths in relation to an unconfirmed life hazard is staggering.

"Fatal Echoes" is the most recent report that came out in December 7, 2016, by the *Kansas City Star*. Investigative journalists Mike Hendricks and Matt Campbell took the state and federal fire reports and analyzed 201 firefighter LODDs. They discovered that 157 of the 201 firefighters died in unoccupied buildings. Only 11 died while trying to rescue civilians trapped in a burning structure. Again, actual civilian rescues are not nationally tracked with any consistency, so such figures are difficult to reference. They included a citation from a 2015 U.S. Fire Administration (USFA) report attributing many LODDs and injuries to a tradition-bound firefighting culture that too often celebrates and awards heroism at the expense of safety:

> Despite improvements in personal protective equipment, apparatus, safety devices, more available training, greater emphasis on firefighter health and wellness, and decreases in the number of fires and dollar loss due to fires, the rate of on-duty firefighter death and injury has remained relatively unchanged in the past four decades.

With the rate of on-duty firefighter fatalities and injuries being nearly as high as it was 40 years ago, the reader is left to conclude that firefighters refuse to change the way they fight fires and fail to learn from their mistakes, although we're finally seeing a slight downward trend in the numbers. NFPA publishes the annual report, *Firefighter Fatalities in the United States,* which provides detailed statistics.

This is not meant to diminish the work of firefighters who have made hundreds of heroic rescues over the decades; that is our job—to rescue and save people. Nor is it meant to diminish the work of those who have made the ultimate sacrifice. These reports shed light on the sobering fact that the number of firefighters getting killed and the number of civilian fatalities in those same fire incidents are extremely disproportionate. What is going on here?

The *Kansas City Star*'s online report includes a video quote from Ron Siarnicki, the executive director for the NFFF, ". . . [I]f we really wanted to improve the profession of firefighting, whether you're career or volunteer, we need to do evidence-based research just like doctors or attorneys and other professions. We *need* that evidence-based research to show us the data of what we are doing and how we can change it to make it better."

The other two intersecting trajectories are lightweight construction components with engineered lumber used in modern residential structures, and the fuel load of modern home furnishings that are primarily made from hydrocarbon-based plastics and synthetic materials. The operational safety window is much smaller now. Before 1980, firefighters may have had a relatively safe operational period of 20 minutes to perform search and rescue and fire extinguishment before flashover. Today, that window might be 4 to 5 minutes after the fire starts. Most fire departments haven't even arrived on scene in that amount of time. The average response time is about six minutes.

Thanks to technological advancements in scientific instrumentation, computer modeling, thermal-imaging resolutions, and digital photography, we are experiencing major changes in our understanding of basic fire behavior and the tactics used in fire attack. If we do not understand the impact of these changes, we ignore the science, and we may be putting ourselves in positions that are so unsafe that they prevent us from accomplishing our primary mission of saving lives (including our own) and property.

Traditionally, fires have been extinguished by removing the heat or fuel component. We tend to think only in terms of water and ventilation in extinguishing fires. Except for food-on-the-stove fires, which can be extinguished by placing a lid over the burning pot, or ship fires where watertight compartments can be sealed off until the fire suffocates from lack of oxygen, removing the air was a somewhat futile attempt. After all, how can you remove air from the atmosphere? We discuss in this chapter the information from the UL/NIST experiments that emphasizes controlling fires by removing or manipulating the air of the fire.

Chapters 2 and 3 covered changes in modern fire dynamics and building construction. In this chapter, we'll review the evidence that proves today's fires burn at higher temperatures when adequate oxygen is present, release energy faster, and reach flashover potential sooner. They are much more likely to become ventilation-limited because of energy-saving construction features than did structure fires of even a few years ago. As a result of these many changes, it is important that our tactics for ventilation, water application, and search and rescue adapt to the recommendations made from the conclusive evidence of these live-fire behavior experiments.

Decades of experience have provided us with our traditional practices to fight fires. But these experience-based techniques did not provide us with the means to measure and understand fully the actual progression of fire and the impact of each action we take at a fire. As a result, we have sometimes drawn inaccurate conclusions. On occasion, these conclusions have resulted in ineffective and counterproductive efforts while we are risking our most valuable resources: firefighters, apparatus, and equipment.

One reason for the increase in our understanding of fire behavior is the findings from live-fire experiments. Over the last 15 years, NIST and UL have conducted controlled fire scenarios in specially constructed laboratories and in acquired structures. The researchers used a variety of instruments to measure the temperatures throughout the structure at various heights within each room, the heat release rate, and heat generated from the burning contents in the room. They measured the air speed and the direction of the currents in and out of the fire compartment as well as from the ventilation points. They noted changes in visibility throughout the structure and measured the chemical makeup of the fire gases within and outside the structure. All the observations and data were carefully recorded so they could be compared and analyzed completely. Each change that was made to the fire scene, such as doors and windows opened or closed, ventilation and water application start times, could be measured. Because the researchers were able to conduct these experiments multiple times, it was possible to determine how each of the actions taken by firefighters affected the growth and extinguishment of the fires. The scientific data also allowed the researchers to predict and speculate the viability of human life (survivability profile) within the structure more accurately.

FIGURE 4-1 Transitional attack. Spraying water through a window from the outside was used during Colonial America firefighting.

The conclusions drawn from these live-fire experiments are not new. The UL/NIST recommendations are actually returning to the strategy and tactics used by our forefathers in the fire service without actually saying it. Bucket brigades and the practice of shooting water through the window from the outside using hand pumpers to knock down the fire (transitional attack) was used in Colonial America **FIGURE 4-1**. Transitional attacks continued through the first half of the 20th century before the advent of SCBA. Before the development of NFPA-compliant SCBA and Nomex® flame-resistant personal protective equipment (PPE), it was rare if firefighters made interior attacks the way we do today. Their helmets, coats, and gloves had limitations. The inability to breathe smoke and keep their ears from burning placed physical limitations on offensive interior attacks. The exception may have been a confirmed trapped occupant or incipient fires, in which case they went in with water pump cans and extinguishers. Otherwise, firefighters started their fire attack from the outside by spraying water from a fire hose through a door or a window to knock the fire down; then they went in. That all changed with the introduction of SCBA, which led to the aggressive interior tactics we use today. NFPA 1981, *Standard on Open-Circuit Self-Contained Breathing Apparatus (SCBA) for Emergency Incidents,* establishes the levels of respiratory protection and functional requirements for SCBA used by emergency services.

If there is any question that firefighters attempted to extinguish fires from the outside, just look at the vintage (antique) nozzles of the past: cellar nozzles, distributor nozzles, piercing nozzles, long fog applicators, and bent nozzles. Even the old solid brass, smoothbore, 30-inch playpipes carried on the horse-drawn steamers were 2½-inch nozzles. These were not interior attack nozzles; they were used from greater distances to deliver large amounts of water with deep penetration into the building to knock the fire down before crews went inside.

Another reason these experiments yield valuable information is because the current number of structure fires is down. Many would-be structure fires remain content fires due to smoke detectors, monitored in-home early detection systems, and the increase of residential sprinkler systems. (*Note:* Sprinkler systems do not reduce the number of fires; they simply keep them from extending.) Public education is much more comprehensive than it was 50 years ago, and it has significantly increased the public's awareness and diligence in fire prevention; for example, NFPA's Fire Prevention Week. Another factor is the reduction of ignition sources. For example, in home décor, many candles (a frequent cause in residential fires when they were open-flame) are now using battery-operated lights that resemble a flickering flame. Smoking, a historically notorious cause of home fires, is also on the decline. According to the Centers for Disease Control and Prevention (CDC), from 1965 to 2015, the percentage of adults who smoke dropped from 42% to 15%.

Fewer structure fires mean firefighters and company officers gain less field experience and therefore must rely more on firefighting theory and computer simulators. Modern fire-resistant PPE structural firefighting ensembles with helmets, ear flaps, hoods, boots, gloves, and a self-contained fresh air source with an impact, heat-resistant facepiece has encapsulated firefighters, giving them a false sense of security and invincibility. Ask yourself: Has modern state-of-the-art PPE with high-intensity flashlights, personal alert safety system (PASS) devices, and portable radios reduced the number of LODDs on the fireground? The answer is no. On average, the fire service is still experiencing between 80 and 90 LODDs annually. Even with fewer structure fires occurring, there is no equivalent drop in the national average of LODDs. They remain at basically the same level, although a downward trend has begun.

In fact, SCBA bottles, which used to be 30-minute capacity, have now been replaced in many fire departments for the more popular 45-minute and even 1-hour air bottles, which in turn means that we can stay inside the fire longer. This has led to strategic changes that enabled firefighters to take exceptional risks when no life or property value existed. When the fire culture gets away with dangerous practices, these practices continue until disaster strikes. In some cases, organizational traditions, intimidating peer pressure, and ego-driven bully politics force firefighters and company officers to continue in dangerous practices even when the evidence shows they are wrong. It takes courage to challenge such institutional resistance to change, even when armed with evidence, knowledge, and science.

What is new is the evidence, which includes temperature measurements, time graphs, and computer modeling regarding the changes in fire behavior with modern construction and furnishings. This chapter introduces you to these findings, conclusions, and recommendations gained from the UL/NIST experiments. This chapter does not present all the information you need to implement changes in every suppression tactic, but it should influence your approach to most tactics, especially the benefit of utilizing a transitional attack strategy. There is certainly sufficient evidence presented to raise your awareness and understanding of these scientific findings to guide you in your size-up, risk-benefit assessments, and decision making on the fireground. This chapter also describes the advances in our knowledge of the principles of fire behavior and compares building construction techniques used in older legacy construction with modern construction techniques used in residential construction of today.

Note that there is still resistance from around the country in accepting the data and conclusions drawn from the UL/NIST experiments because, although the researchers tried to create realistic fire conditions, they are still laboratory test fires or fires in acquired structures within a controlled environment. Real-world fires occur in an uncontrolled environment with many variables. However, the UL/NIST experiments have the most modern, comprehensive data and heat measurements to date and should not be ignored or shrugged off.

Because of increased fuel load and modern lightweight construction techniques, today's fires grow faster and release more heat energy when adequate oxygen is available, creating an interior environment that is much less survivable. These trends have led to greater adoption of residential sprinkler systems. A house with working smoke detectors, heat detectors, and a residential sprinkler system increases the occupants' chance of surviving a fire by over 80%.

Fire Behavior

Firefighting can be broken down into many steps. We'll start by considering the changes in modern fire behavior and the importance of controlling ventilation.

Fire has always followed the same rules of physics and chemistry. A fire is produced when fuel and oxygen are combined with a heat source for ignition. The campfire example combines wood and oxygen with a heat source of ignition to produce a fire. Left unattended, this fire starts in the incipient stage, advances to the growth stage, grows to a fully developed fire, and proceeds to a decay stage when the available fuel is consumed **FIGURE 4-2**. These are the classic stages of fire growth **FIGURE 4-3**. In a **fuel-limited fire**, the decay stage begins when the fuel is depleted. In a structure, what we can call a box, the decay stage can begin when the fuel is depleted, or it can begin when the available oxygen is depleted even though there is still fuel to burn. In this case, the fire becomes a ventilation-limited fire **FIGURE 4-4**.

While the fuel-limited sequence of fire growth holds true for campfires, fires in modern construction buildings do not necessarily follow these stages. The progression from the incipient to the growth stage occurs faster, but before reaching the fully developed stage, the fire can consume all the available oxygen within the enclosed structure and begin to go into a ventilation-limited decay stage. The smoke generated from the incomplete combustion of modern fuels contains aerosols; dust; oils; tar; creosote; and numerous flammable gases like hydrogen cyanide, hydrogen chloride, carbon monoxide, benzene, formaldehyde, and acrolein. The autoignition temperatures of these flammable gases, except for the last three, are over 1,000°F (538°C). Benzene has an ignition temperature of 928°F (498°C), formaldehyde has an ignition temperature of 806°F (430°C), and acrolein has an ignition temperature between 428°F and 532°F (220°C and 278°C) depending on which chemical fact sheet you

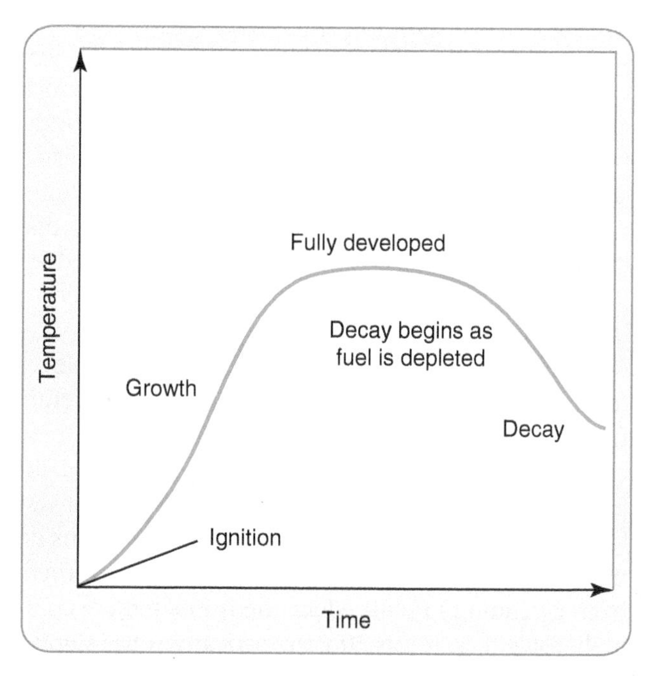

FIGURE 4-3 Graph of a traditional, fuel-limited fire growth curve noting the stages of fire growth.

Courtesy of NIST.

FIGURE 4-2 A campfire is a fuel-limited fire, meaning it continues to burn until it runs out of wood, which serves as the fuel for the fire.

FIGURE 4-4 In a structure fire, the decay stage can begin when the fuel is depleted or when the available oxygen is depleted even though there is still fuel to burn.

Courtesy of NIST.

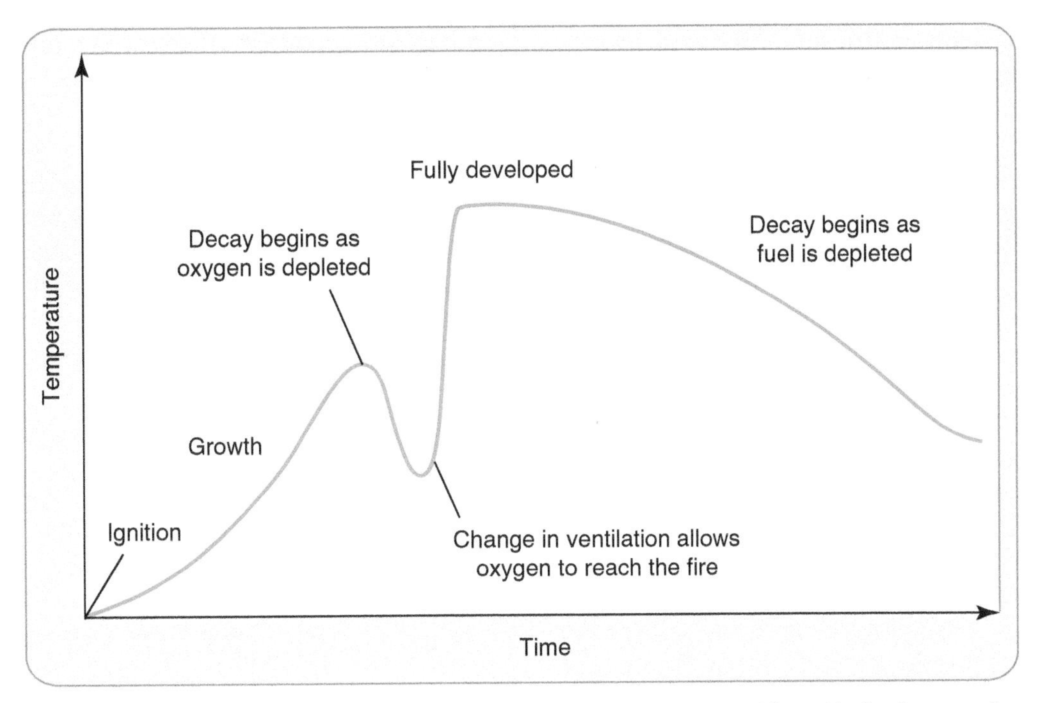

FIGURE 4-5 Graph of the time-temperature curve in a ventilation-limited fire with the fire growth stages noted.
Courtesy of NIST.

consult. An ignition temperature of 428°F (220°C) is an extremely low ignition temperature for a fire gas in the decay stage compared to fires that occurred in the decades between 1900 and 1959, when the primary decay stage smoke was carbon monoxide, which has an ignition temperature of over 1,000°F (1,128°F [609°C]).

The reignition of a fire in the decay stage is no longer dependent on superheated fire gases. In **FIGURE 4-5**, the temperature of a fire in the decay stage quickly starts to drop. It can range from 200°F to 300°F (93°C to 149°C). In other words, this is cool smoke, not superheated smoke. If acrolein is always present in modern fuel smoke, then it stands to reason that, with an ignition temperature as low as 428°F (220°C), it is a pre-igniter or ladder fuel to the other flammable gases in the smoke. When a firefighter opens a door or breaks out a window for ventilation and enters the structure, a flow path is created that introduces fresh oxygen to the oxygen-limited fire. Once the acrolein ignites, the ignition temperatures of the other fire gases become irrelevant. They will all ignite within seconds of the acrolein igniting, causing firefighters to get caught in a smoke explosion (cooler smoke) or a flashover as the rest of the smoke ignites **FIGURE 4-6**.

From a fire-behavior standpoint, there are slight differences between these two fire events, but it's important to understand the differences so you know when to expect them. A flashover, like a backdraft, is a high-temperature event occurring on the upside rise of the fire growth curve when temperature and pressure are increasing, and the fire is progressing

FIGURE 4-6 A. Acrolein is a pre-igniter or ladder fuel to the other flammable gases in the smoke. **B**. A firefighter making entry can introduce fresh air, causing all the gases to ignite within seconds.

from the growth stage to the fully developed stage. It requires the proper air mixture for these two events to occur. A smoke explosion happens on the downward side of the fire growth curve, where the temperatures are decreasing, cooling below 500°F (260°C), and the fire is entering the decay stage **FIGURE 4-7**. The building is already as pressurized as it will get. This pressurization can push smoke some distance from the fire. An explosion has to have an ignition source: The smoke explosion occurs when fresh air mixes with low-temperature smoke and finds an ignition source (for example, the decayed fire or firefighters uncovering burning embers). If no explosion occurs when a new flow path is created by firefighters entering the structure, the preheated smoke can quickly advance to a flashover because the new flow path is intensifying the fire and raising the interior temperature. Therefore, you can have a flashover in the decay stage on the back side of the fire temperature curve. A flashover can follow a smoke explosion in a matter of seconds, and both events can be followed by rapid fire growth.

In recent years, firefighters have encountered a surprising increase in the number of violent and rapid flashovers and/or smoke explosions in residential fires. Between 2010 and 2015, there has been a 28% increase in firefighters caught in flashovers and smoke explosions. In the same time frame, there has been a 13% decrease in structure fires. What is happening here? Firefighters have not completely understood and recognized the conditions that cause these flashovers and smoke explosions. NIST and UL have conducted an extensive series of experiments specifically designed to provide the fire service with the answers, knowledge, and the tools it needs to better understand this phenomenon.

These experiments included component testing of modern furnishings, furnished room experiments, full-scale house burns in the laboratory, and full-scale burns in acquired structures **FIGURE 4-8**. This evidence-based research points to topics that the fire service needs to understand better in order to operate safely in today's fire environment, including:

- Fire dynamics with modern fuels
- Ventilation, fire flow paths, and how fires spread
- Water application and coordinated fire attack
- Search and rescue safety considerations
- Basement fires

Changing Fire Environment

The interior framing of walls, floors, attics, and roofs in older houses (constructed before the 1950s) were built primarily of dimensional lumber, solid wood beams, and plaster. The exteriors were wood; brick; solid masonry walls; and, in some cases, asbestos. They were finished in wood trim and furnished with contents that were made primarily of organic material and natural fibers such as wood, wool, cotton, and paper. The homes had limited insulation in the walls and ceilings, and windows were made with single-pane glass. This type of construction is called **legacy construction**.

For years, firefighters have successfully fought fires in these types of legacy buildings using a defensive

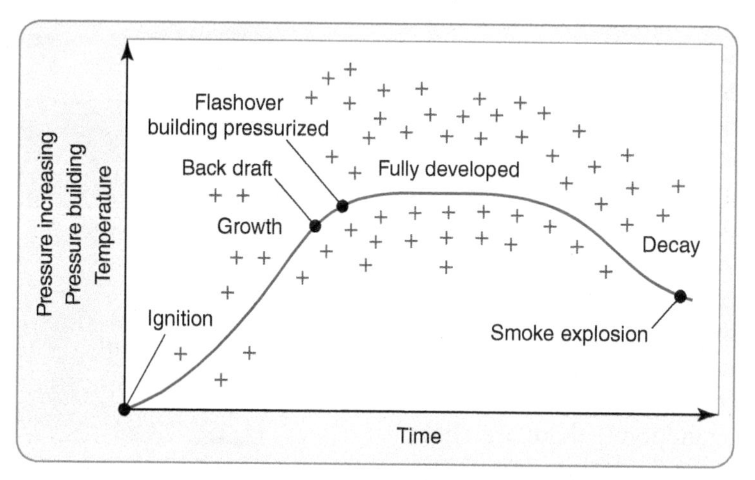

FIGURE 4-7 Flashover and backdraft are high-temperature events occurring on the upside of the fire growth curve, when temperature and pressure are increasing. A smoke explosion typically happens on the downward side of the curve, when temperatures are decreasing but the building is already pressurized, pushing the smoke away from the fire.

Courtesy of the National Institute of Standards and Technology (NIST).

FIGURE 4-8 The legacy room with older furnishings shows little flame 3 minutes and 25 seconds after ignition, while the modern fuel room furnished with petroleum-based products has flashed over in the same time period. (Most fire departments are not even on the scene in this same time period.)

Courtesy of UL.

exterior attack to knock the fire down before continuing with an interior attack. They often had little trouble finding the location of the fire because the single-pane glass would fail from the interior heat, giving away the location of the fire.

Houses built since the mid-1980s are constructed using different construction techniques and materials. Solid dimensional lumber has been replaced with engineered lumber and manufactured wood beams because they are stronger and do not check. (Checking occurs when wood splits along the grain because the wood fibers separate.) Engineered trusses are made with as little wood as possible or from light-gauge steel. They can also be stronger than natural wood and span greater distances. Houses are furnished with a wide variety of plastic-based material derived from petroleum products (hydrocarbons), such as polyurethanes and polystyrenes, and they are filled with petroleum-based foam-filled furniture. Furniture constructed from plastic-based materials is found in all types of buildings and contributes to a greater fuel load with a more rapidly developing fire and a greater heat release rate (producing more energy). Petrochemical materials yield approximately 15,000 British thermal units (BTUs) per pound and legacy materials yield about 8,500 BTUs. Today's society is also more affluent, so modern homes have twice the amount of fuel by weight, which means the modern fire has a BTU potential three to four times greater than homes of the 1950s.

Modern houses are larger with more open spaces and are constructed to be energy-efficient. The result is a house that is more tightly sealed. Windows are often constructed with double or triple panes, which means that, by the time the last pane of glass is exposed to the heat of the fire, the fire may have already moved into the decay stage from a lack of oxygen. As the interior temperature decreases, that last pane of glass has much less chance of failing, making it more difficult for firefighters to locate the fire. The company officer needs to look for and understand the clues given by the windows: Moisture condensation or water vapor on the glass indicates an interior change of temperature from an incipient fire and possibly the room of origin **FIGURE 4-9**.

Cracked glass or crazing (a network of fine cracks along the glass surface) along with blackened windows indicate high-heat areas. Cracked glass indicates that you're getting closer to the fire or that this is the room on fire, especially if smoke is seeping through the glass. Glass cracks when the interior temperature is between 500°F and 600°F (260°C and 316°C). The water vapor has dried, leaving stains, and the window appears to be covered in a gray, brown, or black film. These are indications that the fire is in the growth stage **FIGURE 4-10**. When interior temperatures reach over 600°F (316°C), the glass will fail.

FIGURE 4-9 Condensation on glass indicates a change of interior temperatures from an incipient fire and may be the room of origin.

© Jones & Bartlett Learning. Photographed by Glen E. Ellman.

FIGURE 4-10 Cracked glass indicates that you're getting closer to the fire or that this is the room on fire, especially if smoke is seeping through the glass.

Courtesy of Raul Angulo.

FIGURE 4-11 When interior temperatures reach over 600°F (316°C), the glass will fail.

FIGURE 4-12 The research demonstrated that firefighters making an entry through a front door can introduce enough air into the fire area to produce rapid fire growth and flashover, significantly reducing the opportunity for an effective search within the compartment.

Dark-brown stained glass is due to a creosote film on the glass from incomplete combustion. This indicates that the fire is probably in the decay stage, especially if the window glass is not cracked. Inside temperatures are approximately 250°–300°F. The longer the fire stays in the decay stage, the more creosote layers form on the glass.

Blackened windows are a classic sign for a potential backdraft. Reading the windows is an important part of a proper exterior size-up **FIGURE 4-11**. The changes in modern home construction and furnishings are fine—until a fire occurs. Then they have a great impact on the growth, progression, suppression, and ventilation of fires in modern construction.

NIST and UL have conducted fire experiments comparing the growth of a fire in a room with synthetic furnishings versus one with older contents (modern fuels versus legacy fuels). In these experiments, a room with legacy furnishings took a longer time to flash over or it never reached flashover temperatures. In other words, a room with furnishings produced from natural materials burns much more slowly that a room furnished with synthetic or plastic-based materials. Before the 1960s, legacy fires provided a cooler atmosphere, giving firefighters the opportunity to fight smaller fires with an interior attack, with a greater chance of successfully rescuing an occupant. However, in modern synthetic furnishing test fires, flashover occurred in the room in 3 to 4 minutes. Even if the fire department were to be notified at the time of ignition, it becomes mathematically impossible to arrive in time to launch an offensive interior attack and make a successful rescue within the compartment.

Perhaps the greatest value of these experiments was their ability to demonstrate to firefighters that fires fueled by synthetic furnishings contain a much greater fuel load, develop faster, generate a greater heat release rate (measured in BTUs), and flash over more quickly. This creates a more dangerous condition for both building occupants and firefighters. Again, fires in modern residential occupancies are likely to become ventilation-limited prior to the arrival of the fire department. This means that the introduction of air results in rapid fire growth. The research also demonstrated that firefighters making an entry through a front door can introduce enough air into the fire area to produce flashover, which significantly reduces the opportunity for an effective search within the compartment **FIGURE 4-12**. We cannot operate in a flaming environment above the rating of our structural PPE (500°F, 260°C for 5 minutes).

The experiments documented that a room-and-contents fire in a modern house often flashes over and then enters the decay stage due to decreased oxygen levels in the house. Research has also documented that a typical room-and-contents fire in a standard residential occupancy may use up the available oxygen before it reaches flashover. This produces a ventilation-limited fire, where the fire goes into a dormant state or decay stage. Often there are no visible flames, and the fire no longer appears to be actively burning. In this decay stage, the fire growth and fully developed stage is limited only by the lack of available oxygen. This rapid-fire growth phenomenon has led researchers to identify a second growth stage of a fire: *explosive growth stage*. It was realized that the traditional four-stage fire-growth sequence was based on a fuel-limited model and had to be modified for ventilation-limited fires. The two new stages of fire growth include ventilation-limited and **explosive fire growth**. Therefore, a new six-stage fire growth sequence for ventilation-limited fires includes the following stages **FIGURE 4-13**:

- Incipient stage
- Initial growth stage
- Ventilation-limited stage

- Explosive growth stage
- Fully developed stage
- Decay stage

**Note: Ventilation-limited* (or *vent-limited) fire* is quickly becoming an accepted term, but it is technically incorrect. It is more accurate to call these fires *oxygen-limited*, *oxygen-deficient*, or *oxygen depleted*. What the fire needs for growth is oxygen, not ventilation. Ventilation is merely a mechanism that moves and exchanges the atmosphere to deliver and introduce fresh oxygen to the decaying fire. Without ventilation, the oxygen cannot enter the smoke-pressurized compartment.

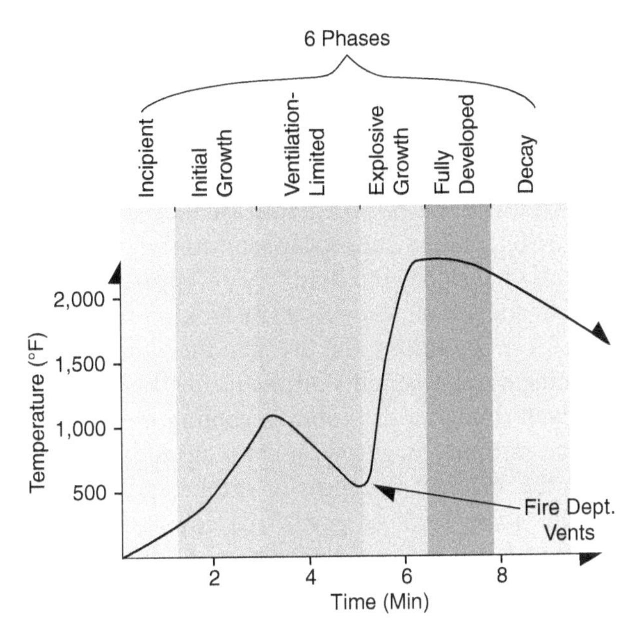

FIGURE 4-13 This time-temperature graph illustrates the six stages of a ventilation-limited fire.

Not all fires follow this sequence. A fire follows either a fuel-limited model or a ventilation-limited model. It depends on the involvement of legacy fuels versus modern fuels, legacy construction versus modern lightweight construction, and which component is needed to complete the combustion process—fuel or air **FIGURE 4-14**.

Survivability Profiles

Marsar defines **survivability profile** as "the educated art of examining a situation and making an intelligent decision of whether to commit firefighters to life saving and/or interior operations." That is a good definition because, as firefighters' knowledge and experience increase in the areas of fire behavior, fire spread, smoke conditions, smoke movement, flow paths, and building construction, they'll make smarter decisions. Safe decision making is rooted in smart decision making. If you want safer firefighters, they must become smarter firefighters. If we consider ourselves professionals, we must not abandon scientific information, knowledge,

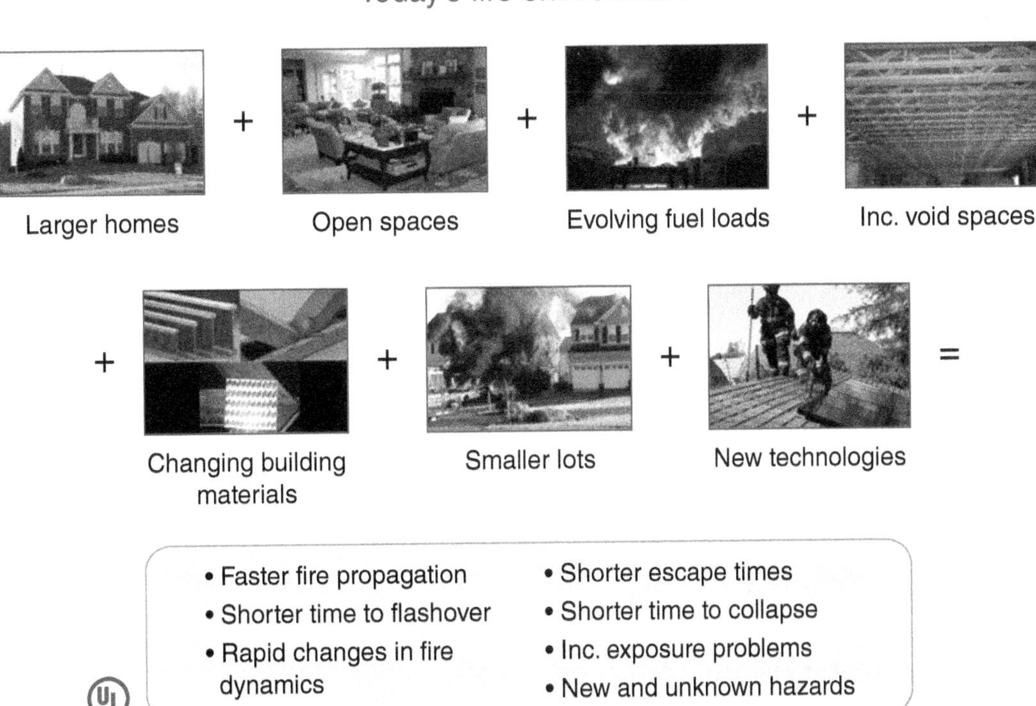

FIGURE 4-14 The current fire environment changes the way fires burn in modern homes.

and experience in exchange for making emotional decisions. Experience level plays a vital role in profiling the survivability of a known or reported trapped occupant.

In a conversation the author had with Retired Fire Chief Mark Wessel, Keokuk, Iowa, Fire Department, he said, referring to the Keokuk fire, "The distraught mother screaming in the front yard, as emotionally painful as that sounds, wasn't going to kill us—but the fire will . . . what you need to do is slow down your operation and think. What will immediately make conditions better for the survival of the occupants and your firefighters? Water." He continued, "We should have focused more on getting a hose line in the building and water on the fire, and less on the mother screaming in the front yard. We didn't spray a drop of water. Doing so would have given everybody more time, and the outcome would have been different. I have to live with that for the rest of my life."

Smoke Toxicity

When considering survivability profiles, we must persuade those who tend to think in terms of flames only and that, if the fire hasn't flashed yet, the survivability profile is still good and there may be a savable life. In fact, in the decay stage fire, we must also consider the immediately dangerous to life or health (IDLH) atmosphere in evaluating the survivability profile. Within the smoke of a decay fire are toxic levels of carbon monoxide, hydrogen cyanide, benzene, and acrolein. The National Institute for Occupational Safety and Health (NIOSH) lists the IDLH for acrolein at 2 parts per million (ppm). The concentration of carbon monoxide can be as high as 35,000 ppm. The IDLH for carbon monoxide is 1,200 ppm. At 12,000 ppm, carbon monoxide can render a person unconscious in two to three breaths, and death can occur in less than 3 minutes. The IDLH for hydrogen cyanide is 50 ppm, and it is 35 times more toxic than carbon monoxide.

Smoke Temperature

There may be a difference of opinion about the temperatures in a decay state fire, but we can say that its lower range is between 200° and 300°F (93°C and 148°C) and that it is less than 500°F (260°C). Consider that pain registers on human skin at 111°F (44°C), and first-degree burns occur at 118°F (48°C). Second-degree burns occur at 131°F (55°C). Pain receptor nerves become numb at 140°F (60°C), and at 162°F (72°C), human tissue is destroyed on contact. Clothing that ignites can cause third-degree burns in a matter of seconds. Therefore, we must use these physiological facts with confidence when making risk-benefit analyses. Otherwise, we will continue to make emotional judgments and risk firefighter lives when no viable life value exists.

Emergency medical services (EMS) is approximately 85% of the fire service workload. We often encounter physician's orders for life-sustaining treatment (POLST) and do not resuscitate (DNR) orders for certain patients. We can choose to ignore these orders and resuscitate the patient because, after all, our primary mission is to save lives at any cost, or we can acknowledge that a judgment call was made by a doctor and the family based on the physiological viability of the patient. In the field, paramedics and emergency medical technicians can recognize and withhold medical treatment based on a traumatic mechanism of injury (MOI) like a gunshot wound to the head or the lack of vital signs with lividity. Firefighters should be able to apply their knowledge by looking at the visual cues from the smoke, the fire, and the building to determine the survivability profile inside the fire room.

Even though a fire room is cooling, the fuel contents continue to pyrolyze, sending additional fuel into the space in the form of smoke and fire gases. Again, some of these gases have low ignition temperatures—all the fire lacks is oxygen. To firefighters, a fire in the decay stage can look like the fire is very small or has gone completely out, but this is a dangerous error in judgment that needs to be reversed through more education until the fire service begins to recognize the dangerous situation that lies in front of them. Experiments have repeatedly shown that when oxygen from the outside air is reintroduced into a ventilation-limited fire compartment in the decay stage, it can rapidly generate a ventilation-induced smoke explosion or flashover in a short period.

Note that some fires can burn in a sealed room devoid of oxygen when the fire involves strong oxidizers like fluorine, chlorine, bromine, and chlorine trifluoride. These oxidizers produce their own oxygen in the combustion process, but these fires are rare and will probably never be encountered in residential fires. It is mentioned to acknowledge the science in this exception to the rule.

Firefighters need to carefully consider what constitutes ventilation. Opening any door, window, skylight, or roof introduces oxygen into a burning building. The fire service has traditionally taught that making entry into a building is part of forcible entry—not ventilation. Yet studies have shown that opening the front door has a profound impact on the growth of the fire. Repeated experiments produced a violent flashover or smoke explosion shortly after the front door is opened. Perhaps gaining access should now be considered part of ventilation because it creates a direct flow path to the fire. At the very least, firefighters should no longer look at forcible entry as an isolated task; there is a cause and effect that immediately affects the ventilation profile. This requires an emphasis on controlling the door until the hose crew is ready to put water on the fire **FIGURE 4-15**.

This research has also demonstrated that our traditional assumptions about the effects of ventilation need to be modified. Most fire service texts define *ventilation* as the systematic removal of heat, smoke, and fire gases from the building and replacing them with cool fresh air, which improves interior conditions and the atmosphere; it can change the direction of smoke and the fire and is a means of preventing flashover. This definition is still technically correct—if the fire is knocked down from the outside. Without putting water on the fire first, any ventilation creates a flow path that draws fresh air toward the seat of the fire, increasing the intensity, interior temperatures, and the products of combustion. What really happens when a vertical vent hole is created is the lifting of the neutral plane. Any cooling effect or improvement of interior conditions produced by ventilation alone is short-lived without the application of water **FIGURE 4-16.**

FIGURE 4-15 Firefighters should no longer look at forcible entry as an isolated task; there is a cause and effect that immediately affects the ventilation profile that increases fire behavior.

What the research has repeatedly demonstrated is that venting a ventilation-limited fire draws the flames to the low-pressure ventilation hole that was just created. That ventilation hole creates a bi-directional flow path where higher pressure hot gases are pushed out the top while low-pressure fresh air is entrained or sucked into the seat of the fire. This increases the intensity of the fire, thereby increasing the temperature, the heat release rate, and smoke production with other products of combustion inside the structure. This puts firefighters and occupants at increased risk of being caught in a flashover or smoke explosion.

Because of the extremely fuel-rich environment found on today's fireground, ventilation that is not preceded by, concurrent with, or immediately followed by effective water application and suppression efforts introduces sufficient oxygen to bring the fire area rapidly to flashover conditions.

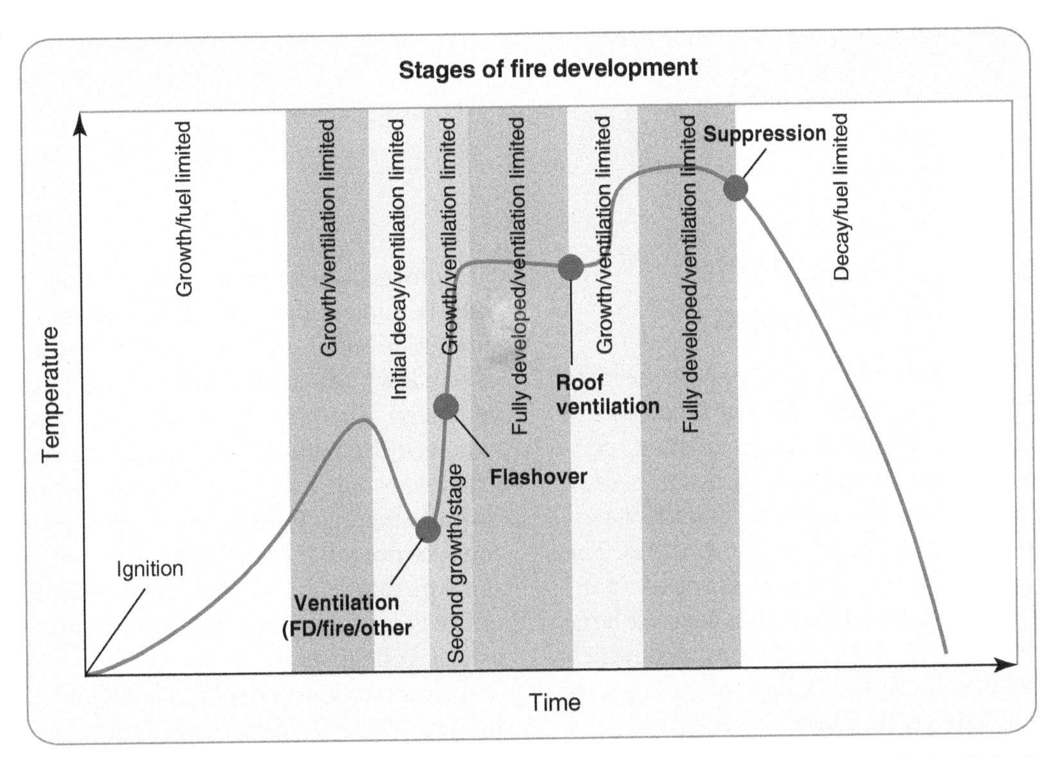

FIGURE 4-16 Fire growth in a well-insulated modern family home often becomes ventilation-limited. This graph illustrates that vertical ventilation without suppression results in fire growth.

RESIDENTIAL FIRE BEHAVIOR

Here is a list of the conclusions for residential fire behavior that were drawn from the fire experiments conducted by NIST, UL, and FDNY:

- Increasing the air flow into a ventilation-limited structure fire by opening doors, windows, or roof openings increases the intensity and temperature of the fire, thus increasing the hazard for occupants and firefighters.
- Increasing the air flow into a ventilation-limited structure fire may lead to a smoke explosion or to rapid fire development and flashover.
- The fire attack must be coordinated and should include the activities of a 360-degree size-up, an exterior fire attack with the application of water through a window, making entry, ventilation, and search and rescue.

Courtesy of Raul Angulo.

KEY POINTS ON FIRE DYNAMICS

Some important points to remember regarding fire dynamics are:

- The stages of fire development change when a fire becomes ventilation-limited. It is common in the modern fire environment to have a decay period prior to flashover and the fully developed stage, which emphasizes the importance of coordinating ventilation with the application of water. It is best to start with the application of water defensively from the outside. This resets the fire and takes the energy out of it. If ventilation is followed by water application, the flow path will not intensify the fire for the interior fire attack crew.
- The absence of visible smoke from the exterior doesn't mean that there is no fire or no danger. A common event noted during the experiments was that, once the fire became ventilation-limited, the smoke being forced out of the gaps of the houses greatly diminished or stopped altogether. This is because interior heat was decreasing and so was the interior pressure that pushes smoke out. However, pyrolysis is still taking place, breaking down the remainder of solid fuels and creating flammable vapors.
- No smoke showing during the size-up should increase awareness of the potential existence of ventilation-limited conditions inside. Check window construction. Are these modern windows with double or triple panes? In a decay stage fire, windows are layered in creosote buildup. Dark-brown stained windows are a strong indicator that the fire is in the decay stage. A creosote-stained window is also a strong indication that the door to the room is open and should not be selected for a vent-enter-isolate search (VEIS). The survivability profile is nonexistent, and opening that window will likely produce a rapid change in fire conditions and possibly a smoke explosion. On the other hand, a clear window indicates that the door to the room is closed from the fire and the survivability profile is favorable. There are lots of variables, but check for visibility before opening or breaking any windows for VEIS.
- Structural collapse should always be a consideration in your size-up. Any involved or exposed residential floor system can collapse in your operational time frame, especially with an unprotected engineered floor system found in lightweight construction. It may be difficult, if not impossible, to determine from the exterior. It must be suspected by trying to determine the age (era) of the building. In other words, is the building modern or legacy?

Courtesy of Raul Angulo.

Ventilation, Flow Paths, and How Fires Spread

In order to extinguish fires effectively, we need to understand how they spread. Chapter 2 covers fire spread based on the principles of conduction, convection, and radiation. The classic fire tetrahedron discussed in that chapter is still helpful in understanding the growth as well as the suppression of fires. The theory of extinguishment is in removing one component of the fire tetrahedron. Content fires generate large volumes of smoke, particulates, and gases, all of which are considered fuel. If this fuel is hot enough, it ignites with the introduction of adequate oxygen.

In addition to these principles of fire spread, we need to add another factor that has not received much attention: fire spread is largely a pressure-driven phenomenon. Fires spread along a flow path. A **flow path** is a lower pressure space starting at the structural inlet portal for fresh air (such as an open door or window), toward the fire where higher pressure is generated. The hot smoke and pressurized gases rise and seek the path of least resistance toward an exit exhaust portal, such as another open window, door, or roof opening. Fires in imperfectly sealed buildings produce pressures that, while very small, still result in pressure differentials that cause significant movement of fire gases throughout the building and out through external openings.

Understanding how pressure differentials and fire flow paths influence the growth of fires leads us to create more effective ventilation and fire control techniques. Hot smoke always rises to the top of the compartment space and escapes through the opening above the neutral plane. Fresh air enters the compartment below the neutral plane. The greater the difference between the air inside and the air outside, or the greater the distance between the top exit opening and the bottom of the intake opening of a structure, the more profound the air movement will be. Air movement can also occur in any direction when there is a high-low pressure differential because high pressure always moves toward low pressure.

Once an opening establishes a flow path, hot gases move away from the seat of the fire, following the flow path back to that opening. The air movement creates a pressure differential that draws any available cooler fresh air toward the seat of the fire, accelerating its growth. If there is no exterior opening, the fire creates a churning, drawing colder air in at the bottom while expelling hotter gases, which eventually return to the fire if there is no opening for escape. In a fire where only the front door is open, the fire creates a flow path of hot gases exiting through the top part of the door, while the bottom half of the door allows the introduction of cool oxygen-rich air into the fire. This type of opening produces a two-way or **bidirectional flow path** through the front door.

Some fire flow paths move in a single direction. Single direction flow paths can occur when there are two openings, one high and one low, creating a pressure differential. This is called a **unidirectional flow path**. For example, an opening at the roof carries all the smoke out of the structure, and a simultaneously open front door brings in fresh oxygen to fill that space. If the fire is not extinguished, it continues to grow and produce smoke volumes beyond the size of the roof vent's ability to exhaust the smoke. The smoke and heat seek another path of least resistance, which could be the front door, and return to a bidirectional flow path. A horizontal unidirectional flow path can also be created on the same level of the fire by having two openings, for example, a door and a window, but the flow path needs a pressure source to create the atmospheric differential, like a strong wind or positive-pressure ventilation (PPV) fan, to determine the direction of the flow path. This is the theory behind positive pressure attack **FIGURE 4-17**.

Fire growth and development are affected by both the amount of air entering the fire area and the amount of fire gases exiting the fire area. Opening a door or a window provides oxygen to a ventilation-limited fire. Providing an opening in the roof creates a unidirectional exit flow path for hot fuel and accelerates air flow intake to the fire as well as spreads the fire along the flow path. This significant increase in oxygen combined with the fuel-rich environment of a ventilation-limited fire results in explosive fire growth

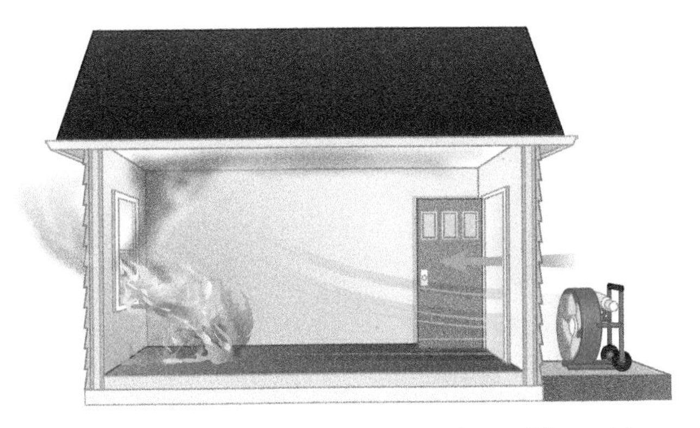

FIGURE 4-17 By introducing an atmospheric differential, a PPV fan (blower) can create a controlled horizontal unidirectional flow path.

FIGURE 4-18. This is why it is important to get water on the fire, preferably right before ventilation takes place.

Wind-Driven Fires

Wind is a powerful force that influences the direction of a flow path. Any time a window or door is opened on the side of a building that faces the wind (windward side), a tremendous amount of oxygen is introduced into the fire **FIGURE 4-19**. The introduction of oxygen may result from the heat of the fire breaking out a glass window, the fire causing a door to fail, or from the intentional actions of firefighters breaking a window for ventilation or forcing a door to make entry. The sudden increase in oxygen combined with hot flammable gases from the fire can result in rapid fire growth. It can produce a sudden change in the direction of the flow path from bidirectional to unidirectional due to the pressure imposed by the wind. The effects of the wind can be more severe in multistory and high-rise buildings, where the upper wind currents are stronger. If you think of the wind as a giant positive-pressure fan forcing huge quantities of oxygen into a ventilation-limited fire, you can imagine the results. This is called **extreme fire behavior**. The hot gases are pushed along the flow path. If these gases move through the building before exiting to the outside, the oxygen-rich air from the intake mixing with the hot gases can turn the entire flow path into a rapid moving wall of flames. Even a firefighter in full PPE and SCBA cannot survive these conditions for more than a few seconds, much less outrun the superheated, pressure-driven air currents.

One means of controlling a fire is to remove the air, which can be done by controlling the door and limiting the amount of oxygen that is available to the fire. In the example in the previous paragraph, some major city fire departments use a fire curtain, a heavy-duty, weighted tarp, in high-rise firefighting. The fire

FIGURE 4-18 A ventilation-limited fire sequence in which the main photo in each panel shows the room where the fire was started. The small photo at the lower right shows the exterior of the house at that same time. **A**. The initial flaming and incipient growth of the fire. **B**. Halfway through the initial growth curve. **C**. Fire peaks at its initial growth stage period. **D**. Fire enters into the initial decay stage just before the fire becomes ventilation-limited. **E**. The fire reignites 5 seconds after the front door is opened, an action that increased the oxygen supply to the fire. **F**. One minute after the front door is opened, the fire enters into the explosive growth stage and the room flashes over, becoming fully involved.

Courtesy of NIST.

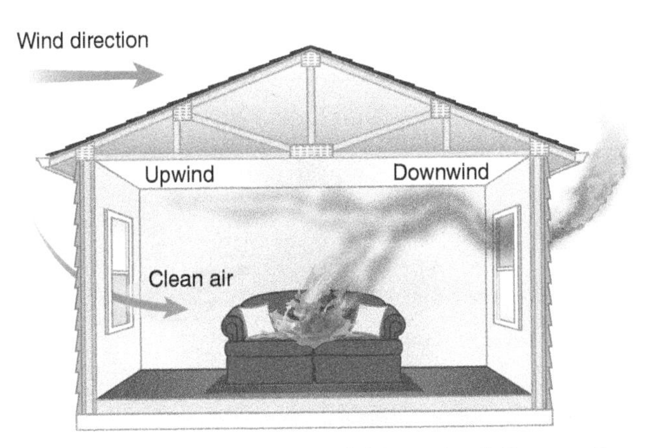

FIGURE 4-19 Wind is a powerful force that influences the direction of a flow path.

FIGURE 4-20 The KO Fire Curtain.

Photo courtesy of Scott Broer, KO Fire Curtain.

FIGURE 4-21 An effective way to control a wind-driven fire is to hit the fire defensively with overwhelming force from the upwind side of the structure so the wind carries the steam wherever the fire is moving.

Courtesy Scott Peterson.

FIGURE 4-22 The interior fire attack must be sustained with a reliable water supply.

Courtesy of John Odegard.

curtain is hung from the window above the fire floor, then dropped directly over the broken-out window on the windward side of the fire floor FIGURE 4-20. The objective is the same as it is with door control: to stop the wind-driven air intake vent from forming a unidirectional flow path that spreads the fire with blow-torch force toward occupants and engine crews in the fire room and hallways. Another way to control a wind-driven fire effectively is to hit the fire defensively with overwhelming force from the upwind side of the structure so the wind carries the steam wherever the fire is moving FIGURE 4-21. This transitional technique may require multiple master streams.

Understanding a flow path aids in understanding the importance of door control. Limiting the available oxygen to the fire until suppression crews are in place with charged hose lines keeps the fire smaller. Because opening the front door of a house supplies the fire with oxygen that may quickly produce a violent flashover or smoke explosion, ventilation needs to be considered as a critical part of the fire suppression attack. It must always be conducted in coordination with charged hose lines in position for effective suppression efforts. The interior fire attack must be sustained with a reliable water supply FIGURE 4-22. Leaving the front door closed as long as possible significantly aids in controlling the fire. Designate a firefighter for door control at the front door. Keep the door partly closed as the hose line is advancing into the fire.

Some fire departments and training organizations now use smoke curtains, which are hung to cover the entry door (usually the front door) with about a 10-inch gap at the bottom for hose advancement. These lightweight curtains are made from fire-resistive material and effectively prevent a flow path from

FIGURE 4-23 Smoke curtains are hung to cover the entry door with about a 10-inch gap at the bottom for hose advancement.

Photo by Joe and Andy Starnes.

developing when firefighters make entry into the structure **FIGURE 4-23**. They also don't hinder emergency egress for firefighters. If we understand the science of fire behavior and flow paths, incorporating the use of fire curtains may become a more popular method of attacking interior fires.

The primary resistance to smoke curtains is the feeling that delaying improvement of the IDLH atmosphere, which traditionally is accomplished by ventilating the structure, decreases the survivability profile of any occupants inside. This technique also requires more firefighters to enter the IDLH atmosphere and creates a psychological challenge to remain oriented while operating in the dark smoke. There's also a feeling that the emergency egress for firefighters is blocked or hidden. The importance of door control is hard to put into practice because there is comfort in seeing daylight or apparatus emergency lights through the open door, and firefighters want the ability to bail out the entry door quickly if extreme fire behavior suddenly develops. Because the majority of residential doors do not open in the direction of egress, many firefighters are reluctant to close the door. Therein lies the great debate. But if scientific fire-growth time lines are any indicators, victim survivability within the fire room is doubtful by the time firefighters arrive on the scene.

Consider the alternative method: One firefighter standing outside the IDLH atmosphere can hit the fire through a window with a defensive-to-offensive transitional attack **FIGURE 4-24**. This action eliminates the need for a smoke curtain and door control because, once the fire is knocked down, the open door allows you to start ventilation immediately (by creating a low-pressure flow path). Crews can then enter the structure to search for victims and finish extinguishing the fire. This method reduces the IDLH atmosphere by cooling it before those firefighters make entry. It's the smarter and safer method because it eliminates the chances for a flashover and backdraft by reducing the interior temperatures within the limits of our gear. This method also puts water on the fire, reduces temperatures, and clears the atmosphere faster than firefighters can working behind the smoke curtain in the dark.

FIGURE 4-24 Firefighters working from outside the IDLH atmosphere can hit the fire through a window with a defensive-to-offensive transitional attack.

Here are some important points to keep in mind about flow paths:

- Think of a fire flow path as a river bed of water. The headwaters are the source of the river; the mouth of the river is where the water empties into the ocean. Barriers, dams, and channels can redirect the river flow. Water temperature from the mountain headwaters is colder than its temperature when it finally makes its way to the mouth. In comparison, the fire flow path starts at an air intake portal into the fire compartment. You influence the direction of the flow path by opening doors, breaking windows, or creating a ventilation opening in the roof. You can also restrict the flow path with door control. The air from the flow path intake portal is cooler, and the air in the flow path exhaust portal is hotter.

- If the fire cannot be knocked down from the outside and an offensive interior attack is made, designating a firefighter at the front door for door control should be strongly considered. This assignment is a critical support function for the success and safety of the interior fire crews. After the attack crew enters the structure, controlling the door is allowing just enough of a gap to allow the hose or hoses to advance. This limits the air entering behind the firefighters. Even though there is a gap from the floor to the top of the door, it is better than leaving the door open, which accelerates the intensity of the fire until firefighters apply water. Another option is to have a firefighter use a chainsaw to cut the bottom eight inches off the door, starting with the side opposite the hinges. Keep the hinges intact: After the crews enter the house with charged hose lines, the door can be completely shut over the hose lines without pinching or kinking the hose while still controlling the intake vent point of the flow path effectively. This accomplishes the same objective as using a smoke curtain. The firefighter assigned to door control should have a bright flashlight and a sounding tool to help the interior attack crew stay oriented to the entry door. As soon as there's water on the fire, ventilation should immediately start. The firefighter should open the door and place a fan once an exhaust port is established close to the seat of the fire.
- Remember to keep the wind at your back during fire attack **FIGURE 4-25**. This may mean accessing the rear door. Don't assume the front door is always the best entry point. Before SCBA, old-time firefighters entered structures after the fire was knocked down with the wind at their backs so they could breathe fresher air. It is the duty of the incident commander to make sure the fire is always going away from the firefighters making the interior attack. Once an exhaust ventilation hole is established, it can be supported and controlled by setting up a PPV fan behind the entry team.

FIGURE 4-25 Try to keep the wind at your back during fire attack.

- Forcing the front door open is ventilation and must be thought of as ventilation. While forcible entry is necessary to fight fires, it must trigger the thought that air is being introduced into the fire unit and thus the clock is ticking before either the fire is extinguished or it grows to create untenable conditions, jeopardizing the safety of everyone in the structure.
- If there isn't a window or door to knock down the fire from the outside, an offensive interior attack is initiated; however, keep in mind that, once the front door is opened, attention should be given to the low-pressure flow path that was just created. You must control the entry door. A rapid rush of air or a tunneling effect could indicate a ventilation-limited fire that can cause a smoke explosion or flashover. The use of a smoke curtain or cutting 8 inches off the bottom of the door allows the door to be completely shut over charged hose lines without pinching or kinking the hose. This action restricts air from entering the intake side of the flow path, preventing the fire from growing.
- Every new ventilation opening provides a new flow path to fire. Each one can create very dangerous conditions when a ventilation-limited fire exists. You never want to be between the fire and where it wants to go without a charged hose line or access to a door that closes. If this is a high-low flow path, where the exhaust portal is high and the air intake portal is low and becomes pressurized, this puts you in a unidirectional flow path—a deadly predicament because you're now in the equivalent of a chimney.
- Fire showing means the fire is venting, but it does not mean the fire is venting safely. Additional ventilation points allow the fire to grow if water is not immediately applied on the fire.
- Vertical ventilation, or "taking off the lid," does not guarantee positive results. Although vertical ventilation is the most effective type of natural ventilation because it allows for hot gases to

exit through the entire roof vent hole, it also allows the fresh air to be entrained into the structure unless the front door is closed. Without immediate water application, this action draws the fire toward the vent hole right into the attic space and roof-truss assemblies supporting firefighters on the roof. Coordination of vertical or horizontal ventilation must occur with fire attack, and it is best to hit the fire from the outside before venting the structure. UL/NIST tests showed that the only way vertical ventilation was effective was when the fire was first hit from the outside; otherwise, the fire always grew in intensity, increasing the interior temperature.

- Only water absorbs the heat and extinguishes the fire during offensive operations. Emphasis needs to be placed on applying water on the fire from the outside, then ventilating the fire gases. Smoke is fuel; do not hesitate to apply as much water as needed into the smoke to cool it as you move through it. Preserving the thermal balance is still beneficial but not as crucial as it used to be. Prior to SCBA, it was critical for better visibility and breathable air below the neutral plane for occupant and firefighter survival. The temperature was also cooler at floor level. However, thermal-imaging cameras allow us to see through smoke effectively. Immediately applying water into the space cools interior temperatures, which gives any trapped occupant a better chance for survival than protecting the thermal layer did in the past. Modern fuel smoke is more volatile than legacy fuel smoke and needs to be cooled. Hesitation of water application because the nozzle firefighter is trying to protect the thermal layer balance can mean missing the critical opportunity to prevent a smoke explosion or flashover. Getting water on the fire also reduces the net smoke production and starts to reduce IDLH conditions **FIGURE 4-26**.

FIGURE 4-26 Prior to making entry, the nozzle firefighter directs a straight stream through the top of the door or window and deflects the stream off the ceiling. Evaluate conditions before transitioning to an interior fire attack.

CONCLUSIONS ABOUT FLOW PATHS

The conclusions that were drawn regarding the fire flow path from the fire experiments conducted by NIST, UL and FDNY are as follows:

- Increasing the size and number of inlets and/or exhaust flow paths results in fire growth and spread.
- Interrupting the fire flow path by limiting or controlling the inlet or controlling the outlet can limit fire growth.
- Controlling the door or doors (keeping them closed) allows less oxygen into the fire and lowers interior temperatures by reducing the intensity and energy of the fire.
- Anyone in the exhaust portion of the flow path, that is, between the fire and the direction of its travel, is basically in a chimney, which is a high hazard and a potentially deadly location.
- Controlling the flow path actually improves victim survivability by preventing interior temperatures from rising to untenable levels.
- Applying water on the fire from a defensive exterior position has the greatest impact on all of the above points.

Water Application and Coordinated Fire Attack

Our primary means of suppressing fires is to apply water using a fire hose and a nozzle. The research conducted by NIST and UL provides valuable knowledge about more effective means of applying water to a fire. It has been widely accepted that attacking a fire from the outside by opening a nozzle and spraying water into a window may "push" the fire through other parts of the building. This research proved that in every test fire, this widely accepted "fact" was not true. The researchers were *not* able to push fire with water; however, it should be clarified that *they were not able to push fire with straight streams.* In the past, what firefighters thought was pushing fire was actually steam following the flow path the fire had already chosen, or they were increasing the pressure within the compartment by entraining air from a combination nozzle.

Since the adoption of SCBA, it has been taught that a residential fire should be attacked through the door from the unburned, uninvolved side, and advance toward the fire. This is often the best tactic to use today and results in a successful fire attack. However, this has also led to the assumption that offensive fire attacks *must* be made exclusively from the inside of the structure and that the fire must *always* be approached from the uninvolved side to effectively "push the fire" away from the uninvolved portion of the building and away from any occupants who might be behind the advancing hose team. This technique requires firefighters to enter a burning building, often with low visibility and without confirmation of where the fire is located. Still, many defend this tactic for its effectiveness based on years of experience with successful fire attacks and it has been in use for decades. However, the UL/NIST studies show that first knocking down the fire from the outside is faster for cooling inside temperatures and most effective for improving interior conditions. It is also safer for firefighters.

When you need to get a hose line between the occupants and the fire, the above exterior tactic is the correct one to use. The science shows that if a fire is hit with a straight stream from the exterior window, the products of combustion will exit that same window, making it much easier for an attack team to enter and place a line between the fire and the occupants. Even in a wind-driven fire, it is best first to hit the fire defensively, then enter from the unburned side of the structure to place a hose line between the fire and the occupants.

We need to believe the scientific evidence. The NIST and UL research has demonstrated that entering the structure fire with flames visible from the windows simply because a company officer feels that applying water through the window to knock the fire down will push the fire throughout the structure is not true. As NIST and UL researchers measured the movements of the fire, they repeatedly determined that, in every experiment and test fire, it was not possible to push the fire with a straight stream of water because it does not entrain air. It can be said with certainty that *straight streams do not push fire or fire gases*. For proof, you can conduct an experiment the next time you have a backyard bonfire. If a garden hose equipped with a little smooth-bore tip is aimed into the flames, the convected currents remain basically undisturbed, and the straight stream shoots out the back side of the flames from the bonfire. On a grander scale, if a house is fully involved and there is a clear path from an open front door to an open back door, a straight stream from a 2½ -inch hose line will shoot right through the house and out to the backyard without having any effect on the fire. The velocity (speed) and mass (the tightness of the water molecules) of the straight stream provides limited surface area to absorb any significant heat to make a difference as it passes through the fire without hitting an obstruction.

Back to the bonfire, if you used your thumb to cover the straight stream of the garden hose, you'd create a spray pattern of water droplets, increasing the surface area and heat-absorbing capability of the water and thus entrain air. The convection currents would immediately be disturbed, the vapors would cool, and the bonfire would be extinguished in seconds. In fact, no stream technically pushes fire. Only the air entrained in the stream pushes the fire. A fog nozzle and fog patterns move significantly more air compared to a straight stream, and that entrained air pressure is what moves the fire gases.

Think back to your liquefied petroleum gas (LPG) training in drill school. Many training facilities utilized a live-fire LPG tank or Christmas tree prop. The officer stood between two attack teams, each with a 1¾-inch hose line. The officer had one hand on the shoulder of the left nozzle firefighter and the other hand on the shoulder of the right nozzle firefighter. Both teams huddled together along their fire hoses and slowly approached the burning LPG tank with wide fog streams fully opened to push the flames away from the tank **FIGURE 4-27**. Once they made it up to the burning tank, they had to hold their position. Massive flames were roaring all around the attack crews, but they were protected from the fire by two giant shields of water. The officer reached through the fog pattern and shut off the valve to the prop and the fire would go out. This was more of a drill in teamwork, coordination, and self-confidence than it was

FIGURE 4-27 Straight streams do not push fire; wide fog streams entrain significantly more air that can push or direct fire gases and flames away from firefighters with considerable force.

about actually fighting an LPG fire. (There are usually other sectional valves away from a burning tank that can shut off the fuel source.) Nevertheless, this demonstration clearly leaves a visual impression, and firefighters can mistakenly conclude that hose streams push fire. We now know it is the air entrained by the fog pattern that is pushing away the flames, not the water. Note that, in this same training evolution, if a firefighter accidently narrowed the fog pattern to a straight stream, the flames came through to the attack team, sometimes causing minor burn injuries. (This is why backup safety teams are also part of this training demonstration.) This painful scenario also confirms that straight streams do not push fire.

Another example is hydraulic ventilation. After a room fire is extinguished, the nozzle firefighter brings the nozzle to the open window to vent the smoke with the hose line. Smooth-bore nozzles can be used for hydraulic ventilation, but the bail needs to be shut halfway to break up the stream into a spray pattern so it can entrain air. A fully open combination fog nozzle is more effective for hydraulic ventilation after a room fire has been extinguished. The nozzle should be 2 to 4 feet (0.6 to 1.2 m) away from the window opening, and the fog pattern should be adjusted so that it takes up about two-thirds of the window space. The entrained air effectively ventilates the smoke from the compartment **FIGURE 4-28**.

It is not the water moving the smoke and gases, but the entrained air created by the fog pattern that creates the venturi air movement. A straight stream has no effect on the smoke at all. A straight stream of water simply flows out the open window because very little entrained air is generated by a straight-stream flow pattern. Flames, smoke, and other fire vapors have to follow the gas laws of physics.

Wide fog patterns can push fire to uninvolved areas of the building if the BTUs produced by the fire are greater than the gallons per minute (gpm) produced by the fog stream can absorb. The entrained air "fans the flames" and spreads the fire. This tactic, although it may result in spreading the fire, is appropriate when smoke and fire need to be pushed away from a known fire victim or a downed firefighter **FIGURE 4-29**. If the fire cannot be extinguished, a fog pattern to "protect in place" is the only way to push back the flames effectively until the rescue can be accomplished. Then the tactics and the streams can be adjusted to fight the fire. Those who would argue that this same tactic can be accomplished with a smooth-bore stream would also have to admit that it can be accomplished only by gating-down the nozzle to create a broken spray pattern to entrain air; however, gating-down a smooth bore also reduces the gpm flow, which is the last thing you want to do if you're trying to make a rescue while the fire is still burning out of control. Whichever nozzle is in place, it needs to be operated effectively.

NIST and UL researchers determined that the movement of the fire depends on the flow path of heated smoke and gases. The fire flow path carries the fire from an area of high pressure to areas of lower pressure. Firefighters often mistake steam in a flow path as pushing fire when the steam is only following the flow path current. After verifying these results with multiple experiments, the researchers determined that if water is introduced into the fire from the outside, it produces amazing cooling effects, not just in the fire room but also in other areas of the fire building.

Significant cooling was achieved even when the water entered the flow path at a distance from the seat of the fire. In these experiments, a straight stream of water flowing at 180 gpm (11 liters per second) was applied for 28 seconds and flowed 84 gallons (318 liters)

A

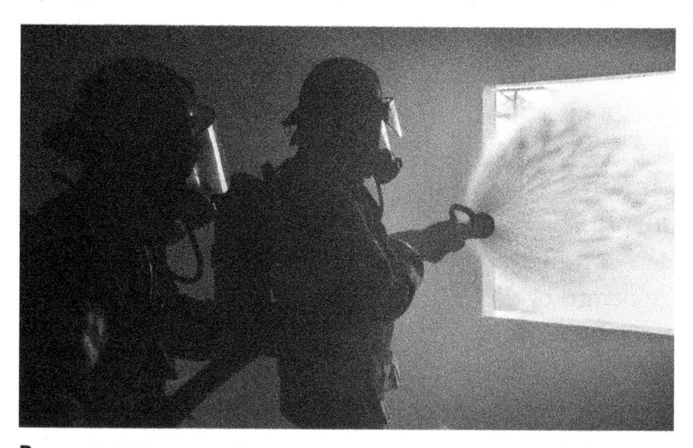

B

FIGURE 4-28 **A**. Smooth-bore nozzles can be used for hydraulic ventilation, but the bail needs to be shut halfway to break up the stream into a spray pattern so it can entrain air. **B**. A combination fog nozzle fully opened 2 to 4 feet from the window and filling two-thirds of the open space is more effective.

FIGURE 4-29 Entrained air from wide fog patterns can fan the flames and push fire to uninvolved areas of the building, but a wide fog pattern also protects and pushes flames away from a trapped occupant.

of water into the house from a distance. The water reduced the temperature in the front of the living room from 1,200°F (649°C) to 300°F (149°C). (*Note:* These numbers apply only to the Modern Fire Behavior experiments conducted at Governor's Island by the FDNY, NIST, and UL) FIGURE 4-30. These experiments proved that introducing water through a window from the outside reduced temperatures in other parts of the house, including those some distance from the fire. It did not completely extinguish the fire, but it took the energy out of the fire and slowed its growth by cooling large quantities of hot gaseous and solid fuels below their ignition temperatures. This defensive-to-offensive exterior attack is sometimes referred to as a blitz attack; a transitional attack; or, using a military metaphor, softening the target. A fire vernacular expression for these terms is "hitting it hard from the yard" FIGURE 4-31.

The effect of this transitional exterior attack is significant but temporary. To prevent the fire from regaining its intensity, the hose team must make a quick aggressive entry to complete final extinguishment. Once the hose team has water on the fire and ventilation has been started, interior conditions greatly improve, and the temperature is significantly reduced. Visibility improves, making it easier to move through the rest of the structure faster for a more efficient search and rescue. The sooner the fire goes out, the better, so put the fire out. Immediately launching into a primary search through a hot smoky building is not a quick task. You can make the argument that extinguishing the fire and ventilating should be part of the overall search and rescue process because it makes the process faster and safer to complete thoroughly.

Here's the takeaway: A defensive-to-offensive attack strategy can often put effective water on the seat of the fire faster than an interior hose team can by forcing entry and crawling on their bellies through thick smoke and high heat trying to find the seat of the fire. Timed evolutions on single room and contents fires showed that a hose stream through the outside window can knock down the fire in about 30 seconds using a single firefighter outside the IDLH area; an interior offensive attack, with firefighters crawling through the smoky house, can take between 90 and 120 seconds, and requires more firefighters on scene to accomplish the same task, including following the two in/two out rule.

Water application from the exterior quickly reduces interior temperatures, making the atmosphere safer and more tenable for any trapped occupants while also reducing the threat of flashover. With the fire knocked down, firefighters enter a safer atmosphere, and

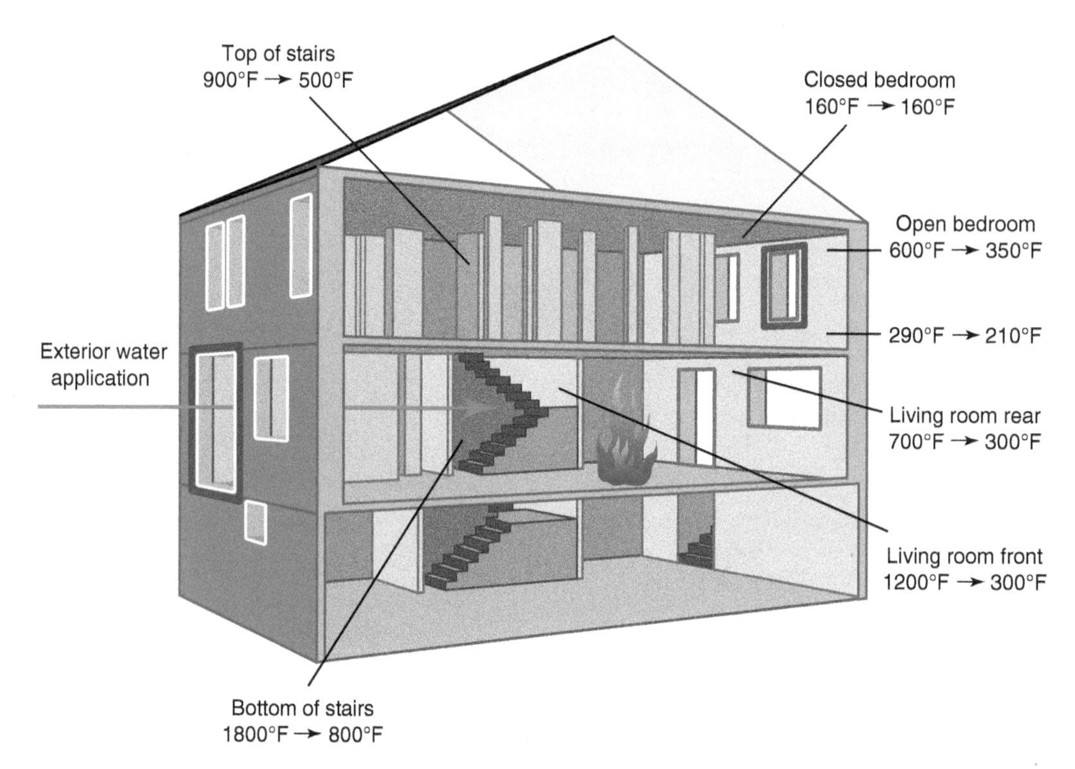

FIGURE 4-30 An offensive exterior attack some distance from the fire reduces temperatures throughout the house and partially suppresses the fire. The temperatures listed are from before and after the exterior application of water.

Courtesy of NIST.

FIGURE 4-31 Introducing water through a window from the outside reduced temperatures in other parts of the house, including those some distance from the fire.

Courtesy of UL.

APPLYING WATER DURING A COORDINATED FIRE ATTACK

Some important points to remember regarding applying water during a coordinated fire attack are:

- You cannot push fire with straight streams of water from smooth-bore or combination nozzles. Air entrained by a fog-cone stream pattern causes air to follow the flow path. A fog pattern entrains more air flow than a straight stream. You need to reduce the amount of air introduced with the stream of water and let the hot gases escape and flow out as you apply the water through an opening. Just like any other tactic, there is a correct way to flow water into an opening, such as a window or door, if you do not want the heat to follow the flow path.
- Cutting ventilation holes of sufficient size no longer passes for effective ventilation. Venting does not lead to cooling unless the fire is out. The sense of cooling is from the lift of smoke, which is actually the raising of the neutral plane; the results for improved conditions are temporary. Ventilation creates the flow path that brings in fresh air, thus intensifying the fire and raising interior temperatures. Ventilation must be coordinated with the hose team and is most successful if preceded by water application from the outside.
- Effective firefighting depends on a coordinated fire attack strategy. This includes a 360° size-up to identify the problems to be addressed with tactics; a risk-benefit analysis; determining the location of the fire; hitting the fire defensively for knockdown; and ventilation, preferably away from occupants and exposures. Then enter with hose lines for final extinguishment, the use of door control to restrict the flow path, search and rescue with a line between the fire and occupants, and overhaul. These are all important sequential critical tasks.

Courtesy of Raul Angulo.

CONCLUSIONS ABOUT EXTERIOR TRANSITIONAL FIRE ATTACK

The conclusions about exterior transitional fire attacks drawn by researchers from the fire experiments conducted by NIST, UL, and FDNY are:

- An offensive exterior fire attack through a window or door, even when that window is the only exterior vent, does not push fire.
- Water application is most effective if a straight stream is aimed through the smoke at the ceiling of the fire room. Water should be applied for approximately 30 seconds, or as long as it takes to knock down the fire. The effects of a transitional attack are significant but temporary. The favorable conditions are short-lived unless additional water can be applied to the seat of the fire for complete extinguishment with an interior attack.
- A wide fog stream can block a ventilation opening. The entrained air can increase interior pressure, effectively changing the flow path. A straight stream from a combination nozzle was not found to be different than a solid-bore straight stream at moving air. The solid-bore straight stream provides an advantage at reaching deeper into a building with less evaporation, which means that a larger area can be cooled by the steam. However, the straight stream still needs to be deflected off a solid object, a wall, or the ceiling to have the optimum effect of absorbing heat. A combination nozzle gives you options because it also provides better protection for firefighters and trapped occupants. You cannot protect a downed firefighter or an occupant with a smooth-bore stream without risk of injury.
- Gating down a smooth-bore tip doesn't provide as effective a spray pattern as a combination fog nozzle. It also reduces the gpm.
- Applying a hose stream through a window or a door into a room involved in a fire results in improved conditions throughout the structure.
- Applying water directly into the compartment as soon as possible and deflecting the stream off a wall or ceiling results in the most effective means of suppressing the fire.
- Even in cases where the front and rear doors are open and windows are vented, application of water with a straight stream through one of the vents improved conditions throughout the structure. Again, the straight stream should be deflected off a wall or ceiling.
- A transitional attack from defensive to offensive is a strategy that should be announced over the radio to avoid conflicting actions by other companies. It occurs just prior to entry, tactical ventilation, and search. This defensive-to-offensive attack is also known as a deck gun blitz attack, a blitz line, softening the target, and hitting it hard from the yard. It should be done before the hose team enters the structure for final extinguishment.
- Shooting water into the fire room from the outside can be accomplished with one firefighter. Because the firefighter is not in an IDLH area, the two in/two out rule does not apply. This is an effective tactic when resources are limited.
- The transitional attack strategy can begin from the outside (defensive), but it is necessary to *transition* to an interior (offensive) attack to complete final extinguishment. It can also begin on the inside (offensive to make a known rescue on a loser building), and then it should transition into defensive operations because there is no longer life or property value. The changes in strategies should be announced to all units over the radio to avoid confusion and maintain control in what should be a coordinated attack.
- Coordinate the fire attack with vertical ventilation. Do not ventilate before an attack team is ready with charged hose lines. It is best to apply water just prior to ventilation from the exterior to knock down the fire. This *resets* the fire, giving more time for firefighters and survivable occupants. It also takes the energy away from the fire and slows the destruction of the building.

Courtesy of Raul Angulo.

aggressive ventilation efforts can follow to improve visibility and exhaust smoke and fire gases without the risk of flashover. A faster and thorough search can be accomplished. The results of these experiments should force us to reconsider our long-standing approaches to water application and ventilation of structure fires.

Search and Rescue Safety Considerations

The same research that increased our understanding of fire flow paths, the effects of ventilation in ventilation-limited fires, and the use of a transitional fire attack strategy also increases our knowledge of rescue techniques. Anytime the door is closed between the fire and other parts of the building, it increases the chance of survival of any trapped building occupants, including firefighters. Isolating a room from the fire can provide a safe environment for a longer period of time. Ideally, this is accomplished by the trapped occupant (and is something we need to add to the public education curriculum). When an occupant is in an isolated area, it is often best for that occupant to shelter in place behind a door or for a firefighter to remove that person from a window with the objective of preventing her or him from being exposed to the deadly products of combustion rather than walking the occupant through them **FIGURE 4-32**. This latter tactic was used many years ago before the use of SCBA was common. A size-up determined if there was a space in the occupancy with a positive survival profile. This would not necessarily be the case if the room door was open.

The rescue technique of vent-enter-search (VES), which requires a firefighter to ventilate a window from the outside, enter through the open window, and search for victims, is a technique that creates a dangerous situation for the firefighter and any victims in that room if the room door is open to the fire. By opening or breaking the window, the firefighter can create a low-pressure vent for the fire to move toward. In other words, the firefighter just created a new flow path that brings the fire right toward the firefighter and the victim. UL/NIST experiments have demonstrated that this rescue technique is much safer if the interior door is closed prior to conducting the room search. This isolates the room from the fire flow path and allows the room to vent naturally, improving visibility and lowering the room temperature. Safety note: The window should never be opened until the firefighter is on air and ready to enter the room. This maximizes the time to find and close the door before a low-pressure flow path starts to develop.

The modified VES process has become **vent-enter-isolate-search (VEIS)**. Those who resist the change in the acronym state that having the door closed is *implied* when the firefighter makes the decision to VES. Not so. Our job as company officers is to teach the firefighting craft

A

B

FIGURE 4-32 **A**. Place the tip of the ladder directly below the window sill. **B**. The objective of a window rescue is to protect the victim from being exposed to heat and the deadly products of combustion from an interior rescue.

to mastery level. Don't assume every firefighter knows all the trade secrets. There should not be any "secret methods"; there's too much at stake. VES depends on isolating the space from the rest of the fire. The emphasis needs to be placed on closing the door and isolating that space.

VEIS consists of systematic steps. First, try to determine if there is a known victim in the burning structure. This can be confirmed visually by the firefighter, a reliable person at the scene, or an occupant who has escaped. Next, try to confirm the location of the victim from the reliable source. If the victim's location cannot be confirmed, the firefighter must identify the location of possible fire victims in a burning building.

Reading the Windows

In VEIS, it's important to read the clues from the fire on the glass before selecting a window. Moisture condensation and water droplets can form on the glass with a change in interior temperatures, indicating an incipient fire. Light gray smoke or a haze can be seen through the window. Flame temperature can be between 700°F and 1,000°F (371°C and 538°C), but the room temperature may be just above ambient. The oxygen level is still between 20% and 21%, and CO production is still low. Therefore, the survivability profile is good **FIGURE 4-33**.

Cracked glass or crazing (a network of fine cracks on the glass surface) along with blackened windows indicate higher temperature. Cracked glass indicates that you're either getting closer to the fire or this is the room on fire. A window that is blackened from smoke and creosote and has cracked glass from the heat of the fire, especially if smoke is seeping through the cracks in the glass, is a strong indication that the door to the room is open or that the fire may be in that room. Window glass cracks when it's exposed to interior air temperatures between 500°F and 600°F (260°C and 316°C). Cracked glass indicates a high-heat area above the human threshold. Not only is this a warning sign for a backdraft, but the survivability profile in this room based on the fire gases alone is zero. Go to the next window. Glass fails at temperatures above 600°F.

Select a window that has some visibility, even if there is gray haze (avoid the black). If you have to make a choice between a clear window and a hazed window, it's a judgment call based on a risk-benefit analysis and a favorable rescue profile. The IDLH conditions in the room with the hazed window is more severe and getting worse, but don't bypass or unselect the clear window simply because it isn't risky enough for a firefighter to VEIS **FIGURE 4-34**. Remember, fire gases like carbon monoxide, hydrogen chloride, and hydrogen cyanide are colorless. Glass lets visible light in and out. Gauge the level of visibility before opening or breaking the window. Note the location of the door and determine if it is closed or will need to be closed. Keep in mind that *glass blocks the infrared spectrum and the thermal image*, so a TIC shows a blank screen until the window is opened. Do not use the tip of the ladder to break out the window until you're on air and ready to enter the room in order to maximize your safety before a low- pressure flow path develops.

Next, the room is quickly accessed by opening or breaking out the window. The entire window should be broken out to accommodate a firefighter with all his or her gear. You'll also need the widest space

FIGURE 4-33 Although the fire room has flashed over and there are no survivable occupants in that space, the survivability profile for the other rooms on Floor 2 is good. Any of these windows can still be selected for VEIS.

Courtesy of Raul Angulo.

FIGURE 4-34 Black smoke has charged the attic and the second floor. The window to the left of the fire is already at smoke level, but it needs closer inspection before the window is broken for VEIS. Creating a low-pressure flow path can bring flames right to the firefighter and victims in a matter of seconds. The other windows are below the fire, and the survivability profile is good. These windows should not be overlooked for VEIS.

Courtesy of UL.

possible to pass an unconscious person out the window. Then the firefighter quickly enters the room, locates the door into that room, and immediately closes it. This isolates the room from the main body of the fire, decreases the chance of the room becoming a new fire flow path, and allows the room to ventilate naturally. A partner can stay at the window and use the TIC from that position to guide the firefighter making the interior systematic search.

The whole purpose of VEIS is to remove a victim through the window **FIGURE 4-35**. This is not an easy task. Immediately call for help, a second or even a third ground ladder, and a rescue rope. The first ladder should have been placed with the top rung just below the windowsill. Once the unconscious victim is lifted to the window, the rescue rope should run from the inside of the ladder over the first rung underneath the windowsill. A snug round turn, two half-hitches should be tied around the chest and underneath the arms of the victim for a belay. While maintaining constant contact with the victim, the rescue firefighter descends the ladder while the firefighter on the ground manages the victim's weight with the belay line. Another option is to use the second or third ladder as a high directional anchor point by extending the tip well above the top header of the window and run the rope down to the unconscious victim. Again, while maintaining full contact and control of the victim, the rescue firefighter descends the ladder, while a firefighter on the ground manages the victim's weight with the belay line.

FIGURE 4-35 A VEIS evolution should always include a second firefighter on the ground belaying a conscious or unconscious victim with a rescue rope, while the rescue firefighter maintains contact with the victim and descends the ladder.

The point is to teach the entire VEIS evolution, which should always include belaying a conscious or unconscious victim with a rope. Plenty of online videos show firefighters who were unable to manage an unconscious victim or lost their grip and the victim fell from the ladder to the ground—sometimes taking the firefighter down, too.

Closing the doors and creating distance from the fire needs to be taught to the public in the workplace and in grade schools, both public and private. Fire education programs need to advance beyond "call 911" and "stop, drop, and roll" and teach these life-saving concepts until they become common knowledge. Members of the public are more likely to find themselves in a situation where they need to know how to shield themselves from the fire flow path and avoid creating one, rather than having their clothes catch on

THE IMPACT OF FIRE ON BUILDING OCCUPANTS

Conclusions from the fire experiments conducted by UL, NIST, and FDNY regarding the impact on building occupants are:

- Suppressing the fire from the exterior as soon as possible improves the potential rescue profile and survival time of building occupants.
- Ventilation that is not immediately followed by effective fire suppression, that is, by water application, reduces the rescue profile and potential survival time.
- Being in the exhaust flow path between the fire and the exit point reduces potential survival time and can actually be deadly.
- Controlling the flow path improves the rescue profile and victim survivability.
- Controlling the door to a room when performing VEIS improves the safety of the firefighter and the building occupant. Once the door to the room has been closed, natural ventilation can occur, lowering the room temperature and improving visibility.
- Compartmentation (being behind a closed interior door) prior to fire department arrival provides increased protection and survivability for the occupant compared to being in a room or area with an open door to the fire.
- Greater distance from the fire improves the chances of survival for the occupant.

RESCUE AND SAFETY CONSIDERATIONS

Some important points to remember regarding rescue and safety considerations are:

- If you add air to the fire and do not apply water in the appropriate time frame, the fire becomes larger and safety decreases. Coordination of the fire attack crews is essential for a positive outcome in today's fire environment involving modern fuels.
- The greatest probability of finding a victim is often behind a closed door.
- During a VEIS operation, primary importance should be given to determining if the room is a survivable space, then isolating the room by closing the door to it before beginning the search. This reduces the impact of the open vent turning into a flow path and increases tenability for potential occupants and firefighters while the smoke ventilates from the now isolated room.
- If firefighters get into trouble and need to escape, closing a door between them and the fire buys valuable time. This may mean forcing a door to an adjacent apartment across the hallway.
- A sagging floor is an indicator of an imminent floor collapse. However, it may be very difficult to determine the amount of deflection while moving through the structure, especially with a thick carpeted floor.
- Sounding with a tool determines the thickness of the floor decking. Simply sounding the floor for stability is not reliable and therefore should be combined with slow advancements across the floor to determine the condition of the floor joists. Utilize flashlights and a thermal-imaging camera.

fire. Knowing their chances of being caught in a fire flow path and learning to get behind a closed door is a better service to the public than simply knowing how to stop, drop, and roll.

Basement Fires

When you first make entry into a structure without an exterior below-grade access to the basement, watch to see what the smoke does. If the smoke thins briefly when the door is opened (because the backup pressure is being released), but then steady smoke flows out through the entire doorway, the door is an exhaust point for a unidirectional flow path. This means that the fire is below the level of the opening **FIGURE 4-36**. The chance of flashover is temporarily reduced due to the lack of oxygen within the exhaust port, but this can quickly change with the failure of a below-grade window by creating a fresh air lower-level intake portal. This type of unidirectional flow path is basically a chimney, indicating that this is a basement fire or below-grade fire—one of the most dangerous fires we face. Singular stairway access, small windows or no windows, low ceilings, and

FIGURE 4-36 Steady smoke filling the entire doorway is the exhaust portal of a unidirectional flow path. This means the fire is below the level of the opening, indicating a basement fire or below-grade fire.

inability to ventilate smoke quickly make basement fires extremely difficult for escape. Other than a visual confirmation, this unidirectional flow path of smoke may be your only indication that you have a basement fire.

Basement fires don't have to be difficult to fight, but the way many fire departments currently access and fight basement fires in residential structures makes them dangerous and difficult to attack and extinguish due to limited access, low ceilings, lack of windows, and limited egress. There is often only one way in and one way out of below-grade levels and basements. Because pressurized superheated fire gases want to rise and move from areas of higher pressure to areas of lower pressure, starting a fire attack from the floor above the basement creates a bidirectional high-low flow path. As firefighters begin to work their way down the basement stairs, the smoke and fire is coming up the stairs. That doesn't put firefighters in the best vantage point; in fact, they're in the middle of the thermal exhaust column, making it a deadly place to be. If a door or window on the low side were to fail from the heat of the fire with a prevailing wind opposite the fire attack, or if a below-grade door were to be opened by a firefighter, the bidirectional flow path changes to a unidirectional flow path with little chance of survival for firefighters caught in the middle.

A tenable condition can become untenable in seconds. A case study example of this type of fire behavior was the Diamond Heights Fire in San Francisco, California, where two firefighters were killed on June 2, 2011.

The traditional tactic of pushing down the interior basement stairs to the seat of the fire should no longer be considered a safe option, period. In fact, it makes as much sense as crawling down a chimney with a log fire burning below—because that is what it is. Besides, one of the findings in the UL/NIST research on basement fires was that water applied via the interior basement stair had a limited effect on cooling the basement or extinguishing the fire. Now that firefighters have a better understanding of flow paths, that reasoning makes sense. If a high-low flow path is created when the door at the top of the basement stairs is opened, the bidirectional flow path causes the smoke and heat to travel back up to the open door on the first floor.

Consider this question: When we have a well-involved room fire on the first floor of a two-story house, do we put a ground ladder up to the second floor, enter through a window, and make our way down the main stairs to the first floor to attack the fire? Of course not! Always consider the safer and more effective approach.

Another serious hazard is the wooden stairs that lead to the basement. They are often steep and there is no way to know the condition of the stair assembly. Basement stairs often have open risers which expose more surface area of the step (tread) to fire. Prefabricated wooden stairs are hung using lightweight steel or aluminum hangers, which fail quickly in fire conditions or under the weight of an advancing firefighting team.

There may be scenarios where fire crews may have to push their way down a basement stairway in Type I fire-resistive construction, or old, beefy Type III ordinary construction, where basement stairs may be made of concrete. This warning about compromised stairways is predominantly for Type V wood frame, especially those of lightweight construction. The tactic of pushing down the stairs is dangerous and extremely risky. In a well-involved basement, it is doubtful that the majority of firefighters would be able to tolerate the superheated flow path. Again, consider the science. A basement fire is under pressure. Opening the door at the top of the stairs of the basement introduces a lower atmospheric pressure. The higher pressure seeks to equalize. This means the smoke, fire, and all the heated gases head up the stairs, taking the path of least resistance. Firefighters on the basement stairs end up being in the flow path between the fire and where it wants to go, regardless of stair construction. This is called the **chimney effect**. We've already stated that this can be deadly, especially when the flow path converts from bidirectional to unidirectional.

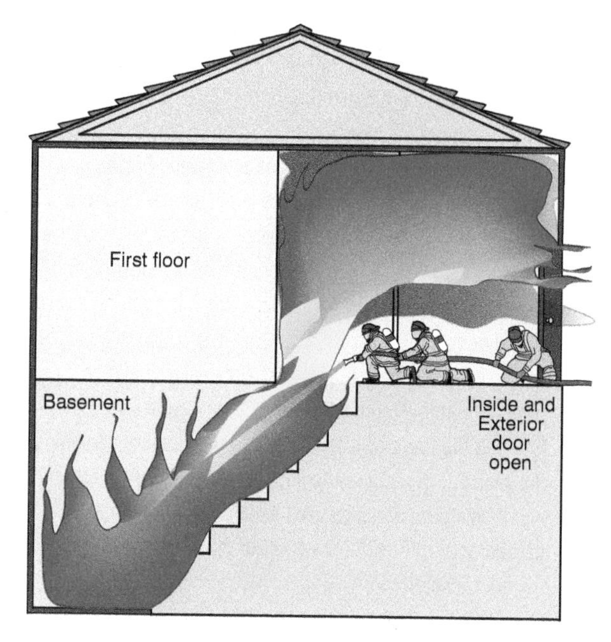

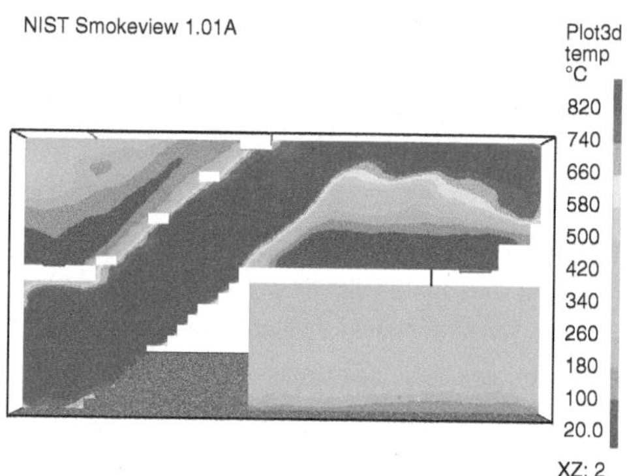

FIGURE 4-37 A NIST illustration and a temperature computer model of a townhouse basement fire in Washington, D.C.

Top: © Jones & Bartlett Learning; **Bottom:** Courtesy of NIST.

FIGURE 4-37 shows an NIST illustration and a temperature computer model of a townhouse basement fire that occurred in Washington, D. C. Note that the temperature color within the flow path in the stairwell is all red, with the temperature ranging between 1,364° F and 1,508°F (740°C and 820°C). Our structural firefighting gear is rated for only 500° F (260°C) for 5 minutes. This rating doesn't mean that you can endure 500° heat for 5 minutes; most firefighters can endure 500° for only a few seconds. These basement stairwell temperatures are three times above the PPE rating. In other words, it is physiologically impossible to survive this event.

The testing methods and specifications of NFPA-compliant PPE is found in Chapter 8 and Annex B of NFPA 1971, *Standard on Protective Ensembles for Structural Fire Fighting and Proximity Fire Fighting*, 2018 edition. This standard protects firefighting personnel by establishing minimum levels of protection from thermal, physical, environmental, and bloodborne pathogen hazards encountered during structural firefighting operations.

Many vertical voids within a building originate or terminate in the basement (pipe chases, trash chutes, laundry chutes), providing numerous channels for fire gases to spread throughout the building. In many parts of the country, especially in newer subdivisions with lightweight construction, basement ceilings are left unfinished when the home is sold. Under these circumstances, fires originating in the basements quickly involve the floor system, resulting in a failure of the floor over the fire and early collapse, with the potential for causing injury or death to firefighters entering the building in conditions of limited visibility.

Basement fires should be and can be one of the easiest fires to knock down and extinguish. They actually provide the firefighter with the least exposure to heat and carcinogenic smoke if they're attacked from the outside. They're only dangerous and difficult because many fire departments don't understand basement tactics. If we examine how extinguishing tools were used before the adoption of SCBA, we'll have a better understanding of how basement fires were successfully extinguished. SCBA and modern, Nomex® heat-resistant gear has allowed us to enter interior IDLH atmospheres that were previously impossible and untenable for firefighters. The new ability to enter the basement allowed for the introduction of new interior tactics—but not necessarily safer tactics. SCBA and modern PPE should not render obsolete earlier tactics that were proven to be safe and effective. New interior SCBA tactics took hold because firefighters were reaching the seat of the fire more quickly and they were extinguishing them with success. But we've already proved, and painstakingly so, that modern fire fuels combined with lightweight construction have outmatched and surpassed the rating of our PPE structural ensemble and thus our ability to fight these fires with unchanged offensive interior attacks with the same number of successes. The number of firefighter fatalities in relation to the number of structure fires bears that out.

In some major cities, older multistory Type III buildings that were constructed between 1900 and 1940, most of which are still in use today, do not have sprinkler systems—except for the basements. Even back then, building code officials knew basement fires threatened the entire structure and were the hardest to access for the fire department. In fact, building code officials also understood that door and window requirements in occupied spaces were not only for entry and egress, allowing natural illumination and ventilation; they also served as a portal so the fire department could put water on the fire. Many of these Type III buildings are only one, two, or three stories in height; in other words, they did not require a standpipe. So, if you find a fire department connection (FDC) on an older building of three stories or less, it's a sure clue the building has a basement or parking garage. Sometimes the FDC is a single intake pipe with a 2 ½-inch female swivel **FIGURE 4-38**. In older buildings, this single intake was a dry pipe system that ran along the ceiling of the basement and had small distributor heads. Much like a deluge system, the fire department would hook into the dry pipe system, and the spinning heads would extinguish the fire. Most of these systems have been upgraded and converted to modern sprinkler systems, but the extinguishment theory was sound.

FIGURE 4-38 A single-intake FDC with a 2½-inch female swivel on an old Type III construction building that is one, two, or three stories in height is a sure clue that the building has a basement or parking garage and a sprinkler system.

Courtesy of Raul Angulo.

As soon as a basement fire was hot enough (approximately 180°F [82.2°C] depending on the fusing temperature), the sprinkler heads would fuse, applying water directly onto the fire even before the fire department arrived on scene. Sprinkler systems aren't worried about spreading the fire or if there may be occupants in the basement. They don't ventilate the basement—they just flow water—and the fire usually didn't get worse. Whenever you get water quickly onto the fire, all the other conditions get better. If you think about it, an operating sprinkler system is what firefighters are actually trying to accomplish with a transitional indirect attack.

The best way to extinguish a basement fire is to use the tools that were designed for such purpose: the New York Baker cellar pipe (its handles and pipe resemble an upside-down periscope), a cellar nozzle, a round nozzle, a Bresnan distributor nozzle, a fog applicator, or a piercing nozzle. All these tools are designed for an **indirect attack**, which basically creates a giant sprinkler head that can flow between 50 and 300 gpm, depending on the nozzle **FIGURE 4-39**. If these nozzles are not readily available, you can accomplish the same effects of the indirect attack with a combination nozzle. Adjust the nozzle to a 45° fog, shove the hose through the window and whip it around. Ideally, the firefighter should cut a hole through the floor decking above the basement or cut a hole horizontally through the siding closest to the fire. The hole should be large enough to accommodate the size of the nozzle and the fire hose so the hose line can be inserted and withdrawn. Using a roof ladder can assist in supporting the weight of a charged hose **FIGURE 4-40**. Doing so creates little or no ventilation, so the issue of creating a flow path that provides fresh oxygen to intensify the fire is eliminated. For an indirect horizontal attack, the firefighter can also insert any of these nozzles through a basement window, which is often much smaller and narrower than a bedroom window. However, if this window is not close to where the fire is, don't hesitate to use power tools to create a window through the siding right where you need it.

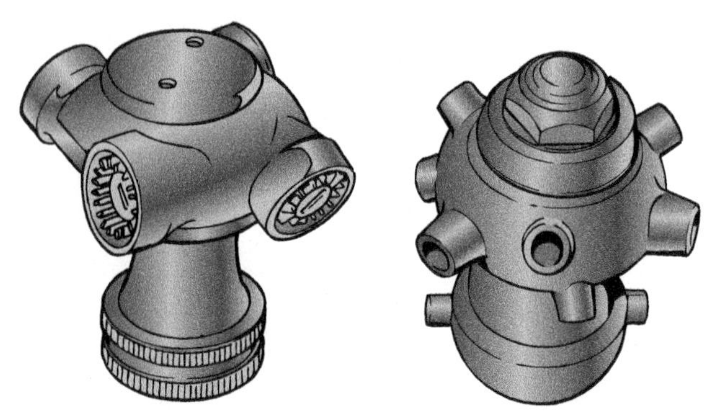

FIGURE 4-39 Prior to the introduction of SCBA, cellar nozzles and Bresnan distributors were the nozzles used most often to attack basement fires.

Left: © Jones & Bartlett Learning; **Right:** Courtesy of Elkart Brass Mfg.Co. Inc.

This tactic is part of a transitional strategy, indirect to direct, because these nozzles effectively knock down a fire, taking away its energy by reducing the heat release rate. Once the fire is indirectly knocked down, the hose team can enter the basement to finish extinguishing the fire, while other crews ventilate and search for victims. Check voids for extension.

One of the findings in the recent research on basement fires was that water applied via the interior basement stairs had a limited effect on cooling the basement or extinguishing the fire. However, water applied through an external window or a door quickly darkened down the fire and reduced temperatures throughout the building, and according to the thermal couple readings, no fire or hot gases were "pushed" up the interior stairs when the door at the top of the stairs remained closed. This effect lasted for several minutes before the fire grew back to the size it was before the application of water. Therefore, the preferred and safest tactic is to attack the fire from the exterior by spraying water through a basement-level window or door. Small basement windows can be on all four sides of the structure just above the foundation, or there may be only one. The basement exterior entry door is typically at the rear C side of the structure. If you have to make a choice between taking the exterior basement door or the window, take the window. Try for the window closest to the fire **FIGURE 4-41**. Breaking out a window is faster at getting water on the fire than forcing a door that may be heavily secured and provides less of an opening for a flow path to develop. If you decide to take the door, use the reach of your stream to shoot water into the basement from a safe exterior position. Once the fire is knocked down, you can enter the basement for final extinguishment while other crews search for victims, start ventilation, and check for extension. It is safer to access the basement from the exterior entry door until the integrity of the interior stairs is verified.

Firefighters in charge of the exposure hose line on the floor above the basement must ensure that the interior door at the top of the basement stairs stays closed. The exposure hose team does not have to huddle behind the door. Again, take advantage of the reach of the hose stream. These firefighters can protect the interior basement door from a distance where the floor decking is firm. After ensuring that the basement interior door is closed, they can place a PPV fan into Floor 1 to pressurize the floor, keeping the

FIGURE 4-40 A–C. In this demonstration, firefighters insert a Bresnan distributor nozzle from the floor above a basement, a hole large enough to accommodate the nozzle and a 2½-inch charged hose line should be made. Smoke will vent through the hole but will change to steam once the nozzle is opened. A roof ladder can be used to support the weight of the hose and to assist in lifting and lowering the nozzle into the basement. D. Note that the radius of the stream flowing 290 gpm is enough to spray beyond the basement window to the outside.

Courtesy of Raul Angulo.

FIGURE 4-41 Firefighters using an offensive exterior attack. A. A basement fire with flames showing at the basement window. B. Water applied through an external basement window darkens the fire and reduces temperatures throughout the building. Application of water does not push the fire into other parts of the building.

Courtesy of NIST.

basement fire and smoke from spreading. The basement should be horizontally ventilated.

In addition to balloon construction, many older buildings have pipe shafts that extend from the basement to the attic, so the attic needs to be checked quickly for extension. If the fire originated in the basement and there is smoke coming from the attic, a round nozzle or fog applicator needs to be inserted into the attic space without opening the roof to extinguish the fire. (An attic charged with hot smoke is an enclosed oxygen-deficient space. There are no flames unless the roof is opened.) PPV fans should not be used on balloon-frame construction until the fire has been extinguished. Using a fan during the attack can spread the fire through every open vertical and horizontal channel.

This research involving basement fires has also provided valuable information that challenges fire departments to reexamine their approach in fighting basement and below-grade fires. The techniques for ventilating a basement will be covered again in Chapter 10.

Tactical Considerations for Basement Fires

One of the initial things the first-in officer needs to do at a fire in a Type V structure is determine whether there is a basement, and if so, whether the basement is on fire. This is a critical part of the 360-degree (walk-around) size-up. A fire in a basement can be a game changer in the initial attack strategy. Many of these structures have an exterior entry door into the basement level. If they do not, there should be at least one, if not more, basement window. Basement fires should be attacked from the basement window or from the exterior basement door. This is the safest and coolest position for the fire attack team.

If there is no door, a basement window should be taken out and the hose stream should be directed into the basement. Depending on the stream pattern, you are creating a sprinkler system. If there are no windows, make one. In fact, if you're going to make one, make two. The first window should be utilized to get water on the fire as quickly as possible. The second window can be cut on the opposite side of the basement to assist in ventilation. Using a PPV fan from the attack side can turn a bidirectional flow path into a unidirectional flow path. It stands to reason that when water expands into steam 1,700 times its volume, it will increase the interior atmospheric pressure. Higher pressures move toward lower pressures to equalize, so making that second window as an exhaust portal will direct any pressurized smoke and gases out of the basement instead of up into vertical void spaces.

Keep in mind that these conclusions are from experiments. Real-world fires can act differently. Pressurized atmospheres still seek the path of least resistance to equalize. Void spaces, walls, and the attic still need to be opened and checked for fire extension when dealing with basement fires. Utilizing a TIC is extremely helpful in locating elevated temperatures and hidden fires.

The Parker Doctrine

"There will always be resistance to organizational change. You will not be able to successfully implement a new change against deep seated, tradition-bound, cultural strongholds until the pain of current behavior and staying the course becomes greater than the pain of making the change." –Larry Parker, International business executive

This is the Parker Doctrine. One of the most gut-wrenching illustrations of the Parker Doctrine is the Sofa Super Store fire that killed nine firefighters of the Charleston, South Carolina, Fire Department on June 18, 2007. During the investigation, media outlets quoted the fire chief as saying, "Our firefighting techniques are not going to change in the City of Charleston Fire Department. . . . We're safe and we've got the best equipment, we've got the best people, and that's the way we fight fires." He also suggested that his firefighters performed just as they were trained, and he would not do anything different if the same fire happened again.

Extinguishing fires is not that complicated. In fact, the UL/NIST recommendations are, in essence, urging us to return to the old ways of extinguishing fires. Prior to the introduction of SCBA, extinguishing methods for interior structure fires hadn't changed much since the days of horse-drawn steamers. The first-in company would spray water into the building from the outside to cool interior conditions before going in for search and rescue and final extinguishment.

The concept of offensive interior attack is relatively new in our history and evolved with the introduction of SCBA. What was once physically impossible became possible with new equipment. With legacy fuels and construction, it was effective 95% of the time. With modern hydrocarbon-based fuels and modern lightweight construction, this tactic has become dangerous and is killing firefighters—primarily because we are still attacking them like legacy fires. When a lightweight construction occupancy furnished with modern hydrocarbon-based plastics and synthetics is

CONCLUSIONS ABOUT BASEMENT FIRES

The conclusions drawn from the fire experiments conducted by NIST, UL, and FDNY regarding basement fires are:

- Fire extension from basement fires developed in locations other than the stairs; the floor assembly above (basement ceiling) often failed close to the location where the fire started.
- Flowing water at the top of the stairs had limited impact on basement fires. The structural integrity of the stairs is unknown—and vulnerable to the fire—so the stairs can be a vulnerable location for firefighters. If a stairway collapses under the weight of an entry team, firefighters cannot make a quick retreat. They will fall down the stairwell into the burning basement. They may lose the hose line, and a fall could cause their SCBA facepiece to become dislodged. A portable radio can also be lost in the fall, eliminating their ability to call for a Mayday.
- An offensive exterior attack through a basement window is effective in cooling the fire compartment. If there isn't a basement window, make one.
- An offensive fire attack through the exterior basement door is effective in cooling the fire compartment. If there isn't a door, make one.
- It is best to extinguish a basement fire with a tool designed for this purpose, such as a cellar pipe, distributor nozzle, fog applicator, or piercing nozzle. An indirect attack utilizing these tools is safer for firefighters and is the fastest way to knock down a fire and lower the interior temperature. The two in/two out rule does not apply when the indirect attack is made from the outside.
- Initiating a fire attack on a basement from the floor above and coming down the interior stairs puts firefighters directly into the thermal flow path, called the chimney effect. This practice, which can turn deadly in seconds, should no longer be considered an accepted practice.
- Thermal-imaging cameras used on the first floor may help indicate that there is a basement fire, but they cannot be used to assess the structural integrity of the floor joists and decking.
- A floor assembly above a basement fire has the potential to collapse due to the fire exposure underneath. This can create a dangerous and deadly scenario for firefighters.
- Attacking a basement fire from the interior stairway places firefighters directly in a high-risk, possibly deadly flow path of hot gases flowing up the stairs, known as the chimney effect. This practice should be abandoned. Attack from the exterior to knock the fire down. Then send the entry team in through the exterior basement door. It is good practice to place a PPV fan at the back of the entry team to protect them when an exhaust portal in the basement has been established opposite the entry team's access point.
- Protecting the exposure floor above the basement fire should be done by ensuring that the interior basement door is closed. Keeping the door closed keeps firefighters out of the potential unidirectional flow path. An open interior door can allow the fire to extend and cause smoke damage, which can be prevented by first commencing with an exterior transitional attack.
- Protect the closed interior basement door by using the reach of the hose line while operating from a safe area with secure floor decking.
- Coordinating ventilation is extremely important. However, engine crews should not delay applying water to the fire if the ventilation team isn't in place yet.
- Ventilating the basement up the interior stairs through the first floor is an option, but it creates a flow path up the stairs and out through the front door of the structure—almost doubling the speed of the hot gases and increasing the temperatures of the gases to levels that could cause injury and death to fully protected firefighters. Ventilating the basement is best accomplished through horizontal ventilation by utilizing basement windows, even if exhaust holes need to be cut through the siding. PPV fans can direct the desired ventilation flow path.
- Electric smoke ejectors or electrical PPV fans can be used to ventilate the basement by using the exterior entry door as the exhaust portal.

Courtesy of Raul Angulo.

on fire, the window of opportunity for a relatively safe operational period (in firefighting terms) is reduced to approximately 5 minutes before flashover and structural collapse can occur. This is about the time most fire departments are arriving on scene. This window of opportunity is lost, in effect, for understaffed rural volunteer fire departments or for any metropolitan fire departments with longer response times.

These UL/NIST experiments conclude that the most effective and fastest way to improve interior conditions throughout the structure is to spray water into the fire from the exterior. This technique knocks down the fire faster than a crew can make an interior offensive attack with a hose line through the front door. Once the fire is knocked down, the energy and the temperature of the fire is reduced, making conditions better for possible occupants who are trapped and safer for firefighters who are now entering an environment below the maximum rating of their structural firefighting PPE. The UL/NIST recommendations are urging a return to a defensive-to-offensive attack strategy. This is not the only strategy for extinguishing a fire; fires at the incipient stage should still be attacked with a pressurized pump can or hose line, and defensive fires should be fought from the outside with master streams. It's all the fires in between where we need to slow down, perform a proper size-up to determine the fire conditions, and evaluate the risks before committing firefighters to the interior of a burning structure.

The UL/NIST results provide valuable information about the growth of fires, fire flow paths, optimal suppression techniques, and effective ventilation techniques. Firefighters still need to make interior attacks for search and rescue, final extinguishment, and overhaul. The issue at hand are the steps that should be taken to improve and stabilize interior conditions before firefighters make entry. The facts are convincing, and numbers don't lie. Firefighters around the country who resist the implementation of these recommendations do so because it's not risky enough, dangerous enough, and exciting enough and that, somehow, without these factors, it cheapens the nature of our noble profession. Professional firefighters understand that nothing could be further from the truth.

Risking your life is a double-edged sword. Saving a life is the greatest reward for a firefighter and often involves extreme risk, but it should be for a life saved. Your life has value, too, especially to your family. Don't risk it simply for heroics. Making a no-go decision isn't losing a life—it's saving yours. Old traditions die hard; when new ideas and methods challenge them, it is natural that the new ideas and methods are met with resistance, especially when fire departments are quick to hand out medals for heroism to firefighters whose actions violate safety policies and procedures. Rewarding risky behavior always encourages more risky behavior. Operating within the rated safety limitations of your gear and taking a calculated risk to rescue another human being deserves a medal for heroism, but risking your life for property that no longer has value or for lives that can no longer be saved is not heroic. It is reckless and should never be rewarded. Neither should fire departments reward body recoveries. We have to stop rushing into burning structures where no human lives are in danger until the fire department arrives and enters the building.

We often make the comparison between the fire service and the military because we're a paramilitary organization—but we're not the military. We're a municipal public safety and emergency services department. There are no "acceptable losses" in the fire service. But let's take an example from the military. The Green Berets (Special Forces) are instructed that one of their missions is to save civilian lives but not by giving their own lives: The US Army has too much invested in them. They need to stay alive so that they may save hundreds of civilian lives throughout their career. When a Special Forces unit is outnumbered, its members do not rush the enemy and attack simply because they're the Green Berets. No, they withdraw to find a better tactical advantage. The end game is for a successful mission, not one that ends in failure.

From day one of the fire academy, we're taught to "get in there and put the fire out!" We're rarely taught to retreat because we take that as a personal defeat. We're losing lives because all our traditional training is based on legacy fires and buildings. We should make entry into the interior environment only when it is within the safety rating of our PPE (which is about 500°F [260°C]). A smoldering combustible environment or the decay stage of a fire is within the rating of our gear; a flaming environment with modern fuels is not. We need to learn when it's time to retreat and look for a better tactical advantage to fight the fire. Simply put, we need to reduce the interior temperature to 500°F or less before making entry into a burning structure.

The information in this chapter attempts to illustrate, clarify, and apply the findings. NIST and UL are planning additional experiments. Although this book is the most up-to-date information at the time that it was printed, it is not possible for textbooks to always contain the most current research. Additional information about these fire experiments can be obtained online at the NIST Firefighting Technology and UL Firefighter Safety Research Institute websites. To keep up with the data, study the information on the NIST and UL websites.

The UL/NIST recommendations are now *evidence-based practices* and, in many cases, they validate the traditional practices of firefighting in the era before SCBA. Our litigious society has forced the fire service to weigh their operating policies against risk management. Risk management and legal liability drive all the areas of services provided by the fire department. These recommendations can become *best practices* and can factor into the decisions and actions we take on the fireground. Fire departments will have to reevaluate and modify some of their approaches to offensive interior fire attacks, and company officers need to develop the expert knowledge and discipline to implement these changes within their fire companies. For the safety, protection, and well-being of all personnel, fire departments are also encouraged to adopt and to comply with NFPA safety standards and recommendations. Don't make the Parker Doctrine ("Change will not happen until the pain of current behavior becomes greater than the pain of making the change") applicable to your department. No doubt, in a court of law, we will be held accountable to the science and standards behind these recommendations.

After-Action REVIEW

IN SUMMARY

- Fire attack should be coordinated with ventilation.
- Ventilation releases pressurized smoke, but it does not cool a fire. It actually intensifies the fire because it creates a flow path bringing oxygen into the fire.
- Controlling the flow path is essential in stopping the fire from intensifying and prevents flashover.
- Water knocks down the fire, diminishes the heat release rate, and resets the fire, giving everyone inside more time.
- Engine companies are typically in place with hose lines long before ladder companies can climb ladders and vent a roof. Applying water to the fire should not be delayed.
- Interior search and rescue should not be commenced without a charge hose line in place between the possible occupants and the fire.
- VES has been upgraded to VEIS, which places emphasis on closing the door to a room, thus isolating it from the fire and preventing the creation of a new low-pressure flow path before the search is commenced.
- The decision to VEIS is easier to make when there is a confirmed trapped occupant whose location is known.
- If there are possible occupants who may be trapped, the firefighter or company officer needs to consider the survivability profile based on the signs from the flames, smoke, and the color of the windows.
- The company officer must consider human thresholds—the victim's exposure to the heat and toxicity of the smoke—before committing firefighters to VEIS.
- The integrity of the first floor and basement stair assemblies may be compromised and unable to support the weight of a firefighting team.
- Accessing the fire via the basement interior stairs places the firefighters directly into the flow path of the fire.
- Basement fires should be attacked from the outside perimeter using an indirect attack with tools that were designed for such purpose.

KEY TERMS

bidirectional flow path A two-way or high-low air current of fresh air and smoke competing for the same opening. The lower air intake is fresh air from the outside being entrained or sucked in toward the seat of the fire. The high point or exhaust path is where the smoke and other heated gases are using the same opening (door or window) to escape, seeking equilibrium.

blitz attack An aggressive fire attack that often utilizes a 2½-inch hand line or deck gun and occurs just prior to entry, search, and tactical ventilation. Also known as *defensive-to-offensive exterior attack, softening the target, transitional attack.*

chimney effect The area between a high-low pressure differential flow path. The fresh air intake portal is low, and the exhaust portal is high. As heat from the fire travels up through this flow path, the fire dynamics are the same as in a chimney. Temperatures can reach over 1,000°F (538°C). The chimney effect happens in stairwells, and under these conditions, the stairs can become a deadly trap for firefighters.

defensive-to-offensive exterior attack A transitional attack that starts from the exterior of the structure to knock down and take the energy away from the fire. This is sometimes referred to as resetting the fire because the application of water cools the interior temperature, thus increasing the time period to flashover. Once the fire is knocked down, an interior offensive attack can begin. Also known as *blitz attack, softening the target, transitional attack.*

explosive fire growth A room-and-contents fire in a modern house often flashes over and then enters the decay stage due to decreased oxygen levels in the house, or it may use up the available oxygen before it reaches flashover. This produces a ventilation-limited fire where the fire goes into a dormant state or decay stage. In this decay stage, fire growth and fully developed stage is limited only by the lack of available oxygen. When air is introduced back into this environment, the fire can grow with explosive force.

extreme fire behavior The sudden increase in oxygen combined with hot flammable gases from the fire that results in rapid fire growth and changes the direction of the flow path from bidirectional to unidirectional due to the pressure imposed by the wind. This is also called a *wind-driven fire.*

flashover A transition phase in the development of a compartment fire in which surfaces exposed to thermal radiation reach ignition temperature more or less simultaneously, and fire spreads rapidly throughout the space, resulting in full room involvement, or total involvement of the compartment or enclosed space. (NFPA 921)

flow path The movement of heat and smoke from the higher pressure within the fire area toward the lower pressure areas accessible via doors, window openings, and roof structures. (NFPA 1410)

fuel-limited fire A fire in which the heat release rate and fire growth are controlled by the characteristics of the fuel because there is adequate oxygen available for combustion. (NFPA 1410)

indirect attack Firefighting operations involving the application of extinguishing agents to reduce the buildup of heat released from a fire without applying the agent directly onto the burning fuel. (NFPA 1145)

legacy construction An older type of construction that used sawn dimensional lumber and was built before the 1970s.

oxygen-limited fire A fire that depends on the amount of oxygen available for combustion.

softening the target An aggressive offensive exterior fire attack that occurs just prior to entry, search, and tactical ventilation. Another expression for this term is "hitting it hard from the yard." Also known as *blitz attack, defensive-to-offensive exterior attack, transitional attack.*

survivability profile An assessment that weighs the risks likely to be taken, versus the benefits of those risks, of the viability and survivability of potential fire victims under the current conditions in a burning structure.

transitional attack An offensive fire attack initiated by an exterior, indirect hand-line operation into the fire compartment to initiate cooling while transitioning to interior direct fire attack in coordination with ventilation operations. Also known as *blitz attack, defensive-to-offensive exterior attack.*

unidirectional flow path A fresh-air intake opening or an exhaust exit port. It simply means the convected air currents are flowing in the same direction through the entire opening and exit port.

vent-enter-isolate-search (VEIS) A method of searching for fire victims that consists of selecting and opening a window to a bedroom or other living space; entering the room; closing the door to isolate the room from the fire, thus preventing a flow path from developing and allowing the room to ventilate naturally; quickly searching for any possible victims; and rescuing them out through the window and down a ladder.

REFERENCES

A Career Lieutenant and Fire Fighter/Paramedic Die in a Hillside Residential House Fire—California, NIOSH Report F2011-13, March 1, 2012.

Marsar, Stephen, *Can They Be Saved? Utilizing Civilian Survivability Profiling to Enhance Size-up and Reducing Firefighter Fatalities in the New York City Fire Department*, Executive Fire Officer Thesis and recipient of 2010 Outstanding Research Award from the National Fire Academy, EFOP, New York, NY. Self-published. 2009.

Marsar, Stephen, *Survivability Profiling: Applying What We've Learned*, Fire Engineering, Fair Lawn, NJ. 2011.

Marsar, Stephen, *Survivability Profiling: Are the Victims Savable?* Fire Engineering, Fair Lawn, NJ. 2009.

Marsar, Stephen, *Survivability Profiling: How Long Can Victims Survive in a Fire?* Fire Engineering, Fair Lawn, NJ. 2010.

National Fire Protection Association, NFPA 1971, *Standard on Protective Ensembles for Structural Fire Fighting and Proximity Fire Fighting*, 2018.

National Fire Protection Association, NFPA 1981, *Standard on Open-Circuit Self-Contained Breathing Apparatus (SCBA) for Emergency Services*, 2019.

Nine Career Fire Fighters Die in Rapid Fire Progression at Commercial Furniture Showroom—South Carolina, NIOSH F2007-18. February 11, 2009.

Rose, Stewart, *Strategy and Tactics MCTO.* Insource Inc., Salem, OR. 2015.

Hendricks, Mike, and Matt Campbell, Fatal Echoes, *Kansas City Start Newspaper*, Kansas City, MO. December 7, 2016.

Schottke, David, *Evidence-Based Practices for Strategic and Tactical Firefighting.* Burlington, MA: Jones & Bartlett Learning, 2019. In conjunction with National Institute of Standards and Technology (Daniel Madrzykowski, NIST Fire Protection Engineer), UL LLC, Firefighter Safety Research Institute (Stephen Kerber, Director).

CHAPTER

Equipment and Initial Hose-Lay Operations

LEARNING OBJECTIVES

- Recognize the advantages and disadvantages of both smooth-bore tip nozzles and combination fog nozzles.
- Be aware of the current technology advancements in nozzle design, tools, and equipment.
- Describe the four basic hose lays used by a one-piece (single-pumper) engine company.
- Describe the advantages and disadvantages of the various hose lays.
- Be familiar with NFPA 1901, *Standard for Automotive Fire Apparatus*.

Introduction

The first section of this chapter deals with equipment. The fire apparatus—the engine company—is an emergency vehicle that must be relied on to transport firefighters safely to and from an incident, and to operate reliably and properly to support the mission of the fire department.

> A piece of fire apparatus that breaks down at any time during an emergency operation not only compromises the success of the operation but might jeopardize the safety of the firefighters relying on that apparatus to support their role in the operation. An old, worn-out, or poorly maintained fire apparatus has no role in providing emergency services to a community.—National Fire Protection Association (NFPA) 1901, Annex D.6, *Standard for Automotive Fire Apparatus*

Company officers and firefighters, no matter how well trained, cannot perform efficiently without proper equipment and adequate staffing. The combination of well-trained firefighters in the knowledge of hose evolutions and competency and confidence in handling their equipment—including the quality of their equipment—determines the outcome on the fireground. Because engine company personnel are called on to engage in combating many different types of fires, their equipment must be chosen for its quality and dependability to ensure adequate performance while covering a wide range of fireground situations. But none of that can happen unless firefighters arrive on the apparatus safely.

Emergency vehicle operators must always follow federal, state, and local laws and regulations governing the movement of apparatus on a roadway. The only time there are exemptions to traffic laws is when the apparatus is operating in emergency mode. Even when responding to an emergency, the driver can be found criminally or civilly liable if the apparatus is involved in a crash. The exemptions do not relieve the driver from the duty to operate the apparatus with reasonable care for all other persons or vehicles using the roadways, and the responsibility for all the passengers on the apparatus. This starts with every member wearing their seat belts any time the apparatus is in motion, regardless of emergency or nonemergency status—it's the law. Every fire department has a regulation stating so, but it is often not enforced. Members who resist wearing seatbelts claim it slows them down from donning their PPE and SCBA. There have been numerous injuries and line-of-duty-deaths associated with firefighters not wearing their seatbelts. They were either ejected, thrown around, struck, or run over by the fire apparatus before they even arrived at the scene to perform their job. Some have occurred when the apparatus is driving around in service. Most of the deaths could have been prevented if the member was wearing a seat belt. If we cannot change the culture to follow this safety rule, we shouldn't be surprised when other safety rules are disregarded. Would a fighter pilot disregard wearing seatbelts in the cockpit on the way to a mission? Of course not. It's unsafe and unprofessional. Wear your seat belt—enough said.

It is not enough to be well trained, adequately staffed, and equipped. Time is the ally of the fire, not of the firefighter. An engine company must deploy equipment quickly to and at the scene of a fire. The first few minutes at the scene of a working fire could mean the difference between a successful operation and a failed one. An old fireground adage is, "As the first attack line goes, so goes the rest of the fire." The initial task of an engine company is to provide fast water onto the fire using a preconnected hand line and tank water if it will be effective, then establish a reliable water supply from a hydrant to the fireground.

The second section of this chapter discusses initial supply hose lays. Departmental standard operating guidelines (SOGs) should dictate the procedures used to deliver an adequate, constant, and uninterrupted supply of water to the fireground, with consideration of the types of apparatus responding, available personnel, arrival time of additional companies, existing water sources and their capacity, building construction, and the availability of mutual aid.

Section 1: Engine Company Equipment

NFPA 1901 is a valuable document that defines the requirements for new automotive fire apparatus designed to be used under emergency conditions to transport personnel and equipment, and to support the suppression of fires and mitigation of hazardous situations. It is a must-read document for anyone assigned to an engine company—especially for the driver/pump operator and the company officer.

The AHJ should be familiar with documents that allow them to make informed decisions before purchasing fire apparatus and equipment. They should also consult the fleet manager of neighboring cities and cities of similar size and population to solicit input before selecting the model and make of a fire engine for purchase. The most important people to consult for valuable input are the firefighters themselves because they will be the ones ultimately using every aspect of the fire apparatus for many years to come. The opinions of senior-level drivers, company officers, and firefighters should influence any decision to purchase fire apparatus and equipment so that a fire department is not surprised when equipment arrives and does not meet the requirements or standards. Apparatus manufacturers can be of assistance to your department. Remember, simply going with the lowest bid may save the department money in the short run, but a fire apparatus that is unreliable and cannot endure the extreme wear and tear of the job will constantly be in the fire garage for repairs. In addition to out-of-service time, it may end up costing the department more in the long run.

The engine itself should be equipped with a pump having a rated pumping capacity of no less than 750 gallons per minute (gpm) and a water tank that carries at least 300 gallons of water. These are absolute minimums for the general operation of engine companies. The capacities should be increased on the basis of a fire department's knowledge of its needs and SOGs. The majority of fire apparatus have pumps rated at 1,500 gpm and have a tank capacity of 500 gallons. The department should choose a water tank size that best supports efficient and effective fireground operations. For example, a rural engine company might find that a minimum 1,000-gallon water tank is required to

NFPA 1901 EQUIPMENT LIST FOR PUMPER FIRE APPARATUS

Suction Hose or Supply Hose

NFPA 1901, 5.8.2

A minimum of 20 feet (6 m) of suction hose (hard suction) or 15 feet (4.5 m) of supply hose (soft suction) shall be carried.

Fire Hose and Nozzles

NFPA 1901, 5.9.3

The following fire hose and nozzles shall be carried on the apparatus:

- 800-feet (240 m) of 2½-inch (65 mm) or larger fire hose
- 400-feet (120 m) of 1½-inch (38 mm), 1¾-inch (45 mm), or 2-inch (52 mm) fire hose
- One hand-line nozzle, 200 gpm (750 L/m) minimum
- Two hand-line nozzles, 95 gpm (360 L/m) minimum
- One smooth-bore or combination nozzle with 2½-inch shutoff that flows a minimum 250 gpm

Equipment for Pumper Fire Apparatus

Ground ladders carried on pumper fire apparatus are listed in NFPA 1901, 5.8.1. All ladders meet the NFPA Standard 1931, *Standard for Manufacturer's Design of Fire Department Ground Ladders*.

- One 12-foot roof ladder (straight ladder with hooks)
- One 24- to 26-foot ground extension ladder
- One folding ladder (attic ladder)
- One 12-foot attic ladder

Miscellaneous Equipment

NFPA 1901, 6.9.4

- One 6-pound (2.7 kg) flathead axe
- One 6-pound (2.7 kg) pickhead axe
- One 6-foot (2 m) pike pole or plaster hook
- One 8-foot (2.4 m) or longer pike pole
- Two portable hand lights or battle lanterns
- 80 B-C dry chemical extinguisher
- One 2½ gallon (9.5 L) or pressurized water extinguisher
- Self-contained breathing apparatus (SCBA) plus spare bottles for every member on the crew or each seated position
- One first aid kit with an automatic external defibrillator (AED)
- Four combination spanner wrenches
- Two hydrant wrenches
- One 2½-inch (65 mm) double-male adapter coupling with National Hose (NH) threads
- One 2½-inch (65 mm) double-female adapter coupling with NH threads
- One rubber mallet, suitable for use on suction hose connections
- Two salvage covers or tarps, 12 feet × 14 feet (3.7 m × 4.3 m)
- Two (or more) wheel chocks
- Five fluorescent orange traffic cones with reflective bands
- Five flares
- High-visibility traffic vests for each seated position

Additional Equipment

NFPA 1901, A.5.9.4

- One fire service claw tool
- One smoke ejector or small positive-pressure ventilation (PPV) fan
- One 3-foot (1 m) crowbar
- Insulated bolt cutters with a 7⁄16-inch (11 mm) minimum cut
- One Halligan tool
- One 2½-inch (65 mm) hydrant valve (screw-type gate)
- One double-gated reducing leader wye sized to fit the hose used in the fire department (FD)
- One scoop shovel
- One pointed shovel
- Four hose straps
- One 125-foot (38 m) utility rope with a breaking strength of at least 5,000 pounds (2,200 kg)
- One 3,000-watt portable generator if the apparatus is not equipped with a fixed-line power source
- Two 500-watt portable lights
- Two cord reels to reach 400 feet (120 m) (Heavy Duty) HD
- One portable pump
- Tool box with assorted hammers, pliers, wrenches, screwdrivers, and so on
- Master stream appliance, 1,000 gpm (4,000 Lpm) minimum
- Foam delivery equipment compatible with onboard foam system
- One hose clamp
- Hose adapter for water supply connections in neighboring communities

maintain an initial attack during the length of time needed to set up its pumper. If so, that engine company should be equipped with a 1,000-gallon water tank.

NFPA 1901 states that the requirements of service in different communities might necessitate additions to the equipment required. The operational objective is to arrive at the scene of the emergency with the necessary equipment for immediate life safety operations and emergency control. NFPA 1901 recommends that pumper fire apparatus carry a minimum list of equipment described in the following sections and listed here. Remember, a pumper is not a tender.

Hose Storage

NFPA 1901 requires a minimum hose storage area of 30 ft^3 for 2½-inch or larger fire hose, and two areas, a minimum of 3.5 ft^3, each to accommodate 1½-inch or larger preconnected hose lines. A divided hose bed is one that is separated into two supply hose compartments running the length of the hose bed. This arrangement permits two separate hose lines to be laid simultaneously **FIGURE 5-1**. It also allows for two different hose setups for two different types of hose lays. For example, one side of the hose bed may be set up for a reverse lay and the other for a forward or straight lay or with the use of the proper couplings, can join both hose beds together to extend the single supply line for a long lay evolution. **FIGURE 5-2**. This setup allows flexible operation at the fireground. A divided hose bed can be used with a 2½-inch, 3½-inch, or large-diameter hose (LDH) of 4- and 5-inch (100 mm and 125 mm) diameter. **FIGURE 5-3**. The 4- and 5-inch LDH are usually connected using Storz-type couplings **FIGURE 5-4**. Today's fires involve modern synthetic fuels with larger household fire loads, and they produce fires that burn hotter and faster, meaning the temperature and heat release rate has greatly increased. To meet the increased gpm required to fight these fires, the majority of fire departments have exchanged 1½-inch hose lines for 1¾-inch hose lines, and now use 4- and 5-inch hose for their supply lines.

FIGURE 5-1 A divided hose bed allows two supply lines to be laid simultaneously. The right hose bed consists of 4-inch LDH, while the left hose bed is deploying 2½-inch hose.

FIGURE 5-2 In the divided hose bed shown, the bed on the left is loaded with 2½-inch hose with an attached combination nozzle and is set up for a reverse lay (fire to water). The bed on the right is loaded with 4-inch LDH set up for a forward lay (water to fire).

FIGURE 5-3 A divided hose bed can be used with a 2½-inch, 3½-inch, or LDH of 4 and 5 inches for supply.

Many fire departments find it more efficient to use 4-inch or 5-inch LDH to move water effectively from a source to the fireground from areas with wide hydrant spacing or areas with no hydrants at all. In these

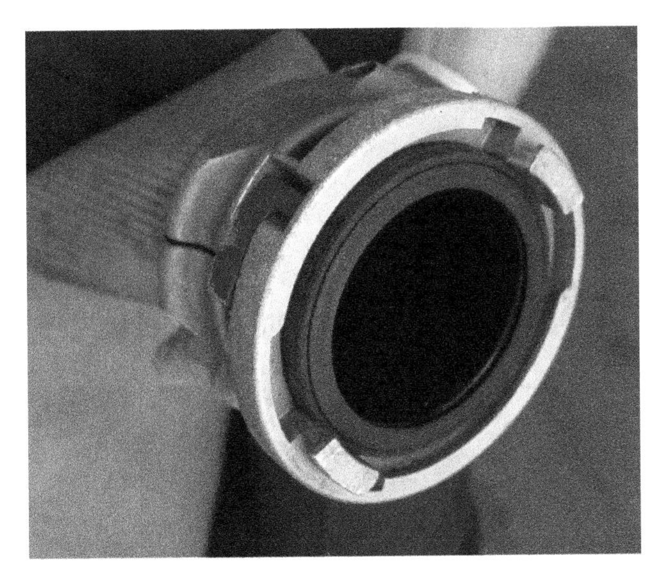

FIGURE 5-4 Four- and 5-inch LDH are usually connected using Storz-type couplings.

cases, it is wise to carry additional hose. A department should evaluate its specific needs and choose the size and amount of hose that best support its operation.

Hose storage areas should be arranged so that the configuration best supports the operational procedures following SOGs. There are many variations as to how supply and attack hose lines are carried on fire apparatus. Most attack loads are department-specific and have been developed over years of trial and error. We will cover only the three most common supply loads and a couple of variations for attack hose. The three most common supply loads are the flat load, the horseshoe load, and the accordion load FIGURE 5-5. The flat load and the accordion load start with the female coupling. The horseshoe load starts with the male coupling. The flat load flakes the hose back and forth the length of the hose bed in a W pattern. The horseshoe load stands the hose on edge and follows the outer wall of the hose bed in a horseshoe pattern, working its way toward the center. The accordion load also weaves the hose on edge back and forth in the hose bed.

Fire Hose

NFPA 1901 requires a minimum of 400 feet of 1½-, 1¾-, or 2-inch fire hose to be carried on pumper fire apparatus. Although 1½- and 2-inch hose are used by fire departments for attack lines, 1¾-inch hose has become the most widely used within the fire service for this purpose. Two-inch hose still uses 1½-inch couplings. The 1½- or 1¾-inch hose, with 1½-inch couplings, usually includes preconnected lines supplied by apparatus tank water and can be quickly advanced for rescue, interior exposure coverage, and direct fire attack FIGURE 5-6. The size and capacity of the fire

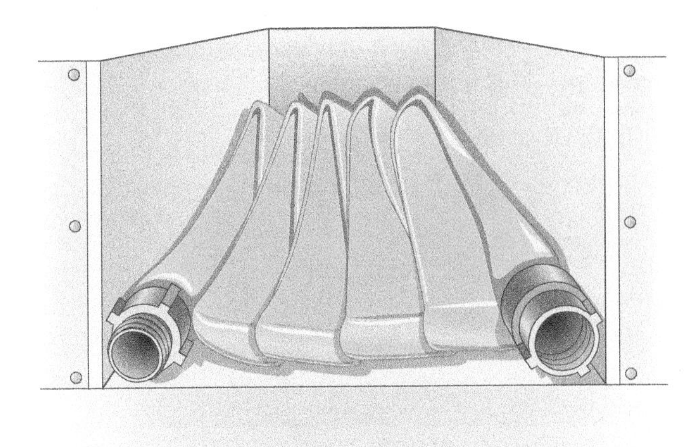

A

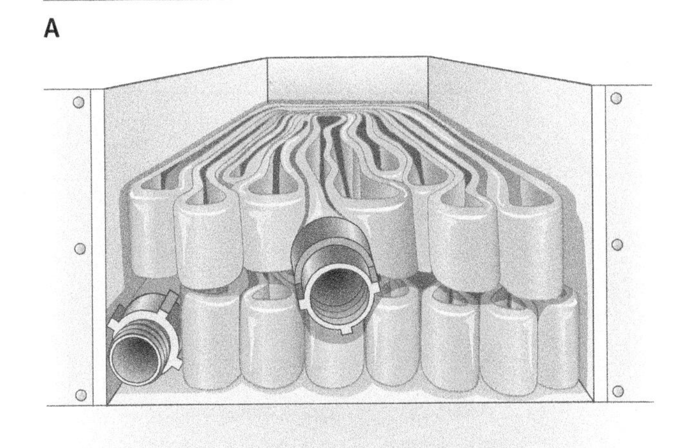

B

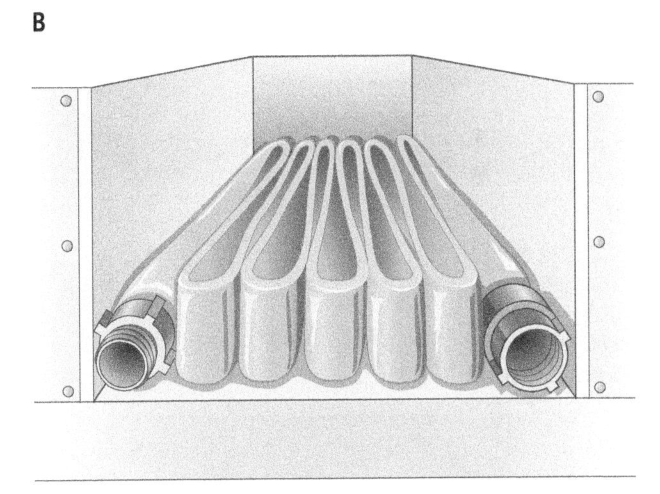

C

FIGURE 5-5 Three basic hose loads are used for supply hose on the apparatus: **A**. flat hose load, **B**. horseshoe hose load, and **C**. accordion hose load.

hose allow fast movement and sufficient delivery of water for most interior fires; however, 1¾-inch hose lines cannot be used on every fire, in spite of their mobility. When the size or intensity of the fire dictates the use of 2-inch or 2½-inch hose lines, for greater reach, penetration, and gallonage, then they must be placed in service. The basic rule of thumb is this: If you have a big fire, you operate a big line FIGURE 5-7.

A

B

FIGURE 5-6 **A**. One firefighter can deploy the highly mobile 1¾-inch hose. **B**. The 1¾-inch preconnected attack line is usually 200 feet and should be no longer than 300 feet due to friction loss.

FIGURE 5-7 When the size and intensity of the fire dictates the use of a hose line for greater reach, penetration, and gallonage, then a 2½-inch hose must be placed in service.

Wherever possible, the 1¾-inch hose lines should be preconnected to pump outlets. **Crosslays**, or transverse hose beds, are usually located behind the crew compartment. This hose lay is also known as a bucket line, a speed lay, or a Mattydale lay (so named because it was designed in 1947 by Chief Burton L. Eno of the Mattydale, New York, Fire Department). This configuration of the hose bed allows preconnected attack lines to be deployed from either side of the engine **FIGURE 5-8**. They are connected in the hose bed with a swivel valve. The hose is flat-loaded with the nozzle on the bottom so that it can be stretched or

FIGURE 5-8 Crosslays allow preconnected attack lines to be deployed from either side of the engine.

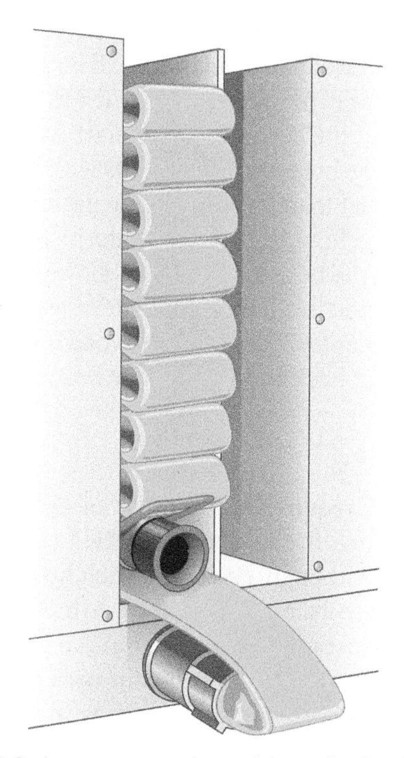

FIGURE 5-9 A preconnected attack hose flat-loaded with the nozzle on the bottom allows the top flakes to pay off the top while it is advanced to the fire.

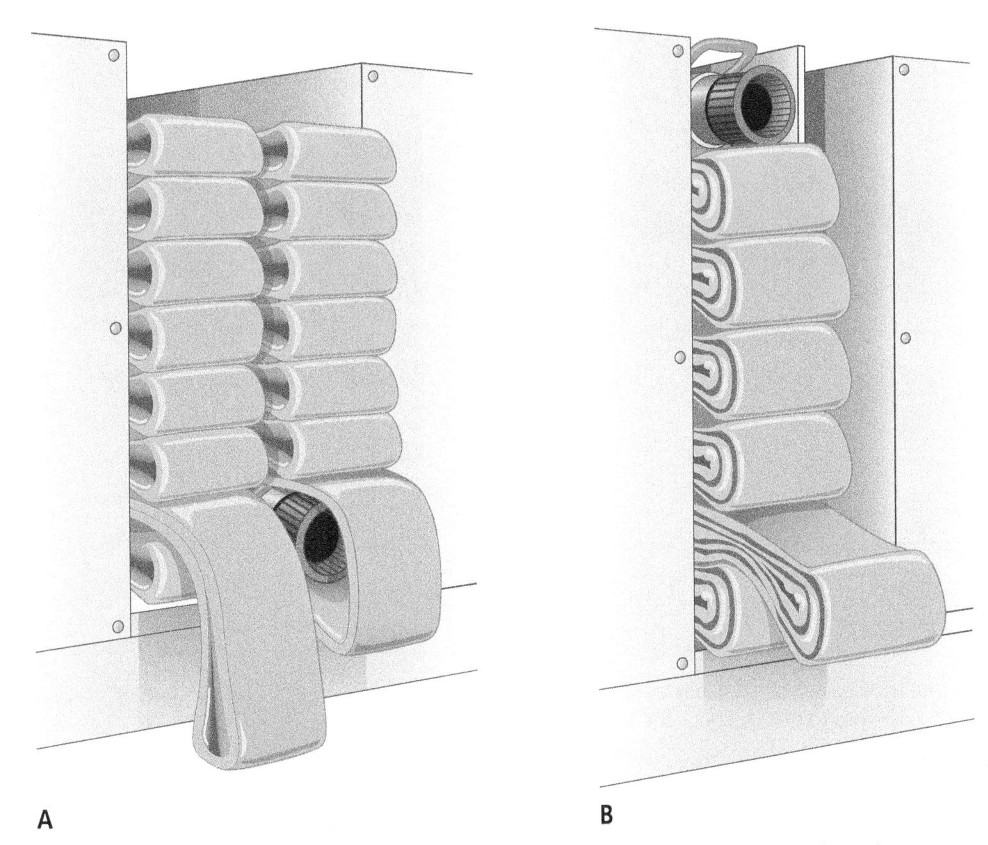

FIGURE 5-10 **A**. The minuteman hose load allows a single firefighter to stretch or drop one fold of the hose at a time from the shoulder while advancing to the fire. **B**. The triple-layer hose load is a method in which the hose is folded back onto itself to reduce the overall length to one-third.

shoulder-carried (shoulder load) quickly upon arrival at the scene. The flakes pay off the top of the shoulder load (some loads have the nozzle on top) **FIGURE 5-9**. Firefighters, with the help of the driver/operator, should be able to get this attack line in position and have it charged and in operation in less than 1 minute. Sometimes this load is referred to as the minuteman hose load **FIGURE 5-10(A)**. The triple-layer hose load is a method in which the hose is folded back onto itself to reduce the overall length to one-third before loading the hose in the bed **FIGURE 5-10(B)**.

If the apparatus doesn't have crosslay hose slots for preconnected lines, it will be equipped with separate hose beds with the 1¾-inch hose lines loaded so that the nozzles are on top. The nozzle and hose line can then be advanced efficiently to the firefighting position. When the crew and a nozzle is in position, the hose can be broken at a convenient coupling and attached to the pump discharge with a 2½- to 1½-inch reducer **FIGURE 5-11**. Although this method works, modern fire apparatus are usually equipped with crosslays for preconnected hose-line operations.

One-and-three-quarter-inch preconnected hose lines should not exceed 300 feet in length because of the excessive friction loss in longer lays. If these hose lines need to be extended a greater distance, they should be connected to a manifold or 2½-inch diameter hose lines. The most common lengths of 1¾-inch hose on preconnected lays are 200 feet. A 1½-inch gated wye is often used and is very advantageous when attached to the end of a 2½-inch diameter hose line to extend the lay because it gives the option for a second line.

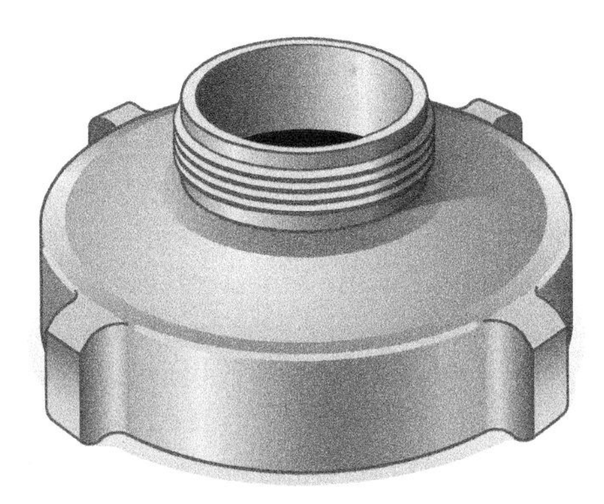

FIGURE 5-11 Reducers are used to connect couplings of different diameters. A 2½-inch to 1½-inch reducer is used to connect a 1¾-inch hose line to a 2½-inch discharge port off the pumper, or when an extended lay of over 300 feet is required.

2½-Inch Fire Hose

NFPA 1901 requires a minimum of 800 feet of 2½-inch or larger fire hose. It also recommends a designated

FIGURE 5-12 Fires that have gained some headway should be attacked with 2½-inch hose lines to quickly knock the fire down.

FIGURE 5-13 When there is a long hose lay from the pumper to the fire, and a 1¾-inch line will be used for attack, a 2½-inch hose line with a gated wye can be used for the first hose length to extend the lay. The wye gives you the option to connect an additional 1¾-inch hand line.

bundle or preconnected hand line of at least 200 feet of 2½ inches with an attached nozzle. Many of these 2½-inch preconnected hand lines come off the rear of the apparatus and are called Blitz lines. The 2½-inch hose line is recommended for fires that cannot be controlled by smaller hose lines **FIGURE 5-12**. Although the 2½-inch line is bulkier and more difficult to maneuver and operate, its water delivery capacity of up to 300 gpm is absolutely necessary for attacking large, intense fires. Probably because of a lack of proper staffing, there is a strong tendency to use the smaller hose lines for the initial attack. Some fire departments assume that the 1¾- or 2-inch hose line is the universal remedy for controlling all fires. Nothing is further from the truth. In fact, since the introduction of this smaller-diameter fire hose, some fire departments have eliminated 2½-inch fire hose for attack lines. This is a big mistake. If the fire is obviously too large for smaller attack lines, they should be left on the pumper.

For efficient operations, a 200-foot, 2½-inch preconnected hose line should be set up in a separate compartment on the pumper, and a combination spray nozzle or solid smooth-bore tip nozzle should be attached. Because of the reduced nozzle pressure and lower nozzle reaction, a line with a solid smooth-bore stream nozzle is easier to maneuver and operate inside the fire building. The nozzle should be a leader line type. This tip is equipped with a screw-down thread at the end. If needed, a smaller-diameter hose with 1½-inch couplings can be attached to the 2½-inch hose line using the screw-down feature on the tip. In this way, a smaller-diameter hose line can be advanced a greater distance than 200 feet from the pumper. The attack line could be used if this smaller diameter line was adequate for the current task. A preconnected 2½-inch hose line with a solid-bore nozzle and a smooth-bore tip leader line can be used with smaller-diameter hose attached to the screw-down tip. Many departments also have an adapter above the shutoff on spray nozzles that allows the tip to be removed and a smaller line extended. The advantage of either of these lines is that the 1½-inch connection is ahead of the shutoff; thus, the 2½-inch line can remain charged while the smaller hand line is extended. (Both types of nozzles are commonly available with the break-apart or leader line to allow a smaller line to be added or used for overhaul.)

Another example in which a 2½-inch hose line is beneficial is when there is a long stretch from the pumper to the fire. If 1½- or 1¾-inch hose lines are sufficient for fighting the fire, the 2½-inch hose line can be stretched from the pumper toward the fire. Two 1½- or 1¾-inch hose lines can be advanced from a double-gated reducing leader wye to combat the fire **FIGURE 5-13**. As with the 1½- and 1¾-inch hose line, the preconnected 2½-inch hose line can be placed into service with a minimum of delay. If discharge outlets are not available for a 2½-inch preconnected hose line, a separate hose compartment can be set up to hold 150 to 300 feet of 2½-inch hose. This hose line, called a skid load, can be quickly removed and hooked up to a standard outlet when needed.

Supply Hose

A supply hose is designed for the movement of water between a pressurized water source and a pump that is supplying attack lines. By definition, a supply hose is an LDH of 3½ inches or larger (although 3½-inch hose is rarely used anymore for a supply). LDH provides the movement of large amounts of water from a water source to the fireground with less friction loss and fewer firefighters to establish a constant uninterrupted water supply. Most LDHs are 4 or 5 inches, with the 4-inch being the most popular. In areas without a municipal water system, or when an existing main is shut off for maintenance or has been damaged by an earthquake or other disaster, LDH can become a temporary aboveground water main from the source to the fireground. LDH offers many advantages, especially with a limited number of personnel. The main advantage of LDH is not only an increase in water flow volume but also a corresponding reduction of friction loss compared with a common 2½-inch hose used for a supply. NFPA 1142, *Standard on Water Supplies for Suburban and Rural Fire Fighting*, states that where delivery rates exceed 500 gpm and water is moved long distances, LDH provides the most efficient means of minimizing friction loss and developing the full potential of both water supplies and pumping capacities.

Another major advantage of LDH is increased productivity because the same number of personnel and apparatus can use LDH to move more water over longer distances. Extended horizontally, a 4-inch LDH flows almost three and a half times the capacity of 2½-inch hose of equal length. Therefore, an LDH supply allows firefighters to get the operation under way faster and more reliably with a larger flow, less friction loss, and less physical stress on firefighters.

Nozzles

The purpose of a nozzle is to shape the stream and convert pressure energy to velocity energy. Many specialized nozzles exist; however, they can be classified in two ways: solid-stream smooth-bore and combination spray or fog nozzles **FIGURE 5-14**. NFPA 1901 requires the following nozzles to be carried on the pumper apparatus:

- One hand-line combination spray nozzle, 200 gpm (750 L/m) minimum
- Two hand-line combination spray nozzles, 95 gpm (360 L/m) minimum
- One playpipe, with shutoff, and 1-, 1⅛-, and 1¼-inch tips
- One smooth-bore or combination nozzle with 2½-inch shutoff that flows a minimum of 250 gpm

Solid-Stream Smooth-Bore Tip Nozzles

Solid-stream smooth-bore tip nozzles are classified according to the nozzle diameter. Nozzles with tips up to 1⅛ inch or perhaps 1¼ inch are generally considered for use on hand lines (i.e., those normally operated by hand and handled by one or two firefighters). A 1¼-inch tip is the cutoff point for hand lines and streams at 50 pounds per square inch (psi) versus 80 psi.

The 1⅛-inch nozzle used with 2½-inch hose produces the so-called standard fire department stream of 250 gpm at about 50 psi nozzle pressure **FIGURE 5-15**. Many fire departments use hand lines with flows exceeding 250 gpm. Department members must train and be thoroughly comfortable with 2½-inch hose lines flowing the maximum number of gallons per minute before using these high-flow nozzles on the fireground. Nozzles with tips larger than 1¼ inch have such strong reaction force that they must be mechanically restrained to a solid object with webbing, straps, or chains. Tips from 1½ to 2 inches are usually used on master stream appliances, such as monitors, deluge sets, or deck guns.

Solid streams are useful when a maximum range is desired and when penetrating capabilities are needed, for example, when the thermal degradation of

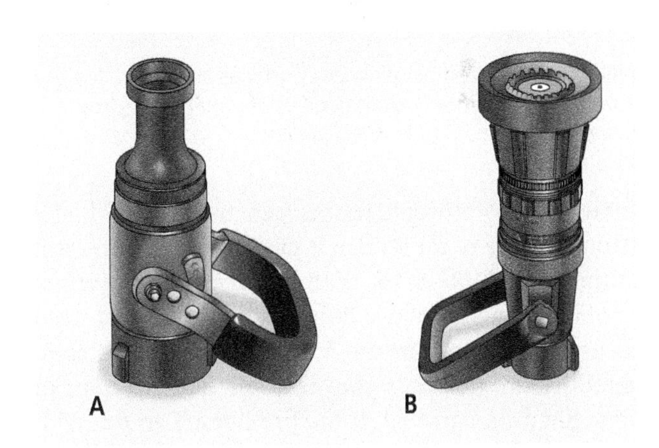

FIGURE 5-14 **A**. Solid-stream smooth-bore tip and **B**. combination spray are the two basic types of nozzles.

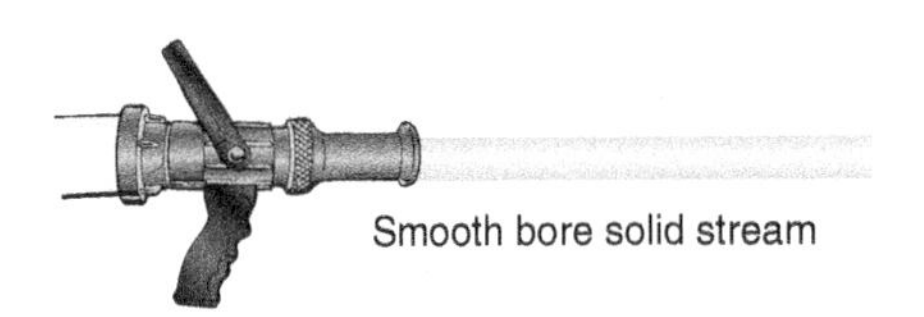

FIGURE 5-15 The 1⅛-inch tip on a smooth bore nozzle used with 2½-inch hose produces 250 gpm at about 50 psi nozzle pressure.

A Operating a solid-stream nozzle

B Operating a spray nozzle in a fog pattern

FIGURE 5-16 Solid-stream smooth-bore and combination spray nozzles produce different types of streams.

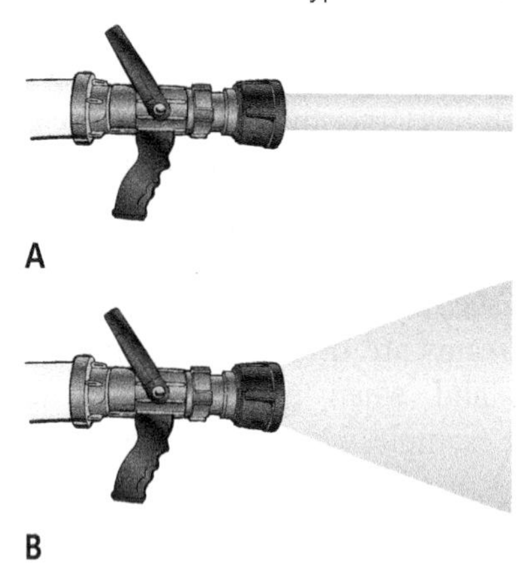

FIGURE 5-17 Combination spray nozzles are also called fog nozzles and produce varying degrees of water spray, from **A**. a straight stream to **B**. a fog stream.

spray streams prevents proper penetration or when a stronger force of the stream is needed to reach the seat of the fire **FIGURE 5-16**. With higher elevation (upper floors) and winds above 30 miles per hour, you need to use a smooth-bore tip for penetration. The solid smooth-bore nozzle is your last "big gun" weapon in your hand-line arsenal. If the fire doesn't go out with the 2½-inch smooth-bore, you have to switch to master streams. Remember, solid streams do not have as good heat-transfer characteristics as do spray nozzle streams; consequently, they are not as effective in absorbing heat, but they are effective in cooling the fuel below its ignition temperature by direct water application.

Combination Spray Nozzles

Combination spray nozzles, also called combination fog nozzles, or simply, combination nozzles produce varying degrees of water spray. **FIGURE 5-17**. They may have a fixed-spray angle, like the old Navy all-purpose nozzle, or they may be adjusted from a straight stream to a large wide-angle spray. Many spray nozzles have predetermined pattern settings, usually a straight stream with 30-, 45-, 60-, and 90-degree (or full fog) spray angles; however, most can also be set at intermediate angles.

The designated flow from a spray nozzle is usually rated at 100-psi nozzle pressure, although there are now low-pressure spray nozzles that operate at lower pressures, like 75 psi. Some spray nozzles have different flows at various angles. Other spray nozzles are constant gallonage; that is, they have the same flow at all spray angles.

For most interior fire attack operations, an effort should be made to knock down the fire first from the outside of the structure with a straight stream into the top of a door or window and deflected off the window header or ceiling for approximately 30 seconds, or as long as it takes to knock down the fire. This is referred to as *softening the target* or *hitting it hard from the yard.* The effects of a transitional attack are significant but temporary. The favorable conditions are short-lived unless additional water can be applied to the seat of the fire for complete extinguishment with an interior attack. Evaluate the effects with a thermal-imaging camera (TIC) before transitioning to an interior direct fire attack. Then, as the attack team enters the structure, the fog pattern should be no greater than 30 degrees **FIGURE 5-18**.

The Advantage of Combination Fog Nozzles. Water from spray nozzles set at a fog pattern effectively absorb heat more quickly than those set at a straight stream pattern—more surface area of the heated environment is covered with a fog pattern. The UL and National Institute of Standards and Technology (NIST) experiments have disproven the belief that water can push the fire. But the entrained air coming from the fog nozzle causes strong turbulent movement of the immediate atmosphere and can direct the movement of fire gases, smoke, and other products of combustion in

FIGURE 5-18 Prior to making entry, the nozzle firefighter directs the straight stream through the top of the door or window and deflects the stream off the ceiling. Evaluate the effects with a TIC before transitioning to an interior direct fire attack. This is called *softening the target* or *hitting it hard from the yard*.

the direction of the stream. It is the entrained air that gives the *appearance* of water pushing fire by creating pressurized air movement. When using spray nozzles for interior firefighting operations, extreme care must be taken not to drive fire and the products of combustion into uninvolved areas of the building. This is accomplished by having a vent portal opposite the fire attack. There are many scenarios in which a fog pattern is essential in holding or redirecting the smoke and flames during an interior attack. The best example is hydraulic ventilation after a room fire is extinguished **FIGURE 5-19(A)**. The nozzle firefighter, standing 2 to 4 feet from the window, narrows the fog pattern and aims it out so that it covers about two-thirds of the window opening. The smoke is drawn into a low-pressure atmosphere behind the nozzle; then the high-pressure atmosphere created by the fog stream produces a venturi effect that drafts in the smoke and ventilates it out the window with the fog spray. The entrained air forcefully ventilates the room quite effectively **FIGURE 5-19(B)** and **(C)**. This cannot be done as efficiently with a smooth-bore stream.

Think back to the liquified petroleum gas (LPG) prop at the recruit fire academy. This training evolution required the company officer to stand between two hose teams, with one on the left and one on the right. As the flames roared from the LPG tank, the radiant heat could be felt over 50 feet away. The evolution started with both streams cooling the tank from

A

B

C

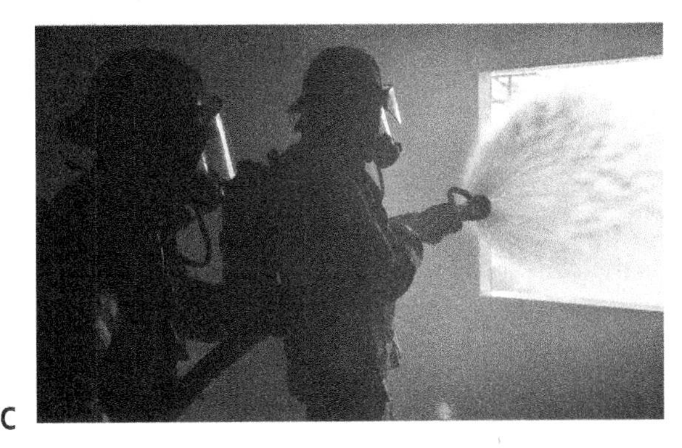

D

FIGURE 5-19 **A**. To ventilate a room hydraulically after a fire, the nozzle firefighter positions 2 to 4 feet from the window, narrows the fog pattern, and aims it out so that it covers about two-thirds of the window opening. **B**. The smoke is drawn into a low-pressure atmosphere behind the nozzle; then the high-pressure atmosphere created by the fog stream produces a venturi effect, drafts in the smoke, and ventilates it out the window with the fog spray. **C**. The entrained air forcefully ventilates the room. **E**. The entrained air generated by fog patterns is effective in redirecting and controlling even the pressurized flame reach of a fire.

a distance using a straight stream. As the two teams cautiously approached the tank in unison, the nozzle firefighters started to slowly widen the fog. The closer the team got to the burning tank, the wider the fog angle that was needed to entrain or "push" the flammable gases away from the officer and firefighters. The objective was to get right up to the tank so the officer, under this protective umbrella of water, could shut off the supply valve to the tank. As happened on more than one occasion, all that was needed for the flames to quickly flank around the protected stream and burn the team was for the nozzle firefighter to narrow the fog pattern. It was quickly resolved by widening the fog stream immediately to push back the flames. Although this training evolution is primarily a drill to develop confidence and teamwork, the drill also demonstrates clearly how fog patterns are extremely effective in redirecting and controlling even the pressurized flame reach of a fire **FIGURE 5-19(D)**. This can't happen with a straight smooth-bore stream.

These same fire behavior conditions make the combination fog nozzle the best tool for getting a line between the occupants and the fire, protecting a hallway from the advances of fire, and protecting the stairway from an encroaching fire. Then firefighters can effect the rescue of occupants through a stairway or work safely on the floor above the fire while a hose team protects their means of egress. For these same reasons, the combination fog nozzle is also the better nozzle choice for rapid intervention teams (RITs). If the RIT needs to protect a trapped firefighter from flames, the combination fog nozzle provides a larger umbrella of water with the full-rated gpm of the nozzle. Trying to protect this same firefighter with a straight stream can cause injury, and trying to get a protective spray from a smooth-bore tip requires the nozzle to be gated down, which produces a less than desirable fog with reduced gpm.

Master Stream Appliance

The **master stream appliance** is a portable or pre-plumbed water cannon mounted on the top of an engine apparatus. It is often called the deck gun, monitor, or deluge, or it is referred to by a model or brand name, like the *Stinger* or the *Stang*. Some portable units can be used either mounted on or detached from the pumper. Other portable monitors need to be removed from the apparatus and are operated on the ground. NFPA 1901, A.5.9.4 recommends a master stream appliance of 1,000 gpm minimum. Portable master streams are supplied by single, dual, or triple 2½-inch supply lines or by a single section of LDH. When the master stream appliance is not preconnected, one, two, or three sections of 2½- or 3-inch hose or an LDH can be connected to the appliance **FIGURE 5-20**. This allows the appliance to be connected to the pump's discharge gates and placed in operation rapidly.

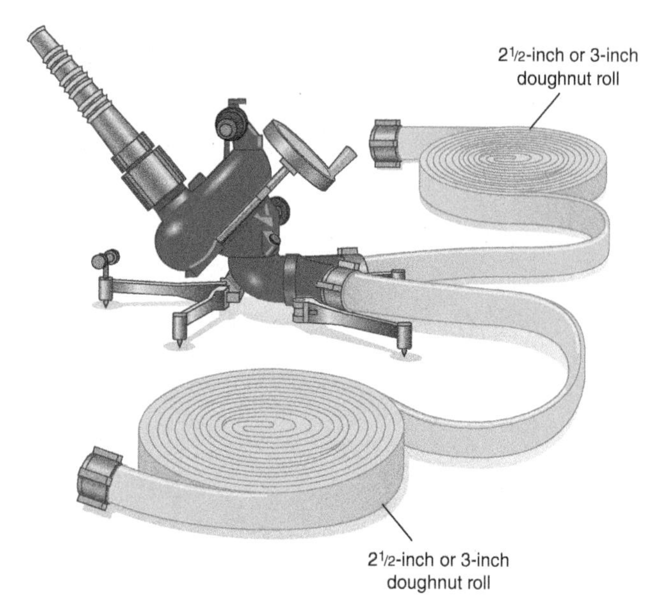

FIGURE 5-20 Two sections of 2½-inch or 3-inch hose can be connected to the master stream appliance for quick hookup to the pump.

Pre-Plumbed Master Stream Appliance

A **pre-plumbed master stream appliance** can be permanently mounted to the fire apparatus, or it can be removed and used as a portable monitor. The main disadvantage to having a permanently mounted master stream appliance is that the apparatus needs to be in the right position for the deck gun to be effective. Once supply lines are connected to the pumper, that rig isn't moving. The advantage to a pre-plumbed master stream is its capability for immediate application of a heavy stream as soon as the pumper is at the fire and the nozzle is positioned **FIGURE 5-21**. Initially, the master stream appliance may be supplied from the water tank and used as a Blitz attack to help knock down the main body of fire, thereby allowing an attack with hose lines. Because the master stream appliance soon empties the water tank in about 30 seconds, the deck gun needs to be aimed carefully before it is opened. Its continued operation requires a quick connection of the pump to the water system to maintain an uninterrupted supply of water.

A pre-plumbed master stream appliance has a separate discharge pipe of adequate diameter that runs from the fire pump to the appliance. Master stream appliances on engine companies swivel 360 degrees, raise up and down, and may have a telescoping feature.

FIGURE 5-21 The advantage to a pre-plumbed master stream is its capability for immediate application of a heavy stream on a fire.

Newer models can now be operated by remote control. Opening a gated discharge outlet allows water to flow into the appliance. Some apparatus allow for the appliance to be detached from its base and operate as a ground monitor supplied by one or more hose lines.

Soft-Suction Hookups

NFPA 1901 requires a minimum of 15 feet (4.6 m) of soft-suction hose, or 20 feet (6.1 m) of hard-suction hose. Twenty to 25 feet (7.7 m) of soft suction is better suited for the driver; unless drafting is a regular evolution, many fire departments don't carry the hard suction unless there is an earthquake where water mains may be out of service. Compartment space is at a premium, and room is needed to store the ground and roof ladders. The soft-suction hose allows a quick, efficient connection of the pump intake to a hydrant or other pressurized water source. Soft-suction hose is generally available in sizes from 2½ to 6 inches, with 4 and 5 inches being the most popular sizes. A hard-suction hose is primarily used for drafting water from a static water source such as a lake, pond, or reservoir. It is also used during tanker shuttle operations for drafting, or siphoning, water from a portable water tank. A hard-suction hose is generally available in sizes from 2½ to 6 inches. Pumpers carry a minimum of two 10-foot sections of hard sleeves (hard suction) **FIGURE 5-22**.

If one end of the soft-suction hose is preconnected to the pump intake, one firefighter can quickly stretch the unattached end, couple it to the hydrant, and charge the supply hose line. A side or rear intake may be used, but many apparatus are equipped with a front soft-suction intake. Side intakes are ideal for narrow streets when cars are parked on both sides. If the pumper can position, side intakes are also better suited for drafting from a static water source because the length of intake pipe from either side goes directly into the pump. Although hard suctions can work on pressurized hydrants, they are unforgiving and difficult for the driver to handle alone. Soft suctions should be used for hydrants and hard suctions for drafting.

Pump Intake Connections

NFPA 1901 requires that a pump have at least the number of intakes required to match an arrangement for the rated capacity of the pump, and the required intakes must be at least equal in size to the size of the suction lines for that arrangement.

Pump intake connections are used in conjunction with soft-suction hose for hydrant operations and with hard-suction hose for drafting evolutions. Front or rear intakes permit better positioning of the pumper close to the curb than do side intakes, but side intakes bring the water directly into the pump and are better suited in densely populated areas when cars are parked on both sides of the street **FIGURE 5-23**.

FIGURE 5-22 A hard-suction hose is generally available in sizes from 2½ to 6 inches.

FIGURE 5-23 Soft suctions on side intakes.

The front intake is particularly effective on cab forward apparatus when connecting directly to a hydrant **FIGURE 5-24**.

There are various apparatus designs and, for cab-forward apparatus, drafting from the front intake permits easier maneuvering and better positioning of the pumper. On cab-forward apparatus, a front intake can be set close to the water while the front wheels remain on firm ground **FIGURE 5-25**. If the pumper can position, drafting from side intakes are closer to the pump. Ultimately the practices are governed by apparatus design and department standard operating procedures (SOPs).

2½-Inch Pump Intake Connections

A pumper located at a fire usually receives water from hydrants or other pumpers. To operate efficiently, the pump must receive the required amount of water through appropriately sized supply lines. Additional intakes are provided for this operation.

Most fire departments now use LDH to supply pumpers on the fireground. If needed, a 2½-inch pump intake connection allows a pump to receive a supplemental supply line using a 2½-inch hose line or greater. In departments using 3½- or 4-inch hose to supply pumpers on the fireground, the 2½-inch pump intake connection is well suited for a second supply.

FIGURE 5-25 A front intake on cab-forward apparatus permits easy and efficient drafting operations. Side intakes draft the water right into the pump.

FIGURE 5-24 Front or rear intakes in combination with an attached soft-suction hose permit good positioning and quick hydrant hookup.

Photo by Martin Grube/FireRescue TV.

On existing units that have but one 2½-inch intake, a two-way Siamese provides an alternate method of providing a supplemental water supply should the need arise. Two 2½-inch hose lines can be combined into one by using a gated Siamese, or a Siamese with a clapper valve **FIGURE 5-26**.

Hydrant Assist Valve

The **hydrant assist valve (HAV)**, also known as a four-way valve, comes in two sizes. The larger fits on the large hydrant discharge opening or steamer port. The smaller HAV fits the 2½-inch hydrant outlets. An HAV is attached to the end of a pumper's supply line and is designed for a forward lay **FIGURE 5-27**. During an incident, the HAV is attached to the hydrant by the first-in engine company. The hydrant firefighter gets off the rig with all the hydrant tools to make the connection and wraps the hydrant with the supply hose. The pumper then lays forward and proceeds to the fireground. The driver sets a hose clamp or connects the uncharged LDH into the pumper, then signals the firefighter (with hand signals or by radio) to open the hydrant and charge the supply line. Once the line is charged, that firefighter returns to the engine and

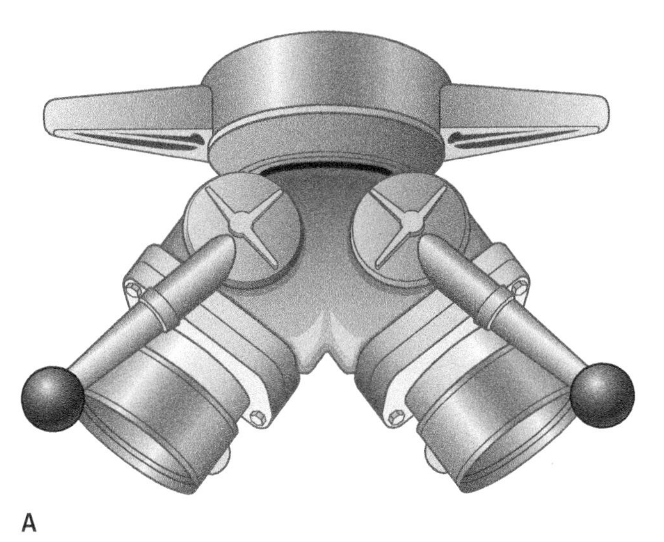

A

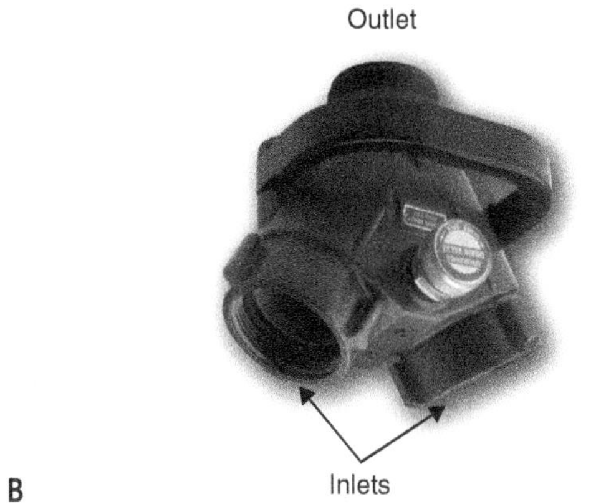

B

FIGURE 5-26 A. A 2½-inch gated Siamese. B. A 2½-inch Siamese with a clapper valve.

B: Courtesy of Akron Brass Company.

FIGURE 5-27 An HAV is attached to the end of an engine's LDH supply line and is designed for a forward lay.

rejoins the crew FIGURE 5-28. If the volume of water from the hydrant needs to be increased, a second pumper can hook up to the HAV without shutting down the hydrant, and increase the pressure and volume through the supply line to the pumper on the fireground. An HAV can be a useful piece of equipment if

FIGURE 5-28 Once the initial supply is established, the second pumper can hook into the HAV without shutting down the hydrant and increase the pressure and volume through the supply line to the forward pumper on the fireground.

© Jones & Bartlett Learning.

it is used properly. Remember that an HAV serves no purpose to a second pumper if the hydrant is unable to discharge a sufficient supply of water needed on the fireground. If the pumper on the fireground needs a supplemental water supply and the hydrant is unable to produce that supply, then the water should come from a different source with a sufficient volume.

Valves and Hydrant Gates

Valves are used to control the flow of water in a standpipe, hose lays, and pipes going into and out of the pumper. FIGURE 5-29 shows the three common types of valves found on engine apparatus: ball valve, gate valve, and butterfly valve. The gate and butterfly valves use control wheels to open and close the valve. They're primarily used on the intake ports of the pumper. A **ball valve** is an in-line valve that has an internal plastic or steel component shaped like a ball and is the valve used for the discharge ports on the pumper FIGURE 5-30. The ball has a hole through its center that allows water to flow through it when the valve is in the "open" position; no water is allowed to flow through it in the "closed" position. When the lever is in-line with the hose, the valve is open and flowing water. When the lever is perpendicular to the hose, the valve is shut, and no water is flowing.

Two ball valves placed on a hydrant allow an additional supply line to be laid without shutting down the hydrant FIGURE 5-31. A ball valve can be placed on the hydrant when the initial supply line is connected to the hydrant whether from the steamer connection or a 2½-inch discharge. If a second ball valve is to be used, both ball valves must be connected at the same time; otherwise, the hydrant must be shut down to complete this evolution. A hydrant gate serves the

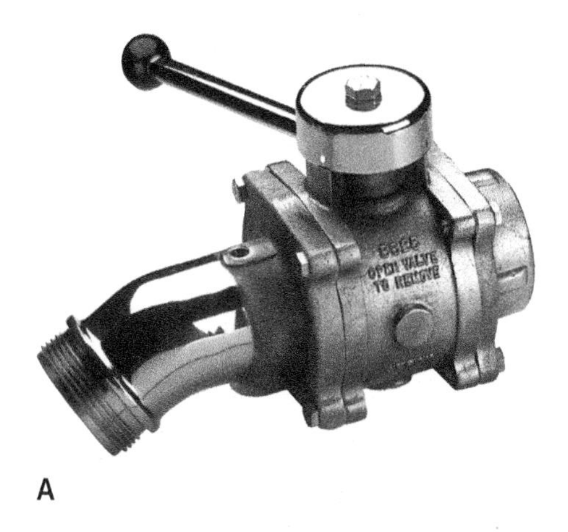

A

B

C

FIGURE 5-29 The three common types of valves found on engine apparatus: **A**. ball valve, **B**. gate valve, **C**. butterfly valve.

Courtesy of Akron Brass Company.

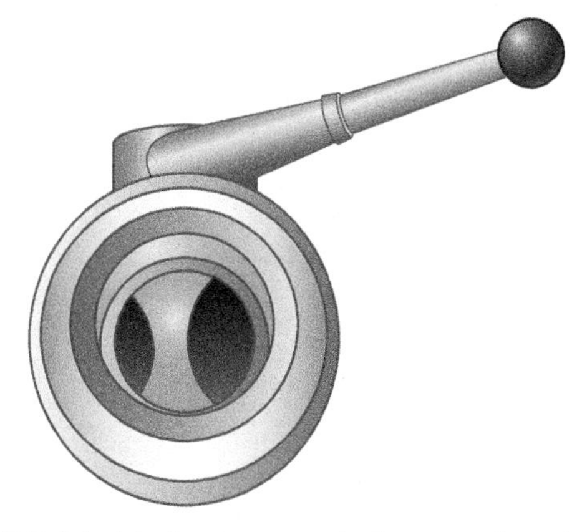

FIGURE 5-30 A ball valve controls the water flow through the hose.

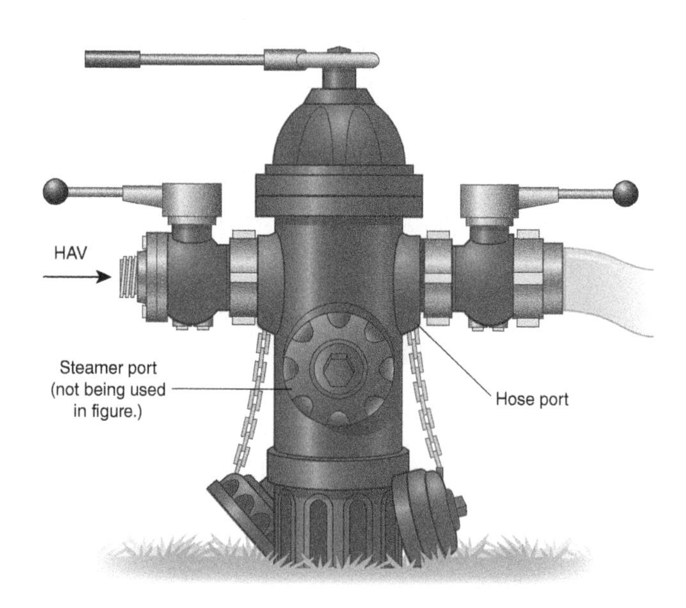

FIGURE 5-31 Ball valves permit connection of a hose line for a second suction into the pump without shutting down the hydrant.

same purpose as a ball valve except it is a gated valve. A lever rotates counterclockwise to open the gate and rotates clockwise to close it. Hydrant gates connect to the 2½-inch hose ports of a hydrant FIGURE 5-32. Water on the fireground is flowing at tremendous pressure and velocity. All valves must be opened and closed *slowly* and *smoothly* to prevent a water hammer, which can burst the hoses or damage the pump.

A gated wye can be used as a Siamese, but a Siamese can never be used as a wye. A Siamese is always two lines into one. A 1½-inch wye has a 2½-inch intake and splits that line into two hand lines of 1½-inch couplings FIGURE 5-33. A 2½-inch wye has one 2½-inch intake and two 2½-inch discharge ports to split the line. A water thief is a tri-gated wye with a 2½-inch intake, a 2½-inch discharge, and two additional 1½-inch discharge ports FIGURE 5-34. This appliance would split a 2½-inch line into one 2½-inch line and two 1¾-inch lines.

FIGURE 5-32 A hydrate gate can be connected to the hose port of a hydrant on a forward or reverse lay and allows for the driver to take a second suction into the pump without shutting down the hydrant.

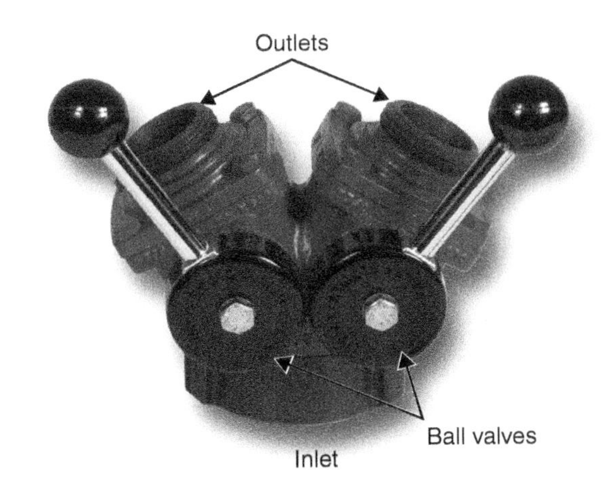

FIGURE 5-33 A 1½-inch gated wye.

Double-Male and Double-Female Couplings

Double-male couplings and double-female couplings are used to connect two threaded connections of the same size and same threads FIGURE 5-35. They are

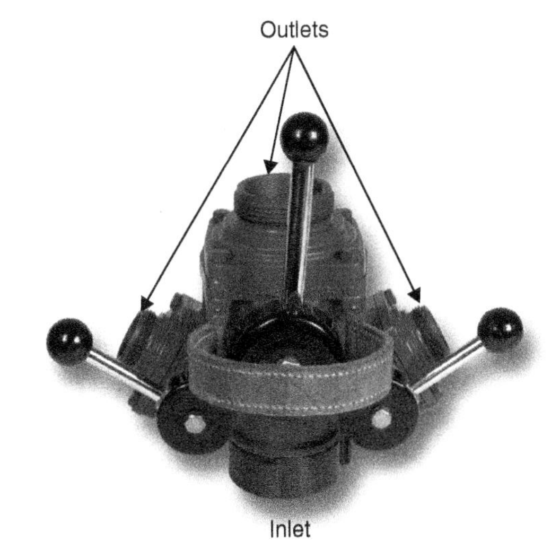

FIGURE 5-34 A water thief.

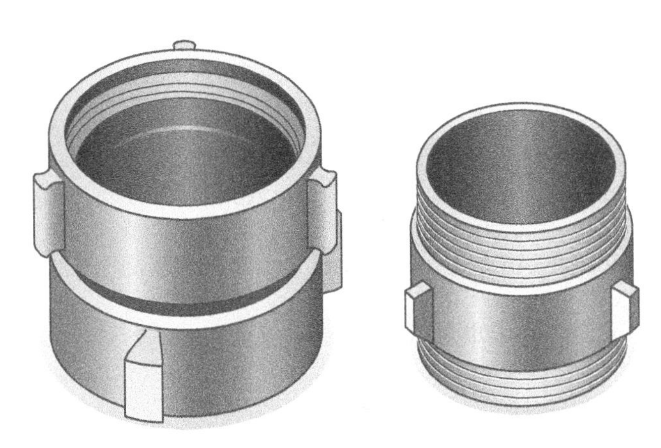

FIGURE 5-35 Double-male and double-female couplings.

needed when a pumper, set up for a forward lay, uses a reverse lay, and vice versa. To make the proper connections at the pumper or the nozzle, a double-male and double-female are needed. NFPA 1901 requires one double-female 2½-inch adapter with National Standard Threads, and one double-male 2½-inch adapter with National Standard Threads be carried on the apparatus. They used to be made of heavy brass, but today, they are made with strong lightweight material.

Ground Ladders

NFPA 1901, 5.8.1.2 requires, at a minimum, that the following ladders be carried on the apparatus:

- One straight roof ladder equipped with hooks (12 feet)
- One extension ladder (24 feet, 25 feet, or 26 feet)
- One folding attic ladder

In addition, a 12-foot, two-section baby ladder is recommended (NFPA 1901, 5.8.1.3).

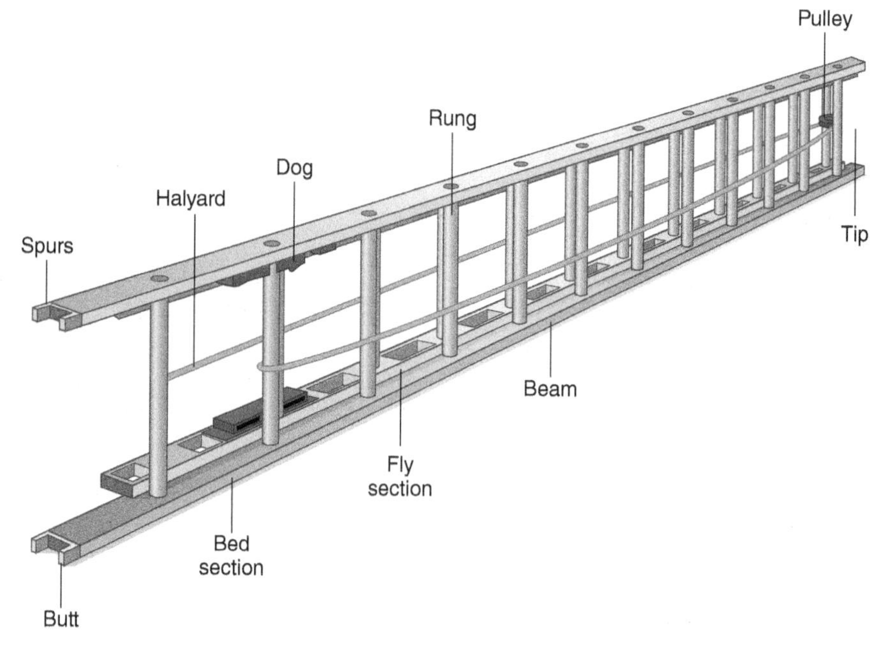

FIGURE 5-36 A 24-feet, 25-feet, or 26-feet ground extension ladder must be carried on the apparatus to reach the second floor of a house.

The majority of single-family residences throughout the country are one- or two-story structures. The required 24-, 25-, or 26-foot ground extension ladder reaches at least the second story of a house. Where there are no ladder trucks in service, pumpers should be equipped with a 35-foot ground extension ladder **FIGURE 5-36**.

An engine company can use a ground or roof ladder to gain entry to upper stories for firefighting or rescue operations. Ladders carried on engine companies can also be used to operate hose lines in a fire building. There are many special circumstances in which ladders can be used, including civilian rescues, rapid intervention rescues, ice rescues, bridging, shoring, hoisting operations, securing 2½-inch hand lines, constructing a makeshift monitor, damming operations, and water catch-all operations. Ladders may be carried on the top or the side of the pumper and may use electric or hydraulic arms to lower the ladders to a safer ergonomic height and increase storage capacity.

Bolt Cutters

Something as simple as a padlock can impede access to a fire with possible rescues. Although not complicated, a gate secured with a padlock can add a significant delay in stretching hose, throwing a ladder, or laying a supply. The best tool for a gate secured with a padlock: bolt cutters. Cutting a lock with bolt cutters is fast, unless the set is too small. Make sure the engine is carrying the largest set available **FIGURE 5-37**. The hockey puck round locks are difficult, if not impossible, to cut with bolt cutters and require a handheld, battery-powered wheel grinder.

FIGURE 5-37 Cutting a padlock with bolt cutters is the fastest way to gain entry. Carry the largest set available.

Modern Advancements in Nozzle Technology and Equipment

Success or failure in firefighting ultimately comes down to the nozzle. That is the last tool in a series of actions that delivers water to the fire by shaping it into an effective stream. Without the nozzle, all other tools for fighting a structure fire are useless. From the days of the horse-drawn steamer until now, the core design of the

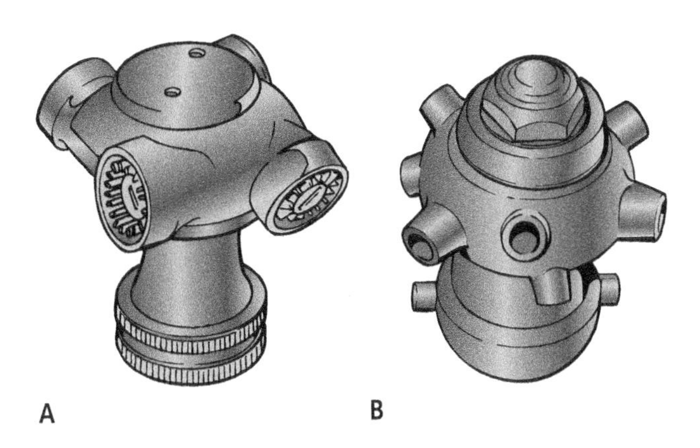

A B

C

FIGURE 5-38 Distributor nozzles: **A**. Cellar nozzle, **B**. Bresnan distributor nozzle, **C**. Various smooth bore and combination nozzles

A: © Jones & Bartlett Learning; B: Courtesy of Elkhart Brass Mfg. Co. Inc. C: Courtesy of Craig Maciuba.

nozzle remains the same. The long smooth-bore playpipe evolved first with brass handles, then with a gate valve shutoff. Over the years, the evolution of nozzle design has come down to two basic types: the smooth-bore nozzle, and the combination fog/straight-stream nozzle. The rest, like piercing and distributor nozzles, are considered specialty nozzles **FIGURE 5-38**. Two major design features that revolutionized the nozzle were the quick ball-valve shutoff and bail, and the pistol grip, both of which enhanced the firefighter's ability to control the nozzle and the nozzle reaction forces. All the other advancements in nozzle technology still fall within the two categories of smooth-bore and combination nozzles.

Although all the nozzle advancements over the years have contributed to more efficient water delivery, nine new nozzle designs are worth reviewing because they specifically deal with the problem fires we regularly face in the fire service. We cannot look at these new nozzles as gimmicks or novelty items. If we're to prepare and train the next generation of firefighters to fight fires more effectively, more efficiently, and more safely, we need to adopt the new technology advancements for nozzles, tools, and equipment instead of teaching the same tactics with the same tools and expecting different results.

TFT Flip-Tip

The Task Force Tips® (TFT) Flip-Tip combines two nozzles into one. The combination nozzle can be adjusted to fixed, selectable, or automatic gallonage and can be adjusted from straight stream to a wide fog pattern. The Flip-Tip also works with firefighting foams **FIGURE 5-39(A)**. If fire conditions require a harder-hitting, deeper-penetrating, straight stream from a smooth-bore tip, the combination tip is flipped down, and the pivot-lock secures it in the down position **FIGURE 5-39(B)**. There are 1¾-inch and 2½-inch nozzles with smooth-bore tips ranging from ⅞, ⅝, 1⅛, 1¼, and 1⅛ inches. This allows for flows of 50 to 200 gpm. If used exclusively as a smooth-bore nozzle, two different-sized smooth-bore tips can be attached. **FIGURE 5-40**.

TFT Vortex

The TFT Vortex modifies and enhances the use of a traditional smooth-bore nozzle. It is attached behind the shut-off and can be used with water or firefighting

A

B

FIGURE 5-39 The Flip-Tip in **A**. combination mode and **B**. smooth-bore tip mode.

Photos from Task Force Tips.

A

B

FIGURE 5-40 When used as a smooth-bore nozzle, two different-sized smooth-bore tips can be selected with a flip. **A**. The ⅝ tip. **B**. The 1⅛ tip.
Photos from Task Force Tips.

FIGURE 5-41 The TFT Vortex appliance.
Photos from Task Force Tips.

FIGURE 5-42 The Blitzfire portable monitor.
Photos from Task Force Tips.

foam. Six fluted vanes in the bore of the appliance reduce the turbulence inside the straight stream. A twist of the stream shaper from "stream" to "Vortex" causes the vanes to pivot proportionally and the nozzle converts from a straight stream to a uniformly dispersed 30- to 40-degree spray pattern without gating down the valve and reducing the gpm fire flow **FIGURE 5-41**. This turns the smooth-bore into a nozzle that is now efficient for hydraulic ventilation, overhaul, car fires, and protecting exposures without causing straight-stream damage. The inability to achieve effective fog patterns when needed has been part of the debate against smooth-bore tips.

The Blitzfire Portable Monitor

The Blitzfire is a 2½-inch, single-operator portable master stream, ideal for initial interior attacks **FIGURE 5-42**. The nozzle can flow up to 500 gpm (2,000 L/min). If there is unintentional tipping or unstable movement during fireground operations, or if there is a sudden increase in pressure, the monitor automatically shuts down to prevent uncontrolled whipping. The monitor is compact and lightweight. The base model is designed for low-angle interior attack. Ten- to 46-degree elevation is ideal for directing a fire stream into any door or window opening during an initial attack. The high-elevation (HE) version provides an even higher 86-degree elevation angle for tactical advantages, such as the ability to place a high-volume stream directly into an overhead. Both monitors can be equipped with a water-driven turbine that drives the oscillating unit in a selectable 20-, 30-, or 40-degree sweeping motion. This nozzle can be used instead of 2½-inch, smooth-bore or other 2½-inch nozzles, which require at least three firefighters because they are difficult to hold onto and operate for an extended period, usually because the nozzles are over-pressurized. There is little, if any, nozzle reaction for the firefighters to brace against, making this portable monitor an excellent alternative nozzle for offensive 2½-inch attacks when there is limited staffing. The appliance can be deployed by a single firefighter and once set, can be left unattended, so it can support and protect hand line crews in fighting commercial and high-rise fires.

The Chimney Snuffer

During the winter months, fire departments respond to chimney fires from fireplaces or woodstove units. Although minor in scope, they are sometimes stubborn fires to extinguish. The Chimney Snuffer nozzle is effective at suppressing chimney fires by using an ultra-fine spray, which reduces potential water damage. The weighted 5-pound (2.3 kg) torpedo-shaped nozzle head (which looks like a mini mortar round) has eight misting ports that discharge water at 40 gallons (150 liters) per hour at 60 psi (4 bar), creating a mini-sprinkler system. It attaches to a 1-inch (25 mm) line or a heavy-duty garden hose and can easily break through soot blockages, creosote buildup, or nests within the chimney without cracking the flue integrity with large amounts of water **FIGURE 5-43**. Cracking the integrity of the chimney flue is what leads to attic fires.

The HydroVent™

HydroVent™ is a hydraulic dual-attack, ventilation and suppression system nozzle. Hydraulic ventilation and suppression in the fire room takes place simultaneously with a single nozzle operating two separate tips simply by opening the bail. No other nozzle does this. One firefighter can place this nozzle into operation. It connects to any 1¾-inch hand line by untwisting the tip of the nozzle and connecting the HydroVent™ tube to the 1½-inch male threads in front of the shutoff. The HydroVent™ wand is just shy of 8 feet in length. It has a pointed wedge spike at the tip, so the tool can be used to break out the window **FIGURE 5-44(A)**. It can easily be placed in a basement, first-floor, or second-floor window. It can even be placed in upper stories by working off a ground ladder, aerial ladder, or elevated platform. The nozzle can be used in

FIGURE 5-43 The TFT Chimney Snuffer.

Photos from Task Force Tips.

A

B

C

FIGURE 5-44 **A**. In these live fire training demonstrations, the firefighter approaches the window with the HydroVent™ wand while remaining safely outside the IDLH zone. **B**. The suppression nozzle is a round nozzle with four smooth-bore discharge holes that flow approximately 95 gpm toward the top of the heated compartment. **C**. The combination fog nozzle is positioned at the windowsill aimed toward the outside for hydraulic ventilation.

Photos by Kevin O'Donnell.

vertical ventilation roof cuts or for attacking a fire at the gabled end of an attic opening. It can also be used effectively for residential high-rise fires by working safely from the floor below. The nozzle is meant to be placed into the fire room from the outside by a single firefighter (outside the IDLH). There is a sawtooth grip plate called a rake that holds the nozzle firmly to a windowsill using the weight of the tool. The wand is designed to redirect the nozzle reaction back into the building instead of away from it, so the nozzle reaction force actually secures the nozzle to the exterior of the building. Once set, the nozzle is left unattended **FIGURE 5-44(B)**.

The suppression nozzle is a round nozzle with four smooth-bore discharge holes that flow approximately 95 gpm toward the top of the heated compartment, where ceiling temperatures can easily reach 1,000°F (538°C) or higher. This creates a sprinkler system where the water converts to steam 1,700 times its volume to significantly knock down the fire and reduce the thermal energy output. This slows the fire growth and fire extension. The ventilation nozzle is a small 95-gpm combination nozzle with variable fog patterns, including straight stream. This nozzle can be positioned at the windowsill and aimed toward the outside for hydraulic ventilation. It is capable of pulling thousands of cubic feet per minute (cfm) of smoke, steam, and other gases of combustion out of the room and away from fire victims **FIGURE 5-44(C)**. Cooling the fire with water while simultaneously removing the heat through hydraulic ventilation drastically reduces interior temperatures and improves smoke conditions and visibility with immediate results. There is no faster method.

This indirect transitional attack immediately starts as soon as the firefighter simply opens the bail on the nozzle. This is done from outside the **immediately dangerous to life or health (IDLH)** area, so it does not require adherence to the two in/two out rule, nor does it require a firefighter to be covered on air with SCBA any more than an officer performing a 360-degree size-up. While working safely from outside the IDLH, it can be set-up and placed into operation by the driver while the officer is performing the size-up and the other firefighters are performing forcible entry and laying lines for the interior fire attack. When firefighters make entry from the unburned side toward the burned, the fresh air is at their back. The entry is much safer because the temperatures are drastically cooler, reducing the chance of flashover. Contaminated air is replaced with fresh air, visibility is improved, and the nozzle directs the flow path out the window from the fire room. These are huge safety advantages for firefighters and possible victims. A primary search can take place with great speed due to favorable visibility.

This nozzle could change the approach in attacking interior fires for the following reasons:

- The single-operator design allows a tremendous amount of work to be accomplished by a single firefighter and drastically improves the current fire conditions. This is a significant advantage for volunteers or any departments who have extremely limited staffing in the first few minutes of the incident.
- Because the nozzle is designed to be placed in service from the exterior of the structure — safely outside the IDLH area—the two in/two out rule isn't being violated. Many fire departments have wrestled with its compliance when staffing is limited.
- This is the only nozzle that addresses all the recommendations from the recent UL/NIST experiments. It allows for fast water application from the outside to knock down and reset the fire and for coordinated ventilation, and it controls the flow path.
- The best-trained engine and ladder companies cannot apply water and initiate horizontal ventilation using the traditional approach from the unburned side for interior attack faster than this nozzle can.

Specialty High-Rise Nozzles

The most difficult, time-consuming, labor-intensive fires we respond to are high-rise fires. A high-rise building is one that is over 75 feet (23 m) from the lowest level of fire department vehicle access. It can also be a building taller than your fire department's tallest ladder. High-rise buildings are no longer exclusive to major cities. High-rise hotels can be found in any community around the country. According to the NFPA's 2016 *High-Rise Building Fires* report, between 2009 and 2013, there was an average of 14,500 high-rise fire responses per year. During those years, the average number of civilian fatalities in high-rise fires was 40 per year. Sixty-two percent of high-rise fires occur in residential high-rise buildings, 4% occur in hotels, and 2% occur in high-rise office buildings. Note that more high-rise office buildings are sprinklered compared to residential high-rise apartments. In fact, of all the high-rise categories, residential apartments have the lowest number of sprinkler systems.

Of the numerous high-rise case studies, four are worth mentioning: the First Interstate Bank Fire on May 4, 1988, in Los Angeles, California; One Meridian Plaza Fire on February 23, 1991, in Philadelphia, Pennsylvania; the Chestnut Fire on December 10,

2009, in Chicago, Illinois; and the Marco Polo Apartment Fire on July 14, 2017, in Honolulu, Hawaii. In these fires, numerous floors were lost to fire because the fire departments were not able to make entry on the fire floor for an offensive attack. Once the fire gains control of an entire floor, temperatures can be well above 2,000°F (1,093°C), making entry impossible for firefighters. Aerial ladder pipes are only effective up to the twelfth floor. They may have some effect up to the sixteenth floor, but the twelfth floor is usually the limit for aerial master streams. This was confirmed at the unprecedented 24-story Grenfell Tower high-rise fire in Kensington, West London, United Kingdom. The fire started on Floor 4 and, because of the polyethylene-core cladding, quickly engulfed the entire building in less than 20 minutes. The aerial master streams of the London Fire Brigade could not reach past the twelfth floor. Once the fire controls the floor, there is little else the fire department can do except to protect the fire doors and wait for the combustible fuel to be consumed, which wasn't even an option in the Grenfell Tower example. Type I fire-resistive construction should be able to withstand the fire for at least 2 hours and even up to 4 hours or longer.

The HEROPipe is an 8- to 15-foot (2.4 to 4.6 mm) telescopic waterway high-rise nozzle that gives firefighters another opportunity for an exterior attack when the fire is well beyond the reach of aerial ladders and elevating platforms. Big fire requires big water. The HEROPipe simply takes the elevation out of the problem of delivering high gpm flow required for a high-rise fire. The HEROPipe is a substantial piece of firefighting equipment. It weighs about 80 pounds (36 kg) and is stored within a 7¾-foot wheel and track base unit. It's wheeled like a wheel barrow and can fit inside any high-rise elevator. This system is designed to be assembled and operated from the safety of the floor below the fire. The adjustable stabilizing arms are secured with a zip nut, and together with the base unit sill-clamps, secure the unit to any Type I fire-resistive construction, whether windows, balconies, or glass-curtain walls that extend from floor to ceiling **FIGURE 5-45(A)**.

Once assembled, the HEROPipe is supplied by a single 2½-inch hose coming off the standpipe. If additional gpm is required, two parallel 2½-inch lines can supply the monitor when a Siamese or a gated wye is attached to the waterway intake. The waterway is made of anodized aluminum, so it is lightweight yet durable. The main waterway is a 3½-inch internal diameter pipe, and the extension pipe has a 3-inch internal diameter. Water is charged to the control valve at the unit. The firefighters slide the wheeled waterway along the tracks, and the counterweight of the charged hose lines helps shoot the waterway up into position to the fire room window above **FIGURE 5-45(B)**.

The tip is equipped with an Elkhart Sidewinder EXM water cannon. This monitor tip is a giant combination nozzle that has varying degrees of fog patterns, including straight stream. The monitor is rated for 700 gpm at 80 psi. If a 1½-inch or 1⅞-inch smoothbore tip is used, the HEROPipe can flow up to 900 gpm—if you can supply it. Six hundred to 700 gpm are more realistic attainable flows for high-rise firefighting. The Sidewinder EXM is also a remote-controlled monitor. It has a vertical sweep range of 60°; 20° below 0 level, and up to 40°. It has a horizontal sweep range of 40°; 20° to the right, and 20° to the left. When the unit is set to automatic mode, it can be left unattended, and the monitor tip oscillates up and down, and right to left simultaneously, within the respective degree ranges mentioned above. A 2-hour lithium ion battery pack runs the oscillating motor; the spare battery can extend it to 4 hours, which is more than enough power for a high-rise operation. The tip assembly can also accommodate a TIC with wireless image transmission back to the operator. After the fire is knocked down, the TIC can detect remaining hot spots, and the monitor can be aimed directly to these locations. If the monitor is put into manual mode, the vertical angle can shoot straight up, 90°. The ability to shoot up 90° can be a significant control measure for the lapping of flames to upper stories.

Many major cities have high-rise companies or squad units that carry special high-rise equipment. A well-trained team can assemble and put the HEROPipe in service in less than 5 minutes. Even if it took 10 minutes to set up, consider the amount of time it takes to run a high-rise standpipe evolution with hose lines for fire attack on Floor 20. With SCBA, and battling heat and smoke conditions, the initial hose team would be lucky if its members made it to the fire unit within 15 minutes. Also consider the amount of time it would take to set up a heavy-duty high-rise fire curtain from the floor above or from the roof to prevent the effects of a wind-driven fire. Setting up the fire curtain takes a few minutes, works only on high-rise buildings with windows that match the dimension of the curtain, and requires a second team below the fire floor to finish anchoring the curtain in place. In the meantime, crews are standing by to wait for the curtain to stop the wind. The attack crew still has to fight through smoke and high heat to get to the fire. In the meantime, no water is being applied to the fire.

In February 2008, the FDNY/NIST tests on Governor's Island in New York deployed the K.O. fire curtain using a large mobile ventilation unit (MVU) aimed into the fire room to create 25 mph wind conditions.

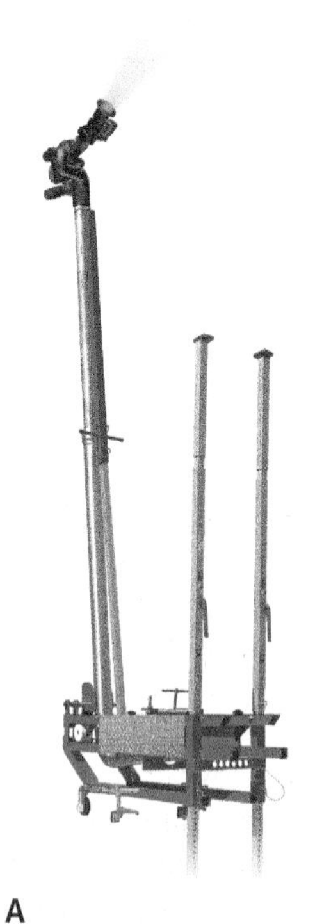

A

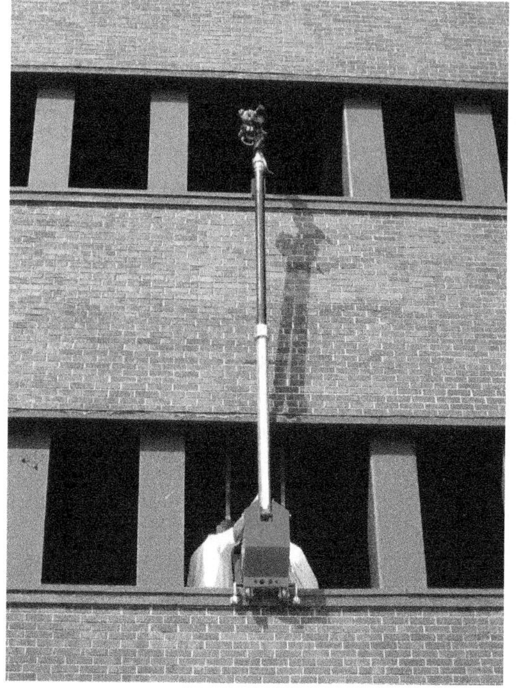

B

C

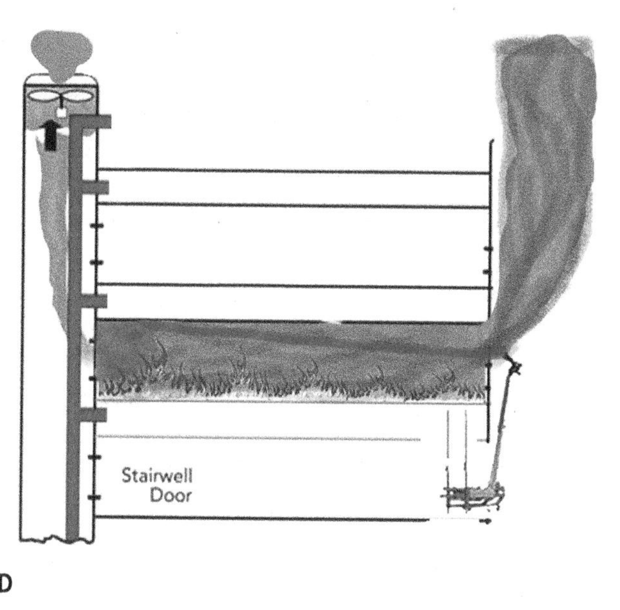

D

FIGURE 5-45 **A**. The HEROPipe. **B**. The HEROPipe is set up from the floor below the fire where temperatures may barely be above ambient, enabling firefighters to set up the unit without being covered and on air. **C**. When placed into the window inlet of the flow path, 600 gpm flow into the fire room to knock down the fire. **D**. The HEROPipe is able to shoot a 145-foot straight stream that reaches deep into the high-rise compartment.

Photos by Michael Wielgat; illustrations from National Institute of Standards and Technology.

The fire room front door was open to the hallway, and a window was open within the hall, completing a unidirectional flow path. The fire came through the hallway with blowtorch force, and hallway temperatures spiked to 1,650°F (899°C). Once the fire curtain was dropped to cover the fire room window and block the fan, the wind speed dropped to 2 mph, the room temperature dropped to 750°F (399°C), and the hallway temperature dropped to 700°F (371°). However, interior crews now have to make entry and battle 700° temperatures. Keep in mind, our fire ensemble personal protective equipment (PPE) is rated only to 500°F

(260°C), and there's still no water being applied to the fire. Add up all the time for these two scenarios with the risk involved to firefighters, and it should make you ask, "Is there a better way?"

Wind-driven fires are the perfect scenarios for the HEROPipe. When a unidirectional flow path is created in through the fire room and out through the hallway due to wind, those 1,600° temperatures make it impossible to enter the hallway. This was the scenario on December 18, 1998, when three Fire Department of the City of New York (FDNY) firefighters were killed at the Vandalia Avenue Fire. They got caught between where the fire was burning and where the fire wanted to go via the flow path. This incident revealed and brought attention to the effects of wind-driven fires.

The HEROPipe takes advantage of the flow path. If placed into the window inlet of the flow path, it flows well over 600 gpm into the fire room and quickly knocks down the fire **FIGURE 5-45(C)**. No other nozzle can deliver this gallonage from the floor below the fire. The wind currents push all the steam and moisture through the flow path and drastically cool and change conditions for the better. Fire curtain tactics have to try and stop the flow path before the traditional interior attack can be resumed. In the meantime, no water is being placed on the fire.

High-rise nozzles are also excellent tools. These 8-foot bent-pipe nozzles are connected to a single 2½-inch hose and are also operated from the floor below the fire. The 68-degree turnback angle allows the nozzle to be aimed only at the ceiling just inside the windowsill. The nozzle relies on the deflection of the stream against the ceiling for it to be effective. It flows between 200 and 250 gpm. That might be sufficient for a residential high-rise fire, but not for an office or commercial high-rise. In comparison, the HEROPipe monitor ends up directly in front of the fire room window, so it is able to shoot a straight stream that reaches and penetrates deep into the compartment, at least 145 feet (44.2 m) **FIGURE 5-45(D)**. This is the effect we are trying to accomplish on any exterior attack regardless of elevation.

The safety advantage of the HEROPipe cannot be overstated. Any operations on the fire floor require firefighters to be on air (that is, using SCBA) and deal with heat and smoke conditions. Setting up the fire curtain from the floor above (since they have to open the window above the fire) also exposes the team to smoke and heat because the gases want to rise. Flames lapping up from auto extension often make direct flame contact with the window above. The curtain team will have to go higher, and maybe all the way to the roof. The HEROPipe is set up from the floor below the fire, where temperatures may barely be above ambient, and the atmosphere should be clear of smoke, enabling firefighters to set up the unit without being covered and on air. There should be no fire conditions hindering their ability to set up this nozzle and start flowing water.

If the fire service is serious about making changes in combating fires involving modern synthetic fuels, there should be no half-measures. They must look seriously at these new nozzles that solve the problem. The alternative is business as usual, which on average, according to the NFPA, still results in 40 civilian fatalities and 520 civilian injuries per year in high-rise fires.

The HEROPipe is a major investment for a fire department. If you look at any major city, the number of high-rise buildings under construction is increasing. Suburban fire departments must also be ready for this challenge. Fire chiefs should be looking at grants, or propose that insurance companies and building owners partner with them and invest in this portable monitor that will help protect their property. The financial burden should not rest solely with the fire department. The HEROPipe and other high-rise nozzles could be an extension of the required fire protection systems that go into high-rise designs.

Flamefighter and Transformer Piercing Nozzles

Piercing nozzles and round nozzles have been around for a long time and are one of the most effective yet underutilized tools we carry. The fire service needs to start making better use of them. The Flamefighter original battering ram nozzle is a workhorse that delivers 300 gpm of water at 150 psi, which is a far superior flow than other piercing nozzles of years past. A 6-foot lightweight piercing nozzle of steel and aircraft aluminum, with the pistol grip and shutoff attached to the nozzle for better control is now available **FIGURE 5-46**. The lightweight piercing nozzle still delivers 300 gpm at 150 psi. These nozzles are very effective for extinguishing attic and basement fires. In a residential high-rise fire, when wind-driven conditions make entry into the fire unit dangerous, the fire can be attacked or flanked through the wall of the adjoining unit. From a safe position, these piercing nozzles can go right through the wall separation and introduce a 300 gpm sprinkler system into the fire unit. These piercing nozzles can be used with any type of foam concentration delivered through the fire hose.

The TFT Transformer Piercing nozzle is a different kind of piercing nozzle. With the two extension tubes and transition block that serves as the striking head, the piercing point-flow tube can be assembled

FIGURE 5-46 The Flamefighter piercing nozzle.
Courtesy of Flamefighter Corporation.

FIGURE 5-47 The TFT Transformer Piercing nozzle.
Photos from Task Force Tips.

in a variety of configurations and lengths. The straight configuration can be short or long, depending on how many extension tubes are inserted. Each extension tube measures 19 inches (483 mm), and the piercing point-flow tube measures 14 inches (356 mm). The straight configuration can give you a piercing nozzle that is 56 inches (1.4 m) long. The hammer configuration, using the transition block, puts a 90-degree turn in the nozzle **FIGURE 5-47**. The piercing depth depends on how many extension tubes are inserted, making it an excellent nozzle for attic fires with deep rafters and for car fires. The third configuration is the distributor nozzle configuration, where the piercing point is switched out for a three-spinner round nozzle. All the tips flow a spherical spray pattern with a 10- to 15-foot reach, flowing 150 gpm at 100 psi (570 L/min. at 7 bar). An in-line twist-grip rotational valve controls the water flow, so it doesn't need a separate breakaway ball valve nozzle. The selection options make it a versatile piercing nozzle for sealed compartment fires. The only drawback is the nozzle has to be assembled from the carrying case.

Thermal-Imaging Camera (TIC)

Very few tools revolutionize firefighting tactics, but the TIC is one of them. The TIC has become an invaluable

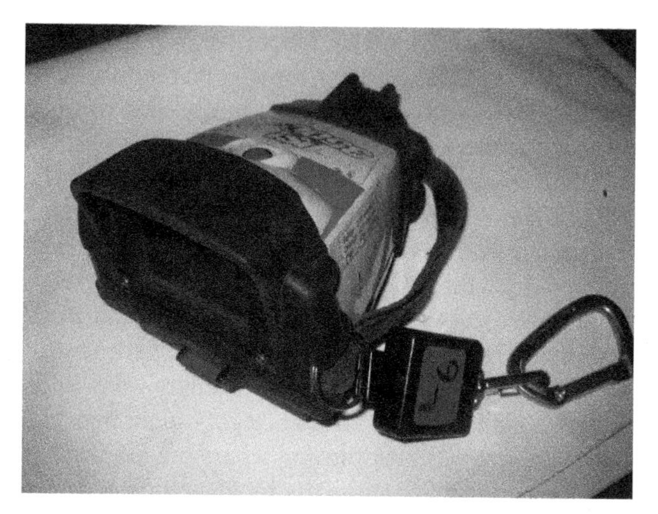

FIGURE 5-48 International Safety Instruments (ISI) 3500 TIC. In the colored temperature scale, yellow is 275°F (135°C), orange is 400°F (204°C), and red is 600°F (316°C).

tool for use on the fireground **FIGURE 5-48**. By capturing heat images, the device is able to show the relative temperature of different objects on a display screen. With its ability to see human figures in pitch black darkness and through smoke, the TIC has given firefighters the ability to perform swift primary searches by scanning the room in zero visibility. The TIC does not eliminate the need to perform a thorough physical search, but open areas that are obviously clear can be skipped, thus speeding up the search for a possible trapped occupant. It is also the tool most relied on in finding a downed firefighter during rapid intervention rescue operations.

Another advantage of the TIC is that it can locate fire and heat in concealed spaces. It can locate the source of fire behind walls, ceilings, and in other void spaces, and it can also indicate the direction of fire spread. For ladder companies, the TIC can pinpoint the location of an attic fire as well as register the direction of the roof rafters and joists in preparation for vertical ventilation cuts. TICs also provide temperature readouts of the surrounding area—a major safety feature in warning company officers that it may be time to back out their crews.

Hose Alert Hose Restraint Safety System

NFPA 1901, 15.10.5 states:

> Any hose storage area shall be equipped with a positive means to prevent unintentional deployment of the hose from the top, sides, front, and rear of the hose storage area while the apparatus is underway in normal operations **FIGURE 5-49(A)**.

A

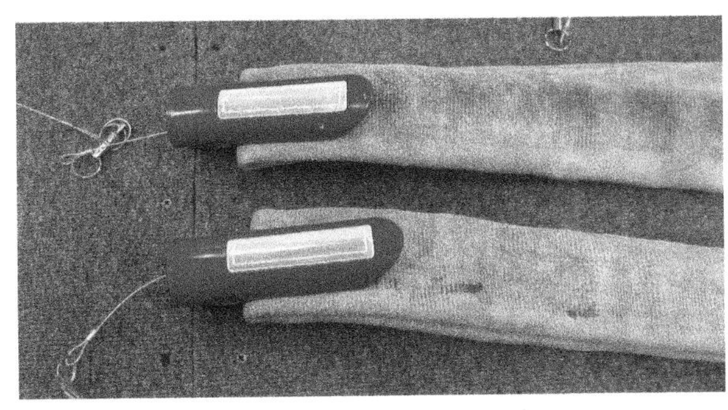

B

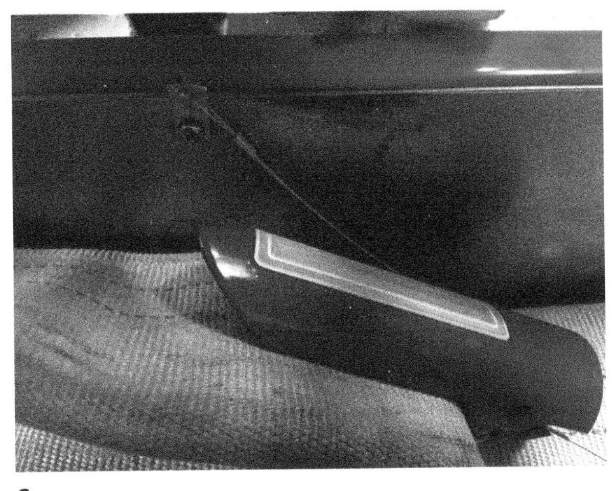

C

FIGURE 5-49 **A**. According to NFPA 1901: Any hose storage area shall be equipped with a positive means to prevent unintentional deployment of the hose from the top, sides, front, and rear of the apparatus. **B**. The Hose Alert clamp. **C**. The Hose Alert clamp secured to the top flake of the LDH bed.

NFPA 1901, A 15.10.5 also states:

> Many fire departments have experienced fire hose inadvertently coming off apparatus traveling to and from incidents. Several incidents have resulted in injuries, damage to property, and death. Fire departments and manufacturers have developed various methods of preventing inadvertent deployment, including fully enclosed hose bed covers, buckled straps, hook-and-loop straps (Velcro), fabric covers, webbing mesh, wind deflectors, and other material restraints or combination of restraints. It is also important that fire departments develop methods of storing hose and appliances in a manner that does not promote the inadvertent deployment of the hose and appliances.

The problem with the wording in both parts of the NFPA directive is that it still gives fire departments a lot of leeway to interpret what works best for their operations, and the proper function of the restraint system depends on firefighters actually securing them as designed. Procedures, actions, and safety devices are often ignored in cases where firefighters feel it delays or hinders their ability to respond or to deploy equipment quickly. In addition, the wording does not take into account that, in many of these accidental deployments, the driver and crew were not aware any hose had fallen off the rig. The sirens, air horns, and the roar of the diesel engines are overwhelming stimuli, and it's almost impossible to feel when equipment has fallen off the apparatus. The driver is busy watching the road, the officer is looking at the dispatch information on the mobile data computer, and the firefighters are securing their PPE and SCBA. Restraint systems may hold the hose loads in place, but if the hose does fall out, even partially, none of these systems alert the driver in the cab that a section of hose has fallen off. Therein lies the deadly chink in the system. The following cases illustrate this point:

- 2002, Tualatin Valley, Oregon, Fire and Rescue: A fire hose that accidentally fell off the apparatus caused a crash, killing a 41-year-old man.
- August 21, 2004, Coraopolis, Pennsylvania, Volunteer Fire Department: As an engine responded to an emergency, a 30-foot section of the 150-foot attack line, still attached, fell off the apparatus. A 10-year-old girl was killed after being struck in the head with the 6-pound nozzle, and her 10-year-old friend was severely disfigured after being struck by loose fire hose and nozzle trailing behind the responding apparatus.

- January 1, 2010, Cambridge, Massachusetts, Fire Department: An 82-year-old woman was struck and killed while waiting at a traffic island for the fire engine to pass, which was dragging a long tail of hose with an attached nozzle.
- October 18, 2014, Toledo, Ohio, Fire Department: A 58-year-old man was killed while riding his bicycle after being struck by 150 feet of fire hose with an attached nozzle.
- April 24, 2016, Walpole, Massachusetts: At a house fire, two police officers grabbed the LDH supply from an engine that was stopped to wrap a hydrant—without the driver's knowledge. The engine was directed to take a different hydrant. As the engine drove away, it tensioned the slack in the deployed supply hose, forcefully throwing both officers up in the air. One officer landed on his head. He was air-lifted to the hospital, where it was discovered he suffered a severe concussion, a back injury, and a separated shoulder. The other officer was transported via ambulance with injuries to both arms.
- May 23, 2017, Macon County, Georgia: Firefighter Darrell Plank, a 10-year veteran was struck in the head while masking up at a fire. Other firefighters pulled a crosslay hand line without the driver's knowledge. The driver pulled away to take a hydrant, yanking the hose line out of the hands of the attack team. The wildly swinging hose line trailed the moving apparatus and struck Firefighter Plank in the head. He suffered a skull fracture with traumatic brain injury. The following day, Firefighter Plank died from his injuries, leaving behind a wife and five children.

These incidents do not include the numerous accidental hose deployments that did not cause any injuries but did cause hundreds of thousands of dollars in property damage. Our mission statement is to save civilian lives and property from fire, not injure and kill civilians and destroy property on the way to an emergency to carry out our mission. Each jury award for the above incidents was in the millions.

Hose Alert is a fail-safe, simple concept to alert the driver when any hose is detached or deployed from the engine, either intentionally or unintentionally. The red clamp, which looks like a giant clothespin, is actually a spring-loaded hose-gripping unit that clamps on to the top flake of the supply hose bed or attack line hose slots. The Hose Alert clamp is tethered and anchored to the apparatus with a thin, nonobtrusive steel cable, which doesn't interfere or get caught when pulling off hose lines **FIGURE 5-49(B)**. When the hose lines are deployed or removed for any reason, the gripping unit is pulled away from the hose and the contact sensors connect, which sends an electronic signal to the dash unit control screen inside the cab. The alarm sounds, immediately notifying the driver that a hose load has accidentally been deployed. The driver can stop safely and in a timely manner before hundreds of feet of hose are laid out in the street or before a bystander is struck and killed by the wild swinging tail and nozzle.

The 3-inch × 6-inch display screen is about the size of a large smart phone. Up to 20 separate clamps can be programmed into the system, more than enough for any apparatus. When all the clamps are properly connected **FIGURE 5-49(C)**, a green screen displays "All Hose Secured," letting the driver know it's safe to proceed. When the hose line is intentionally pulled off by a firefighter, the clamp snaps shut, allowing the contacts to touch. The screen turns red and displays the specific hose slot that was deployed, for example, "Hose 3 Deployed." If the hose load is intentionally pulled, the driver can silence the display unit with the push of a button. If unintentional, the driver can quickly stop the apparatus. This system allows fire departments to also comply with the intent of NFPA 1901, 15.10.5.

Firefighting Foam

Because of its cooling effect and ready availability, water has always been the extinguishing agent of choice for fire. But no text would be complete without a review and understanding of the benefits of firefighting foam. The various types of foam are:

- Class A foam: Used on fires involving ordinary Class A fuels like paper; wood; cardboard; textiles; and organic vegetation such as grass, brush, trees, hay, and straw.
- Class B foam: Used on fires involving Class B flammable fuel like gasoline, diesel, and other hydrocarbon-petrochemical flammable liquids. There are various types of Class B foam:
  - Protein foam
  - Fluoroprotein foam
  - Film-forming Fluoroprotein foam (FFFP)
  - Aqueous film-forming foam (AFFF)
  - Alcohol-resistant foam
- Universal extinguishing foam: A modern formula that works on both Class A and B fires. It is especially suited for Class A fires, but it is still effective on Class B flammable liquids, although not as effective as foams specifically designed for Class B fires.

Low, medium, and high-expansion foams, which are sometimes mentioned in battling commercial basement fires, use various Class A and B foams, but their quality comes from the foam application through the use of special nozzles and equipment that introduce air into the stream. Low-expansion is the typical foam used on most fire apparatus to combat Class A fires and some Class B fires. Low-expansion foam has an expansion ratio of less than 20:1, and no specialized air-aspirated nozzles are required.

Medium-expansion foam has an expansion ratio between 20:1 and 200:1. It requires specialized air-aspirated nozzles to produce this characteristic of foam. It is typically used in fixed-fire protection systems designed to protect three-dimensional structures, although some fire departments use it in manual firefighting.

Hi-expansion foam has an expansion ratio between 200:1 and 1,000:1. It contains more air within the foam. It produces large bubbles and is designed to fill an entire space. It requires special equipment like a generator, a blower, and a delivery sock to dump the foam into a space. It is excellent for extinguishing stubborn or hard-to-get-to basement fires. There must be enough foam to fill the space, and the room or the space must remain sealed. The foam blanket can be several feet high, floor to ceiling. The foam displaces the oxygen and absorbs the heat so combustion can no longer occur. Hi-expansion is found in commercial and industrial fire protection systems and is also used aboard ships in marine firefighting operations.

History, Science, and Development of Foam

Ron Rochna and Paul Schiobohm, two of the most experienced experts on wildland firefighting and Class A foam, said in 1987, "Foam will replace all current water applications and present new suppression opportunities to the fire management communities." Apparently, they were right; in Seattle, Washington, all 33 first-line engines have built-in 10-gallon tanks for Universal Extinguishing foam (Novacool), in addition to six 5-gallon containers of Class A foam carried on each apparatus.

To understand the evolution and development of foam's use in the fire service, it helps to review the history of Class A foam, which started in the mid-1970s within the Texas Forest Service. Soap-skim is a dark brown, sticky material resembling axle grease. It is a by-product of paper manufacturing. This thick residue was skimmed from kraft paper liquor vats that came from paper mills. The mills were looking for ways to dispose of this concentrated residue, and the Texas Forest Service was experimenting in techniques and products to improve rural and wildland firefighting. It was discovered that, by adding water to the soap-skim in an 8% to 9% concentration, similar to the mixtures for Class B foams, it produced a wetting agent that penetrated and soaked through brush, wood, and charred surfaces. If compressed air was injected into the hose lines, the wetting agent would foam up and discharge from the nozzle as snow-like foam.

Even though soap-skim from the kraft paper process was biodegradable, soap-skim from white-paper operations contained chemicals that made the mixture toxic to the environment. Therefore, it was not an acceptable extinguishing agent for wildland fires. These early tests with soap solutions, detergent additives, and equipment led to the development of the current generation of Class A foams.

Although initial test results were encouraging, there were some logistical drawbacks to delivering product efficiently for a sustained fire attack. One of the drawbacks with compressed air foam systems (CAFSs) is that large-capacity air compressors were needed to support structural firefighting. Fire apparatus without these air compressors had to return to quarters to recharge the system, similar to recharging a pressurized fire extinguisher. Unfortunately, fire departments still had a hard time justifying the expense for this equipment when water was readily available to extinguish fires in the way it's been done for years.

Foam manufacturers continued to experiment with mixtures throughout the 1980s, substituting detergents for soap and thus increasing water's ability to penetrate through, and be absorbed by, Class A fuels. Foam solutions were also developed that did not depend on compressed air to make the foam. Air-aspirated venturi nozzles and eductors can also agitate foam concentrations to make thick, effective blankets of foam that can be incorporated into existing structural firefighting water application. The new solutions were also biodegradable, nontoxic, and safe for the environment.

Difference between Class A and Class B Foams

The difference between Class A and Class B foams is in the way these two foams extinguish fires. In Class B mixtures, water is needed to mix the proper foam solution ratio, and water is also the vehicle to deliver the product into the fire. For example, with AFFF, as the foam solution is applied to a flammable liquid like gasoline, a film is formed over the surface of the burning gasoline to stop and prevent further release of the flammable vapors. Thus, the fire goes out as long as the film barrier remains intact. With protein foam and high-expansion foam, the foam blanket is the actual barrier sealing the vapors. Water does have some cooling effect on the flammable fuel.

With Class A foam, water is the dominant extinguishing agent, not the foam. The foam enhances the effectiveness of water by creating bubbles, thus increasing the heat-absorbing surface area of the water droplets. The water is fortified with detergent wetting agents called surfactants, which reduce the surface tension of water, thus allowing it to penetrate fuels and inhibit combustion. Without foam, surface tension, a natural characteristic of water, holds a fire stream together in relatively large droplets, limiting the water's heat absorption capabilities. A small percentage of the droplets, (the outer 10%) actually removes heat, while most of the droplets (the inner 90%) run off the fuel source in water runoff.

When a Class A concentrate is added to water, the surfactant molecules spread apart the water molecules. As air is introduced into the process, foaming agents in the mixture allow the combination of the water molecules and surfactant molecules to form bubbles, increasing the surface area of the water available to absorb the heat. The foam covering contains water to cool the fire; a surfactant to allow the water to penetrate burning and unburned material; and air, which makes the mass lighter in weight and provides an effective heat-insulating cover.

Research by NIST has determined that droplets measuring between 250 and 350 microns were more efficient in absorbing heat than were larger, heavier droplets. An example would be like cooling a glass of water with a single ice cube rather than crushed ice of the same volume. The crushed ice cools the water faster.

Tests conducted by the Department of the Interior, the US Forest Service, and NIST have proven that the use of Class A foam can increase the effectiveness of water from 3 to 15 times over that of plain water by reducing the natural surface tension of water, which increases its ability to penetrate through fuels and increases its capability to absorb heat. These tests have shown that using a Class A suppression agent provides the avenue for delivering a more efficient droplet size into the flame and fuel surface area without having the droplets evaporate or run off during the application, as would be the case with plain water.

Emulsifiers

Emulsifiers are an important chemical component of Class A foams. An emulsifying agent is one that can render a fuel nonflammable by encapsulating the hydrocarbon molecules. When added to surfactants, these agents emulsify grease, petroleum hydrocarbons, paints, and other barriers to water penetration. This is the reason Class B foams do not contain emulsifiers. If they did, the foam blanket would sink to the bottom of the flammable liquid instead of floating on the surface.

Experiment 1

You can try a couple of simple experiments at the fire station to demonstrate to your crew how Class A foam works. For the first experiment, take two small pieces of cardboard (about 3 inch × 5 inch) and lay them side by side. Place a drop of water on one piece of cardboard **FIGURE 5-50(A)**. Notice how the surface tension holds the drop of water in the shape of a dome. The water does not immediately soak through

A

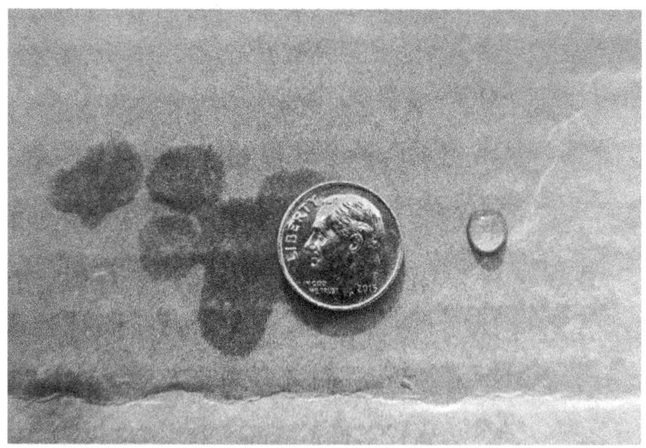

B

FIGURE 5-50 **A**. The surface tension of water holds the drop in the shape of a dome. The water does not immediately soak through the cardboard and sits for 30 minutes unchanged. **B**. A drop of water with premixed foam solution immediately absorbs into and saturates the same cardboard.

the cardboard. In fact, the drop will sit there for 30 minutes unchanged. This illustrates that simple water does not get absorbed into the fuel to extinguish deep-seated fires. In fact, if you tipped this cardboard, the water droplet would run off the surface (water runoff). Next, place another drop of water on the other piece of cardboard. The same thing happens: The surface tension of the water holds the droplet together. Then, add a drop of Class A foam solution on top of this second droplet. The water droplet will immediately be absorbed into the cardboard and saturate it.

Another version of the experiment would be to have a dropper of premixed foam solution. When you add the second drop, it immediately will absorb into and saturate the cardboard **FIGURE 5-50(B)**. This graphically demonstrates how Class A foam breaks down the surface tension of water, making it more effective in extinguishing fuels.

Experiment 2

For the second experiment, fill two large clear glass bowls to midline with water. Add 5 to 10 drops of Class A foam solution into one of the bowls and stir. Leave the other bowl with plain water. Next, fill the two bowls to just under the brim with equal amounts of dry peat moss and let it sit. Do not stir either bowl. You will notice that the peat moss floats on the water **FIGURE 5-51(A)**. As the foam solution breaks down the surface tension of the water, it starts to penetrate through and is absorbed by the peat moss (a Class A fuel) **FIGURE 5-51(B)**. The foam solution took 10 minutes to saturate all the peat moss. In the bowl with plain water, the peat moss floated on the surface for more than 30 minutes, unchanged **FIGURE 5-51(C)**.

Using Class A foam (or not using it) has been compared to washing dishes. We don't wash dishes in plain cold water. We add soap, a surfactant detergent, which reduces the surface tension of water to break down grease and cut through the dirt, thus making the water more effective in getting the dishes clean.

Other Advantages to Class A Foam

There are other advantages to Class A foam. The foam and bubbles make its mass lighter in weight, allowing the solution to resist gravity. This makes it possible for the water and foam solution to stick to three-dimensional surfaces for an extended period of time, which allows time for the foam to penetrate and cool the surfaces of the fuels **FIGURE 5-52**.

A

B

C

FIGURE 5-51 **A**. The natural surface tension of water is strong enough to cause the peat moss to float on the surface. **B**. Water mixed with Class A foam solution breaks down the surface tension of water and starts to penetrate. **C**. In the bowl with Class A foam solution, the peat moss is totally emulsified. In the bowl with plain water, the peat moss floats on the surface for more than 30 minutes, unchanged.

FIGURE 5-52 The foam and bubbles of Class A foam make its mass lighter in weight, allowing the solution to resist gravity and stick to three-dimensional surfaces for an extended period of time.

Troy Carothers and W.S. Darley Co.

The foam provides a thicker barrier of air and water to help protect exposures from radiant heat. Since each bubble has a hollow center, it absorbs heat faster than a water droplet with a liquid center. The combination of thick foam, water, and surfactant forms a barrier at the flame/fuel interface that suppresses the release of flammable vapors from burning material. With this combination of barrier formation and the ability to stick to three-dimensional surfaces, Class A foam can prep and protect threatened structures in wildland/urban interface fires and within the structural wildland interzone. These two terms (*wildland/urban interface fire* and *structural wildland interzone*) refer to a wildland fire that is moving out of the wildland vegetation into a developed and inhabited area, usually residential.

The transition has been slow from wildland/urban interface firefighting to big-city structural firefighting with regard to using Class A foam during the initial offensive attack. Water has always worked in the past. Unless the foam can be pre-plumbed into the apparatus for a hose line to be stretched as quickly as a regular preconnected attack line, the company officer is not going to take the time to set up an eductor with 5-gallon containers simply so the crew can use Class A foam on a house fire when seconds count. Newer apparatus now have pre-plumbed foam systems for Class A and B foams as a standard feature. With the benefits to firefighter safety, and the scientific proof of faster extinguishment, big-city fire departments are taking a closer look at Class A foam for initial company strategy and tactics.

Firefighter Safety and Foam

Firefighter safety is a fundamental reason why the next generation of fire apparatus will have built-in 10-gallon tanks for Class A foam concentrate in addition to 500-gallon booster tanks (capacities vary). It will still be at the company officer's discretion on when to use it, but let's look at some reasons to consider always using foam.

In 1995, tests were performed by the Los Angeles County Fire Department, a long-time pioneer in implementing Class A foam as an extinguishing agent for structure fires. These tests were done primarily to evaluate CAFS. The results yielded valuable information to the superior effectiveness of Class A foam compared to plain water application. The test structures were three identical 1,105-square-foot, one-story, wood-frame, single-family residences. Each structure was furnished with identical home furnishings to simulate the normal fire load found in a typical home. Here are the results in similar-sized fires:

- Knocking down the fire with Class A foam took only 25 seconds compared to plain water application, which took 50 seconds—twice as long.
- In terms of gallons of water used, the attack team using Class A foam used 44 gallons of water to knock down the fire. The team using plain water used 75 gallons.
- The fire rooms were rigged with thermocouples, which recorded the heat-absorbing properties of Class A foam. The Class A foam reduced the average interior temperatures significantly faster than did plain water. The fire room extinguished with Class A foam took one minute and 45 seconds for the average interior temperature to drop from 600°F to 200°F. The fire room extinguished with plain water took six minutes and three seconds for the temperature to drop down to 200°F.

These results are significant when it comes to making the interior environments safer for firefighters and possible occupants. Ask yourself these questions: "How long does it take for a backdraft to occur?" "How fast can a flashover happen?" "What is the speed of a smoke explosion?" The answer to these three questions is simple: fast! All these violent events happen in a split second, and they can be deadly to firefighters. They all happen when the fire is allowed to continue to burn without introducing water to interrupt the combustion process. The unknown factor is, Where is this fire on the fire curve? Is it on its way to flashover? Or is it on its way to backdraft? We don't know because we do not know how long the fire has been burning before it was discovered and reported to the fire department.

Here's what we do know. With today's modern fuel household furnishings (furniture, carpeting, and appliances) being constructed with plastics, synthetics, and glues, fires using these fuels burn hotter and faster. Their heat-release rate and British thermal unit (BTU) output are greater than the ordinary legacy combustibles of wood, paper, and cotton. By the time a fire company arrives at a modern house fire with an average 4-minute response time (without a delay in alarm transmission), its members should be making entry right about the time flashover is ready to occur, if it hasn't already. Why do you think so many more firefighters are experiencing flashover events, and more flashovers are being caught on video and posted on the Internet? It's because the chemical makeup of the modern fire load fuels accelerates the heat release burn rate. The shaving off of 30 to 50 seconds to extinguish a fire isn't insignificant anymore. Dropping the interior room temperature from 600° to 200° in one minute and 45 seconds dramatically increases the safety margin for trapped occupants and firefighters.

There is a safety benefit in quickly reducing interior temperatures, in addition to knocking the fire down, and if Class A foam can help accomplish that faster than plain water can, why wouldn't we use it? Class A foam allows the water to penetrate burning material quickly, and the fuel becomes thoroughly wet. That means water won't evaporate or run off the burning surfaces. As temperatures are reduced, less steam and smoke are produced. A reduction in the flammable vapors being emitted from the fuel surface means a leaner atmospheric environment for rekindles, flashovers, or smoke explosions. That translates into a safer environment for firefighters.

For a variety of reasons, ventilation is sometimes delayed before the fire attack begins. It is quicker to lay a hose line than it is to set up proper PPV or open a roof. Have you ever noticed that, once PPV takes effect after a fire is knocked down, the sudden introduction of fresh air flow is often all that is needed to reignite the fire or hot spots? You still have sufficient heat to sustain the emission of flammable vapors from the fuel. With air completing the fire triangle, you have fire. A company officer who elects to use a preconnected Class A foam attack line when it's known that ventilation may be delayed is making a wise move. By thoroughly wetting the burning fuel with Class A foam, the fire triangle is quickly broken. Less smoke and heat are being generated, which greatly reduces the amount of time fire crews must spend operating in hostile and dangerous atmospheres.

Reducing the gallons of water used may again seem insignificant (44 gallons compared to 75 gallons), but it's more than just preventing excessive water damage for the homeowner. Some areas have long lays between hydrants. In rural fire departments, tender operations and attaining water supplies requires time and personnel. Stretching water usage and making that booster tank last by using Class A foam may be all that is necessary for a quick knockdown or to make a successful rescue in an offensive-to-defensive transitional attack. Again, this translates into a greater safety margin for firefighters and occupants **FIGURE 5-53**.

FIGURE 5-53 Using a preconnected attack line with Class A foam thoroughly wets the burning fuel, and the fire tetrahedron is quickly broken. Flames are extinguished faster. Less smoke and heat are generated, which greatly reduces the amount of time fire crews must spend operating in hostile and dangerous atmospheres.

Troy Carothers and W.S. Darley Co.

Overhaul and Rekindles

Primary damage refers to the damage that is caused directly by the fire. *Secondary damage* is the damage caused directly by firefighters during firefighting efforts, including leaving the property inadequately secured after a fire, where inclement weather or vandalism can result in secondary damage.

Overhaul is the complete extinguishment of fires and usually starts after the initial fire is knocked down. It involves looking for hidden fires in walls, ceilings, and void spaces that are still causing primary damage. Some fire officers, under the guise of "property conservation," elect to use Class A foam only during overhaul in an attempt to limit secondary damage. This is where Class A foam can be a doubled-edged sword: It can give the company officer a false sense of security. Simply "foaming it" and leaving the scene does not prevent rekindles. A thorough overhaul must be conducted even if foam was used on the initial attack. Many items in a house as well as the building construction do not allow for the penetration of any liquid. Items like tile, concrete, metal flashing, skylights, glass, triple-paned windows, unburned plastics, and metal appliances can collapse onto and cover hot spots. Without digging these hidden fires out by hand, they're left to smolder and generate heat until conditions are right for a rekindle. And the foam blanket does not last indefinitely.

Returning to a rekindle is embarrassing and unprofessional for you and your department. In addition, the rekindle makes the efforts of the previous attack useless, and now, the second attack is launched on a weakened, fire-damaged structure, making it even more dangerous for firefighters. Fire chiefs who are concerned with the costs of "needlessly using foam" must look at this from a firefighter safety perspective.

CAFS

CAFSs are still specialty systems, although select fire departments have CAFS engines. CAFS is an advanced method of making Class A foam solution. Compressed air foam is produced by injecting compressed air from an onboard compressor into the water stream that has been mixed with 0.1% to 1.0% foam. This results in a highly compacted foam blanket of small bubbles. Compressed air foam has excellent surface-adherence properties, so it is very popular in the wildland interface firefighting operations. Wildland firefighters apply the compressed air foam to exposure homes endangered by the fire. Runoff is minimal, and the structures look like they're covered with snow **FIGURE 5-54**. Compressed air foam is also an excellent surfactant for use during overhaul and deep-seated Class A fires.

FIGURE 5-54 Compressed air foam has excellent surface-adherence properties.

Troy Carothers and W.S. Darley Co.

Section II: Initial Hose Operations

Safety First

It's important to start this section with a safety reminder. Emergency vehicle operators must always follow federal, state, and local laws and regulations governing the movement of apparatus on roadways. The only time there are exemptions to traffic laws is when the apparatus is operating in emergency mode. Even when responding to an emergency, the driver can be found criminally or civilly liable if the apparatus is involved in a crash. The exemptions do not relieve the driver from the duty to operate the apparatus with reasonable care or the responsibility for all the passengers on the apparatus. This responsibility starts with all members wearing their seat belts any time the apparatus is in motion, regardless of emergency or nonemergency status. Every fire department has a regulation stating so, but it is often not enforced. Members who resist wearing seat belts claim it slows them down from donning their PPE and SCBA. Remember that there have been numerous injuries and line-of-duty-deaths (LODDs) associated with firefighters not wearing their seat belts.

CAFS

By Alan M. Petrillo

A CAFS uses a fire-retardant foam mixed with water and compressed air designed to extinguish a fire rapidly. Its components include a water source, centrifugal pump, foam tank, foam-proportioning system, rotary air compressor, and control system to mix the proper quantities of foam concentrate, which allows for greater reach of compressed air foam than that of either aspirated or standard water firefighting nozzles.

A CAFS attacks three sides of the fire tetrahedron. By covering the fuel with foam, it reduces its capacity to consume oxygen. By adhering to walls and ceilings, the foam aids in reducing heat. Third, it shields other fuel sources from radiant energy.

Foam consistencies range from very dry (type 1) to wet (type 5). The pump operator controls the air-to-solution ratio and the concentrate-to-water percentage. The bubbles that make up dry foams don't readily burst to give up water, which increases cover duration compared to wet foams that break up more quickly in heat. The benefits of CAFSs are:

- **Initial attack capability is improved.** The success of a transitional attack from defensive to offensive is accomplished by darkening the main body of a large fire.
- **Reduced water use.** The number of gallons required to extinguish a fire is reduced to as little as one-third when compared to plain water. With the ability of compressed air foam to cling to vertical surfaces, less water is wasted from runoff protecting exposures.
- **Increased firefighter safety.** Increased reach equals increased safety. CAFS streams have greater reach and penetration into structures than water or nozzle-aspirated foam fire streams.
- **Less firefighter stress and fatigue.** Hoses filled with compressed air foam are approximately 30 percent air by volume, making those lines lighter and easier to advance through a structure.
- **Reduced smoke and steam.** CAFS leaves minimal smoke and steam in the atmosphere, giving the fire attack team greater interior visibility. It doesn't drive them to the floor with steam penetrating their turnout gear, which is often experienced during a conventional water attack.
- **Reduction of wasted resources.** In wildland and urban interface fires, the application of compressed air foam on trees, brush, and exposure buildings is very visible, preventing excess application of agent and water wastage. The need for reapplication becomes evident.

Equipment and Initial Hose Lay Operations Compressed Air Foam Systems (CAFS) by Alan M. Petrillo.

Single Engine Hose Lays

A Type 1 fire engine, often referred to as an engine company, a fire engine, a pumper truck, or a structural firefighting engine, is the most common type of fire engine in use today. Type 1 fire engines are specifically designed to support urban, rural, and suburban fire departments for structural firefighting. Per NFPA 1901, Standard for Automotive Fire Apparatus, every Type 1 engine is required to have a pump with a minimum tank capacity of 300 gallons, although most Type 1 engines in urban and metropolitan cities feature a 500-gallon water tank. In suburban and rural areas without hydrants, it is not unusual to find Type 1 engines that carry between 750 and 1,000 gallons of water, and sometimes even more. This section discusses the initial hose lays of the first two engine companies arriving at a fire. There are four basic hose lays for establishing a water supply:

- Forward lay
- Split lay
- Direct-to-engine: no line laid; second-in engine lays reverse from first-in engine
- Reverse lay

Forward lay and *reverse lay* are standard nomenclature. The other variations use different terms depending on the region. No matter how you get there, in the end, all hose lays end up with a forward engine in front of the fire building, supplied by another engine hooked up to a hydrant.

Forward Lay—Wet

The most advantageous hose lay is the forward lay from a hydrant to the fire, with the immediate charging of the supply line to the engine **FIGURE 5-55**. This enables the first engine company to function independently with a constant and uninterrupted supply of water from the beginning of the operation once the hydrant is charged. The engine is hooked up to the closest hydrant, lays forward, and can quickly take up its position at the front of the fire building, while leaving room for the ladder company. Most engines have a water tank capacity of 500 gallons. The forward lay allows the deployment of a preconnected attack line to apply fast water onto the fire through a door or window from the outside while the hydrant is being charged. If the fire continues to gain headway after the initial attack, other pumpers can supply water to this first engine, which is already in position. This engine basically becomes a giant manifold for the rest of the operation. The forward lay is also the only one that allows the first-in engine to comply with the two in/two out rule. In addition, this is your toolbox; all the extra hose, nozzles, master streams, ladders, pike poles, forcible entry tools, other tools, extra SCBA bottles, and emergency medical services (EMS) equipment is readily available, instead of at the hydrant down the street. You want your toolbox close to your work. The forward lay permits the use of 1½-, 1¾-, or 2½-inch attack lines, or a master stream if

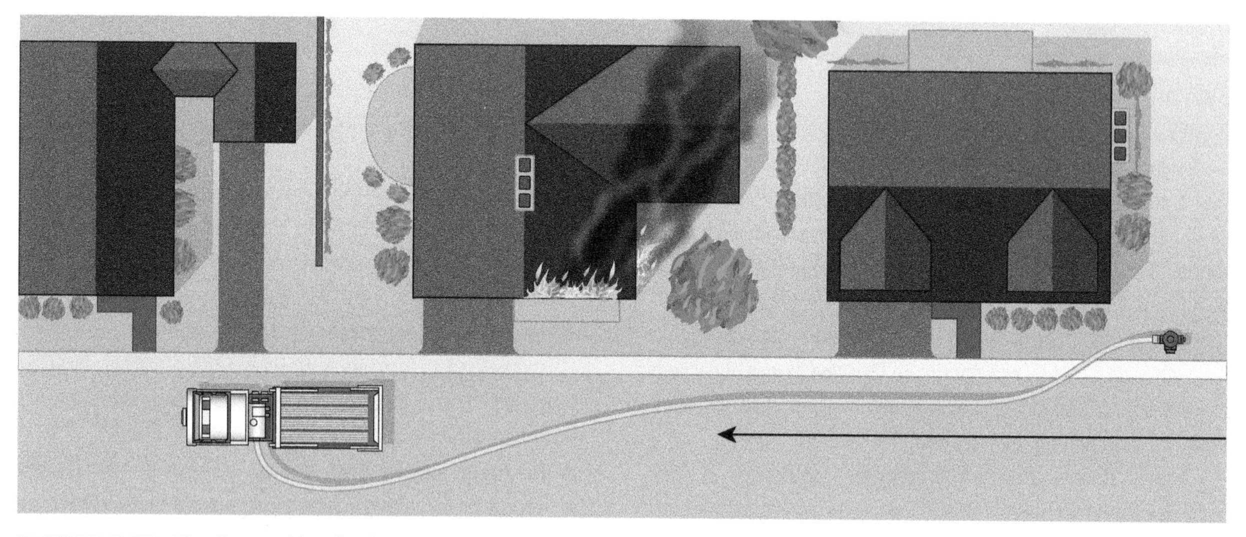

FIGURE 5-55 The forward lay (hydrant to fire) gives the ability of the forward engine to stretch a preconnected attack line using the tank for fast water and the firefighters to comply with the two in/two out rule, and it turns the forward engine into a manifold for additional hand lines once the hydrant is charged.

the hydrant is adequate, and the proper diameter supply lines have been laid to the pumper.

The following are advantages of a forward lay:

- The company can deploy 1¾-inch or 2½-inch preconnected hand lines for fast water.
- The two in/two out rule can be adhered to.
- The engine company is self-sufficient in terms of a constant uninterrupted water supply once the hydrant is charged.
- The company is free to take up any position in the front, rear, or side of the burning structure and begin operations with an uninterrupted water supply.
- The engine becomes a manifold so other responding engine companies can take hose lines off the first engine.
- With a sufficient supply, the company can utilize a pre-plumbed deck gun or master stream for a Blitz attack to knock down the fire.
- All extra hose, nozzles, ladders, tools and EMS equipment is readily available in a forward position.

The following are disadvantages of a forward lay:

- There is a slight delay while the engine stops and drops off a firefighter to wrap the hydrant with the supply line.
- There is a temporary loss of this firefighter, who must remain at the hydrant until the supply line is charged.
- If any necessary hydrant tools are accidentally left on the engine, the hydrant firefighter must run back to the engine to retrieve them, thus delaying the water supply.
- Supply lines being laid may hinder access of ladder company or aerial apparatus.
- Depending on staffing, the two in/two out rule may be delayed until the hydrant firefighter charges the supply line and rejoins the company.
- There is no redundancy for safety. Should a supply hose rupture or a jacket pull away from a coupling, the supply is lost.
- The company could unknowingly have taken a dead or out-of-service hydrant.

Forward Lay—Dry

In this approach, a supply line is laid from the hydrant to the fire, but the line is left uncharged for another company to hook up to the hydrant and charge the supply **FIGURE 5-56**. The firefighter who wraps the supply line around the hydrant must make sure enough tail remains (about 15 feet) to reach the intake port of the second engine. The hydrant bag with the proper tools and couplings should be left at the hydrant, including the hydrant wrench. This allows any knowledgeable person who may show up before the second engine to hook up the hydrant and charge the supply line. Then the firefighter reboards the pumper and immediately proceeds with the rest of the crew to the fireground. Now the entire crew of the first pumper can lay hand lines while the officer performs a 360-degree size-up and risk assessment. Then they can proceed with the initial fire attack using tank water.

Another company is now responsible for going to the hydrant, hooking up the supply hose, and charging the line. If the area is known to have high-pressure hydrants, or if a four-way HAV hydrant valve is utilized, anyone (an ambulance crew or the battalion chief) can

FIGURE 5-56 This dry forward lay (hydrant to fire) has the supply line uncharged until the arrival of the second company.

hook up the hydrant in order to start the initial supply. Otherwise, the second engine company can finish the hookup and charge the water supply line on its arrival. The evolution proceeds faster if all members of the second engine help in hooking up the hydrant, but the driver should be able to handle the hookup alone to allow the rest of the crew to grab their equipment and head toward the fire.

Department SOPs may require a third engine apparatus to provide the first-in engine with a second LDH water supply to the fireground to provide a backup safety system of redundancy. Should one of the supply lines burst, there will still be a constant uninterrupted water supply.

This operation is smooth and efficient when the second or later-arriving engines can be expected to arrive soon after the first engine. SOGs and radio communications between engine companies can ensure that someone on the second engine is aware that someone on the first engine laid a dry forward lay. If the forward lay isn't obvious, radio communications also make certain that the second engine goes to the right hydrant; for example, "E2 from E8, take the hydrant on the corner of Warren Avenue and Lee Street."

If the engines do not arrive in fairly rapid succession, the first pumper, working off its water tank, will run into a problem when it runs out of water. The driver of the first engine must run back to the hydrant and charge the supply line, and that decision should be made before the tank runs dry. If there is a delay in water, the driver must notify the officer and attack crew so they can back out of the structure until a constant uninterrupted water supply is established.

It is assumed that a 360-degree size-up and risk-benefit analysis have been made if the company officer decides to make the initial attack with tank water only. The two in/two out rule is adhered to, and the incident is a room and contents fire in the incipient or growth stage that can easily be extinguished with less than 500 gallons of water. Most room and contents fires in the incipient or early growth stages are extinguished with less than 100 gallons of water. If the first-arriving officer sees a large column of smoke showing or determines that the scene is a working fire, the company should perform a standard forward lay, leaving a firefighter at the hydrant to charge its own supply line unless extenuating circumstances dictate otherwise.

The advantages of a dry forward lay supply line that is left at the hydrant to be charged by a second pumper are:

- After laying the supply line at the hydrant, the entire crew can proceed to the fire scene to fight the fire.
- The company is free to take up any position in the front, rear, or side of the burning structure and begin operations with tank water.
- The company can deploy1¾-inch or 2½-inch preconnected hand lines for fast water.
- The two in/two out rule can be adhered to.
- The engine becomes a manifold so other responding engine companies can drop off its crew and use hose lines and equipment taken

off the first engine company after the hydrant is charged.

- All extra hose, nozzles, ladders, tools, and EMS equipment are readily available in a forward position.

The disadvantages of a dry forward lay supply line and leaving it at the hydrant to be charged by a second pumper are:

- Two engine companies are required to provide one engine company with an uninterrupted water supply.
- Another company must respond from the same direction for this task to be completed in a timely manner.
- At least two engine companies must arrive at the fireground in quick succession if this procedure is to be efficient.
- If radio communications are not monitored or SOGs are not followed, there is the possibility for confusion or a delay in starting the water supply if the second engine goes to the wrong hydrant.
- If there is a problem with obtaining a constant uninterrupted supply (i.e., a second engine pump failure, taking a dead hydrant) the first-in engine will run out of water.

Split Lay

The split lay is a dry forward lay by the first-arriving engine. The only difference is there's no hydrant to wrap, or no water source at the T-intersection. A firefighter anchors the LDH supply while the driver starts the forward lay. Two sections of LDH should be enough to weight the lay so the rest of the supply hose can freely pay out of the rear hose bed. The driver stops, and the firefighter gets back on the apparatus. The pumper then proceeds to the fire and their actions thereafter would be the same as those for a dry forward lay. The LDH coupling should be a 3½-inch or 4-inch female or a Storz coupling. When the second engine arrives at the T-intersection, the crew uses whatever couplings or adapters are needed to connect their LDH to that first section of hose from the initial engine. The second engine then lays reverse to the hydrant, connects, and charges the supply line. Because the first engine lays forward and the second engine lays reverse, it's called a split lay **FIGURE 5-57**. This operation is useful when long narrow driveways, limited access roads without hydrants, or dead ends hamper operations at the fire building. The advantages and disadvantages of a split lay are the same as those above for a dry forward lay supply line that is left at the hydrant to be charged by a second pumper.

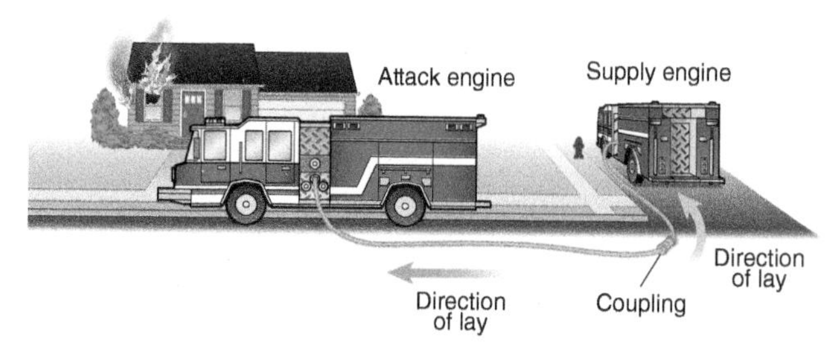

FIGURE 5-57 A split lay.

Direct-to-Engine—No-Line-Laid Approach

In this approach, the first pumper to arrive at the fireground lays no supply line. It proceeds directly to the fire and begins working off the water in its tank for the initial fire attack **FIGURE 5-58**. Using fast water from the tank without a supply can be a safe and effective procedure only when:

- The fire is in the incipient stage or early in the growth stage and can be easily extinguished with 500 gallons or less. Most room and contents fires in the incipient stage or early growth stage can easily be extinguished with less than 500 gallons of water. Use the National Fire Academy (NFA) fire flow formula: L × W ÷ 3.
- At least two engine companies are sure of arriving close together or at the same time. Better still is when they both can see each other arriving.

When both engines arrive at the same time, the first-listed engine on the response should be allowed to deploy the preconnected hand lines to avoid confusion, and the second engine gets the water supply. However, if the second engine is better positioned, the first-listed engine should yield and start establishing the water supply. Use good radio communications and follow your SOGs.

There are a few options for how the second pumper can establish a supply:

1. If the second pumper is approaching from the opposite direction, it stops next to the first pumper and hands the driver the end of the LDH. The driver anchors the LDH with enough slack to reach and make the connection into the pump. The second pumper drives to the hydrant—this is a reverse lay using the first-in pumper like a manifold **FIGURE 5-59**. If the area is known for high-pressure hydrants, the second pumper can connect the LDH directly to the hydrant and supply the first pumper. This frees the second engine to cover another position and establish a second supply.

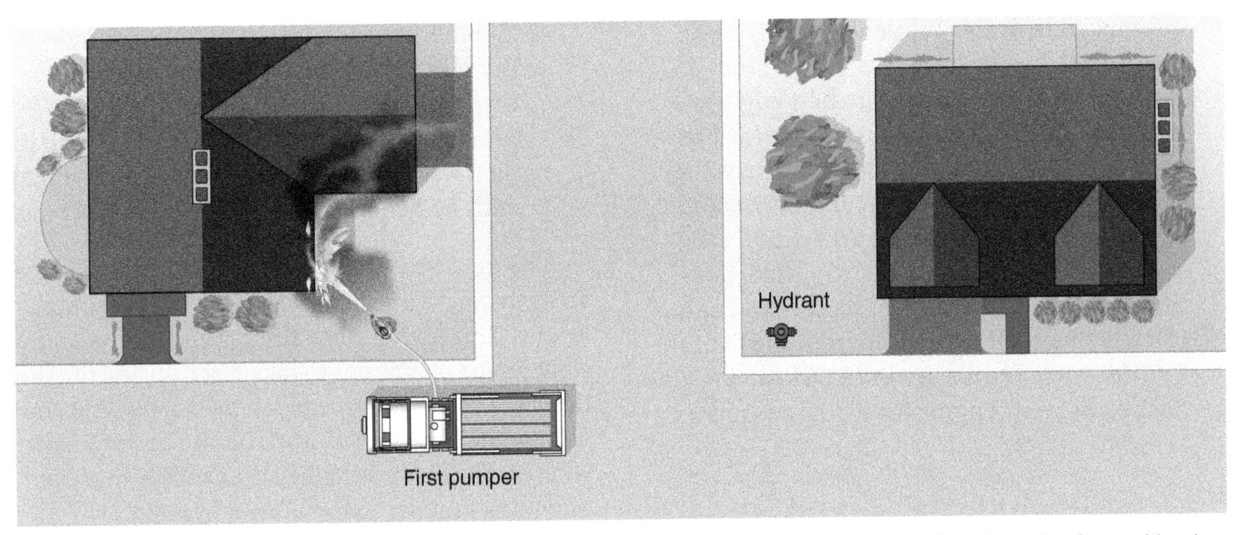

FIGURE 5-58 In the direct-to-engine—no-line-laid approach, the first-in pumper proceeds directly to the fire and begins working off tank water for the initial fire attack.

FIGURE 5-59 The supply hose is handed to the first-in driver. This driver anchors the LDH with enough slack to reach and make the connection into the pump. The second pumper drives to the hydrant—this is a reverse lay using the first-in pumper like a manifold.

2. A better and safer option is to follow the procedures in number 1 above except that the second engine connects to the hydrant and supplies the first engine. A second LDH can be overhauled by hand back to the first engine for a second supply with parallel lines.
3. If the second engine is responding from the same direction, it should spot on the

designated hydrant. While the driver connects to the hydrant, the crew of the second engine runs up to the first engine, takes the end of the LDH from the rear hose bed, cross-lots, shoulder-loads, or drags the supply line back to the second pumper, and connects it to a discharge port **FIGURE 5-60**. This takes some effort and extra time, but the end game is the same: The second engine is supplying the first engine from the hydrant **FIGURE 5-61**. Then a second LDH can be laid back to the first engine for a dual secondary supply.

FIGURE 5-60 The crew of the second engine runs up to the first engine, takes the end of the LDH from the rear hose bed, shoulder-loads or drags the supply line back to the second pumper, and connects it to a discharge port.

4. Another alternative is having the second pumper lay a supply line from a hydrant to the first pumper for a quick water supply. If needed, the second pump can then obtain its own water supply. This is risky if the hydrant ends up being a low-pressure, low-flow plug.

The direct-to-engine approach is quite effective in areas that have a reliable water system with ample hydrants. This allows some firefighting efforts and possibly rescue operations to take place during the time it takes the second engine to lay the supply line and get water. One unit goes to work on the fire immediately, whereas the second engine lays a supply line from the operating pumper to the nearest hydrant.

This strategy is riskier in areas without hydrants because you run the risk of running out of water, but the officer may not have any other choice but to operate on tank water alone until a water supply is established from another source. The officer needs to do a proper 360-degree size-up and risk-benefit analysis, and utilize the NFA fire flow formula, or Royer/Nelson to calculate if there is enough water in the tank. The officer needs to decide very carefully how to use those 500 gallons. Perhaps the only alternative is to hit the fire from the outside through a window or a door to knock down and reset the fire. Then the remaining water can be used to cover the search team for a quick primary search before backing out all personnel from the building until a reliable uninterrupted water supply is established.

If there are no hydrants, the water needs to come from a tender or be taken from a reservoir, lake, pond, or stream. At this location, the second pumper hooks

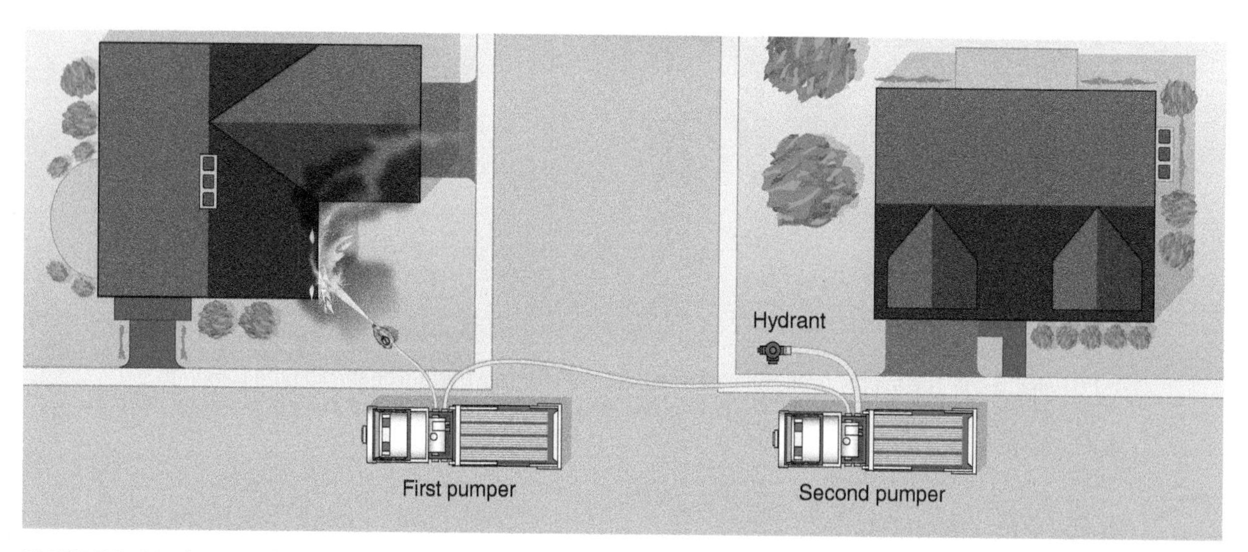

FIGURE 5-61 The crew from the second engine takes the uncharged supply line from the first engine, drags it back to the second engine, and connects it to a discharge port. In the end, the second engine is supplying the first engine from the hydrant.

up the hard suctions for drafting. If the second pumper cannot reach the water supply due to distance, the other engine companies must set up a relay pumping operation from the water supply to the pumper at the fire. Another option is to have the remaining responding engines come in and give their tank water to the first pumper or use it in their own hose lines, as the situation demands. If four engines are assigned to the alarm, each having a 500-gallon tank capacity, that's 2,000 gallons of water to work with when no hydrants are available. Two thousand gallons is a lot of water.

The advantages of the direct-to-engine—no-line-laid approach are:

- The first-arriving pumper and its entire crew go directly to the fire to begin firefighting and rescue operations.
- This allows the first-arriving pumper to utilize preconnected hand lines to apply fast water to knock down and reset the fire.
- Knocking down the fire with tank water lowers the temperatures inside the structure, allowing for a quick primary search of savable victims. It also increases the safety margin for the interior operational period against flashover.

The disadvantages of the direct-to-engine—no-line-laid approach are:

- The operation does not work unless companies arrive close together, communicate with each other, and follow SOGs.
- Two engine companies are required to provide one engine company with a continuous uninterrupted water supply.
- There is little or no margin of safety for interior attack crews.
- If a supply cannot be established or will take an extended period of time, the water in the tank will run out and crews will have to stand back and let the fire go.

Reverse Lay Using a Charged Line

A reverse lay can be carried out in many variations, but they all include the engine laying from the fire to the hydrant. In the reverse lay, the engine lays firefighting hose lines; a trigate, a portable hydrant, or a manifold is connected to an LDH supply line; then the engine lays from the fire to the hydrant **FIGURE 5-62**. The pumper is then hooked up to the hydrant; when it is ready, pumping operations can begin. **FIGURE 5-63** shows the actions of one engine company using the reverse lay.

FIGURE 5-62 For a reverse lay, the LDH can be connected to a manifold.

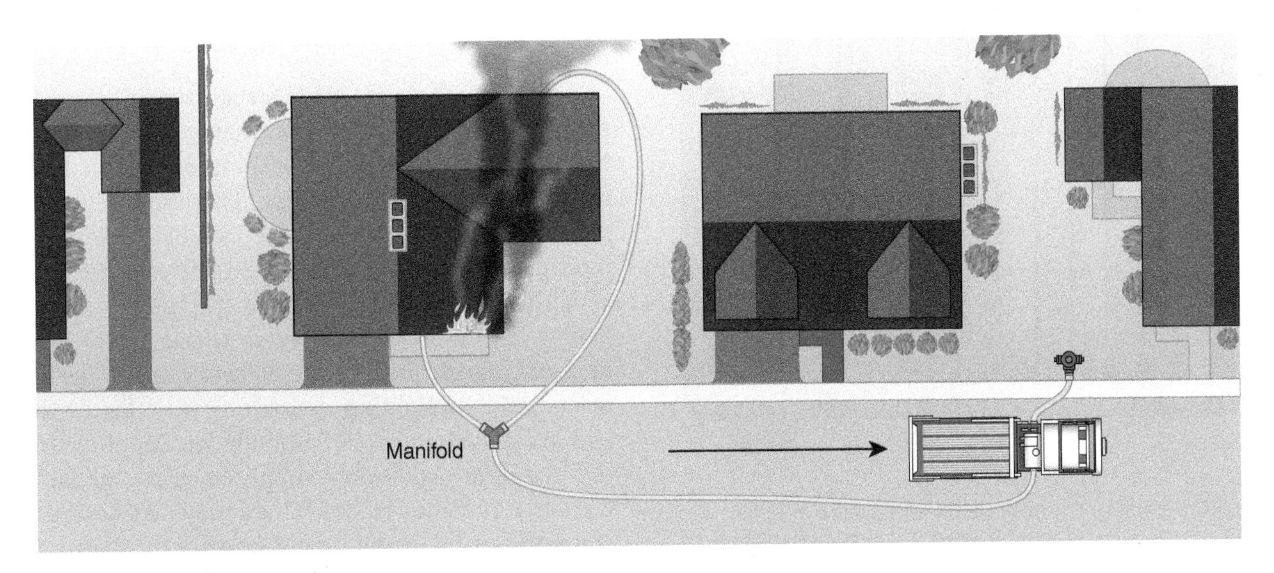

FIGURE 5-63 The reverse lay.

In the past, a 1½-inch reverse lay was used, and it consisted of one or two bundles of 1½-inch hose with nozzles connected to a 1½-inch wye, which was connected to the 2½-inch reverse hose bed. There could easily be 1,000 feet of 2½-inch hose available for the supply. This was a good evolution for long reverse lays, but it limited fire attack options to the two hand lines. Although this evolution is good for small brush fires and brush trucks, it is not a good option for today's structure fires. Most metropolitan, city, and suburban fire departments do not use this load anymore on their engine apparatus.

The 2½-inch reverse lay was a similar evolution. A 2½-inch nozzle was connected to the 2½-inch reverse bed **FIGURE 6-64**. A section of 2½-inch hose was pulled to the ground, and the engine would lay reverse. This allowed the operation of only a single 2½-inch hand line unless a 2½-inch wye was connected before the line was charged; then you had the option of running two 2½-inch hand lines, or two 1½-inch hand lines off the wye, and a 2½-inch line. Although there were a few more options, the friction loss associated with this lay along with the limited gpm capability now make it impractical for today's structure fires. This lay can sometimes come in handy for grass and small brush fires, so it's good to know about it.

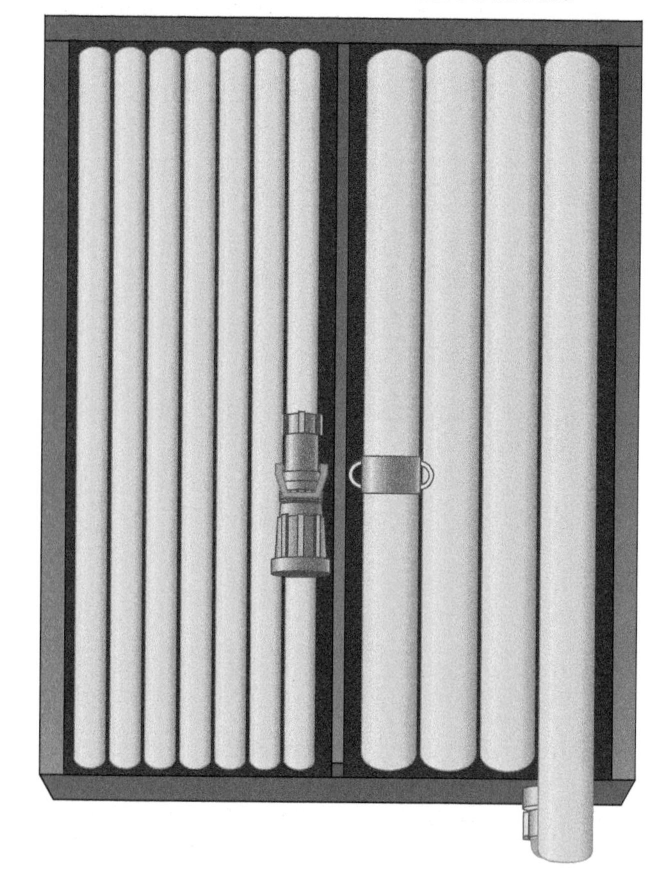

FIGURE 5-64 The left hose bed is set up for a 2½-inch attack line, but it is also set up for a 2½-inch reverse lay.

The most efficient way to lay reverse is using a trigate, a portable hydrant, or a manifold connected to an LDH supply line. The LDH provides more volume and less friction loss to the manifold. Hand lines and supply lines for other engines, or hose for supplying an aerial ladder pipe, can be taken off the manifold.

When the engine pulls up to the fire for a reverse lay, the officer and the crew members exit the cab and grab all the equipment they anticipate will be needed. This includes 1¾-inch hand lines with nozzles and a 1½-inch wye, 2½-inch hand lines with nozzles and a 2½-inch wye, ground ladders, pike poles, forcible entry tools, and the first aid kit because their toolbox (the engine) will soon lay to the hydrant. Finally, they drop the manifold and physically anchor it. The signal is given for the driver to lay reverse toward the hydrant. Once the LDH hose is sufficiently weighted and is paying out freely from the reverse bed, the officer and crew members can connect lines to the manifold. The manifold is gated, so the driver can charge the LDH line as soon as the pump is hooked up to the hydrant. The officer can charge the hand lines directly from the manifold when they are in place.

This type of operation has three advantages not mentioned in the discussions of other hose lays. First, if the first-arriving engine drops a manifold and lays out to the hydrant, it puts a portable water source directly in front of the building. If the second-in engine is close behind, the crew members can position in front of the fire building, utilize their preconnected attack lines, and join the first company to attack the fire. The driver can take the initial supply from the manifold. Additional supply lines are laid to the second engine, so there are essentially two manifolds in front of the fire building.

Second, if there are other structures threatened by the fire, a manifold can be overhauled or cross-lotted to the rear of the building or another specified location without apparatus access. This establishes a remote pressurized water source capable of supplying multiple hand lines or a master stream appliance.

Third, the engine is away from the fire building. It may be important to position aerial ladder trucks or elevating platforms near the building to perform rescue and firefighting operations. The ladder pipe and master streams can initially be supplied from the manifold. Additional supply lines can be laid to the aerial or tower ladder. There may be a severe fire condition or the possibility of structural collapse that would preclude the fire apparatus being close to the building.

The advantages of the reverse lay are:

- The engine company is self-sufficient in terms of its own water supply.
- Other arriving pumpers can take their supply from the manifold.
- Aerial ladder pipes and elevated platforms can initially be supplied by the manifold.
- Engines are not positioned close to the fire and thus do not block the approach and operation of aerial ladder trucks or elevating platforms.
- The engine apparatus is out of the collapse zone.

The disadvantages of the reverse lay are:

- A reverse lay does not get water onto the fire as fast as does a forward lay, where fire attack can start with preconnected hose lines.
- Equipment needed on the fireground must be removed from the engine before it proceeds to the hydrant. SOGs should list the equipment that firefighters should remove from the engine on a reverse lay.
- The trigate, portable hydrant, and manifold have limited intake and discharge ports, and do not provide the same options that a pumper does. If more water volume is needed, a second manifold must be laid using another engine apparatus.

After-Action REVIEW

IN SUMMARY

- The tools for fighting fire and providing emergency services include the fire apparatus and the equipment carried onboard the apparatus.
- Pumpers and equipment should be purchased following recognized standards, federal requirements, and procurement policies set forth by the AHJ.
- The purchaser should ascertain design, function, and performance requirements to ensure that equipment received will function effectively, efficiently, and safely.
- The selection of fire apparatus and equipment should not be left to one or two individuals.
- Sometimes fire chiefs or high-ranking officials lose touch with the everyday operation of the department.
- Senior firefighters and company officers should be queried about their thoughts and ideas for the types of tools and equipment that best serve the needs of the department.
- A concerted effort should be made to purchase equipment that will serve the department for years to come.
- SOGs should be established for all engine company tasks that need to be performed on the fireground. They include, but are not limited to the following:
  - Water supply
  - Hose lays
  - Initial fire attack
  - Backup lines
  - Relay pumping
  - Tanker shuttle operations
  - The use of private fire protection systems
  - Property conservation
  - Overhaul
- With SOGs in place, an engine company can commence operations that they as well as other firefighters are familiar with.
- Because of the size and complexity of different fire departments and demographic considerations encountered, SOGs differ from jurisdiction to jurisdiction, reflecting the different needs of each jurisdiction.
- For an engine company to be effective, it must have sufficient personnel that are well trained, properly equipped, and led by competent staff who operate using an incident management system.

KEY TERMS

ball valve Valve used on nozzles, gated wyes, and engine discharge gates. It consists of a ball with a hole in the middle of the ball.

crosslays Traverse hose beds. Also known as *bucket line, speed lay*, or *Mattydale lay*.

divided hose bed A hose bed that is separated into two or more supply hose compartments running the length of the hose bed.

double-female coupling A hose adapter that is used to join two male hose couplings.

double-male coupling A hose adapter that is used to join two female hose couplings.

hydrant assist valve (HAV) A specialized type of valve that can be placed on a hydrant and allows another engine to increase the supply pressure without interrupting flow. Also known as *four-way valve*.

immediately dangerous to life and health (IDLH) Any condition that would pose an immediate or delayed threat to life, cause irreversible and/or adverse health effects, or interfere with an individual's ability to escape unaided from a hazardous environment. (NFPA 1670)

master stream appliance Devices used to produce high-volume water streams for large fires. Most master stream appliances discharge between 350 gpm (1,325 L/min) and 1,500 gpm (5,678 L/min), although much larger capacities are available. Also known as *deck gun, monitor, deluge*. Also referred to by a model or brand name, like the *Stinger* or the *Stang*.

pre-plumbed master stream appliance A master stream appliance that has a separate discharge pipe of adequate diameter that runs from the fire pump to the appliance.

pumper fire apparatus Fire apparatus with a permanently mounted fire pump of at least 750-gpm (3,000 L/min) capacity, a water tank, and hose body whose primary purpose is to combat structural and associated fires. (NFPA 1901)

skid load One hundred fifty to 300 feet of prebundled 2½-inch hose carried in a hose slot. This hose is not preconnected, but it can be deployed quickly by connecting to any 2½-inch discharge port. The skid load can be used as a 2½-inch attack line, a backup line, an exposure line, or a quick supply to a master stream appliance, standpipe, or sprinkler connection.

supply hose Hose designed for the purpose of moving water between a pressurized water source and a pump that is supplying attack lines. (NFPA 1961)

REFERENCES

Cavette, Chris, "Bubbles Beat Water," *Fire Chief Magazine*, July 1, 2001.

Cotner, Jeff, "Our Department Has Class A Foam; Now What?," *Fire Engineering Magazine*, Volume 160, Issue 7, July 2007.

Fornell, David P., *Fire Stream Management Handbook*, Fire Engineering/Pennwell, 1991.

National Fire Protection Association, NFPA 1901, *Standard for Automotive Fire Apparatus*, Quincy, MA: 2016.

National Fire Protection Association, NFPA *High-Rise Building Fires Report*, Quincy, MA: 2016.

National Institute of Standards and Technology, Report 1629, *Fire Fighting Tactics under Wind Driven Conditions*, Gaithersburg, MD: National Institute of Standards and Technology, 2009.

Plunkett, Kurt, *Firefighting Foam Operations, Apparatus, and Equipment*, Seattle Fire Department Training Guide #1-1, September 2009.

CHAPTER 

Apparatus Positioning

LEARNING OBJECTIVES

- Recognize the general aspects of complete coverage of fire buildings.
- Explain why the first and most basic step in coverage is the positioning of the apparatus.
- Describe the concept of developing a pre-incident plan to determine resources and actions necessary to mitigate anticipated emergencies at a specific facility.
- Recognize that basic apparatus positioning assignments are often defined in written SOGs.
- List the eight phonetic geographic indicators for the fire building.
- Explain how certain buildings may present positioning and coverage problems because of size, construction, location, and use.
- Describe the challenges with covering shopping malls, mercantile buildings, garden apartments, and other problem buildings.

Introduction

The success of a fire emergency incident depends on how engine company apparatus is positioned on the fireground. To a great extent, the spotting positions taken up by the engines as they arrive at the fire building are based on standard operating guidelines (SOGs), past experience, and anticipating the problems that may be encountered before the fire occurs. This requires knowledge of the company's district, which can be acquired through pre-incident planning; building inspections; district driving drills; and paying attention to details such as construction features, hydrant locations, other sources of water, and access routes. The first part of this chapter discusses pre-incident planning, including the benefits of planning. The second part of this chapter deals with basic coverage responsibility. The last two sections deal with coverage in more detail and types of buildings that can pose coverage problems. Remember that SOGs should mandate apparatus placement; it takes the guess work out of routine decision making and lets other incoming units know which initial actions to expect.

Pre-Incident Planning

A pre-incident plan is a document developed by gathering general and detailed data used by responding personnel to determine the resources and actions necessary to mitigate anticipated emergencies at a specific facility. NFPA 1620, *Standard for Pre-Incident Planning*, was developed with the primary purpose to aid in the development of a pre-incident plan and thus help responding personnel manage emergencies with available resources effectively. The pre-incident plan should not be confused with fire inspections, which monitor code compliance.

The pre-incident plan should be coordinated with the incident management system. The ultimate goal of pre-incident planning is to know the problems involved in every potential fire building in a company's response area. If every building had to be pre-planned, however, the process would be time-consuming; therefore, emphasis should be placed on the larger commercial and multi-residential occupancies within the district.

Building versus Structure

Fortunately, most of the buildings in an area are usually similar in many respects; for example, a neighborhood might consist mainly of two-family semidetached dwellings, with access to the rear of each building available through a driveway, and a hydrant within 200 ft on either side. Inspection of one such building is in effect pre-incident planning for them all. Although it might be helpful to examine each building in detail, the return might not be worth the effort. Too much information can be overwhelming and defeat the purpose.

Buildings that must be examined and pre-planned are those that are unusual in some respect. Unusual construction features in terms of dimensions or materials warrant special consideration. The following factors should be considered when assessing the potential situations that could affect a facility during emergency conditions:

- Construction types and classifications
- Access to buildings
- Occupancy characteristics (life safety considerations)
- Protection systems, for example, standpipes, sprinkler systems, and fire department connections (FDCs)
- Capabilities of responding personnel and equipment
- Availability of mutual aid
- Water supply, for example, hydrant locations, water main sizes, and reservoirs
- Exposure factors

When developing criteria for pre-incident plans, strong consideration should be given to items such as the following:

- Potential life safety hazard
- Structural size and complexity
- Value
- Importance to the community
- Location
- Presence of chemicals (hazmat potential)
- Susceptibility to natural disasters

Unusual buildings present special positioning and coverage problems. Proper positioning of a pumper at a fire depends on several factors:

- Physical characteristics of the structure
- Location of the structure
- Width of street
- Dead ends and other street characteristics
- Setbacks
- Overhead electrical wires
- Availability or lack of water supply
- Normal alarm response routes with companies in quarters

- Size and location of the fire
- Topography
- Need for aerial fire apparatus

Basic Coverage Responsibility

Apparatus positioning and coverage are often defined in a set of SOGs, sometimes called standard operating procedures (SOPs) or policy and operating guidelines (POGs). A policy can be rigid, but guidelines give you leeway for variables in decision making. Department SOGs may differ from jurisdiction to jurisdiction. For example, a department may mandate that the first-in engine go directly to the fire building and operate on tank water, while the second-in company establishes a water supply. Generally, the company that is expected to arrive first on the fireground from its station (the first due) is usually assigned to the front of the building. The first-due company officer assumes command and assigns apparatus according to the initial size-up. Additional responding apparatus may be assigned to cover a side of the building or a task, or it may be sent to an initial staging area (base). Companies should not be assigned a task unless there is a task for them to perform. Engine companies should not be allowed to freelance (i.e., decide independently what needs to be done or act outside the incident commander's [IC's] orders or the initial action plan). The IC should assign apparatus to positions that will support the attack. A company should be assigned to a position that allows it to perform the assigned task while ensuring its safety outside the collapse zone and, if possible, that other apparatus have access to the area. In most cases, the first-arriving pumper should pull past the fire building. This accomplishes two objectives:

- Allows officers to see three sides of the building for size-up
- Leaves room for the ladder truck in front of the building

If the first-in ladder company approaches from the opposite direction on a narrow street, the first-in engine should stop short.

Coverage Assignments

Coverage assignments are usually based on the proximity of companies to the fireground, but it is important to realize that they are only a guide for normal operations. No policy or guideline should be used as a substitute for initiative and the exercise of good judgment by the company officer. If the fire situation requires that the first-arriving company take a position at the C side rear of the burning structure, that is where it should be positioned; however, the officer should notify the other responding companies via radio of any deviation from assigned positions. The second engine and additional responding companies will then be prepared to adapt to the situation. In this example, the second-in pumper would be advised that it should cover the A side front.

Any time an engine company is out of quarters ("on the air") and out of vicinity ("out of position"), situated so that they will not arrive at the fireground when expected, the officer should immediately relay this fact to the dispatcher upon receiving the alarm. With this information, dispatch assignments can be modified; for example, "E1 to dispatch, E1 is responding from the very south end of the city limits, E2 will be first-in."

Whether an SOG is followed to the letter or is modified because of special circumstances, the end results are that alarm assignment responsibilities are distributed among responding engine companies and that each crew knows its own coverage responsibility.

This section discusses some general aspects of the complete coverage of fire buildings. The word *coverage*, as used here, means the assignment of companies to a particular side of the fire building or the fireground for size-up, with the objective of accomplishing any or all of the objectives of a firefighting operation. The first and most basic step in coverage is the positioning of the apparatus. The C side rear of the fire building may indicate an entirely different situation than the A side front or the sides **FIGURE 6-1**. This is why a 360-degree size-up is so important. The C side needs to be checked.

Every officer and firefighter should always maintain geographical awareness. There are eight directional indicators: N, S, E, W, NW, NE, SW, and SE. Without a compass or visible landmark, however, it might be difficult to determine where north is in the middle of the night. To avoid confusion, the phonetic alphabetical indicators, **A**lpha, **B**ravo, **C**harlie, and **D**elta, are now used, with the "Alpha" side typically being the front of the structure or the address side of the building, but the "Alpha" side can also be determined by the IC. The letters continue clockwise to identify the sides around the building. On a two-story single-family residence, the front side is obvious. On a downtown high-rise building, it may not be so obvious. Therefore, eight phonetic directional indicators also exist: Sides, A, B, C, D, and AB corner, BC corner, CD corner, and AD corner. When transmitting assignments over the radio and any radio update messages, Alpha, Bravo, Charlie, and Delta should be used, or Alpha-Bravo corner, Charlie-Delta corner, and so on, should be used. (For the purpose of brevity in this text, only the letters will

FIGURE 6-1 Fire conditions vary from the A side front (left) and to the C side rear (right) of an involved structure.

be used.) Both front and rear (A and C) and the sides, (B and D) must be covered quickly to establish effective control of the fire, for example:

- First engine: A side (front) of the building
- Second engine: water supply/backup line
- Third engine: C side (rear) of the building
- Fourth engine: water supply for rapid intervention team (RIT)
- First ladder: A side of the building
- Second ladder: C side of the building

The examples described here provide a review of apparatus positioning at some of the most common types of buildings that a fire department may encounter. These are general recommendations. Depending on the type of building and any extenuating circumstances, command must be able to develop an incident action plan to establish an overall operational approach.

Positioning: Alpha Side Front

The first-arriving company is normally assigned to cover the front of the building. Again, in most situations, the front of a building is obvious; sometimes it isn't. Buildings may have more than one main entrance, or the entrance may be away from a street or parking area. When command designates the front of the building as the A side, all personnel must be notified. This divides the building into areas that are easy for command to manage. If resources are needed at the right side of the building in the rear, they would be directed to the CD corner of the building. The position that the first-in engine company takes may depend on the building type, as described in the following sections.

Detached Buildings

On arriving at a single-family detached dwelling, the first engine should proceed just past the building.

FIGURE 6-2 As the first engine arrives at the A side front of a single-family residence, it should proceed just past the structure, observing one side, the front, and finally the other side during its approach. The officer needs to conduct a 360-degree walk-around size-up to check the C side rear of the house.

This permits the officer and crew to observe one side as they approach, then the front, and then the far side when they stop **FIGURE 6-2**. Only the C side rear will not have been observed, and that will be covered when the company officer gets out and performs a 360-degree size-up. If the initial IC cannot physically make a 360-degree walk-around survey, an officer from the next arriving company, even if it is a ladder or rescue company, should be assigned to check the C side. In Figure 6-2, the engine company, positioned in this manner, leaves room for a later-arriving ladder company to get into position in front of the building. This approach and position are effective even when the building is relatively tall. The important point is that the first-arriving crew obtains a quick view of at least three sides of the structure, but someone has to check the back. A situation on the C side rear can change the entire fire attack plan.

Another important observation is to determine if the structure has a basement or a below-grade level. You can add a B to the acronym RECEO VS (rescue, exposures, confinement, extinguishment, overhaul/ventilation and salvage): BRECEO VS (basement [or below-grade] rescue, exposures, confinement, extinguishment, overhaul/ventilation and salvage). Having the B at the start of this acronym helps to put the determination of a basement at the forefront so it isn't overlooked. The existence of a basement must be transmitted to all units on the fireground, especially if it is on fire. A basement fire is another situation that can immediately change the fire attack plan.

Wide-Frontage Buildings

Structures such as warehouses, garden apartments, large stores, and factories may be detached, but they usually have wide street frontage. In such cases, the first-arriving company can at least observe the approach side of the building and the front. The engine should be positioned so that entrances to the building can be used in attacking the fire **FIGURE 6-3**. It is best to try for the entrance closest to the fire because it will require the shortest hose lay. If the fire can be reached with the preconnected hand lines, this approach is best for applying fast water to control the situation. Longer hose lays take more time, require more staffing to get a single line into position, and increase the friction loss to the nozzle.

Depending on the need for assistance with the water supply or with laying initial attack lines, the second- or third-arriving engine company can approach from the rear or cover the C side. This decision is based on the strategic goals of the quick initial action plan (QIAP).

These buildings are instances where a 360-degree walk-around size-up by the first-in company officer is not possible. The C side rear still needs to be checked. One effective way to get "eyes on the back side" is to

FIGURE 6-3 In approaching a building with wide frontage, the engine should be positioned at the entrance that is most accessible to the fire. This figure depicts the perfect scenario to apply fast water onto the visible fire through the window using a preconnected hand line with tank water, thus knocking down and resetting the fire while cooling interior temperatures for any survivable victims.

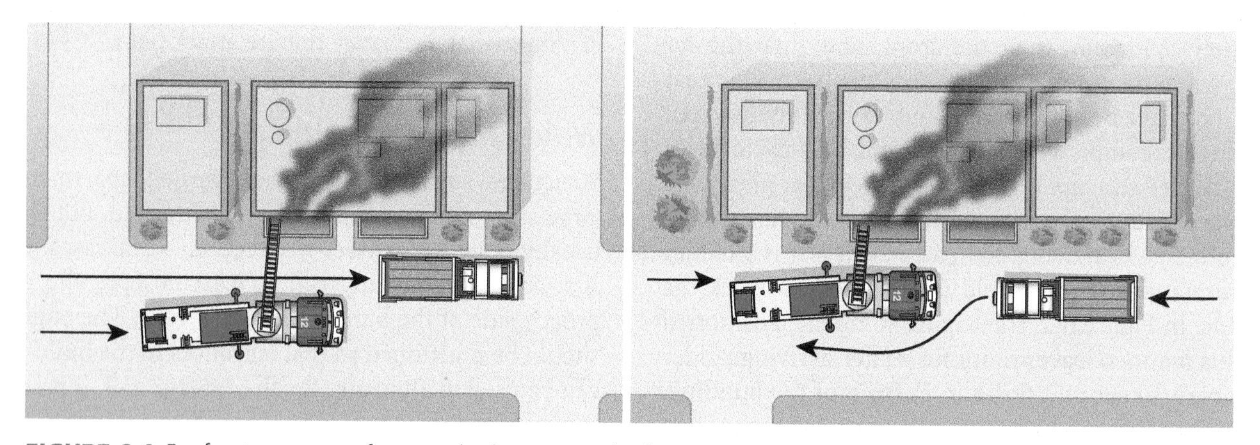

FIGURE 6-4 For front coverage of an attached structure, the first-arriving engine should be positioned according to the approach of the first-arriving ladder truck. Priority should be given to positioning of the aerial apparatus.

send a firefighter or an officer through an uninvolved adjacent building to get to the C side. Another way is to have a crew member throw a ladder to the roof (if it is safe) or the roof of an adjacent building, traverse the roof, and look down onto the C side rear to report conditions.

Attached Buildings

Fire in a building that is part of a continuing complex of attached structures, like row houses, presents a more acute problem. The sides of all but the end buildings are hidden, and the first-arriving company has a limited view of the situation. When the ladder company is approaching the fire building from the same direction, the engine company should position itself slightly past the front of the fire building. This will keep the front clear for ladder company operations. When the ladder company is approaching from the opposite direction, the engine company should stop short of the front of the building **FIGURE 6-4**. They can always drive around the ladder to get into position. Remember, you have more fire hose than aerial ladder. The ladder has to be in position to reach the roof and be effective for window rescues. It's usually easier to ladder a building at the front. Once the ladder is stopped, the engine can easily maneuver around the truck to view the third side of the structure, lay hose to the hydrant or the standpipe connection, or position wherever it needs to set up.

Companies at the front of an attached building usually have easy access to its interior. They should select the entrance closest to the fire to lay the least amount of hose. The front entrance is the best access for locating the fire alarm panel, the elevators, interior halls, and stairways in multistory buildings. In buildings equipped with standpipes, the FDC intakes are usually near the front entrance. If they are not, they are in the rear alley or next to a stairway exit door. In some older buildings—those constructed between 1900 and 1950—the standpipe is adjacent to the exterior fire escape.

Depending on the type of structure involved, the current fire situation, the number of engine companies on scene, and personnel at the fireground, positioning on the A side front of the building usually gives you the best options for advancing hose lines to attack the fire, getting hose lines to the floor above the fire, and stretching backup lines as required.

Positioning: Charlie Side Rear

As just described, covering the front of a building is usually not a problem; the first-arriving engine is positioned on the street that the building faces. Covering the C side rear may not be so easy because positioning is affected by the layout of the building, the block, or the area in which the fire building is located. Many times, vehicle access is only in the front. Nevertheless, an engine crew must get a line to the rear of the building. If there is vehicle access to the C side rear, it's good practice to have the second- or third-in engine positioned there with a supply. If the second engine is needed in front to help establish a water supply, then the third-due engine should take the C side rear. Often there are accessible fire hydrants from the rear of the building. Most modern fire apparatus have an in-cab mobile data computer where the map features show the locations of all the hydrants in the area. If this information is not available, every engine should have a district map notebook with the location of all the hydrants plotted on the map.

There are many neighborhoods in which the backyard of a house is separated only by a fence from the backyard of the house one block over. Sometimes the engine has to get a supply and position one block over, lay hose lines through the yard of the exposure house, and take out a section of the fence in order to access the C side rear of the fire building. Do not hesitate to take the fence out. The fire needs to be contained; fences are easy to repair.

Detached Buildings

Companies assigned to rear coverage should use alleys, service roads, or driveways to get into position quickly. Engine companies should be positioned to allow room for ladder companies to get to the rear of the building. This may mean moving beyond the rear of the building or stopping short of it, depending on the direction from which the ladder truck would normally approach.

When one engine company has allowed space for a ladder company to position, other drivers should take care not to block this space. If there is room for more than one apparatus, an effort should be made not to block narrow alleys or other restricted passages that are the only available access routes. An engine should avoid parking directly behind a ladder truck so that it prevents the ladder company from removing their equipment or from getting to the rear of the building. An example is when an engine parks within 25 feet of the rear of the ladder truck. The rear compartment is where all the ground extension ladders and long pike poles come off the rig. A two-section 35-foot ladder needs about 25 feet of rear clearance to be removed from the truck, as do the Bangor ladders. The truck crew also needs extra room to swing them around **FIGURE 6-5**. This is a major error on the part of the engine driver. Care should also be taken with parking next to ladder trucks as well. Newer apparatus have roll-up compartment doors, but the majority still have compartment doors that open out or swing up. Make sure room is left for truck crews to access their long tools that are stored traverse inside the equipment compartments.

Sometimes there is room only for one apparatus on a narrow road. Other crews must adapt to the situation. In some residential, light commercial, or industrial areas with detached buildings, rear accessways

FIGURE 6-5 Engines should never park directly behind a ladder truck. Stowed ground ladders ladders require 25 to 35 feet of clearance to be removed.

may not be wide enough for a pumper. This means that the company or companies assigned to cover the C side rear must position their apparatus on the street and carry their hose lines, ladders, and equipment to the rear of the building **FIGURE 6-6**.

In some areas, driveways and parking lots permit positioning of the engine close to the rear or alongside the fire building **FIGURE 6-7**. This position can be taken if the driveway is not too close to the building and if fire and wind conditions allow such action.

FIGURE 6-6 When no rear access to a building is available, engines need to be positioned in front of the building or on a side street so that hose lines, ladders, and equipment can be carried to the C side rear.

FIGURE 6-7 Where alleys or other access routes allow, engines may be positioned at the rear of the building.

Pay special attention to underground garages. Some accessways that go over the underground garage may not be able to support the weight of fire apparatus. Weight limit signs will be posted. These situations should be identified during pre-incident planning.

Attached Buildings

Attached buildings make it unpractical for the initial IC—the first-arriving company officer—to perform a 360-degree walk-around size-up. In these situations, the IC can ask the second- or third-in engine, or even the ladder truck, to drive around the block (or as close as possible) to size up and report conditions on the C side rear before being assigned. The first-in battalion chief should definitely drive around to get a 360-degree perspective before reporting to the front and assuming command.

Rear coverage should not be downplayed because nothing is showing. C side coverage is important in attached buildings because the backs of buildings on adjacent parallel streets are often fairly close to each other, especially in apartment buildings. In such buildings, the C side rear windows, porches, and balconies must be checked for victims in need of rescue.

When all the buildings on a street are attached, there is usually an alleyway behind them. The second-arriving engine company can proceed up the alley from the cross street to the C side rear of the fire building **FIGURE 6-8**. Again, this company officer provides the eyes for the initial IC and should perform and report a C side rear size-up of the building and existing conditions. The third arriving engine can assist with the water supply to the first-in engine.

If the pumper cannot be driven to the rear of an attached building, it must be positioned at the front or the end of the row according to the location of the building on fire. Firefighters can also get to the C side rear by going through the fire building (when it is safe), an adjoining exposure building, or perhaps a breezeway or walkway. On commercial structures, the C side rear doors may be difficult to access. To prevent burglaries, the owner will have heavy-duty steel doors, in addition to substantial locking systems and other deterrents. Anticipate that forcible entry may be required.

If the fire is on the upper floors, it should be easy for crews to pass to the rear through the fire building; however, this might require positioning the pumper where it would add to the congestion on the A side front of the fire. It is better for firefighters to use an adjoining building or one on the next parallel street that is back-to-back with the fire building to accomplish their objective **FIGURE 6-9**.

Engine companies at the rear might not find it as easy to get hose lines into position as they would at the front of the building, unless there are ample rear alleys. Actions necessary to get hose lines into position

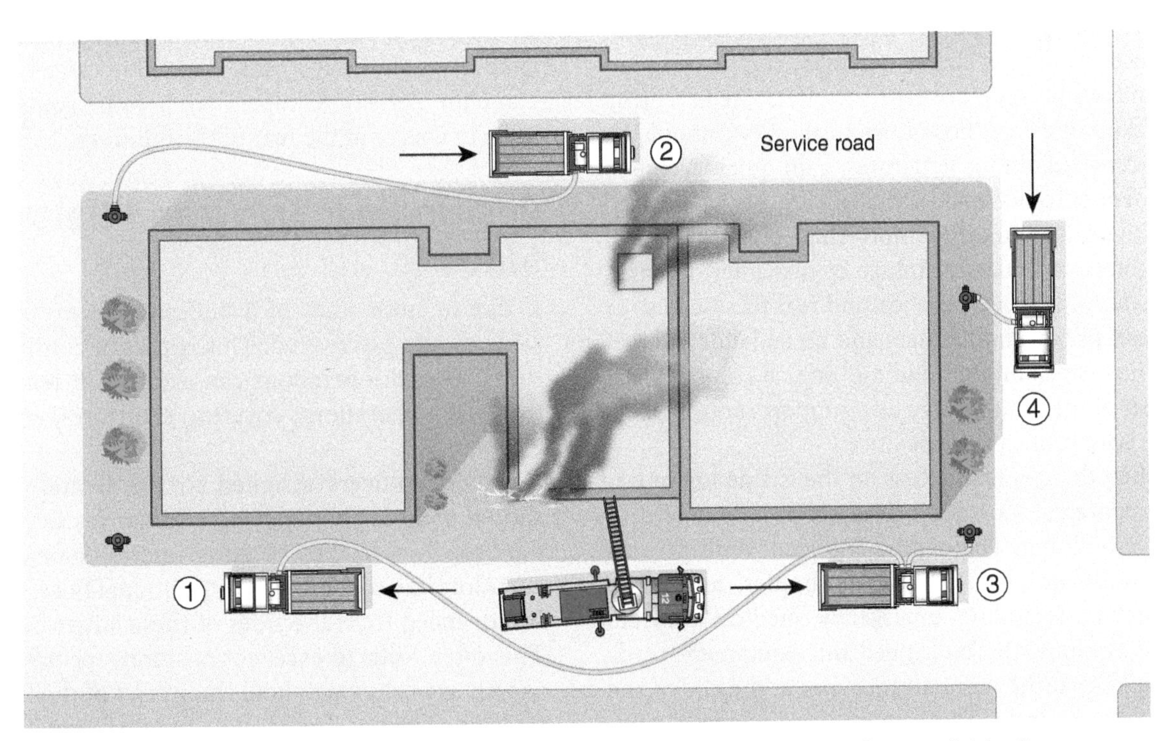

FIGURE 6-8 Positioning at the C side rear of an attached building should be made from available alleys or service roads.

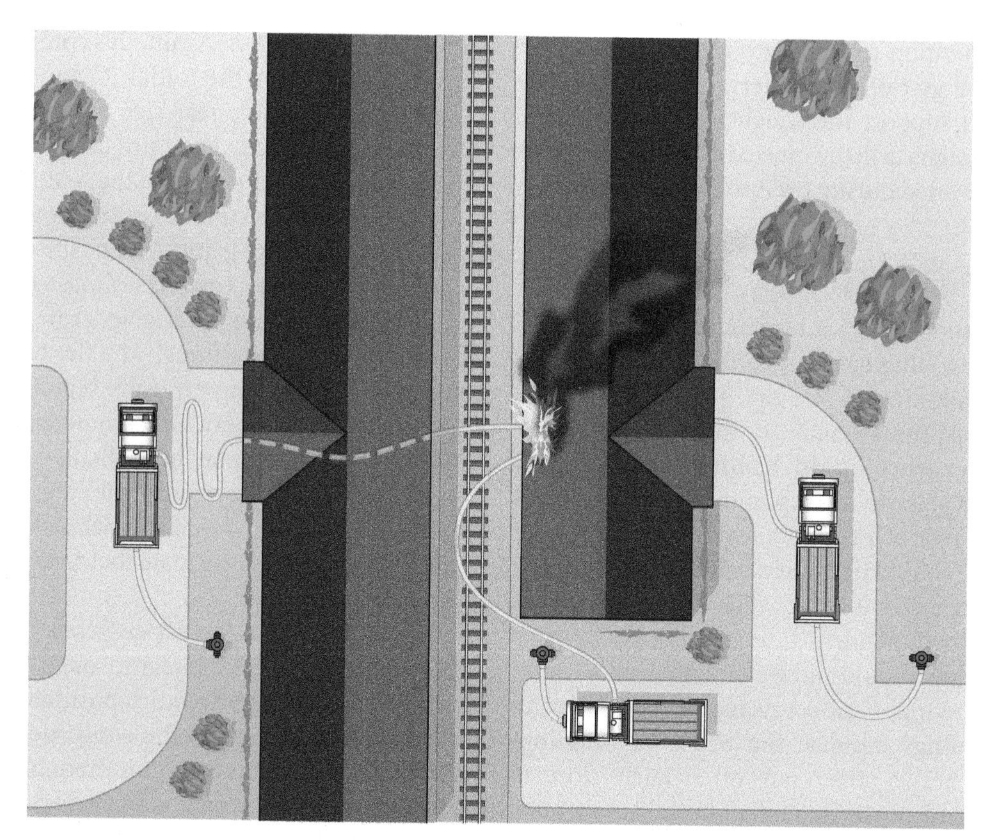

FIGURE 6-9 Where there is no street access to the rear of the fire building, hose lines can be stretched through the inside of the exposure building. Fire dispatch must notify railway dispatch to stop the trains when fire hoses are laid over train tracks.

vary with the situation. Note: Any time fire hoses are laid across railroad tracks, the initial IC must radio fire dispatch to immediately contact the railway dispatcher to stop all train traffic on the affected tracks.

Once C side rear access is gained, engine companies can advance their hose lines to attack the fire, bring in backup lines, get hose lines to the floor above the fire, access adjoining structures, gain quick access to basement entrances, and carry out any other required operations. In buildings more than one story high, firefighters can take advantage of rear interior stairs, rear exterior stairs to porches, and rear fire escapes, or they can use ground ladders and aerial ladders to get their hose lines into a building. Engine companies at the rear of the building can also support rooftop operations with protective hose lines.

When there is rear access on the C side for an engine apparatus, it's a good decision to lay forward to the fire building. The engine is a giant toolbox with extra hose, nozzles, ladders, air bottles, and other equipment, including emergency medical services (EMS) equipment. You'll need this equipment readily available at the rear entrance, not at the end of the alley at the hydrant. This engine also becomes a giant manifold for additional hose lines and master streams that may need to be put into service. A third engine can hook up to the hydrant to increase the water supply pressure to the second engine. The engine officer at the rear should report rescue profiles and/or fire conditions to the IC. Most likely this officer will become the Division C supervisor because someone has to be in charge at the rear of the building.

Positioning: Bravo and Delta Sides

If one or both sides of a building are exposed, the sides should be covered. This is part of confining the fire, and such operations can also aid in rescue, initial attack operations, covering exposures, and rapid intervention.

Engine officers assigned to the B and D sides should consider options and alternatives for advancing hose lines to the fire. After interior stairways and corridors have been covered, additional hose lines can be advanced from the sides of these larger buildings. This often leads to excellent positions for fire control as well as positions to hold the spread of the fire. Side interior stairways, fire escapes, porches, and balconies can be used to advance hose lines if interior crews are having trouble advancing and if such actions do not hinder occupants from escaping the building. However, the side access should not be used if this will be the exhaust portal for horizontal positive-pressure ventilation. Coordinate the ventilation plan with the ladder company.

Ladders at the sides of the building can be used to get hose lines around and above the fire and thus reduce the number of hose lines that can clutter stairways. Three or four 1 ¾-inch hose lines inside a stairway make it nearly impossible to advance any hose except the last one laid. Two 2 ½-inch hose lines inside a stairway are about all that firefighters can effectively advance. This is called hose layering. Hoses on the bottom become buried by the weight of the other charged lines, and the friction created between the jackets makes it extremely difficult for the hoses to slide over each other. Hose lines stretched in from the side might find a better vantage point to knock down the fire ahead of those crews advancing up the stairs and down corridors, thereby lessening the hazardous conditions faced by the crew on those lines. If this tactic is used, companies must coordinate the fire attack, communicate by radio, and be cognizant of operating opposing hose lines against each other. If opposing hose lines occur, the crews coming in from the sides should shut down and back out. The other precaution that must be taken when accessing the fire from ladders placed on the sides is to ensure that firefighters do not end up in a high–low flow path. The company officer must recognize this situation.

Engine company ground ladders are one of the most underutilized tools on the fireground; firefighters often leave that assignment for the ladder company. However, ladder trucks do not carry many 24- or 26-foot ladders; they tend to carry the longer ladders. A 24-foot to 26-foot ladder is the best ladder for fires in single-family residences, and most of those sizes are carried on the engines. If possible, aerial and ground ladders should be placed on all four sides of the structure for access and egress from the building. Ladders hanging on the side of a parked engine serve no purpose **FIGURE 6-10**. Once the pump is set, the driver should remove the ladders and get them in a forward position in case they are needed. If staffing allows, place ladders at every room window on the fire floor. In a sudden flashover, any window can become an emergency exit point for a trapped firefighter trying to escape. The next time you're on a fire scene, count how many ground ladders are left on the engines, serving no purpose other than decoration. When you need a ground ladder, you need it now! Valuable time is wasted if firefighters have to run all the way back to wherever the engine is parked to retrieve a ladder.

FIGURE 6-10 A 24 to 26-foot ladder is the best ladder for fires in single-family residences. Don't leave them on the apparatus.

Courtesy of John Odegard.

In one-story structures, especially large well-involved structures, hose lines from the sides can be used to bring additional streams to bear on the fire. Again, these activities must be coordinated with the crews attacking the fire from the front and rear to promote efficiency and safety.

General Front, Rear, and Side Operations

Emphasis in this section has been on the need for complete coverage around the fire building and the positioning of apparatus for the best strategic advantage. Complete coverage ensures that exposures are protected by hose lines placed for maximum effect **FIGURE 6-11**. In terms of rescue and firefighting operations, complete coverage provides the following benefits:

- The IC has eyes on all four sides of the building.
- The location of the fire is determined.
- The full extent of the fire and where it's headed is known.
- Safety hazards and rescue situations can be identified or forecasted with few, if any, surprises.
- Severe exposure problems are discovered and protected.
- All companies surrounding the fire building can work together, assisting each other for maximum efficiency.
- Firefighters are already in place on all sides of the building to assist should a rapid intervention situation develop.

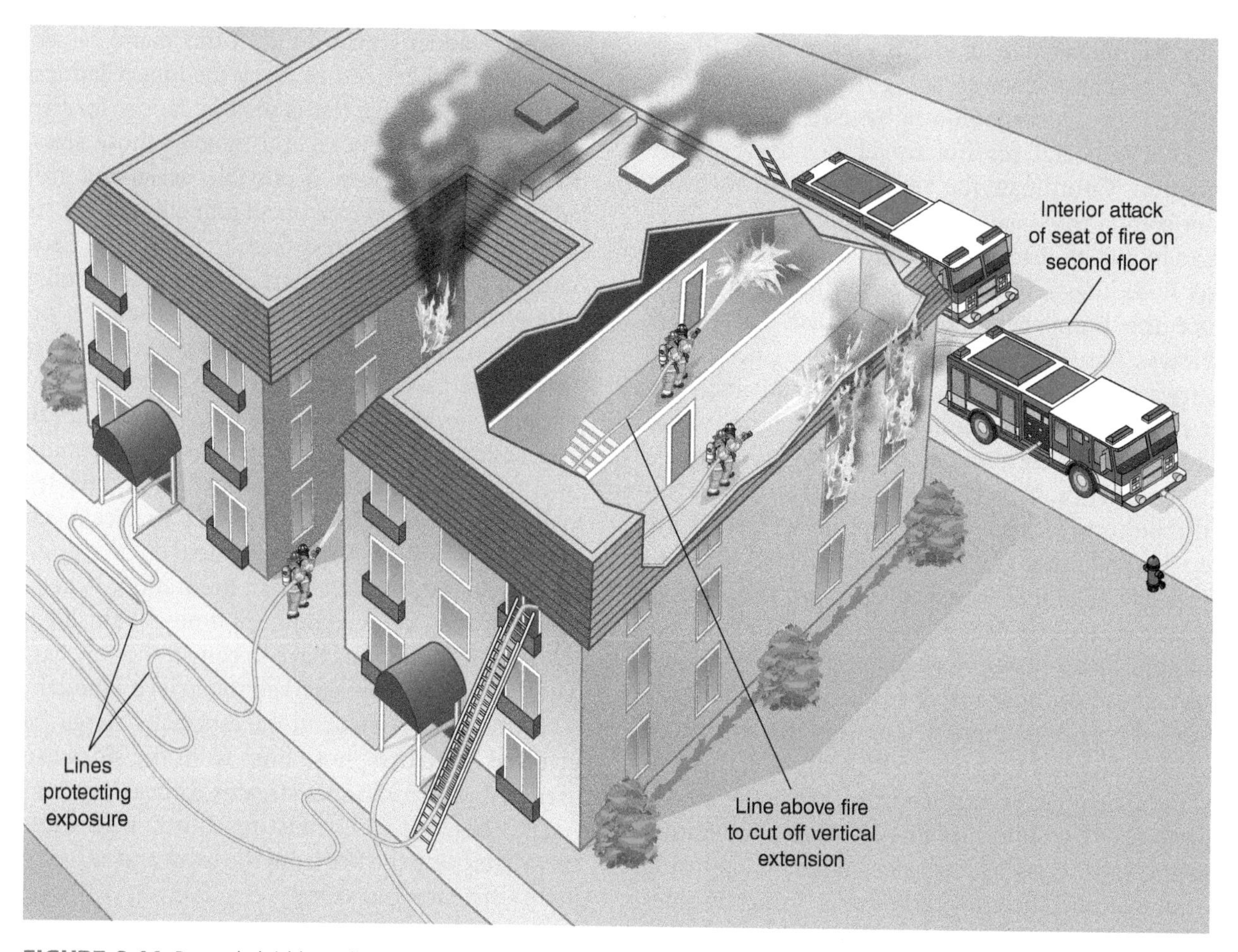

FIGURE 6-11 Properly laid hose lines protect exterior and interior exposures.

Problem Buildings

Some buildings might fit into one or more of the categories discussed in the last section, but they may present additional positioning and coverage problems because of their size, construction, location, or use.

Setback Buildings

A detached structure might be set far off the street, with a long drive leading to it and no room in which to maneuver apparatus **FIGURE 6-12**. There may be other barriers or hindrances such as trees, a wall, or fence that block the proper positioning of apparatus. Many long private driveways have security gates; if passkeys or fobs are not provided in a Knox-Box, forcible entry may be required. In such cases, the first engine company to arrive should take the hydrant and lay forward into the property. If no hydrant is within 100 feet of the entrance, then lay a split lay. Approach as close to the building as fire and wind conditions permit. Use preconnected attack lines and tank water to knock down the fire with fast water. If the fire is small enough and a size-up and risk-benefit analysis is done, the officer may go offensive if the fire can be extinguished with less than 500 gallons. Because they laid forward, the officer can comply with the two in/two out rule. If the fire requires more than 500 gallons, the officer must wait for a supply and remain in a defensive (noninterior) mode.

The second engine company should connect their large-diameter hose (LDH) to the split lay. The officer and crew should grab two bundled sections of 1¾-inch hand lines and report to the first-in engine officer. The driver should continue the lay, hook up to the hydrant, and supply the first engine. The crew of the second engine can be assigned to the C side rear or B and D sides for coverage. The second engine officer thus has a chance to size up the far sides and the rear if this could not be done from the road when she or he arrived or if the first-in officer hasn't yet performed a 360-degree walk-around size-up.

Shopping Malls

Large enclosed major shopping malls are usually long rectangular shapes. They can take up an entire city

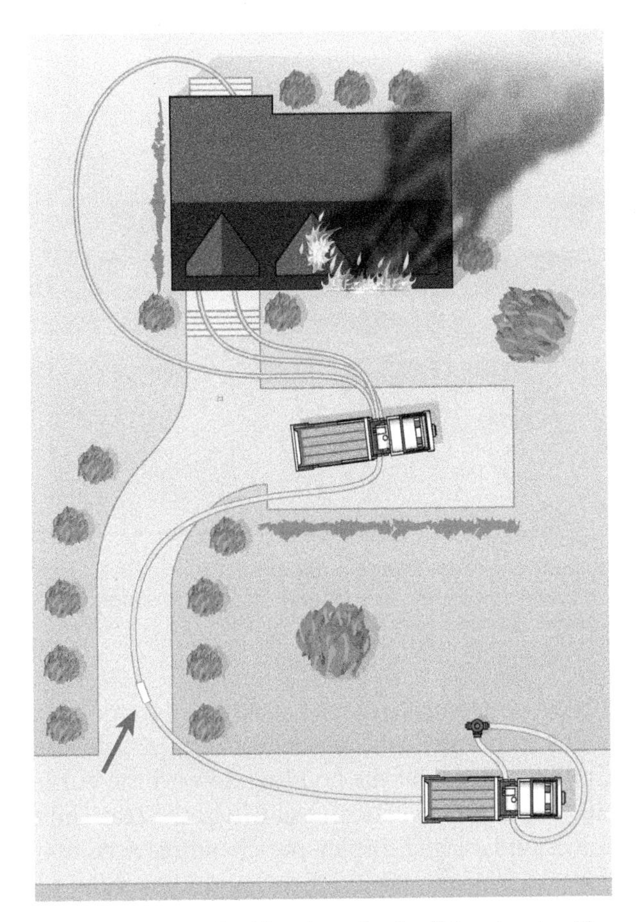

FIGURE 6-12 This building is set back off the street, with a long driveway and little room for an apparatus to maneuver. In this situation, the first-in engine should lay forward or perform a blind alley or split lay.

block or even two, and may have two and even three interior levels. Most are designed to allow direct entrance into the central mall and court area through large doorways on all sides of the structure. Food courts have numerous side-by-side flame grills and kitchens. With millions of lights, electrical displays, and tons of modern synthetic fuels, the fire load is tremendous. Hose lines can be carried through the doors with little problem, but in case of a fire, you may be battling the crowds who are trying to escape using the same doors you're trying to enter. All large shopping malls have a horizontal standpipe system throughout and are fully sprinklered. Even so, long hose lays are required. Each hose line needs to be between 200 and 300 feet. Two-and-one-half-inch and 1½-inch wyes need to be part of the equipment taken in. As engine companies arrive, there needs to be a *minimum* of four hand lines taken in, but more will be required, especially if the mall has multiple levels, including:

- An attack line
- A backup line
- An exposure line for the right occupancy
- An exposure line for the left occupancy
- An exposure line for the rear interior service corridor
- Exposure lines for the upper levels
- Two-and-one-half-inch lines for portable master streams
- An exposure line for rooftop operations

This allows coverage on the front of the individual stores and offices inside the mall. The rear of individual stores and offices is covered from the sides of the structure in which they are located FIGURE 6-13.

Malls are usually surrounded by parking lots or roads with numerous hydrants. Thus, there is ready access to all sides and entrances of the mall for front, rear, and side coverage of a fire in a store or in the central court itself. When standpipes are not provided, unusually long attack lines are necessary. Companies should already be aware of such situations and should prepare and practice for this operation. Dry hose drills can be set up with mall management after operating hours where actual hose lays can be performed. Such a drill should include positioning of 1¾-inch and 2½-inch hose lines and master stream appliances. Pre-incident planning and live multicompany drills certainly benefit firefighters if a fire should occur in a mall.

Standard Shopping Centers

Standard shopping centers usually consist of attached blocks of buildings. Front and rear coverage here is as important as in the situations discussed earlier. Most shopping centers have rear alleys or roadways for delivery trucks, so there should be plenty of room for apparatus to maneuver. Loading docks have a heavy fire load with high piles of merchandise or products. Sometimes rear doors and windows permit ready access to the back of the building. In most cases, rear windows are limited in size, are barred, and/or are located high on the wall; doors are made of steel and are securely locked from the inside to prevent theft. However, most windows can be used for fire attack or ventilation. Ground ladders off the engine are essential access tools. Commercial doors on the C side rear are beefy and fortified. Forcible entry could be time-consuming and labor-intensive. As a last resort and if conditions warrant, walls may be breached to gain entry, perform rescue, or attack the fire. Again, breaching a commercial wall, unless it's made of wood, is labor-intensive and time-consuming. In addition, the side and rear walls are bearing walls; breaching

FIGURE 6-13 In covered malls, engine companies should be positioned to take advantage of the best entrances in relation to the fire. An engine should connect to the FDC for the fire protection system as soon as possible and supply it.

a concrete or masonry bearing wall can have catastrophic consequences if the firefighters choose the wrong area to breach. It is extremely unlikely any member will have the structural engineering expertise to select the portion of a wall that is safest to breach. Anyone who does have that structural engineering expertise probably doesn't work for the fire department. Unless it's for a confirmed rescue where someone is trapped in a corner room without exterior doors or windows, breaching a bearing wall should not be attempted.

Mercantile Areas

Rear coverage is vital in mercantile areas in general. Many of these fires start in the work/storage sections behind the sales floors (i.e., in the rear of the store). Utilities (water, gas, and electricity) pass from store to store at the rear and pierce the walls. Fire can spread rapidly through these utility openings. Fire companies positioned at the rear may be the first to notice this fire spread and take action to stop it with fast water application.

Remember, big fire needs big water. Due to the fire load and large square footage associated with commercial occupancies, the company officer should be thinking "2½-inch" for the initial hose line pulled. In the Sofa Super Store Fire in Charleston, South Carolina, on June 18, 2007, the showroom square footage ranged between 42,000 $ft.^2$ and 51,000 $ft.^2$ (12,802 m^2 and 15,545 m^2). The loading dock was approximately 2,250 $ft.^2$ (686 m^2). (Refer to the Sofa Super Store Fire case study in Chapter 10.) In this type of fire, the main body of fire in the rear of the store needs to be attacked head-on from the exterior and knocked down with overwhelming gallons per minute (gpm). Crews should then enter and attack from the A side front or unburned side of the building, toward the burned side. This tends to contain the fire in the rear and to protect undamaged display merchandise. A combination of 1¾-inch and 2½-inch hand lines should be deployed in the interior offensive attack. You need the maneuverability of the 1¾-inch lines, but you still need the high-caliber water streams of the 2½-inch lines for reach and penetration.

Garden Apartments

Front and rear coverage are most important in the multifamily structures commonly known as garden apartments. Older garden apartments are typically Type III construction, but newer one are actually Type V wood-frame buildings with wood, brick, or masonry veneer, and they use many lightweight components and connectors. Many share common attic or cockloft spaces, and there is little to slow the spread of fire once it gets into the void spaces within the building. A mansard roof is a popular style on garden apartments and, like a common cockloft, it consists of lightweight truss framing that wraps all around the building. Once the fire enters this space, it can spread horizontally 360 degrees around the structure, outflanking the fire attack, with the potential for a ribbon or a domino-effect collapse **FIGURE 6-14**.

The manner in which these buildings are laid out is a challenge for apparatus positioning. In the usual design, half of the apartments can be seen only from the A side front, and half only from the C side rear. Thus, the first-arriving company may not even notice a fire in a rear apartment **FIGURE 6-15**. If the interior design does not permit passage through the building from front to rear, occupants in trouble in rear apartments cannot be seen or are not reachable from the front. For this reason, it is important that the initial IC assigns the second- or third-arriving engine to the C side rear and that a size-up is quickly made of this area.

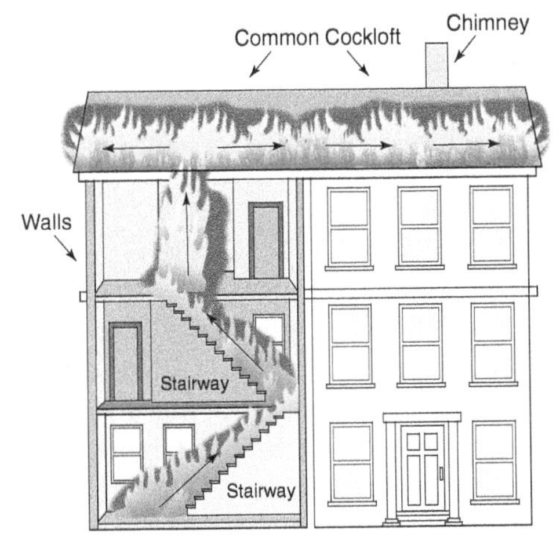

FIGURE 6-14 Mansard roof assemblies often have a common cockloft where fire can spread horizontally 360-degrees around the structure, outflanking firefighters with a potential for a ribbon or domino-effect collapse.

Another problem is the varying height of some garden apartments. Building codes usually limit these buildings to three and sometimes four stories; however, this is often measured from the street-side front entrance. If such a building is set into a steep grade, additional stories may be added at the rear. This means that firefighters on the A side front may be looking at a three- or four-story building, whereas crews on the C side rear must deal with a five- or six-story structure.

A garden apartment building that is extremely long is sometimes divided into two sections lengthwise, with a courtyard in between, blocking access to the rear through the building. Also, the outside accessway to the rear may be too narrow for an engine. If the fire is in a rear apartment near the BC or CD corner of the building, there's not a major access problem, but when it is located 200 or 300 feet or more toward the C side center, crews must find a way to get hose lines into action quickly. In some cases, lines can be stretched around to the back from pumpers positioned at the ends of the building. These are

FIGURE 6-15 Because of the interior design of garden apartments, fire in a rear unit might not be readily visible by the first-arriving engine company.

long lays, and friction loss is a consideration. LDH supply hose may have to be cross-lotted by hand with a manifold attached to the end. The hand lines can be taken off the manifold.

If extremely long hose stretches are required, another option might be to drop a crew of firefighters at the end of the building so that they can proceed on foot to the fire area on the C side rear. The engine is then located on the A side front and quickly ladders the roof. Connected uncharged hand lines are carried up the ladder, passed over the roof, and then lowered down to the waiting hose teams before charging the lines **FIGURE 6-16**. Many fire departments have had success with this evolution, but only after pre-planning and hands-on, on-site training.

Central Corridor Construction

Other apartment houses and most motels, hotels, and office buildings are laid out differently, but with designs presenting problems similar to garden apartment buildings. In these buildings, a central corridor on each floor divides the floor in two. Again, the result is that half the rooms can be seen only from the A side front, and half only from the C side rear. Because the rear apartment units cannot be covered from the front, both A and C side coverage must be put into operation immediately **FIGURE 6-17**. This means that the C side rear of the building must be checked as quickly as possible. Depending on the situation, the B and D sides of the building may also need to be covered. The first-arriving company is usually positioned at the A side front entrance because that is the major access point to all parts of the building. Clear communication between companies is essential to prevent the impact on firefighters from opposing attack lines.

High-Rise Buildings

The first-arriving company should be positioned so that crew members have ready access to the front entrance and standpipe FDC inlets. The second-in engine may be required for tandem pumping if higher pressures are anticipated. Water supplies should be established from the closest hydrants. The ladder company covers the corners of the high-rise. Other responding engine companies may be positioned on the opposite side of the high-rise where additional standpipe FDCs may be present. The rest of the responding engines should stage 200 feet away, which is known as "base." Engine company high-rise operations are covered in Chapter 15.

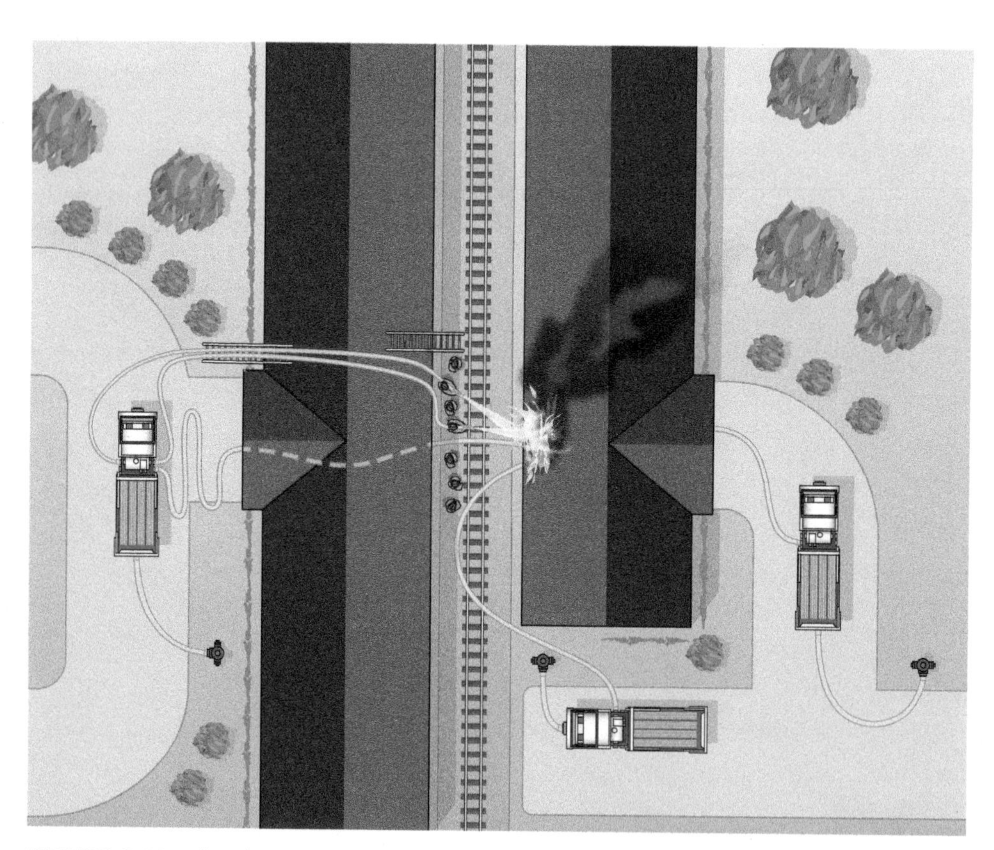

FIGURE 6-16 When long hose lays are required, uncharged hand lines can be carried up the ladder, passed over the roof, and then lowered down to the hose teams on the other side before charging the lines. Railway dispatch should be contacted immediately when hose lines are laid over train tracks.

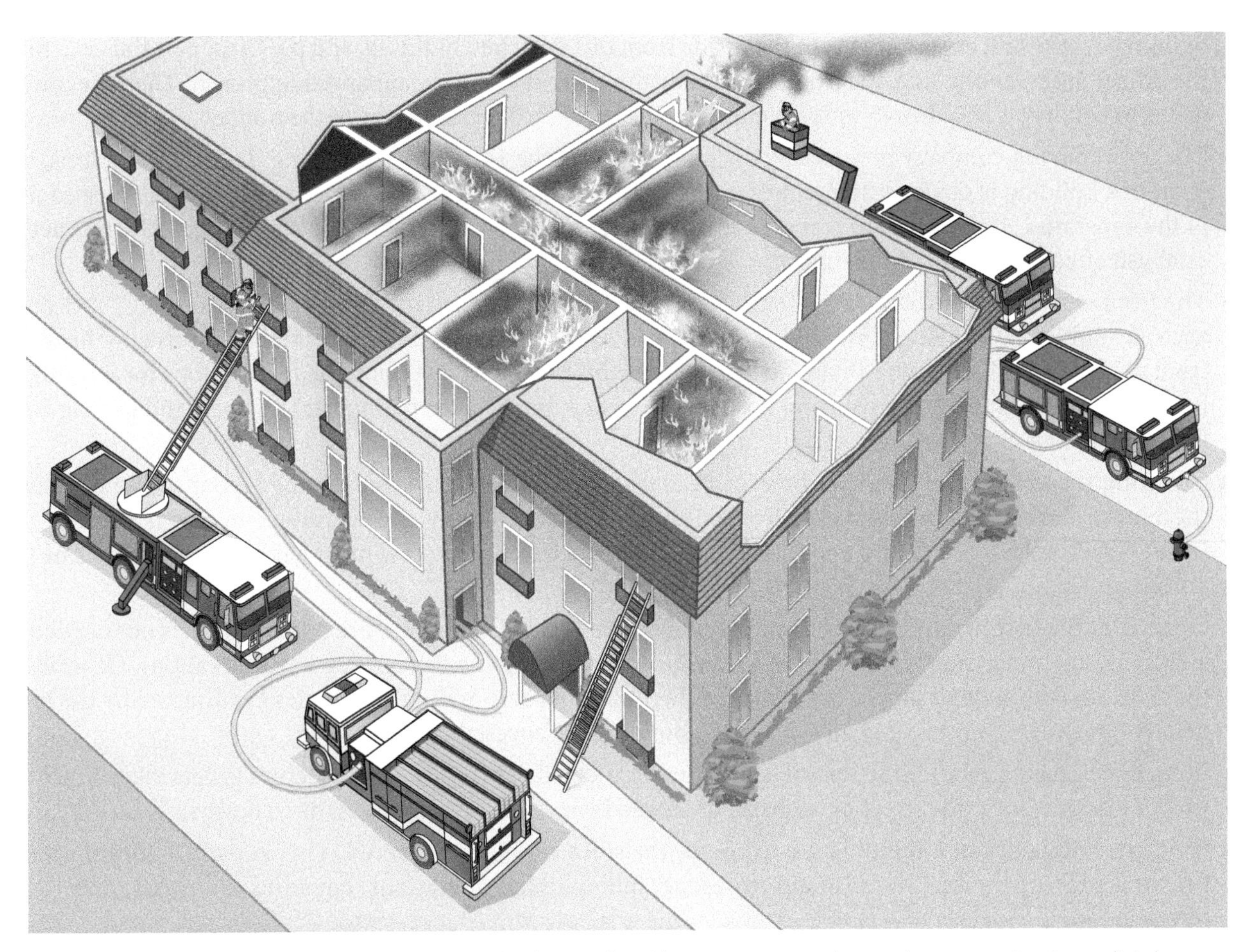

FIGURE 6-17 In central corridor construction, both A and C side coverage must be put into operation immediately.

After-Action REVIEW

IN SUMMARY

- The first and most basic step in coverage is the positioning of the apparatus.
- There are eight alphabetic directional indicators: Sides A, B, C, D, and corners AB, BC, CD, and AD. When transmitting assignments over the radio, Alpha, Bravo, Charlie, and Delta and Alpha-Bravo corner, Charlie-Delta corner, and so on, should be used.
- Development of an SOG for the positioning of apparatus and for coverage of a fire building begins with a good working knowledge of the district and pre-incident planning of the area of responsibility.
- The SOG should clearly spell out the responsibilities for the first- and second-in companies for coverage of a particular part of a building. The spotting positions taken up by the engines are based on SOGs, past experience, and anticipating the problems that may be encountered before the fire occurs.
- SOGs mandate apparatus placement; it takes the guesswork out of routine decision making and lets other incoming units know which initial actions to expect.
- No policy or guideline should be used as a substitute for initiative and the exercising of good judgment by the company officer. The officer should notify the other responding companies via radio of any deviation from assigned positions.
- The first-in company officer, who is the initial IC, must have the confidence and capability to modify or change the SOGs when the situation warrants it.
- The first-arriving pumper should pull past the fire building, accomplishing two objectives: It allows the officer and crew to see three sides of the building for size-up, and leaves room for the ladder truck in front of the

building. As the first engine arrives at the A side front of a detached building, and pulls just past the structure, the officer observes one side, the front, and finally the other side during apparatus approach. Then the officer needs to conduct a 360-degree walk-around size-up to check the C side rear of the house.

- The first-arriving company is normally assigned to cover the front of the building. In most situations, the front of a building is obvious; sometimes it isn't. The first and most basic step in coverage is the positioning of the apparatus. Both front and rear (A and C) and the sides (B and D) must be covered quickly in order to establish effective control of the fire.
- The front entrance is the best access for locating the fire alarm panel, the elevators, interior halls, and stairways in multistory buildings. In buildings equipped with standpipes, the FDC intakes are usually near the front entrance. Positioning on the A side front of the building also gives you the best options for advancing hose lines to attack the fire, advancing hose lines to attack the fire, stretching backup lines, and getting hose lines to the floor above the fire as required.
- Covering the C side rear may not be easy because positioning is affected by the layout of the building, the block, or the area in which the fire building is located. Many times, vehicle access is only in the front. Nevertheless, the C side needs to be checked, and an engine crew must get a line to the rear of the building.
- Ordinarily, the first-arriving engine company covers the A side front of the fire building. If it is not needed to assist with the water supply, the second-in company covers the C side rear in certain situations. Otherwise, the third-arriving engine should cover the C side rear. Other responding companies are directed by the incident commander to ensure that all sides of the building are covered.
- Immediate access to the C side rear of a structure is affected by whether the structure is detached, attached, wide, or narrow; near the street or set back; or served by rear alleys, driveways, or walkways.
- If one or both sides of a building are exposed, the sides should be covered. This is part of confining the fire, and such operations can also aid in rescue, initial attack operations, covering exposures, and rapid intervention.
- Access to the C side rear and the B and D sides may be impeded by obstacles of construction, security fences, or landscaping. These obstacles tend to delay the full coverage of a fire building, making it all the more imperative.
- An engine should avoid parking directly behind a ladder truck so that it prevents the ladders from being removed from the rear compartment. Ladder crews need about 25 feet of rear clearance to remove all the ground extension ladders.

KEY TERMS

coverage The assignment of companies to a particular side of the fire building or the fireground for size-up so that they accomplish any or all of the objectives of a firefighting operation.

pre-incident plan A document developed by gathering general and detailed data that is used by responding personnel in effectively managing emergencies for the protection of occupants, responding personnel, property, and the environment. (NFPA 1620)

REFERENCE

National Fire Protection Association. NFPA 1620, *Standard for Pre-Incident Planning*, 2020.

CHAPTER 7

Water Supply

LEARNING OBJECTIVES

- Explain the importance of establishing a continuous and uninterrupted water supply and the various methods to attain one.
- Recognize the inherent risks associated with laying supply lines, connecting the pumper to water sources, supplying water to the fireground, water tender shuttle operations, and relay pumping.
- Recognize the four possible sources of water at the fireground.
- Describe the components and function of an engine, whose primary purpose is to combat structural and associated fires.
- List the various factors that influence the flow of water through the fire hose.
- Describe the benefits and drawbacks to the different types of hose, diameter, length, and carrying capacity used on the fireground.
- Explain the importance of setting up a shuttle apparatus corridor.
- Describe a typical water tender shuttle operation.
- Explain the reasons for a pumper relay and describe the procedures that must be followed for proper operations.

Introduction

Obtaining an adequate and continuous water supply on the fireground is a basic firefighting task. Without water, firefighters are unable to attack, control, and extinguish a fire. Even without trapped occupants in a burning building, relying on tank water alone without obtaining a continuous supply of water is putting the firefighters at risk. The initial crew should remain in a defensive mode until a continuous supply of water is available or enough personnel and equipment are at the scene to sustain an interior offensive attack with sufficient gallonage. That means that members of the initial crew should use the available water from the tank and spray water from the outside. The UL and National Institute of Standards and Technology (NIST) studies have proven that this single action reduces interior temperatures throughout the fire area by knocking down the thermal energy put out by the fire. Firefighters depend on water to be available when needed and in sufficient quantities (a continuous supply of water) to accomplish their objectives of search and rescue and putting out the fire with the greatest margin of safety.

To achieve these goals, water supply operations provided by engine companies must be carried out quickly and safely. Standard operating guidelines (SOGs) as well as approved safe work practices should be followed. Water sources, such as hydrant systems, static water sources like lakes and reservoirs, pumpers, and mobile water supply apparatus (water tenders), as well as the supply hose, should all be addressed when considering safety issues.

Three factors influence the movement of water at fires: water source, the engine, and fire hose. This chapter discusses each of these in turn, always with the objective of obtaining maximum utilization with minimum personnel. Water delivery limitations are noted, as well as some operations that can be effective despite these limitations. Also included is a discussion of the relay operations necessitated by long supply line lays and a section on pump performance.

Inherent Risks

Obtaining a water supply is sometimes viewed as a task for those who are inexperienced or too old for battle. On the contrary, it requires a great deal of skill and experience. Although the task may appear benign, it is deceivingly dangerous. The following tasks have inherent risks regarding safety:

- Laying of supply lines
- Connecting an engine to a hydrant
- Charging the hydrant
- Supplying water to the fireground
- The usage of hose clamps
- Replacing burst sections of fire hose
- Obtaining a suction from a static water source
- Driving a tender apparatus
- Tender shuttle operations
- Relay pumping operations

Firefighters must pay attention when performing these tasks and recognize the fundamental danger of an activity that may appear to be routine but could cause serious injury or death. An example would be driving a pumper or mobile water supply apparatus during a tender shuttle operation. The driver/operator is responsible for following state law and department SOGs and getting the fire apparatus to the scene safely. This starts with wearing a seat belt **FIGURE 7-1**. According to the annual National Fire Protection Association (NFPA) report of line-of-duty deaths (LODDs), there were 69 firefighter fatalities in 2016. Overexertion, stress, or medical reasons were the leading cause of LODDs, making up 42%. Vehicle accidents were the second leading cause of firefighter deaths; 19 firefighters died in vehicle accidents, making up 25% of the LODDs **FIGURE 7-2** (with **TABLE 7-1**), **FIGURE 7-3**. Unfortunately, there are far too many incidents each

FIGURE 7-1 A professional safety attitude starts with wwwearing your seat belt any time the apparatus is in motion— during emergency or non-emergency transit.

year in which the sudden weight shift of water inside a 1,000- or 3,500-gallon tank causes the water tender or even a pumper to flip over. Many times, due to inexperience, inattention, or poor judgment by fire department personnel, the apparatus flips while taking a curve too fast, or by the rear duals going off the road FIGURE 7-4. Firefighters must consider safety as a top priority in all aspects of firefighting, including those involving water supply operations.

Engine companies exist because of their ability to supply water at the fireground. Their apparatus is designed and equipped for that job. Personnel are trained to provide a continuous, uninterrupted supply of water quickly and efficiently. The first action for the driver of an engine company arriving at the fireground is to obtain water for firefighting and rescue, and this action often starts by using tank water. When the initial action is laying supply lines, this operation should involve only two firefighters: the driver and the hydrant position. This leaves more firefighters available for the initial fire attack and rescue.

FIGURE 7-3 In 2016, vehicle accidents were the second leading cause of firefighter deaths.

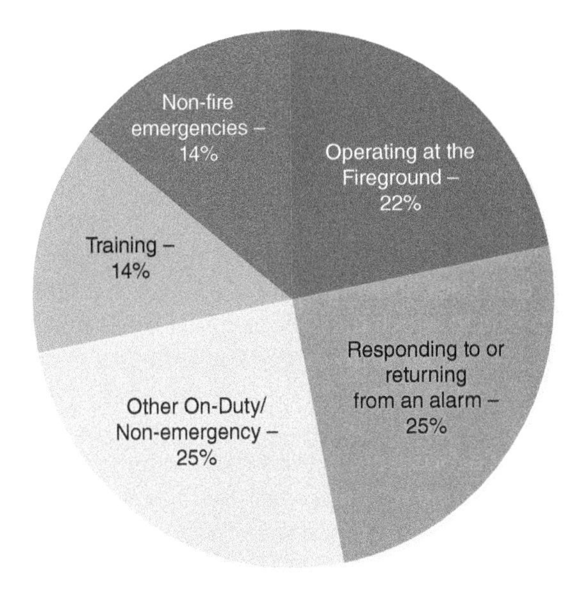

FIGURE 7-2 Firefighter deaths in the United States by type of duty. A total of 69 firefighters were killed in the line of duty.

National Fire Protection Association, Firefighter Fatalities in the United States, 2017.

FIGURE 7-4 Taking a curve too fast can cause the front wheels or rear duals to leave the roadway. Note the front wheel buried into the soft shoulder.

TABLE 7-1 Firefighter Deaths by Cause of Injury

Cause	Percentage
Overexertion, stress, medical	42%
Vehicle accidents	25%
Falls	10%
Struck by objects	6%
Other	6%
Fatal assault	4%
Structural collapse	4%
Struck by vehicles	3%

National Fire Protection Association, Firefighter Fatalities in the United States, 2017.

The Water Source

The first factor that influences the movement of water is the source. There are four possible sources of water at the fireground:

- Hydrant and water main systems, that is, pressurized water sources
- Static water sources
- Apparatus water tanks
- Mobile water supply apparatus

Hydrants and Water Main Systems

Hydrants on water main systems are the major source of water for most metropolitan city and suburban fire departments **FIGURE 7-5**; however, many fire departments may have little control over them. An engine company must accept what is available from a water main system and make the best use of it. It is important, therefore, to know the flow rates of the water system, the parts of the system servicing particular areas, and individual hydrants. Flow-test data can be used to determine these rates. The fire department should either run its own tests or be part of the process if testing is done by the water department or another agency.

There are three types of hydrants:

- Dry-barrel hydrant
- Wet-barrel hydrant
- Dry hydrant

A **dry-barrel hydrant**, or frost-proof hydrant, is installed by the municipality when temperatures can drop below freezing. The main stem valve is located below the frost line and can be several feet below the hydrant **FIGURE 7-6**. Once the top nut is opened, the entire barrel is charged with water. Hydrant gates and hydrant assist valves (HAVs) have to be connected before the hydrant is opened.

A note on safety: During emergency or nonemergency incidents, a firefighter should always stand behind the dry hydrant when charging the barrel, keeping the knees away from the steamer and hose ports. The dry barrel is not under pressure until charged. The nipples of the hose ports have been known to break off under the introduced water pressure and shoot out like a projectile with tremendous velocity. Standing in front of a hydrant port while it is being charged can cause a career-ending injury.

Wet-barrel hydrants can be used when temperatures do not drop below freezing. The barrel is always charged. Each hose and steamer port has separate valves to open or close the flow of water, and each port can be operated independently of the other **FIGURE 7-7**. And, another safety note: Although wet hydrants are always under pressure, it is still good practice to stand behind the hydrant and away from any hydrant ports when the valves are opened.

A **dry hydrant** is a large pipe with one end submerged into a static water source for the intake; the other end rises above ground level with a discharge connection to which the hard-suction hose is attached. It's like a giant straw to extend the hard-suction hose into the engine. Once it's hooked up, drafting operations can begin. The intake end has a protective screen to prevent debris from being sucked into the pump. It provides fast and reliable access to a static water supply like a reservoir, lake, or a stream **FIGURE 7-8**.

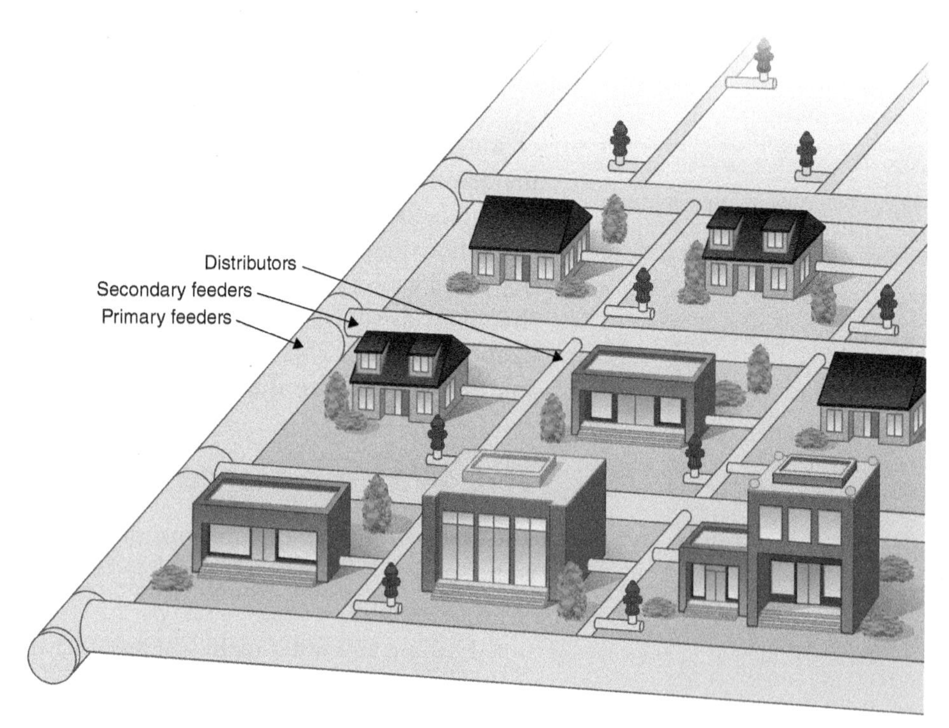

FIGURE 7-5 Hydrants on water main systems are the major source of water for most metropolitan and suburban fire departments. Underground distribution systems include primary and secondary feeders narrowing down in diameter and pressure to the distributor mains, which are connected to the hydrants.

A

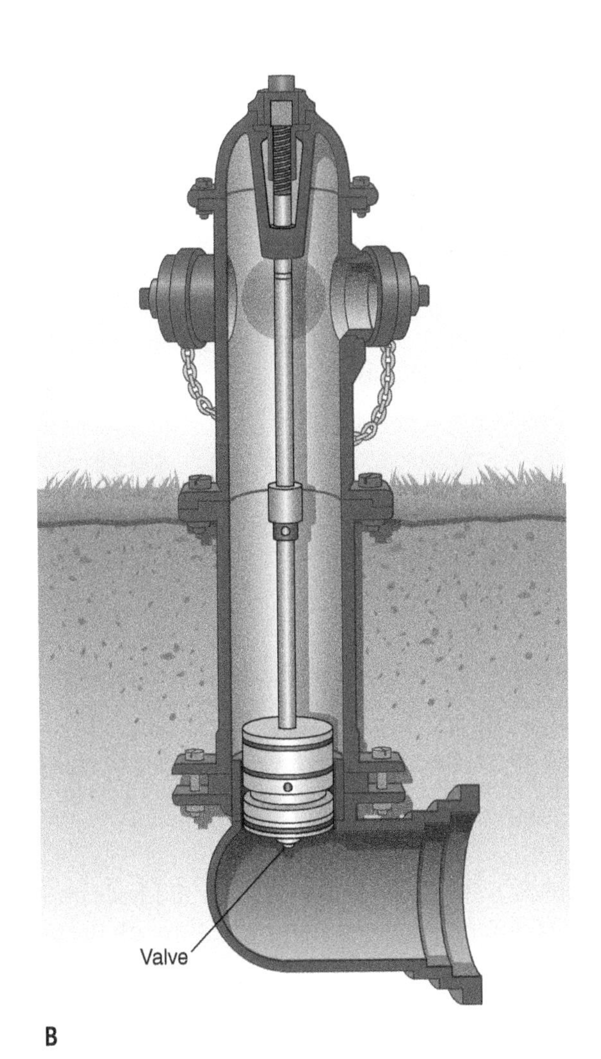

B

FIGURE 7-6 A. A dry-barrel hydrant. B. A dry-barrel hydrant is controlled by an underground valve that may be a few feet below the hydrant to avoid freezing.

A: Courtesy of American AVK Company; B: © Jones & Bartlett Learning.

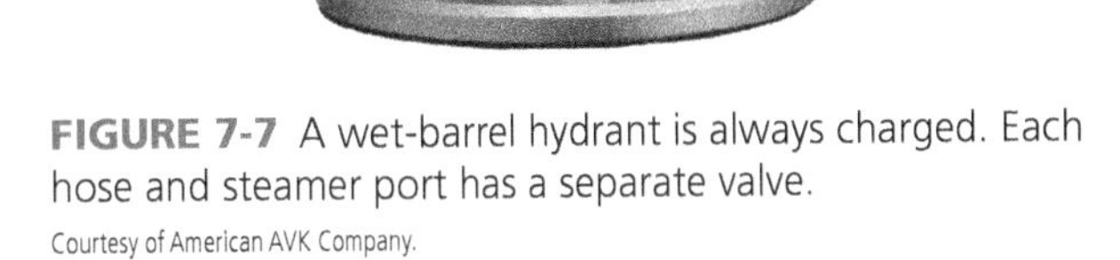

FIGURE 7-7 A wet-barrel hydrant is always charged. Each hose and steamer port has a separate valve.

Courtesy of American AVK Company.

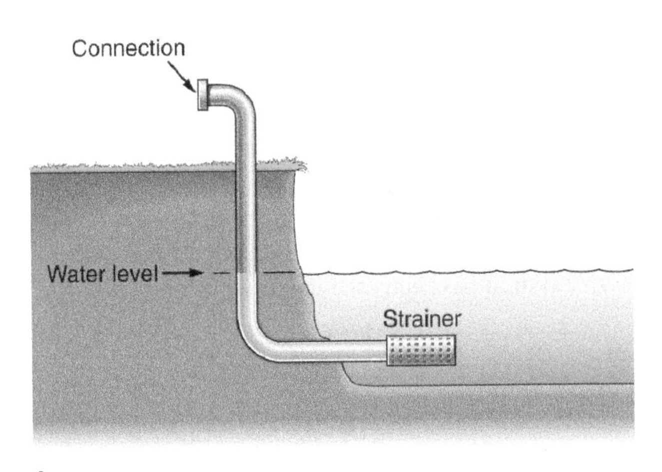

A

B

FIGURE 7-8 A. A dry hydrant is a permanent extension pipe for a hard suction. B. A dry hydrant has one outlet to connect to the hard suction for drafting.

A: © Jones & Bartlett Learning; B: Courtesy of Bill Larkin.

If a hydrant or group of hydrants provides a less-than-adequate supply of water and water pressure, additional sources must be used. Sources used to augment the basic water supply should be identified by pre-incident planning before a major incident occurs. These might simply consist of additional hose lays to more distant hydrants known to have ample water, or they can include the use of static water sources and/or pumper relays. Areas with dead-end mains and small mains are especially vulnerable to low water pressure supply problems. The use of large-diameter hose (LDH), if not already used routinely, should be considered in these areas. Increasing the number of engines responding into these low-pressure areas serves a twofold purpose. First, each engine carries around 500 gallons of water. Adding just two additional engines gives the incident commander 1,500 gallons of water for immediate use. Taking risk management into consideration, most house fires are extinguished with less than 500 gallons of water and, some would argue, with less than 100 gallons. Second, if long lays are required to attach to hydrants on larger mains, the incident commander (IC) will need the extra bodies to make that happen. Setting up long lays and **overhauling** hose by hand (also referred to as **cross-lotting**) can be labor-intensive **FIGURE 7-9**.

Among the factors that affect the flow rate of a water main system are the sizes of mains, the capacities and locations of elevated reservoirs, and the capacities of pumps if used in the system. These are all design features that have been in place for years. It would be difficult and costly to change them once the permanent water system has been established within a jurisdiction.

Reading Pressure Levels

The two master pump gauges are the pump intake, or compound gauge, and the pump discharge gauge **FIGURE 7-10**. **Static pressure** is the pressure of the hydrant water at rest, that is, with the hydrant open to the pump and no water flowing through the pump. **Residual pressure** is the pressure in the hydrant with water flowing from the hydrant through the pump. Both are read on the compound (intake) gauge located on the pump panel **FIGURE 7-11**. Together, they can be used by the pump operator to obtain an accurate indication of the available water supply from the

FIGURE 7-9 Setting up long lays and overhauling hose by hand (cross-lotting) can be labor-intensive.

FIGURE 7-10 The two master pump gauges are the pump intake, or compound gauge, and the pump discharge gauge.

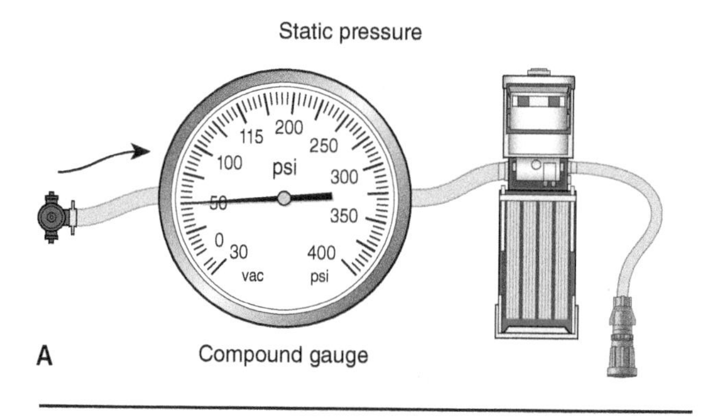

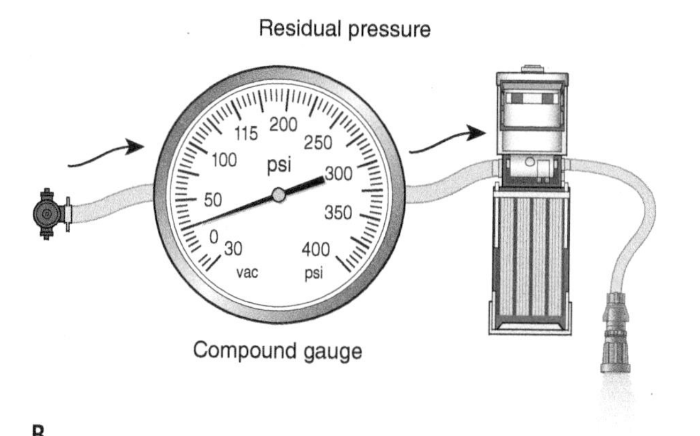

FIGURE 7-11 A compound gauge is used to measure **A**. the static pressure with the hydrant charged to the pumper and all discharge gates closed (no water flowing) and **B**. the residual pressure with at least one discharge gate open and the proper pressure being pumped to the attack line (water flowing).

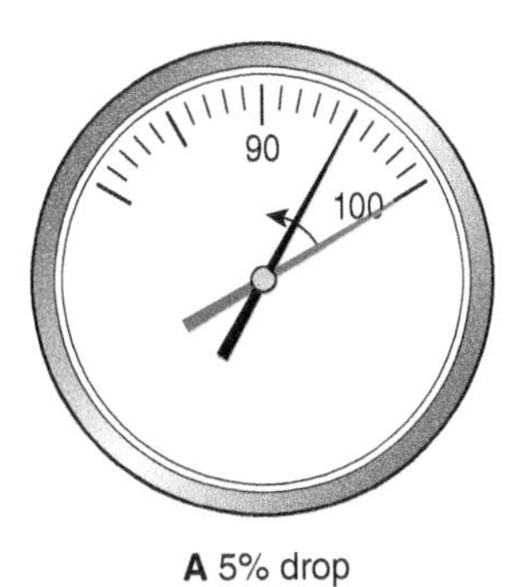

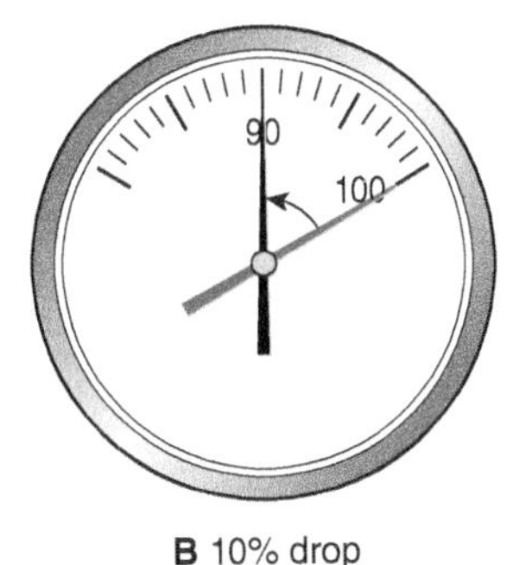

FIGURE 7-12 Pressure readings indicate how many more attack lines can be supplied by the water main system. **A.** A pressure drop of about 5%; **B.** A drop of about 10%; **C.** A drop of about 20%.

hydrant. Any hose connected from the hydrant into the pump is technically considered a suction hose, although the term is rarely used anymore. Any hose leaving the pump under pressure through a discharge port is technically called a supply. Because most cities have pressurized hydrants, the term *supply* is used almost exclusively, and "getting a supply" is understood to mean connecting to a water source.

At the beginning of the operation, the pump operator reads the compound gauge with the hydrant open and all discharge gates closed. This is the static pressure. With one attack line charged to the proper discharge pressure, the compound gauge is read again. This is the residual pressure. Static pressure by itself provides no indication of available water. The difference between static and residual pressure is what gives the true measurement.

Estimating Attack Lines

The difference in the readings allows the pump operator to estimate how many more attack lines can be supplied by the water main system. As shown in FIGURE 7-12(A), a drop of about 5% from the static pressure to the residual pressure indicates that three equal parts, equivalent to the amount being delivered, can be supplied by the water main system. FIGURE 7-12(B) shows a drop of about 10%, which indicates that two more equal parts can be supplied. FIGURE 7-12(C) shows a drop of about 20%, indicating that only one more equal part can be delivered from the water main system. Even after all attack lines are charged, the operator must continually watch the compound gauge closely to ensure that immediate action can be taken if other pumpers operating nearby cause the residual pressure to decrease below the minimum operating level.

Static Water Sources

Drafting operations are typically common in rural areas without hydrants. But they can happen anywhere. Such was the case in San Francisco, California, on October 10, 1989, when the Loma Prieta earthquake hit.

FIGURE 7-13 An earthquake can damage and put a water main out of service. In this case, supply hose must be overhauled to reestablish a water supply to residential or commercial areas.

Courtesy of John Odegard.

A natural gas main ruptured, causing a major structure fire in the Marina District. The earthquake damaged the water main system in the area, and supply hose needed to be overhauled by hand several blocks to San Francisco Bay FIGURE 7-13. Supply lines were connected to the Fire Boat Phoenix and to other engines positioned for drafting operations FIGURE 7-14. This massive effort required the help of civilians willing to carry heavy supply hose. In the end, four civilian lives were lost, four buildings were destroyed by the fire, seven buildings collapsed, and 63 buildings were severely damaged by the earthquake in the Marina District alone.

When drafting is the only way to ensure an adequate water supply, pre-incident planning is a must. Locations of static water sources, like lakes, ponds, and reservoirs; the volume of water available; and their distances from the structures they could serve should be known in advance of a fire. Alternative and supplementary static water sources should be determined as part of pre-incident planning FIGURE 7-15. When the static water source is inaccessible to the fire apparatus, a portable pump can be used FIGURE 7-16.

FIGURE 7-14 Long-lay supply lines may have to be overhauled to a fireboat or other engines positioned for drafting operations.

FIGURE 7-15 A static water source is used to get water.

Often a static water source or hydrant is located some distance from the fire structure. A pumper relay may be necessary to move water efficiently from such a source to the fireground. Water for the initial attack will most likely come from a pumper's water tank and/or a water tender.

Apparatus Water Tanks

When possible, given the size of the fire and after a risk-benefit analysis, a determination of the ability to deliver the required gallonage, and the initial hose lay, the first-arriving engine should begin fire attack with water from its water tank—that's its purpose. Most fire engines have one, if not two, preconnected 1¾-inch hand lines that are supplied by tank water (500 gallons). This procedure eliminates the need to wait until supply lines are charged or a relay is set up. Water is delivered to the fire immediately, which is called **fast water**.

FIGURE 7-16 When the static water source is inaccessible to the fire apparatus, a portable pump can be used.

After a pumper is supplying tank water to attack lines, its tank must not be allowed to run dry. The consequences to firefighters, occupants, and the fire structure could be disastrous. In smoke-visible or flames-showing situations, the engine company should establish a supply line from a water source to the fire either by a forward or a reverse lay. The company's members would have the option of either charging this supply line themselves or notifying another arriving engine company to finish hooking up to the hydrant. Fire departments have many options to accomplish this task, depending on their individual circumstances, but a continuous, uninterrupted supply of water to the fireground is paramount, and it is the objective.

Mobile Water Supply Apparatus (Tankers and Tenders)

Depending on what part of the country you're from, there are two main types of mobile water supply apparatus available: tankers and tenders **FIGURE 7-17**. *Note:* There is still resistance to and debate about using one term over the other: *tanker* or *tender*. The

FIGURE 7-17 A mobile water supply fire apparatus—a tender—can deliver hundreds and even thousands of gallons of water to the scene of a fire.

FIGURE 7-18 A mobile water supply apparatus with a 1500 gpm rated pump can serve as a fire engine.

official incident management system (IMS)/incident command system) (ICS) term for a mobile water supply apparatus is *tender.* Many areas around the country still prefer to use *tanker.* Some departments have special fire engines with one hose bed loaded entirely with LDH for extremely long lays or for those times when a temporary water main may need to be laid out for extended periods of time. Two examples are a main closure due to maintenance and when a main is damaged from an earthquake. These rigs are called *hose tenders* or simply *tenders*. Whichever term you prefer, suffice it to say that, in this text, we're not talking about fixed-wing aircraft.

A mobile water supply apparatus is a vehicle designed primarily for picking up, transporting, and delivering water to fire emergency scenes to supply other pumpers or pumping equipment. Depending on the tank size, mobile water supply apparatus generally carries between 1,000 and 3,500 gallons of water. Some fire departments operate tractor-drawn tanks capable of carrying 4,000 or more gallons of water. These new apparatus are called supertankers. A major factor that limits the size of a mobile water supply apparatus for a fire department is the maximum weight loads of bridges and overpasses within its jurisdiction.

A mobile water supply apparatus may or may not be equipped with a fire pump. Most of them are. Many are equipped with a permanently mounted fire pump of 750 gallons per minute (gpm), and the larger apparatus can have a 1,500 gpm pump, making them the equivalent of a fire engine **FIGURE 7-18**. Others may have a smaller capacity pump, or a power takeoff (PTO) drive system to dump the water. Pump capacities are predicated on a fire department's needs.

Vehicles carrying 1,000 gallons of water or more with a fire pump are generally referred to as tankers. Pumper-tankers have a standard fire pump and a large water tank. This type of pumper should be used for fire attack or pumping water from a water source to pumpers on the fireground. It should not be used to shuttle water from a water source to the fireground. Vehicles carrying 1,000 gallons of water or more without a fire pump or using a small PTO-driven pump are called tenders. These vehicles are specifically used to carry water to a fire.

A mobile water supply apparatus can be used for the initial water supply until a continuous water source is obtained from a static water supply or from a hydrant system. In this case, the mobile water supply apparatus provides a pumper directly by supplementing tank water until the continuous water source is obtained through the use of supply lines.

In rural settings, a tender shuttle operation may be required to get water. When a mobile water supply apparatus is used for developing a continuous water supply, it can either supply the pumpers directly or dump its water into a portable water tank from which the pumper drafts. The number of mobile water supply apparatus needed for a tender shuttle operation depends on the distance from the fire to the water source; the size of the fire; how long it takes to fill the water tank at the source; how long it takes to dump the water at the fireground; and traffic and road conditions, time of day, time of year, and weather conditions. Operations during inclement weather or snow conditions add more time.

Large modern dump valves allow water to be offloaded with minimal time loss, thus enabling the truck to return quickly to reload **FIGURE 7-19**. Additional methods might be desired to improve the offloading rate of gravity dumps. These methods include a jet-assist or a pneumatic pump. Basically, a jet is a pressurized water stream used to increase the velocity of a larger volume of water that is flowing by gravity through a given size dump valve. A pneumatic system can be used to pressurize a tank and assist in expelling the water. To make the overall tender shuttle operation more effective, a pumper or pumpers should be located at the water source to fill the mobile water supply apparatus **FIGURE 7-20**. This method is much faster than having these vehicles fill from small pumps or portable pumps.

FIGURE 7-19 A dump valve discharges water into a portable water tank.

Water Shuttle Apparatus Corridor

To maintain the highest level of safety during the water delivery operation, every effort should be made to establish a **water shuttle apparatus corridor** **FIGURE 7-21**. This corridor is basically a circular driveway. The goal is to allow the tender to bring in and dump its water without the necessity of backing up the apparatus or performing a three-point turn. According to the current edition of the *Washington State Emergency Vehicle Incident Prevention (EVIP)* training manual, statistics between 1990 and 2001 cited that, throughout the United States, water tenders made up 2% of the fire apparatus purchased, but they accounted for 25% of fire apparatus accidents. During those 11 years, 38 fatality accidents involved water tenders, but there was actually a total of 42 firefighter deaths. Twenty-five of these accidents were caused by the front wheels or rear duals leaving the roadway (Figure 7-4); 21 of the 38 incidents involved excessive speed. And in 21 of the 38 accidents, the driver wasn't wearing a seat belt. Between 2001 and 2016, there were 13 fatal water tender accidents.

What makes water tenders potentially dangerous to drive is the fact that they are so heavy. A gallon of water weighs 8.34 pounds. For a 1,000-gallon tender, that's 8,340 pounds added to the weight of the apparatus. For a 3,500-gallon tender, that's 29,225 pounds added to the weight of the apparatus. Many of these water shuttle operations happen in rural or suburban areas where the roads are dirt, gravel, or asphalt, and almost all of them are narrow. Combine that with the shifting weight of a heavy live load, driving one

FIGURE 7-20 A pumper drafting water from a portable water tank.

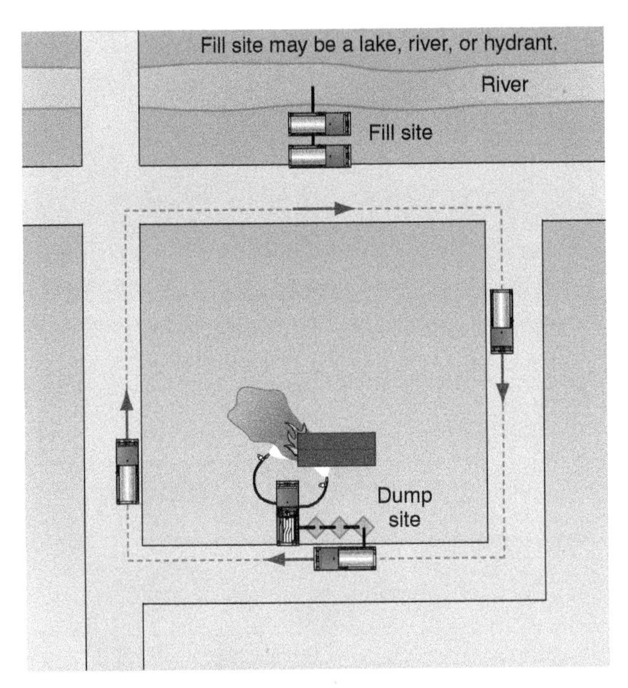

FIGURE 7-21 A water shuttle apparatus corridor.

direction with a heavy load, in a few minutes driving back 29,000 pounds lighter, then in another few minutes, taking on another 29,000 pounds. That's a significant change in driving dynamics, even with baffles. The risks associated with fireground activity, including interior fire attack, may seem greater than shuttling water to the scene. Not so. These are complicated apparatus to operate in less than ideal road conditions. Excessive speed, inexperience, or inattentiveness while being caught up in the urgency, combined with the sudden shifting of water, can lead to a rapid change in the center of gravity, which makes the apparatus easy to roll. In darkness or severe weather, this is a low-frequency, high-risk operation. Often, this apparatus is staffed with a single firefighter; therefore, the water tender apparatus should always be operated by an experienced driver.

Typical Water Supply Evolution

Let's imagine a typical scenario in which two pumpers and mobile water supply apparatus are used. The first pumper arrives at the fire scene and begins the initial attack with water from its water tank. The second pumper arrives and supplies the first pumper with water from its water tank. The second pumper also sets up to receive water from a mobile water supply apparatus. This can be accomplished by receiving water directly into the pump or setting up a portable water tank for drafting operations. (Sometimes the portable tanks are referred to as portable ponds, or port-a-ponds.) The mobile water supply apparatus arrives and provides water to the second pumper, which in turn pumps water to the first pumper. This ensures that the initial attack is maintained, there is minimal movement of apparatus on the fireground, and the water supply is located at a relatively safe and accessible distance from the fire.

After the second pumper is set up, firefighters can assist members of the first pumper in operating attack lines or performing other firefighting tasks. The second pumper can then be set up to supply another pumper located on the fireground. The number of attack lines to be placed in service is determined by the supply of water that can be maintained by the mobile water supply apparatus FIGURE 7-22.

Pre-incident planning, multicompany drills, and mutual aid drills are the key to establishing SOGs about water supply requirements. A fire department must recognize its capability, as well as those of other mutual aid agencies, and be able to work within its parameters to provide a continuous water supply of adequate volume for a particular incident.

Engines and Pumpers

The second factor that influences the movement of water is the engine. Depending on what part of the country you're from, the fire apparatus used to carry fire hose and water to extinguish the fire is called an engine or a pumper. There seems to be agreement that these two terms can be used interchangeably. The terms *engine* and *pumper* apply to the same apparatus without confusion.

An engine or a pumper is a piece of fire apparatus with a permanently mounted fire pump of at least 750 gpm (3,000 L/min) capacity, a water tank, and a hose body whose primary purpose is to combat structural and associated fires. The water delivery capacity of a pumper is limited by the capacity of the pump and by the number of suction intakes and discharges with which it is equipped. In addition to these mechanical factors, operation of the pumper is limited by engine speed and residual pressure.

Rated Capacity

The flow rate at which the pump manufacturer certifies compliance of the pump when it is new is known as rated capacity. The pumping system provided must be capable of delivering the following:

- One hundred percent of rated capacity at 150 psi (1,000 kPa) net pump pressure
- Seventy percent of rated capacity at 200 psi (1,400 kPa) net pump pressure
- Fifty percent of rated capacity at 250 psi (1,700 kPa) net pump pressure

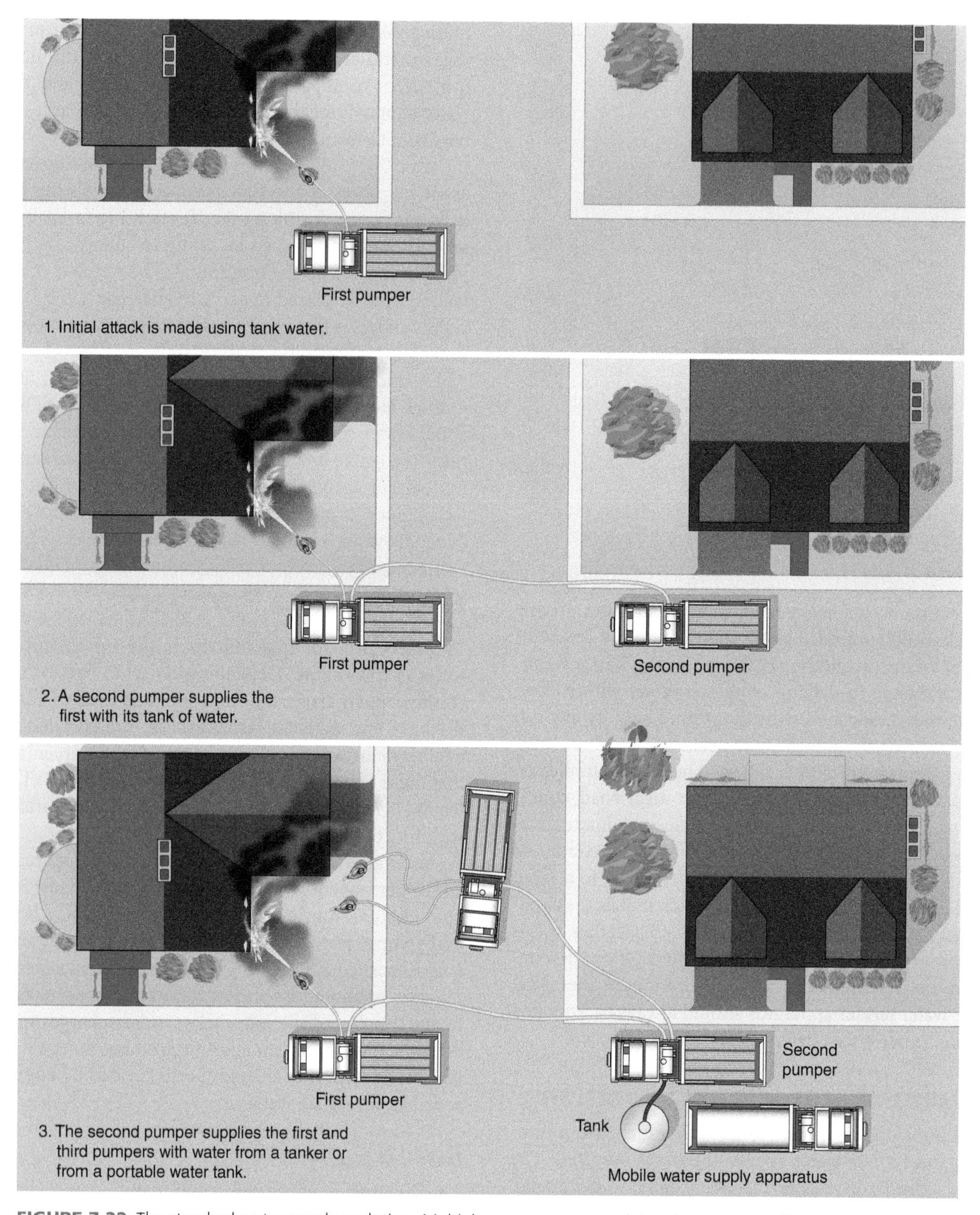

FIGURE 7-22 The standard water supply evolution. Multiple pumpers are supplying the water at a fire.

Most fire engines in major city fire departments are rated at 1,250 gpm (4,738 L/min) at 150 psi (1,000 kPa) intake pressure, or they are rated at 1,500 gpm (5,685 L/min) at 150 psi (1,000 kPa) intake pressure, making them a Class A pumper.

All pumpers are rated from draft. Thus, a 1,500 gpm pumper can draft and discharge 1,500 gallons of water per minute from a static water source. Pumpers exceed their rated capacity if they receive water under positive pressure; thus, if the pumper is at the fireground and is being supplied by another pumper at a hydrant, or directly by a pressurized hydrant, and if the hydrant delivers sufficient water, the pumper can pump more than 1,500 gpm. The limiting factors include the number and sizes of intakes and discharges.

Intakes

The pump must have at least the number of intakes required to match the appropriate arrangement for the rated capacity of the pump, and the required intakes must be at least equal in size to the size of the suction lines for that arrangement. The main pump intakes must have male National Hose (NH) threads if the apparatus is to be used in the United States. This is to accommodate hard suction hoses. However, most if not all pumpers have an intake control valve that is either a large female threaded, or a Storz connection with a ball or butterfly control valve **FIGURE 7-23(A)**.

Suction intakes are usually located on the sides, front, or rear of the pumper. When a pumper is working directly from a hydrant, the common procedure is to hook up with standard soft suction hose, generally a length of 20-50 feet (6.1-15.2 M) of LDH. This supply line is connected to one of the suction intakes **FIGURE 7-23(B)**.

When drafting, the side intakes on the pumper are closest to the pump itself **FIGURE 7-24(A)**. Intakes at the front or rear of the apparatus, or otherwise specially situated, might not allow drafting rated capacities at rated pressures due to the distance away from the pump. Using a ground ladder to support the hard-suction hose keeps it off the shore line and helps lift the hose to the proper angle for an easier connection into the pump. Tying the strainer section to the ladder secures the hard suction in moving water. The ground ladder also makes retrieving the hose much easier after the incident is over **FIGURE 7-24(B)**.

When a pumper stretches supply hose to the fire using a forward lay, it is usually done with an LDH (4 to 5 inches). Many fire departments now find it useful to use large-diameter supply hose to move water effectively from the source to the fire scene **FIGURE 7-25**. The department should evaluate its needs and choose the size and amount of hose that best support its operation.

If the hydrant does not have a steamer port but only 2½-inch discharge ports, it is necessary to hook up with 2½-inch or larger hose. At least one valve intake must be provided on the pumper that can be controlled from the pump operator's position. The valve and piping must be a minimum of 2½ inches. The intake must be equipped with a female swivel coupling

A

B

FIGURE 7-23 **A**. Most engine apparatus use a pump intake control valve, which is essentially a large, double-female appliance with a ball or butterfly control valve. **B**. A 50-feet section of large diameter soft suction hose is connected to a front intake.

B: Courtesy of Kevin O'Donnell.

with NH threads. A gated suction Siamese can also be used **FIGURE 7-26**. This adapter is placed on a suction intake of the pumper and allows for two additional supply lines, a minimum of a 2½-inch hose, to feed the pumper.

Discharges

Discharge ports of 2½ inches or larger must be provided to discharge the rated capacity of the pump. All 2½-inch or larger diameter outlets must be equipped with male NH threads. The usual practice is to install one 2½-inch discharge for each 250 gpm of rated capacity.

Discharges may be located on either side of the pumper or in the front or rear. They may be capped,

A

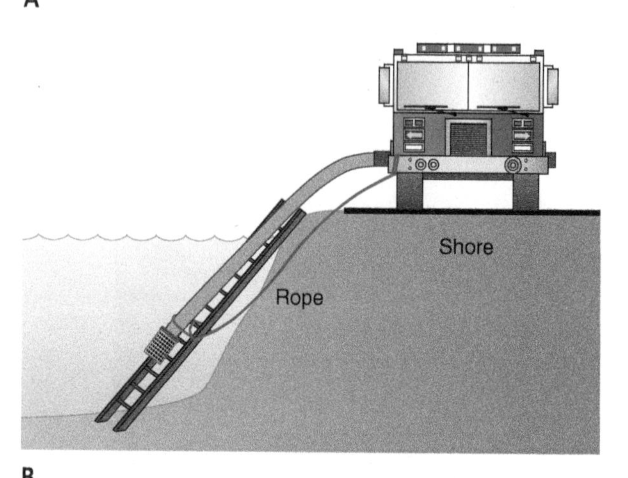

B

FIGURE 7-24 **A**. When drafting, it is preferable to use the side intakes on the pumper because they are closest to the pump itself. **B**. Using a ground ladder to support the hard-suction hose helps lift the hose to the proper angle for an easier connection into the pump. Tying it to the ladder secures the hard suction in moving water.

as at the pump panel, or they may be preconnected to 1¾- or 2½-inch attack lines. A discharge may feed a master stream appliance located on top of the pumper.

Many fire departments miss the opportunity to get the most from a new pumper. They order a large, powerful diesel engine for road performance, but they do not make full use of the engine in driving the pump. Pumps should take advantage of their full pumping capacity. Adequate discharge ports should be provided on pumpers to complement their rated capacity.

Pump Speed and Capacity

Although a pumper can deliver more than its rated capacity under certain conditions, there is a limit to how far it can be pushed. This limit may be imposed either by pump speed or true pump capacity, depending on the situation.

The pump must not be operated above 80% of its rated peak speed for any length of time. If it is, the engine, drive train, or pump may be damaged. The pump operator can tell when the pump is operating at its peak capacity by watching the discharge pressure gauge. If an increase in engine pump speed (revolutions per minute [rpm]) is not accompanied by an increase in discharge pressure, the pump is moving as much water as it can. It has reached its true capacity or the limit of the water supply **FIGURE 7-27**.

Residual Pressure

As noted earlier, residual pressure is the pressure remaining on the intake side of the pump while water is flowing through the pump. It is, a measure of the reserve capacity of the hydrant. When it reaches zero, there is no flow because there is no water left. For this reason, the pump operator must continually monitor the pumper's residual pressure. This is especially important when several pumpers are drawing water from the same hydrant system. The more pumpers drawing water, the more acute the water supply problem can be and the more quickly it can develop.

Pump operators should maintain a residual pressure of 20 to 25 pounds per square inch (psi) or more if at all possible. The residual pressure should not be allowed to drop below 10 psi, except under extreme conditions. When it looks as if this might happen, the pump operator must notify the company officer immediately while attempting to maintain at least enough residual pressure to avoid losing water completely. This is called pump cavitation or the pump "running away from water." The proper procedure is to lower the discharge pressure of the pump. This reduces the pressure on all the hose lines operating off the pump.

For safety reasons, the pump operator must never gate down a hose line to reduce the flow without the knowledge and consent of personnel operating that hose line. If it is necessary to limit the flow of water to increase residual pressure, the IC should determine what actions should be taken to correct the problem. Increasing residual pressure can be accomplished by shutting down unnecessary hose lines, half-gating or gating down hose lines, or reducing tip sizes to decrease flows. A supplemental supply line should be placed in service as quickly as possible to reestablish an adequate water supply to the pumper, which increases the residual pressure.

The residual pressure at a hydrant is decreased by increases in the pumper discharge pressure, discharge volume (gpm), and speed (rpm). Each of

FIGURE 7-25 A large-diameter supply hose is used to move water to the fire scene.

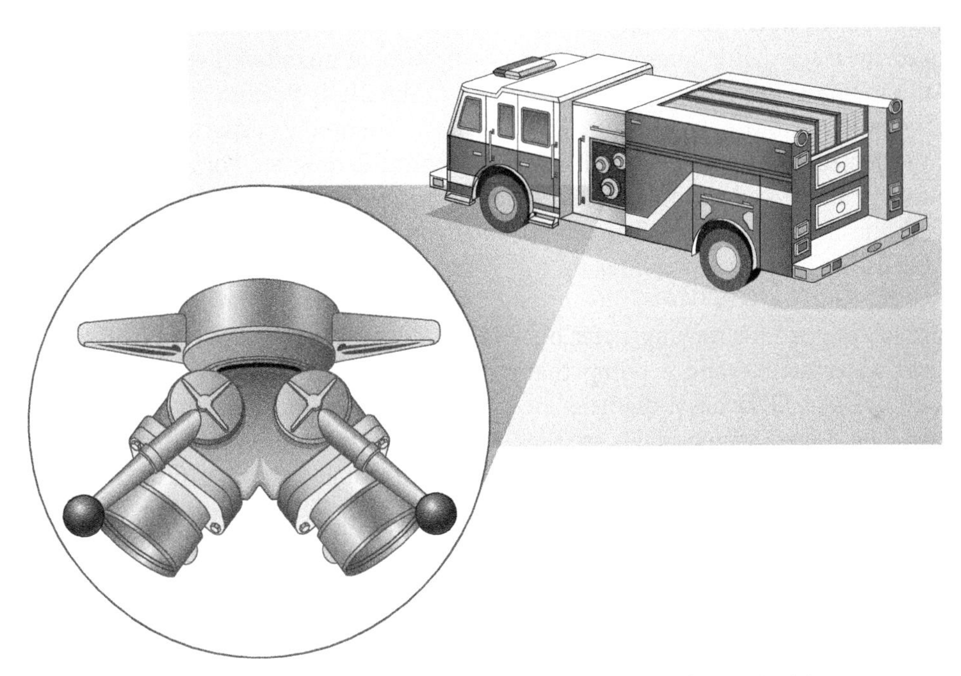

FIGURE 7-26 A gated suction Siamese is used to allow water to be received from two different supply lines. Note all couplings are female swivels.

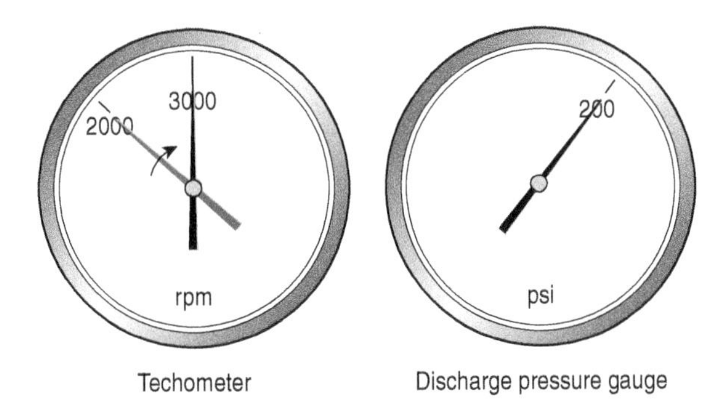

FIGURE 7-27 An increase in engine pump speed (revolutions per minute [rpm]) that does not result in an increase in discharge pressure indicates that the pumper has reached its pumping capacity or the limit of the water supply.

these factors has the effect of increasing the rate at which water is used from the hydrant. The more water taken from the hydrant, the less water remains as potential capacity. One way in which the engine company can increase the efficiency of its operations is to minimize friction losses in the supply hose and attack lines.

The friction loss characteristics of fire hose are an important consideration in the selection of hose. Friction loss varies considerably depending on the construction and design of the hose, the roughness of the lining, and its internal diameter. The type of couplings can also affect friction loss.

The Hose

The third factor that influences the movement of water is fire hose. Fire hose is the flexible conduit used to deliver water to an objective. There must be enough hose lines, with sufficiently large diameters, to carry the water. Because of friction loss in the hose, the diameter of the hose and the length of the hose line from the pumper to the fire stream directly affect the water-moving capability of the system. The larger the diameter of the hose, the less friction loss there is for a given flow rate. Conversely, the greater the flow rate in a given size hose, the greater the friction loss. **FIGURE 7-28** illustrates various fire hose sizes.

To provide proper nozzle pressures, pumpers must overcome the friction losses. Effective use of hose lines can reduce required pump pressures and permit more efficient movement of water. An attack hose is designed to be used by trained firefighters to combat fires beyond the incipient stage. The attack hose is designed to convey water to hose-line nozzles, distributor nozzles, master stream appliances, portable hydrants, manifolds, standpipe and sprinkler systems, and pumps used by the fire department. The attack hose is designed for use at operating pressures up to at least 275 psi.

The supply hose is designed for moving water between a pressurized water source and a pump that is supplying attack lines. An LDH has a diameter of 3½ inches or larger. The 4- and 5-inch LDHs are the most common sizes used in the fire service. An LDH is effective in moving large amounts of water from pumpers at a water source to other pumpers at the fire scene. An LDH also can be used by a pumper to lay a supply line from a water source to the fire scene for a continuous, uninterrupted water supply. An LDH is also used to supply attack lines, master stream appliances, portable hydrants, manifolds, and standpipe and sprinkler systems. Whenever an LDH is used for these applications, a pressure relief device with a maximum setting of 200 psi should be used.

2½-Inch Hose Line (Single Supply Line)

The standard hose of the fire service is the 2½-inch diameter line. It is the work horse of the fire department. It can be used for attack lines; supplying master stream appliances, standpipes, and sprinkler systems; and water supply. Other sizes may fill some of these needs, but not all of them.

The 2½-inch hose is not perfect, however; it does have limitations, especially when used to move large quantities of water. For example, a pumper at a fire receiving water from a hydrant, or from a pumper at a hydrant, might be attempting to supply two 1¾-inch attack lines along with a 2½-inch attack line equipped with combination nozzles. A flow rate of about 550 gpm would be required. In working directly from the hydrant, it must be known in advance that the hydrant is capable of maintaining such a flow and of providing an acceptable residual pressure at the same time. There would be a limit of an operation with a single 2½-inch hose line between two pumpers; the flow

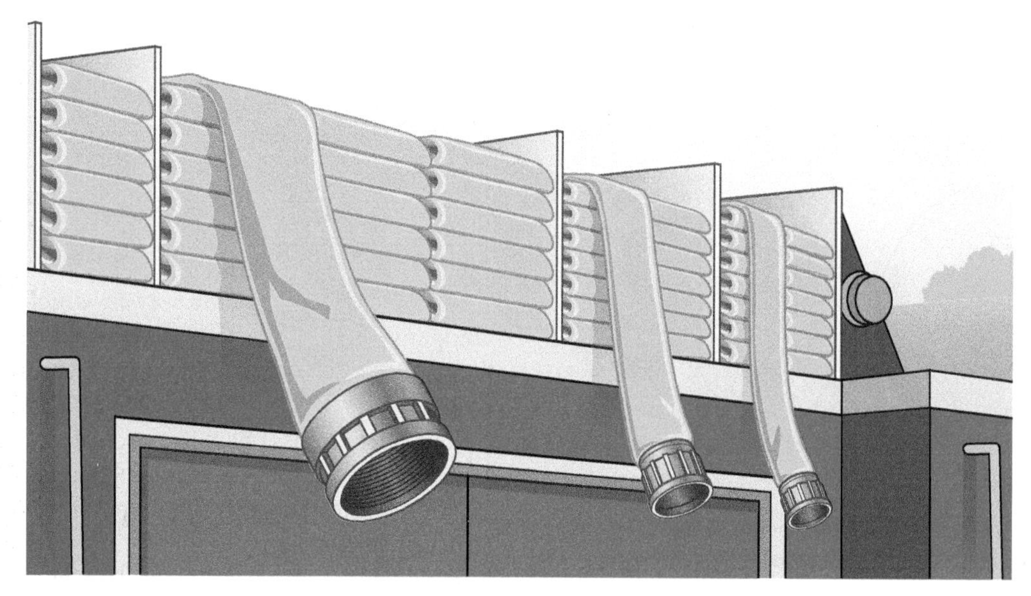

FIGURE 7-28 Fire hoses have a range of sizes for different situations.

could be maintained only if the pumpers were comparatively close to each other. Thus, a 2½-inch supply line should not be used to deliver much more than 350 gpm. A 2½-inch supply line is well suited for car fires, dumpster fires, shed fires with no exposures, and small grass or brush fires. The 2½-inch hose is often used as a second supply (a second suction) into the pump, with LDH being the primary supply (suction).

FIGURE 7-29 shows the maximum available flow using 1,000 ft as a standard supply line at a pump discharge of 175 psi. In this illustration, a single 2½-inch supply line can flow only 275 gpm.

Two 2½-Inch Hose Lines (Dual Supply Lines)

The friction loss in a single section of 2½-inch hose is 20 psi per 100 ft of hose. It would be far better to use two 2½-inch hose lines for a supply or a single LDH. Using two 2½-inch dual supply lines instead of one to move a given amount of water can reduce friction loss substantially. Suppose two 2½-inch supply lines are used to move 500 gpm. Now, because only 250 gpm are passing through each hose line, the friction loss per 100 ft is reduced to about 15 psi. Dual 2½-inch

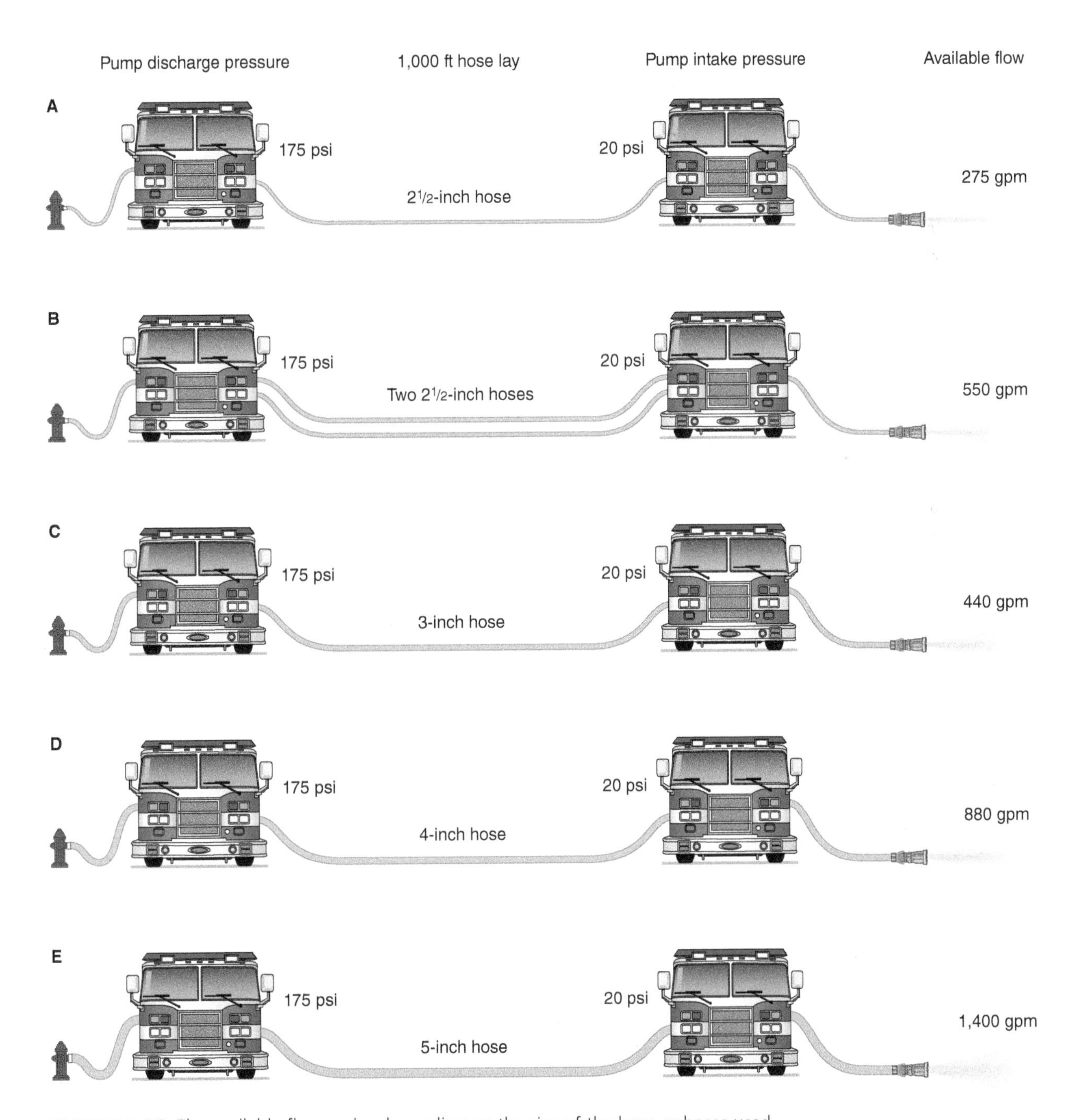

FIGURE 7-29 The available flow varies depending on the size of the hose or hoses used.

hoses are rarely used anymore to supply a pumper at a structure fire. The dual 2½-inch lines are used primarily to supply sprinkler systems, standpipes, and portable monitors.

Nevertheless, 2½-inch hose may sometimes be all that is available for a water supply. Remember, when working directly from the hydrant, the use of two 2½-inch supply lines permits a more efficient operation that allows movement of more water to the pumper than the use of a single 2½-inch line. As shown in Figure 7-29, two 2½-inch supply lines are capable of flowing 550 gpm, whereas a single 2½-inch supply line is capable of flowing 275 gpm.

As long as the total flow of water and the hose-line length remain the same, an increase in the number of supply lines of a given diameter decreases the required pump pressure. Whether hose lines are carrying water to a single point or to separate positions, the friction loss is the same in each line if the flows are the same and the hose lines are the same diameter. There is always a practical limit to the number of 2½-inch hose lines that can be connected between pumpers.

The new standard is to use 4-inch and 5-inch LDH. As noted previously, increasing the diameter of the hose decreases the friction loss as well as the number of hose lines needed to move a given amount of water.

3-inch and 3½-inch Hose Lines

The next standard diameter size of hose after 2½ inches is 3 inches. However, using 3-inch hose for a supply line has been phased out. The demand for larger flows introduced the practice of using 3½-inch hose, which was very popular for many years. With the realization that modern synthetic fuels, combined with lightweight construction, produced more rapid-burning fires with greater heat release rates, the demand for more water flow to overwhelm the British thermal units (BTUs) produced by modern fuel fires made 4-inch and 5-inch LDH the new standard for supply line. Most metropolitan fire departments around the country now use 4-inch LDH for water supply. If 3-inch hose is still being used, fire departments use it to supply fire protection systems like sprinklers and standpipes or for ladder pipe operations on aerial ladders that do not have a pre-plumbed waterway.

Large Diameter Hose (LDH)

For practical purposes, NFPA 1961, *Standard on Fire Hose*, defines an LDH as a hose with an inside diameter of 3½ inches (89 mm) or larger. Generally, an LDH measures 4 inches or greater. An LDH has made an impact on the fire service as a tool for moving large amounts of water over long distances. The hose is basically an aboveground water main for transporting water from a source to the fireground. It is also used to supply attack lines, distributors, master stream appliances, portable hydrants, standpipe, and sprinkler systems **FIGURE 7-30**. Four-inch and 5-inch LDH is used by fire departments throughout the United States in rural and suburban areas and in cities with great success. The decision to select 5-inch LDH over 4-inch is regional. There are trade-offs. More water can flow through 5-inch with less friction loss, which is ideal for low-pressure water sources, but the hose is extremely heavy and difficult to manage once it's charged. Where it's laid is where it stays. Many cities with excellent municipal water systems having sufficient flow and pressure find that 4-inch LDH is sufficient. In industrial facilities, 7¼- and even 10-inch LDH is not uncommon.

An LDH provides the most efficient means of minimizing friction loss and developing the full potential of water supplies and pumping capacities. An

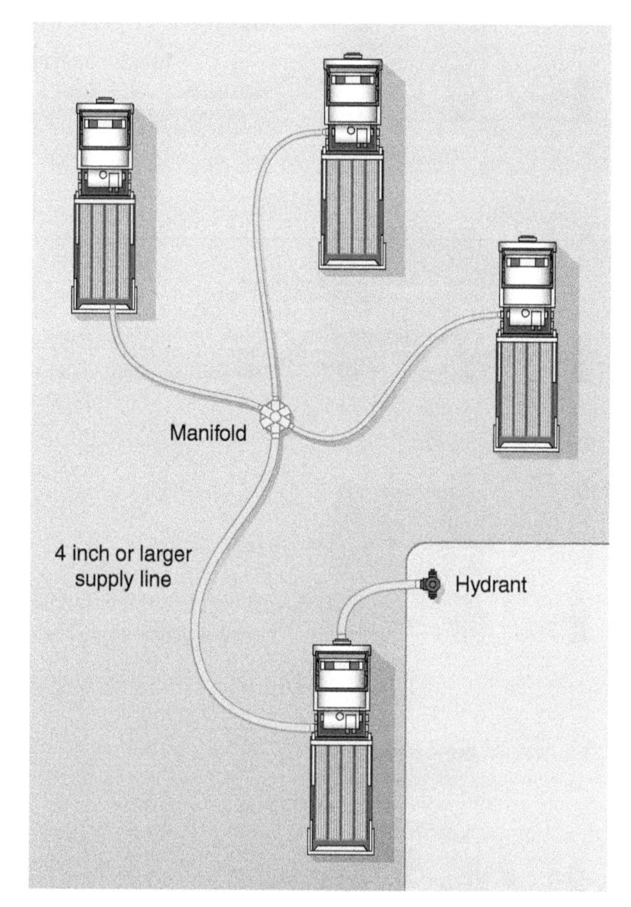

FIGURE 7-30 An LDH may be used to supply a manifold, which in turn can effectively supply pumpers at the fire or the firefighting lines.

BIG WATER OPERATIONS

By Paul Shapiro, Director, Fire Flow Technology

Big water operations refers to large-flow operations from large-caliber master streams and large hand lines. They depend on water supply and water discharge. Without the proper amount of water coming in from the supply, the streams cannot be developed. The key to setting up a successful big-stream operation is to be proactive and anticipate the need. Don't wait. Start getting the water.

Large-diameter hose (LDH) is a key piece of equipment for moving big water because of its low friction loss characteristics. The most common sizes are 4 inches and 5 inches. The 5-inch hose delivers twice as much water as the 4-inch hose. Both sizes work well for moving big water and getting the maximum flow from the hydrant. A single LDH supply line coming directly off hydrant pressure may be good enough for low-flow fires, such as house fires, but when it comes to large-flow fires, more needs to be done. Relay pump operations are needed to boost flow volume and pressure in the supply line. Even LDH has friction loss that eats up the pressure as water moves through the supply line. Often, water is available in the hydrant system, but there's not enough residual pressure to move it through the supply line into the engine. By setting up a relay pump operation, the pressure and flow can be maximized from the water that's available within the system. The engine pumping from the hydrant should maximize the available water from the hydrant by hooking up to at least two ports, if not three.

Issues that will limit the source engine include:

- Running out of water from the hydrant
- Reaching the maximum operating pressure of the supply hose
- Running out of throttle travel on the throttle mechanism itself (the knob won't twist anymore)
- Reaching the maximum rpms, limited by the governor on the powerplant of the apparatus

Large-flow fires usually mean that there are multiple engines pumping, which in turn can create complex supply line operations that need to be managed. Two and even three supply lines to a single engine can make a world of difference to its pump capabilities. It is important to have a water supply officer designated as soon as possible when a large-flow operation starts. The duty of the water supply officer is to develop the water supply system to meet the demands of the large streams.

There are a couple of sayings that I like to use for large-flow water supply operations: "Where there's a will, there's a way" and "The water's out there; you just have to go get it."

LDH increases the distance between the water system and the fireground because of its lower friction loss characteristics.

The carrying capacity of an LDH versus standard fire hose is striking. A 4-inch hose line delivers a volume of water approximately equivalent to three and a half 2½-inch hose lines at any given pressure or distance. A 5-inch hose line is approximately equivalent to six 2½-inch hose lines.

As Figure 7-29 showed, a 4-inch supply line can flow 880 gpm, whereas a 5-inch supply line can flow 1,400 gpm. For proper operation, 5-inch LDH requires hose fill-time, valves, and adapters. The case study entitled *Big Water Operations* depicts another LDH perspective.

Supply Line Procedures

LDH of 4 or 5 inches should always be used for water supply at structure fires. There's an old fire saying that makes a lot of sense: "When in doubt, lay it out." There is no sense in being caught without water at a fire that "looked insignificant on arrival." A supply line should be laid from the initial pumper if there is any indication of a fire on arrival. Prepare for what the fire may become.

Forward and Reverse Lays

A good driver should know the first-alarm district and the locations of hydrants and other reliable water sources, especially when an incident occurs in an area where the hydrants are scarce. Never pass up the last reliable source of water.

An engine approaching the fire from a reliable water source should perform a forward lay. Keep in mind, the firefighters on the first-arriving engine can abide by the two in/two out rule only if they lay forward. The first-in engine cannot comply with this rule if they lay reverse. If the engine arrives at the fire first, then a reverse lay from the fire to the hydrant should be initiated. The crew and the officer should drop a manifold or trigate and all the necessary hose and equipment for initial operations; the driver lays out to the hydrant. The officer should radio for the second-in engine to pull up in front of the burning structure, use tank water for preconnected hose lines, and take the supply from the manifold. Now the two in/two out rule can be implemented. You need to have an engine in front of the house or building that's on fire. This engine is your toolbox. It has all the additional hose, nozzles, ladders, air bottles, power tools, master streams, and emergency medical services (EMS) equipment that may be needed at a moment's notice. Once the supply is established, this engine also becomes a giant manifold for additional hand lines and appliances **FIGURE 7-31**.

The supply line should be charged to provide the second-in engine with a continuous water supply. If the

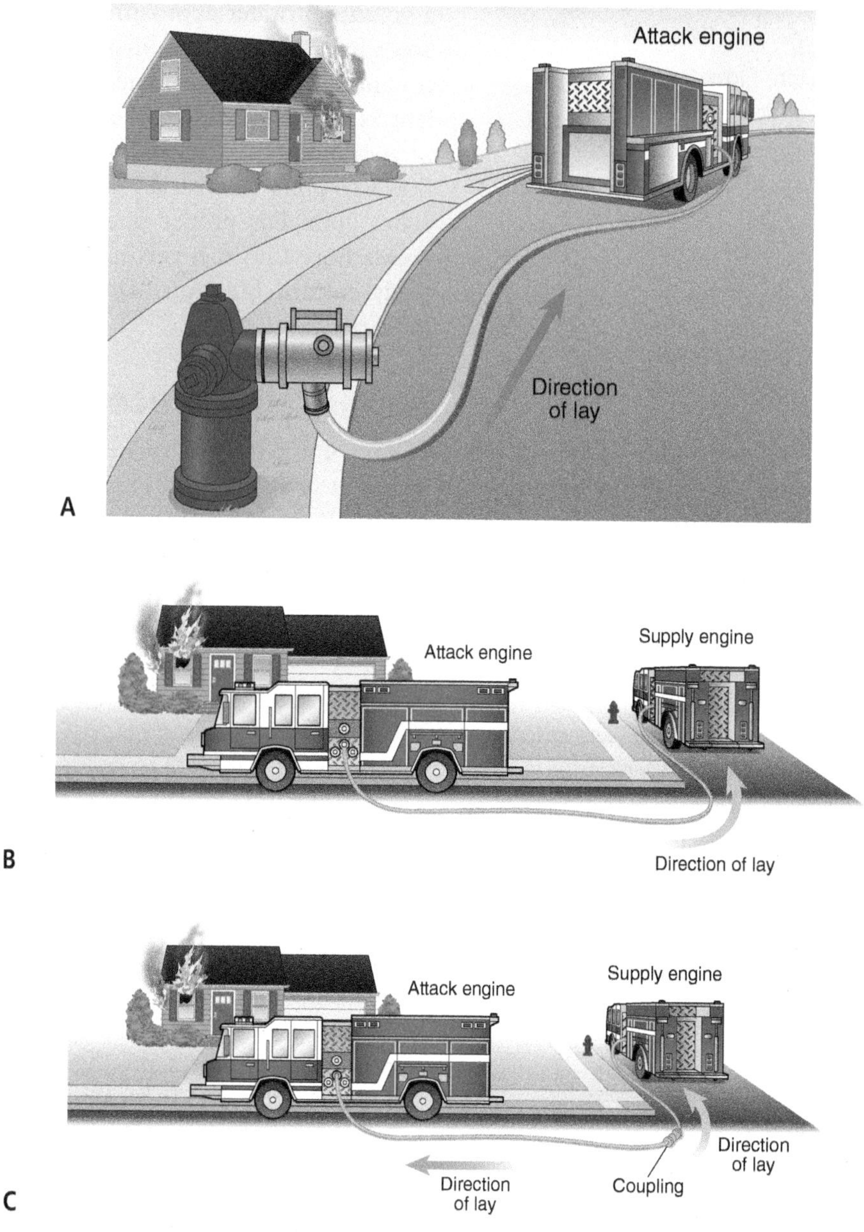

FIGURE 7-31 **A**. A forward lay. **B**. A reverse lay. **C**. A split lay.

officer of this second engine decides to use tank water on the initial attack while the two drivers are establishing the supply, this is a sound tactic. There should be at least 500 gallons to work with, and most house fires can be put out with 500 gallons or less. Once the driver of this second engine has an uninterrupted water supply, the engine officer inside the structure should be notified via radio that a supply has been established. If there are problems at the hydrant or the water supply cannot be obtained for any reason, the interior officer should immediately be notified via radio that only tank water is available. The driver should announce how much water is left in the tank, that is, three-quarters, one-half, or one-quarter of a tank. If there is a possibility that this water supply could be depleted before the fire is controlled, the engine officer should withdraw all interior crews. The situation could become disastrous unless a third engine arrives to supplement the initial pumper's water supply.

A forward or reverse lay can be accomplished using many methods. Departmental procedures may allow for the initial engine to lay a forward supply line but leave it uncharged. The second engine would be assigned to finish the lay by hooking up to the hydrant and charging the supply line. Hose can also be overhauled or cross-lotted to lay the supply lines. Whether laying forward or reverse, the result should be the same. You need an engine on the hydrant pumping to the forward engine, positioned in front of the fire building.

As a starting guideline, the engine on the hydrant should initially begin pumping to the engine at the fire at a pump pressure of 100 psi. Under normal conditions, this provides the water and residual pressure required by the pumper at the fire. As the situation changes, or if 100 psi is found to be insufficient, the pressure can be increased. For long lays, the initial pressure may need to be substantially higher to compensate for the increased friction loss depending on the length of the lay, the carrying capacity of the supply hose, and the amount of water being used. Drivers need to make sure the residual pressure is kept above the minimum. When water supply systems are unavailable, the task of supplying water to initial pumpers should be assigned to other engines or mobile water supply apparatus. The IC needs to call for this additional support at the beginning of the incident.

Pumper Relays (Tandem Pumping)

Excessively long lays—those exceeding 1,000 ft and up to 2,000 ft—require tandem or relay-pumping operations to counteract the effects of friction loss and elevation pressure, even when LDH is used. These pump-relay operations are simple if they are handled properly because they involve merely pumping water through two or more pumpers stationed at intervals between an adequate water source and the fireground **FIGURE 7-32**. The number of pumpers and supply hoses available are the only limitations to the distance over which water can be relayed. It is important to know before the incident occurs where such operations might be required so they can be put into effect as engines arrive at the fire scene.

FIGURE 7-32 Excessively long lays require relay-pumping operations to counteract the effects of friction loss and elevation pressure in supply lines.

Fire department personnel should be trained in the proper procedures to be applied in a relay-pumping operation, and they should drill on this evolution at least once every six months. A fire is not the time for on-the-job training. Personnel must know how to set up this system so that it can be placed in service as quickly as possible, and a water supply officer should be designated. The maximum capacity of a relay operation is determined by the smallest pumper and/or the smallest supply hose used in the relay.

Setting Up the Relay

The actual positions taken by pumpers in the relay line and the movements involved in getting to these positions depend on the relative locations of the fire and the water source, as well as the availability of roadways and room in which to maneuver pumpers at the fire. There are two ways in which most relay pumping operations are set up. In the first method, the first-arriving pumper is positioned at the fire, and its crew begins attacking the fire with tank water **FIGURE 7-33**. The fire should be attacked from the outside first by spraying water into a window or door to knock down and reset the fire. This action immediately lowers interior temperatures, prevents flashover, and thus provides a favorable survivability profile. It also makes the best use of the available tank water while maintaining firefighter safety. Additional responding pumpers lay supply lines from the first pumper toward the water supply. Each pumper takes up its position in the relay line after it has laid the entire LDH hose bed. Then the end of the LDH is connected to the rear discharge port. The driver at this position awaits the arrival of the next engine, receives the end of its LDH, and connects it into the pump intake port. This process continues until the final pumper reaches the water source. In essence, this is a series of reverse lays.

In the second method, the first-arriving engine lays its supply line from the property entrance to the fire building and then begins initial attack **FIGURE 7-34**. This is referred to as a **split lay** or a **blind alley lay**, but it is essentially an uncharged forward lay. Again, UL/NIST initial attack recommendations should be applied.

The next-in engine needs the proper couplings to connect the initial hose lay to its LDH bed and continue the lay (now a reverse lay) toward the water source until the LDH hose bed is laid out. Each pumper takes up its position in the relay line when it has laid out its LDH bed. Then the process continues as before until the water source is reached.

In the rare case that LDH is not being used for the initial supply line, and if conditions warrant (for

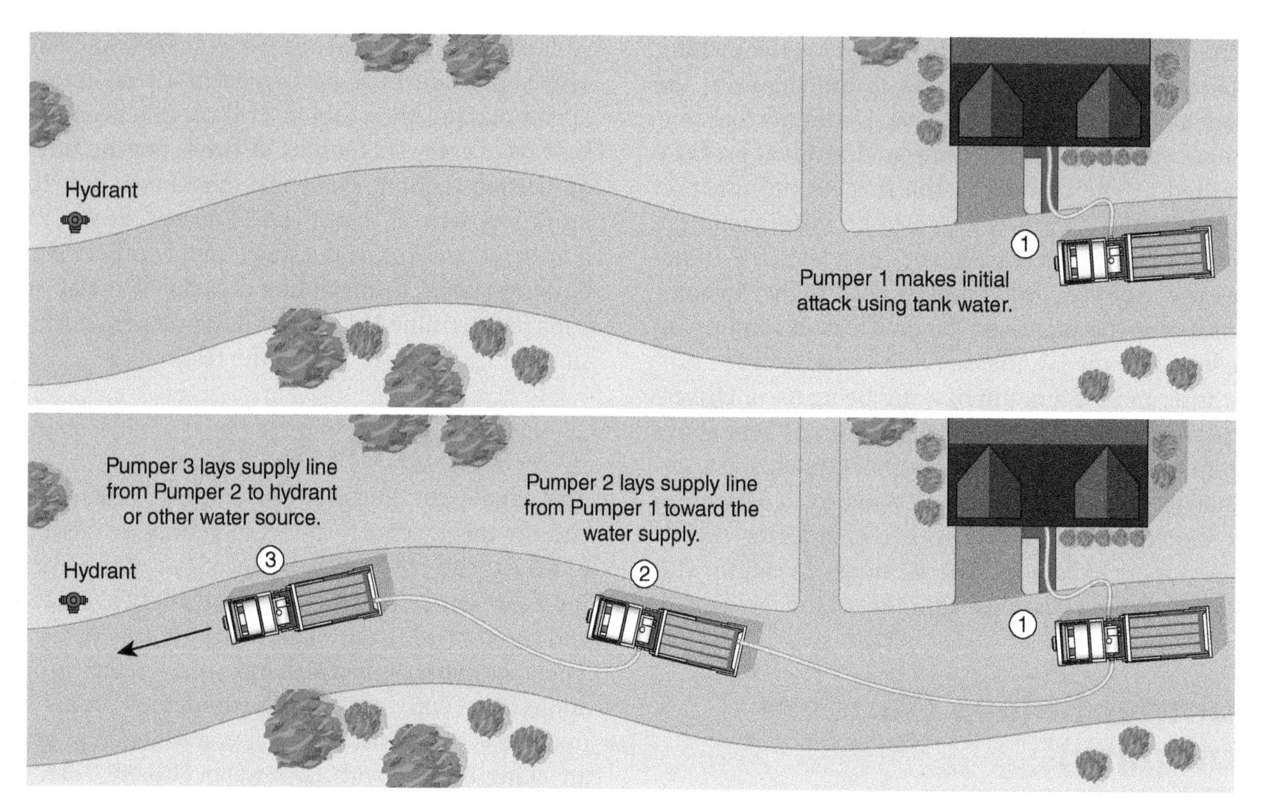

FIGURE 7-33 In this first method of laying a relay supply line, the first pumper begins the initial attack and does not lay a supply line. The second pumper and each successive engine in essence lays reverse until the water source is reached.

FIGURE 7-34 In this second method of laying a relay line, the first pumper lays a dry forward supply line before it begins the initial attack. The second engine can connect to the hose and lay toward the water source or take its position in the relay and have the third engine lay to the water source.

example, in the case of a grass or brush fire), a 2½-inch line can be laid for an initial supply. A second parallel 2½-inch supply line will have to be made at some point. It's best if these two supply lines can be laid out simultaneously.

Relay Pumping

After all the relay pumpers are in position and the supply lines are hooked up, pumping should begin. Supply lines should be charged from the source to the fire. The pump operator at the water source first charges the supply line to the next pumper. When the water reaches this second pumper, the operator opens the intake gate to charge the pump. The operator then opens the discharge to build up pressure to the next pumper. This pumper-by-pumper charging of the supply line continues until the pumper at the fire is receiving water.

After a pumper is discharging water, the pump operator must keep the discharge gates fully opened. The operator must also watch the pump discharge gauge because discharge pressure falls when the next pumper up the line begins taking water. As soon as it does, the operator should begin to adjust the throttle smoothly to bring the discharge pressure up to the starting pressure. The pump operator should continue to adjust the throttle until the pressure remains steady. This keeps fluctuations in pressures between pumpers to a minimum.

At the start of the relay operation, all pumpers in the line should be set to pump 150 psi to the first-in pumper at the fire; the first-in pumper should pump at the pressure required by the attack lines in use. The 150-psi pressure is easy to remember. It ensures a good initial flow and sufficient residual pressure to the pumper at the fire. Conversely, incoming pressure to the next pumper in the relay should never drop below 20 psi.

After the relay line is in operation, discharge pressures can be increased or decreased as necessary to meet firefighting requirements. The basic principle is to have all pumpers within the relay pump at the same discharge pressure. Good radio communications between the pumpers is essential for an efficient relay operation. Each pumper must be aware of the actions of the others so the operation can be coordinated.

Increasing Water Flow

If additional water is needed after the relay has been established, a second supply line should be manually laid parallel to the original supply between the operating pumpers. A second separate relay operation is not recommended. It is much more efficient to build up the flow of water to the original attack lines **FIGURE 7-35**. After this is accomplished, if resources are available, and if conditions still require additional water, a second relay operation can begin.

If an engine needs to be inserted in the middle of the lay to boost the volume and pressure of the supply, here is one way to do it:

1. Position the engine at the desired location parallel to the supply line, facing the water source (away from the fire).
2. If the driver plans to use the rear discharge port, position the rig so the coupling of the supply is at mid-engine.
3. If the driver is going to use a side discharge port, position the rig so the coupling of the supply is just clear of the front bumper.
4. Lay out the required couplings if needed, along with sufficient sections of LDH to reach the coupling of the supply and the intake port of the pumper. Connect the uncharged hose to the intake, and lay the other end next to the existing supply hose coupling to be broken.
5. The water supply officer radios to the first pump operator at the fire to make sure the apparatus tank is full and prepare to switch to tank water.
6. Then the water supply officer radios the pump operator toward the water source from the engine to be inserted to throttle down the pump and close the valve to the supply. The first pump operator at the fire switches to tank water and continues to supply the attack hand lines.
7. The crew of the new engine breaks the coupling of the supply line and connects the new section of uncharged LDH.

FIGURE 7-35 Once the relay is established, a second supply line should be laid parallel to the original supply between the relay pumpers.

FIGURE 7-36 When a reliable water supply cannot be established, the last resort option is to have all the pumpers proceed to the fire to use tank water.

8. The original supply line to the fire has residual water but is no longer under pressure. This supply hose is connected to the rear or side discharge port. As long as the new engine is properly positioned, there should be enough slack in the supply hose to reach the respective discharge ports easily.
9. Once all the hoses are reconnected, the new engine can use tank water to start refilling the supply line while the water supply officer radios the next engine toward the water source to open the discharge valve and throttle up the supply.
10. Once the supply is reestablished, the water supply officer should immediately radio the first pump operator that this is so.

Because the new engine and the new LDH are already pre-positioned and connected, this whole evolution should take as long as it takes for the crew to break one coupling and reconnect two LDH sections.

The last resort option when a reliable water supply cannot be established is to have these pumpers proceed to the fire to use tank water. Several pumpers or mobile water supply apparatus with full tanks represent a fairly large supply of water **FIGURE 7-36**. Three engines can carry 1,500 gallons of water. Four engines could give you 2,000 gallons of water. That's a lot of water for a house fire. There may be enough to allow additional attack lines to be used on the fire or at least to support rescue operations. This is an especially important tactical option if hose lines are needed in locations that are difficult to reach from the pumper being supplied by the relay.

Clearly, many different types of situations may be encountered on the fireground. A continuous size-up and risk-benefit analysis needs to be applied for unusual situations. A combination of department training, SOGs, the IMS, and a trained water supply officer dictates which course of action the IC will use to mitigate the incident successfully.

After-Action REVIEW

IN SUMMARY

- It is the task of the engine company to set up a continuous, uninterrupted supply of water at the fireground. Obtaining an adequate and continuous water supply on the fireground is a basic firefighting task. Without water, firefighters are unable to attack, control, and extinguish a fire.

- The crew must be familiar with the sources of water that constitute an adequate and reliable water supply.
- Three factors influence the movement of water at fires: the water source, the engine, and fire hose.
- There are four possible sources of water at the fireground: hydrant and water main systems, static water sources, apparatus water tanks, and mobile water supply apparatus.
- Static pressure is the pressure of the hydrant water at rest, with the hydrant open to the pump and no water flowing through the pump. Residual pressure is the pressure in the hydrant with water flowing from the hydrant through the pump. Both are read on the compound (intake) gauge.
- Any hose connected from the hydrant to the pump is technically considered a suction. Any hose leaving the pump under pressure through a discharge port is technically called a supply. However, the term *supply* is used almost exclusively, and "getting a supply" is understood to mean connecting to a water source.
- A mobile water supply apparatus is a vehicle designed primarily for picking up, transporting, and delivering water to fire emergency scenes to supply other pumpers or pumping equipment.
- When mobile water supply apparatus are used for developing a continuous water supply, they can either supply the pumpers directly or dump their water into a portable water tank from which the pumpers draft.
- For safety, a water shuttle apparatus corridor should be established. This corridor is a circular, drive-through roadway that allows the tender to bring in water without backing up or performing a three-point turn.
- Most fire departments now use 4-inch and 5-inch LDH for water supply. Increasing the diameter of the hose reduces the friction loss as well as the number of hose lines needed to move large volumes of water.
- Excessively long lays—those exceeding 1,000 ft and up to 2,000 ft—require relay-pumping operations to counteract the effects of friction loss and elevation pressure, even when LDH is used.
- Relay pumping is pumping water through two or more pumpers stationed at intervals between an adequate water source and the fireground. The number of pumpers and supply hose available are the only limitations to the distance water can be relayed. The maximum capacity of a relay operation is determined by the smallest pumper and/or the smallest supply hose used in the relay.

KEY TERMS

blind alley lay A scenario in which the attack engine forward lays a supply line from an intersection to the fire, and the supply engine reverse lays a supply line from the hose left by the attack engine to the water source. Also known as *split lay*.

cross-lotting A fire department term referring to when an apparatus cannot lay supply hose and an LDH supply line needs to be shoulder-carried or hand-jacked several hundred feet to a water source. It requires four to six firefighters (or more) to be spread out equally along the length of the connected, but uncharged, supply line to reach the distant water supply. Also known as *hand jacking*.

dry-barrel hydrant A type of hydrant used in areas subject to freezing weather. The valve that allows water to flow into the hydrant is located underground below the frost line, and the barrel of the hydrant is normally dry.

dry hydrant An arrangement of pipe permanently connected to a water source other than a piped, pressurized water supply system that provides a ready means of water for firefighting purposes and that utilizes the drafting (suction) capability of a fire department pump. (NFPA 1142)

fast water Water delivered to the fire immediately through a preconnected hand line using tank water from the engine.

overhauling As opposed to post-fire operations, the term can also be used to manually attain a water supply by physically dragging or shoulder-loading LDH.

residual pressure The pressure that exists in the distribution system, measured at the residual hydrant at the time the flow readings are taken at the flow hydrants. (NFPA 24)

split lay A scenario in which the attack engine forward lays a supply line from an intersection to the fire, and the supply engine reverse lays a supply line from the hose left by the attack engine to the water source. Also known as *blind alley lay*.

static pressure The pressure that exists at a given point under normal distribution system conditions measured at the residual hydrant with no hydrants flowing. (NFPA 24)

suction hose A short section of supply hose used to supply water to the suction side of the fire pump.

supply Any water leaving the pump under pressure through a hose connected to a discharge port.

water shuttle apparatus corridor A circular, drive-through roadway that allows the tender to bring in and dump its water without backing up the apparatus or performing a three-point turn.

wet-barrel hydrant A type of hydrant that is intended for use where there is no danger of freezing weather and where each outlet is provided with a valve and an outlet. (NFPA 24)

REFERENCES

Fire Apparatus Driver/Operator, Third Edition, Jones & Bartlett Learning, Burlington, MA. 2019.

Fundamentals of Fire Fighter Skills and Hazardous Materials Response, Fourth Edition, Jones & Bartlett Learning, Burlington, MA. 2019.

National Fire Protection Association. Firefighter Fatalities in the United States, 2017.

CHAPTER 

Initial Size-Up and Developing a Quick Incident Action Plan

LEARNING OBJECTIVES

- Describe the considerations that go into a size-up.
- List and explain the three phases of size-up.
- Explain what a risk-benefit analysis is.
- List and explain the minimum criteria included in a risk management plan.
- Explain what a QIAP is.
- Learn how to create a value-time-size worksheet, including PSTR.
- List the typical problems found on the fireground.
- Demonstrate how to make tactical assignments on a QIAP tactical worksheet.
- Describe how a QIAP can help determine the need for multiple alarms.
- List the criteria of a functional firefighter accountability system.
- Demonstrate how to operate a firefighter accountability system while operating a tactical worksheet.
- Test the effectiveness of an accountability system by selecting the name of a firefighter and locating his or her position, assignment, and task at any given moment.

Introduction

The study of size-up and risk assessment is often glossed over, simply defined, or left to a series of acronyms to check a box—size-up by acronym. An inadequate size-up and failure to perform a risk assessment consistently ranks in the top five contributing factors in fireground line-of-duty-deaths (LODDs).

The first-in company officer becomes the initial incident commander (IC) and is responsible for the overall management of the incident and the safety of all members involved in the fire operation. Before you accept this tremendous responsibility, you must understand fire science, modern fire behavior, building construction, the most recent scientific evidence–based practices, size-up, risk assessment, and firefighter accountability. Developing size-up skills requires practice and continuous study. Peer discussion, studying research material and case studies, and reviewing pre-incident plans and videos are all excellent ways to help you master the art of size-up.

In the first four chapters of this text, we talked about the enemy, the battlefield, and the munitions the enemy uses against us. In this chapter, we will connect the dots and put all the information together to create the battle plan—the **quick incident action plan (QIAP)**. The QIAP is for the first-arriving officer who takes command of the fire. The chief officer assuming command builds upon these initial assignments and creates a more detailed incident action plan, but all plans start with size-up.

The IC

NFPA 1500, *Standard on Fire Department Occupational Safety, Health, and Wellness Program*, 8.1.8, lists the responsibilities of the IC at an emergency incident. The following items are specific to the first-in company officer:

- Arrive on scene before assuming command.
- Assume and confirm command of an incident and take an effective command position.
- Perform situational evaluation that includes risk assessment.
- Initiate, maintain, and control incident communications.
- Develop an overall strategy and an incident action plan, and assign companies and members consistent with standard operating procedures.
- Initiate an accountability and inventory worksheet.
- Develop an effective incident organization by managing resources, maintaining an effective span of control, maintaining direct supervision over the entire incident, designating supervisors in charge of the entire incident, and designating supervisors in charge of specific areas or functions.
- Review, evaluate, and revise the incident action plan as required.

Size-Up

Size-up is the observation and systematic evaluation of the existing conditions in order to identify the problems presenting themselves; determine what has occurred, what is occurring, and what is about to occur; develop strategic objectives; and select the tactics for rescue and firefighting. It is not a speech describing current conditions; size-up is often confused with the initial radio report. They are not the same thing. There are three phases of size-up: pre-incident size-up, initial size-up, and ongoing size-up. **Risk-benefit analysis** is a component of each size-up. Size-up is a dynamic, ongoing, and continual process throughout the incident.

Unless it's impossible, every size-up should include a 360-degree walk-around survey; however, remember that you're also performing a **six-sided size-up**, which includes areas above and below the fire. Floors above the fire, basements, and below-grade occupancies must be identified or ruled out. If basements are present, this information should immediately be announced over the radio, for example, "All units, be advised this structure has a basement." Address the following questions: Where is the basement access? Is it a walkout basement? Is there a rear exterior entrance? Are there emergency access windows or will access have to be made from the interior? Is it on fire? These questions can be answered in seconds, and this information should be transmitted via radio so responding units can get a mental image about what they may soon have to handle.

Exposures above the fire present a separate set of problems, and they are more obvious. Size-up must be performed at every fire and emergency incident. Size-up should determine the following:

- occupancy type
- size of the occupancy
- type of construction
- whether the structure is occupied or unoccupied
- whether there are any people trapped
- survivability profile
- whether to initiate a search and rescue

- building contents
- what is burning
- fire location
- size of the fire
- current fire conditions
- where the fire is headed
- what the fire will look like in 5 minutes
- whether the required resources are responding

FIGURE 8-1 contains a thorough size-up checklist of everything that should be considered at some point during the incident. Although the first-in company officer serves as the initial IC, not everything on this list can be addressed by the first-in officer. The main size-up points for the first-in company officer will be covered later in this chapter. Don't forget: *Everything* in the modern fire service is driven by risk management.

Risk Management

Firefighting is not without risk. The risks and dangers are inherent. Risk management is probably the most overused phrase at the administrative level, and the most misunderstood phrase at the company level of the fire service. We cannot integrate risk management with incident management by assuming everyone knows what it is. Firefighters and chief officers may believe they're on the same page when, in fact, they may not be. A quick review of effective risk management is essential. Risk is the possibility of suffering harm, an injury, or a loss from the dangers and hazards of fire and other emergency and non-emergency operations. Risk management identifies risks, then utilizes methods to reduce exposure to those risks. Simply put, risk management is showing you how to avoid getting hurt or killed in the first place.

The Risk Management Plan

To address any problems in life effectively, you have to develop a plan. Integrating risk management with incident management is no different. A risk management plan serves as documentation that risks have been identified and considered, and that all reasonable control efforts have been applied and adhered to by the firefighters. Most fire departments have a designated health and safety officer (HSO) and/or incident safety officer (ISO) who are charged with the responsibility of being risk managers. However, if you have a department without these full-time designated positions, the fire chief or the IC is still responsible for addressing risk management within the incident action plan in order to prevent or reduce the possibility of injuries and LODDs.

Having a risk management plan increases operational safety and effectiveness for the firefighters and helps protect fire departments from liability. These are just some of the benefits. Risk management plans also have a positive effect on department morale, a reduction in personnel injuries, and an increase in firefighter longevity, as long as fire department members are familiar with and committed to the plan. If we are going to integrate risk management into incident management at all levels, we cannot view this as a function of the administration or "a chief thing." Every line firefighter has to be on board because they are on the receiving end of fireground injuries and fatalities.

At a minimum, a risk management plan should include:

- Risk identification
- Risk evaluation
- Risk prioritization
- Risk control techniques
- Risk monitoring

Risk Identification

Risks are activities that can lead to:

- Injuries to personnel or civilians
- Deaths of personnel or civilians
- Property damage, for both the fire department and public, private, or environmental property
- Lawsuits

Every identifiable risk is a manageable risk. Foresee or forecast worst-case scenarios or potential cause-and-effect consequences. Predicting and anticipating the end result provides the greatest margin of safety. You're looking for things that can go wrong. Fireground injuries that are predictable can be preventable.

Risk Evaluation

Once the risks and hazards have been identified, evaluate the likelihood of occurrence.

- Evaluate whether the risk involves trapped occupants.
- What is the survivability profile? Is there someone we can save?
- Does the risk involve property? Is there something we can save?
- Do I have time to save this property?
- If the benefits are worth the risk, for how long?

Time is a critical factor and plays a vital role in risk assessment because the structural integrity of modern

SIZE-UP/PRE-PLAN CHECKLIST LIFE SAFETY/FIREFIGHTER SAFETY

- Smoke and fire conditions
  - Fire location
  - Direction of travel
- Ventilation status
- Occupancy type*
- Occupant status
  - Estimated number of occupants*
  - Evacuation status
  - Occupant proximity to fire
  - Awareness of occupants*
  - Mobility of occupants*
  - Occupant familiarity with building*
  - Primary and alternative egress routes*
  - Medical status of occupants*
- Operational Status
  - Adherence to SOPs
  - Fire zone/perimeter
  - Accountability*
  - Rapid intervention*
  - Organization and coordination*
  - Rescue options*
  - Staffing needed to conduct primary search
  - Staffing needed to conduct secondary search
  - Staffing needed to assist in interior rescue/evacuation
  - Staffing needed for exterior rescue/evacuation
  - Apparatus and equipment needed for evacuation
  - Access to building exterior*
  - Access to building interior* (forcible entry)
- Structure
  - Signs of collapse
  - Collapse zone*
  - Construction type*
  - Roof construction*
  - Condition*
  - Live and dead loads*
  - Water load
  - Enclosures and fire separations*
  - Extension probability*
  - Concealed spaces*
  - Age*
  - Height and area*
  - Complexity and layout*

Extinguishment

- Probability of extinguishment*
- Offensive/defensive/non-attack
- Ventilation status
- External exposures*
- Internal exposures*
- Manual extinguishment
  - Fuel load*
  - Calculated rate of flow requirement*
  - Number and size hose lines needed for extinguishment*
  - Additional hose lines needed*
  - Staffing needed for hose lines*
  - Water supply*
  - Apparatus pump capacity*
  - Manual fire suppression system*
- Automatic fire suppression equipment*

Property Conservation

- Salvageable property*
- Location of salvageable property*
- Water damage
  - Probability of water damage*
  - Susceptibility of contents to water damage*
  - Water pathways to salvageable property*
  - Water removal methods available*
  - Water protective methods available*
- Smoke damage
  - Ventilation status
  - Probability of smoke damage*
- Susceptibility of contents to smoke damage*
- Damage from forcible entry and ventilation*

General Factors

- Total staffing available versus staffing needed*
- Total apparatus available versus apparatus needed*
- Staging/tactical reserve*
- Utilities (water, gas, electric)*
- Special resource needs*
- Time
  - Time of day
  - Day of week
  - Time of year
  - Special (e.g., holiday season)
- Weather
  - Temperature
  - Humidity
  - Precipitation
  - Winds

* This factor can be at least partially known in advance of the fire through pre-planning.

FIGURE 8-1 A size-up/preplan checklist.

lightweight construction is weakened and fails quickly under heavy fire conditions. We often do not know how long a fire has been burning or how intensely it has been burning prior to the arrival of the fire department. A calculated risk for an interior offensive attack may therefore be limited to a very small window of time. The IC must ask, "Do I have the available resources and adequate water supply to control this fire?

Do I have the resources to take a calculated risk?" *Don't confuse calculated risk with taking chances.* Insufficient resources or inadequate water supply to deliver the required gallons per minute (gpm) to control and extinguish the fire may mean that the risks outweigh the benefits. Although it may be one of the hardest decisions an IC may have to make, there may be situations when attempting to make a rescue or an interior offensive attack may be too great a risk to the safety and survival of the firefighters, and that it is not worth the benefits. A defensive strategy must be implemented.

Risk Prioritization

Based on the evaluation process, establish the risk priority. Any risk that has a high probability of occurrence and a serious degree of consequence deserves immediate consideration. Various color-coded risk matrices exist to illustrate the levels of risk, but risk priority generally falls into three categories **FIGURE 8-2**:

- High risk (significant risk)
- Medium risk (elevated risk)
- Low risk (standard risk)

Actions that present a significant risk to the safety of firefighters shall be limited to situations where there is a potential to save lives that are in immediate danger. Actions that are routinely employed to protect property are inherent risks to the safety of firefighters, and preventable measures should be taken to reduce or avoid these risks. No risk to the safety of firefighters is acceptable when there is no chance of saving lives or property.

Risk Control Techniques

- ***Risk avoidance.*** Simply put, avoid the activity that creates the risk. Obviously since we are emergency services, there are some risks we simply cannot avoid, so we need to look at control methods to minimize the risks. However, policies can be put in place that are very effective at avoiding risks. Some examples include defensive fire attacks on vacant and dilapidated buildings once the life hazard has been mitigated. With such a policy, any risks that normally occur during offensive interior attacks are now avoided.
- ***Risk transfer.*** This term means physically transferring the risk to someone else. An example is a department overhaul policy to include final extinguishment, but the actual removal of cold fire debris is contracted to a private salvage company, thereby eliminating the risks of post-fire injury and contamination of your fire crews.
- ***Control measures.*** Risk reduction is the most common and most visible method utilized by risk managers. Risk reduction includes implementation of an incident management system (IMS) on all incidents; a functional accountability system; following standard operating procedures and guidelines; wearing all your personal protective equipment (PPE), including self-contained breathing apparatus (SCBA); adhering to the two in/two out rule; and establishing a rapid intervention team. Implementation of control measures is selecting

Risk assessment/rules of engagement			
Firefighter injury/ life safety risk	High probability of success	Marginal probability of success	Low probability of success
Low risk	Initiate offensive operations. Continue to monitor risk factors.	Initiate offensive operations. Continue to monitor risk factors.	Initiate offensive operations. Continue to monitor risk factors.
Medium risk	Initiate offensive operations. Continue to monitor risk factors. Employ all available risk control operations.	Initiate offensive operations. Continue to monitor risk factors. Be prepared to go defensive if risk increases.	Do not initiate offensive operations. Reduce risk to firefighters and actively pursue risk control operations
High risk	Initiate offensive operations only with confirmation of realistic potential to save endangered lives.	Do not initiate offensive operations that will put firefighters at risk for injury or fatality.	Initiate defensive operations only.

FIGURE 8-2 Risk priority generally falls into three categories: high risk, medium risk, and low risk.

the best option for taking a calculated risk with the highest margin of safety. Analysis weighs the pros and cons of the solution. Firefighters must follow orders on each incident, and company officers and chief officers must have the courage to enforce the safety policies and hold members accountable for deviations from the policy.

Risk Monitoring

As with any applied solution, there needs to be a comparison between the desired results versus the actual results. Evaluating the potential risk to firefighters compared to the benefits to be gained requires constant monitoring. For example, the IC should constantly reevaluate the effectiveness of the strategy and, if necessary, modify the tactics. If no significant progress is made on the fire where fire crews are working inside the structure, the IC should consider changing the strategy because the ongoing fire is weakening the building's structure, thus increasing the risk to firefighters. The IC may determine that the risk is too great, withdraw all firefighters from the interior, and change to a defensive strategy.

A risk-benefit analysis needs to be applied to each of the tactical objectives. A risk-benefit analysis is the process of weighing the predicted risk to firefighters against the potential benefits for property owners and occupants, and making decisions based on the outcome of that analysis. Like size-up, risk-benefit analysis does not stop once the QIAP is implemented and firefighters are taking action. It must be ongoing throughout the incident. **FIGURE 8-3** is a detailed flowchart for operational risk assessments.

The most important question an IC has to ask after the civilian life hazards have been mitigated is: "At what point does this building no longer have any value?" Once the building no longer has any value (meaning the owners are probably going to knock it down anyway), write it off and don't risk any more firefighter lives. Be willing to take the criticism for creating parking lots. Take a defensive posture.

The National Fire Academy Command Sequence Model

The National Fire Academy (NFA) Command Sequence model, based on the study of successful military commanders and soldiers, is still one of the best logical decision-making models for fireground strategy and tactics. In the center of the model are the four incident priorities: life safety, incident stabilization, extinguishment, and property conservation. All emergency incident activities and tactics are centered around these incident priorities and form the basis of the command sequence **FIGURE 8-4**. All firefighters would agree our mission at any fire is to save lives and property, which is best accomplished by putting the fire out. The command sequence is think, plan, act:

- Think: The thinking phase consists of size-up, problem identification, and risk-benefit analysis. Before we can act, we must think. Before any actions are taken, we must first identify the problems so our actions actually address or solve them. There are many examples of fireground operations that have gone awry because firefighters rushed into a situation (moth to flame) performing random, uncoordinated tactics before anyone performed a thorough size-up to determine what the problems were or developed a plan.
- Plan: The planning stage consists of determining the strategy and tactics and developing the QIAP (offensive, defensive, or transitional).
- Act: The action phase is implementing the plan by making tactical assignments that support the strategy.

As with any plan, there needs to be an evaluation process to ensure the desired outcomes. If there is no progress, the plan must be modified.

The Three Phases of Size-Up

Phase 1: Pre-Incident Size-Up

Pre-incident size-up means just that: analysis before the incident occurs. Pre-incident surveys, whether pages in a binder or the apparatus mobile computer screen, contain information sheets with a site map for most occupancies within your first-alarm district. Mobile computer software also lets you access any pre-incident plan for any occupancy within your jurisdiction if it is entered into the system. Many pre-incident plan templates are available and contain a list of crucial information for decision making, including **FIGURE 8-5**:

- Name and address of the occupancy
- Type of occupancy
- Type of construction
- Roof construction
- Area dimensions
- Number of stories
- Existence of a basement
- Location of hydrants
- Calculated fire flow requirements (in gpm) (liters/min)

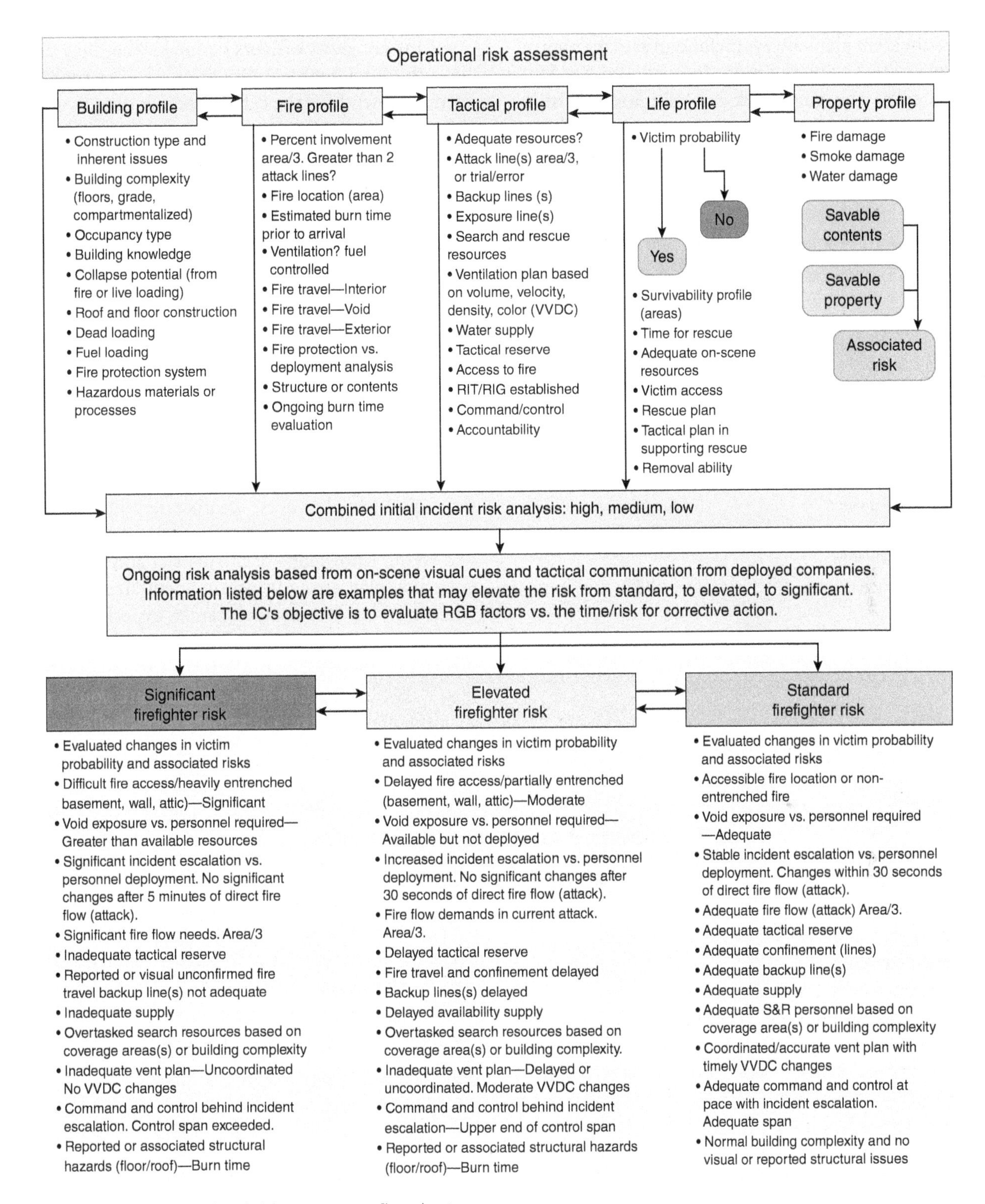

FIGURE 8-3 Operational risk assessment flowchart.

- Topography and terrain
- Building entrances and exits
- Access considerations, including fences, security gates, lockboxes, and access keys
- Location of fire alarm panel
- Location and number of stairwells
- Which stairwells have roof access
- Location of standpipe and sprinkler connections (fire department connections [FDCs])
- Location of elevators
- Location of machinery rooms
- Location of utility shutoffs
- Other specific building information or special considerations

Pre-incident plans always include an area map and a floor plan identifying the above features. Pre-incident size-up takes the guesswork out of initial decision making and allows for standard operating procedures to be applied so that companies know what to expect from the first-in units.

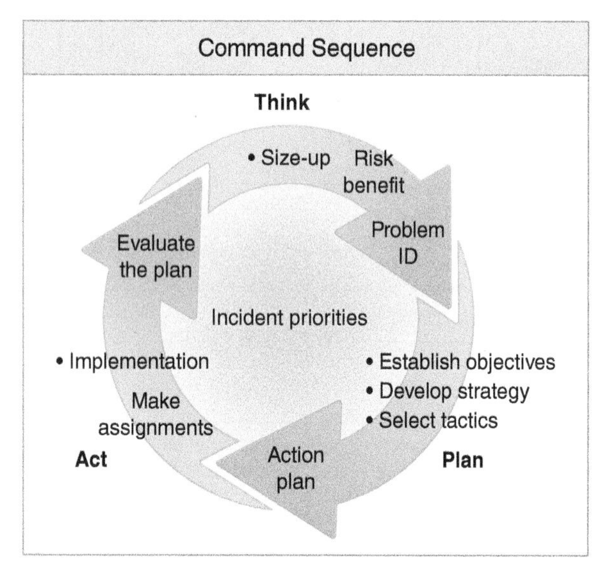

FIGURE 8-4 NFA Command Sequence model.

NFA Command Sequence model, State of Indiana, Retrieved from http://www.state.in.us/dhs/files/Decision_Module1_SM.pdf

Many company officers complain that they do not have the time or the ability to review the prefire information while in the cab and responding to an incident. They are right. Pre-incident plans are designed to be reviewed with the company during a drill or classroom session. The more the plans are reviewed, the more information is committed to memory so that, at minimum, a mental picture of familiarity is developed, and the company officer isn't caught off guard. This helps the first-in officer with quick problem identification and decision making. Once the battalion chief and other command staff arrive at the command post, pre-incident plans and maps can be reviewed in more detail to ascertain pertinent information.

Other size-up factors can be noted before an emergency, for example, time of day, day of the week, holidays, special events, weather conditions, traffic patterns, typical response times for incoming units, and so on. Knowledge of the district and past experience also play a role in pre-incident size-up. This is when all your fire service career experience comes into play (your mind is a database). For example, if you're responding to an alarm in the industrial district at two in the morning, chances are you're going to immediately have a forcible entry challenge on a

MEMORY LANE APARTMENTS

26 MEMORY LANE

Life Safety/Firefighter Safety

Occupancy Type

- 89-unit apartment building
- Most occupants are elderly, some with severe handicaps
- 7 small retail stores on first floor facing Main Street
- Video Rental Store (50 ft × 45 ft [15.2 m × 13.7 m])
- Joe's Barber Shop (24 ft × 45 ft [7.3 m × 13.7 m])
- Hair by Gloria Beauty Parlor (24 ft × 45 ft [7.3 m × 13.7 m])
- Old Tyme Antiques (12 ft × 45 ft [3.6 m × 13.7 m])
- Nail Boutique Manicure Salon (12 ft × 45 ft [3.6 m × 13.7 m])
- Song Shop CDs (24 ft × 45 ft [7.3 m × 13.7 m])
- Thrift Shop (24 ft × 45 ft [7.3 m × 13.7 m])
- Basement
  - Bingo Hall (180 ft × 45 ft [55 m × 17.3 m]) posted for 500 maximum occupants
  - Exits for Bingo Hall stairs to Memory Lane and Main Street sides

Estimated Number of Occupants

- Bingo Hall could have 500 on Tuesday and Thursday evenings
- Occasionally used as a party hall
- Most apartments occupied by a single resident
- Shops vary, seldom more than 10 per shop

Primary and Alternative Egress Routes

- Three stairways and fire escape
- One-story roof to rear could provide emergency or secondary egress from second floor

Access to Building Exterior

- Access streets to front (Memory Lane) 50 ft (15 m) setback
- Right side (Main Street)
  - Driveway on left side, narrow, but apparatus accessible
  - Too close for aerial use
  - U-shaped driveway to front (may not support apparatus without damage)
  - Rear – one-story roof – good access to second floor

FIGURE 8-5 A pre-incident plan sample.

Construction Type

- Ordinary construction (brick exterior wooden floor/ceiling assembly)

Roof Construction

- Built-up flat tar roof supported by 2² x 12²

Condition

- Well maintained – no noted damage

Live and Dead Loads

- Average load for residential property

Enclosures and Fire Separations

- Only open areas in basement and stores
- Metal fire doors from apartments to hallways
- Hallway no separated, open from end to end

Extension Probability

- Stairways are not enclosed
- Utility openings penetrate floors

Concealed Spaces

- Inaccessible crawl space between top floor and roof

Sample Preincident Plan Narrative

Age

- Built in 1932

Height and Area

- Four-story building
- Basement to rear of building above grade level
- 220 ft × 180 ft (67 m × 55 m) U-shaped

Complexity and Layout

- U-shaped hallway with center hallway
- Access to basement down stairs
- Stores all face Main Street with direct access to street

Extinguishment

Probability of Extinguishment

- Basement is only significant problem

External Exposures

- College of Design on left side (Side B) 15 feet (4.6 m)

Internal Exposures

- Johnny's Restaurant attached to rear (Side C)
- Open stairways and floor to floor via walls/ceilings

Fuel Load

- Average residential
- Stores and Bingo Hall average

Calculated Rate of Flow Requirement

- Rate of flow for apartments and stores within the capacity of standard pre-connect and back-up line
- Bingo Hall 810 gpm (15 L/sec) required

Number and Size Hose Lines Needed for Extinguishment

- Standard pre-connect except Bingo Hall where 21/2-inch (64 mm) lines may be required

Additional Hose Lines Needed

- Nothing special (SOP calls for back-up and line on floor above the fire)

Water Supply

- Hydrant located to front of building and on Main Street
- Additional hydrants on Main Street
- Hydrant flow on Memory Lane 1,000 gpm (63 L/sec) per hydrant
- Hydrant flow on Main Street 1,500 gpm (95 L/sec) per hydrant
- Water supply on grid system 8" (203 mm) main on Memory Lane; 20-inch (508 mm) main on Main Street
- System flow in area is approximately 12,000 gpm (757 L/sec)
- System can be cross-tied to two other supply systems for a total 30,000 gpm (1,892 L/sec)

Property Conservation

- Ordinary household items in apartments
- Stores have property of moderate value

General Factors

Staffing/Apparatus on Alarm Card

Alarm	Engine Companies	Truck Companies	Others
First	23, 19, 32	23, 19	District Chief 1
Second	3, 9, 5	32, 3	District Chief 4
Third	46, 34, 29	29	H. Rescue 9, ALS 31, Deputy Chief
Fourth	14, 31, 12	None	Chief of Department
Fifth	21, 28, 20	None	Training Chief, Prevention Chief

Additional Companies and Command Staff Available

- 11 engine companies
- 8 truck companies
- 1 heavy rescue company
- 4 ALS, 6 BLS units
- Mutual aid

Staging/Tactical Reserve

- Suggested staging area – parking lot off Main Street and Taft Road

Utilities: Water, Gas, Electric

Gas, water, and electric shut-offs in basement utility room just left of Bingo Hall

FIGURE 8-5 *(Continued)*

building of Type II or Type IV construction. This is a commercial fire with possible hazardous materials involved. Based on the address alone, the problems you'll face are quite different than what you would face at an address in a quiet neighborhood of single-family dwellings. Many specific problems can be anticipated even before the company arrives on scene based on pre-incident size-up and advance knowledge of risks. Standard operating procedures (SOPs) become part of the QIAP and give the first-in company officer a head start on size-up and decision making.

Known factors at the beginning of the shift include which shift is working, day of the week, holidays, special events, and weather forecast. A list of water shutoff notices and out-of-service hydrants and/or building fire protection systems should be listed in the watch office. Be alert for these notices throughout the day and make sure everyone is aware of them, especially the apparatus drivers. Department-wide training may have certain units or apparatus out of service or out of position. Other specialized apparatus, like aerial ladders, may be in the fire garage for maintenance.

Tracking the Movement of Apparatus

Although many fire departments have a member assigned to the watch office to monitor emergency radio traffic in addition to other responsibilities, good company officers have a department radio turned on in the engine office or carries one with them to monitor the activity in surrounding districts. For example, if the engine company of the next district is out of service on an emergency medical services (EMS) alarm, your second-due engine for a structure fire may be coming from another part of the city with a longer response time; therefore, you may be alone for a while before you get support from another engine company. The same is true for ladder companies; all cities have more engine companies than ladder companies. Particular attention needs to be paid in tracking the movement and out-of-service times for ladder companies. If your first-due ladder is out of service for any reason, the next truck will be responding from another part of the city with an increased response time. That means you do not have the option for immediate aerial operations like rescue and ladder pipe master streams; your ground ladder cache is limited to those carried on the engine; and any specialized forcible entry tools, rescue tools, or positive pressure ventilation fans (blowers) are not immediately available. In order to effect certain rescues, your primary strategy may have to be to extinguish the fire.

Apparatus Move-Ups

Most fire departments have a policy in place that empowers the fire alarm center to move apparatus around the city in order to maintain balanced city-wide coverage during busy hours when many companies are out of service on active alarms. Some companies can be tempted to use this official interruption as a legitimate reason to cancel the training day for the crew. A move-up to another station means just that—remain in quarters until an alarm comes in or you are sent back to your home station. Not so. Whether you're on an engine or a ladder company, responsible company officers take advantage of move-ups to drive the district and pre-fire buildings to learn more about the area and occupancies in what is now their first-alarm district.

UL/NIST Factors

Applying the latest scientific findings from the UL/NIST experiments also comes into play during the pre-incident size-up. Because modern fuels release more energy faster than do legacy fuels, we know to expect rapid fire growth. This leaves less time to get hose lines into position. We can also expect the possibility of extreme fire behavior once crews enter the structure if they have not cooled the fire from the outside or controlled the flow path.

Visible and nonvisible indicators in reading smoke allow us to better predict what is happening inside the structure based on what we're seeing from the outside. There are two sides of the fire tetrahedron we can quickly affect: heat and oxygen. With fast water application from the outside, we can reduce the interior temperature, improving the environment for possible victims. By limiting the available air through closing doors and windows, or using smoke curtains, we block the flow path, reducing the intensity of the fire. At least conditions will not worsen and, to a lesser extent, temperatures are reduced. A smaller fire, or one that is ventilation-limited, puts out less energy. Look for these clues during the 360-degree size-up to control these aspects early. For example, if you have smoke and fire showing from a window, note the exhaust portal: Is it bi-directional or unidirectional? If it is unidirectional, the fire is getting its air supply from an intake portal located at another location around the structure, so be attentive for an air-inlet during your size-up. You may have the opportunity to shut down or cover the inlet flow path and limit or eliminate the air available to the fire.

If you have a large volume of fire showing, consider a rapid exterior attack on the fire to control the heat and energy momentum from building up. These tactical actions should be at the forefront of the company officer's pre-incident size-up and should be implemented if these situations are encountered. Occupant survival time is of the essence, and rapid exterior water application has consistently proven to make interior conditions better. Firefighters can accomplish this without entering an immediately dangerous to life or health (IDLH) environment; therefore, there should be no delay in getting water on to the fire by the fastest means possible. Firefighters can take these actions while the officer is performing the 360-degree size-up FIGURE 8-6.

Phase 2: Initial Size-Up

The initial size-up is the quick mental process of receiving rapid-fire information that starts as soon as the bell hits in the station or the alerts go off, but also includes the on-scene, 360-degree walk-around size-up. The initial size-up factors are:

- **Time of day.** All alarm dispatches start by announcing the time.
- **Location, including the address of the incident (which may include the name of a local landmark).** The address can indicate a residential, rural, industrial, or commercial area. It can also give you clues of topography and possible road conditions, that is, steep hills, narrow roads. These can become a major issue during snow or heavy rains.
- **Size and type of the initial response, including which company is listed first-in.** A larger than normal initial response indicates a confirmed fire from information that was obtained from the dispatchers. The dispatch usually includes the type of response, for example, dumpster fire, car fire, house fire, structure fire, commercial fire, high-rise fire, hazmat fire, ship fire, numerous calls, report of flames visible, people trapped, across from this address, caller on the phone, meet the caller, and so on.
- **Traffic, temperature, and current weather conditions, including wind direction.** You should note these factors when leaving the station. Wind speed and direction can be determined by flags flying in the area, leaves and branches blowing in the wind, or smoke from chimneys. These can be assessed within seconds of responding out of the station.

If your company is listed first-in, you have limited time to prepare yourself mentally. If you see any visual confirmation of a fire from a distance, announce it over the radio to give the responding units a heads

A

B

C

FIGURE 8-6 A,B. This defensive-to-offensive exterior attack is the fastest means of getting water on the fire. It is referred to as a blitz attack, a transitional attack, softening the target, or "hitting it hard from the yard." **C.** Fast water quickly knocks down the fire, reducing interior temperatures for possible victims.

Courtesy of John Odegard.

up, for example, "We have flames visible, we see a large column of black smoke, this appears to be a working fire," and so on.

The alarm assignments also give you an indication if additional resources will arrive on scene as expected. If your usual next-in units are out of service on another alarm, your engine company may be by themselves until other units from a more distant part of your jurisdiction arrive on scene. For example, E1 is first-in, and E2 and E3 should follow respectively along with L1, but instead the alarm assignment is E1, E5, E8, and L6. This means that your closest units are out of service. The response times for E5, E8, and L6 will be longer, which means you, as the officer of Engine 1, must perform a risk-benefit analysis and determine which tasks you can reasonably accomplish with calculated risks until additional units arrive on scene. Other delays for responding units can leave you to manage the initial stage of the fire by yourself. Stay calm and focused. Don't take unnecessary risks with the lives of your crew. Remember, this is not your emergency. This is your chance to get into a proactive role instead of getting excited and becoming reactive.

Having limited staffing while waiting for additional units to arrive on scene does not mean you can't take decisive action. Consider the following scenario, if an EI was staffed with a company officer, pump operator, and firefighter: You are on the scene of a house fire with flame showing from a window and a known rescue. Once the driver charges the hose line, the officer can shoot water through the window from the outside to knock down the fire, cool the interior temperatures, and remove the thermal threat. The firefighter can now vent-enter-isolate-search (VEIS). This is a fast evolution without breaking the rules.

It's still a good practice to drive the apparatus just past the structure. This allows you to see three sides of the building upon arrival and leaves the A side and a corner of the building clear for an aerial apparatus **FIGURE 8-7**. After the apparatus stops, get out and put on all your protective PPE, including SCBA with the facepiece in the standby position. Set a good example for your crew. Just because you may be standing outside does not mean you shouldn't be ready for any unforeseen emergency or rescue—you may be part of the initial two in/two out.

Remember the overall four incident priorities: life safety, incident stabilization, extinguishment, and property conservation, along with the strategic and tactical objectives of rescue, exposures, confinement, extinguishment, overhaul/ventilation and salvage (RECEO/VS). They need to be in the forefront of your mind because they help you maintain a systematic organized thought process when evaluating the situation and prioritizing your actions. According to the IMS, the first-in company officer becomes the initial IC. You must establish command and a command post, perform the 360-degree size-up, identify the problems, perform a risk-benefit analysis, establish your strategy, make tactical assignments, and maintain firefighter accountability.

FIGURE 8-7 Driving the apparatus just past the structure allows you to see three sides of the building upon arrival and leaves the A side and corner of the building clear for an aerial apparatus.

ACRONYMS

COAL WAS WEALTH: 13-Point Size-Up

COAL WAS WEALTH is probably one of the oldest and most popular acronyms used for size-up. (There are some variations on this acronym, like WALLACE WAS HOT.) This 13-point memory aid is based on experience and historical information to coordinate the fireground by imposing consistency:

Construction
Occupancy
Apparatus and personnel
Life Hazard

Water supply
Auxiliary appliances
Street conditions or **S**tructural conditions

Weather
Exposures
Area
Location and extent of the fire
Time
Height

Each category of the acronyms listed is a critical part of size-up and must be considered at some point during the emergency incident. The two potential issues with these acronyms are that they do not follow a logical sequence, and it is hard to recall what the letters stand for when the initial IC is under pressure on an uncontrolled dynamic fireground. For example, there are three **A**s in the acronym, COAL WAS WEALTH, but which one comes first? Now it has been modified to include hazardous materials. *Area* and *height* are combined under the third **A**, and **H** now stands for *hazmat*. WALLACE WAS HOT has two **W**s, two **A**s, and two **L**s. The 13-point considerations are better suited for the pre-incident phase of size-up, or they should be part of a tactical worksheet posted at the command post. The emphasis must be on learning to identify problems and selecting the appropriate strategy and tactics to solve the problems rather than on performing size-up by acronym.

SLICE-RS

The International Society of Fire Service Instructors (ISFSI) along with Hanover Fire and Rescue in Richmond, Virginia, created another acronym to guide the first-in engine officer who serves as the initial IC in fire attack operations: SLICE-RS. It incorporates the recent UL/NIST recommendations of early water application from the outside. It emphasizes identifying and controlling the flow path. SLICE-RS claims to improve firefighter safety and improves the efficiency of fire attack.

SLICE-RS is a seven-step process. The first five letters represent steps that should occur in a sequential manner, and two that may occur at any point in the operation:

Size-up: In addition to all the 13 points listed above, pay close attention to the wind. Strong winds can create severe fire conditions or a wind-driven fire.

Locate the fire: Perform the 360-degree survey and utilize the TIC. Fires are usually in a ventilation-limited state. Look for pressurized smoke.

Identify and control the flow path: Closing the door at the inlet or the outlet can decrease the intensity of the fire. Keep windows closed until hose lines are in place. This also helps limit fire growth. Set up the smoke curtain at the front door if your department uses such equipment. Don't create an opening until hose lines are in place. If you find yourself in the flow path, work safely and quickly to change conditions. Remember, it's better to work on the intake side of the flow path rather than on the exhaust side.

Cool from a safe location: Cool superheated compartments from a safe location to reduce thermal threat to firefighters from a flashover. Utilize the reach of the hose stream.

Extinguish the fire: After "hitting it hard from the yard," move in quickly while the energy of the fire has been knocked down. Traditional fire attack sequences can continue until final extinguishment is achieved.

Rescue
Salvage
} Actions of opportunity

According to ISFSI, SLICE-RS is considered the initial attack sequence for the first-arriving engine company at a structure fire. ISFSI claims that the SLICE-RS predecessor, RECEO-VS, now serves as command priorities, guiding the command thought process of the IC. (There are many who disagree with this change, asserting that the first-in officer is the IC.)

SLAB SAVERS

Safety profile
Life profile
Air track
Building
} 360-degree size-up; Determines strategy; Offensive, defensive, transitional

Search and rescue
Attack
Ventilation
Extension/exposures
Rapid intervention
Salvage
} Tactics to support the strategy

Performing the On-scene Size-up

Consequences of an Inadequate Size-Up

In his book, *I Can't Save You But I'll Die Trying—The American Fire Culture,* Dr. Burt Clark from the National Fire Academy (ret.) cites a 2010 study by research doctors Kunadharaju, Smith, and DeJoy, from the College of Public Health at the University of Georgia where they studied the contributing factors of 213 firefighter LODDs. They concluded that operating with too few resources, compromising certain roles and functions, skipping or short-changing operational steps and safeguards, and relying on extreme individual efforts and heroics may reflect the cultural mindset of firefighting.

An inadequate size-up, or a lack of one, is always in the top five contributing factors in line of duty deaths. If there has not been a significant reduction of firefighter occupational fatalities (LODDs) on the fireground, it only stands to reason that we must examine how current size-up criteria is being taught. It is unrealistic to expect company officers to perform a task they do not entirely understand or cannot remember; otherwise, it wouldn't be a repetitive contributing factor to occupational deaths. The 13-point size-up list along with the various versions are too hard to memorize and too hard to recall under the pressure of "smoke and flames visible" with incoming units right on your heels waiting for an assignment. Rather than teach size-up by using acronyms, we need to teach company officers how to identify problems. The rapid mental evaluation of the current situation and the identification of problems starts with the 360-degree walk-around size-up.

360-Degree Walk-Around Size-Up

Problems identified in the 360-degree size-up will typically fall into seven categories: visible or verifiable occupants; possible occupants; access; exposures; smoke; fire; and the possibility of hazardous materials being involved. Not all problems will present themselves, but any actions we take will end up addressing one or more of these problems. The root of all other side considerations will still be within the seven categories.

Whether it's you walking the 360, a company or chief officer assigned to Division C, or firefighters assigned to reconnaissance, the initial IC must know what is happening on all sides of the building to identify problems and assess risks accurately, especially the rear C side of the building. For most residential and light commercial structures, the 360-degree walk-around is easy to accomplish very quickly. For small industrial buildings or superblock rowhouses, a firefighter can climb a ladder and traverse the roof (roof man) to the C side rear and report conditions via portable radio to the IC. Another quick way to survey the C side of a long-block row house is to enter through an exposure occupancy to access the rear of the building. For larger commercial facilities or high-rise buildings, consider requesting an incoming engine company or command staff vehicle to drive around the block for a survey before reporting to your command post. The reason: You must know what's happening on the C side rear of the building or critical problems and hazards will be missed.

Thermal-Imaging Cameras (TICs)

During the size-up, note visible and nonvisible indicators. Clues from a ventilation-limited fire are hard to read. The TIC is a critical tool now because it helps us identify and evaluate what we cannot see. Thermal imagery can help you immediately locate the fire by registering high-heat-source temperature readings in color or by giving you the ability to see convective heat currents radiating from the structure. The TIC is probably the most valuable tool to carry during the 360-degree survey. TICs used to be as large as a barbell. Now they're being integrated into the firefighter's SCBA facepiece **FIGURE 8-8**.

Like any tool, not all TICs are the same, but they basically fall into two categories: situational awareness TICs and decision-making TICs. They vary with features such as field of view and temperature modes, and in levels of resolution (high resolution). Situational awareness TICs are being integrated into the next-generation SCBA units and are designed primarily to help a firefighter escape in zero-visibility smoke or when temperatures suddenly escalate. Decision-making cameras are handheld high-end units with larger screens, high resolution, and extremely high sensitivity to temperature changes. They are best suited for size-up and search and rescue.

TICs enhance the fundamental tactical skills and experience we already possess by adding a thermal perspective, scientifically confirming signs that we're estimating during size-up but that are difficult to actually see. For example, as mentioned in Chapter 3, triple-paned modern windows can provide superior insulation that resists exposure to heat, often without failing. Therefore, the fire may have consumed all the available oxygen inside the structure and may now be

FIGURE 8-8 A TIC. Situational awareness TICs are now being integrated into the firefighter's SCBA facepiece.
Photo of TIC/facepiece, Scott.

ventilation-limited (oxygen deficient) but maintain high interior temperatures. There may be no smoke showing from the house. The situation is ripe for a backdraft. A high-resolution decision-making TIC detects higher temperature readings from exterior siding panels closest to the fire.

We've been taught to read smoke, but we cannot read temperature. Although TICs are not thermometers, they can be extremely beneficial in identifying the flow path and register its temperature differentials above and below the neutral plane. Heat always travels from hot to cold, and the TIC also helps you visually note the speed and direction the fire is headed by reading the radiated convected heat currents. Contemporary synthetic fuel and glues have a high heat-release rate, making the modern fireground one of rapid progression with dramatic changes, which can occur in seconds. With training, we can benefit from TIC data and diagnosis to find the location of the fire and its severity before we launch an interior offensive fire attack.

The TIC gives us the ability to see through smoke, which enhances our search methods to locate victims and firefighters rapidly. For example, with the help of the TIC, the first-in officer may quickly check the front door prior to starting the 360-degree survey. Occupants trying to escape from a fire usually take the most common path of egress. If the door is unlocked or easily forced, the officer can aim the TIC inside for a quick sweep of the area and scan for signs of life. If there is someone who registers on the TIC screen, the officer can quickly call for a rescue or do it him- or herself before conditions get worse. It also gives the officer a peek at the interior floor plan, which will give him or her a better feel for the entire layout while continuing the 360-degree size-up. Scan for handprints on the walls; they do register on the TIC. Look for areas where an occupant is most likely to survive. An occupant behind a closed door may find a very survivable atmosphere, while an occupant unprotected from the flow path has little chance for survival. The benefits and methods of using the TIC for various tactics in search, firefighting, ventilation, and overhaul will be referred to throughout this book.

Continue to determine the current situation with the following questions:

- What has happened? What do I have? What type of structure is involved?
- What is happening? What is burning?
- Where is the fire? This is where you apply the skills of reading smoke. Look for signs of high-pressure smoke. Note volume, velocity, density, and color. Black, turbulent, pressurized smoke is closest to the fire. Pressurized smoke coming through cracks indicates an area where water application will have the greatest effect. In the absence of pressure and heat, you may be dealing with an incipient fire, or a fire starved for oxygen (ventilation-limited) as long as it has lost the heat necessary to support combustion. Utilize the TIC to confirm unseen indicators. Both incipient and decayed fires have light gray smoke.
- What will happen? Where is the fire going? Exposures. Heat travels from hot to cold.
- What is my life hazard?
- What is the survivability profile?
- What effect is the fire having on the building construction?
- What is the risk versus benefit?
- Check the windows. Look for handprints or smears. Note curtains or blinds that are in disarray. Depending on the moisture level, glass cracks at 500°F to 600°F (260°C to 315°C). Dark brown stained glass indicates a creosote film on the glass from incomplete combustion, which in turn indicates that the fire is probably in the decay stage, especially if the window glass is not cracked. Inside temperatures are approximately 250°F to 300°F (121°C to 149°C). The longer the fire stays in the decay stage, the more creosote layers form on the glass.

- Cracked glass or crazing (a network of fine cracks along the glass surface) along with blackened windows indicate higher temperature. The water vapor has dried, leaving stains, and the window appears to be covered in a gray, brown, or black film. Cracked glass indicates you're getting closer to the fire or that this is the room on fire. If you find a window that is blackened from smoke by a layer of creosote and cracked glass from the heat of the fire, and especially if smoke is seeping through the cracks in the glass, you have strong indications that the door to the room is open or the fire may be in that room. Window glass cracks when it's exposed to interior air temperatures between 500°F and 600°F (260°C and 316°C). Cracked glass indicates a high-heat area above the human threshold. This a warning sign for a backdraft, and the survivability profile in this room based on the fire gases alone is zero.
- What resources do I have to put this fire out?
- Do I have the water to put this fire out?

Basements

Determine whether there is a basement or a below-grade occupancy and if it is involved in the fire. Either way, announce "basement" or "no basement" on the radio. By their nature, basement fires do not always present themselves like fires on the ground level or upper levels. Failure to identify the existence of a basement has led to firefighter fatalities. One trick is to add a B in front of the strategic and tactical objectives of RECEO/VS, changing it to BRECEO/VS. The B is a prompt to confirm or eliminate the existence of a basement at the start of a 360-degree survey. Remember, the important sequence for basement fires is to hit it from the outside with water from a safe location (door or window) and let it vent. This can be done with a single firefighter operating outside the IDLH while the officer is still performing the size-up. Delaying the application of water only allows the fire to get worse.

Setting Up the Fireground

Throughout the rest of this chapter, we will refer to a series of photographs taken of the same two-story residential house fire from different vantage points and stages of the fire to illustrate the components of size-up, identifying problems, and developing a QIAP.

The first-in company officer must do a 360° size-up, perform a risk-benefit analysis, identify the problems, and formulate a QIAP. Until then, you are in a defensive mode. *Defensive* doesn't always mean setting up master streams for a surround and drown. It's perhaps better stated that firefighters must remain outside the structure or outside the IDLH atmosphere until the officer finishes the size-up and shares the attack plan. In the same way, *offensive* doesn't always mean performing interior firefighting activities. Many offensive actions can take place from the outside. For example, placing a ladder at a window for a rescue, or a firefighter standing outside to operate a hose line through a window into the burning structure are forms of offensive actions.

The point is, firefighters don't have to sit and wait on the apparatus until the officer returns. There are many tasks that can take place to set up the initial fireground while the officer performs the size-up. Obvious access problems can be addressed with forcible entry (softening the building). Hose lines that will most likely be used can be stretched to the front door and charged. Ladders can be placed, and positive pressure fans can be started and set up at the entrance in a ready position. Firefighters can also start connecting to the hydrant for an uninterrupted water source. The firefighters cannot go inside until value, time, and size (VTS) have been determined by the officer. VTS is the IC's initial documented risk-benefit analysis and the first indication about whether your immediate strategy will be offensive or defensive.

Value

The first question you need to ask is "Do I have any value here?" *Value* falls into two categories: life value and property value. "Do I have life value here?" "Is there someone savable?" This is where the survivability profile comes into play. "Do I have property value here?" "Is there something I can save?" A yes answer to these questions calls for some level of offensive strategy involving calculated risks. A no answer—where there are no savable lives or property—warrants a defensive operation where no firefighter lives are put at risk. Let's look at each one a little more closely.

Life safety is the first incident priority in any emergency incident. In asking the question, "Do I have life value here?" there are two likely answers: visible or verifiable occupants (VOs) and possible occupants (POs). If there are VOs who are trapped and in immediate danger from the fire, rescue efforts should commence immediately. Immediate rescue efforts still require managing risks. For example, an occupant at a second-floor window with heavy smoke requires the rapid placement of a ground ladder but may also require the quick deployment of a hose line to protect the occupant and the rescuer. Remember to think! Our job is to rescue the victim and continue the firefight.

There are plenty of other critical tasks to accomplish. Conversely, not every visible occupant at a window needs to be rescued. If occupants are not in immediate danger, perhaps the best action to take is to instruct them to shelter in place. State that they will be evacuated after the fire is knocked down or extinguished. This was the case at the Villa Plaza Fire on September 21,1991, in Seattle, Washington. The first-in engine company made the decision to perform a ladder rescue instead of applying fast water to knock back the advancement of the fire. The victim was successfully rescued, but by the time efforts were refocused on attacking the fire, it had progressed well beyond the firefighters' capability to fight it with 2½-inch hand lines. The fire ended up destroying 66 residential units: one-third of the entire complex.

The second likely answer is *possible occupants*. An example would be a single-family residence at two in the morning. Although there may not be any visible occupants, it is likely there are trapped occupants, and a primary search and rescue operation needs to be initiated—but not without evaluating the risks. This is where survivability profile comes into play (survivability profile is covered in Chapter 4). Survivability profile is based on:

- The stage of the fire (how long has it been burning?)
- The length of exposure to heat and smoke
- The materials burning (synthetic modern fuels versus legacy fuels)
- The occupant's proximity to the fire
- Access to effect rescue

Keep in mind that the upper limit of temperature tenability for humans is 212°F (100°C). At these temperatures, burns to the nasal mucus membranes are almost always fatal, and when oxygen levels drop to 6% to 7%, death occurs within a few minutes. Historically, carbon monoxide (CO) is the most toxic fire gas and prevents the blood from assimilating oxygen. The human body has a 250:1 ratio affinity to CO over oxygen (CO has 250 times the affinity for hemoglobin than oxygen). Physical exertion (like someone trying to escape through a smoky structure) increases the effects of CO poisoning. When CO concentration is at .02%, collapse can occur in 20 minutes, and death can occur in 45 minutes. At .06%, death can occur in 10 to 15 minutes. When CO is at 1.28%, a person can become unconscious with two to three breaths, and death can occur within 3 minutes.

Chapter 2 addresses modern fuel toxicity. Cyanide is present in modern fuel smoke, and recent studies show 87% of fire victims died from toxic cyanide levels. Cyanide incapacitates the major organs and at 3,400 parts per million (ppm, levels found in most structure fires), cyanide can kill a person in less than 60 seconds. Cyanide exceeds 3,400 ppm when flashover occurs. Therefore, from the time a fire starts, the survivability time for a trapped occupant in a fire is about 10 minutes.

Remember to consider the alarm sequence, which starts when the fire alarm center is notified of a fire. Companies are dispatched, there are turnout times and response travel times, arrival, setting up the fireground, accessing the structure, and initiating fire attack. In the meantime, structural stability is deteriorating, and civilian survivability is diminishing. In reality, once the fire department arrives, occupant survival times are better measured in seconds, not minutes.

Finally, there's reading the smoke. Is there visible fire? Where is the smoke coming from? What color is it? What is its density? Is it under pressure? What about the windows? Is the glass cracked or covered with creosote? These criteria will indicate positive or negative survivability profiles. If the smoke indicates a possible backdraft condition, the survivability profile is negative. Keep in mind that a negative survivability profile for one part of the structure does not mean the entire structure has a negative survivability profile. There may be many survivable spaces that need to be searched **FIGURE 8-9**.

Although all these questions seem quite detailed and thus require serious contemplation, making the initial decision takes only seconds.

The second category to the value question is property: "Is there savable property here?" This question is more straightforward and easier to answer yes or no. In Figure 8-9, there is significant property that can be saved. Saving property is part of every fire department's

FIGURE 8-9 No one is alive in the fire room shown here; however, the adjacent bedroom and first floor have positive survivability profiles. The car in the driveway indicates possible occupants, and primary searches need to be made immediately.

mission statement, but it requires limited to no risk to firefighters to accomplish. Remember, no structure is worth the life of a firefighter. Also keep in mind the new term in the industry: **disposable buildings**. Modern, lightweight construction is cheaper to rebuild than to repair after it has been severely damaged in fire. A sobering example is the McDonald's restaurant fire that occurred on February 14, 2000, which claimed the lives of two Houston, Texas, firefighters. A similar early-morning fire occurred in a McDonald's restaurant in Seattle, Washington, although there were no firefighter injuries or fatalities. From a firefighting perspective, it was a great stop on what appeared to be a spectacular fire. The crew thought the restaurant was saved, but the McDonald's Corporation came in a few weeks later and demolished the entire structure to build a brand-new restaurant. In both fires, there were no civilian lives at risk, the two restaurants were demolished and rebuilt, but the Houston Fire Department paid an extreme price.

Time

Time relates to the fire resistance of the building construction. Ask the following questions: "What type of building construction do I have here?" "How much time do I have to perform interior operations before I can expect structural compromise, failure, or collapse" (based on collapse time estimates)? The first thing to determine is whether this is a contents fire or a structure fire. For example, if the fire is limited to a garbage can in a laundry room or a sofa in the living room, you have a contents fire. Although there may be significant smoke and flames, the structure is not yet involved, and its integrity is intact. However, once the fire spreads to the structure and building components become fuel for the fire, you have a structure fire and the clock starts ticking. Elapsed time in the alarm sequence comes into play again in determining collapse times based on the type of building construction exposed to the fire, and whether it is a contents fire or a structure fire. Some fires start as structure fires.

As stated in Chapter 3, there are no exact collapse times for the fire ratings of the different construction types because these numbers are laboratory times. These times are merely indicator guidelines of the fire resistance qualities in each type of classification. The objective is for the IC to understand that certain construction types endure the effects of fire for a longer period than do others, thus allowing for a longer operational period within manageable risks. The opposite is also true: Some construction types can fail quickly when exposed to flames, so the operational time is shorter before crews need to be pulled out in order to switch to a defensive strategy.

Type I fire-resistive considerations are as follows:

- The structure will not contribute fuel to the fire.
- The load-bearing structural members are protected with fireproofing material.
- Columns have a 4-hour rating.
- Beams have a 3-hour rating.
- Floors and shafts have a 2-hour rating.

Type I construction, for example, high-rise buildings in major cities, is designed to contain contents fires to the floor of origin. They usually have large open areas and a fire load that produces a considerable amount of British thermal units (BTUs). Standpipes are limited to 500- to 1,000-gpm capacity flow. Type I allows the most time for interior operations. Because the floors have the lowest rating, and that is what firefighters will be standing on, 2 hours should be the operational time reference.

Type II noncombustible considerations are as follows:

- The structure does not contribute fuel to the fire.
- Load-bearing structural members are unprotected or have minimal protection: Steel columns, steel bar joists and trusses, and suspended loads can be directly exposed to high-temperature and flames.

Steel can fail at temperatures between 700°F (cables) and 1,200°F (structural) (371°C and 649°C). Ordinary fires can easily produce ceiling temperatures above 1,000°F (538°C). A 100-foot (30.5 m) steel beam can elongate 9 inches (23 cm) at 1,000°F. Elongating steel I-beams and trusses can push down surrounding walls. When the temperature is above 1,000°F (538°C), steel softens and starts to fail, depending on the load. Structural failure is driven by high temperatures, so the time for interior tactical operations could be near zero. For this reason, fire attack in Type II structures should commence from the outside **FIGURE 8-10**.

Type III ordinary considerations are as follows:

- They contain brick masonry exteriors with dimensional lumber inside.
- The brick masonry exterior does not contribute fuel to fire.
- Floors and roofs can be 1 inch thick.
- Sand-lime mortar was used in pre-1920 construction, and it can crumble away or wash away with heavy water streams, affecting building integrity.
- Nonbearing walls and partitions are combustible and contribute to fire load.
- Renovations may have added lightweight wood components and unprotected steel.

FIGURE 8-10 Fire attack in Type II structures should commence from the outside.
Courtesy of John Odegard.

The guideline for interior firefighting is 2 to 20 minutes.

Type IV heavy timber considerations are as follows:

- Main structural members have a minimum dimension of 8 × 8 inches.
- Floors and roofs are 2 inches thick.
- Interior load-bearing members contribute considerably to the fire load.
- Timber trusses can burn 10% to 15% of their mass before they can fail.
- Structures include unprotected steel connectors.

The guideline for interior firefighting is 1 hour.

Type V wood-frame considerations are as follows:

- Dimensional lumber supports the wood floors and roof.
- The entire structure can burn.
- Structures include 1-inch-thick floors and roof.
- Structures include lightweight floors and roofs.
- Each exposed surface burns ⅛ inch every 5 minutes.
- Genuine 2-inch × 4-inch lumber has been replaced with 1.5-inch × 3.5-inch lumber.
- Trusses can fail in 3 to 7 minutes when exposed to heavy flame.
- Trus Joist (TJI) I-beams can fail in 2 to 3 minutes when exposed to heavy flame.
- Lightweight construction can use engineered lumber and panel consisting of glues and fiberboard, which can delaminate and accelerate burn time.

Lightweight wood construction is not a "type" of construction classification, nor is it defined by the International or Uniform Building Codes. Generally, the wood is lighter, with less mass, than conventional construction. A genuine 2-inch × 4-inch piece of lumber is now 1.5 inches × 3.5 inches. The member is unsafe at a 1-inch thickness. Where a 2-inch × 4-inch would take 20 minutes to burn through, a modern-milled 2-inch × 4-inch burns through in half the time. The weakness in lightweight construction are its lack of mass and the connectors. Gang nails or gusset plate teeth penetrate 5⁄6 inches into the wood—if they're properly installed. The construction is normally very strong if the connectors are not compromised or exposed to fire; if compromised, then they can fail very quickly. Engineered lumber and panels, which uses glues and adhesives, isn't lumber at all. Therefore, the guideline for interior firefighting is 2 to 20 minutes. Again, although the above considerations seem lengthy, the first-in company officer should be a field expert in recognizing the type of building construction, and these considerations for *time* can be mentally evaluated in seconds.

Size

The last part of the risk-benefit analysis is *size*. Size refers to the size of the fire and the amount of water needed to put it out. The final decision to go offensive or defensive hinges on whether you can supply the sufficient number of gallons per minute to exceed the BTUs generated by the fire. If the gallonage water flow isn't sufficient to overwhelm and absorb the heat, the fire continues to burn. Remember, big fire = big water, little fire = little water. The hose lines must match the size of the fire or the fire will continue to burn. This is determined by using the fire flow formulas, and there are two: the Royer-Nelson Iowa State University fire flow formula and the NFA fire flow formula.

$$\textbf{Royer/Nelson:}\ \frac{\text{Volume in ft}^3}{100} = \text{GPM}$$

$$\frac{\text{Volume in m}^3}{0.748} = \text{L/min}$$

$$\textbf{National Fire Academy:}\ \frac{\text{Area in ft}^2}{3} = \text{GPM}$$

$$\frac{\text{Area in m}^2}{0.074} = \text{L/min}$$

The Royer-Nelson formula is more accurate than the NFA formula, which makes it better suited for pre-incident fire flow planning versus rapid fireground calculations. The NFA formula is much more generous in determining the required gpm than Royer-Nelson, and it is better suited for quick fireground calculations (more water is better than less). The gpm requirement is per floor at 100% involvement. For example, a one-story structure measuring 30 feet × 40 feet has an area of 1,200 square feet. When this amount is divided by 3, you determine that the structure requires 400 gpm to extinguish the fire if the structure is fully involved. If it is a two-story, fully involved building, it would require 800 gpm. If the single-story structure was 25% involved, it would require 100 gpm; if it was 50% involved, it would require 200 gpm; and if it was 75% involved, it would require 300 gpm. In theory, if the structure was only 25% involved and thus required 100 gpm, an engine apparatus with a 500-gallon tank would be able to put this fire out five times without a supply.

Knowing the gpm requirement also determines how many hose lines you'll need to pull to achieve the required fire flow. For example, a single 2½-inch hose flowing 300 gpm should easily handle a fire requiring 200 gpm. If a fire required 400 gpm to extinguish and two 1¾-inch hand lines were operating at 150 gpm each (300 gpm), the fire would not go out; a third 1¾-inch line would have to be pulled, or a 2½-inch hand line would have to be deployed to achieve (and exceed) the 400 gpm requirement.

After years of experience in applying this formula, well-seasoned company officers can quickly determine if a fire can be handled with less than 500 gpm, a flow requirement that can easily be accomplished with one or two engine companies. Beyond 500 gpm, the fire most likely requires a defensive operation.

With practice and experience, the VTS risk-benefit analysis can be determined in seconds. Here's an example:

- Do I have life value? *Yes.*
- Do I have property value? *Yes.*
- What type of construction do I have? *Wood-frame structure, 2 to 20 minutes of interior firefighting depending on the amount of fire. More time if it is only a contents fire.* So, do I have time? *Yes.*
- What's the size of this fire? *300 gpm.* Do I have enough water and personnel to deliver 300 gpm? *Yes.*

Three yes answers for the VTS questions means that you can take reasonable risks for interior firefighting operations and can make the decision for an offensive strategy.

If any of the VTS questions has a no answer, you must remain in a defensive strategy until some alternative action resolves it. For example, if you do not have the water or personnel to flow 300 gpm, an offensive interior attack would be dangerous and ill-advised; however, if three more engines arrived on scene with 12 firefighters, it is now doable. Or if you do not have the time margin on your side and you're concerned about structural collapse due to fire, you should fight the fire from the outside. Once the fire is knocked down, the fire intensity attacking the structural members is reduced, if not eliminated. It may now be reasonably safe to enter the structure for final extinguishment. This is how risk is managed; however, it doesn't stop here. Risk-benefit analysis continues throughout the incident to determine if the risks taken are worth the benefits gained.

A NOTE ON AUTOMATIC SPRINKLER SYSTEMS

During the 360-degree size-up, the company officer may note that the automatic sprinkler system is operating. The sound of the water gong along with water flowing from the drain valve is clear confirmation that water is flowing through a fused head within the system. Operating sprinklers after regular business hours indicate a working fire, so you might as well connect to the sprinkler Siamese FDC and support it. An automatic sprinkler system activates when the fire temperatures fuse the sprinkler heads; they do not depend on a decision or an order given by the IC and they don't hesitate applying water to the fire. The sprinkler heads are already at the seat of the fire. They're like little firefighters with nozzles impervious to smoke, heat, fear, and fatigue. They don't have access problems or run the risk of running out of air or becoming disoriented.

Supplementing the sprinklers with a water supply utilizes the fire protection system the way it was designed, thus reducing the risk to your firefighters—especially in basement fires. Robbing water from the sprinkler system works against you by having your firefighters take a beating instead of a sprinkler head. Unless the IC specifically wants the sprinkler system shut down during a defensive fire attack, it is always a good decision to support an operating sprinkler system. This can be done upon arrival while the size-up is being conducted by the company officer and doesn't detract from an offensive attack.

Declare Your Strategy

If problems and risk-benefit analysis allow an offensive interior attack, announce over the radio, "We are initiating an offensive attack." All the crews on the fireground must be on the same page. If the IC announces, "This is a defensive fire," crews must refrain from going inside the structure. Switching strategy from offensive to defensive must be clearly announced over the fireground. All interior companies must acknowledge the change in strategy and exit the structure, and then follow with a personnel accountability report (PAR) over the radio.

Identify Your Problems

If you do not identify the problems, you cannot develop an action plan. Without a plan, the IC and firefighters react impulsively to the incident. This results in inefficient use of valuable resources, including firefighters, and can lead to freelancing, which can jeopardize safety and result in increased property damage. The fire ends up running the incident instead of the IC running the fire. Remember, every fire has an IC; make sure it's you.

Identifying problems and decision making are guided by the incident priorities of life safety, incident stabilization, extinguishment, and property conservation, as well as Lloyd Layman's strategic and tactical objectives, RECEO/VS. As stated, there are typically seven we will encounter on the fireground. Get used to using the abbreviations as shown when listing your problems and developing the QIAP:

- Visible occupants or verified occupants (VOs)
- Possible occupants (POs)
- Access (A)
- Exposures (Exp)
- Smoke (S)
- Fire (F)
- Hazardous materials (HM)

HM are not a typical problem found at residential fires, but many attached garages double as workshops and can have a significant array of oils, aerosols, solvents, propane tanks, and other household products that are flammable, and significantly contribute to a fire. This discussion of HM is also a reminder of the increasing appearance of illegal methamphetamine labs throughout the country. They can be encountered anywhere: in hotel rooms, garages, and single-family residences. They're always a surprise, and stumbling onto an illegal meth lab is an immediate game-changer. Crews need to back out and the situation needs to be reassessed according to your department policy regarding hazardous materials.

FIGURE 8-11 is an excellent example of a commercial-made QIAP worksheet that has check boxes for VTS and lists the typical problems that present themselves at a fire, as well as possible strategy and tactics to address the problems. However, once you understand the system, you can use the back of a tactical white board or even any sheet of paper **FIGURE 8-12**. The three top boxes are for VTS and any helpful notations or calculations. The rest of the sheet is divided into four columns with the headings P, S, T, R: problems, strategies, tactics, and resources. The last column could also be a T or a U for team or unit. R is good because some tasks outside the IDLH only require one firefighter. For example, if you listed as a problem secure the gas utility, or set up a PPV fan, these tasks could be accomplished with a single firefighter, so the number 1 would go under the R column. Listing the resources is a count of personnel required for specific tasks and gives you a visual reminder when you're running out of firefighters and need to call for additional alarms.

Let's look at the rest of the categories. The solutions are quite simple, but executing the tasks may be difficult and challenging:

- Are VOs a problem you need to solve? Yes. How? By rescue. The crew can make the rescue by helping someone down the stairs, taking someone down a ground ladder, or using an aerial ladder or tower ladder.
- Are POs a problem you need to solve? Yes. How? By conducting a primary and secondary search. If the possibility of victims exists, crews need to be assigned to conduct a search and rescue (SAR).
- Is access (A) a problem, yes or no? If yes, access needs to be gained by forcible entry of a door or window using irons, a master key, or a key card; raising a ground ladder; or cutting the lock on a fence, etc.
- Are exposures (Exp) a problem, yes or no? If yes, they need to be protected. How? With hand lines, covered with salvage tarps, or removed. If no, exposures are not a concern at this time and can be ignored.
- Is smoke (S) a problem you need to solve, yes or no. If yes, then a crew needs to ventilate it (vertical, horizontal, positive-pressure ventilation [PPV], or hydraulic). If no, and smoke isn't currently a concern, you can ignore it or allow natural ventilation to occur.
- Is fire (F) a problem you need to solve, yes or no? If yes, crews need to extinguish it. How? Offensively or defensively (portable extinguishers, 1¾-inch hand lines, 2½-inch hand lines, deck guns, master streams, ladder pipes, etc.)? If no, let the fire burn itself out while protecting the exposures.

These are the problems that the fire department typically addresses at each fire that require immediate action; however, they must be prioritized. For example, if you have a visible occupant who's trapped by the fire on Floor 2 of a residence, but there is a security gate surrounding the premises, gaining access becomes the first problem that must be solved before a single section of fire hose can be stretched. Life safety, rescue, and extinguishment cannot happen unless access is gained. Solve that access problem by assigning crews to forcible entry—they need to force the gate. If you have a trapped, visible occupant on Floor 2, but

Insource action planning worksheet

Problems **Strategy** **Tactics**	**Value** **Time** **Size** **Problem**	**V** Yes/No **Strategy**	**T** of Value **Tactic**	**S** GPM **Teams**
Fire Ignore Offensive Defensive Transition Hose line Master stream	Command and water supply			
	RIT			
Smoke Ignore Vertical Horizontal Shaft, Sky light Natural PPV				
Possible occupants Locate Direct Determine Primary search Secondary search Intercom, Yell, PA Ask, Look, Listen				
Verified occupants Rescue Direct Shelter Ladder, Rope Yell, PA, Radio Assist				
Exposure Ignore, Cool Remove, Pressurize Monitor Hose line Master stream Drive, Fly, Tnt Dozer, Burn, PPV				
Access Provide, Deny Ladder Force entry Tape, Police line				

FIGURE 8-11 Example of a QIAP VTS/PSTR tactical worksheet. Insource Training and Consulting.

there's limited staffing, the best move may be to assign a firefighter to take a hand line and shoot water into the structure from the outside, knocking the flames down and taking the energy away from the fire, rather than attempting to make the rescue while the fire continues to grow in intensity. The key is identifying your problems, which gives you the ability to weigh which actions will accomplish the most good with the resources you have.

List Only Those Problems You Can Identify

Here's an example. If there are no visible occupants trapped, you do not need to make rescue assignments. If there are no access problems, there is no need to assign a unit to forcible entry. If there are possible occupants inside the structure, you need to assign crews for SAR.

Risk benefit analysis V (Value)		Collapse T (Time)	Fire flow GPM S (Size)
P Problems	S Strategy	T Tactics	R Resources

FIGURE 8-12 The QIAP boxes, columns, and categories can be drawn on the back of a whiteboard or a sheet of paper.

Use good character judgment but don't disregard information from bystanders if it seems reliable. For example, if a father is out on the front lawn and tells you his entire family and the pet are safe and accounted for, do not start by sending crews in for a primary search. Address the problems of smoke and fire. Don't get into far-fetched what-ifs: What if Uncle Joe snuck in the back door in the middle of the night unannounced so as not to wake anybody? What if James came home from college early to surprise the family? Primary and secondary searches will always take place at some point during an incident, but use reliable bystander or caller information as a basis to use your resources wisely.

These are the first-tier problems that need to be addressed immediately by the first-in company officer serving as the initial IC. Second-tier problems will occur and can be addressed in the same manner after the first-tier assignments have been completed, or if additional resources arrive on scene. Examples of second-tier problems include salvage, overhaul, securing the utilities, patient care, rehab, crowd and traffic control, and so on. These are all important tasks that need to be accomplished, but not before the first-tier problems, which are the most important. Focus priorities on first-tier problems. Don't get distracted by the second tier. If you don't handle the first-tier problems, you won't need to worry about the second tier.

There are only so many first-tier assignments the first-in engine officer can make before a chief officer takes over command. These first-tier assignments include:

- Exterior attack line through an opening like a door or window (UL/NIST recommended practices).
- Identify and control the flow path.
- Forcible entry.
- Search and rescue.
- Interior attack line.
- Establish a rapid intervention team (two in/two out minimum).
- Backup line.
- Exposure line.
- Water supply.
- Place ground ladders.
- Ventilation.

The first-in officer needs to make these assignments, and that is about all the first-alarm assignment can reasonably handle. When you break down the tactical objectives, you can see that there are simply not a lot of variables on the fireground. The list of variables is short and simple. These assignments cover the basics. If the fire cannot be extinguished, then control efforts should be made to stop the forward progress of the fire so it can be brought under control (contained). Then you switch to a defensive strategy.

Initial Radio Report

Initial radio reports can be either too short, giving no information at all, or too long. For example, how many times have you heard the first-in unit announce that they're on scene, followed by silence? Or you can get a company officer who loves to talk and fills the airwaves with superfluous information, making it hard to decipher the situation at hand. Radio communications need to be short and concise, and this takes practice. Keep your radio report centered around all the problems you just finished identifying. You're verbally reporting what you just observed during your 360-degree size-up.

Using **FIGURE 8-13**, verbally paint the picture of current conditions and give an initial radio report; for example: "E1 on scene. We have flames coming from the second floor center window on the A side of a two-story, wood-frame single-family residence,

FIGURE 8-13 A fire in the growth stage. Still a room and contents fire, it will quickly involve the structure.

50 feet × 60 feet with an attached garage. No basement. We have smoke on Floor 2 with interior exposures. Floor 1 is not involved at this time. No exterior exposures. E1 is command." That's it. What else needs to be said for an initial report? Does the radio report match the picture in Figure 8-13?

VTS and Risk-Benefit Analysis

Do we have life value? *Yes.* Do we still have property value? *Yes.*

How much time do we have? This is a wood-frame construction, so it's 2 to 20 minutes. Even with intense flames (from a side view), Figure 8-13 still appears to be a room and contents fire venting from the window, so we still have a solid 20 minutes. Note that the smoke is still light gray, although it is beginning to darken at the top of the photograph. In **FIGURE 8-14**, the fire has grown in intensity and is now starting to involve the structure. Note that the smoke column above the flames has turned to black. This is now a structure fire and the 20-minute time clock as started. But do we still have time? *Yes.*

What is the size of the fire and how much water do we need? Using the NFA fire flow formula: 50 feet × 60 feet = 3,000 square feet ÷ 3 = 1,000 gpm (per floor). At 25% involvement, the fire requires 250 gpm to extinguish. This fire is less than 25% involved. It is still confined to the room of origin, so 250 gpm should easily extinguish this fire. With experience, we know that this room fire can easily be extinguished with a 1¾-inch handline flowing between 120 and 200 gpm. An easier determination while using the NFA fire flow formula is asking, "Can I put this fire out with 500 gallons (the tank capacity) or less?" Answer: *Yes*

We have three *yes* answers to the VTS questions; therefore, there is a reasonable risk and benefit to launch an interior offensive attack on this fire.

FIGURE 8-14 The fire has spread beyond the room of origin and is now a structure fire. Using the NFA fire flow formula and experience, this fire can still be extinguished with 500 gallons (the tank capacity) or less.

Listing the Problems and Developing Your QIAP

Refer again to Figure 8-13 to develop your QIAP. Start by listing the problems:

- Possible occupants on Floor 2
- Possible occupants on Floor 1
- Access at front door
- Fire on Floor 2
- Smoke on Floor 2
- Interior exposures on Floor 2
- Interior exposures on Floor 1, including garage

Use a VTS tactical worksheet to list your problems **FIGURE 8-15**. For speed, abbreviate the problems using a shorthand that makes sense to you. For the list above, you could write:

- POs Fl-2
- POs Fl-1
- A (access) front door
- F Fl-2
- S Fl-2
- Exp. Fl-2
- Exp. Fl-1 and G (garage)

V	T 20'	50' x 60' ÷ 3 = 1,000 x .25 = S 250 gpm	
P	**S**	**T**	**R**
PO, Fl. 2 PO, Fl. 1 A, Fr. Door F, Fl. 2 S, Fl. 2 Exp. Fl. 2 Exp. Fl. 1 Exp. G			

FIGURE 8-15 VTS tactical worksheet listing only the problems using abbreviations.

Selecting Strategy and Tactics

Declaring an offensive strategy means that there are viable rescues to be made, you can deliver enough gallonage to extinguish the fire, you have enough resources available to meet the incident demands, and the risks to firefighters are worth the benefits gained. Declaring a defensive strategy means that there are no more lives that can be saved, you do not have enough resources on scene, the building is already a loss, the fire is too big to fight with hand lines, or the risks to firefighters are not worth the benefits gained. Remember, a *strategy* is a broad general goal. For example, offensive attack, defensive attack, offensive-to-defensive or defensive-to-offensive transitional attack, and ventilation are strategies.

The *tactic* is how the strategy is accomplished. For example, an offensive strategy can be accomplished using portable fire extinguishers, or 1¾-inch or 2½-inch hand lines. A defensive strategy can be accomplished using a deck gun, monitor, or ladder pipe. Ventilation can be vertical or horizontal. It can be accomplished by opening the roof or using PPV fans. See **TABLE 8-1** for a list of strategies and matching tactics.

A task is the action an individual firefighter takes to carry out the tactic. A firefighter extending a ground ladder, swinging an ax, operating a chainsaw, or opening a nozzle and adjusting the stream pattern are examples of tasks. Listing the number of firefighters needed to perform each tactical assignment helps determine if you have enough resources on scene or whether you need to call for additional alarms for more apparatus and firefighters **FIGURE 8-16**.

TABLE 8-1 Strategies and Tactics

Strategies	Tactics
Offensive attack	Pump can, 1¾-inch hand line, 2½-inch hand line
Defensive attack	Master streams, monitors, deck gun, ladder pipe
Transitional attack	2½-inch hose line, deck gun
Ventilation	Vertical, horizonal, PPV
SAR	Primary search, ground ladder, aerial ladder
Protecting exposures	2½-inch hose line, monitor, deck gun, ladder pipe

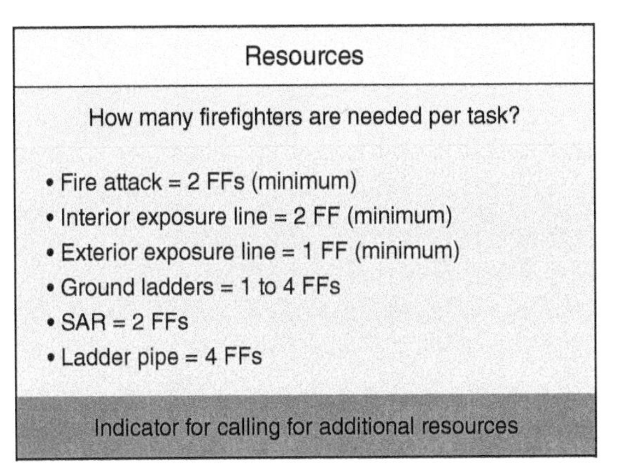

FIGURE 8-16 Examples of required resources.

Now that you've identified and listed your problems, selecting a strategy and tactic to solve each problem is straightforward and simple. Staying with our fire example, we can determine the following:

- The strategy for gaining access through the front door is forcible entry. The tactic is the use of irons or a Halligan tool. One firefighter should be able to force the door.
- The strategy for possible occupants on Floor 2 and Floor 1 is search and rescue; the tactic is a primary search using a TIC. Two firefighters will search Floor 2, and two firefighters will search Floor 1.
- The strategy for the fire on Floor 2 is offensive. The tactic is the use of a 1¾-inch hand line with two firefighters on the attack team.
- The strategy for dealing with the smoke on Floor 2 is horizontal ventilation. The tactic is to place a PPV fan at the front door. The flow path exhaust portal has already been created by the fire self-venting out the window, so one firefighter should be able to set the fan.
- The strategy for the interior exposures on Floor 2 is to protect them. The tactic is to use a 1¾-inch hand line, which also serves as a backup line for the attack team. A team of two firefighters is required.
- The strategy for the interior exposures on Floor 1 is to protect them. The tactic is to use a 1¾-inch hand line. This line also protects the search team and requires an additional two firefighters.
- The strategy for the garage exposure is to search and protect it. The tactic is a primary search using a 1¾-inch hand line. Two firefighters are required.
- A RIT needs to be established.

That's it. You have addressed all the problems you identified. Your VTS strategic and tactical worksheet

V	T	S
	20'	50' x 60' ÷ 3 = 1,000 x .25 = 250 gpm

P	S	T	R
PO, Fl. 2	SAR	Primary/TIC	2
PO, Fl. 1	SAR	Primary/TIC	2
A, Fr. Door	FE	Irons	1
F, Fl. 2	Off.	FA 1¾" HL	2
S, Fl. 2	Hor. V	PPV A1 to A2	1
Exp. Fl. 2	Protect	1¾ " HL/BU	2
Exp. Fl. 1	Prot.	1¾ " HL	2
Exp. G	Prot.	1¾ " HL	2

FIGURE 8-17 An example of a completed QIAP.

FIGURE 8-18 The fire attack (FA) team making entry based on the QIAP.

should look something like the one shown in **FIGURE 8-17** This is your QIAP.

You're now ready to announce your strategy. The strategy must be clearly defined and announced on the radio for all units to hear, for example, "All units from command, this will be an offensive interior attack from side A Floor 1 to side A Floor 2 **FIGURE 8-18**. Ventilation will be horizontal PPV from A-1 to A-2." In other words, the PPV flow path will commence on the A side at the front door and be directed out the second-floor window on the A side, where the fire has already taken the path of least resistance and is currently self-venting out the window.

Making Tactical Assignments

Now it's time to make your tactical assignments. Regardless of the system you follow, use shorthand abbreviations for tactical assignments **FIGURE 8-19**. Many commercial tactical worksheets are available that are functional and quite detailed, but be advised: A tactical assignment worksheet is not a QIAP. It is a checklist of what needs to be done to support your strategy **FIGURE 8-20**. Again, you cannot develop a strategy unless you've identified your problems. Most tactical worksheets do not provide these specific grids or boxes; that's why it is important to learn how to draw the VTS/PSTR grid quickly on any surface.

Let's assume that your first-alarm assignment is four engines, two ladder companies, and a battalion chief. Each unit has an officer and three firefighters. Your assignments can be made face-to-face or on the radio. If done by radio, the unit should repeat your orders, so they can be acknowledged for accuracy and not misunderstood. Your assignments can go something like this:

E1, preconnect. Use forcible entry at front door. Fire attack on Floor 2. E1 driver get a supply (Figure 8-18).

E2, RIT. Side Alpha. E2 driver assist E1 driver with water supply. Establish a secondary supply.

L1, split your crew. Team A: SAR on Floor 2. Team B: set up PPV at front door. Horizontal ventilation from Alpha 1 to Alpha 2. SAR Floor 1.

E3, split your crew. Team A: Backup line to Floor 2 and protect exposures. Team B, exposure line to floor 1.

L2, ladder the roof from Charlie side with vertical ventilation equipment. Do not open the roof yet. If fire is in the attic, initiate indirect attack with a piercing nozzle. I'm sending you E4. (**FIGURE 8-21**)

E4, split your crew. Team A, take a hand line with a piercing nozzle to the roof from side Charlie with L2. Team B, check the garage. SAR and protect exposures.

B1, take Division Charlie.

Assignment Abbreviations

SAR	=	Search and rescue
FA	=	Fire attack
BU	=	Back up line
EXP	=	Exposure
V	=	Ventilation
PPV	=	Positive pressure ventilation
GL	=	Ground Ladder
FE	=	Forcible entry
RIT	=	Rapid intervention team

FIGURE 8-19 Suggested assignment abbreviations.

ASSIGNMENTS MULTI-STORY ON BACK UNIT ASSIGNED		
	ATTACK LINE	
	SUPPLY	
	BACK-UP LINE	
	SEARCH	
VENTILATION PLAN: *ANNOUNCE OVER RADIO*		
		VERTICAL
		PPV
	RIT/RIG	

COMMAND DESIGNATOR:		FIRE	
E N G		L A D	
TACTICAL PRIORITIES CHECK OFF			
	360 COMPLETE		
	WATER ON FIRE		
	SUPPLY ESTABLISHED		
	VENTILATION ESTABLISHED		
	PRIMARY SEARCH COMPLETE FIRE ROOM		
	PRIMARY SEARCH COMPLETE FIRE FLOOR		
	FIRE UNDER CONTROL		
	PRIMARY SEARCH COMPLETE		
	TAPPED FIRE		
	SECONDARY SEARCH COMPLETE		

FIGURE 8-20 A tactical worksheet is a checklist of what needs to be done to support your strategy.

Courtesy of Raul Angulo.

FIGURE 8-21 L2 is carrying out its assignment based on the QIAP.

That's it! What other assignments do you need to make based on the problems you've identified? What information would convince you that your judgment is wrong? Uncertainty is not caused by insufficient information. You can gather enough information in a good size-up to make sound decisions. We need better analysis, not more data. Company officers who make poor decisions do so because they're uncertain and wait too long to make a decision. In the meantime, the fire is allowed to get worse.

As units complete their tasks, they can be reassigned to support units still working on active assignments. For example, Ladder 1 Team B was assigned to set up a PPV fan at the front door, then search Floor 1. Because Floor 1 is essentially clear of smoke, the search should be completed rather quickly. Once a "primary search all clear" is transmitted to the IC, L1 Team B can be reassigned to Floor 2 to pull the ceilings and check the attic for extension. E4 Team B was assigned to check the garage; if there is no life hazard or fire extension, they can be reassigned to place a ground ladder on side A and overhaul the exterior siding and soffit above the fire window, or they can be assigned the secondary search on Floor 1.

A Functional Accountability System

According to the National Institute for Occupational Safety and Health (NIOSH), the lack of firefighter accountability ranks in the top five contributing factors leading to a firefighter occupational fatality (LODD) or severe injury. On January 22, 2004, Greenwich, Connecticut, Fire Department was fined by the state's Division of Occupational Safety and Health for violating regulations when tracking the whereabouts of firefighters during a fire in which three firefighters were seriously injured. Many fire departments think they have an accountability system because they have a series of name tags. Personnel Accountability During Emergency Operations is covered in National Fire Protection Association (NFPA) 1500, 8.5, and here is what the initial IC needs to know:

- The accountability system shall be used on all emergency incidents.
- The IC shall maintain an awareness of the location and function of all companies or crews at the scene of the incident.
- Officers assigned the responsibility for a special tactical level management component at an incident shall directly supervise and account for all companies and/or crew operating in their specific area of responsibility.
- Company officers shall maintain an ongoing awareness of the location and condition of all company members.

For the IC to meet the NFPA 1500 requirements for an effective and functional accountability system, it needs to accomplish five objectives:

1. The names of firefighters (ID number or radio)
2. Where they are assigned (company or unit)
3. Where they are located (on the fireground)
4. What are they doing (task assignment)
5. What time did they enter the structure (IDLH, air management)

If an IC cannot track the movement of fire companies on the fireground, and she or he simply has a roster of firefighters and companies assigned to the incident, the IC does not have a functional accountability system. The following quote is a sobering admonishment that is still relevant today.

> *"Dallas (TX) Fire Dept. uses a Passport accountability system. Getting used to it was the toughest part. During my 30 plus years with Dallas, we have lost six firefighters in structure fires. I firmly believe they died because we did not know where they were."*
>
> Deputy Chief Larry Anderson, DFD (ret.)

We're going to spend considerable time on this subject because firefighter lives depend on your understanding of accounting for and ability to account for firefighters on the fireground. And it's the law. A lack of understanding of the gravity of having a functional accountability system is seen in the case of the Sofa Super Store Fire that took the lives of nine career firefighters in Charleston, South Carolina, on June 18, 2007. In this incident, the only accountability system used was the daily work roster. Several off-duty firefighters and mutual aid companies responded without being dispatched. Not all firefighters entering the structure had their designated handheld radios, and there was no way to track them. Firefighters were known to be trapped inside the burning building, but the number of missing firefighters was not determined—and at least one of the identities was not discovered until the body was recovered after the massive fire was extinguished.

Firefighter accountability is not difficult; it is a very teachable skill. Let's return to our house fire scenario. **FIGURE 8-22** is an example of an accountability status board that reflects the assignments made on the tactical worksheet. It shows exactly where every firefighter is by name, what unit each one is assigned to, the location of each on the fireground, and the task each is performing. If any firefighter on this status board calls a Mayday, the IC knows exactly where the firefighter is and can deploy the RIT to that location.

Technology is evolving for embedding electronic components into SCBA units to track the movement of firefighters within a burning building, but it is still far from being the norm for every fire department in the country. The cost alone for an entire fire department to upgrade or purchase new SCBA with the ability to track firefighters will be cost-prohibitive for many years to come. We cannot engineer our way out of bad behavior to solve the problem of firefighter accountability. It is like learning how to use a calculator before we're taught to add, subtract, multiply, or divide by hand. Simply pushing buttons on a calculator does not teach you how to do math; likewise, watching little moving dots on a computer screen isn't going to teach an IC how to maintain firefighter accountability. Every tactical worksheet should reflect the strategy that was implemented, but not every tactical worksheet is an accountability system. It must include the five criteria listed above. However, every accountability system should mirror or can double as a tactical worksheet.

Exercise 1

Look at **FIGURE 8-23**. The number 1 and 2 at the top of the first two columns indicate that this is a two-story structure. The C is the C side rear of the building, and the G is an attached garage. Can you already picture it? Starting with fire attack (FA), you can see this fire is on Floor 2.

- E33 was first-in. They performed forcible entry and took a line up to Floor 2 for fire attack and search and rescue.
- L8 Team A also went up to Floor 2 to force any doors and searched the rest of the second floor. E18 Team A took a backup line to Floor 2 and also protected interior exposures.

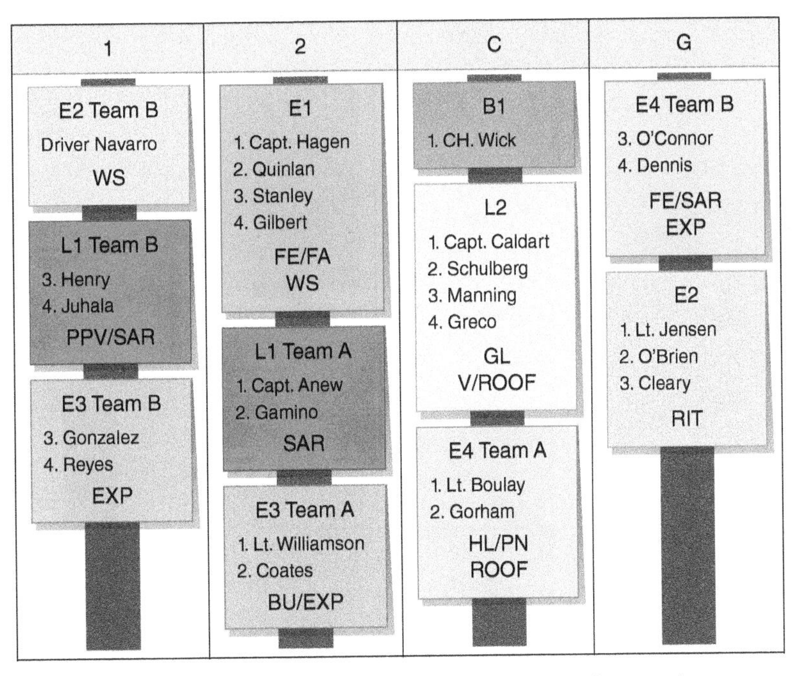

FIGURE 8-22 An accurate accountability status board reflecting the assignments that would be made for the house fire shown in Figure 8-13.

1	2	C	G
E18/B WS	E33 FE / SAR FA	L4/B GL V / Roof	E6 Acc / SAR Exp /HM?
L8/B PPV / SAR	L8/A FE / SAR		
L4/A GL	E18/A BU / Exp		E20 RIT

FIGURE 8-23 Exercise 1 accountability worksheet.

FIGURE 8-24 A two-story, wood-frame, single-family residence with an attic and a basement. There is a fire on Floor 1 and Floor 2.

- Under the first-floor or ground-level column, E18 Team B is getting a water supply, probably for Engine 18 and E33.
- L8 Team B initiated PPV and is searching the first floor.
- L4 Team A is throwing ground ladders on side A.
- At the C side rear, L4 Team B is placing a ground ladder and climbing up to the roof for vertical ventilation.
- E6 is gaining access to the garage, searching it, checking for household hazardous materials, and protecting exposures.
- E20 is the RIT.

This simple exercise is an incomplete accountability system, but it shows you the idea of the accountability concept. It can be performed on a napkin or a piece of paper. The only thing missing are the names of the crew members that can be crossed referenced with a roster printout or compared to name tags, and the entry times—then you would have a complete system.

Exercise 2

Refer to the fire in **FIGURE 8-24** and study **FIGURE 8-25**. The column headings are B, 1, 2, and R which means this structure is a two-story, single-family residence with a basement. R is the roof.

- Starting with Floor 1, E9 was first-in.
- E9 Team A initiated fire attack.

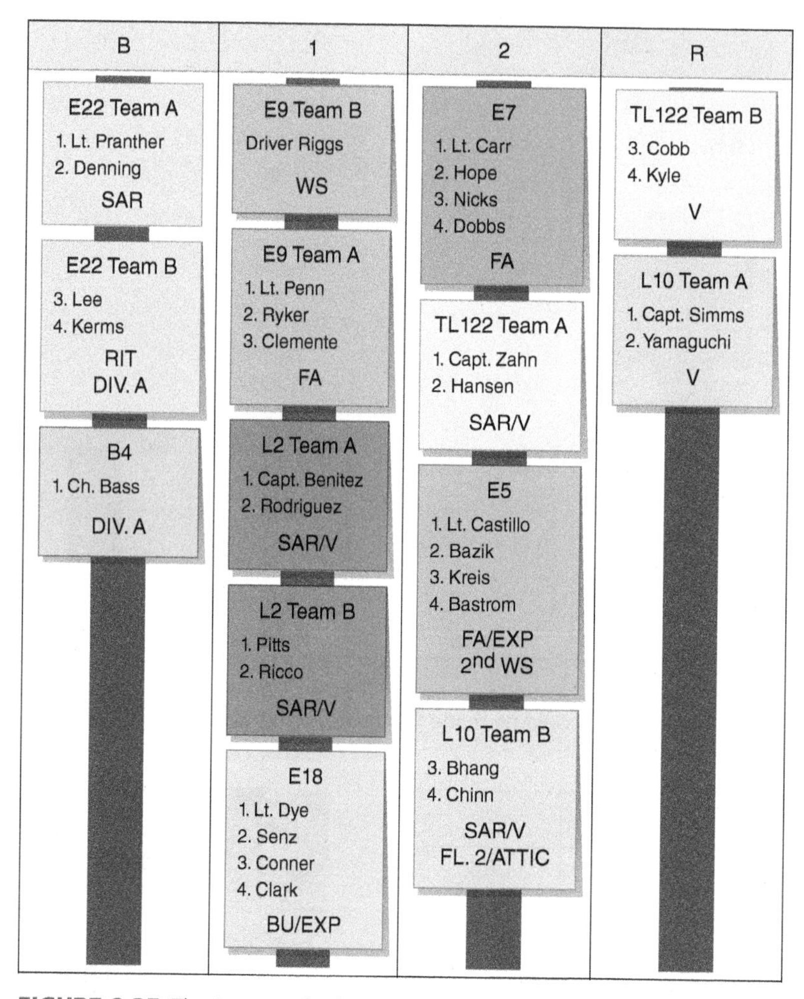

FIGURE 8-25 The Passport firefighter accountability status board accurately reflects the strategy and tactics for Figure 8-24. It lists the names of every company, every firefighter, every assignment, and their respective locations on the fireground. Entry times should be included.

- E9 Team B is getting a water supply.
- L2 Team A and Team B were assigned search and rescue and ventilation.
- E18 laid a backup line and an exposure line.
- On Floor 2, E7 laid a line up the stairs for fire attack.
- Tower Ladder (TL) 122 Team A is performing search and rescue along with ventilation.
- E5 laid a second fire attack line as well as an exposure line. The driver is getting a second water supply.
- Upon arrival, Ladder 10 Team B was assigned to Floor 2 to assist TL 122 with the search and ventilation.
- Ladder 10 Team B is also searching the attic.
- Under the R column, TL 122 Team B was sent to the roof for vertical ventilation. Upon arrival, L10 Team A was also sent to the roof to assist TL 122 with vertical ventilation.
- Under the B column, E22 Team A was assigned search and rescue in the basement. Since the basement isn't involved, the SAR should be quick. If no occupants are found, E22 Team A can rejoin E22 Team B and bring the RIT staffing up to four.
- E22 Team B is the RIT on Side A.
- Battalion 4 is Division A at the front of the building.

Exercise 3

Refer to Figure 8-25 and pick any name and call a Mayday. For example, the IC receives a Mayday from Bastrom. After a quick scan of your accountability board, you can see that Bastrom is part of E5 on Floor 2 assigned to fire attack, or is the exposure line. Deploy E22 Team B (RIT) to Floor 2 with instructions to search for and rescue FF Bastrom. Raise Lt. Castillo on the radio to enquire about Bastrom. Reassign TL 122 Team A and L10 Team B, who are already on the second floor, to search for FF Bastrom.

In another example, say a panicked firefighter yelled "Help! This is Rodriguez!" and that was the last transmission you heard. Where is FF Rodriguez? After scanning your accountability board, you can see that Rodriguez is assigned to L2 Team A. The crew was performing search and rescue and ventilation on Floor 1. Deploy E22 Team B (RIT) to Floor 1 with instructions to search and rescue FF Rodriguez. Contact Captain Benitez on the radio to ask what's going on. Notify Engine 9, Engine 18 and Ladder 2 that FF Rodriguez is missing. Redirect L10 Team B (in the attic or Floor 2) to Floor 1 to join E22 Team B.

Exercise 4

Whether your accountability system uses name tags or passports to track firefighters, it is important to note the times when fire crews enter the IDLH; this helps the IC with air management, helps keep track of operational times based on the building construction, prompts the IC to ask for a progress report, and indicates when it's time to rotate crews out for air replenishment. NFPA, 1500 8.9.1.1 states that personnel shall not be permitted to use more than two SCBA cylinders before they are sent to rehab.

Refer to **FIGURE 8-26**. If you received a Mayday and all you heard was a panicked firefighter yelling "Mayday! I'm out of air!" who would it be? Based on the time of entry into the IDLH, Engine 1 entered the burning house at 04:33. Rescue 1 entered at 04:37. Engine 1 also had to climb another set of stairs to the third floor with a charged hose line for fire attack, a more strenuous job than going to the second floor for search and rescue. In other words, the hose team will use more air than the search team. Let's also add zero visibility smoke conditions on Floor 3. Chances are, the Mayday is from a member of Engine 1 because that group of firefighters has been on air longer than Rescue 1. Call Captain Hilliard on the radio to ask what's going on and notify that a Mayday has been called, possibly from Engine 1. If Captain Hilliard doesn't acknowledge the radio, the Mayday is probably from him. Through the process of elimination, each member on Engine 1 should give a personal status report. Whoever does not answer is the firefighter in trouble. If Captain Hilliard, FF Dennis, and FF Hoefner answer, but Gauweiler doesn't, then he's the one who ran out of air and is probably down. Immediately deploy the RIT to Floor 3 and assign additional units to assist in the rescue effort. If Hoefner and Gauweiler did not answer, then that means you probably have two firefighters down and need to double the assigned units for the rapid intervention rescue. Even in this scenario, the IC still knows immediately who's missing, what unit he or she is assigned to, where he or she is located in the structure, and the assignment he or she was conducting. More than likely, if two members are out of air, Captain Hilliard and FF Dennis do not have much air left either. Once they make contact with the RIT, they have to leave the structure immediately before they run out of air.

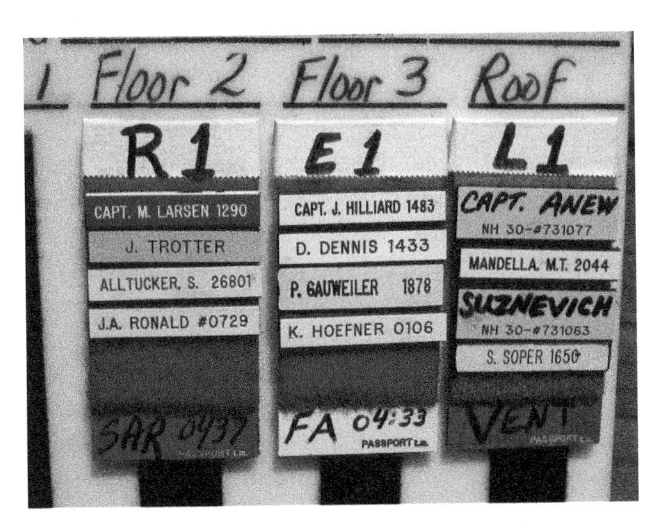

FIGURE 8-26 It is important to note the times when fire crews enter the IDLH area. This helps the IC with air management, and helps keep track of operational times based on the building construction, prompts the IC to ask for progress reports, and indicates when it's time to rotate crews out for air replenishment.

That's all there is to firefighter accountability. It's not hard; in fact, it's rather simple. It just isn't taught correctly. In order for any accountability system to work, no unit can self-deploy. Every unit must check with the IC in person or via radio that it is at another division on the fireground. Only the IC should make tactical assignments unless divisions or sectors are set up. The unit tags need to be placed on the accountability status board, and they can also be incorporated within a tactical worksheet—*at the time the assignments are made*. That is the key to keeping track of firefighters. With proper training and practice, any company officer can manage a tactical worksheet, make fireground assignments, and work the accountability status board simultaneously **FIGURE 8-27**. **FIGURE 8-28** is a different accountability status board than the one shown in Figure 8-27 but if you look carefully, they reflect the same units and tactical assignments at the same fire. Most of the activity is happening on Floor 2, so two columns are used to accommodate all the passports. Most important, you can verify the location of every firefighter on the fireground. Once you learn how to place accountability tags or passports while making fireground assignments, the process is the same, no matter what type of accountability system you're using.

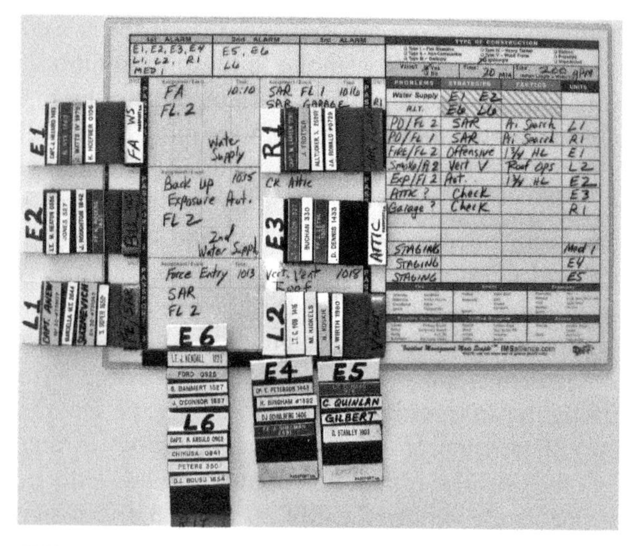

FIGURE 8-27 This accountability status board contains the VTS/PSTR format and can double as the tactical worksheet. Note that E6 and L6 are the RIT, and E4 and E5 are in staging. The passports are placed at the edges of the status board.

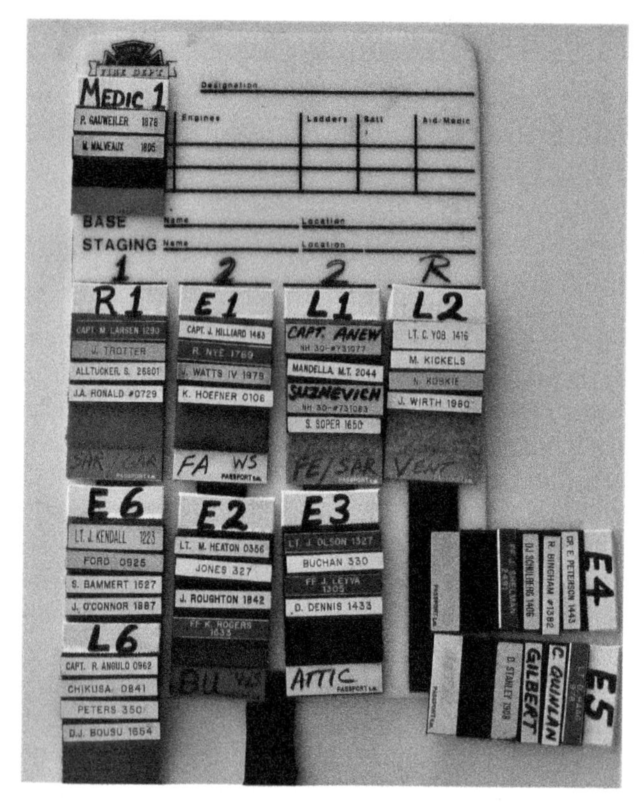

FIGURE 8-28 This accountability status board is simpler but reflects the same fireground assignments as those in Figure 8-27. E6 and L6 are the RIT at ground level. E4, E5, and Medic 1 are in staging and placed on the edges of the status board.

Phase 3: Ongoing Size-Up

Size-up and risk-benefit analysis are ongoing throughout the fire incident. As more interior status and progress reports are transmitted to command, the IC gains more information and has a better picture of whether the current strategy and tactics are effective in controlling the fire. One of the ongoing assessments that is critical to watch for are changes in the flow path and the level of the neutral plane. This is a key indicator about whether interior conditions are getting better or worse. If the fire has been knocked down and contained, the IC can now identify new problems, for example, checking for extension, salvage, overhaul, and securing utilities. These are considered second-tier problems. If the fire has been knocked down, the risk to firefighters, apart from structural instability, is greatly reduced. By using the same decision-making process as above, units that are now freed up or waiting in staging can be assigned to second-tier problems until the incident is terminated.

If the effectiveness of tactical operations does not contain and extinguish the fire (that is, if you do not have incident stabilization), the IC must ask, "At what point does this building no longer have any value?" If the life hazard has been addressed and the building no longer has any value, the firefighters should be withdrawn to set up for defensive operations. The change of strategy from offensive to defensive must be announced on the radio so there is no confusion on the fireground. The lives of firefighters should never be risked for buildings that no longer have any value. Overhaul should not be performed on buildings that are already destroyed. Smoke explosions, which tend to happen in the overhaul phase, have killed or seriously injured firefighters. Risk-benefit analysis continues until the incident is concluded. There is no justification to rush a decision at this point of the fire. Better to have a unit stay on-scene and assigned to fire watch than to have firefighters overhaul a structure that is destroyed with nothing to save. The bulldozers will be there soon enough.

As stated previously, this is not an incident command system (ICS) textbook. There are many components that need to be implemented under the incident management system (IMS). It is assumed throughout this text that all the initial companies are operating under the IMS and following the ICS. The above scenario is used as a template to illustrate how a first-in company officer should set up the fireground by taking command, considering value-time-size (which is actually a documented risk versus benefit assessment), performing a 360-degree size-up, identifying the problems, determining the best strategy, making tactical assignments, and tracking the movement of assigned companies and firefighters by using a functional accountability system. Every company officer should have the ability to run a fire to this point of the incident and effectively assign the first-alarm companies.

After-Action REVIEW

IN SUMMARY

- All emergency incidents need a plan to identify the objective, keep things organized, prevent confusion, eliminate freelancing, and provide for safe and coordinated activities for firefighters and trapped occupants.
- The QIAP is for the first-in company officer. The rest of the incident is based on these initial actions.
- All action plans start with size-up.
- Size-up is the systematic observation of current conditions to identify problems and determine what has happened, what is happening, and what is about to happen.
- There are three phases of size-up: pre-incident size-up, initial size-up, and ongoing size-up.
- Size-up is a dynamic, ongoing, and continual process throughout the incident.
- An adequate risk-benefit analysis is a component in every phase of size-up and is ongoing throughout the incident.
- Risk management has to be at the core of every decision that is made on the fireground.
- A risk-benefit analysis is the process of weighing the predicted risks to firefighters against the potential benefits gained to protect lives and property.
- A risk-benefit analysis needs to be applied to every tactical objective.
- Every initial size-up is a six-sided consideration and should include a 360-degree walk-around survey when possible. It includes areas and floors above and below the fire.
- The presence of basements or below-grade structures should be identified or ruled out and should be announced to the fireground over the radio.
- All emergency command activities are centered around the classic four incident priorities: life safety, incident stabilization, extinguishment, and property conservation.
- RECEO-VS is an abbreviation for strategic and tactical objectives that help the first-in company officer maintain a systematic, prioritized thought process.
- BRECEO/VS adds a B to the acronym to prompt the company officer to first check for the presence of basements or below-grade structures.
- During the 360-degree size-up, the company officer needs to ask him- or herself about value, time, and size. If there is life value and property value, if there is enough time to fight the fire before structural failure, and if you have enough water and firefighters to fight the fire (if there are three yeses to VTS), then the risk for an interior offensive attack can be justified. If there is a no answer to any VTS category, the strategy should be defensive until conditions change to favorable.
- Problems on the fireground need to be identified. You cannot take any action to solve a problem unless the problem is first identified.
- Problems on the fireground typically fall into seven categories: visible or verified occupants, possible occupants, access, exposures, smoke, fire, and hazardous materials. Once problems are identified, tactical assignments can be made to solve them.
- A strategy is a broad general goal for fire attack, that is, offensive, defensive, and transitional.
- A tactic is how the strategy is accomplished.
- A task is the actual process of how the tactic is carried out.
- Tactical assignments are best made face-to-face but can also be made over the radio. Radio assignments must be repeated back to the IC to ensure that the message is accurate and there is no confusion.
- A functional firefighter accountability system is required by law and is more than a roster of who is on scene.
- As tactical assignments are given, the accountability system status board needs to reflect the name of each firefighter, each company, their tactical assignment, and their location on the fireground and entry times.
- Entry times should be noted on the accountability status board when assigned companies enter the IDLH area.
- When a Mayday is transmitted, locating the firefighter on the status board should be quick and easy. The status board works in conjunction with the tactical worksheet.

KEY TERMS

disposable buildings Modern, lightweight construction that is cheaper to knock down and rebuild than to repair after it has been severely damaged by fire.

quick incident action plan (QIAP) The initial action plan by the first-in company officer, who acts as the initial IC. This plan is the foundation of the incident action plan that will be expanded upon or modified after command is transferred to a chief officer.

risk A combination of the probability and the degree of possible injury or damage to health in a hazardous situation. (NFPA 79)

risk-benefit analysis An assessment of the risk to rescuers versus the benefits that can be derived from their intended actions. (NFPA 1006)

six-sided size-up *See* size-up. *Six-sided* refers to the four adjacent sides of a burning structure or compartment, plus the top level and bottom level.

size-up The process of gathering and analyzing information to help fire officers make decisions regarding the deployment of resources and the implementation of tactics. (NFPA 1410)

REFERENCES

Marsar, Stephen, *Survivability Profile: How Long Can Victims Survive in a Fire?*, fireengineering.com, July 1, 2010.

National Fire Protection Association, NFPA 1500, *Standard on Fire Department Occupational Safety, Health, and Wellness Program*, Quincy, MA: 2018.

National Institute for Occupational Safety and Health, *Nine Career Firefighters Die in Rapid Fire Progression at Commercial Furniture Showroom, South Carolina,* NIOSH F2007-18: National Institute for Occupational Safety and Health, February 11, 2009.

Rose, Stewart, *Strategy and Tactics—Multiple Company Tactical Operations*, InSource Inc. Salem, Oregon, 2015.

Dr. Burton A. Clark, EFO, *I Can't Save You, But I'll Die Trying—The American Fire Culture*, Premium Press America, Nashville, TN, 2015.

Starnes, Andy J., *Introduction to Tactical Thermal Data*, FDICI 2019.

CHAPTER 9

Search and Rescue

LEARNING OBJECTIVES

- Recognize that fire extinguishment and ventilation are also part of rescue operations.
- Describe how a survivability profile is applied to search and rescue.
- Explain how forcible entry affects search and rescue.
- Describe how ventilation without prior water application on the fire affects search and rescue.
- Comprehend the chronology of rescue.
- Describe the two in/two out rule and how it plays into RIC.
- Compare the benefits of a combination nozzle to a smooth bore nozzle during search and rescue and rapid intervention.
- Articulate the difference between a primary search and a secondary search.
- Describe standard search patterns.
- Understand and explain VEIS.

Introduction

Every firefighter, from the newest recruit to the seasoned veteran, knows that the rescue of endangered persons is the primary objective at the scene of a fire; however, not every firefighter realizes how broad the scope of rescue operations actually can be. Carrying a victim out of a burning building or carrying a victim down a ladder to safety is rescue work in the purest sense. Directing people to evacuate a fire building is also a rescue action, as is searching a building for victims. Although some of these types of instances are more dramatic than others, each is a rescue operation because it immediately reduces the danger to human life.

Proper placement of the first hose lines can keep a fire away from people in the building. Rapid ventilation removes accumulations of smoke and gas and prevents their further buildup. Both operations reduce the danger to people inside and extend the time they have to evacuate the building. Quickly extinguishing the fire removes the danger to human life. In a very real sense, these are also rescue operations.

Residential fires, especially in single- and two-family dwellings, very often require rescue operations. More firefighters are injured in single-family dwellings than in any other type of occupancy. National statistics show that civilian injuries and deaths in residential fires far outnumber those in other occupancies, such as hospitals, schools, and nursing homes.

Every structure fire will require a search to one degree or another, but not every search results in a rescue. There are some who believe search and rescue are two separate assignments. But in this text, it is implied that if a search results in the discovery of a victim, you will rescue the victim; hence search and rescue (SAR) are used together to address the first incident priority of life safety. This chapter discusses rescue operations, beginning with survivability profiles, pre-incident planning, the primary search, the secondary search, certain rescue drags and carries, fire attack for rescue, rapid intervention teams (RITs) or crews (RICs), search patterns, and vent-enter-isolate-search (VEIS). Each topic will be considered in detail. There are many specific procedures and tool methodologies that can be used to accomplish search and rescue, including the various rescue drags and carries. These are specific to the task level and should be taught within your department.

Risk Management

Today, the entire fire service is driven by risk management. Operating at emergency incidents poses an inherent risk of injury and even death. All firefighters operating at incidents must function in a safe manner. Each member must maintain a level of awareness for her or his well-being as well as that of others. Toward that goal, all members are expected to operate under a risk management profile within the incident command system (ICS). Every fire department should incorporate standard operating guidelines (SOGs) that describe the operating policy regarding risk assessment and safety management at all emergency incidents. SOGs vary nationwide, but they should, at a minimum, incorporate and state the following:

- Firefighters will risk their lives a lot in a calculated manner to rescue savable lives.
- Firefighters will risk their lives a little in a calculated manner to rescue savable property.
- Firefighters will not risk their lives at all for lives or property that is already lost.

To ensure that this guideline is in place at emergency incidents, the following must take place:

- Command must be established.
- A 360-degree size-up must be performed.
- Problems must be identified.
- A risk-benefit analysis must be performed.
- A safety officer must be designated.
- Firefighters need to be in full personal protective equipment (PPE), including self-contained breathing apparatus (SCBA), compliant with NFPA standards.
- A functional accountability system must be established.
- Safety procedures must be in place and adhered to, including the two in/two out rule, followed by the implementation of a RIC.
- A continuous size-up and risk assessment must be conducted throughout the incident.

Survivability Profile

The term **survivability profile** is a relatively new term to the US fire service. Scientific data has allowed researchers to predict with confidence the viability of human life (survivability profile) inside a burning structure composed of modern fuels with more accuracy **FIGURE 9-1**. Determining the survivability profile is the educated art of examining a situation and making an intelligent decision about committing firefighters to lifesaving and/or interior operations. It is based on a company officer's knowledge and understanding in the areas of fire behavior, fire spread, modern fuels, the chemical makeup of smoke, smoke conditions, smoke movement, flow paths, and building construction. If we consider ourselves professionals, we must not abandon

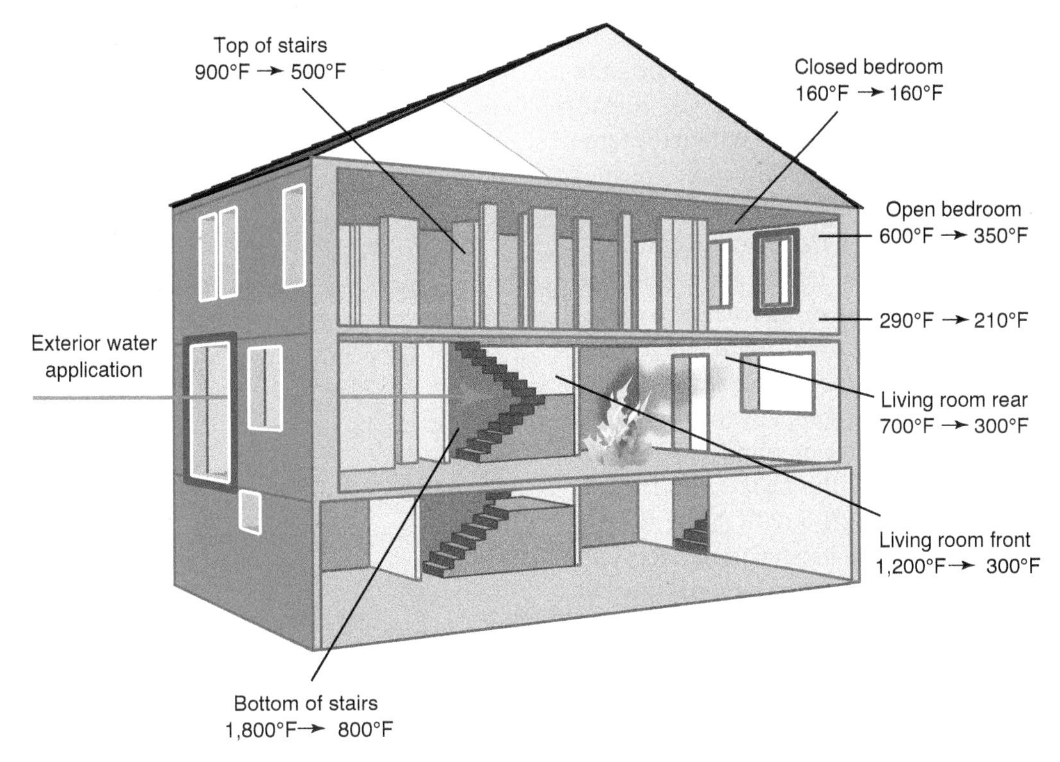

FIGURE 9-1 The temperatures listed were taken before and after the exterior application of fast water. These test-based temperatures help develop the survivability profile. What this diagram doesn't show is the concentrations of smoke—the other significant factor.

Courtesy of NIST.

scientific information, knowledge, and experience in exchange for making emotional decisions. The Underwriters Laboratories (UL) and National Institute of Standards and Technology (NIST) experiments conclude that getting water immediately on the fire by any means possible reduces interior temperatures and slows down the fire, thus improving the survivability profile of building occupants. Fast water, or immediate water application to cool the interior temperatures, makes the atmosphere less volatile for firefighters by preventing flashover and allowing them to enter the structure for search and rescue. Fast-water duration is approximately 30 seconds of straight-stream water application, or as long as it takes to knock down the fire. Fast water knocks down the fire and takes the developing thermal energy away, thus slowing down rapid fire growth. This is known as **resetting the fire**. Resetting the fire buys you more operational time inside the structure that is on fire to extinguish it before the heat has the chance to regain momentum toward flashover. Resetting the fire is like taking the energy of a fire in the fully developed stage and knocking it down to the energy put out by a fire in the growth stage. However, resetting the fire does not reset the rated components of the building construction. The fire may have already damaged the structural integrity of the building, so apply risk management.

Smoke and Room Temperatures

The fire department is rarely lucky enough to arrive on scene and perform a search in a fire room during the incipient-growth stage and before flashover. More than likely, the fire department shows up in the growth stage or right after the fire room has flashed. Temperatures can range from 500°F (427°C) at the floor to 1,500°F (815°C) at the ceiling and certainly over 500°F (260°C) in the fire room, which is the maximum safety rating of our firefighting PPE. In other words, we shouldn't be in the fire room without cooling the interior temperature. Our structural firefighting equipment is designed to protect us from sudden thermal assault atmospheres, not to allow us to work in them. Our PPE firefighting structural ensemble are not aircraft proximity suits. A human being cannot survive in a room that has flashed over. That survivability profile is zero. Our search efforts need to be concentrated on the exposure rooms where temperatures may still be tenable. Temperatures will quickly rise throughout the house unless water is applied immediately on the fire.

There may be a difference in opinion about the temperatures in a decay state fire, but we can say generally that it's between 200°F and 300°F (93°C and 149°C), in other words, less than 500°F. Consider that pain registers on human skin at 111°F (44°C), and first-degree burns occur at 118°F (48°C). Second-degree burns occur at 131°F (55°C). Pain receptor nerves become numb at 140°F (60°C), and at 162°F (72°C), human tissue is destroyed on contact. That's less than the temperature used for baking cookies. (Most cookie recipes call for baking at 350°F [177°C] Clothing that ignites inflicts third-degree burns in a matter of seconds.

It is not simply a matter of temperature tolerance on the exterior skin. The human internal airway, including the trachea and lungs, which are much more sensitive, cannot tolerate these temperatures or the deadly smoke products that accompany them. The entire range of burn temperatures on human skin, from the first sensation of pain to the instant destruction of tissue, falls well below 212°F (100°C), the boiling point of water. We must understand these facts and apply them with confidence when making risk-benefit judgments; otherwise, we will continue to risk firefighter lives when no viable life value exists.

FIGURE 9-2 Repeated experiments produced rapid fire growth or violent flashover shortly after the front door was opened unless water was quickly applied.

Smoke Toxicity

When considering survivability profiles, you cannot think in terms of flames alone; many firefighters think that if the fire hasn't flashed yet, the survivability profile is still good and it is possible to save lives. However, in a decay stage fire, we must also consider the immediately dangerous to life and health (IDLH) atmosphere in evaluating the survivability profile. The smoke of a decayed fire contains toxic levels of carbon monoxide, hydrogen cyanide, benzene, and acrolein. The National Institute for Occupational Safety and Health (NIOSH) lists the IDLH for acrolein at 2 parts per million (ppm). The concentration of carbon monoxide can be as high as 35,000 ppm. The IDLH for CO is 1,200 ppm. At 12,000 ppm, CO can render a person unconscious in two to three breaths and death can occur in less than three minutes. The IDLH for hydrogen cyanide is 50 ppm, and it is 35 times more toxic than CO. Firefighters must apply this knowledge by looking at the visual cues from the smoke, the fire, and the building to determine the survivability profile inside a burning room.

FIGURE 9-3 To prevent rapid fire growth after forcible entry, the door must be controlled until the hose crew is ready to put water on the fire.

Forcible Entry Effects on Search and Rescue

A traditional belief is that making forcible entry into a building is part of gaining access, not part of ventilation, yet studies have shown that opening the front door or a window has a significant impact on the growth of the fire **FIGURE 9-2**. Repeated experiments produced rapid fire growth or violent flashover shortly after the front door was opened. Perhaps forcible entry and access should now be considered part of ventilation because it creates a direct flow path to the fire. At the very least, firefighters should no longer look at forcible entry access as an isolated task; it as an immediate effect on the ventilation profile. This shift in perception requires an emphasis on controlling the door until the hose crew is ready to put water on the fire **FIGURE 9-3**.

In the past, there has been an emphasis for ladder companies (or whoever is assigned to ventilation) to ventilate the building quickly to release the smoke and heat, thus improving interior conditions

FIGURE 9-4 Ventilating a ventilation-limited fire before the attack team is ready to make entry creates a high-low flow path that increases the intensity of the fire, thereby increasing the temperature, heat release rate, and smoke production with other products of combustion inside the structure.

for possible victims. The UL/NIST research has repeatedly demonstrated that ventilating a ventilation-limited fire draws the heat and flames to the ventilation hole that was just created. That ventilation hole creates a high-low flow path that increases the intensity of the fire, thereby increasing the temperature, heat release rate, and smoke production with other products of combustion inside the structure. This puts firefighters and occupants at increased risk of being caught in a flashover or smoke explosion FIGURE 9-4.

Because of the extremely fuel-rich environment found on today's fireground, ventilation that is not preceded by, concurrent with, or immediately followed by effective water application and suppression efforts introduces sufficient oxygen to bring the fire area to flashover conditions rapidly. Water application should come before ventilation and before the primary search begins.

The absence of visible smoke doesn't mean there is no fire or no danger. A common event noted during the UL/NIST experiments was that, once the fire became ventilation-limited, the smoke being forced out of the gaps of the houses greatly diminished or stopped altogether. Interior heat was decreasing and so was the interior pressure that pushes smoke out.

The stages of fire development change when a fire becomes ventilation-limited. It is common with the modern fire environment to have a decay period prior to flashover and the fully developed stage. This emphasizes the importance of coordinating ventilation with the application of water. It is best to start with the application of fast water defensively from the outside to reset the fire and take the energy release out of it. If ventilation is followed by water application from the attack crew, the flow path does not intensify the fire in front of the interior search team. This is what is referred to as a coordinated fire attack.

The Chronology of Rescue

Before the Alarm

Preparation for rescue begins well before the alarm is received. It begins with a thorough knowledge of the company's area of responsibility, including the occupancies, hazards, and potential rescue problems. This knowledge is based on building inspections, pre-incident planning of particular buildings, and learning about the changes that occur in the company's district. The objective is to know beforehand the approximate type and extent of rescue operations and equipment that might be required in any fire incident.

Receipt of the Alarm

The alarm itself can be an indication of the potential rescue situation. The type of occupancy and the time of day, for example, are clues to the need for and possible extent of rescue work. Fires in residential properties, from single-family homes to large apartment houses, always include the possibility of a rescue situation. The possibility is much greater and the problem perhaps more acute at night and in the early morning hours, when most people are at home and asleep FIGURE 9-5.

Buildings such as offices and large stores used to present a reverse situation; in other words, they were normally empty during the night. However, many businesses are now open 24 hours a day, and many people work at home, so the life safety concern is present regardless when during the night the alarm comes in. Be aware of factories that have graveyard shifts within your district.

Even the information given with a verbal dispatch can be important. Phrases such as "across from," "at the rear of," "next door to," "near the intersection of," and "numerous calls" may indicate that the fire alarm was not turned in by an occupant of the involved building but rather by someone outside the building who only saw smoke or fire and does not know anything about the situation inside. Such phrases frequently indicate a confirmed fire. These important clues need to be recognized. When the engine company arrives at the fireground, the clues can be used as part of the information on which problems are identified during the initial size-up.

At the Fire Scene

Even before the engine apparatus has stopped, the officer should begin a careful size-up of the fire scene. The following questions can help to determine what will be required for the rescue operation:

- Is the fire structure a locked-up house with heavy smoke showing?

FIGURE 9-5 The type of occupancy and time of day are two of the most important clues related to rescue operations.

FIGURE 9-6 Cars parked near the house, a swing set, a bicycle and wagon nearby, or a bystander pointing toward the house are all clues that indicate people may be in the building.

- Are there lights on, or is the house completely dark?
- Is it a two-story house that will require ground extension ladders?
- Are there toys, bikes, or swing sets in the yard?
- Are cars parked in the driveway, indicating that an entire family might be inside?
- Is there a wheelchair ramp, indicating a disabled occupant?
- Are people at the windows, or are other occupants calling for help?
- Are there indications that other victims might be inside, for example, in a basement, and that they are unable to get out of the building?
- Are there multiple stories requiring an aerial ladder or an elevated platform?

FIGURE 9-6 and **FIGURE 9-7** illustrate what a firefighter should look for to determine whether people are still inside the fire building.

FIGURE 9-7 Victims at windows often indicate that others may be trapped inside the fire building.

In addition, the extent of the fire; type of occupancy; and size, age, and apparent occupancy load of the building are important in ascertaining what rescue operations are needed. Some of this information should have been gathered during pre-incident planning or building inspections. It can also be gathered on medical alarms. Companies often respond to the same occupancies and residences for medical emergencies. Once the medical alarm is resolved, the company officer can discuss fire attack and rescue challenges with the crew should they return to the same location on an actual fire in the future.

At the fire scene, information that is useful in sizing up the situation might be volunteered by neighbors or solicited from occupants who have escaped the fire building. Of special urgency is any report that people are still inside. On the other hand, unofficial reports that "everyone has escaped" should not delay or halt the primary search, especially in multifamily residences. In an effort to help, bystanders may offer information that may or may not be reliable. Factual information like "a family of four lives there" or "an elderly couple lives there" may be accurate information. Whether the neighbors know if the occupants are home or not is another story; this information requires confirmation by your firefighters. An occupant and family member coming up to a first-arriving officer with reliable information should be acted on. Statements like "my children are still in there!" or "my elderly mother is in there!" have genuine emotion *and* credibility behind them.

Because there is a limit to the time available for size-up, it must be accomplished quickly and efficiently if it is to be of any use for rescue. Again, the more information available before the alarm, the more accurate the initial size-up will be. Initial size-up indicates where firefighters should begin their search and rescue and fire attack operations. These operations, coordinated by the incident commander (IC), should begin immediately.

Occupants attempting to jump from upper floors would constitute an extreme case of immediate rescue. In such a situation, even a crew with limited staffing could make a rescue with the greatest margin for safety; for example, the first-in engine company staffed with two or three should split the crew. One firefighter can raise a ground ladder and effect the rescue, while the other firefighter stretches the line, charges the line with tank water, and shoots water through the door or window from the outside. This can all be done without the crew entering the IDLH. Nothing should delay getting fast water on the fire to knock it down and reset the fire. Additional responding pumpers would be responsible for establishing a water supply and stretching hose lines into the structure for fire attack.

Every situation is different, but immediate rescue by the first-in engine company, without trying first to knock down the fire, should not be attempted except in extreme cases. And even then, the risk is so extreme, the gamble may not pay off. There are plenty of examples where delaying water application on the fire in favor of rescue led to worse fire conditions that resulted in civilian and firefighter fatalities. The worst-case scenario happened in Keokuk, Iowa, on December 22, 1999, when three children and three firefighters were killed—the emotional intensity and stress doesn't get any worse.

The Keokuk Fire

The Keokuk fire in Iowa involved a 130-year-old home of wood balloon-frame construction that was converted into three apartments. The fire started in the kitchen on floor 1, C-D corner. White and dark brown smoke was coming from the apartment. Upon arrival, two apparatus were already on scene with four firefighters. The chief would arrive shortly with one extra firefighter. A woman and child were trapped on the second-story porch roof screaming that there were three more children trapped inside the house. A police officer grabbed a ground ladder and rescued the woman and child from the porch roof. Three of the firefighters entered the structure to search for the three children. Two children were found and passed off to personnel outside. Although a hose line was advanced inside the front door and charged, no one operated it to apply water to the fire. With the limited staffing, all the attention was placed on the rescue of the children. At this time the fire on floor 1 flashed over, and the intense thermal energy, heat, and flames killed the firefighter on floor 1 and two firefighters on floor 2. The third child, who was found in the arms of one of the deceased firefighters was also killed in the flashover. The two children who were rescued and transported to the hospital were pronounced dead on arrival. As heroic as these firefighters were, they were racing against fire behavior and rapid fire growth that were unchecked by water.

Defensive Search Tactics

If firefighters enter a room to perform a search, and extreme high temperatures force them to crawl to tolerate the heat and smoke, this is a warning sign that flashover is imminent. If rollover—the intermittent flaming of fire gases and smoke—is happening above them, this is another sign that flashover is imminent. Rollover is your last warning sign before flashover occurs. The crew must back out of the room. If firefighters cannot tolerate the extreme heat in their PPE, the survivability profile for civilians is zero. A hose line is needed immediately to cool the atmosphere. Any search at this point without the protection of a hose line should be a **defensive search**.

There are four self-limiting defensive search tactics:

1. From the door, yell "Fire Department! Is anyone in there?" Use your light and thermal-imaging camera (TIC) to scan the floor and call anyone who responds to come to your location.
2. Sweep behind the door with your tool, then sweep out in front of you. Do not go beyond the reach of your tool without a hose line **FIGURE 9-8**.
3. Do not enter the room beyond the *point of no return*. This term is often used to describe the reaction time of a firefighter who is suddenly

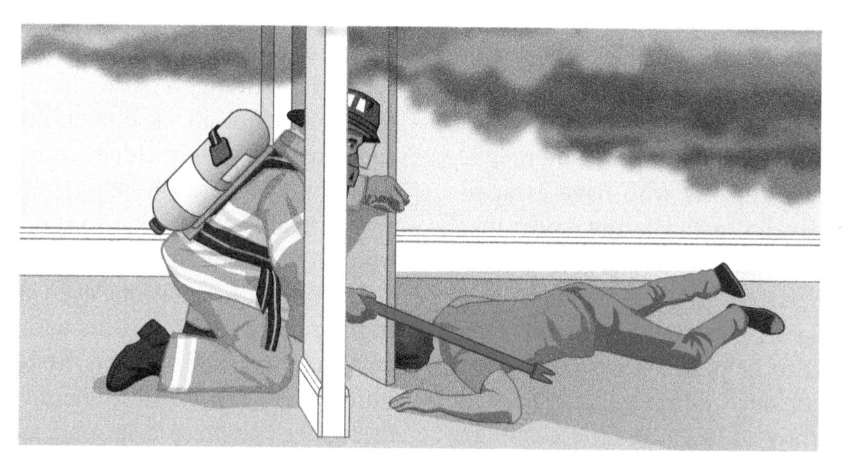

FIGURE 9-8 When rollover is occurring, flashover is imminent. Sweeping behind the door with your tool is a defensive search tactic. Do not go beyond the reach of your tool without a hose line.

caught in a flashover, turns 180°, and survives by making it safely to the door. The point of no return is estimated at 6 to 10 feet to the door, basically, the height of an average firefighter falling forward with enough momentum to crawl past the threshold of a door to safety. Beyond 10 feet into the burning compartment, it is doubtful a firefighter caught in a flashover will survive, or escape, without suffering critical thermal injuries. Remember, our PPE is not rated to withstand flashover temperatures.

4. In extreme heat conditions on upper floors, search the floor area below the window from the ladder by reaching over the windowsill. Stay on the ladder. Use your tool for reach and sweep the floor. If there is a victim below the window, reach over and try to lift the victim up over the windowsill onto the ladder without entering the room. If this is impossible, call for help and a second ladder. If you have to enter the room to lift the victim, do not go beyond the reach of the ladder. If flashover occurs, this is a scenario for a head-first ladder bailout.

Getting caught in the exhaust flow path between the fire and the exit point can be deadly for you and the victim. You need to take immediate action to escape by creating a barrier between you and the fire. This may be closing a door, breaching the wall into an adjacent room, or forcing your way into another apartment across the hallway and then closing the door. Increasing the distance between you and the fire improves your chances of survival. This is a Mayday situation. Call for help.

If victims are on upper floors at windows, additional firefighters may assist them by raising ladders and/or attempting to talk to the victims to calm them until they can be brought down **FIGURE 9-9**. Handheld megaphones or the public address (PA) system on the apparatus are useful in such situations. The victims that you see may not be the ones in the most distress. It may also be an indication there are more people trapped inside the structure.

Engine companies can aid rescue by placing protective streams between the victims and the fire. The entrained air from a fog stream can push the smoke and fire away from occupants in burning structure. Any positive action on the part of arriving firefighters usually has a calming effect on those trapped in the building.

Two In/Two Out Rule

Firefighters must be aware of the two in/two out rule as directed by the Occupational Safety and Health

FIGURE 9-9 Victims attempting to jump from upper floors indicate an immediate rescue situation along with the possibility of additional trapped occupants inside the burning structure.

Administration (OSHA) and National Fire Protection Association (NFPA) 1500. During the initial stages of an incident, firefighters working inside the IDLH area must work with a crew of at least two members. They must be backed up by at least two members outside the hazard area who are fully protected with PPE and SCBA and who are able to go to the aid of the two members inside if the need arises. The two-out firefighters should not stand around waiting for a Mayday to occur. They should be performing a 360-degree size-up: familiarizing themselves with the layout of the house, locating additional access points, determining if there is a basement or below-grade area, and determining what size ground ladders are needed around the perimeter—all while additional companies are being assigned fireground tasks.

The presence of a basement, as well as any change in fire conditions that involve the basement, should be relayed immediately by radio to the interior attack team and the IC. Fires involving a basement can require immediate adjustment to the attack strategy. According to the *Project Mayday* website, the number one reason firefighters call a Mayday is because they fell into a basement.

The ability to identify the presence of a basement quickly is a critical skill for any firefighter assigned to search and rescue, two in/two out, or the RIC. The five pictures described next are examples for exterior identification of a basement or below-grade area. **FIGURE 9-10(A)** shows a window close to the walkway and foundation of a residential home. Such windows are sometimes hidden by landscaping, but the window itself is a clear sign that there is a basement. In both residential and commercial occupancies, windows at the sidewalk level or stairs that go down from ground level are obvious signs of a basement.

Many Type V wood-frame homes built before 1960 have two brick chimneys. One is usually larger and rectangular, and is located on a perimeter wall, close to the A-B or A-D corner of the house. This is obviously the chimney to the fireplace in the living room. The other brick chimney, which is often located at the peak in the center of the roof, is much smaller and is built in a square, and often has a galvanized steel cap. This type of chimney is the ventilation flue for an old oil furnace or wood-burning stove. The firebox is in the basement **FIGURE 9-10(B)**. In Type V wood-frame residential homes of the same era, two flues might share the same brick chimney space, and this chimney is much wider. One flue is for the fireplace; the other is the vent for an oil or woodstove furnace, with the firebox in the basement **FIGURE 9-10(C)**.

In older cities with downtown areas that were around in the first half of the twentieth century, many of the buildings are Type III ordinary and have basements. Sidewalk skylights, which are also called vault lights, consist of thick glass blocks cemented into a steel grid. The grid is laid into the sidewalk to bring in natural light. Any building with vault lights in the sidewalk in front of it has a basement **FIGURE 9-10(D)**.

In these same buildings with vault lights, there is often a stairway or elevator to access the basement from the sidewalk. These sidewalk openings were and, in many cities, still are used for deliveries. Metal doors span the sidewalk. If the floor doors have hinges, the access is still used. If the hinges were removed and welded shut **FIGURE 9-10(E)**, the doors are no longer in use, but they still indicate the presence of a commercial basement. These commercial structures, many of which do not require a standpipe, usually have a Siamese fire department connection (FDC) or even a single-intake FDC for a dry sprinkler system close to the sidewalk, the vault lights, or close to the sidewalk delivery doors, which is another indication that this building has a basement **FIGURE 9-11**. For search and rescue, any of these basement access points may have to be used for making entry, applying water, or for ventilation.

The two in/two out team should anticipate and prepare for the arrival of the RIC and start to compile the necessary extra equipment off the first-in engine, all while maintaining a state of readiness in case a Mayday is transmitted over the radio.

Rapid Intervention Crew

A rapid intervention crew (RIC) must be provided from the initial stages of an incident to its conclusion. Whether the crew or company is called a RIC, RIG, RIT, or Fast Truck, the purpose is to have a dedicated team in a forward position for the sole purpose of launching a rescue for a firefighter who gets into a Mayday situation and needs assistance. An RIC is established when members are engaged in active firefighting activities or other incidents where firefighters are subject to hazards that would be immediately dangerous to life and/or health in the event of an equipment failure, a sudden change in fireground conditions, or a mishap.

On a full response, the first three engine crews are probably assigned the fire attack line, the backup line, and the exposure line, respectively. In this sequence of assignments, designating the fourth-in engine as the RIC is a good practice. The objective of the RIC is to have a fully-equipped rescue team with a dedicated hose line ready on site to react and respond immediately to the rescue of injured or trapped firefighters

FIGURE 9-10 A. Residential basement windows. B. The chimney on the left is for the fireplace. The chimney on the right is for an oil or woodstove. The firebox is in the basement. C. Two separate flues sharing the same chimney. One is for the fireplace; the other is for an oil or woodstove, with the firebox in the basement. D. Vault light or sidewalk skylights. E. These old sidewalk elevator doors were once used for deliveries into the basement, which still exists.

and civilians. If possible, the RIC should be a crew of four firefighters, one of whom must be an officer.

This process begins with the first-arriving companies using the two in/two out rule. When the RIC arrives and is set up, the crew members can absorb the two firefighters assigned to two in/two out and bring the company strength of the RIC up to six. If the incident becomes complex, additional companies should be assigned. RIC members should be in full PPE with the following equipment:

FIGURE 9-11 Many older Type III buildings that do not require a standpipe have a Siamese FDC connection close to the sidewalk, or they have a single-intake FDC for a dry sprinkler system for the basement. The single-intake FDC with a drain pipe outlet is indicative of a basement.

- One-hour air bottles on SCBAs
- Thermal-imaging camera
- Fully charged, high-power battle lanterns
- Search rope to serve as a lead line or for hauling the Stokes litter
- **Rescue air kit (RAK)**, which is an air replacement for the downed firefighter
- Stokes litter, which provides fast, easy, and efficient removal of a downed firefighter
- Tools (i.e., axes, irons, Halligan tool, pry bars)
- A dedicated hose line equipped with a combination nozzle

Other rescue equipment staged for the RIC team includes: hydraulic rescue tools, chainsaws, rescue saws, other power tools, jacks, cribbing, ventilation equipment, and ground ladders. The separate hose line means that if the team is deployed and needs to protect the downed firefighter in place while others try to free him, this hose line will be on top of all the other hose lines laid (hose layering); thus, it will be free to advance and position where needed without becoming entangled with or weighted underneath other hose lines. The combination nozzle allows for efficient fog patterns that can entrain air and push smoke and flames away from a trapped firefighter with full gpm capability. This is more difficult to accomplish with a smooth-bore tip nozzle, which has to be gated down, reducing the gpm flow. The combination nozzle provides better protection for the entire RIC. It's the right tool to use.

The practice of having the initial two in/two out crew and RIC stage next to the IC or command post should be discontinued. It's more of a comfort for the IC than it is for the firefighter needing rescue. The command post can be some distance away from the action; in some cases, it can be located a block or more away. The RIC officer should anticipate where trouble can happen and stage the team in a forward position to reduce the entry distance and maximize the deployment options. For example, if the fire attack is in the basement, the RIC should be staged where it can make the quickest entry into the basement. If the fire attack takes place on floor 10 of a multistory apartment building, the RIC should stage on floor 9.

Setting Up the RIC

During a fire incident, while the RIC is en route, fireground assignments are made over the radio, and interior crews give radio report updates. Therefore, the RIC company should only announce that it is on scene and start setting up the equipment in a forward staging position. There is no need for the RIC officer to check in with the IC at this moment. Besides, you already know your assignment, which is rapid intervention. The person you want the information from is the lead member of the two out crew. The officer should know simply by listening to the radio what the company assignments are, along with current fireground conditions. For example, if a two-story residence is on fire and you heard, "E1, fire attack on floor 1; E2, back up E1; E3, take an exposure line to floor 2; L1, split your crew; Team A, PPV horizontal ventilation on floor 1, and Team B, search floor 1, L2 search floor 2," you know exactly where every company on the fireground is located. Whatever the IC hears on the radio, the RIC officer will hear on the radio, including Maydays. It is more important for the officer and the crew to set up their equipment cache, survey the building, soften the building, get ground ladders in place, and set up their dedicated hose line. Softening the building is the act of performing a 360-degree survey of the fire building to force any locked doors or windows, or removing any obstacles like bars, gates, or fences without changing the ventilation profile or creating a dangerous flow path. Softening the building is any resolution of any obstacle that can hinder or slow down the deployment of the RIC.

At this time, the IC needs to hear that you're set up and ready to go. Then, if you want to check in with the IC face-to-face for an update, go ahead and do so—except now, you'll have pertinent information on what you have done to set up the RIC, the actions taken, and where the team is currently staged. The IC will appreciate that information.

RIC members should closely monitor the assigned fireground radio channel at all times. They should be listening for companies changing locations, making progress, or not making progress; changes in fire conditions; changes in building conditions; and, most important, for a Mayday distress call. The RIC team should listen for firefighters getting excited on the radio and anticipate the order to deploy. A rise in emotions can indicate something is happening or is about to happen. Because seconds count with one of our own, an SOP must be created ahead so the RIC officer can deploy the team as soon as a Mayday is transmitted over the radio. The policy should state that the IC's permission to deploy the RIC is implied rather than having to wait for an actual order, which always has a way of being delayed for one reason or another.

Any company can be assigned RIC, so the officer should train all crews for RIC-specific tasks, like practicing the quick application of an RAK on a firefighter, and discuss possible Mayday scenarios. The following Mayday facts are beneficial to know:

- According to the Project Mayday website, the number one reason firefighters call a Mayday is from falling into a basement, number two is falling through a roof, and number three is becoming disoriented. In the number three category, firefighters who become lost or separated from their hose line is the number one reason for calling a Mayday.
- Fifty-five percent of Maydays are transmitted by the first-in unit. Engine companies call 52% of the Maydays, and ladder companies account for 46%. Rescue companies or other specialized units make up the remaining 2%.
- Most Maydays occur around the 20-minute mark of the incident. This is why it is important for the RIC officer to stay with the team and set up the equipment cache for rescue as soon as possible. The disoriented firefighter in zero or low visibility is often within 15 feet of the front door before turning the wrong way from the initial point of entry. Reports from NIOSH also list disorientation as one of the most serious hazards a firefighter faces while conducting interior firefighting operations.
- At the 20-minute mark, what the distressed firefighter needs most is air.
- If a distressed firefighter is out of air, it is likely the partner and the rest of the initial crew will also be running out of air very shortly. One Mayday can quickly turn into two, or three.

Water Supply and Hose Stream Placement

Water is as essential to the primary search and rescue operations as it is to extinguishment. Water from attack lines is used to:

- Provide fast water by operating the first hose line from the outside, aiming the stream into a window or door to knock down and reset the fire, reducing interior temperatures, and preventing flashover.
- Separate the fire from the occupants closest to it by placing hose lines in strategic locations.
- Control interior stairways and corridors for evacuating occupants and advancing firefighters **FIGURE 9-12**.
- Protect firefighters performing the primary search and ventilation.
- Extinguish around, above, and below the fire.

Engine companies at the scene must therefore set up quickly to provide the necessary hose lines and provide a continuous supply of water to sustain the attack. The 500 gallons or more of water carried in the average-size water tank may not be sufficient for rescue operations. Supply lines should be laid by first-arriving pumpers at the fire scene to ensure an uninterrupted supply and continuous flow.

If no hydrant system is available, water from the water tanks of later-arriving pumpers or mobile water supply apparatus (tenders) should be fed to the first-arriving pumper **FIGURE 9-13**. If later-arriving pumpers run their own attack lines, the first pumper, whose crew might be engaged in rescue operations, will probably run out of water, and their operation and safety will deteriorate. It is important to keep the first-arriving pumper supplied with water so its crew can maintain its position and continue rescue and fire attack operations. Once the first-in pumper is supplied, additional lines can be taken off that apparatus for backup and exposure lines.

Hose Stream Placement

Hose streams should be put into service as fast as possible to attack the main body of the fire. The first actions should be fast water: spraying water through a window or door from the outside. The primary function of the engine company in a rescue situation is to support the primary search by getting water on the fire: extinguishing it or knocking it down. If it cannot be extinguished, knocking it down "resets the fire" and stops the momentum of the energy and heat release rate. Then the fire can be contained so that it doesn't

FIGURE 9-12 Control of interior stairways and hallways must be established.

FIGURE 9-13 If no hydrants are available, the initial attack should be made by the first-arriving pumper (at right) with hose lines supplied by the tank water. The second-arriving pumper (at left) should supply the first pumper from its water tank.

jeopardize anyone within the fire building. For the first-arriving company, this may mean ignoring the rescues at hand and concentrating on knocking down or containing the fire. When the fire is extinguished, the danger to occupants is greatly reduced if not eliminated.

This does not mean that crews shouldn't be assigned search and rescue; it means the assignment for the first-in engine is to control the fire. The next arriving unit can be assigned search and rescue. To ignore the structure fire and immediately go into hands-on rescue mode without applying water to the fire can have disastrous results. This was the case in the multi-fatality fire in Keokuk, Iowa, and at the Villa Plaza Fire on September 21, 1991, in Seattle, Washington.

Hands-on rescue efforts at the expense of water application to the fire only allows the fire to get bigger. This should never be an either/or decision. The increased intensity of radiant heat from a growing fire due to a delay in fast water can make offensive positions untenable. Trapped occupants need to be rescued from smoke, heat, and flames, but those efforts need to coincide with extinguishing the fire. Once the fire is knocked down or extinguished, all these by-products of combustion that are dangerous to humans are greatly reduced, if not eliminated. Therefore, extinguishment should be considered an effective strategy for rescue.

FIGURE 9-14(A) illustrates hose-line positioning that establishes an effective stream between the fire and trapped victims. This positioning is the best rescue procedure for engine companies. When hose lines are placed between the trapped occupants and the fire, combination nozzles are best suited for the job **FIGURE 9-14(B)**. The combination nozzle's ability to provide varying-angle fog patterns and to entrain air makes them the most efficient nozzles for rescue by pushing smoke, gases, and flames away from occupants.

The Villa Plaza Fire

The Villa Plaza Apartments fire occurred on September 21, 1991, in Seattle, Washington. Built in 1968, the U-shaped, wood-frame apartment complex consisted of five buildings designated A, B, C, D, and E, with a total of 96 residential units. Each building was four stories with flat hot tar roofs. The construction consisted of dimensional lumber, and there were no trusses or cocklofts in the roof system. There were dry standpipes in each stairwell but no sprinkler systems. Fire extinguishers, but no hose cabinets, were mounted at the designated intervals along the exterior walkways. The complex had a local alarm with manual pull stations, but the fire alarm was not a monitored system. There were decorative facades made of cedar wood lattice that covered the front four-story exterior walkways of each building. These dry 40-feet high lattice facades were the major factor in the rapid spread of the fire.

The fire started in unit 114 on floor 1 of the C building, which was in the middle of the U-complex configuration and was the largest of the five buildings, measuring 26,100 square feet per floor. The fire occurred at 21:15 hours on a hot Saturday night, so many of the residents were home. There had been no rainfall in the last 30 days, and the cedar wood lattice and siding was very dry. Each unit had three exits: the front door, the rear slider to the balcony, and a large slider window from the bedroom to the front walkway. These ample exits and open balcony walkways allowed many of the occupants to escape. There were no fatalities in this major five-alarm fire.

A

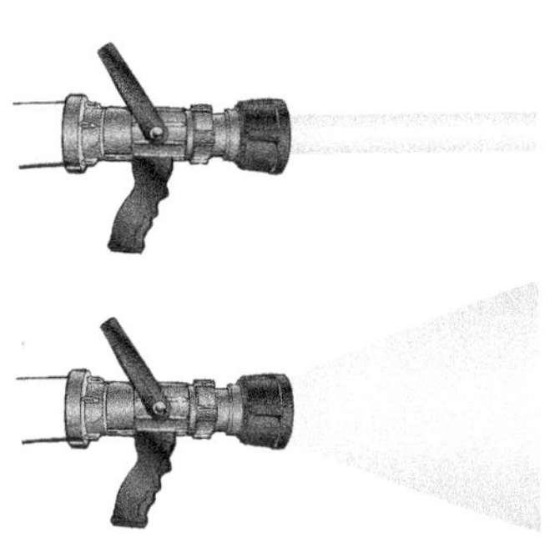
B

FIGURE 9-14 **A**. The correct hose-line position ensures that an effective stream is established. **B**. The use of a combination fog nozzle for rescue is highly recommended for its ability to entrain air that pushes the smoke, gases, and flames away from the trapped occupants.

The occupant, who had her power shut off due to lack of payment, was using candles for light. The fire, which was ruled accidental, was started by the open flame. The occupant attempted to extinguish the fire but was unsuccessful. She never called the fire department because she assumed the neighbors had called. As the fire grew, the occupant panicked and fled the scene with her three children. It is unclear whether she closed her front door, but the single pane windows (installed throughout the complex) quickly failed and allowed the fire to spread. There were many neighbors who noticed the fire but assumed someone else had called the fire department. (In 1991, personal cell phones were not common devices.) A resident pulled the fire alarm, but neighbors did not react due to the numerous malicious false alarm activations in the past. This caused a delay of 15 to 20 minutes before the fire department received the 911 call reporting the structure fire.

As the fire spread rapidly to uninvolved units, many trapped occupants climbed down the exterior balconies, then jumped to the ground or on top of parked cars. Some children were dropped from the second- and third-floor balconies to neighbors and bystanders prepared to catch them. The cooperation and willingness of all the residents to help their trapped neighbors escape the building were other major factors in preventing any fatalities. The fire department successfully rescued eight occupants.

The Villa Plaza was five blocks from Station 33, the farthest station on the southeast border of Seattle. All other initial responding and multiple alarm units came in from the north, increasing normal response times. The initial alarm was dispatched at 21:38. E33 arrived on scene 2 minutes later and parked on a hydrant at the southeast corner of the complex. The firefighters saw a spectacular 40-foot wall of fire, which was actually the vertical fire spread of the cedar lattice facade. They laid a 2½-inch manifold into the parking lot with a single 2½-inch handline coming off the manifold. That is all two firefighters can initially handle. In the middle of their hose lay, they saw a woman about to jump from a second-story balcony. They yelled at the woman to wait, abandoned the hose lay, and ran back to the apparatus to get the 24-foot ground extension ladder. They raised the ladder and rescued the woman and her dog. By the time they got back to hooking up the hose line, there was an 8-minute delay in getting water onto the fire. By now, the 2½-inch hose jacket was smoking, and the outer jacket of the supply hose to the manifold split. The rope handles of the manifold had burned off, and the rubber thread protectors on the manifold were melting. The fire was well beyond the control of a single 2½-inch handline, and the radiant heat made their position untenable. The E33 crew retreated back to the apparatus to grab and set up the portable monitor: an evolution that took these two (fatigued) firefighters another 10 minutes before the monitor was operational. Shortly after, the southwest corner of the building collapsed. The fire had gained so much headway that its radiant heat prevented all the fire crews on scene from getting ahead of it.

Sixty-six apartment units were destroyed, Building C was completely destroyed, and Buildings B and D were severely damaged and could not be reoccupied. The fire department was on scene for six days putting out hot spots. Not long after, the decision was made by the owner to demolish the entire complex—half a city block.

Numerous situational and existing conditions, including dry weather and a steady wind, led to the extremely rapid spread of the fire. This fire was instrumental in requiring four-person staffing on all engine companies. In considering the initial actions of E33, having one extra firefighter could have changed the outcome of the entire incident. One firefighter could have handled the ladder rescue of the woman and her dog, while the other two members finished the hose lay to get fast water on the fire.

This case study challenges our traditional perspective on rescue and extinguishment being separate priorities and tasks. Life safety is our first incident priority, and rescue is the first strategic and tactical component of rescue, exposures, confinement, extinguishment, overhaul/ventilation and salvage (RECEO/VS), but the best way to accomplish life safety is to put the fire out. A fire doesn't get worse because water is being applied. Conversely, a structure fire left unchecked, that is, without any water, only gets worse.

Rescue Profile Considerations

The company officer in charge of the rescue must make a quick mental evaluation or rescue size-up to determine the scope of the search and rescue operation. Can rescue be accomplished with a single crew or will additional companies need to assigned to SAR? These considerations include:

- What is the number, location, and physical condition of trapped occupants?
- Are they ambulatory or incapacitated?
- What will the effects of radiant heat and toxic smoke have on the victims?
- Will they have to be taken out through the IDLH?
- Can non-IDLH escape and rescue routes be established or used?
- Will addition ladders, tools, and personnel be needed?

Be mindful that rescue of occupants transitions into an EMS incident.

Anticipate the need for multiple EMS companies and move them into a forward position to receive the rescued occupants.

Primary Search

The primary search is the first search. If there is any indication at all that victims might be trapped or overcome within the fire building, search operations should begin as soon after arrival as possible. This should coincide with the first hose lines being stretched and positioned. Water must be applied into the structure first to cool the temperature and prevent flashover conditions from developing. Otherwise, making entry creates a flow path, accelerating the fire.

The search for victims is normally the function of ladder or rescue companies. The engine company advances its hose lines and places streams into operation, first to support the primary search; however, if a ladder or rescue company is not available or has not yet arrived, engine company personnel must begin the search. (Search patterns and VEIS are discussed in detail in the last section of this chapter.) No matter who performs the primary search, it is extremely important that every firefighter at the scene be aware that a search is in progress. All activity should be directed toward supporting the efforts of firefighters engaged in the search and toward providing protection for them and for any victims they might find.

Secondary Search

A primary search is a quick systematic search of an area to determine whether victims are still in the building. A soon as possible, a secondary search needs to be conducted after the primary search to determine that no one was overlooked during the original search. Both searches should be conducted in a systematic pattern so that no area is overlooked. The secondary search should be assigned to a different company who did not participate in the primary search. Covering the same area with a fresh set of eyes is good practice. The secondary search is not a rapid search. The emphasis should be placed on thoroughness, not speed.

Ventilation

The building should be ventilated as soon as possible after the application of water on the fire to allow smoke, heat, and gases to move away from occupants who might be trapped inside. Ventilation must be performed and coordinated with the fire attack operation,

FIGURE 9-15 When using PPV, there must be an exhaust opening in the fire room to allow the heat and smoke to escape.

When using positive pressure ventilation (PPV) there must be a exhaust opening in the fire room to allow the heat and smoke to escape. If not, ventilation should begin after the initial attack begins.

Now that we understand how dangerous flow paths are created, door control during the primary search needs to be emphasized. Keeping the doors closed during and after the primary search prevents a flow path from developing before the fire is knocked down or extinguished. After water hits the fire, ventilation should begin, and having the doors already closed actually allows the structure to be horizontally vented much faster using PPV. It takes more time to pressurize an entire house or building with a single PPV fan than it does to systematically vent smaller compartmented areas in sequence.

If the fire is smoldering but there's indication of pressurized smoke and high heat, the building must be vertically ventilated before crews enter it to prevent a backdraft. A TIC can help determine whether a possible backdraft situation exists.

Fire Attack for Rescue

The construction of the fire building, its size, its occupancy, and its layout affect rescue operations because these factors affect the number of people inside and the paths that can be used to reach them. This section discusses fire attack in conjunction with search and rescue in several sizes of occupancies and in fire-resistant construction. In any type of structure, the main thrust of operations is determined by the location and severity of the fire, and the location of the occupants most endangered by it. The goal is to get water on the fire through any means possible to knock down the fire and reset it. This accomplishes many objectives in getting the situation under control as quickly as possible.

Single-Family Dwellings

In a typical two-story, single-family dwelling with two or three rooms on fire on the first floor, the occupants in most danger are those close to the fire on the first floor and those directly over the fire on the second floor. The first floor will be affected by radiant heat; the second floor will be affected by the convected smoke, hot air, and gases.

The main body of fire should be attacked immediately with the proper size attack line. At the same time, firefighters should be sent to the area over the fire to begin ventilating and searching for possible victims. No time should be lost in getting firefighters to the upper floor, but apply water to the fire first. If an attack line is immediately available, it should be taken upstairs. If not, the primary search should begin without it.

FIGURE 9-16A illustrates the exterior of a two-story, single-family dwelling with a fire on the first floor. The fire should be attacked and cut off from traveling to

FIGURE 9-16 **A**. The fire should be attacked and cut off from extending to the second floor. A search of the second floor should begin with the rooms directly above the fire. **B**. Firefighters attack the fire on the first floor in a single-family dwelling, preventing the fire from traveling to the second floor.

A

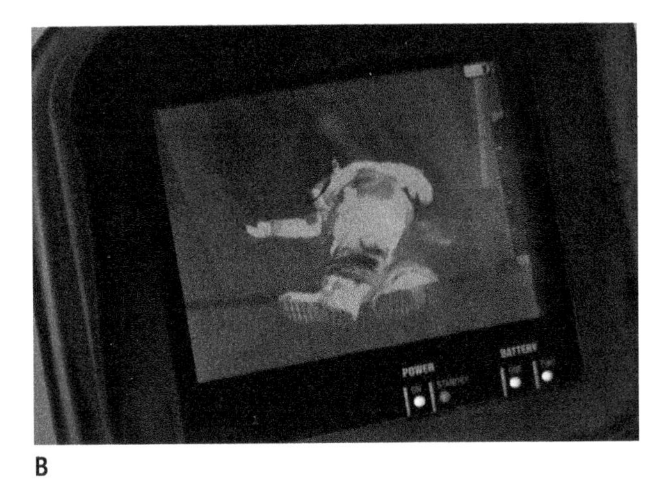

B

FIGURE 9-17 A. If fire is discovered in any upstairs room, the area behind the door should be checked for victims. Use a TIC to scan the room at floor level. B. If there is a victim, it should show clearly on the screen of the TIC. Try for the rescue.

the upper floor, and a search of the upper floor should also be conducted, beginning with the rooms directly above the fire.

Using the example in FIGURE 9-16B, as the attack on the fire progresses on the first floor, the areas around the fire should be searched thoroughly. The second floor fire conditions should still be tenable since the hose line is protecting the stairs from the intensity of the fire on the first floor. If fire is discovered in any upstairs room, the area behind the door should be checked for victims. Use a TIC to scan the room at floor level FIGURE 9-17A. If there is a victim, try for the rescue FIGURE 9-17B. Perform the primary search. If the room is clear, close the door. Keep the rooms isolated until a hose line can be advanced to the second floor. The doors of every room should be kept closed while they are being searched to prevent a flow path from developing. The windows of the uninvolved rooms can be left open, providing the door to the room remains shut. Opened windows dissipate some of the heat and smoke and thus allow more efficient search and fire attack operations. After searching the uninvolved rooms, close the doors. If it is suspected that the open windows may contribute to the spread of the fire, close them. Leave them closed until the fire is extinguished.

Multiple-Family Residences

For any large, occupied residential building, the location of the fire and the smoke above it should be carefully noted during size-up. The smoke indicates the area into which the fire will most likely spread, the path it will take, and the location of occupants who will be in the most danger if the fire does spread. Hose lines should be stretched and placed to hit the main body of the fire, cut off its spread, and cover areas into which the fire will most likely extend.

In a large, occupied residential building, the location of the fire and direction of the most smoke indicate the greatest danger areas for occupants FIGURE 9-18. To aid in the primary search and the evacuation of victims, open stairways must be protected. The fire must be driven away from them or knocked down if one or more stairways are already on fire. Corridors must also be completely controlled, both as paths of safety and to keep the fire, smoke, and gases from penetrating into rooms or apartments.

Most victims of fire are overcome by carbon monoxide gas rather than by burns; therefore, while stairways and corridors are being controlled, every effort should be made to advance hose lines into the upper areas and to ventilate the building. A search of floors above the fire should be started as soon as possible to ensure that all occupants are located and removed from exposure to the products of combustion. Here, as in all building fires, search and rescue must be coordinated with a properly mounted attack on the fire.

Protection of occupants trapped inside the building and control of the fire depend on the number of properly sized hose lines and nozzles positioned in key locations. If the fire has gained considerable headway, streams from a 2½-inch hose line should be used on the main body of the fire FIGURE 9-19(A), whereas 1¾-inch hose lines should be used to cut off the spread by getting around and/or above the fire FIGURE 9-19(B). Small attack lines probably will not control or extinguish a fire that has gained considerable headway. Two-and-a-half-inch attack lines must be placed into operation to control and extinguish a big fire.

Hospitals, Schools, Institutions

Fires in hospitals, schools, nursing homes, and similar institutions are attacked in essentially the same

FIGURE 9-18 The victim on floor 6 is in greater danger than the woman to the far right because he is directly above the fire and needs the reach of the aerial. The victim at the left is also in danger but can be rescued with a 35-foot ground extension ladder.

way as fires in multiple-family residences. The search and rescue problem may be compounded by the larger number of people and their age and physical condition. First-arriving engine companies should resist the temptation to immediately assist and direct people from the building. The officer should not hesitate to call for help in such cases, but remain focused on performing necessary fire attack duties. The second engine company, even though it may eventually be free to put its hose lines in place, should be assigned to SAR efforts. The fire must be controlled. Once the fire is knocked down, the situation will get better.

In all cases, it is the job of the engine company to contain the fire, open passageways, and extinguish the fire. If fire protection systems, such as sprinklers and standpipes are installed, they should be used to the advantage of the fire department.

Schools should be completely evacuated. Whether or not it is necessary to evacuate a hospital, nursing home, or other such building depends on the type of construction, the size of the building, the location of fire doors, the capability to isolate and compartmentalize floor wings, and the location and severity of the smoke and fire. Protect in place is the preferred method. However, patients can be moved to locations within the building that are isolated from the fire area if it is evident that it will take some time before the fire companies will be able to control the fire **FIGURE 9-20**. Often, lowering occupants to the floor may be all that is necessary. This might be preferable to evacuation for patients in poor physical condition, the continuing care that some must receive, and possibly adverse weather conditions; however, if there is any doubt regarding control of the fire, complete evacuation plans must be put into effect. The evacuation of such facilities should be detailed in prefire plans.

Fire-Resistant Construction

Fire-resistant structures, built to resist the spread of fire, also tend to hold in heat rather than let it escape. Fire in such buildings forces firefighters to combat excessive concentrations of heat as they attack the fire or search for victims. These buildings also hold the large volumes of smoke given off by their burning contents. The smoke can quickly overcome

A

B

FIGURE 9-19 Proper hose sizes with combination nozzles are necessary to control a fire situation and establish protection for trapped occupants. A. Two teams of two firefighters attacking the fire with two 2½-inch hose lines. B. Two firefighters attacking the fire with a 1¾-inch hose line.

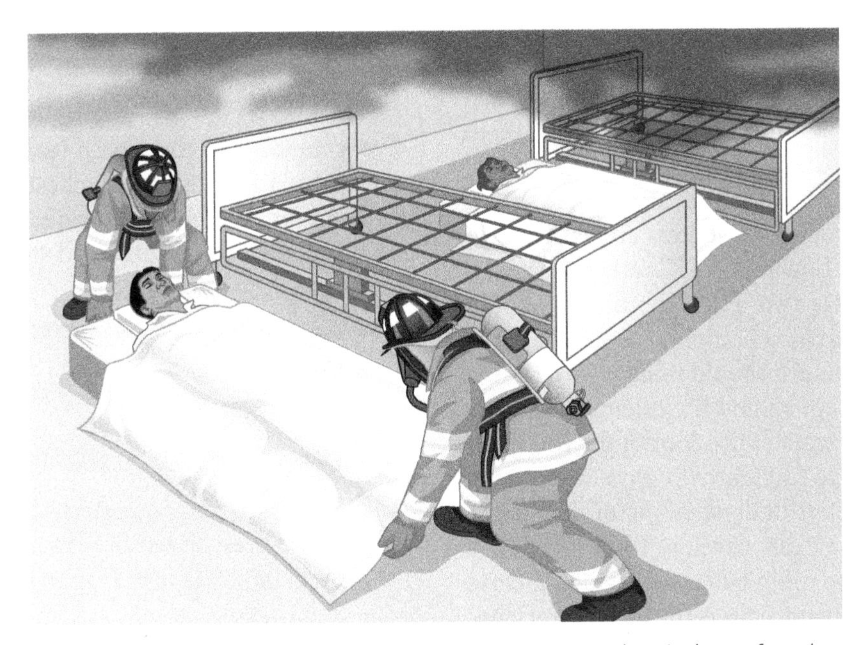

FIGURE 9-20 In hospitals and nursing homes, protect in place is the preferred method.

occupants, who may collapse in just about any location inside.

Fire-resistant buildings generally have sprinkler systems. Their stairways are enclosed, smoke doors and fire doors are used to divide long corridors into smaller sections, and building assemblies are generally less apt to burn; however, the intense heat associated with fires in these structures may require large volumes of water to ensure control. Fire companies should be prepared to use 2½-inch hose lines at fires in a fire-resistant building. The heat buildup near the fire can become so intense that 1¾-inch hose lines may not deliver water fast enough to absorb the heat and control the fire. In fires in these structures, rescues could be made within the apartment, office, or room in which the fire originated; however, the fire might have spread into the corridors, or smoke might have begun to spread through some of the floors. Frightened occupants, who would have been better off staying in their apartments or offices, might attempt to reach an exit and collapse en route. Therefore, engine company members advancing on the fire must search every area through which they pass. These areas include lobbies, elevator alcoves, corridors, and the like. Ladder and rescue companies should be assigned to thorough search operations.

Search

Oriented Search

A thorough, systematic primary search should be conducted at every fire where it is safe for firefighters to enter the building. All firefighters, no matter what type of company to which they are assigned, should be able to conduct a primary search if needed. All firefighters should realize that the safety of each member performing a primary search is their responsibility, and they must be constantly aware of the presence and position of each member. This is known as an oriented search. An oriented search means just that: using methods and tools to keep the team oriented to their location within the building and to each other.

There are many ways to perform an oriented search depending on how many firefighters are assigned to the search team and how many TICs are at their disposal. If the team has two TICs, the officer should have one, and the searching firefighter should have the other. If there's only one, it should remain with the officer. The basic concept behind the oriented search is to have the officer remain at the door of the room being searched while the firefighter(s) physically search the room. With a better understanding of flow paths and how quickly they can develop, the officer may now have to enter the room being searched and close the door behind the team. This is the same theory behind VEIS, only in reverse; whether the officer stays in the hallway or enters the room with the search team, the door should remain closed during the search.

Using a TIC, the officer can identify the type of room if it isn't otherwise obvious. With a TIC, the officer can monitor the search and call out areas that the firefighter might have missed, for example, "There's an alcove area behind the sofa, check behind the sofa." While remaining at the door, the company officer should use the TIC to take temperature readings and check any fire conditions in the immediate area of the search team, all the while remaining oriented to the way out should interior conditions quickly turn for the worse.

If the officer has two firefighters and the search is taking place in a hallway of a hotel, with units on both sides, the officer remains in the hallway with the TIC while one firefighter searches the room to the right, and another firefighter searches the room to the left. The officer can scan the rooms with the TIC while each firefighter performs the manual search, then returns to the hallway and closes the door. Again, there needs to be an emphasis on door control of the rooms being searched to avoid creating an undesirable flow path. The officer can scan the hallways with the TIC and watch for any increases in temperature.

If time allows, a master key, card, or fob should be obtained from the manager of the building to access locked units. All motel and hotel rooms are locked unless propped open. Getting a master key saves time and a lot of unnecessary property damage. Using the key is always faster and safer than performing any forcible entry techniques in low visibility or smoke to open a locked door. Try before you pry can be expanded to mean try using readily available master keys before you pry. Unnecessary property damage is unprofessional.

Required Tools

Every member of the search team should have a helmet light; a flashlight or a chest-mounted flashlight; and one of the following tools, depending on what each can reasonably carry and still perform an effective search:

- Ax
- Halligan tool
- Set of irons
- Pike pole
- TIC
- Water-pressurized fire extinguisher
- RAK
- Search rope and webbing
- Battle lanterns
- Door chocks
- Search markers (industrial chalk or crayon, tape, marker flags, glow sticks, spring-loaded Jeromeo clamp)

The RAK consists of an SCBA 60-minute air bottle, an attached air regulator, and a facepiece and is carried inside a specially designed carrying case. It was originally designed for rapid intervention rescue of a downed firefighter. What a downed firefighter needs is air. The RAK is an independent, reliable source of fresh air that is placed on the down firefighter. It bypasses his or her existing SCBA unit, which may have malfunctioned or its bottle depleted of air. If there is only one RAK available, it should be reserved for the RIC. If more than one RAK is available at the fire scene, it should be taken in with the search team.

Like a downed firefighter, the first thing a victim needs is air. In the past, the RAK has been highly underutilized for civilian searches because it was considered a RIC tool. Every fire department should invest in more RAKs for the purpose of civilian search. The RAK can be carried by the officer as the firefighters perform the manual search and advance through the hallway.

The officer is the anchor point of the oriented search, and he or she sometimes stays in one place while the firefighters perform the search manually. Thus, the officer can carry the RAK while the team advances. If possible, the officer should also move the fire extinguisher and keep it in the immediate area. In addition to the TIC, the officer should carry a 6-foot (1.8 m) pike pole. The officer needs the reach to control the doors and check the overhead ceiling for fire. It is the officer's responsibility to make sure the fire doesn't get above the rescue team or come at them from behind. The officer should also have a powerful battle lantern to illuminate the exit doors for the firefighters. Modern battle lanterns are extremely bright and have different colored lenses and strobe lights to assist with illuminated orientation.

The last thing the officer is responsible for is the air management of the team. The critical air reading is the member with the least amount of air. Ideally, depending on the complexity of the structure and the depth of the search, a good practice is to use one-third of the SCBA air supply for search, one-third of the air supply to exit the building, and one-third for emergency reserve.

Controlling the Door

The UL/NIST studies have provided us with a better understanding of flow path. Door control is essential for preventing the development of undesirable flow paths that could entrap firefighters and occupants caught within them. Traditionally, oriented search is taught by having the officer remain at the door while the firefighter searches the room. If there is an open window in the room, it could possibly create the same situation we are trying to prevent in the VEIS method, only in reverse because we're entering the room through the door. A high-low flow path could be created, thus bringing the fire right up the stairs into the room that is being searched.

Thus, traditional oriented search needs to be modified in one of two ways. First, if the officer chooses to remain outside the room being searched or chooses to remain in the hallway, the firefighter needs to enter the room and close the door behind her or him before conducting a primary room search. In this scenario, the officer cannot supervise or assist the searching firefighter, and it would be best if the firefighter conducting the search had the TIC. The second option is having the firefighter and officer both enter the room and close the door behind them. Then the officer can supervise and assist the firefighter conducting the search in that room. This option is a challenge in, for example, a hotel where one firefighter searches the room on the right of the hallway and the other firefighter searches the room on the left. To maintain door control, the officer must remain in the hallway while the searches are being conducted behind closed doors, or all team members must stay together unless an additional firefighter or company is assigned to the team. Always assume that there may be occupants in a building until a search is completed, even in vacant buildings. Vacant building searches are dependent on fire conditions and structural stability.

Because performing a primary search is dangerous, and the safety of the crew is shared by everyone involved, it should be performed according to SOGs. This guideline should provide for an oriented and systematic search coordinated by the IC, but executed by a company officer or an experienced firefighter. The guideline should require that the search begin where there is the most danger to occupants. This is determined by a quick 360-degree size-up using the TIC. It should be simple and straightforward so that one firefighter can substitute for another at any point in the search.

Typical Search

Consider a fire in a large kitchen and dining area on the first floor of a two-story, single-family dwelling. Engine companies arriving at the scene should immediately perform a 360-degree size-up of the fire situation and obtain a continuous water supply. New UL/NIST procedures have attack lines initially spray water into the house from the outside to knock down and reset the fire, then advance into the house, position to cut off the fire and prevent it from spreading, and attack the main body of fire directly. Hose lines should be positioned between the occupants and the fire. Occupants closest to the fire are in the most danger, whether they are on the fire floor or the floor above.

Search begins immediately after water application. Firefighters on the hose lines, by getting low to the floor, can probably see some clear area above the floor as they search for victims near the fire. Staying low to the floor provides the coolest temperature and the best visibility. One member on the attack team should have a TIC. In fact, each company should have at least one TIC. The stairway and the upper floor will likely be full of smoke and hot gases. An immediate attempt should be made to advance firefighters to the upper floor. If the area is tenable, they can begin searching for victims. If the area is untenable because of the intense heat, ventilation should be conducted from the outside **FIGURE 9-21**, and a hose line should be directed to the upper floor. The attack should be coordinated, with firefighters inside the building attacking the fire and performing the primary search, and firefighters outside the building performing ventilation.

Performing proper ventilation depends on the number of firefighters available for this task. Vertical ventilation is more labor intensive and time consuming than horizontal ventilation. Vertical ventilation should be considered only when the area or the fire room is directly below the roof, a cockloft, or attic; the building is over-pressurized with smoke from a lower unit; or the fire unit entrance is from a common area with an open staircase all the way to the top floor. Otherwise, you have to go with horizontal ventilation. As soon as the upper floor is tenable, the search can begin.

Many fire departments use PPV to force fresh air into a building. This powerful fan can displace a contaminated atmosphere by pushing heat and the products of combustion out of the building. This allows the search team and attack team to enter into the structure in front of this fresh air to conduct a primary search (fresh air at their backs). PPV has advantages as well as disadvantages. Firefighters must be trained to recognize when this tactic can be used safely and effectively. There should always be a charged exposure line in the area being vented. If there is any hidden fire, PPV causes it to flare up. This isn't necessarily bad when the hose line is in place. It actually helps to locate hidden fires so they can be extinguished. If PPV is creating undesirable conditions, simply have a firefighter turn the fan away from the door 90° or shut off the fan. The forced-in air will immediately stop.

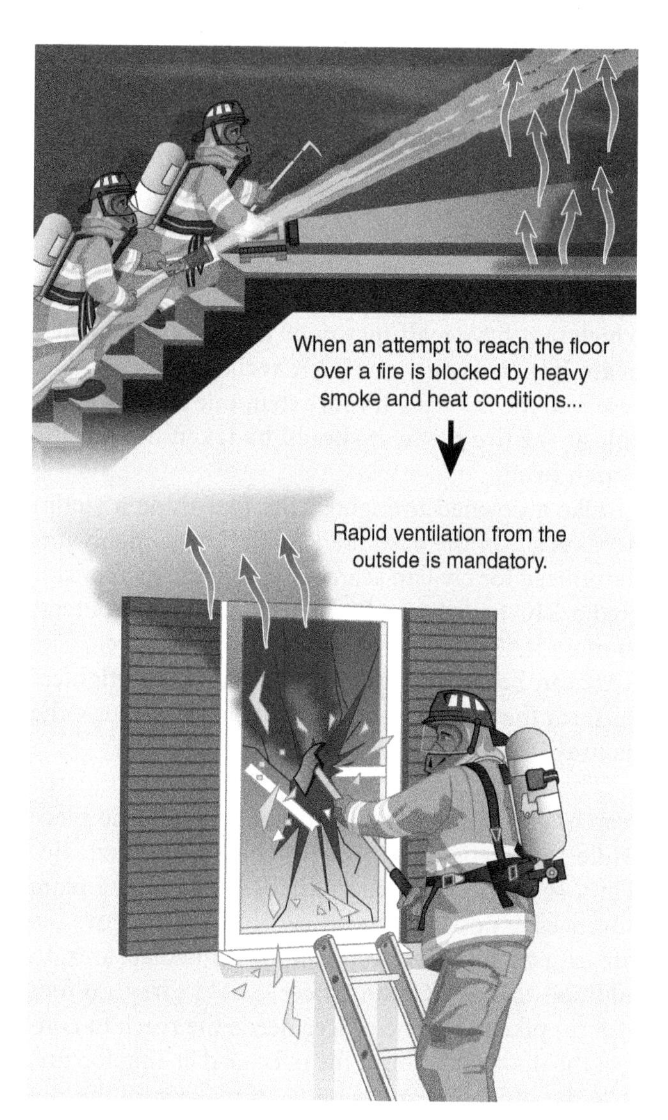

FIGURE 9-21 Rapid ventilation allows the search for victims on the second floor. A hose line needs to be in place to protect the stairway and prevent an undesirable flow path from developing.

Value of the Standard Search Procedure

The important point is that everyone at the fireground knows that an immediate attempt will be made to reach the second floor or the floor above the fire. Firefighters on the hose lines and those performing ventilation tasks must be aware that a primary search is being conducted on the upper floors.

Suppose the firefighters on the hose lines must retreat from the building because of deteriorating conditions. Following SOGs, they know they must notify firefighters conducting the primary search on the second floor or the floor above the fire, and there should be no hesitation about doing the notification. When those assigned to laddering and ventilation duties

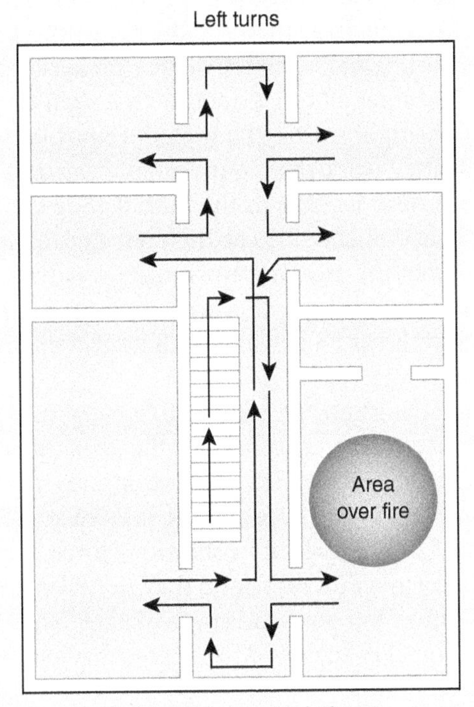

FIGURE 9-22 Search of the floor above the fire follows a pattern of either all left-hand turns or all right-hand turns, based on the first turn into the room over the fire.

know that a search is being conducted, they should place ladders to the second floor and the floors above the fire as a second means of egress for search personnel and possible victims. Hose lines should be in place to protect the stairs and assist search personnel who may be exiting the building. The RIC should be standing by to assist in the rescue of fellow firefighters if the need arises.

Standard Search Patterns

As soon as the fire floor and the floor above the fire are tenable, the primary search begins. Following department SOGs, the first area to be searched after the fire room is the room and area directly above the fire. After reaching the top of the stairway, firefighters turn in one direction or the other to get to the room over the fire. They begin the search by turning either right or left into that room. This turn sets up the basic search pattern for the entire floor. A search behind the door is the first action when entering a room. As the search proceeds, firefighters performing the primary search keep turning in the original direction as they go in and out of rooms. If the first turn into the room over the fire is a right-hand turn, then all turns into subsequent rooms on the floor must be right-hand turns **FIGURE 9-22**. If the first turn into the room over the fire is a left-hand turn, then all turns into subsequent rooms must be left-hand turns.

Use the TIC to expedite the primary search. When the screen clearly shows no victims in the room, there is no need to feel around the perimeter walls. If it shows furniture and other items in the room, they must be physically searched by the firefighter.

Let's assume that firefighters engaged in the search turn right at the top of the stairs and turn right again to move down the hallway to get to the room over the fire. Kicking off the search pattern for the floor, the firefighters turn right to get into the room over the fire, as shown in Figure 9-22. Remember to control the doors to prevent a dangerous flow path from developing. They start by searching behind the door. When they have finished searching that room, they turn right out of the room and begin the search of the hallway until they reach the next room to be searched. Then, according to the search pattern, the firefighters turn right again to enter a room and, after coming out of that room, turn right again and move along the hallway to the next room. The right-turning path is continued until the firefighters have worked all the way around the hall and are back at the stairs. Use the TIC. It is the best search tool we carry.

If the firefighters need to turn back, reversing direction and following a left-left pattern leads them back to the starting point. Firefighters should try to keep in mind where they are in the building and in which direction they are heading. Landmarks inside the search area should be noted, including windows or other areas that may be used for escape. These landmarks can be used to guide a firefighter out of the building. Search rope has been used by firefighters for many years. A search rope or taglines of a predetermined length allow firefighters to tie off and then proceed a prescribed distance. By following the rope back, they have ensured their return to the starting point. Other tools like battle lanterns with strobe light features, and chemical glow sticks can be used to mark and illuminate windows, hallway passages, stairways, and exit doors.

Other Structures

The standard search procedures and the search pattern described above also apply to other buildings, including apartment houses and office buildings. When the firefighters turn off a corridor into an apartment or office, the team engaged in the search should follow the same pattern within the room or area. When they leave the area, they should retain the pattern in the corridor and when they enter the next room or area. If they enter all office areas with a right turn, they place their right side to the wall and must keep to the right as they work their way through the area and back to the door leading to the hall. If they enter with a left turn, they should keep their left side to the wall as they search the area and work their way back to the door. When they leave this area, they must retain this pattern in the hall and when entering the next office, as shown in **FIGURE 9-23**.

It is important that search personnel leave an office or work area through the same doorway used to enter it. Otherwise, part of an area that could possibly contain trapped victims could be overlooked **FIGURE 9-24**. If fire conditions force search teams to leave by a different door, they should report this information to command immediately. Hose lines should be quickly advanced to the area that has been searched thoroughly.

Large structures or buildings with complex floor plans require the use of additional resources. Large office areas with cubicles are relatively easy to search when conditions are clear, but searching them properly is labor intensive under heavy smoke conditions. Obstructions of any kind in unfamiliar surroundings

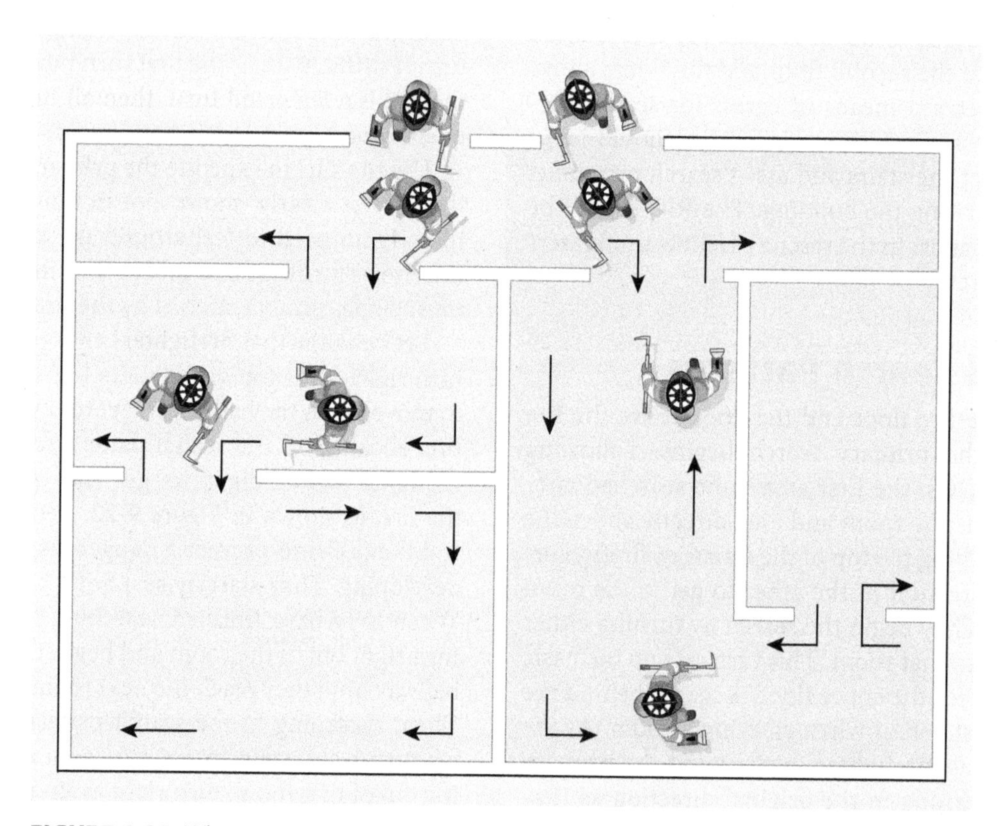

FIGURE 9-23 When turning into an office, apartment, or similar area, firefighters should keep one shoulder to the wall: left if a left turn is made upon entering, right if a right turn is made upon entering.

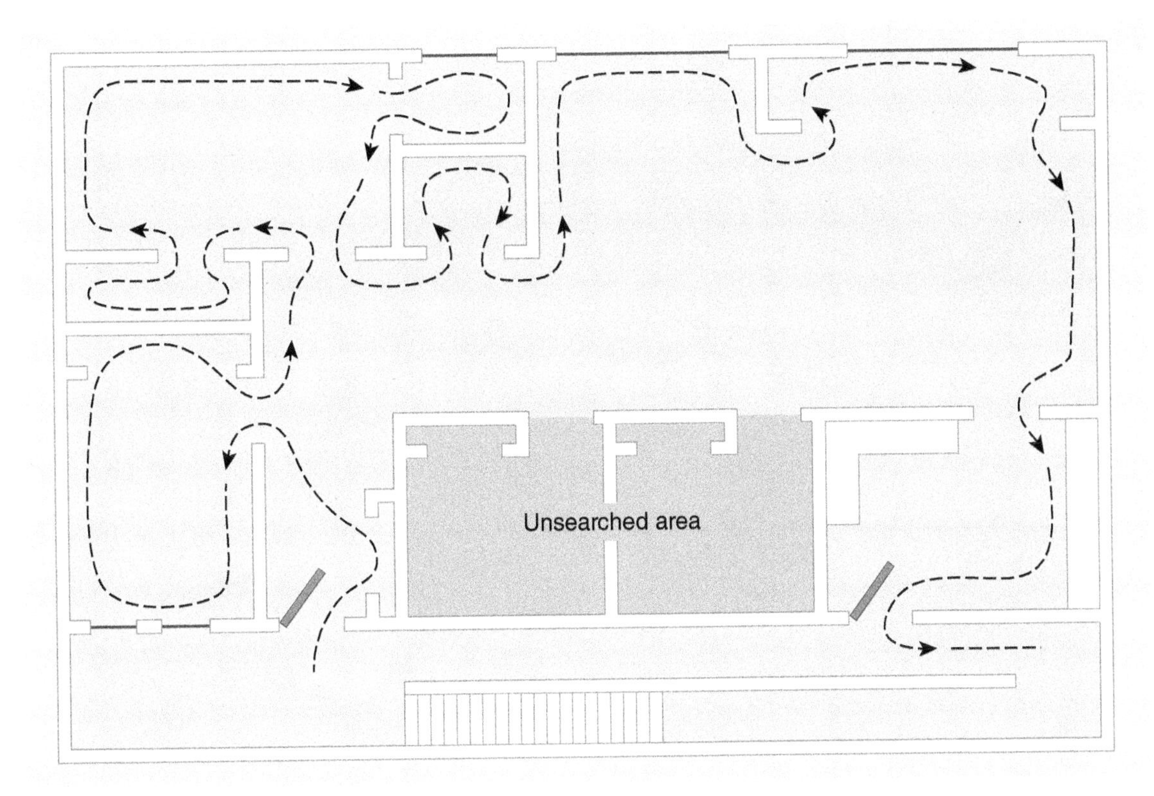

FIGURE 9-24 During an office search operation, the same door should be used to enter and exit an area in order to prevent missing some of the rooms.

under fire conditions create an additional burden for firefighters conducting the primary search. Command must ensure that adequate personnel are on hand to perform all fireground operations. Resources must be available to perform the primary search with deliberate purpose and in a safe manner. If members are running out of air, they must be relieved and additional firefighters must be assigned to continue the search.

Search and rescue are ladder company duties on the fireground. Engine companies should be free to direct all their efforts to advancing hose lines in support of rescue operations; however, when ladder companies are overburdened, do not have adequate personnel, or do not respond in sufficient numbers, engine company personnel must then be assigned to perform search and rescue operations. With SOGs and proper training in place, assignments can be adjusted to cover any situation.

What to Check

The normal path of egress is where most people try to escape from fire. Before entering any room for a search, check behind the door first. Many unconscious victims are right behind the door. In older homes, opening the front door can sometimes block and hide the access to the second-floor stairwell. It may go the entire fire before it is discovered, so the area behind the door must be checked first.

The corridor or hallway, as well as the open areas of each room, should be checked thoroughly. In addition, closets; spaces behind large chairs, furniture, and large appliances; areas beside and under beds; bathrooms, including tubs and showers; and any other area in which a person would seek refuge should be checked **FIGURE 9-25**. People often seek protection in such places.

Every firefighter engaged in the search should carry a fire ax, Halligan tool, or claw tool with which to sound the floor, probe areas, and reach out into open areas of the room. These tools can also be used for forcible entry and ventilation tasks. The officer or one member of the search team should have a TIC **FIGURE 9-26**.

Depending on the overall stability of the structure and smoke conditions, the search team may have to sound the floor as they go. Sounding with a tool determines the thickness of the floor decking. Simply sounding the floor for stability is not always reliable and therefore should be combined with slow advancements across the floor to determine the condition of the floor joists. A sagging floor is a good indicator of potential floor collapse; however, it may be very difficult to determine based solely on

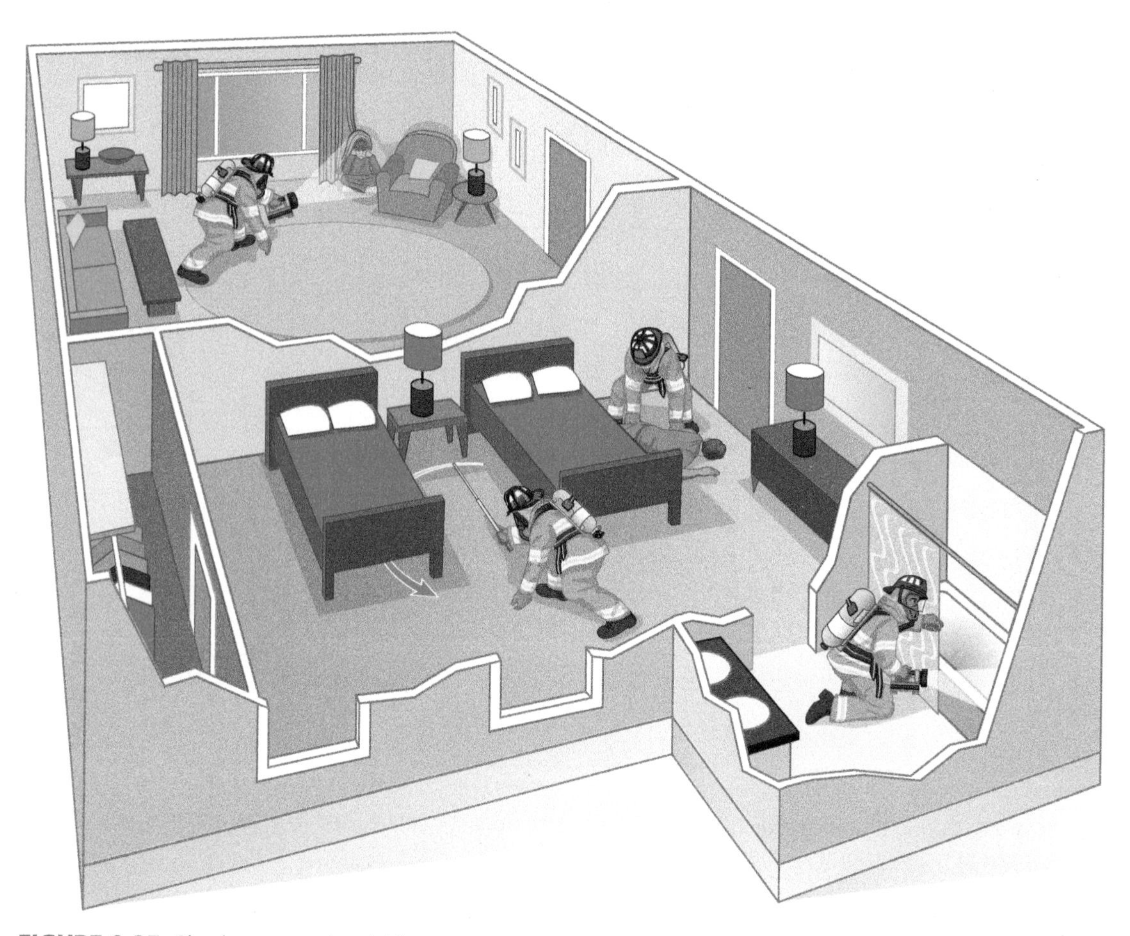

FIGURE 9-25 Check any area in which a person would seek refuge.

FIGURE 9-26 Every firefighter engaged in the search should carry a fire ax, Halligan tool, or claw tool to probe areas and reach out into open areas of the room. The officer or one member of the search team should have a TIC.

the amount of deflection while moving through the structure, especially on a floor with thick carpeting. It is best to stop and utilize flashlights and a TIC from a safe position.

Rooms that the fire has entered should also be searched, then the door should be shut to isolate the fire. Standard guidelines should include an engine company advancing a hose line over the fire for this exact situation. The windows of rooms that are not involved with fire and hallway windows should be opened or removed in order to vent as much of the upper story as possible.

Indicating That a Room Has Been Searched

Different methods are used to indicate that a room has been searched so that there is no duplication of effort, at least in the initial search operation. *The practice of placing a piece of light furniture in the doorway, like a chair or a lamp table, should be discontinued.* With modern-fuel fires, door control must be emphasized. During and after the primary search,

FIGURE 9-27 Marking the door with an X indicates the room has been searched.

doors must remain closed to limit oxygen to the seat of the fire or to prevent a dangerous flow path from developing. If a window is open, propping the door open with a chair can create a flow path, drawing the fire into the room and spreading it. It can also allow smoke to contaminate a room that may have been uninvolved.

Any firefighter can be tasked with performing a search, so all should carry a roll of first-aid tape, a piece of industrial chalk, or a heavy-duty marker in their coat pocket so they have the ability to mark a door after it's been searched. A firefighter should not have to enter a room to find out that it has already been searched—a waste of valuable time. Many departments have had success with placing tags or innertube rubber straps on the doors or doorknobs of inspected rooms, apartments, and offices. These devices can be purchased from a vendor or fabricated by department members **FIGURE 9-27**. The *Jeromeo clamp* is a simple handheld, heavy-duty, plastic, spring-loaded clamp or quick-grip that can be purchased in any hardware store **FIGURE 9-28(A)**. It is lightweight and inexpensive. It can be clipped onto the door handle to indicate that the room has been searched, and it can also be used as a door wedge to keep it open or closed for door control **FIGURE 9-28(B)** and **(C)**. Whichever method is selected, it should be recognized throughout the department that a clamp or tag means a room has been searched.

Vent-Enter-Isolate-Search (VEIS)

Search and Rescue Safety Considerations

The tactic of VEIS is discussed briefly in Chapter 4. The rescue technique of vent-enter-search (VES), requiring a firefighter to ventilate a bedroom window from the outside, enter a bedroom through the open window, and search for victims, is a technique that creates a dangerous situation for the firefighter and any victims in that room if the bedroom door is open to the fire. By opening or breaking the bedroom window, the firefighter can create a *low-pressure exhaust vent* for the fire to move toward. In other words, the firefighter just created a chimney and a flow path that is going to bring the fire toward the firefighter and the victim. Recent experiments have demonstrated that this rescue technique is much safer if the interior door is closed prior to conducting the room search. This isolates the room with a barrier to the fire flow path and allows the room to vent naturally, improving visibility and lowering the room temperature. The modified VES process becomes a safer tactic called **vent-enter-isolate-search (VEIS)**.

First, try to determine if there is a known victim in the burning structure. This can be confirmed visually by the firefighter, a reliable person at the scene like another occupant who escaped, or the dispatcher.

A

B

C

FIGURE 9-28 **A**. A Jeromeo clamp is a handheld, heavy-duty, plastic, spring-loaded clamp that clips onto the door handle to indicate the room has been searched. **B**. The Jeromeo clamp can be used as a wedge to keep a door propped open when the clamp is clipped to the hinge stile side of the door. **C**. The clamp can be clipped to the lock stile side of the door to maintain access while providing door control.

Next, try to confirm the last known location of the victim from the reliable source. (Don't forget the use of cell phones; almost everyone has one and not all victims are unconscious.) If exact locations of victims cannot be confirmed, the firefighter must identify the location of possible fire victims in a burning building. For example, for a fire that occurs at night, a bedroom is likely to contain a fire victim compared to a kitchen.

Reading the Windows

Reading windows is discussed in Chapter 4, but because it is critical to this chapter, it is addressed here, as well. In VEIS, read the clues from the fire on the glass before selecting a window. Moisture condensation and water droplets can form on the glass with a change in interior temperatures, which indicates an ignition stage incipient fire. Light gray smoke or a haze can be seen through the window. The flame temperature can be between 700°F and 1,000°F (371°C and 538°C), but the room temperature may be just above ambient. The oxygen level is still between 20% and 21%, and CO production is still low. Therefore, the survivability profile is good.

Cracked glass or crazing, along with blackened windows indicate higher temperatures. The water vapor has dried, leaving stains, and the window appears to be covered in a gray, brown, or black film. Cracked glass indicates that you're getting closer to the fire or that this is the fire room. A window that is blackened from smoke with a layer of creosote, and cracked glass from the heat of the fire, especially if smoke is seeping through the cracks in the glass, is a strong indication that the door to the bedroom is open or that the fire may be in that room. Cracked glass indicates high heat, above the human threshold. Not only is this a warning sign for a backdraft, but the survivability profile in this room based on the fire gases alone is zero. You need to go to the next window. Remember, you're trying to make a successful rescue, not a body recovery. Don't break out blackened windows. Use the scientific evidence to make a smart judgment call; you want your efforts to be successful, and your family wants you to survive this event.

Check window construction. Are these modern windows with double or triple panes? In a decay stage fire, windows are layered in creosote buildup. Dark brown stained windows are strong indicators that the fire is in a ventilation-limited or the decay stage. A creosote-stained window is also a strong indication that this may be the fire room or that the door to the room is open to the fire and should not be selected for VEIS. The survivability profile is nonexistent, and opening that window will likely

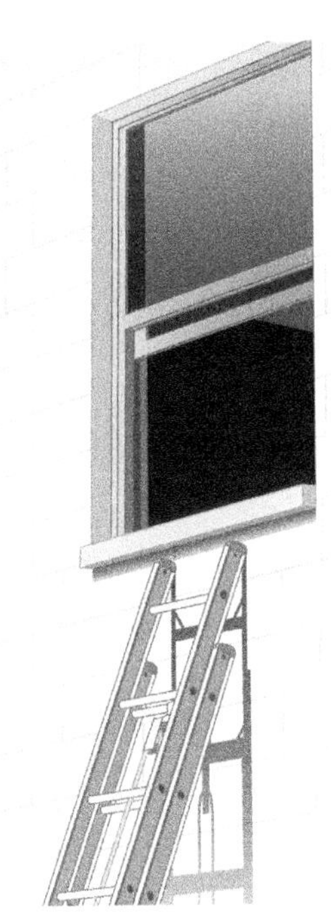

FIGURE 9-29 The tip of the ladder should be placed just below the windowsill. Open or take out the window. The opening should be as wide as possible to accommodate a firefighter with all his or her gear and an unconscious victim.

produce a rapid change in fire conditions and possibly a smoke explosion. On the other hand, a window that is still clear indicates that the door to the room is closed from the fire and that the survivability profile is favorable.

Select a window that is clear or has some visibility, even if there is gray haze (avoid the black). Place the ladder with the tip right below the windowsill FIGURE 9-29. If you have to make a choice between a clear window and a hazed window, it's a judgment call based on a risk-benefit analysis and a favorable rescue profile. Obviously the IDLH conditions in the room with the hazed window are more severe and getting worse, but don't bypass or not select the clear window simply because it isn't risky enough for a firefighter to VEIS. Remember, fire gases like carbon monoxide, hydrogen chloride, and hydrogen cyanide are colorless.

Take advantage of any light to gauge the level of visibility before opening or breaking the window. Notice the door: Is it closed or will it need to be closed? *Remember: Glass blocks the infrared spectrum and thermal images, so a TIC will show a blank screen until the window is opened.*

Next, the room is quickly accessed by opening or breaking out the window. The entire window should be broken out to accommodate a firefighter with all his or her gear. (You'll also need the widest space possible to pass an unconscious or semiconscious person out the window.) Then the firefighter quickly enters the room, locates the door, and immediately closes it FIGURE 9-30. Once the door is closed, the room can naturally ventilate through the window. Improved visibility will help with the room search.

FIGURE 9-30 The firefighter quickly enters the room, locates the door, and immediately closes it. This isolates the room from the main body of the fire and decreases the chance of introducing a new flow path for the fire. Once the door is closed, the room can naturally ventilate from the window.

Because no firefighter should attempt entering the structure alone, the second firefighter can stay at the window and use the TIC from that position to guide the first firefighter making the interior systematic search. Another option is that the interior firefighter can use the TIC to search the room for victims.

The purpose of VEIS is to remove a victim through the window instead of taking them down through the IDLH FIGURE 9-31. This is a challenge, especially if you have an unconscious adult who needs to go down a ladder FIGURE 9-32. Immediately call for help, a second or even a third ground ladder, and a rescue rope. The second ground ladder can be raised parallel with the first ladder, thereby increasing the work area between the beams FIGURE 9-33. Once the unconscious victim is lifted to the window, the rescue

FIGURE 9-31 No firefighter should attempt VEIS in a burning structure alone. The partner stays at the tip of the ladder to guide the search and keep the exit point of reference, while the rescue firefighter performs VEIS. The firefighter at the tip is in a ready position to receive and help the victim down the ladder.

FIGURE 9-32 Taking an unconscious adult down a ladder is not an easy job. Use a rope and call for help.

FIGURE 9-33 A second ground ladder can be raised parallel with the first ladder, thereby increasing the work area between the beams.

rope should run from the inside of the ladder over the first rung underneath the windowsill. A snug round turn, two half hitches should be tied around the chest and underneath the arms of the victim for a belay. The Sling-Link, (described below) is a fast evolution that can also be used as a patient harness attached to the rescue rope. Another option is to use the second or third ladder as a high-directional anchor point by extending the tip well above the top header of the window and running the rope down to the unconscious victim.

The rationale of VEIS should be taught to all firefighters, and it should always include belaying a conscious or unconscious victim with a rescue rope. Plenty of videos show firefighters who were unable to manage an unconscious victim, or they lost their grip and the victim fell from the ladder to the ground, sometimes taking the firefighter down with them. What's the point of risking a VEIS rescue attempt in IDLH conditions, only to lose the victim when he or she is on the ladder?

Commercial Structures and Large-Area Search

More firefighters are needed to search larger and more complicated structures. In commercial, manufacturing, and storage facilities, there may be large open areas that need to be searched. Large-area search is similar to an oriented search, only on a grander scale. It's usually an assignment for ladder companies. It involves using rescue rope up to 300 feet in length that is tied to an exterior anchor point and used as the main line. As the search team enters the structure, the firefighters clip onto the main line with a tag line or personal webbing. The firefighters are now tethered to the main line and can go off in perpendicular directions to perform the search.

There are various methods for performing a large-area search, but the TIC dramatically speeds up the process. The basic concept is to extend the reach of firefighters to cover large areas without becoming lost or disoriented themselves. While firefighters are familiar with residential living areas, they can easily become lost or disoriented in a factory or commercial structure.

Unfortunately, if there is not an adequate firefighting force on hand to commit to the large area search, it may become a difficult task to complete in a reasonable amount of time. Additional firefighters will be needed, or the IC will have to make adjustments to the incident action plan and strategic goals.

Public Assemblies

Search and rescue assignments in public assembly occupancies are more often coordinated evacuations. The same search patterns and methods apply except in extreme cases. The worst-case scenario happened on February 20, 2003, at The Station nightclub in West Warwick, Rhode Island. This fire killed 100 people and injured 230 when a pyrotechnic show set fire to the stage. Flashover occurred within a minute, and the entire club was choked with smoke in 5½ minutes. Due to the lack of exits, the main entrance became a bottleneck with patrons trying to flee. As they succumbed to the effects of the thick toxic smoke, they collapsed, blocking the exit with unconscious bodies.

Firefighters must recognize when a similar scenario presents itself and make as many doors and openings as possible to give perhaps hundreds of patrons a chance to escape **FIGURE 9-34**. Two-and-a-half-inch hand lines or master stream appliances would be needed immediately due to the wide open portals of fresh air that will greatly accelerate this fire.

Rescue Drags and Carries

If the primary search reveals one or more fire victims, you need to remove them from the burning structure. There are many types of rescue drags and carries, and these should be practiced during company drills. You will have to figure out, through trial and error, which one works best for you. There are some points to consider in making an actual rescue.

The traditional firefighter carry of throwing a victim over the shoulder and carrying him or her out of a burning house looks dramatic in the movies and, in some cases, it may be the quickest method of rescue, but you need to consider the typical circumstances that you will encounter when performing this maneuver. Most searches are conducted by firefighters crouching or crawling on hands and knees. The visibility is better when they are low, and the temperature is lowest at the floor. The smoke and fire gases rise, so they are above the neutral plane and toward the ceiling. The most breathable air is at floor level. The *firefighter carry* places the victim's head and her or his airway at a higher level in the IDLH atmosphere and defeats the purpose of trying to save a life.

In extreme fire behavior, a rescue firefighter with a victim may not be able to make it back to the ladder or point of egress. Breaching a wall into the next unit may be your only option to find shelter in a tenable space to shield yourself and the victim from the fire **FIGURE 9-35**. This is another Mayday situation—call for help.

FIGURE 9-34 In public assemblies, firefighters must recognize extreme scenarios, in which making as many doors and openings as possible will give perhaps hundreds of patrons a chance to escape.

FIGURE 9-35 In extreme fire behavior, breaching a wall into the next unit may be your only option to find shelter in a tenable space to shield yourself and the victim from the fire.

During sleeping hours, many people wear skimpy pajamas or may be naked. In other words, there is nothing substantial to grab. In these cases, it is safer to use a blanket or a sheet to drag a victim. It keeps the victim on the ground, where the best air and the coolest temperatures are located. It provides you the best control and can accommodate a RAK if one is applied. Consider using a Sling-Link, which is a five-ring, interconnected, one-inch webbing rescue system **FIGURE 9-36(A)**. This allows you to quickly create a harness by slipping the loops around the arms and legs of a victim without tying any knots. The victim can be dragged or hoisted, regardless of what she or he is wearing. The Sling-Link can also serve as a harness to drag or hoist a downed firefighter. Again, a drag keeps the victim close to the floor: in the lowest heat and the cleanest air **FIGURE 9-36(B)**.

Another tool of choice is the 6-foot, D-handle pike pole. Hooking the SCBA backpack assembly of a downed firefighter with the pike pole and dragging him or her using the D-handle—similar to pulling a

A

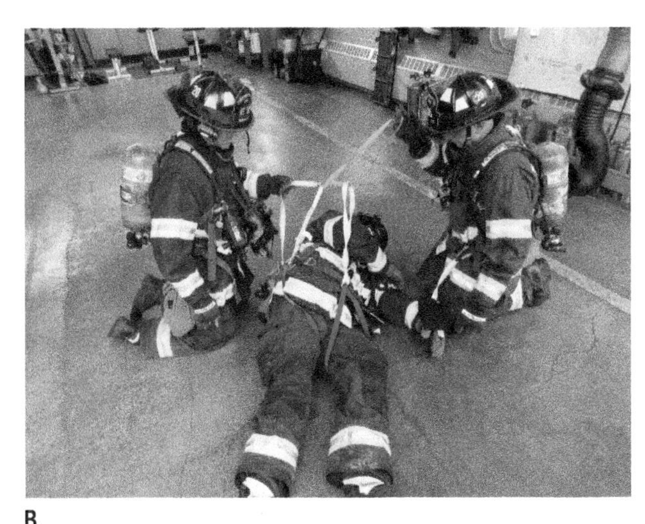

B

FIGURE 9-36 **A**. The Sling-Link is five-interconnected rings of 1-inch webbing. **B**. The Sling-Link creates a harness because you can slip the loops around the arms and legs without tying any knots or lifting the victim.

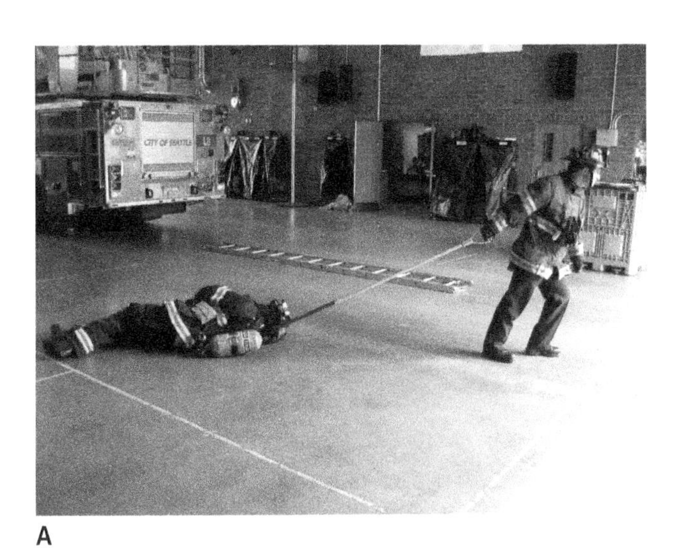

A

B

FIGURE 9-37 **A**. One firefighter can hook the backpack assembly of the SCBA with a D-handle pike pole and drag a firefighter to safety with little effort. **B**. One firefighter wraps webbing underneath the knees, while the second firefighter drags the downed firefighter with the D-handle pike pole—similar to pulling a wagon.

wagon—is a quick and effective way to drag a firefighter out of a building over any surface with very little effort **FIGURE 9-37**.

A baby ladder or a roof ladder can also be used as a sled to carry an unconscious victim or downed firefighter to safety. The ladder carries and controls all the dead weight of the victim, which makes it fast and easy for the rescue firefighters **FIGURE 9-38**.

Roof ladders can be inserted into a basement or a first-floor window to remove a victim quickly. Placing a victim on the roof ladder and then pulling the ladder back out of the window makes for a very fast rescue. The ladder acts as a large platform for the unconscious victim, making the rescue fast and easy because the ladder supports and controls all the dead weight of a victim **FIGURE 9-39**. It can also be used to move a victim or a downed firefighter over tall obstacles with very little effort **FIGURE 9-40**.

Companies need to practice different carries and drags to figure out which method works best for them. The same goes for bringing a victim down a ground ladder. Don't hesitate to raise two ladders side-by-side. This widens the work area for two firefighters to assist in the rescue. There are various methods to carry a victim down a ladder, but always use a rescue rope to tie off the victim. A round turn, two half hitches under the arms and around the chest helps you control the victim being carried down a ladder. No one wants a victim to fall off a ladder after he or she has been rescued from a burning building. No doubt that error would be caught on camera and end up on the evening news!

A B

FIGURE 9-38 A. Firefighter rescue drag using a baby ladder. B. Firefighter rescue drag using a roof ladder.

A B

FIGURE 9-39 A. A roof ladder can be used for rescuing a victim or a firefighter from a first-floor window or from a basement. B. Using a roof ladder through a window is the fastest way of rescuing a downed firefighter from a basement. The ladder supports and controls all the dead weight of the firefighter.

FIGURE 9-40 A roof ladder can be used to move a victim or a downed firefighter over tall obstacles with very little effort.

Emergency Medical Services (EMS)

Every rescue evolves into an EMS incident. When the crew finds one or more victims and brings them out, the officer of the search team should let the IC know as much information as possible. For example, is the victim an adult or a child, male or female? Is there more than one? Are they conscious and breathing, or unconscious and not breathing? Is the victim burned? This kind of information is important for the EMS crew, who are probably listening to the fireground radio channel anyway. Once the IC hears that a victim has been found, EMS crews need to be moved to a forward position, as close as safety allows, to receive the patient. If the victim is a downed firefighter, the scene will be chaotic, but there needs to be sufficient personnel to assist the EMS crew to remove all the bunker gear and assist with advanced life support and cardiopulmonary resuscitation (CPR).

After-Action REVIEW

IN SUMMARY

- Every fire department should incorporate a risk management SOG for all emergency incidents, and it should state: Firefighters will risk their lives a lot in a calculated manner to rescue savable lives, and they will risk their lives a little in a calculated manner to rescue savable property. Firefighters will not risk their lives at all for lives or property that are already lost.
- Proper placement of the first hose lines can keep a fire away from people in the building. Rapid ventilation removes accumulations of smoke and gas and prevents further buildup of either one. Both operations reduce the danger to people inside and extend the time they have to get out of the building. Quickly extinguishing the fire removes the danger to human life. In a very real sense, these too are rescue operations.
- Suppressing the fire from the exterior as soon as possible improves the potential rescue profile and survival time of building occupants.
- Ventilation that is not immediately followed by effective fire suppression using water reduces the rescue profile potential survival time.
- Preparation for rescue begins before the alarm is received. It begins with a thorough knowledge of the area of responsibility, including the occupancies, the hazards, and the potential rescue problems.
- Initial size-up indicates where firefighters should begin their search and rescue and fire attack operations.
- Immediate rescue by the first-in engine company, without trying to knock down the fire first, should not be attempted except in extreme cases. Even then, the risk is extreme, and the gamble may not pay off. Delaying water application on the fire in favor of rescue has led to worse fire conditions that resulted in civilian and firefighter fatalities.
- The primary function of the engine company in a rescue situation is to support the primary search. Engine companies are responsible for the proper placement of hose lines between the fire and victims. Engine companies are also responsible for protecting firefighters conducting the search. This is best done by putting out the fire.
- Extreme high temperatures that force firefighters to the ground, where they have to crawl to tolerate the heat and smoke, is a warning sign that flashover is imminent. Rollover is the last warning sign before flashover occurs. Any search at this point without the protection of a hose line should be a defensive search.
- The two in/two out rule is directed by OSHA and NFPA 1500, which states that, during the initial stages of an incident, firefighters working inside the IDLH hazard area must work with a crew of at least two members. They must be backed up by at least two members outside the hazard area who are fully protected with PPE and SCBA and who are able to go to the aid of the two members inside if the need arises.
- A RIC must be provided from the initial stages of an incident to its conclusion. The RIC is a fully equipped, designated rescue team on site that is ready to react and respond immediately for the rescue of injured or trapped firefighters. The RIC should be a crew of four firefighters, one of whom must be an officer.
- Water is as essential to the primary search and rescue operations as it is to extinguishment. Water from attack lines is used to provide fast water by operating the first hose line from the outside, aiming the stream into a window or door to knock down and reset the fire, reducing interior temperatures, and preventing flashover.
- The primary function of the engine company in a rescue situation is to support the primary search by getting water on the fire. Extinguish it or knock it down. If it cannot be extinguished, knocking it down resets the fire and stops the momentum of the energy and heat release rate.
- The primary search is the first search. If there is any indication that victims might be trapped within the fire building, search operations should begin as soon after arrival as possible. This should coincide with the first hose lines being stretched and positioned. Water must be applied into the structure first to cool the temperature and prevent flashover conditions from developing; otherwise, making entry creates a flow path and accelerates the fire.

- The secondary search should be conducted after the primary search to determine that no one was overlooked during the original search. The secondary search should be assigned to a different company who did not participate in the primary search. The secondary search is not a rapid search. The emphasis should be placed on thoroughness, not speed.
- The building should be ventilated as soon as possible after the application of water on the fire to allow smoke, heat, and gases to move away from occupants who might be trapped inside. Water first, then vent. A dangerous flow path is less likely to develop during ventilation if water is applied to the fire.
- Keeping the doors closed during and after the primary search prevents a flow path from developing before the fire is knocked down or extinguished.
- Compartmentation (being behind a closed interior door) prior to fire department arrival provides increased protection and survivability compared to being in a room or area with an open door to the fire.
- A thorough, systematic primary search should be conducted at every fire where it is safe for firefighters to enter the building. An oriented search is one in which firefighters use methods and tools to keep the team oriented to their location within the building and to each other.
- Using standard search patterns, the first area to be searched is directly over the fire. After reaching the top of the stairway, firefighters turn in one direction or the other to get to the room over the fire. They begin the search by turning either right or left into that room. This turn sets up the basic search pattern for the entire floor.
- Every rescue evolves into an EMS incident. The IC needs to move EMS crews to a forward position, as close as safety allows, to receive victims. If a victim is a downed firefighter, sufficient personnel must be available to assist the EMS crew with removing all the PPE gear and performing CPR.

KEY TERMS

defensive search A type of search conducted in the immediate vicinity of the door; used when heat conditions in a room indicate an imminent flashover and a hose line is not in place to cool the atmosphere.

Mayday A verbal declaration indicating that a firefighter is lost, missing, or trapped and requires immediate assistance.

primary search An immediate and quick search of the structures likely to contain survivors. (NFPA 1670)

rapid intervention crew (RIC) A minimum of two fully equipped personnel on site, ready for immediate rescue of disoriented, injured, lost, or trapped rescue personnel. (NFPA 1006)

rescue air kit (RAK) A prepackaged SCBA unit with a facepiece and 60-minute air bottle. The backpack assembly is omitted. It is a reliable, independent air source used in rapid intervention rescue and primary search.

resetting the fire Applying fast water on a fire that results in knockdown. Knocking down the fire takes away the heat and thermal energy of the fire and slows down rapid fire growth. The goal is to reset the time to flashover.

secondary search A detailed, systematic search of an area conducted after the fire has been suppressed. (NFPA 1670)

survivability profile An assessment that weighs the risks likely to be taken versus the benefits of those risks for the viability and survivability of potential fire victims under the current conditions in the structure.

vent-enter-isolate-search (VEIS) A method of searching for fire victims that consists of selecting and opening a window to a bedroom or other living space; entering the room; closing the door to isolate the room from the fire, thus preventing a flow path from developing and allowing the room to ventilate naturally; quickly searching for any possible victims; and rescuing them out a window and down a ladder.

REFERENCES

National Institute for Occupational Safety and Health, Report F-2000-04, April 11, 2001, "Structure Fire Claims the Lives of Three Career Firefighters and Three Children." Cincinnati, OH.

NFPA 1971, *Standard on Protective Ensembles for Structural Fire Fighting and Proximity Fire Fighting*, 2018 edition.

NFPA 1981, *Standard on Open-Circuit Self-Contained Breathing Apparatus (SCBA) for Emergency Services*, 2019 edition.

CHAPTER 10

Initial Fire Attack and Ventilation

LEARNING OBJECTIVES

- Analyze the concept of using a direct attack during offensive fire attack scenarios.
- Analyze the advantages and disadvantages of using an indirect attack during offensive fire scenarios.
- Analyze the concept of using a combination attack during offensive fire scenarios.
- Examine the factors involved in determining the choice of the initial attack lines and nozzles used on the fireground.
- Recognize the conditions that indicate the presence of a ventilation-limited fire.
- Understand all the hazards associated with fighting a basement fire.
- Describe the challenges in standpipe operations.
- Describe the various smoke conditions and ventilation challenges for a fire attack that is initiated from a stairwell in a multistory building.
- Explain the proper methods for ventilating a ventilation-limited fire or one with potential backdraft conditions.
- Describe the various methods of ventilation.

Introduction

Major changes have occurred in the fire service since the terrorist attacks of September 11, 2001. Responding to a large column of smoke, indicating a working fire, or arriving on scene with flames visible doesn't always mean an accidental fire. High-rise fires, plane crashes, explosions, and mass casualty incidents (MCIs) are viewed with greater suspicion today. The other changes over the last two decades coincide with the heightened threats of terrorism but are unrelated: the increase in the use of modern fuels of plastic and synthetic household furnishings combined with the ever-changing methods and materials found in lightweight construction.

Since 2001, UL and the National Institute of Standards and Technology (NIST) have been conducting controlled fire scenarios in specially constructed laboratories and in actual houses that are completely furnished: some with modern fuels, others with legacy fuels to draw comparisons. The researchers used a variety of sophisticated instrumentation and computer modeling to measure the temperatures throughout the structures at various heights within each room, the heat release rates, the heat generated from the room contents, the airspeed and air current directions coming into and out of the fire compartment as well as from ventilation points, and the visibility and chemical makeup of the smoke and fire gases within and outside the structure. All of the data and observations were carefully recorded and completely analyzed. These experiments were repeated numerous times to confirm how each action taken by firefighters in the areas of water application and ventilation affects the growth and extinguishment of fires.

All the previous chapters in this text have been building a solid foundation to bring us to this point: initial fire attack. If you have not read the first four chapters on modern fire behavior, building construction, and the UL/NIST experiments, it is highly recommended you do so now before continuing. It is important to connect the dots to see the importance of all these factors before even pulling a hose line off the engine.

Evidence-based practices allow us to make tactical adjustments in how we initiate the interior fire attack for residential structures and basements and withhold ventilation until hose lines are in place to cool and control the changing environment that results.

Three Strategic and Tactical Areas for Initial Fire Attack

To incorporate the findings of the UL/NIST experiments for initial fire attack, three major strategies and tactics need to be modified or implemented: transitional attack (defensive to offensive), our approach to basement fires, and coordinated ventilation.

1. When the fire has self-vented through the window or a door, an indirect attack or a **transitional attack** should precede the interior attack. Working from the outside of the structure (outside the immediately dangerous to life or health [IDLH] area), the firefighter should spray water into the burning compartment, either through the window or the door for approximately 30 seconds or as long as it takes to knock down the fire. The nozzle should be set to straight stream for maximum reach and penetration. The straight stream will not push fire into the rest of the structure, and it can be bounced off the ceiling, deflected off the top window header or off the walls, or aimed at the seat of the fire if the angle permits. This technique is often referred to as softening the target, or hitting it hard from the yard **FIGURE 10-1**. Another important reason the straight stream must be used, is it allows the natural ventilation process to occur through the same opening. The heat generated from the fire room causes fire gases to expand, thus pressurizing the room. In the case of a self-vented window, the pressurized smoke and gases seek the path of least resistance from high pressure to low pressure, always trying to equalize. A straight stream takes up very little surface area of the open window, so fire gases are allowed to escape. A fog pattern through the window would have the opposite effect. The main objective is to get **fast water** into the fire room by the quickest means possible to knock down and reset the fire before entering the IDLH. Resetting the fire means stopping the fire growth momentum and pushing it back to where it was in the incipient stage or early growth stage. This fast water application immediately reduces interior temperatures, slows down the heat release rate, and decreases the production of toxic smoke and fire gases.
2. In residential basement fires, the practice of making entry through the first floor and down the stairs into the basement should be discontinued except as a last resort—and even then, it may be extremely dangerous if not impossible. Going down the basement stairs is always risky because it requires the door to remain open at the top. Smoke and heat immediately start rising up against the

FIGURE 10-1 Transitional attack: defensive to offensive. Also referred to as softening the target, hitting it hard from the yard, and resetting the fire.

Photos by Rick McClure.

attack crew and into the first floor. The attack crew is literally in a chimney flow path. If a basement door or window fails or is opened for any reason, a high-low flow path is created, and the firefighters will be caught between where the fire is and where the fire wants to go. The convected high-temperature currents with this high-low flow path can travel with great speed and generate temperatures well over 1,000°F (538°C). Firefighters do not have time to react in these conditions, nor can they survive these temperatures. More than one firefighter has been killed by thermal assault in this below-grade scenario, and this was the exact scenario that killed two San Francisco, California, firefighters at the Diamond Heights Fire on June 2, 2011.

Every effort should be made to use any exterior opening to the basement for fire attack to knock down the fire. A rear door entrance is preferred to entering from the floor above. Basement-level entries put the fire attack team on the same level as the fire, which is a more advantageous position **FIGURE 10-2**. If fire is blowing out a basement window, a hose line should be aimed from the outside, and a straight stream or a narrow fog should be applied into the burning basement. If there is no window, a chainsaw can easily make one near the seat of the fire, and it only has to be large enough to get the nozzle stream into the basement. The same science applies. Knocking down the fire with fast water immediately reduces temperatures within the compartment.

With concern for possible occupants, we have to rely on the scientific facts to determine a survivability profile. If the basement fire is hot enough that the incident commander (IC) opts for a transitional attack from defensive to offensive, it's unlikely an unprotected human would be able to survive the smoke toxicity and temperatures inside the fire compartment. When water is applied, the situation always gets better.

3. The final strategic area concerns coordinated ventilation. Coordinated ventilation is easier said than done. It's just a fact that engine companies usually arrive on the fire scene before the ladder company, and the engine crew is usually in place and ready to go before the ladder company is ready to ventilate the fire. Ventilation should occur immediately before the attack team makes entry **FIGURE 10-3**. This may be possible with horizontal ventilation, but not with vertical ventilation. Raising ground ladders, positioning aerial apparatus, sounding the roof, making the roof cuts, louvering, and punching through the ceiling is labor intensive

FIGURE 10-2 In a basement fire, a rear door entrance is preferred to entering from the floor above. Basement-level entries put the fire attack team on the same level as the fire, a more advantageous position.

Courtesy of John Odegard.

and time consuming. It is unreasonable to expect the hose team to wait until vertical ventilation is complete before making entry **FIGURE 10-4**.

The next ideal situation would be to ventilate while the engine company is making entry. Realistically, ventilation usually ends up taking place after the engine company has started attacking the fire. The engine company has often knocked down the fire and initiated hydraulic ventilation through a window in the fire room before the ladder company has established ventilation. The UL/NIST test results have already proven that the immediate application of water reduces interior temperatures, and fast water without ventilation has positive effects on the inside environment. These same tests have also proven that when ventilation takes place before hose lines are in position for quick water application, there is an immediate release of heat, but the newly established flow path creates air movement, increasing the intensity of the fire. Remember, ventilation always creates a flow path, and the movement of air currents always intensifies the fire. This results in higher temperatures, increased production of toxic smoke and fire gases, and fire spread. If water isn't applied, the fire will be drawn to the ventilation exhaust hole. Therefore, water first, then vent.

Perhaps a better way to understand the adjustments can be as follows: in the past, we taught that ventilation supports the fire attack; now we understand that having hose lines in place and ready to go supports effective ventilation. Ventilation without water application makes the fire worse. Immediate water application into the fire through a self-vented window or door, by the fastest means possible, improves the situation. After that, interior operations can continue in the normal fashion.

What You Need to Know Before You Even Pull a Hose Line

Several extreme fire behavior events can occur during a structure fire that firefighters must know and be able to anticipate, including thermal layering, rapid fire growth, rollover, flashover, backdraft, the behavior of ventilation-limited fires, and smoke explosions. Though they seem out of sequence, flashover and rapid fire growth are being introduced at the beginning of this chapter because with modern fuels, flashovers are happening with more frequency before the fire department arrives, or as the fire department arrives on scene. With double- and triple-paned windows, a well-insulated house can deprive the fire of oxygen, putting it into a ventilation-limited state. The sudden introduction of fresh air, often caused by the initial attack team making entry into the structure, may cause rapid fire growth, a sudden backdraft, or flashover. Being met with extreme fire behavior is an unexpected event when the initial crew is anticipating a routine fire attack.

Flashover

A flashover is a high-temperature event. As a fire continues to burn in a room, additional heat and smoke

FIGURE 10-3 Ideally, ventilation should occur immediately before the attack team makes entry.

are generated. The hottest layers of fire gases, those at the ceiling, begin to bank down toward the floor. These superheated gases raise the temperature of all the surrounding unburned contents within the room. When the contents in the room have absorbed all the heat energy they can withstand from the fire, they reach their ignition temperature, and flashover occurs. Flashover is the rapid transition from the growth

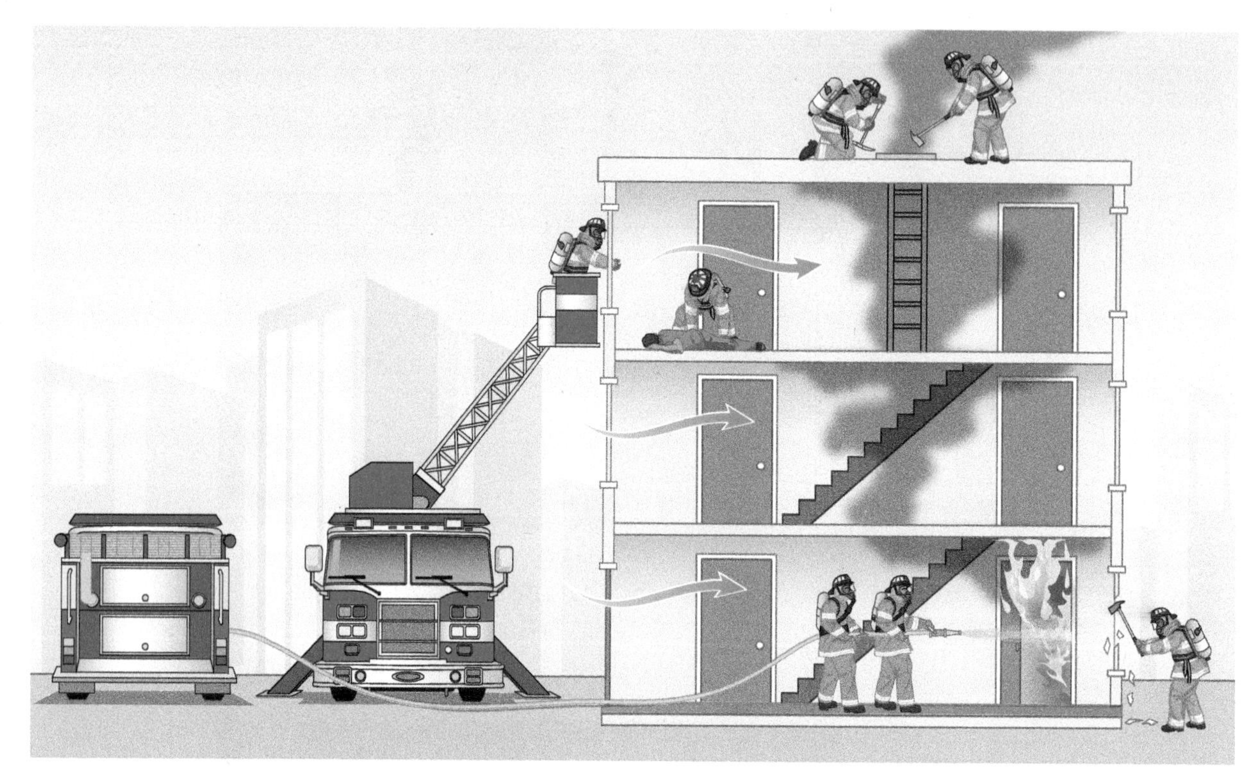

FIGURE 10-4 In a coordinated fire attack, ventilation and fire attack should happen simultaneously.

stage to the fully developed stage. In a flashover, the entire contents of a room ignite almost simultaneously, releasing a tremendous amount of energy and generating intense heat and flames. This rapid change in the compartment occurs in seconds. The threshold temperature for a flashover to occur is approximately 1,000°F. Once this temperature is reached, all the fuels in the room are involved in fire, including the floor coverings, so the temperature at floor level may be as high as the temperature of the ceiling when flashover occurs.

In certain fire conditions, flashover can be delayed, or may not occur at all. For example, flashovers are less likely to occur if the fire is in a large room with high ceilings, a condition that supplies more air during fire development. Larger areas with high ceilings entrain more air in the fire plume, which tends to cool the plume, thus reducing the radiation and heat convection toward other contents in the room. With high ceilings, it takes longer for accumulated heat to radiate back down to other fuels at floor level. This doesn't mean the fire isn't going to continue to grow; it simply means a flashover may not occur.

Safety and Survival

Firefighters, even with all their structural personal protective equipment (PPE) rated to withstand temperatures up to 500°F (260°C), cannot survive for more than perhaps 10 to 15 seconds in a flashover. To account for reaction time, a firefighter would have to be between 6 and 10 feet from the exit door for any chance of surviving this very dangerous and deadly occurrence. There is no chance for human civilian survivability in a room that has flashed over. Although the primary objective of initial attack lines and direct attack is to extinguish the fire, the first priority should be to prevent flashover from occurring by using water streams effectively.

Flashovers can be prevented by cooling the fuel with water; cooling the ceiling atmosphere with water, which reduces the temperature of the gases; removing the fuel; and/or reducing the amount of oxygen present by controlling ventilation. Keeping doors and windows closed in the fire room helps to reduce the amount of oxygen available to the fire and may help prevent flashover until hose lines are ready to move in. Preventing flashover allows firefighters to continue to work in an IDLH environment within a reasonable threshold of safety to search for and rescue any possible occupants and to completely extinguish the fire. If we cannot prevent or delay flashover, it is deadly for any occupant or firefighters caught in that room.

Rapid Fire Growth

Introducing air into a ventilation-limited (oxygen-limited) fire can result in explosive fire growth. Research has demonstrated that firefighters making entry through the front door of a structure can introduce enough air into the fire area to produce rapid fire growth and flashover.

FIGURE 10-5 Remember, smoke and fire gases are fuel. Repeated experiments have shown that opening the front door has a significant impact on the growth of the fire and can produce violent flashovers or backdrafts.

This change can occur within 2 minutes and can progress so fast that it is impossible for firefighters to escape.

We need to consider carefully what constitutes ventilation. Opening any door, window, skylight, or roof introduces oxygen into a burning building. Studies have shown that opening the front door has a significant impact on the growth of the fire. Repeated experiments have produced violent flashovers or backdrafts after the front door is opened. Remember, smoke and fire gases are fuel **FIGURE 10-5**. Rapid fire growth can be prevented by cooling the superheated ceiling environment, cooling the solid fuels, removing the fuel, and/or controlling the amount of oxygen entering the structure. During rapid fire growth, a fire can progress to its maximum heat release rate unless immediate actions are taken to limit the flow path of oxygen or to apply water on the fire. Hot smoke flowing quickly through hallways, corridors, and stairways rapidly spreads the fire. Smoke ignition causes this rapid fire spread. Know the difference between heat-pushed smoke versus volume-pushed smoke. Note and be forewarned of smoke that enters and moves in faster than firefighters can; this indicates a ventilation-limited fire.

Introduction to Initial Attack Lines

At a working structure fire, the placement and operation of initial attack lines protect occupants and firefighters, and also provide the first water on the fire for extinguishment. The IC should conduct a 360-degree size-up and risk-benefit analysis; determine value, time, and size (VTS); and consider if enough water and personnel (resources) are available before the decision is made to allow firefighters to enter a building to conduct an aggressive interior attack. If firefighters cannot enter, a defensive strategy of operation should be used. Firefighters must take up safe positions to protect exposures and operate on the fire building from outside the collapse zone. Firefighters must decide the type of initial attack line and nozzle that will best manage the situation. Choosing the correct hose line protects firefighters and victims and enables crews to control the fire. Randomly selecting smaller, ineffective hose lines for initial attack usually does not control the fire and creates negative results until the proper hose lines are used. This potentially places everyone on the fireground in danger while the structure continues to burn.

Most fire departments operate pumpers with preconnected hose lines and designated hose sizes, lengths of hose lines, and specific types of nozzles. These hose lines are suitable for initial attack lines most of the time. Unfortunately, many fire departments have become complacent and stretch these hose lines at every incident because they have proven successful in the past. This was the case at the tragic Sofa Super Store Fire on June 18, 2007, in Charleston, South Carolina, where nine firefighters were killed. A company officer must be able to choose the correct hose line and nozzle for the intended application. This is done by calculating the rate of flow and using either the Royer/Nelson formula or the National Fire Academy formula. Experience plays a role, but sometimes the IC must deviate from the norm and operate a larger-diameter hose line that will provide safety to firefighters and victims while controlling and extinguishing the fire.

Firefighters must operate within the incident management system (IMS) and coordinate efforts between other engine and ladder company personnel. The fireground must be managed so that engine company personnel are prepared to enter the building after the fire has been knocked down from the outside. If this is not possible, for example, if a fire is in a closet or an interior room with no windows to the outside, personnel must be prepared to make entry in conjunction with or immediately after ventilation. Ventilation must be coordinated with suppression efforts; ideally, a coordinated fire attack ensures that both tasks take place simultaneously. Any delay in ventilation should not delay the application of water on the fire. Every effort should be made to apply water on the fire from a safe position by any means possible. Water application slows the growth of the fire and reduces interior temperatures, whereas a delay in water application, especially after ventilation, allows the fire to intensify and spread.

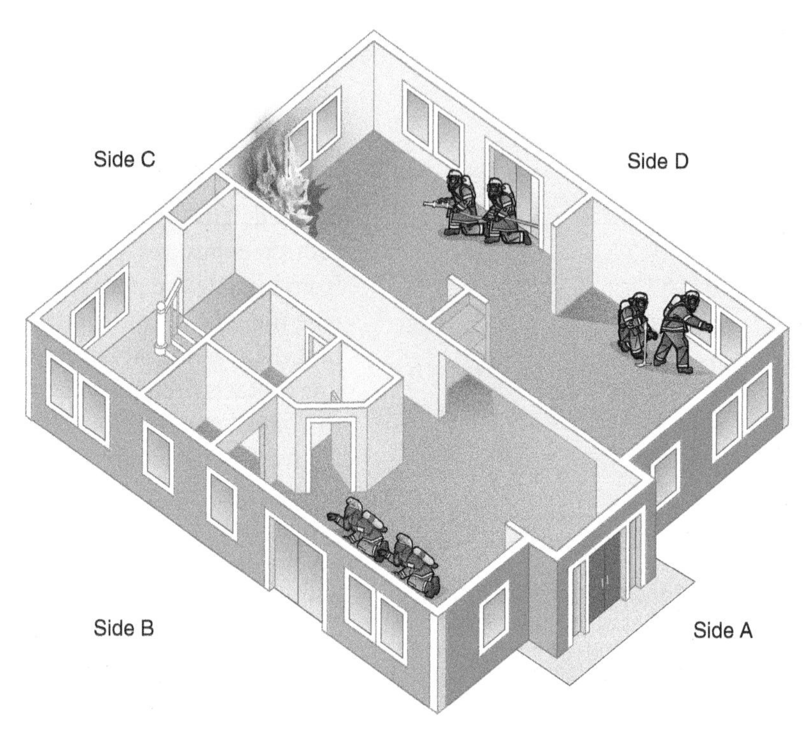

FIGURE 10-6 Firefighters need to get inside to attack the seat of the fire, place attack lines between the victims and the fire, figure out where the fire is extending, place interior exposure lines, conduct a primary search for victims, and ventilate promptly.

With pumpers well positioned and an adequate water supply established, the outcome at the fireground depends, to a great extent, on the effectiveness of the initial attack. The success of the initial attack most often centers on the effectiveness of the company officers' decisions regarding the size, number, and placement of attack lines, as well as the types of nozzles to be used against the fire.

An interior direct attack is an offensive operation in which firefighters enter a building with an attack line to control and extinguish the fire. If there is any chance that a fire building is occupied and an interior attack is possible, the fire should be knocked down from the outside, then fought from within. Firefighters need to figure out where the fire is and get inside to attack the seat of the fire, place attack lines between the victims and the fire, figure out where the fire is extending, place interior exposure lines, conduct a primary search for victims, and ventilate promptly **FIGURE 10-6**. An interior attack can be conducted using a direct, indirect, or a transitional combination method.

Locating the Fire

Determining what's on fire, where is the fire located, and where the fire is headed should be determined during the size-up. The tactical objectives of Rescue, Exposures, Confinement, Extinguishment, Overhaul/ Ventilation and Salvage (RECEO/VS) starts with laying the first hose line. Rescue can be accomplished by extinguishing the fire. Exposures in this case refers to interior exposures. To cover exposures, the company officer has to figure out what is burning and where it is burning. Sometimes the location of the fire is obvious; other times it is not. Determining the location of the fire is important so the amount of hose required to reach it can be calculated. Once the fire is located, it can be determined where it is headed. Getting an attack line and an exposure line ahead of the fire confines it, and then it can be extinguished.

These actions are not always separate, carried out by separate crews. Sometimes these tactical objectives can happen almost simultaneously with the first attack hose line. There are five basic fire attack strategies: direct attack, indirect attack, combination attack, transitional attack, and a defensive attack.

Direct Attack

A **direct attack** should be used during the incipient stage and growth stage of a fire. It can also be used during most free-burning stage fires, but not all of them. The simplest example of a direct attack is using a pressurized 2½-gallon water fire extinguisher

FIGURE 10-7 Using a 2½-gallon water pressurized fire extinguisher is the simplest example of a direct attack.
Courtesy of John Odegard.

FIGURE 10-8 A direct attack is one in which a preconnected 1¾-inch, 2-inch, or 2½-inch hose line with a solid smooth-bore or straight-stream combination nozzle is used.

to put out a trash fire inside or outside a structure FIGURE 10-7. Small fires in a laundry room; beauty bark fires; burning food on a stove; stuffed-chair, sofa, and even mattress fires can be extinguished with one or two 2½-gallon fire extinguishers. For any burning solid fuels that are larger or if the fire is in the early growth stage, a direct attack using a 1¾-inch hand line should be used.

In most offensive fire attack situations, a direct attack is the preferred method. A 1¾-inch preconnected hand line provides enough water flow that the majority of room and content fires can be controlled and extinguished by this first line. A direct attack is the surest way of controlling the fire and minimizing the danger to occupants. A direct attack is one in which a preconnected 1¾-inch, 2-inch, or 2½-inch hose line with a solid smooth-bore or straight-stream combination nozzle is used to attack the fire. The fire attack team enters the structure from the unburned side and works toward the burning side (uninvolved toward the involved) to deliver water directly onto the seat of the fire FIGURE 10-8.

Using the reach of the hose stream, the nozzle firefighter should work the nozzle at all the areas involved in fire using a steady, deliberate Z-, T-, or circular O-pattern to cover the entire area of the room. Smoke is fuel, and the fire gases are already in a vapor state for immediate combustion once the ignition temperature is reached. Hitting the ceiling area cools the fire gases and burning material below their ignition temperature and prevents flashover. Hitting the center area of the room from side-to-side using a steady, straight stream breaks out any windows in the room, allowing for immediate ventilation. If a fire occurs during the day, sunlight will eventually be detected from the window and the officer can confirm that the window is open. If a fire occurs at night, the nozzle firefighter can hear when the stream is hitting the wall and when the stream is passing through a broken-out window. The change in sound confirms that the window is open for ventilation. Most solid fuel loads are at floor level. Home and office furniture, for example, chairs, tables, desks, sofas, beds, and carpeting, are all below waist level. When these Class A solid fuels are heated by the radiant heat of the fire, they produce combustible vapors in a process known as pyrolysis. Hitting all the solid fuels, including the base of the fire, reduces the temperature of the solid fuels and halts the vapor-producing process. This is when the fire is extinguished.

The best way to save the maximum number of lives is to put the fire out with a quick, aggressive direct attack. A direct attack works best if the fire attack is coordinated with other engine companies advancing and operating backup and exposure lines, truck companies performing search and rescue, and ventilation.

Indirect Attack

A free-burning fire may also be attacked indirectly. Obstacles, obstructions, furnishings, or stacked material often prevent a direct attack on the seat of the fire. An indirect attack is one in which a solid, straight, or narrow fog stream is used to direct water at the ceiling to cool superheated gases in the upper levels of the room FIGURE 10-9. The objective of the indirect attack is to prevent flashover by removing heat from the upper atmosphere. This method injects a stream of water into the superheated upper levels of the room. As the water is deflected and bounces off the ceiling into droplets, it is converted to steam and expands 1,700 times its volume and thus absorbs a much

FIGURE 10-9 In an indirect attack, a solid, straight, or narrow fog stream is directed at and bounced off the ceiling or the window threshold of the intensely heated area in order to break up the spray pattern to create steam.

greater amount of heat. This process quickly drops the temperature in the atmosphere, helping to extinguish most of the fire.

The disadvantage of an indirect attack is that the water-to-steam expansion ratio is capable of causing steam burns to firefighters and victims. Because of this, an indirect attack isn't the best choice in areas where victims may be located. It certainly makes rescue attempts more difficult after this method is applied unless ventilation occurs immediately before, simultaneously, or immediately after the indirect attack. Nevertheless, the opening of this paragraph could read "The advantages of an indirect attack . . ." because indirect attacks are quite effective for extinguishment.

Choosing an indirect attack should not pose a moral dilemma or create any indecision for the company officer. We must consider the science and evidence available in making a judgment call on a favorable or poor survivability profile within the burning compartment. It is no longer justified to launch an interior direct attack simply because there *might* be someone inside. The IC must perform a risk-benefit analysis, apply all the information obtained at the scene, and weigh it against the scientific facts about the probability of human survival within a fire compartment involving modern fuels.

Consider the following points:

- If a direct attack is doable and has a great chance for success, the company officer should opt for it. If an indirect attack was implemented, a direct attack was probably not possible.
- The fire needs to be extinguished. The longer it burns, the higher the temperatures will climb until the room flashes over. In the meantime, hot, toxic, deadly smoke and fire gases that are unbreathable continue to be generated. Hesitating to apply water due to the concern for creating steam around trapped occupants only allows the conditions to worsen.
- If the engine officer is concerned with creating steam from an indirect attack, then the room and ceiling temperatures are well over 212°F (100°C). If we have proven that water application and steam conversion absorb the heat and quickly lower the interior temperature, then the room should now be cooler than it was before the nozzle was opened, whether or not a direct or indirect attack was used.
- Steam burns are extremely painful and can cause serious burn injuries, but they are not fatal 100% of the time. Just a few breaths of hot toxic smoke can be fatal. An unprotected occupant caught in a room that flashes over will certainly die 100% of the time, and it is unlikely that firefighters would survive the same event because their PPE ensemble has a rated protection ceiling of 500°F (260°C) for 5 minutes. Water must be applied to the fire by any means possible to prevent flashover from occurring.
- The entire range of burn temperatures from when humans first feel pain to complete tissue destruction falls well below 212°F, the boiling point of water.
- Deciding that there is a zero survivability profile in the fire compartment does not mean there is a zero survivability profile throughout the structure. Using an indirect attack has a positive effect on and improves the survivability profiles of trapped occupants outside the fire room if it results in extinguishing the fire.

Firefighters should work the nozzle all across the ceiling and the upper walls to deflect the water stream, then shut down the attack line as soon as enough water has been discharged to knock down the fire and cool the area. If too much water is converted to steam, the steam and hot gases are pushed down toward the floor, where they can injure or kill unless the space is quickly ventilated. When the area has been cooled down below 212°F, steam production stops, and firefighters feel and hear the water falling on the floor. Once ventilation is taking place, the operation should be changed from an indirect attack to a direct attack for final extinguishment.

Using a piercing nozzle, fog applicator, or distributor nozzle is another excellent method to carry out

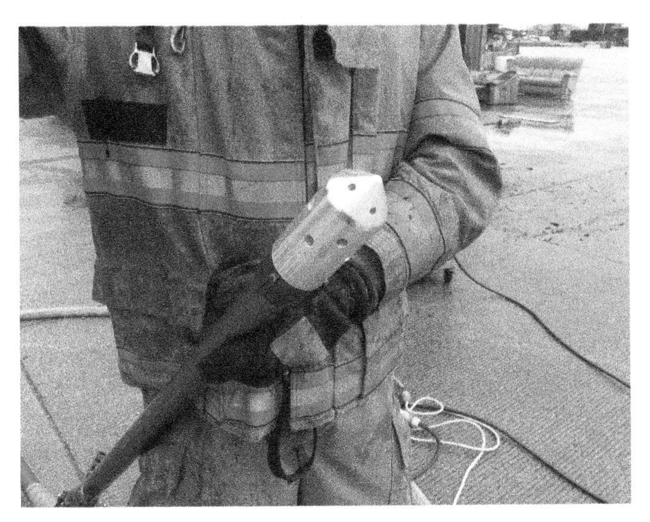

FIGURE 10-10 An indirect attack using a piercing nozzle is best suited for unoccupied areas.

Photo by Raul Angulo.

an indirect attack. The indirect attack using these specialty nozzles is best suited for unoccupied areas such as storage areas, basements, attics, cocklofts, wall spaces, and other concealed or sealed-up compartments **FIGURE 10-10**. It can also be used in flanking attacks from adjacent uninvolved units. In attacking basement fires, the indirect attack creates a sprinkler system. The water converts to steam to absorb heat and knock down the fire. A shift to a direct attack will still be necessary for final extinguishment.

Indirect Attack for Attic and Cockloft Fires

Attic and cockloft fires are common because all the smoke and heated gases rise to the highest point of the structure. Once they hit the vertical barrier (roof decking), the smoke and gases start to spread horizontally until they reach the lateral barriers (walls), then they mushroom and work their way down to the floor. Consider the fire science: if the products of combustion from a room fire rise into the attic space, the smoke and gases eventually fill that space and then escape from the gable-end vents or from the soffits. At this time, the attic space is oxygen-deficient and cannot support combustion. There is heat and there is fuel (the smoke and the trusses), but there is insufficient oxygen to complete the fire triangle, so no fire is present. If a piercing nozzle is inserted through the roof into the attic space or a small hole is cut into the roof to accommodate a round nozzle or fog applicator, the water entering the attic converts to steam that is 1,700 times the volume of the water and cools the combustible gases well below their ignition temperature. The fire won't extend into the attic and thus will be contained to the room of origin. As long as the roof is not opened, the fire room can be horizontally vented (this would be true no matter what floor it is on). The indirect attack into the attic can also be accomplished by inserting a piercing nozzle into the attic space from the room below the attic or by accessing from the outside through the soffits or from the gabled ends.

The key to success in this type of indirect attack is *not* opening the roof for vertical ventilation. If the roof is vertically vented, a high-low flow path is created. Once fresh air enters the flow path, the fire triangle is complete, and the combustible vapors will ignite, bringing the fire right into the attic space, and exposing the trusses and roof assembly to fire. The air mixture at the vent hole is also sufficient to ignite the escaping gases, which produce spectacular flames for the cover of the trade magazines. Opening the roof to vertically ventilate an attic or cockloft fire ultimately causes more structural damage to the building than the indirect attack. Also, because the fire is now in the attic space, it requires a direct attack to control. If the hose lines are not in place for a direct attack, the fire runs the entire length of the attic or cockloft space. Apartment complexes and strip malls that share a common attic space do not have the protection of a fire wall, and the entire attic and roof structure becomes involved with fire after vertical ventilation. Water first, then vent.

Indirect Attack in Residential High-Rise Fires

There are three types of high-rise buildings: commercial offices, hotels, and residential apartments or condominiums. Of the three, according to the National Fire Protection Association (NFPA) report on high-rise fires, residential high-rises are sprinklered the least often. A wind-driven fire can be deadly in a residential high-rise. A wind-driven fire scenario is created when the fire self-vents by breaking out a window on the windward side. The wind along with the fire pressurizes the unit. As long as the front door of the unit remains closed, the flames, smoke, and gases try to escape against the wind through the open window. As soon as firefighters attempt to make a direct attack through the front door, a high-pressure to low-pressure flow path is created. Remember, high pressure always seeks the path of least resistance and moves toward the low pressure in an attempt to equalize. But in this scenario, the wind drives the flow path currents with extreme speed and at temperatures well over 1,000°F (538°C). Firefighters cannot survive this event. This is the scenario that killed three New York City (FDNY) firefighters at the Vandalia Avenue Fire on December 18, 1998, in Brooklyn, New York.

The safest method to attack this type of fire successfully is an indirect attack from the adjoining residential units. Depending on the wall construction, an indirect attack using a piercing nozzle could also be inserted through the hallway interior wall to the unit or even through the front door of the unit on fire. The first task is to ensure that the front door to the fire unit remains closed. Attack lines should be laid into both adjoining exposure units—to the right and to the left of the fire unit. A TIC can help determine which wall has the highest heat readings, and the indirect attack should be concentrated on that particular wall. Then, using piercing nozzles or distributor nozzles, an indirect attack can commence into the fire unit. Another indirect method is to breach the walls into the fire unit, just large enough to accommodate combination nozzles set to a medium fog. Then, the indirect attack can begin. This indirect attack is like introducing a giant sprinkler system into the unit **FIGURE 10-11**.

Backup lines should be laid to the exposure units to support the attack team because the nozzles will be through the wall, leaving the firefighters vulnerable should something happen. The two factors that make this type of fire difficult to fight is the wind and the elevation. The fire load and the residential square footage require water flow rates similar to a one-story apartment or condominium. Therefore, multiple flanking hand lines operating from the adjoining units should be able to flow the required gallons per minute (gpm) easily to extinguish this fire. Firefighters serving as observers from adjacent buildings or adjacent rooftops can let the attack crews know when the indirect attack has successfully knocked down the fire. Then attack lines and backup lines can be repositioned for a direct attack and final extinguishment. If the wind is less than 25 mph, multiple positive-pressure fans can also be set up in the hallway or at the entry of the fire unit to keep the wind at the back of the fire crew while its members overhaul the fire.

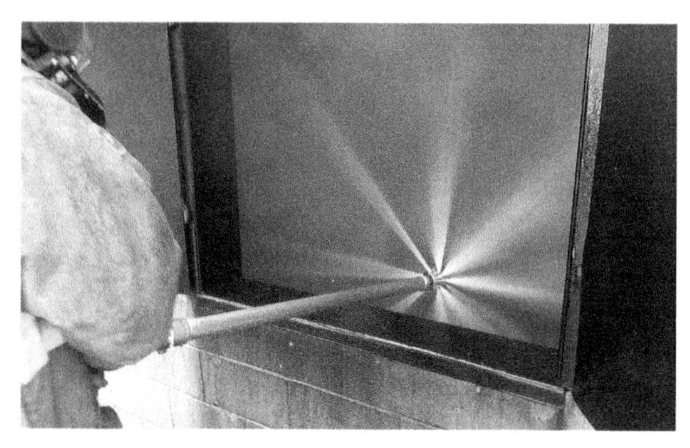

FIGURE 10-11 This training drill simulates how a firefighter would indirectly attack the fire room from the wall of an adjacent unit using a piercing nozzle.
Photo by Raul Angulo.

Depending on the floor construction, a piercing nozzle or Bresnan distributor nozzle can also be inserted into the fire unit from the floor above using the same tactics for an indirect attack on a basement fire, a cellar fire, or a pier fire. A bent pipe high-rise nozzle may also be used for an indirect attack from a window on the floor below the fire unit.

Combination Attack

The **combination attack** uses both the direct and indirect methods, one after the other. If the room is extremely hot and nearing flashover conditions, the survivability profile in that area is probably zero. The stream reach of an indirect attack is used to bring down the temperature. A sufficient amount of water should be used during the indirect attack to reduce the threat of flashover. If the firefighters operate from a safe position and utilize the reach of the stream, the amount of steam produced shouldn't force firefighters out of the area. (A stream from a 1¾-inch hand line can easily reach 70 to 80 feet.) After the threat of flashover has been reduced, the direct attack is used to extinguish the main body of fire.

Transitional Attack

There are two types of transitional attack: defensive to offensive, and offensive to defensive. Each one will be discussed separately in the next subsections.

Transitional Attack—Defensive to Offensive

The defensive-to-offensive transitional attack is the one most frequently used. The new terms in response to the UL/NIST recommendations, *softening the target* and *hitting it hard from the yard* are referring to this type of transitional attack. The decision to use a transitional attack, defensive to offensive, starts with a risk-benefit analysis and determining if there is life and property value still worth saving (the VTS strategy). This strategy implies that a direct attack into the structure will immediately follow after the fire is knocked down from the outside. If there is no occupant life

FIGURE 10-12 Steam from a quick burst of water from the nozzle onto the entry door to a fire can indicate high heat conditions.

FIGURE 10-13 Prior to making entry, the nozzle firefighter sprays water into the burning compartment through the window or the door for approximately 30 seconds or as long as it takes to knock down the fire.

hazard at risk and if no property still has value, there is nothing left to save. There is no justified reason to attack offensively, and the strategy should be defensive.

A trick of the trade to check for high heat behind a door before making entry is to apply a quick burst of water onto the surface of the entry door to the fire. If the water turns to steam, it is an indication of high heat conditions on the opposite side of the door **FIGURE 10-12**.

This modified transitional attack, defensive to offensive, is used when the fire has self-vented through a window or a door and should precede the interior direct attack. Working from the outside of the structure, the nozzle firefighter sprays water into the burning compartment through the window or the door for approximately 30 seconds or as long as it takes to knock down the fire. The nozzle should be set to straight stream for maximum reach and penetration. The straight stream will not push fire into the rest of the structure and can be bounced off the ceiling, bounced off the walls, or aimed at the seat of the fire if the angle permits **FIGURE 10-13**. The deflected stream breaks into droplets and converts to steam. A straight stream takes up very little surface area of the open window, so the steam, smoke, and fire gases are allowed to escape by venting out the same window. A fog pattern through the window would have the opposite effect.

The main objective of this transitional attack is to get fast water into the fire room to knock down and reset the fire. Resetting the fire means stopping the fire growth momentum and pushing the fire back to the incipient stage or early growth stage. This immediately reduces interior temperatures, slows down the heat release rate, and decreases the production of toxic smoke and fire gases. As the temperature decreases, fire gases contract and pressure is reduced, slowing the spread of fire by convection. This helps to confine the fire. The immediate reduction of interior temperatures also interrupts the countdown to flashover. All these changes in conditions, which happen simultaneously with the application of fast water, increase the survivability and rescue profile of trapped occupants while making the interior environment safer for firefighters.

Modern nozzle technology advancement has produced the HydroVent, a nozzle like no other. The HydroVent is a dual-attack, dual-nozzle suppression and ventilation system that accomplishes simultaneously all the tactical objectives mentioned above **FIGURE 10-14**. The nozzle is meant to be placed into the window of the fire room from the outside (outside the IDLH) by a single firefighter and, once set and charged, it can be left unattended **FIGURE 10-15**. The suppression part of the nozzle is a round nozzle with multiple smooth-bore discharge holes that flow approximately 95 gpm toward the top of the heated compartment. This creates a sprinkler system (the water converts to steam 1,700 times its volume) that knocks down the fire and reduces the interior temperatures

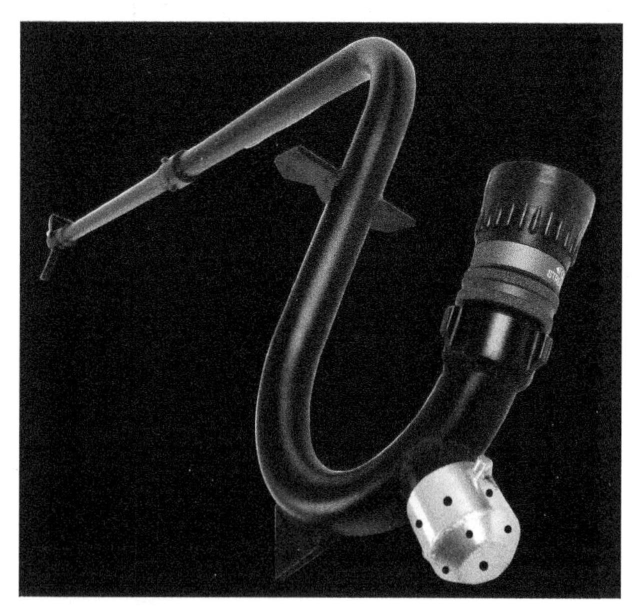

FIGURE 10-14 The HydroVent dual-attack nozzle.
Courtesy of Kevin O'Donnell, HydroVent.

FIGURE 10-15 The HydroVent nozzle is meant to be placed into the window of the fire room from outside the IDLH by a single firefighter. Once set and charged, it can be left unattended.
Courtesy of Kevin O'Donnell, HydroVent.

and thermal energy output, thus preventing flashover. This also slows the fire growth and fire extension. Operating simultaneously is the ventilation 95 gpm combination fog tip. By design, this part of the nozzle is already positioned at the windowsill aimed toward the outside for immediate hydraulic ventilation. It is capable of pulling thousands of cubic feet per minute (cfm) of smoke, steam, and other combustion gases out of the room and away from any fire victims **FIGURE 10-16**. Cooling the fire with water while simultaneously removing the heat through hydraulic ventilation drastically reduces interior temperatures and improves smoke conditions and visibility. The immediate results are that the interior environment is safer for firefighters and the survivability profile of possible occupants is improved. This is truly a transitional attack nozzle. There is currently no faster method to attack and ventilate a room fire.

This nozzle can also be placed in service quickly into the second-floor window of a two-story house faster than any hose team can stretch a line to the seat of the fire and before ladder crews can ventilate the structure **FIGURE 10-17**. After knocking down and resetting the fire, the initial fire attack team can reposition and make entry for a direct attack for final extinguishment, search and rescue, ventilation, a check for extension, salvage, and overhaul.

This new technology allows rural and volunteer fire departments with limited staffing to implement the recommendations made from the UL/NIST experiments. The nozzle significantly improves, if not eliminates, the hostile thermal environment to trapped occupants in residential fires by the actions of a single firefighter working outside the IDLH area, all within 30 seconds of opening the nozzle.

The defensive-to-offensive transitional attack can also be accomplished using master streams. This tactic is often used by the FDNY when a fire occurs on the upper floors of a tenement building involving one or more units. You can find several video examples of this type of attack on the Internet. In these examples, the fire has already self-vented and flames are visible from two or three windows. The fire flow requirement could be well over 600 gpm, more than two hand lines can deliver. Room is left in front of the building for the tower ladder to position. The tower ladder raises the bucket and, once supplied, opens up the smooth-bore master stream. This is the same technique as *softening the target*, only with master streams. After the fire is knocked down in one window, the bucket is moved to the next window, and the same procedure is followed. In the meantime, engine crews have laid lines up the stairwell and are in position for a direct attack. Once all the fire is knocked down, the tower ladder shuts off the master stream and the engine crews take over.

Transitional Attack—Offensive to Defensive

An offensive-to-defensive transitional attack is rarely used. This attack can be selected when all indications point to a defensive strategy except there is a known life hazard, or maybe a corner of the building can still be searched before it is time to abandon the building. An example would be a gas explosion and fire in an

FIGURE 10-16 By design, the fog nozzle is positioned at the windowsill aimed toward the outside for immediate hydraulic ventilation. It is capable of pulling thousands of cfm of smoke, steam, and other combustion gases out of the fire room and away from any trapped victims.

Courtesy of Kevin O'Donnell, HydroVent.

FIGURE 10-17 The HydroVent nozzle can be placed in service quickly into the second-floor window of a two-story house faster than any hose team can stretch a line to the seat of the fire and before ladder crew ventilate the structure.

Courtesy of Kevin O'Donnell, HydroVent.

occupied building. The majority of the building may be so damaged by the explosion that it isn't worth saving, and the survivability profile in the destruction zone is probably zero. The ensuing fire may be so large that it has to be fought as a defensive fire using master streams. Because the building was occupied before the explosion, there is a chance that there may be some survivors in the areas not yet affected by the fire; thus, the IC should make an aggressive interior direct attack—not to extinguish the fire but to hold back the encroaching flames while search and rescue is performed. Once savable lives are rescued, all crews back out of the building, and the strategy changes to defensive operations.

In another scenario, a warehouse that is fully involved may have an attached office area that still has savable property. The office area may contain business files, records, inventories, payroll, and computers that contain the entire administrative portion of the business. With the proper risk assessment, an offensive attack could be made into the office area to salvage all the records and computers and then crews would be pulled out to transition into a defensive attack. This single salvage action may allow the business owner to rebuild, even though the structure may end up being a total loss.

Defensive Attack

A defensive attack is used when the building no longer has any savable value, or when the gpm requirement to extinguish the fire is more than hand lines can supply. Keep in mind that an aggressive interior attack should not be made in buildings that are in various stages of demolition, are vacant and/or dilapidated, have been abandoned for long periods, have been burning for a predetermined period of time, contain construction features detrimental to safe interior operations, have had previous fires, or are under construction. These buildings no longer have any savable value, so the lives of firefighters should not be risked to preserve used building material. The IC should conduct a risk-benefit analysis before the decision is made to allow firefighters into the building for interior firefighting activities, including a primary search. The life safety possibility needs to be addressed, but keep in mind that occupants in vacant buildings are squatters and trespassing. In other words, they know they shouldn't be there. If a fire is accidentally started, usually from cooking or warming fires, these illegal occupants are quick to leave and are long gone before the fire department shows up—unless they are trapped. If any interior operations are to be made in these buildings, including the primary search, they should be carried out only after the fire has been knocked down from the outside to reduce the uncontrolled thermal threat. A careful check must be made on the condition, stability, and relative safety of the structure on the inside. The exception would be when there are confirmed or visible trapped occupants. Then every effort should be made to effect rescue.

The normal direct interior attack should be made on those buildings that are in use, especially those that are occupied at the time of the fire. Even in this case, if a large intense fire is encountered, it may be necessary to knock down or control the fire from the outside using solid or straight streams from 2½-inch lines or using master stream appliances before an interior attack can be made. Remember that occupants benefit most when the fire is extinguished. If the fire is so intense that an interior attack cannot control or extinguish the fire with a sufficient amount of appropriately sized attack lines, then a defensive attack should be conducted from the outside.

Choosing Attack Lines

One size of hose or one type of stream is certainly not the answer to every fire situation that confronts an engine company on arrival at the fireground. The choice of initial attack lines and nozzles depends on the strategy of the attack, whether it is a holding action; an exposure protection; a defensive operation; or an offensive operation starting with an indirect attack, then continuing with an interior attack on the main body of the fire. The choice is also determined by the ability to deliver the water: the rate of flow required to extinguish the size of the fire.

Factors that affect this choice include the size and location of the fire, how the attack lines are to be used against the fire, available equipment, personnel available for fire attack, and water supply. Based on these factors, the following decisions must be made:

- What size hose diameter is appropriate?
- What type of nozzle should be used?
- How many hose lines are needed?
- Where will hose lines be positioned?
- How many firefighters are ready and available to handle 2½-inch hose lines?
- In what type of operation, offensive or defensive, will the hose lines be used?
- Do I have the ability to flow the required gpm of water to overwhelm the fire and sustain the attack?
- Do I have a reliable water supply source?

Sizes of Attack Lines

Choosing the size or sizes of attack lines depends on the stage of the fire: the incipient stage, the growth stage, or the fully developed stage. Coupled with the square footage or area of the building that is on fire, each stage requires a different amount of water to extinguish it. For example, a free-burning fire in the fully developed stage requires a tremendous amount of water compared to a small fire in the incipient stage.

The number of gpm required to put out the fire is determined by using one of the two fire flow formulas—the Royer/Nelson Iowa State University fire flow formula:

$$\text{Length} \times \text{width} \times \text{height} = \text{volume} \div 100 = \text{gpm}$$

or the National Fire Academy (NFA) fire flow formula:

$$\text{Length} \times \text{width} \div 3 = \text{gpm, multiplied by the percentage of structural involvement, for example: 25\%, 50\%, 75\%, or 100\%.}$$

Fires in the incipient stage or growth stage can be extinguished easily with a single 1¾-inch hand line, which can flow up to 250 gpm with the proper nozzle, tip size, and pressure. However, if a fire is in the fully developed stage and requires a flow rate of 500 gpm, a single 1¾-inch line will be ineffective. This fire requires, at minimum, a 2½-inch line that can flow up to 300 gpm and the 1¾-inch line to get 550 gpm, or two 1¾-inch lines to get 600 gpm. This is a simple example, but the sizes of the hose lines are determined by their ability to deliver the required rate of flow to extinguish the fire.

Choosing the size of the hose also depends on the location of the fire, how it will be fought, and the number of personnel on scene to effectively manage and advance hose lines. Most fire departments use 1¾-inch and 2½-inch hose as attack lines for an interior attack. If a 2½-inch line is required to be stretched to upper floors, however, and sufficient personnel are unavailable to maneuver the charged line, multiple 1¾-inch lines will have to be laid to match the delivery of the required gpm. If a 2½-inch line is required for an exterior exposure, one firefighter can manage it by tying it off to a stationary object, or making an 18-foot loop in the hose.

The 1¾-inch hose line is easy to maneuver and to advance with limited personnel. After the attack line is in position and charged to the proper nozzle pressure, one firefighter generally can handle the nozzle. (It is dangerous to assign only one person to an attack line for interior firefighting, and this practice should not be allowed. The buddy system must be maintained with either another firefighter or the company officer.) The typical 1¾-inch attack line equipped with a combination spray nozzle can discharge 120 to 250 gpm with the proper volume and pressure.

The mobility of the 1¾-inch attack line has made it popular among firefighters. It can be advanced quickly and easily to the fire area and can be very effective against small fires and residential fires. There is a tendency to use it against larger commercial fires, where it is much less efficient or completely ineffective. It wastes water, time, and effort; risks avoidable injuries to firefighters; and could allow the building to burn down. This tendency to operate in a residential mind-set when combating commercial structure fires should be avoided.

The 1¾-inch attack lines, which are as mobile and as easy to handle as the 1½-inch lines with the proper pressure and volume, can discharge 120 to 260 gpm with either a combination spray nozzle or a smooth-bore nozzle, respectively. This permits the attack line to be used safely and efficiently on somewhat larger fires than the 1½-inch attack line. A 1½-inch hand line flows considerably less water; therefore, it is recommended that the use of a 1½-inch hose be eliminated for structural firefighting. The 1¾-inch hose line has become the standard initial attack line in the fire service. The additional number of gallons is extremely effective in extinguishing fires with modern fuels combined with lightweight construction. These fires burn faster and hotter than legacy fuels, and the increased heat release rate has to be combatted with more gallons per minute. The theory is simple: If you have bigger fires, you need bigger hoses.

Many fire department apparatus have one, two, and even three preconnected 200-foot bundles of 1¾-inch hose for fire attack, and 200-feet of 2½-inch hose for a blitz line. With larger homes being built, three-story townhouses, or properties set back from the street, some of these preconnected bundles have to be extended with an extra 100 feet of 1¾-inch hose, or extended with a 2½-inch supply attached with a wye.

Here again, firefighters should avoid the tendency to use a 1¾-inch hose against larger commercial fires where it may not be effective. Firefighters need to stop pulling one type of attack line off the engine for every fire situation simply because it has worked in the past. A classic case study that every firefighter must read is the report on the Sofa Super Store Fire that occurred on June 18, 2007, in Charleston, South Carolina. If a fire cannot be knocked down, controlled, or extinguished within 60 seconds with the hose lines that are being operated, then additional, larger hose lines are needed. The Royer/Nelson formula is based on the premise that the best flow rate of water application is one that results in controlling the fire within 60 seconds.

Because the 2½-inch attack line is heavier than the 1¾-inch attack line, it is not as easily manipulated. However, before the line is charged, it too can be readily advanced and maneuvered; practice in handling this hose line develops confidence and overcomes difficulties in using it. An adequate number of personnel can get the 2½-inch hose line into service more quickly.

Although more effort is required to get a 2½-inch attack line into position, the payoff, in terms of water delivery in gpm for fire control, is worth the effort. This is especially true if the fire has advanced beyond the penetrating capability of a 1¾-inch attack line. The 2½-inch attack line, when equipped with a spray nozzle operating at a nozzle pressure of 100 psi, or a solid-bore nozzle with a 1⅛-inch tip operating at a nozzle pressure of 50 psi, discharges at least 250 gpm when operated properly. The larger stream absorbs more heat and reaches farther, providing better fire control and extinguishment in less time.

Attack Lines and Nozzles

The shape of the stream and diameter of the hose should be selected to provide the greatest reach and penetration into the fire area. It is preferable that the hose line flow as much water as can be reasonably handled by the number of firefighters on the attack team. *A full-out fire attack actually will use less water because it will extinguish the fire faster.* Smooth-bore nozzles producing solid streams have greater reach and more water delivery capability than combination spray nozzles producing straight streams. The smooth-bore nozzle most often used on 2½-inch attack lines is equipped with the 1⅛-inch tip with a 265-gpm discharge rate at a nozzle pressure of 50 psi. A 2½-inch line with a 1³⁄₁₆-inch smooth bore tip will flow 300 gpm at 50 psi, and the 1¼-inch smooth-bore tip will flow 325 gpm at 50 psi. Even with their greater reach and water delivery, these solid-stream tips require only half the nozzle pressure of the 2½-inch combination spray nozzle; however, they are more difficult to handle than the spray nozzle because of the increased water flow. Practice with 2½-inch hose lines and nozzles increases operating efficiency and overcomes any reluctance to use them.

With 1¾-inch hose lines equipped with a smooth-bore nozzle, a ⅞-inch tip will flow 160 gpm at 50 psi, and a 15/16-inch tip will flow 185 gpm at 50 psi. When pushed to 60 psi, the 15/16-inch nozzle will flow 200 gpm but the nozzle reaction is tough to handle.

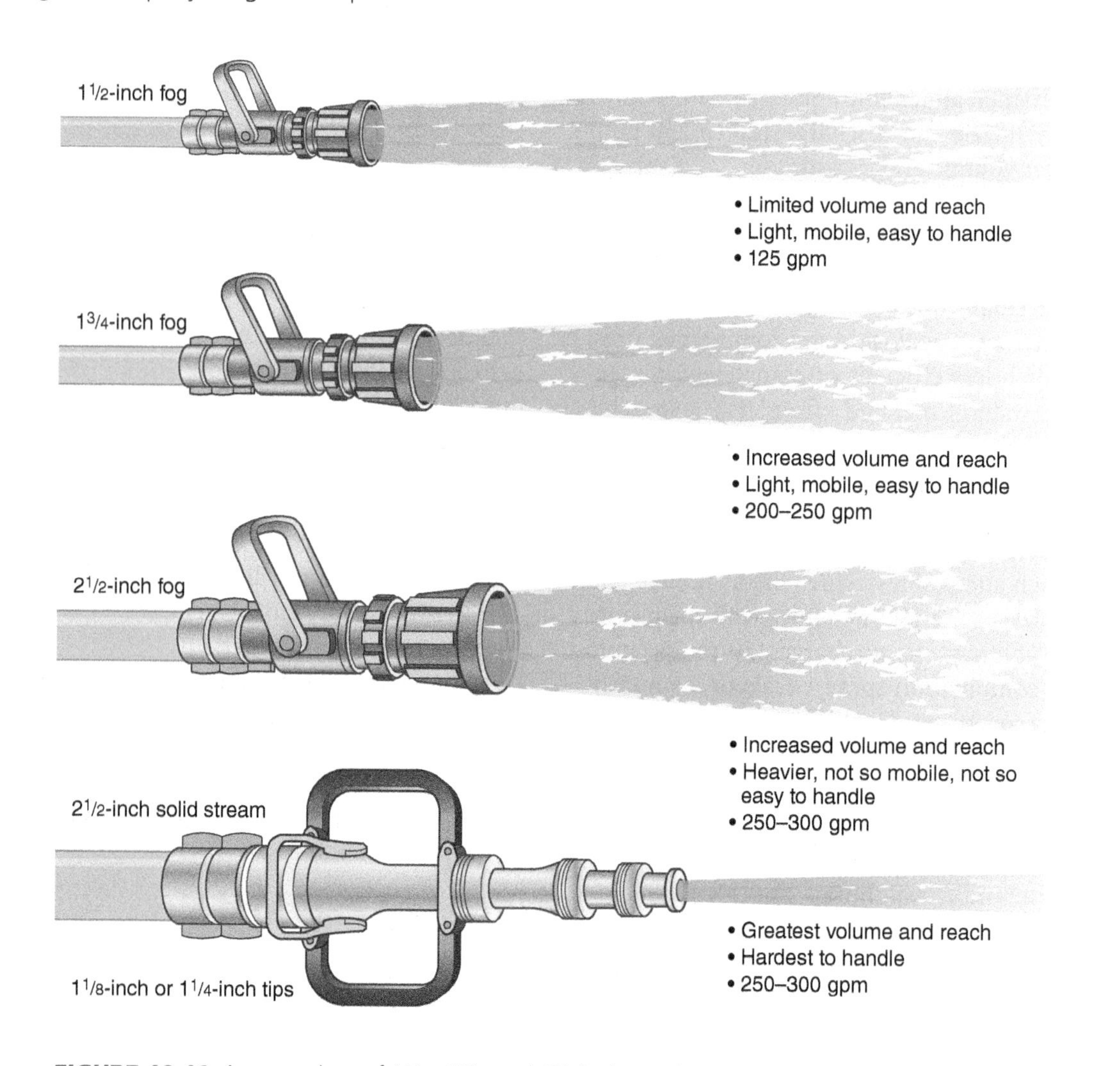

FIGURE 10-18 A comparison of 1½-, 1¾-, and 2½-inch nozzles and streams, showing the advantages and disadvantages of each.

FIGURE 10-18 shows a comparison of nozzles and streams.

Solid Streams Versus Straight Streams

For the safest and most effective operations for interior firefighting where civilians or firefighters will be in the area, smooth-bore nozzles producing a solid stream, or combination spray nozzles adjusted to the straight stream pattern should be used. Either option aids rescue and fire control because these stream patterns are less likely to disturb the thermal layer compared to a fog pattern. They also help with visibility and limit the water converting to steam, which could scald both firefighters and fire victims. However, don't worry about creating steam when conditions warrant the cooling of the atmosphere to prevent flashover.

Solid streams and straight streams move little air compared with a fog stream. The use of combination spray nozzles adjusted to a fog pattern inside a building provides the company officer with options for specific fire situations, but they must be used correctly: the results can be advantageous or disastrous. A wide-angle fog stream does not penetrate a fire. Instead, the entrained air from a wide fog pattern can push the smoke and fire. Without a horizontal ventilation exhaust hole on the opposite side of the fire attack, a fog stream can have the detrimental effect of pushing fire, heated gases, and smoke into uninvolved areas of the building as well as increasing atmospheric pressure within the compartment, thus contributing to the spread of fire. It can also push the fire toward trapped occupants or back on to the attack team **FIGURE 10-19**.

Conversely, the entrained air of a wide-angle fog stream can push smoke and fire away from trapped occupants or downed firefighters if the nozzle is

FIGURE 10-19 Without a horizontal ventilation exhaust hole on the opposite side of the fire attack, the entrained air from a wide-angle fog stream can push fire, heated gases, and smoke into uninvolved areas of the building as well as increase the atmospheric pressure within the compartment, thus contributing to the spread of fire.

positioned between the person needing rescue and the encroaching fire **FIGURE 10-20**. A fog nozzle is better suited to protect and support the rescue efforts than a smooth-bore straight stream, which can also cause injury to victims. The only way to get a fog or broken pattern from a smooth bore is to gate the nozzle down by partially closing the bale, which reduces the gpm flow—exactly the last thing you want when a rescue is in progress during a fire. A trick of the trade memory aid to remember which way to twist the adjustable tip of a combination nozzle is "Right for reach, Left for life."

Combination nozzle fog patterns are best suited for unoccupied confined spaces, such as unoccupied basements, attics, or storage areas. With heavy fire involvement in the attic, a scuttle can be removed, an attic stairway can be pulled down, or a hole can be made in the ceiling below with a pike pole to access the attic and attack the fire. The fog stream can be placed into the attic space to let the cooling and steam conversion do their work. This should be followed by venting the area and checking for spot fires.

When a horizontal ventilation exhaust hole has been made opposite to the attack from which the spray nozzle is being advanced, a fog stream can be very effective. In this case, however, it should be

FIGURE 10-20 A fog nozzle is better suited to protect and support rescue efforts than a smooth-bore straight stream.

operated at no more than a 30° angle. The 30° angle produces a combination of reach and fog pattern that entrains the air and pushes the fire, smoke, and gases out of the exhaust portal. The moisture pushed from the entrained air extinguishes fire along the way. A positive-pressure ventilation (PPV) fan set behind the attack team keeps the wind at their back and creates a controlled flow path that pushes the products of combustion away from the attack team and out the ventilation exhaust hole. Increasing the angle to produce a wider fog pattern gives firefighters more protection, but it decreases the reach of the stream, diminishes firefighting effectiveness, creates steam, and can unintentionally push the fire through the building if the exhaust hole isn't big enough to handle the volume of smoke being horizontally ventilated. If the entrained air from the wider fog pattern pushes the fire into uninvolved areas, narrow the fog pattern and turn off the PPV fan until control is regained. The dimensions of the exhaust portal will have to be increased. Many firefighters have found, however, that even the 30° pattern cannot control a hot interior fire and that the simple action of adjusting the angle to a straight stream can result in better reach and penetration.

A fog stream may be used during an indirect attack to absorb high levels of heat. Once this is accomplished, a direct attack should then be used. A fog stream should not be used unless there is adequate ventilation. If a fog stream is used incorrectly, it is possible that the fire and heated gases, smoke, and steam can outflank and endanger the firefighters operating the attack line.

Smooth-bore nozzles should be used where solid streams are necessary to control the fire. Remember that a straight stream from a spray nozzle is still a broken stream, a stream of water separated into coarsely divided droplets.

The controversy over solid streams versus fog streams has raged on for many years. It is not the intention here to resolve this argument, only to point out the differences between the two stream patterns. However, modern nozzle technology has produced two nozzles that should help settle the debate. The TFT *Flip-Tip* nozzle combines the smooth-bore and combination spray tip into a single nozzle **FIGURE 10-21**. The combination of the two provides a single nozzle that offers fixed, selectable, or automatic gallonage into the tip that can be adjusted from a straight-stream to a wide-angle fog pattern. It performs all the functions and features of any other combination nozzle because it is a genuine combination nozzle. If the fire situation warrants a harder-hitting, deeper-penetrating straight stream from a smooth-bore

FIGURE 10-21 The TFT *Flip-Tip* nozzle.

tip, the locking ring can release the combination tip so that it can be flipped down and locked, exposing the smooth-bore tip. There are nozzles for 1¾-inch and 2½-inch hose lines. Smooth-bore tips range from ⅞ to 5⁄16, 1⅛, 1¼, and 1⅜ inches. This allows for flows of 50 to 300 gpm.

The TFT *Vortex* is an appliance that modifies and enhances the use of a traditional smooth-bore nozzle and is installed between the tip and the shutoff bale. Six short vanes in the bore of the *Vortex* reduce turbulence inside the straight stream. A single twist of the stream shaper from "STREAM" to "VORTEX" causes the vanes to pivot proportionally and converts the hard-hitting straight stream to a uniformly dispersed 30° to 40° spray pattern without gating down the valve and reducing the gpm fire flow **FIGURE 10-22**. This makes the smooth-bore nozzle versatile for hydraulic ventilation, supporting rescue efforts, and protecting exposures without causing straight-stream damage. The inability to achieve effective fog patterns when needed has been part of the great objection against smooth-bore tips.

Unfortunately, many fire departments have abandoned the use of smooth-bore nozzles and 2½-inch hose for attack line operations. Some have opted for 2-inch hose, but eliminating 2½-inch attack lines is a big miscalculation. Solid streams provide reach, penetration, and lower nozzle pressure, and they should be considered for both offensive and defensive operations. Fire departments using only 1¾-inch hose for attack lines are at a disadvantage when confronted with a large fire because they will be unable to apply large quantities of water through a hose line. They need the ability to deploy 2½-inch attack lines.

On the 1¾-inch attack lines, the smooth-bore 15⁄16-inch tip (180 gpm at 50 psi nozzle pressure) or the 1-inch tip (flowing 200 gpm at 50 pounds per square inch [psi] nozzle pressure) is now used with much

FIGURE 10-22 The TFT *Vortex* stream shaper appliance.

success over comparable combination spray nozzles, which operate at a higher nozzle pressure.

Combination Spray Nozzles

Most combination spray nozzles are designed to deliver their rated flow capacity at 100 psi nozzle pressure. "Low-pressure" combination spray nozzles are also available that operate at 75 psi nozzle pressure. Four types of combination spray nozzles are commonly available and in use within the fire service: the basic combination spray nozzle, constant gallonage spray nozzle, constant pressure (automatic) spray nozzle, and the constant/select gallonage spray nozzle.

The basic combination spray nozzle is an adjustable-pattern spray nozzle in which the rated discharge is delivered at a designed nozzle pressure and nozzle setting. The constant gallonage spray nozzle is an adjustable-pattern spray nozzle that delivers a constant discharge rate throughout the range of patterns, from a straight stream to a wide spray at a designed nozzle pressure. The constant pressure (automatic) spray nozzle is an adjustable-pattern spray nozzle in which the pressure remains relatively constant through a range of discharge rates. This nozzle is designed to maintain 100 psi at the tip regardless of the flow. For instance, if 200 gpm is flowing at 100 psi and the flow is reduced to 125 gpm, the nozzle adjusts to produce 100 psi nozzle pressure and maintain a "good-looking stream." Firefighters must realize that this stream still might be ineffective on the fire. The gpm flow controls a fire, not the nozzle pressure; therefore, if 200 gpm is required to control the fire, the 125 gpm "good-looking stream" will not do the job. With the basic combination spray nozzle and the constant gallonage combination spray nozzle, a weak stream is readily apparent. The constant/select gallonage spray nozzle is a constant-discharge-rate spray nozzle with a feature that allows manual adjustment of the orifice to obtain a predetermined discharge while the nozzle is flowing.

Effective Stream Operations During Initial Attack

Guidelines and the sequence for safe and efficient stream operations during the initial attack are listed here:

- Conduct a risk-benefit analysis using VTS before entering a burning building.
- Before entering the building, have an attack team plan for tactical objectives: direct attack using a straight stream; confining the fire and cutting off extension; deciding in which direction the smoke, combustible gases, and steam will be pushed; checking the overhead along the way; and deciding who will remove the victim if one is discovered.
- A full-out fire attack actually uses less water because it puts the fire out faster. A fire attack is not simply flowing gpm of water into a fire; it is the effective application of water on the fire that extinguishes the flames.
- Flake out excess hose using the method that best allows for the efficient advancement of a charged hose line.
- Never enter the fire compartment with an uncharged hose line.
- Crack the nozzle to bleed the air out of the hose line ahead of the water and check the stream pattern before making entry.
- Before the door to a fire area is opened, all firefighters at that location should be positioned on the same side of the hose and the same side of the door while keeping low.
- Spray a short burst of water on the surface of the entry door. Water quickly steaming off indicates high heat on the other side of the door.
- Stay low when entering the fire area to let the heat and gases vent out the door before moving in.
- Use a TIC to scan for victims in the immediate area of entry/egress.
- Watch the smoke and see how the flow path develops. Smoke filling the entire doorway and exhausting to the outside is an indication of a chimney flow path and the fire is below the ground floor—most likely a lower-level or

basement fire. If a neutral plane develops, with smoke venting out the top half of the doorway and fresh air entering the bottom half of the doorway creating a clear tunnel, the fire is on the same level as the attack team. If the smoke clears and stops venting out the door altogether, with fresh air rushing in, the fire is on the floor above the attack team.

- Use a direct attack to deliver water directly onto the fire.
- Adjust 1¾-inch and 2½-inch combination spray nozzles attached to preconnected attack lines to straight stream for attacking the fire.
- The use of smooth-bore nozzles on 1¾- and 2½-inch preconnected attack lines is extremely effective in controlling larger fires.
- If fire shows at the top of the door as it is opened, the ceiling should be hit with a solid stream to cool it and control fire gases. Short bursts of 3 to 4 seconds directly into the overhead should darken the flames and cool the combustible gases.
- The officer must use a pike pole and a TIC to check above the ceiling to ensure that there is no fire overhead. Firefighters should not advance if fire is overhead. The fire will travel behind them and outflank the attack team. Fire in the trusses can collapse on top of the attack team.
- Unless firefighter safety is a factor, do not open the nozzle until fire can be seen. However, hot temperatures and smoke banking down toward the floor indicate that the environment may be reaching flashover. Water has to be directed into the smoke to cool the temperatures of the fire gases below their ignition point.
- If possible, take care to prevent the thermal imbalance of the atmosphere. Unfortunately, water directed at smoke also makes visibility and smoke conditions worse and creates unwanted steam, which may make the area untenable for the crew; a fog stream greatly increases this problem. Remember that firefighters are much more tolerant of dry heat than they are of wet heat, but no one can tolerate high heat—or flashover.
- As the advance is made, the angle of the stream should be lowered and an attempt made to attack the main body of fire. The faster the line advances to the seat of the fire, the faster the fire can be extinguished and the less time the attack crew has to endure the beating and exposure from heat and other products of combustion.
- Direct the straight stream at the base of the fire, which lowers the temperature of the fuel and extinguishes the fire. Once the heat source and energy release are taken away, flame spreading up the walls and across the ceiling will decrease or disappear altogether.
- If the area is heavily involved, direct the solid stream forward and upward at the ceiling; apply it in a side-to-side motion; or rotate the stream using a Z-, T-, or O-pattern so all areas of the room are covered. Avoid quick, whipping movements.
- Remember that the majority of the fire load is in the lower half of the room.
- A horizontal sweep across the room with a straight stream, midway between the floor and the ceiling will likely take out any windows for immediate ventilation of the fire room. Use this technique early when attacking the fire in the room to improve conditions. Once the fire is out, continue to ventilate the room hydraulically using the hose stream.
- Sweep the floor with the stream to cool burning debris and hot surfaces, including the carpet if there is one. The synthetic fibers of carpet add to the modern fuel load. Cooling the carpet helps to prevent burns to hands, knees, and to the attack line as it is advanced.
- If the fire isn't knocked down or controlled within 60 seconds, the current line may be too small. A larger hose line that can deliver more gpm of water is needed with additional backup lines to augment the attack.
- In fully involved rooms where the survivability profile is zero, a narrow fog pattern can be used. Shut down the line and close the door. Allow the steam to do its job.
- In long hallways that have significant fire, a straight stream or narrow fog can be used to create steam while the crew backs down the stairs, well below the floor level, for protection.
- When a horizontal ventilation hole has been created on the fire side of the attack (or on the opposite side of the attack), a fog stream can be used to entrain air and push the smoke and fire gases out toward the ventilation exhaust hole.
- Fog streams are necessary to push back flames and smoke from hallways where rooms are being searched or where rescues are in progress.
- Fog streams must be used to entrain air and push away smoke and flames to protect

stairways and landings in order to maintain an egress route for attack teams and search and rescue teams operating on the fire floor.

- Fog streams may be needed to protect trapped or incapacitated civilians and firefighters from encroaching smoke and flames until they can be rescued.
- Protect-in-place tactics are best carried out using fog streams.
- When the main body of the fire is knocked down, shut down the stream, and let the area "blow" (allow smoke and gases to rise and vent). Listen for crackling sounds, and look for areas that "light up" so that extinguishment can be completed. Use a TIC.
- When the fire is knocked down, shut down the attack line. This helps control water damage and the weight of the water on the floor.
- Steam may have saturated the atmosphere, making the smoke and gases less buoyant. Effective ventilation requires the use of PPV fans as well as hydraulic ventilation from the hose line.
- Immediately pull the ceilings and extinguish any overhead fires using a fog pattern.
- If no headway has been gained within 20 minutes, or the fire has been out of control for 20 minutes, withdraw from the fire building. Collapse is inevitable. Switch to a defensive strategy.

After entering an area that is very hot and finding no fire, check the area below; adjacent areas, walls, and ceilings; and other vertical and horizontal voids for fire extension. Use a TIC. If the fire cannot be located, but heat and smoke conditions are getting worse, the fire is below you. Suspect a basement fire. Firefighters must quickly move out of this area if the fire can't be found. The IC should be notified of the conditions within this area. From an adjacent location, an attack line could protect the building or means of egress from the building while firefighters withdraw.

Sizes and Number of Hose Lines

The four sizes of hose lines allow a variety of tactics in fire attack. Speed is important, and the 1½- and 1¾-inch attack lines are very mobile and allow an attack team to advance quickly to the seat of the fire. With normal pressures, the 1½-inch line can flow up to 125 gpm of water, and the 1¾-inch can flow around 200 gpm. They have excellent reach capability and water delivery capacity for incipient fires, growth stage fires, and many fully developed fires in residential structures. The 2-inch hose line is becoming more popular and is a good in-between line that delivers more water than a 1¾-inch line but isn't as difficult to handle as a 2½-inch line. The 2-inch hose can easily flow 220 gpm or more with less friction loss than a 1¾-inch. However, both the 1¾-inch and the 2-inch hose are still connected by 1½-inch couplings; thus, there is a bit of a restriction because the water needs to flow through a reducer. As noted previously, a fire department using 1½-inch hose should consider switching to 1¾-inch hose as a replacement for structural firefighting. The water-carrying capacity can be doubled using a 1¾-inch hose with little difference in weight and ease of operation.

The 2½-inch attack line is the largest and heaviest hand line we use. It is the workhorse of the fire service because it can be used as a hand line, a master stream, and a supply line. It is not easy to move or handle when it is charged, but it delivers greater water capacity with power and reach. Two-and-a-half-inch hose testing has shown flows up to 470 gpm using a 1½-inch smooth-bore tip on the nozzle.

The choice of which attack lines to use and how they are operated depends on the fire situation, including the size and location of the fire. It is still based primarily on the required gpm to overwhelm the fire, which is determined by applying one of the two fire flow formulas: Royer/Nelson or NFA. Any fire requiring 300 gpm or more of water is a 2½-inch operation until the fire is knocked down to a smaller size or completely extinguished. All defensive operations that utilize hand lines should be 2½-inch.

An incipient stage fire can easily be extinguished with a single, preconnected 1¾-inch hose line. Most residential house fires are usually controlled and extinguished with the first line. The placement of the first hose line is critical. After the fire is knocked down from the outside (if fire is showing), the first line should be placed between the occupants and the fire so search and rescue can be accomplished. This line also protects the means of egress. Use the reach of the hose stream to attack the seat of the fire. A second attack line may be needed to attack the main body of the fire, while exposure lines are stretched to areas above and below the fire as well as into adjoining areas of the building. Hose lines may also be needed to protect exposure buildings.

At any structure fire, at least five 1¾-inch hose lines are laid:

- One fire attack line
- One backup line

- One interior exposure line
- One exterior exposure line
- One rapid intervention team line

The above list is the *minimum* to consider. Some fires may require additional attack lines and interior exposure lines, while other fires may not need an exterior exposure line. It all depends on what's burning, where is it burning, where the fire is headed, and how many gpm of water are needed to overwhelm the fire.

If 1¾-inch lines can accomplish these tasks, they should be used **FIGURE 10-23**. If any doubt exists about the effectiveness of these smaller attack lines or if significant progress and knockdown doesn't occur within 30 to 60 seconds, the hose diameters are too small. A 2½-inch attack line should be placed in service immediately. Often a combination of the two different attack lines works well. For example, a 1¾-inch attack line can be used for the initial attack on the main body of the fire. This attack line can be backed up by a 2½-inch hose line. A 1¾-inch attack line can be taken to the floor above the fire to check for extension and protect the interior exposures **FIGURE 10-24**. Backup lines should be at least the same size as the attack line or larger. Two-and-a-half-inch attack lines should be backed up with another 2½-inch line.

In a building with several floors, 1¾-inch hose lines are ideal for covering interior exposures and getting above the fire to stop vertical spread from floor to floor by way of the windows, whereas 1¾-inch or 2½-inch hose lines can be used to attack the main body of fire.

If there is a large fire in an open area inside a department store or supermarket and an offensive operation is under way, the main body of the fire can be attacked with 2½-inch attack lines; 1¾-inch attack lines may be useful in keeping the fire from spreading horizontally through doors or other openings into unaffected areas. These smaller attack lines should be backed up by additional 2½-inch hose lines.

Master Streams for Initial Fire Attack

Occasionally, a fire is so large that master stream appliances must be used for initial attack. This decision may be based on experience, but it is still determined by using one of the two fire flow formulas to determine the required gpm of water. There would be little sense in trying to control such a fire with handheld 1¾-inch attack lines.

Modern nozzle technology has produced a lightweight, low-profile, compact, easy-to-deploy portable monitor. The TFT *Blitzfire* nozzle is a 2½-inch

FIGURE 10-23 On small fires and on fires in the incipient and growth stages, 1¾-inch attack lines are effective and efficient.

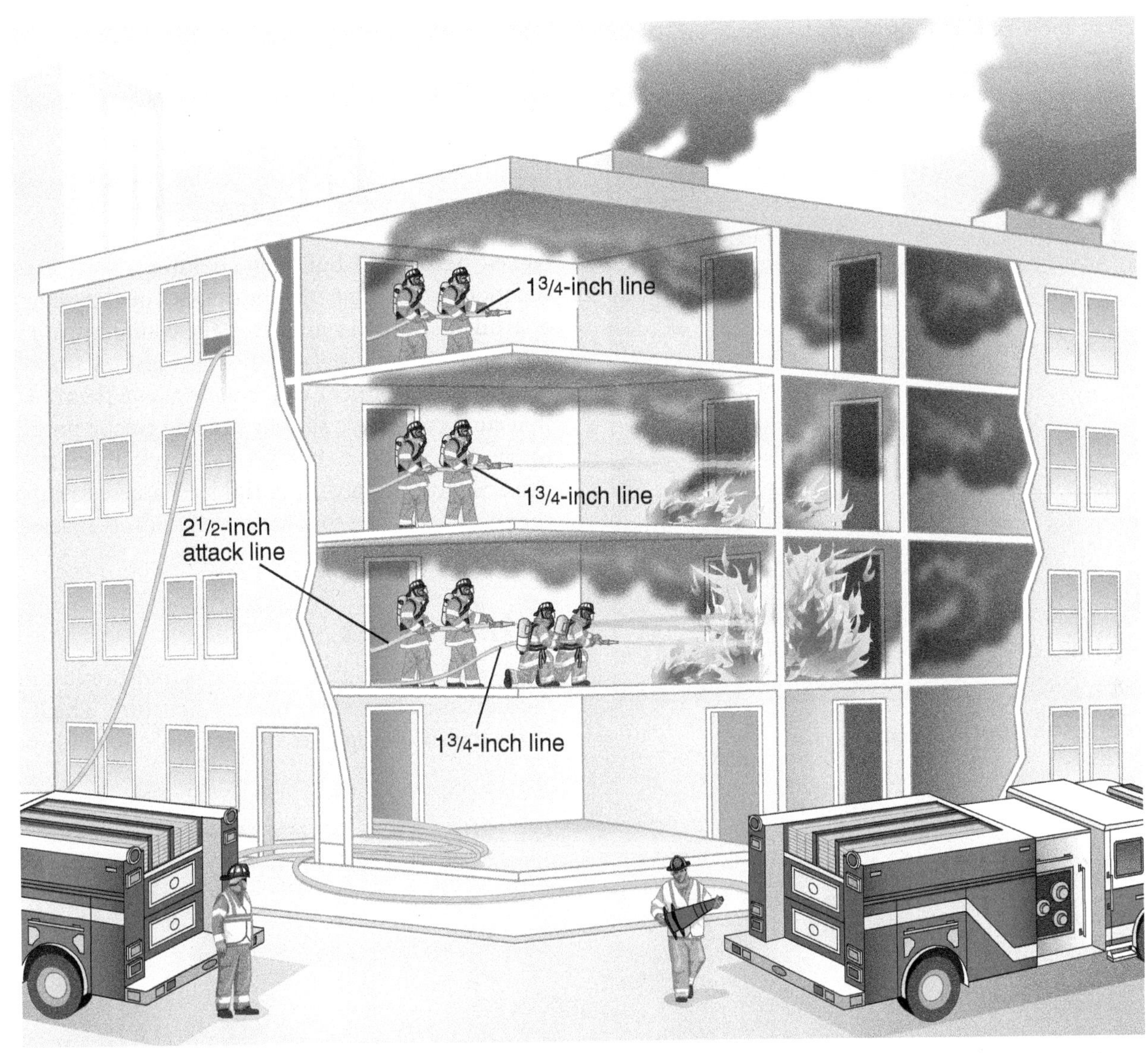

FIGURE 10-24 A combination of 1¾- and 2½-inch attack lines is usually effective and efficient in fighting a larger fire.

single-operator portable master stream that is ideal for initial interior attack. The monitor nozzle, which deploys like a 2½-inch attack line, can flow up to 500 gpm (2,000 l/min). It can be equipped with an automatic combination tip or with a smooth-bore tip. It is designed for a 10° to 46° low-angle interior attack and is capable of directing a fire stream into any door or window. It can also sweep 20° to the right and 20° to the left. The high-elevation model is capable of aiming the stream up to 86° to place a high-volume stream directly into the overhead. Firefighters often shy away from pulling the 2½-inch line because it is difficult to hold on to and operate for an extended period—usually because the nozzles are over-pressurized. The anchor spikes of the monitor absorb the 250-pound nozzle reaction while flowing 500 gpm. There is little, if any, nozzle reaction for the firefighter to brace against, making this portable monitor an excellent tool for offensive interior attacks that require 2½-inch hand lines, like high-rise and commercial fires. One firefighter can carry and place this nozzle in service, and once it is set, the monitor can be left unattended. Many fire departments have this portable monitor preconnected to a 2½-inch hose load for immediate deployment for a direct attack or transitional blitz attack **FIGURE 10-25**.

If an exterior attack is warranted, master streams should be placed in service as soon as possible. During an exterior defensive attack, interior offensive operations must be curtailed, and firefighters should not be allowed to enter the structure.

Master streams are the safest and most effective option for a defensive attack. A master stream appliance mounted on a pumper or portable monitor operated from the ground should be placed in service. This appliance can be operated initially with tank water as soon as it is in position to knock down the fire, but it requires careful aim before the deck gun is opened because the tank runs dry within 60 seconds. A continuous water supply should then be established as soon as possible. Master stream appliances can also be operated from aerial fire apparatus **FIGURE 10-26** and **FIGURE 10-27**.

FIGURE 10-25 The TFT *Blitzfire* nozzle.

Initial Attack Operations

Mixed occupancy buildings, buildings with a combination of commercial businesses and residential apartments, exist in almost every community. These structures usually have stores on the first floor and one or more floors of apartments above them. A fire in a store can create serious fire and rescue problems in the floors above. If a sprinkler system exists, the driver should supply the sprinkler Siamese to support the fire attack. The following are typical scenarios involving mixed occupancy structures.

Attack Lines

Let's say that firefighters arrive to find fire in one of the stores of a mixed occupancy structure. The IC realizes that it is imperative to control the main body

FIGURE 10-26 When a large fire is encountered, a pre-plumbed master stream appliance can be placed in service quickly to attack and knock down the fire.

of fire as quickly as possible. If entry can be made and the store is small, 1¾-inch attack lines might be adequate for controlling the fire. If the store is

FIGURE 10-27 In a transitional attack, master stream appliances can be operated from aerial fire apparatus for the initial fire attack to knock the fire down.

Couresy of Scott M. Peterson.

well-involved (perhaps fire is extending out the front of the store), 2½-inch attack lines should be used on the store fire, and 1¾-inch attack lines should be advanced immediately above the fire to protect occupants of the apartments, assist in rescue, and cut off any vertical fire spread. With 50% involvement of a commercial space, use 2½-inch hose lines as your interior attack line.

Attacking the well-involved store fire with 1¾-inch hose lines would be a waste of time and effort in a situation in which both are at a premium. The heat and force of the extending fire would turn away smaller streams, allowing little penetration into the fire and resulting in insufficient fire control **FIGURE 10-28**. The 1¾-inch attack lines should be shut down, and a sufficient number of 2½-inch attack lines should be placed in operation on the main body of the fire.

The 2½-inch solid stream will penetrate into the heated area. With its greater reach and volume, it has a better chance of controlling the fire. A 2½-inch preconnected attack line should not take much more time

A B

FIGURE 10-28 When a 1¾-inch attack line cannot penetrate a large, hot fire **A**, a 2½-inch attack line should be brought in to gain control **B**.

to be placed into operation than a smaller attack line if a sufficient number of firefighters are present. The point to remember is that the main body of the fire has to be knocked down and extinguished to protect the residential apartment units above (Figure 10-28).

Fires in the lower floors of an occupied structure must be knocked down as quickly as possible. Engine crews assigned to controlling the main body of the fire should not be assigned to other areas of the building until this task is accomplished. Occupants of the building benefit as much from the control of the main body of the fire as from any other search and rescue actions.

In modern garden apartments, fires are often encountered in basement storage rooms. These are usually large open areas divided into small storage cubicles for individual tenants. The same problem exists here as in the storefront apartment house: a fire under the occupied apartments above. Conditions on upper floors will deteriorate if the fire isn't knocked down and controlled.

If the fire in the storage area has gained considerable headway, it should be attacked with 1¾- or 2½-inch hose lines and a 2½-inch backup line. Access should be made from an exterior entry door, usually found in the rear of the building. Firefighters should be assigned to the floors above to conduct the primary search and perform rescue operations if needed. The 1¾-inch attack lines should be advanced to these areas to stop any vertical fire spread and to protect searching firefighters and any possible victims. The floor directly above the basement fire is a collapse hazard area and should be avoided. Find other areas of access to search the first floor and the floors above. A first-floor exposure line should be positioned in a safe area, and the reach of the hose stream should be used to cover the interior door to the basement.

Because it may be difficult to ventilate a basement thoroughly, solid or straight streams from 1¾- or 2½-inch attack lines should be used. This is especially true in basements built like concrete vaults, with concrete walls, floors, and ceilings and thus with no means to vent quickly to the outside. In such cases, the safest and most efficient attack is with 2½-inch smooth-bore nozzles from an outside doorway, bulkhead, or window opening. This stream allows greater reach into the area, provides heavy water flow into the fire, and keeps the firefighters from being subjected to high concentrations of heated gases and steam while accomplishing fire control.

Knockdown should be accomplished before an interior attack is made. Firefighters must protect the opening and the stairway to the first floor. The basement may be difficult to enter unless ventilation is accomplished. Firefighters must be well supervised and be aware of the danger that this situation presents. Fires in similarly constructed areas other than basements should be fought in the same way.

The general attack procedures given in these two examples above can be used in almost any kind of structure and with the fire in any location. For instance, where fire has gained control of a good portion of a large floor area, 2½-inch attack lines should be used on that floor. The 1¾-inch attack lines should be used on the floors above, around, and under the fire to prevent fire extension and, if necessary, to assist in rescue and evacuation. This usually results in the most efficient firefighting operation accomplished with the fewest personnel.

The important point is to use the proper sizes of hose and types of nozzles to confine and extinguish the fire at all points within the building. If, after 30 seconds to a minute, it becomes obvious that a hose stream is not accomplishing its objective, get a bigger one. If more are hose lines are needed, get them.

Advancing the Hose Line

There are several methods to advance a charged hose line into a building. The placement of firefighters on the hose line depends on the available resources of the responding agencies. Large fire departments may have a crew of four or more firefighters to advance a hose line, while other volunteer departments are lucky if they can get two firefighters to a line. Experiment and practice with the various techniques using 1¾- and 2½-inch hose. The best technique is the one that works best for you. There are 10 things to remember with any hose-advancing technique within a burning structure:

1. Flake out as much excess uncharged hose as safety allows to get the nozzle into position. This limits the amount of weight firefighters have to wrestle with once the line is charged.
2. Limit the amount of bends to ensure easy advancement with the least amount of friction.
3. Use gravity to your advantage with hose in stairways.
4. Be aware of door control so that doors don't close against the hose.
5. Ensure that the uncharged hose doesn't pass under the door threshold. The door can act as a hose clamp, stopping the flow of water to the nozzle when the line is charged and thus endangering the attack team. The charged line can also wedge the door closed, preventing firefighters from escaping from the hallway.

FIGURE 10-29 Coiling the hose is the best technique to use in stairway landings where space is limited, and it's the best method with limited firefighter staffing.
Photo by Raul Angulo.

6. Never take an uncharged line inside the fire compartment.
7. Sharp turns and bends may require a firefighter to be posted at these choke points to assist in advancing the hose.
8. Coiling the hose is the best technique to use in stairway landings **FIGURE 10-29**.
9. Coiling the hose is the best method with limited staffing.
10. Be aware of hose layering. The attack line is the first line laid. It can get buried under the weight of backup lines and exposure lines and then stall the fire attack.

Basement Fires

This section is a comprehensive study on the strategy and tactics of basement fires. Often, this subject is glossed over but many line of duty deaths and injuries to firefighters have resulted from the failure to identify the presence of a basement, or from firefighters falling through the floor into a burning basement. The new UL/NIST studies include numerous experiments on basement fires. New construction plays a big factor here. Many spec homes around the country leave the garages and basement ceilings unfinished. A fire that originates in the basement can quickly involve the floor system above it, leading to early collapse of the first floor. This situation has been deadly for unsuspecting firefighters making entry on floor 1. It has happened enough that we need to reexamine how we attack basement fires.

Basement fires are one of the most difficult and dangerous types of incidents that firefighters encounter—but they don't have to be. They can be one of the safest if we choose to implement the new tactics of transitional and indirect attack to combat them. These so-called new tactics are not even new. It is more accurate to say that we should be willing to return to how basement fires were attacked in the past.

Basement fires are challenging and dangerous because there may be only one or two entry points into a basement area from either the inside or the outside of the building. In addition to limited access and limited egress, the following factors also make basement fires dangerous:

- Narrow stairs
- Poorly maintained stairs
- Stairs with chutes and conveyor rollers
- Maze or labyrinth floor plan
- Limited number of or no windows
- Small or narrow windows
- The nature of stock, product, and other items stored in basements
- Low ceilings
- Rapid heat and smoke buildup
- Trapped heat and smoke
- High concentrations of carbon monoxide

This list, plus the limited access and egress, elicits a natural response of heightened awareness about location because none of them are normal for long-term or permanent human occupancy. Unless designed as a lower-level or below-grade habitation, bedrooms and living spaces that were converted from basements still have a cavernous feel to them due to the lack of windows, and people rarely have peace of mind regarding their level of safety. When something seems wrong and occupants in a residential basement or employees in a commercial basement feel threatened, or if smoke and fire starts to develop, their first reaction is to get out quick!

If a basement fire does exist in a bona fide living area, the IC must consider the survivability profile during the size-up and risk-benefit analysis in relation to the severity of the fire. A residential basement fire doesn't automatically warrant an interior attack. One of the realities facing the fire service today is the illegal marijuana grow labs that are often discovered in the basements of residential-type homes because of a fire. Some operations convert every room of the house, including the garage and the basement, into growing rooms. One San Francisco, California, firefighter became disoriented and then lost fighting a basement fire due to the labyrinth setup the growers created. He called a Mayday and luckily, he was found by the rest of the crew. However, two Philadelphia, Pennsylvania, firefighters were killed in 2004 fighting a basement fire that contained an illegal marijuana grow operation when they became entangled in overhead electrical wires that fell from the low ceiling. In a news report covering the major Bay Area cities in California, there were 25 structure fires between 2013 and 2016 involving basements and illegal marijuana grow operations that caused many departments to examine their approach to basement fire strategy and tactics. Reports of basement fires involving marijuana grow operations occurring in Seattle, Washington; Urbana, Illinois; Graham, Washington; and Bronx, New York, indicate that the problem is nationwide. A search on the Internet will reveal numerous examples of this now common scenario.

The UL/NIST research findings on the chemical composition of smoke and other toxic gases released from modern-fuel fires need to be included in our risk-benefit analysis when determining the survivability profile of possible occupants in a burning basement. These facts give us the moral basis to decide not to send firefighters down into a burning basement before the fire is knocked down or extinguished by other than an interior direct attack. We need to utilize these scientific facts and say, "There's no one alive in there," instead of blindly sending firefighters inside a burning basement to their death, only to confirm that there was no one alive in there.

Fires in unfinished basements pose another dangerous hazard. Unfinished basements allow the fire to attack the floor above because the floor joists are often constructed with truss-joist I-beams (TJI or TGI), lightweight prefabricated wooden I-beams. In addition, the lightweight galvanized steel hangers and gusset plates that support the floor assembly are exposed to the heat of the fire. These assemblies can collapse in as little as 2 to 5 minutes of direct flame contact. Some texts and manuals advise firefighters to be aware of such conditions, check the floor for a spongy feeling, and proceed cautiously with the interior attack. *Spongy* is a relative term; nevertheless, this term describes a condition that the floor has weakened and is in danger of collapse.

The Bricelyn Street Fire

The Bricelyn Street Fire is referred to throughout this section because it is a perfect example of a worst-case scenario regarding the dangers of basement fires and why we should no longer consider the interior stairs as our first choice of access for attack.

The Homewood neighborhood of Pittsburgh is a densely populated residential area with narrow streets of one- and two-family, closely spaced, wood-frame dwellings that are at least 100 years old. From the A side front at street level, the house at 8361 Bricelyn Street appeared to be two stories, but many of the houses on the street are built on sloping elevation and have additional floor levels that cannot be seen from the street. From the C side rear of 8361 Bricelyn Street, the house was equivalent to four stories, including the basement.

On February 14, 1995, three Pittsburgh, Pennsylvania, firefighters died at 8361 Bricelyn Street when they ran out of air and were unable to escape from the interior of a lower level, below-grade house fire. The three victims were all assigned to Engine Company 17 (E17) and had advanced the first hose line into the house to attack what was later ruled an arson fire in the basement. When the three firefighters were found, all three were together in one room and had exhausted their air supplies. Three other firefighters had been rescued from the same room, which caused confusion over the status of the initial attack team.

This tragic incident serves as another example where the identification of a basement or below-grade level was missed due to an improper or total lack of a 360-degree size-up—or at least looking at the C-side rear of the building—a critical error that has occurred in several similar firefighter fatality incidents involving basements. When one of the Pittsburgh crews discovered that the fire was in the basement, it was not immediately reported via portable radio to the IC or the other units on the fireground.

To honor the deaths of these three brave firefighters, Captain Thomas A. Brooks, FF. Patricia A. Conroy, and FF. Marc R. Kolenda, we must learn the lessons from their ultimate sacrifice.

Identifying the Presence of a Basement

It is paramount that the first-in engine officer determines immediately whether or not there is a basement fire and announces this game-changing information over the radio to incoming units. Basements are the most dangerous part of any structure due to limited and reduced access. It is a good idea to announce on the radio the existence of a basement or below-grade levels, whether they're involved in fire or not, for example, "All units, be advised this structure has a basement" or "All units, be advised this is a basement fire."

Basement fires can present like fires on the ground level or in upper levels. A basement fire in a house with balloon-frame construction allows the fire to spread up the vertical channels of the walls that run from the basement all the way up into the attic unimpeded by any fire stops or horizontal barriers. This was the case at Bricelyn Street. When E17 made entry, the officer reported "smoke from the first floor." Heavy smoke was already showing from the third floor and from the attic. Truck 17 used ground ladders to vent the windows on the second floor above the porch and then went to the roof for vertical ventilation.

You should always suspect and expect that vintage or historical wood-frame homes that are 100 years old or older have balloon-frame construction. Balloon-frame construction started in 1833 and remained a popular construction method for multistory wood-frame buildings until the late 1950s. Thus, hundreds of thousands, if not millions, of these structures still exist throughout the United States and Canada, and will remain so for many years to come.

Basement fires are not always obvious, and smoke conditions along with other signs can lead the company officer to conclude that the fire is on an upper level. Thus, the initial IC can lose focus on the fact that a basement exists. As a reminder to check for the presence of a basement, add a B to the strategic and tactical objectives of RECEO/VS so it becomes BRECEO/VS. RECEO/VS is so well-engrained into the fire service, that adding the B for *basement* or *below-grade* at the beginning of the acronym is easy to remember and easy to say: linguistically it fits. In addition, if there is a basement fire, it puts it at the front of the list for initial fireground actions. Company officers should be thinking BRECEO/VS as they approach any structure to perform a 360-degree size-up.

Smoke Behavior When Making Entry

When you first open the door to make entry into a structure, watch and see what the smoke does. If the smoke thins briefly when the door is opened (because the backup pressure is being released), a flow path begins to develop. If smoke exits through the top half of the door and a tunnel of clean air enters through the bottom half of the door below the neutral plane, then the fire is probably on the same level as the attack team. This phenomenon is a bidirectional flow path and is sometimes referred to as "smoke that has found balance" **FIGURE 10-30**.

If the smoke rises and the door opening clears out and then begins to pull fresh air through the entire doorway back into the building, then this is an intake portal of a unidirectional flow path (high-low flow path). An intake of a unidirectional flow path indicates that the fire is probably on the floor above you or on an upper level of the structure **FIGURE 10-31**.

If steady smoke begins to flow out through the *entire* doorway, from the foot plate to the top header,

FIGURE 10-30 If smoke exits through the top half of the door and a tunnel of clean air enters through the bottom half of the door below the neutral plane, then the fire is probably on the same level as the attack team. This phenomenon is a bidirectional flow path and is sometimes referred to as "smoke that has found balance."

Courtesy of John Odegard.

FIGURE 10-31 Note that no smoke is exiting from the front door. This is a low intake portal of a unidirectional flow path (high-low flow path), which indicates that the fire is on the floor above.
Photo by Raul Angulo.

FIGURE 10-32 If steady smoke begins to flow out through the *entire* doorway, from the foot plate to the top header, this is an exhaust portal for a unidirectional flow path. This means the fire is below you.

this is an exhaust portal for a unidirectional flow path. This means the fire is below you or on another level below the opening. This type of unidirectional flow path is the chimney of a basement fire or below-grade fire. It is an extremely dangerous position to be in; the pressure behind the flow path is generated by the fire, but if a window or a door is accidentally or deliberately opened at the basement level, a low-entry portal will increase the temperature and speed of the convected air currents to inescapable and unsurvivable conditions. Other than a visual confirmation, this smoke behavior may be your only indication that you have a basement fire **FIGURE 10-32**.

At Bricelyn Street, E17 was first to arrive, just ahead of Truck 17. The homeowner was standing outside and directed E17 to the correct address, 8361. The captain went to the front porch to investigate and met with heavy smoke coming from the front door when it was opened. The captain briefly entered the house, then returned to the front porch and signaled his crew to lay a 1¾-inch preconnected line. He used his portable radio to report that E17 was on the scene with "smoke from the first floor."

If visibility at the entry door is bad, stop! Don't enter the structure blind. Use a TIC or a flashlight to see what the smoke is doing. Establish the direction of the smoke current or flow path. If there is no movement, you are likely in a dead-end corridor or below the fire floor.

Making Entry Down the Stairs

In residential basement fires, the practice of making entry through the first floor and down the stairs into the basement should be discontinued except as a last resort when civilian lives are at stake—and even then, it may be too dangerous if not impossible. Going down the basement stairs is always risky because it requires the door to remain open at the top. Smoke and heat immediately start rising up against the attack crew and into the first floor **FIGURE 10-33**. Those who advocate for this method of entry stress that this advancement down the stairs needs to be rapid. The sooner the crews make it down a hot smoky stairway, the less thermal assault that has to be endured and the sooner a firm floor is reached.

In theory, this sounds doable, but two realistic questions need to be asked. First, what is the condition of the underside of the stairs, and can they handle the weight of an attack crew scurrying down the steps? Answer: It is impossible to know. Second, if visibility is zero to poor, do firefighters rush forward into unknown areas when they can't see? Or is it human nature to exercise caution and slowly work your way through smoke and darkness. The firefighter who takes the brunt of the heat will be in the nozzle position. This firefighter shields the radiated heat from the rest of the attack team, so they each sense different levels of heat. If the nozzle firefighter reaches his or her ceiling threshold to tolerate the heat, this firefighter

FIGURE 10-33 In residential basement fires, the practice of making entry through the first floor and down the stairs into the basement should be discontinued except as a last resort—and even then, it may be too dangerous if not impossible.

will bail. But how do you put a team of two or three into reverse going down a stairway? It won't happen soon enough, and this is when panic sets in.

The attack crew is literally in a chimney flow path. If a basement door or window fails, or one is opened for any reason, a high-low flow path is created. Fresh air intensifies the fire and the firefighters will be caught between where the fire is and where the fire wants to go. The convected high-temperature currents with this high-low flow path can travel with great speed and can generate temperatures well over 1,000°F (538°C). Firefighters do not have time to react **FIGURE 10-34**.

At Bricelyn Street, E18 was the second-in engine, establishing a supply for E17. E19 was the third-in engine and arrived immediately after E18. E19 took a second 1¾-inch line to back up E17 because conditions were getting worse. They were going to follow E17's line but when they reached the front door, the heat and smoke were so intense, it prevented them from advancing any farther. Although it wasn't confirmed in the report, the fire behavior is consistent with conditions that develop with a high-low flow path.

Some instructors teach that, when descending stairs, your boot placement should be spread out on the tread between the outer stringer and the wall stringer. This will supposedly distribute your weight evenly along the tread. Again, there is no way of determining the underside stability of the stairs. Basement stair assemblies in modern lightweight construction often use lightweight steel hangers and gusset plates. You're essentially rolling the dice. If the hangers or connectors fail, it doesn't matter where you're standing on the tread: gravity will take over. Rather than trying to figure out how to outsmart the stair assembly for a direct attack, we should be teaching the effectiveness of an indirect and transitional attack from other access points for basement fires.

At Bricelyn Street, the captain of Truck 17 was on the roof. When he realized the fire was coming up into the attic from the open wall channels, he wanted to check fire conditions on the lower floors and notify the captain of E17 that the fire was extending vertically through the void spaces into the attic. He masked-up, grabbed a pickhead axe, and entered the house through the front door, following E17's line. As stated earlier, the street-side front entrance to the house is actually floor 2. He followed the charged hose line down the hallway, then made the right turn down the stairs to floor 1. As he descended toward the family room, the stairs collapsed under his weight because the fire had weakened the stair supports. He fell approximately 10 feet and landed on the basement stairs, close to the seat of the fire. It is a miracle the captain

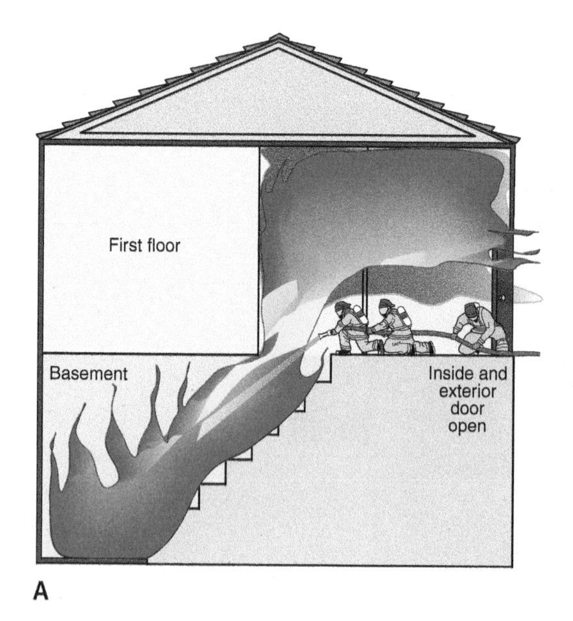

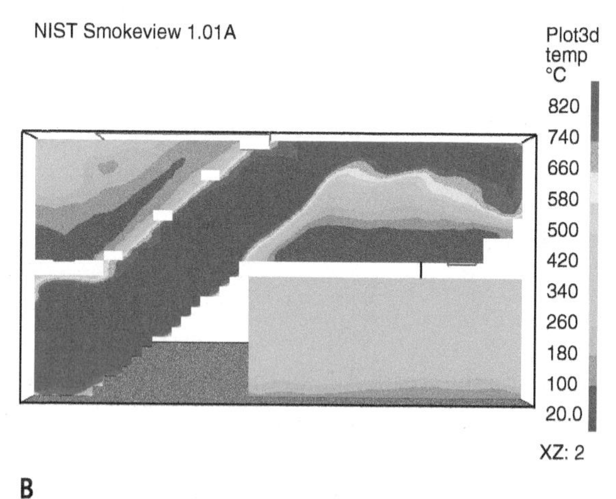

FIGURE 10-34 **A**. A basement fire venting via a stairway and open door leading to the exterior. **B**. NIST fire model showing the flow path from the basement to the first floor at a Washington, DC, townhouse fire.
From the National Institute of Standard and technology (NIST).

survived the fall. He managed to climb back up the burned-out stairs to the family room on floor 1 and met the crew of E17. There was thick smoke and high heat. He wasn't the only one who would fall through these burned-out stairs. Two other firefighters would fall through the same hole.

The reason the Truck 17 captain was able to make it to the hallway and the E19 crew could not was because, in the time sequence, the T17 captain went down first. It was the new hole in the stairs created by the captain's fall that allowed a new flow path to be created from the basement, thus allowing the high heat and smoke to increase and preventing the entry of E19's crew on floor 2. Because of the hole in the stairs, the fire now had a new vertical space to spread to and would eventually burn through E17's line—the only operating hose line inside the structure. Keep in mind that E17's line never made it down to the basement; the crew made it down only one level, to floor 1. When the hose burned through directly over the hole in the stairs, it is possible the water showering into the basement had a positive effect on the fire, confirming the science that fast water into the space, by any means possible, improves conditions. Even pressurized stream from a split in the hose, operating like an indirect attack from above, can have a cooling effect on the fire.

After a self-contained breathing apparatus (SCBA) bottle exchange, two members from E19 made a second attempt through the front door to back up E17. This time, they noticed that heat and smoke conditions had greatly improved, allowing entry. As the nozzle firefighter followed E17's line down the stairs to floor 1, he fell through the stairs into the basement. Shortly after, his partner reached the same point and also fell 10 feet into the basement. Miraculously, both survived the fall; although they were injured and roughed up, both firefighters managed to climb out of the basement using the burned-out stairs to floor 1. Now, all three members of E17, who are barely conscious; the T17 captain; and the two firefighters from E19 are all trapped in the family room on floor 1 (one level below grade) without a hose line. The stairs back to floor 2 (the street level entrance) have burned out, and the basement fire is below them. There were never any civilians in this structure. The only life hazard in this house are these six firefighters.

First-Floor Collapse

Because basements typically have low ceilings and limited access, either by windows or doors, rapid heat buildup occurs. Many newly constructed residential homes often leave the garage and basement areas unfinished, which means that a fire in a basement can easily attack the lightweight TJI beams and connectors of the floor assembly above. There have been numerous fires where the initial attack team has made entry on the first floor, only to have the floor collapse into the basement.

Sounding the floor, checking for sponginess or sagging, is not a reliable way to test for the integrity of the floor. Sponginess and sagging can be late indicators that the floor assembly is ready to collapse. The use of a TIC to check for a basement fire from the floor above is also not reliable, especially when the first floor is carpeted.

Some would argue that some builders use dimensional lumber for basement stairs and floor assemblies over basements, but for the firefighter on the nozzle, there is no way to determine beefier construction methods were used, so why take the chance? Look for an alternative access point or create one.

Exterior Window Indirect or Transitional Attack

Every effort should be made to utilize any exterior opening to the basement for fire attack to knock down the fire using an indirect or transitional attack. A rear door entrance is preferable to entering from the floor above. Basement-level entries put the fire attack team on the same level as the fire—a more advantageous position. If fire is blowing out a basement window, a hose line should be aimed from the outside, and a straight stream or medium fog should be applied into the burning basement **FIGURE 10-35**. If there isn't a window, a chainsaw can easily make one near the seat of the fire. The opening only has to be large enough to get the nozzle stream into the basement. The same science applies. Knocking down the fire with fast water immediately reduces temperatures within the compartment, and conditions will begin to improve.

FIGURE 10-35 If fire is blowing out a basement window, a hose line should be aimed from the outside, and a straight stream or medium fog should be applied into the burning basement.

From the National Institute of Standard and Technology (NIST).

At Bricelyn Street, E18 took a hydrant supply and laid a 5-inch supply line to E17. After that, they lowered a ground ladder down a retaining wall to get to the below-grade, C-side rear of the building. They advanced a 1¾-inch attack line from their apparatus down the ladder, through the side yard, and around to the rear of the involved house. They observed that the fire was in the basement and operated the hose line through the rear basement window. They never reported this on the radio. (Remember to think BRECEO/VS).

After a couple of minutes of water application, E18's line was successful in knocking down most of the visible fire in the basement from the outside. The crew was preparing to advance into the basement through the rear door for complete extinguishment and sent one firefighter to break out the side window for additional ventilation. This window happened to be the window of the family room on floor 1. When the window was broken out, he heard the taps from the low-pressure alarm bell on an SCBA and a moan from a downed firefighter. He entered the room through the window and discovered the semiconscious T17 captain down on the floor, close to the window.

The firefighter radioed that he had a downed firefighter and needed assistance. Another firefighter and the emergency medical services (EMS) personnel from Rescue 1 helped him pull the captain out the window and lower him to the ground. The report was given that the firefighter was rescued. This was the start of the radio confusion. As the EMS personnel were treating the T17 captain, who was barely conscious, he gasped that there were more firefighters inside and that he thought they were dead. Several firefighters returned to the side of the house where they encountered the two firefighters from E19 who had just crawled out through the same window. They thought these were the other firefighters the captain was referring to. The three members of E17 were missed completely and weren't discovered until sometime later in the fire. They all died in that room.

Cellar Pipes and Distributor Nozzles

Another option for basement attacks is inserting round nozzles, distributor nozzles, or cellar pipes into the compartment. They can be inserted through a window or from the floor above **FIGURE 10-36**. If

A

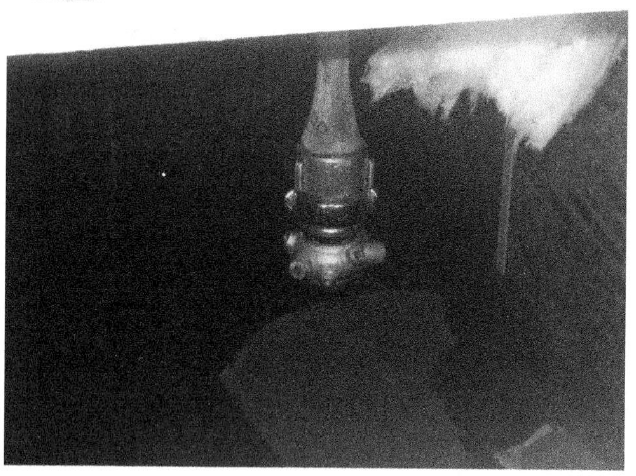
B

C

FIGURE 10-36 **A**. In this demonstration, firefighters used an axe to cut a hole in the floor above the basement. They're using a roof ladder to maintain the vertical angle and weight of the hose line when it's charged. **B**. The Bresnan distributor nozzle hangs like a giant sprinkler into the basement space. **C**. Once charged, it will flow 290 gpm 360°, thus creating a powerful sphere spray pattern.

Photos by Raul Angulo.

cellar pipes are used from the floor above, a backup line should accompany firefighters for protection. Making the hole for the cellar pipe introduces some smoke and heat, and possibly flames, into the first floor, but this will subside after water application.

This type of indirect attack is very effective and highly underutilized. With this indirect attack, you're basically introducing a sprinkler system into the basement. Any concerns about water application should be matched to any concerns you would have with an existing sprinkler system. In a sprinklered basement, the heads fuse at about 140°F to 165°F (60°C to 73°C) regardless of what's going on and who's in the basement, and whether the space can be ventilated. With concern for possible occupants, we have to rely on the scientific facts to determine the survivability profile. If the basement fire is hot enough that the IC opts for a transitional attack from defensive to offensive, it's unlikely an unprotected human would be able to survive the smoke toxicity and temperatures inside the narrow fire compartment. When water is applied, the situation always gets better.

Opponents of an indirect attack immediately stress that the steam created will severely burn, if not kill, possible civilian occupants, and the excess steam will make the atmosphere untenable for firefighters. Let's consider the science. The interior temperature would have to be at least 212°F (100°C) for the water to convert to steam. Humans cannot survive in 212°F heat. Opponents say that those are ceiling temperatures; the trapped occupants can be lying on the floor, where temperatures may still be slightly above ambient. This might be true: steam burns are extremely painful and can cause serious burn injuries, but they may not be fatal 100% of the time. An unprotected occupant caught in a flashover will certainly die 100% of the time. Because firefighters' PPE ensembles have a rated protection ceiling of 500°F (260°C), it is unlikely they will survive the flashover either. The water must be applied to the fire by any means possible to prevent flashover from occurring.

Steam Burns

Humans begin to feel burning pain on their skin when the temperature reaches 111°F (44°C). The skin develops superficial burns at 118°F (48°C). Partial-thickness burns are sustained when the temperature reaches 131°F (55°C). The skin's pain receptor nerves become overloaded and numb at 140°F (60°C), the temperature at which certain low-temperature sprinklers begin to fuse. At 162°F (72°C), human tissue is destroyed on contact. The entire range of burn temperatures from when humans first feel pain to complete tissue destruction falls well below 212°F (100°C), the boiling point of water.

Deciding there is a zero survivability profile in the fire compartment does not mean there is a zero survivability profile throughout the structure. Using an indirect attack has a positive effect and improves the survivability profiles of trapped occupants outside the fire room if it results in extinguishing the fire.

Carbon Monoxide

In the Bricelyn Street findings and autopsy report, the SCBAs used by the members of E17 were found with the air supplies fully exhausted. The SCBA worn by the T17 captain still had 300 psi of air pressure remaining in the cylinder. All three firefighters died from asphyxiation. Two of the three victims had high carboxyhemoglobin levels, in the 40% to 50% range, which indicates that they died primarily from smoke inhalation. The third deceased firefighter had a relatively low carboxyhemoglobin level, approximately 10%, and appears to have died primarily from hypoxia (oxygen deficiency). The captain from Truck 17 also had a carboxyhemoglobin level in the 40% to 50% range, which was incapacitating and potentially fatal.

The high-level carboxyhemoglobin levels (which are consistent with unconsciousness, coma, or death, depending on the individual and the length of exposure) suggest that three of the four individuals were exposed to the fire atmosphere for some time without the protection of their breathing apparatus. High-level exposure to carbon monoxide is known to cause disorientation and compromise motor skills quickly, which would have made it more difficult for the firefighters to find a means of escape from the dwelling. The elevated temperatures inside the house may have also contributed to their rapid incapacitation and disorientation. These are the products of combustion we should be worried and concerned about; not creating steam. The Pittsburgh, Pennsylvania, firefighters did not suffer thermal assault burns or steam burns. They died of carbon monoxide and smoke inhalation. Any civilian would have suffered the same fate.

Introducing a Basement Sprinkler System Using an Indirect Attack

Again, let's look at the theory and design behind an automatic sprinkler system. If a residential or commercial basement is equipped with sprinklers, the heads will fuse at approximately 165°F at the ceiling

FIGURE 10-37 A single-intake FDC on a turn-of-the-twentieth-century Type III ordinary construction building of three stories or less is a clue that the building has a sprinklered basement.

Photo by Raul Angulo.

and discharge a water fog spray at 8 to 24 gpm to knock down, if not extinguish the fires—regardless of the survivability profile. It doesn't matter if firefighters are on scene, there are TJI floor assemblies, or there's limited access or egress—or if there are windows. It doesn't matter if the stairway is narrow, or what type of stock, products, or furnishings are stored in the basement. It doesn't matter if it is too hot or too smoky. By the time the fire department arrived, this fire likely would be contained if not extinguished. If you look at old buildings in many major cities that were built between 1900 and 1950, many have sprinkler systems to cover the basement and sub-basements, even though the rest of the building is not sprinklered **FIGURE 10-37**. They knew basement fires would be difficult for the fire departments, and if a basement fire wasn't quickly contained, the fire spread would jeopardize the rest of the building—so they sprinklered the basement.

Every effort must be taken to create a "sprinkler system" in the basement by using hose lines, fog applicators, round nozzles, Bresnan distributor nozzles, or cellar nozzles (There's a reason they are called cellar nozzles: They should be used in cellars and basements!) This is the atmosphere we need to create when attacking or protecting exposures during a basement fire. Many of these tools can be inserted into the basement area from the outside. That means firefighters are outside the IDLH area. That means firefighters are not taking a beating when the only life hazard would be them, by going into the IDLH area or going down stairs when there is no way to even determine if the stairs are strong enough to support the weight of a firefighter or the weight of the entire crew.

Once the fire is knocked down, firefighters can make an interior entry to the basement for search and rescue, and final extinguishment with a much wider margin for safety. Ventilation openings may be scarce, which makes ventilation challenging but not impossible. Water first, then vent.

The higher in a building that a fire is occurring, the better off firefighters and occupants are in terms of fire and smoke spread. A fire in the basement of an apartment house is much more dangerous to occupants throughout the building than a fire on an upper floor. Heat, smoke, gases (especially carbon monoxide), and hot embers traveling vertically through the building can ignite secondary fires and overcome occupants on all floors.

Controlling Basement Fires

The first-arriving company officer will be the initial IC and must perform a thorough size-up to identify problems and determine the location of the fire within the basement, the stage of the basement fire, and where it is likely to spread based on the construction features of the building. The IC needs to determine life and property value, time to fight the fire based on the building construction, and the size of the fire to determine how much water will be needed to control it (value-time-size [VTS]). The IC needs to list all the problems and identify doors or windows for points of entry and points to ventilate the fire. Then the officer must perform a risk-benefit analysis to determine whether the basement is safe to enter or whether the attack should start from the outside.

Incipient or Early Growth Stage Basement Fires

Part of size-up and risk management is gathering information. Certainly, if tenants or employees give you accurate information on what's burning, where it's burning, how long it has been burning, and how to access the basement, and smoke and heat conditions indicate that the fire is in the incipient or early growth stage, taking a preconnected attack line down the interior stairs and into the basement to extinguish the fire may be an appropriate tactic and the right decision to make. An inexperienced firefighter or company officer who meets with success with this tactic may think basement fires aren't that tough to extinguish.

These aren't the basement fires we're talking about here. We're lucky if we can catch a basement fire this early. We're talking about the more typical basement fires: the ones that take time before they are even discovered. Add to that the amount of time it takes for the alarm to be transmitted, turnout and response, and setup time for making entry. Sometimes the exact location of these fires and their travel direction may be difficult to ascertain because smoke conditions are ugly.

Growth Stage and Fully Developed Stage Basement Fires

Small basement windows may be visible on the A side front or on any side of the structure. The exterior entrance to a basement is usually on the C side rear, but it could be on any side of the structure. In a lot of these incidents, the fire, heat, and smoke are venting out a door or window, and it's pretty obvious you have a basement fire, but you may not see this unless you make it to the C side rear. That is why the 360-degree walk-around survey is important.

When fire is blowing out of basement windows in residential or commercial buildings that have basement doors to the outside, it is best to use these doors for an aggressive interior attack. Work from the unburned side toward the burning side. The fire is already taking the path of least resistance and venting from the window closest to the fire, so utilize the reach of the hose stream and continue to advance toward the fire. A PPV fan can be set up at the entry door, keeping cool fresh air at your back and moving the smoke toward the ventilation window. If there are other windows close to the ventilation window on the fire side of the structure, they should be broken to ventilate the basement more quickly and effectively.

The PPV fan may accelerate the flame out the ventilation window with blowtorch force, so an exterior exposure line should be positioned on that side of the structure. If the air pressure entering the basement is too strong for the volume of smoke venting out the window, the smoke will look for another path of least resistance and start to exit the entry door. Lower the engine revolutions per minute (rpms), move the fan farther away from the door, or shut off the fan until the air currents reverse. Ventilation is a key element in controlling the fire by permitting the heat and smoke to escape while allowing firefighters to advance the interior attack into the basement.

Firefighters must keep ahead of the fire. If fire is burning up stairways or other vertical openings into the first floor, an exposure line must be positioned to protect the integrity of the stairway door. Utilize the reach of the stream and operate from a safe position, avoiding the floor area directly above the fire. Don't hesitate to flow water on the floor to cool it, like you would any exposure. Water seeping through the floor may also help extinguish any fire moving up the vertical cracks. Sufficient exposure lines must be provided with an adequate number of firefighters to accomplish the objective of protecting the first floor. Once exposure lines are in place, and an adequate number of hose lines are attacking the fire, another PPV fan can be set up to pressurize the first floor. The pressurization will be stronger than the smoke and flames trying to extend vertically and should reverse their direction back toward the ventilation portal.

If the IC suspects a basement fire in a balloon-frame construction building, the other objective is to keep the fire from extending to upper floors, including the attic and cockloft areas. *No PPV fans should be used in balloon-frame construction during the fire attack.* The vertical and horizontal open channels will spread the fire everywhere. PPV is very effective when you can control the exhaust portal, and it is disastrous if you can't. Firefighters need to be assigned above the fire to open up any vertical or horizontal space that is suspected of containing fire. Use a TIC. Hose lines must be positioned and operated by engine companies if these conditions are encountered.

When the fire is raging at the very bottom of the structure and forcing its way up into the building, but there are no exterior doors available to make access, it is important to control the main body of the fire as quickly as possible. It is not a quick evolution to try and make the interior stairs, not just yet. All openings into the basement should be located and considered as points of attack. Windows, access chutes, trapdoors, and any other openings should be used for positioning hose lines to attack the fire. The best option is the windows. Standard attack lines of the proper size should be used where indicated. Straight streams should be aimed into the windows from the outside to knock down and extinguish the fire. A 2½-inch hose line with a Bresnan distributor or Bulldozer nozzle can be secured with webbing on top of the rungs of a 12-foot roof ladder. The hooks support the coupling, and the round nozzle can spin freely 360° **FIGURE 10-38**. The ladder can now be inserted into the window for deeper penetration into the basement, creating a 290-gpm sprinkler **FIGURE 10-39**. One firefighter can handle the positioning of the roof ladder with the charged Bresnan distributer **FIGURE 10-40**. Cellar pipes and Bresnan distributor nozzles can be placed in service from the floor above into positions that permit them to knock down or assist in extinguishing the fire. A backup line should be laid to protect the firefighters carrying out this tactic. Once the fire is knocked down, engine crews can enter the basement for search and rescue, and final extinguishment. Clearly, outside attack lines are not to be used from openings into the basement, such as windows and doors, if firefighters have already made entry and are working inside the basement.

FIGURE 10-38 In this demonstration, a 2½-inch hose line with a Bresnan distributor or Bulldozer nozzle is secured with webbing on top of the rungs of a 12-foot roof ladder. The hooks support the coupling and the round nozzle can freely spin 360°.

Photo by Raul Angulo.

FIGURE 10-39 In this demonstration, the ladder can be inserted into the window for deeper penetration into the basement, creating a 290-gpm sprinkler.

Photo by Raul Angulo.

Along with the application of sufficient streams, a basement fire requires full ventilation of both the basement and the first floor. Basement windows, preferably opposite those used for fire attack, should be used for ventilation. The first floor should be ventilated completely to allow efficient and safe operations. Forcing the products of combustion out of the building through the basement and first floor reduces the chance that they will create problems on upper floors **FIGURE 10-41**.

FIGURE 10-40 One firefighter can handle the positioning of the roof ladder with the charged Bresnan distributer. The ladder supports the weight of the charged hose line and there is no nozzle reaction.

Photo by Raul Angulo.

Mercantile and Commercial Store Basement Fires

Basements serving mercantile buildings can be heavily loaded with product, often stacked up to the ceiling. This limits the effectiveness of streams. In such situations, additional hose lines will be required to control the fire. Firefighters should avoid operating streams from opposite directions, an action called **opposing lines**. A coordinated fire attack must be ensured. Thorough ventilation of the basement and first floor is extremely important.

If there are outside entrances but no windows to the basement, the situation becomes more difficult. Solid streams from 2½-inch hand lines or portable monitors are needed. The fire may travel vertically more

FIGURE 10-41 Available openings to the basement can be used for attack from one direction while ventilation is being conducted from the opposite side of the attack.

FIGURE 10-42 Sometimes the only way to ventilate a basement is by cutting an exhaust portal through the first floor just inside the windows on the same wall above the basement fire; then the windows can be removed or opened.

quickly, and the first floor could be in greater peril. Observing the effect of initial attack streams is especially important. If the attack appears to have a quick effect on the fire, ventilation of the first floor may suffice. If the attack is not effective, openings should be made in the first floor just inside the windows on the same wall above the basement fire; then the windows should be removed or opened to provide ventilation of the basement to thus improve fire attack effectiveness and lessen interior exposure problems. A PPV fan and an exposure line should be set up inside the room facing the window to assist and direct the smoke to the outside with positive pressure and hydraulic ventilation FIGURE 10-42.

In fighting a basement fire in a store, the storefront and display window area can be used for fire attack or ventilation according to the fire situation and the location of other basement openings. If there is a rear door but no other openings to the basement, the attack can begin from the rear door, and ventilation efforts should be conducted through openings at or near areas already burned FIGURE 10-43. If there are other openings such as windows or sidewalk doors, opening of the storefront may not be necessary. Conversely, if the basement has no outside opening but only an interior stairway or a trapdoor entrance, the display window will have to be broken out for ventilation, and the fire can be attacked with hose lines through interior openings, or with Bresnan distributor or cellar nozzles from the floor above. PPV fans can be used to direct the smoke toward the storefront display window, and exposure lines must be laid on the floor above the basement.

Finally, if interior attacks are untenable, and there are no advantageous exterior openings to fight the fire, High-Expansion® (HI-EX®) foam can be used to smother large commercial basement fires. HI-EX® foam was developed in England during the 1950s to fight fires in coal mines (the ultimate basement fire). The foam expands over 200 times its original volume and is designed to fill the entire space, from floor to ceiling; absorb the heat; and displace the oxygen. There must be a sufficient supply of foam available to sustain the application, and all parts of the basement

FIGURE 10-43 In a store with only a rear door to the basement, the attack can begin from the rear, and the wall or floor under the front display window can be opened for ventilation. The display window needs to be taken out.

must be adequately sealed. Unfortunately, many fire departments are not equipped with HI-EX® foam or have discontinued using it; however, with modern fuels, which put out a faster and higher energy heat release rate, combined with lightweight construction, HI-EX® foam units are making a comeback.

Protecting Exposures

Stairways and other openings from the basement to the first floor are major channels for the vertical spread of a basement fire. These openings must be covered as soon as possible. Attack lines should be brought in to first-floor landings to knock down any fire that may have spread from the basement; walls and partitions should be opened and checked with a TIC for vertical fire spread, and baseboards should be removed in suspect areas.

These first-floor operations are extremely important, but the safety of firefighters and occupants is even more essential. When a basement fire has gained considerable headway or fully involves the basement, the IC must assess the risks versus benefits and make a decision about allowing firefighters to enter the building. If a decision has been made to allow entry into the building, firefighters must pay particular attention to conditions within the structure. The floor should be checked for a soft or spongy feeling. Smoke coming through the floor should not be ignored, especially at outside entrances, along sidewalks, at cracks in the walls, and at other places that may involve main structural members. These signs confirm a significant basement fire because pressurized smoke is pushing through the cracks of the foundation or the sidewalk. It also indicates the deterioration and weakening of the floor, so firefighters should be withdrawn from the building immediately.

Because of substantial live load and dead load weight combinations, first floor collapses can be sudden and happen without warning. This was the case at the Mary Pang Fire on January 5, 1995, in Seattle, Washington. Four firefighters were killed when the first floor collapsed into a burning basement. All of the abovementioned warning signs were present at this fire but weren't recognized in time. In addition, five interesting firefighter statements in the Federal Emergency Management Agency (FEMA)/U.S. Fire

Administration (USFA) report are worth noting. The original attack was on floor 1; firefighters did not know they were dealing with a commercial basement fire yet, much less an arson fire. (Remember BRECEO/VS.)

- During the initial attack on the floor above the basement, firefighters were operating in zero-visibility smoke. The heat level was such that attack crews needed to stay crouched down, but it wasn't so intense that it caused unusual concern.
- They noticed numerous spot fires at floor level. These were quickly extinguished, but the seat of the fire couldn't be located.
- Once the ladder company opened the roof, the first-floor atmosphere was noticeably cooler. Firefighters could now stand up, but it was still extremely smoky. (Heat and smoke don't ventilate at the same rate.) Floor-level spot fires were still being discovered in the office area.
- Even after 25 minutes of operating on the first floor, interior companies could not find any significant fire on this level to match the amount of smoke and flames that were clearly visible from the outside.
- The first-floor decking was concrete because it was a frozen food processing facility. Even after the roof was ventilated and the heat lifted, the concrete floor was so hot from the radiated heat of the fire that firefighters could not kneel down. It was even too hot to walk on. The heat was radiating right through their fire boots.

These additional warning signs must be added to the list of indicators that a significant fire is occurring in the basement. The entire report should be required reading for firefighters of every rank. Except for Type I noncombustible buildings, a 25-minute interior operational period without progress is too long to go without a careful analysis of fire and structural conditions. A good guideline is to keep an interior operation to 20 minutes unless progress is made. Do not proceed any further over these floors. If occupants still need to be evacuated or firefighters are working on upper floors, ground ladders or aerial ladders should be placed to the second story for emergency egress. Once firefighting efforts on the basement and first floor are abandoned, the decision should be made to change to a defensive strategy. A structural collapse is likely to follow.

Vertical Spread

Many of the vertical openings in a building originate at the basement level, and there is nothing to keep fire from entering the openings and spreading through them. Vertical openings should be opened at the first floor if there are any signs of fire spread there. Confirm any fire spread using a TIC. In any case, an exposure line needs to be laid to the floor above the fire, and these openings should be opened at roof level to allow the heat, smoke, and gases to leave the building *after* water has been applied to the spaces. Otherwise, a high-low flow path is created, and you may end up with a fire in the basement, one at the top of the building, and fire in between. Water first, then vent.

Horizontal Spread

Adjoining basements must be checked for horizontal fire spread. Hose lines must be taken to those locations to extinguish extending fire. This is particularly important in older structures, where deterioration, remodeling, and additions might have destroyed the integrity of the fire walls between buildings. Party walls, those that support beams or joists from two buildings, often have openings where the beams rest on the walls. Fire can easily spread from basement to basement through these openings. Regardless of their construction, adjoining basements must be checked.

The same is true for adjoining attics, especially if the fire building was not vented quickly. Whether the fire originated in the basement or in some other area of the building, the tops of adjoining buildings must be checked and, if necessary, vented.

Basement fires are typically viewed as one of the most dangerous and difficult fires we face, but they don't have to be. In fact, they can be one of the safest—*if* we start our fire attack by first knocking down the fire from the outside.

Advancing Attack Lines to Upper Floors

In the previous section, fires were on the first floor or in the basement of a particular building; however, a fire can start on any floor of a multistory building, and attack lines must be advanced up to and above the fire floor. Even if the fire is on the first floor, hose lines must be advanced to the floors above the fire to protect the primary search and rescue efforts; protect interior exposures; and confine and extinguish the vertical spread of the fire. Vertical fire extension can occur when the fire jumps from floor to floor through the failure of exterior windows. This is called auto-exposure **FIGURE 10-44**. Hose lines must be advanced to upper floors of the fire building to combat extension and auto-exposure **FIGURE 10-45**. Engine

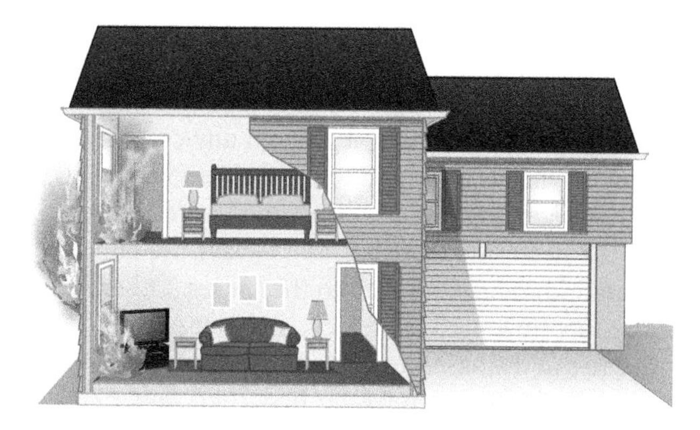

FIGURE 10-44 Vertical fire extension can occur when the fire jumps from floor to floor through the failure of exterior windows. This is called auto-exposure.

FIGURE 10-46 An engine crew controls the main body of the fire and protects the stairway, whereas a ladder crew performs search and rescue operations.

FIGURE 10-45 A hose line must be advanced to the floor above the fire to combat extension from auto-exposure.

company attack lines for high-rise fires are covered in Chapter 15.

On fires above the first floor, the first-arriving engine companies usually use stairways for getting attack lines up to and above the fire because stairways give them quick access to the building. As they advance their attack lines, the firefighters must extinguish any fire that has moved into a stairway. Control of stairways is important—both to eliminate them as channels for the vertical spread of fire and to keep them open for the rescue of building occupants and as escape routes for firefighters FIGURE 10-46. If there are multiple stairways, use one for fire attack and one for evacuation.

Additional arriving companies can also use the stairways if they are safe and wide enough; however, the stairs may be narrow or clogged with hose lines or poorly located in relationship to the position that firefighters must take. Other means of advancing hose lines to upper floors are available and should be used when necessary. Hose lines to upper floors may be advanced using the following methods:

- Standpipes
- Ground ladders, aerial ladders, or aerial platforms
- Hoisting with ropes
- Carrying them into the building and then connecting them to a standpipe
- An outside hose line through a window, balcony, or porch
- Passing them up to a window with a pike pole or roof or rubbish hook

Such alternative methods are often detailed in training manuals, but they are rarely practiced during training sessions and then they are forgotten on the fireground. These alternative methods may require less hose than advancing the line up stairways. When hose lines are taken up the face of the building, only one section of hose is needed for every four or five stories. In a stairway, one 50-foot section must be allotted for each floor because so much length is taken up in winding around stairs and through hallways.

An aerial ladder or the waterway of a tower ladder may sometimes be used to advance a hose line to upper floors or even the roof, for example, when a

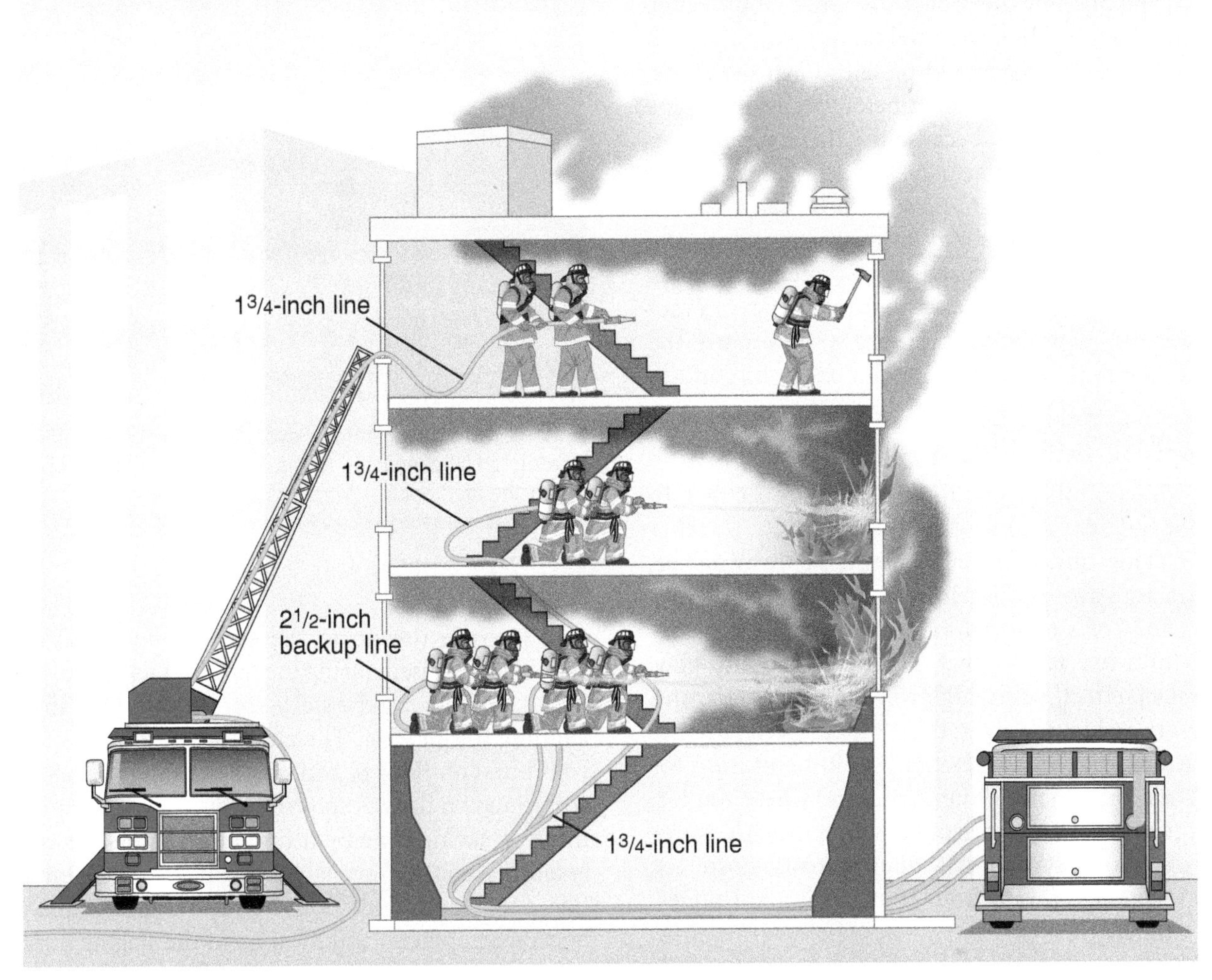

FIGURE 10-47 An aerial ladder or the waterway of a tower ladder may be used to advance a hose line to upper floors or even the roof.

hand line needs to be advanced to an upper floor and there are already numerous hose lines in the stairwell **FIGURE 10-47**. However, this tactic should not become a preferred method or used in lieu of a perfectly functional standpipe. Aerial ladders are a limited specialized resource. Some departments only have one or two, and some departments don't have any. Using an expensive piece of apparatus to establish a temporary standpipe commits the ladder to supporting the attack and renders it useless for aerial rescues and elevated master streams, which are the designed purposes of this fire truck.

Standpipe Operations

Standpipes provide a significant tactical advantage for quick and efficient direct attack evolutions and thus get fast water on an incipient or a growth stage fire, as well as some fully developed fires using 1¾- and 2½-inch hand lines. Other fully developed fires that require more than 300 gpm to knock down and extinguish also require the use of a lightweight portable master stream monitor or the TFT *Blitzfire* nozzle, both of which need to be supplied from the standpipe. An additional supply line may have to be taken from a second outlet from the floor below.

Every firefighter should expect a standpipe to be present in any building that is four stories or higher from ground level—this is required by the International Fire and Building Codes. If a building is only three stories high, it will not have a standpipe; in this instance, hose lines must be laid up the stairs or by some other means. Every standpipe must have the top discharge valve on the roof or just inside the top stair landing at roof level. In a four-story building, the standpipe should be used if the fire is on floor 3 or 4, or on the roof. If the fire is on floor 1 or floor 2, the hose lines should be laid as if it were a two-story building. However, the driver should still charge the standpipe because exposure lines must be laid to floors 3 and 4, and possibly the roof. Those lines should come off the standpipe. In lower-level floors below the main entrance, hose lines should be laid down the stairs, even though there may be a below-grade standpipe discharge connection, because the fire is already below you; heat and smoke may be pushing up as you're working your way down. If conditions change for the worse, you need the ability to back out and up the stairs with a charged hose line to protect the attack team, and protect the stairway in case rescues or evacuations are in progress. Clearly, any fires above the third floor require standpipe evolutions for fire attack.

Low-Rise Buildings

Low-rise buildings are those less than 75 feet in height and exist in almost every community. By coming in at just under 75 feet, sometimes only by inches, contractors avoid the stricter fire protection regulations and costs required for high-rise buildings. Except for a few factors that may not come into play, like stack effect, wind, high elevation pressures, and pressure-reducing valves, these buildings require the same strategy and tactics that would be utilized for high-rise operations. Other structures where you can expect vertical and horizontal standpipes are warehouses, sports stadiums, shopping malls, hospitals, parking garages, waterfront piers, freeways, bridges, and tunnels.

The tactical objectives on the fire floor of a multistory building remain the same as in a one-story building:

- Forcible entry
- Fire attack line
- Rapid intervention line
- Backup line
- Exposure line to the floor above
- Search and rescue
- Ventilation
- Salvage and overhaul

At a minimum, four hose lines need to be deployed. Additional attack lines and interior exposure lines may be needed on the fire floor, and additional exposure lines may be needed on the floor or floors above the fire. Speed and maneuverability are extremely important for controlling fires on upper floors. The size of the hand lines and nozzles still depend on the application of one of the two fire flow formulas; however, achieving rooftop pressures in a low-rise building isn't usually a problem, so 1¾-inch attack lines, backed up by a 2½-inch line, should be sufficient.

FIGURE 10-48 In standpipe evolutions, the 1½- or 2½-inch wye should be connected to the discharge outlet one floor below the fire floor.

Elevators

In a low-rise fire up to the seventh floor, elevators should not be used during the initial attack setup. Firefighters should use the stairs so a member isn't lost in running the elevator. The attack line still needs to be connected on floor 6, and other lines may have to be connected on floor 5, so you're not really gaining anything or saving time by using the elevators. As more resources become available, the IC can still initiate Phase II elevator operations to transfer firefighters, air bottles, equipment, and EMS up to the staging floor.

1½-inch and 2½-inch Gated Wyes

The attack line and backup line should be connected to the standpipe on the floor below the fire using a 1½-inch gated wye or a 2½-inch gated wye **FIGURE 10-48**. This gives the company officers options. Depending on the occupancy, fire load, size of the fire, and whether the building is fully sprinklered, a 1¾-inch attack line with a 1¾-inch backup line should be sufficient. For example, consider the fire load and square footage of a fire in a unit in a low-rise motel or hotel that is sprinklered. There's usually one room with one or two beds, a nightstand, lamps, a TV, an easy chair, a desk or small table with a chair, chest of drawers, curtains, and carpet. This fire does not need a 2½-inch line for the initial attack. One company can manage a 1¾-inch attack line and backup line coming off a 1½-inch gated wye and handle this fire quickly.

Using a 2½-inch gated wye can yield a number of combinations:

- Two 2½-inch lines
- One 2½-inch line and two 1¾-inch lines coming off a 1½-inch gated wye

- Four 1¾-inch lines coming off two 1½-inch gated wyes for exposure protection

These wye configurations add a tremendous amount of cantilevered weight to the standpipe discharge valve and should be supported with webbing tied to the standpipe bracket. A better solution, which many fire departments have implemented, is to have a pigtail: a 10- to 12-foot section of 2½-inch hose that connects to the standpipe discharge valve and is then laid on the floor. Wyes, the in-line pressure gauge, and attack hose lines are connected to the male end of the pigtail. This relieves practically all the weight on the standpipe discharge valve and transfers it to the floor **FIGURE 10-49**. The pigtail also allows the connections to be moved to a corner of the floor landing so they aren't blocking the door and aren't in the way of firefighters.

Opponents of using any wyes in standpipe evolutions opt for a single 2½-inch line per floor. They claim that the friction loss created through these appliances diminishes the needed flow, and accidentally bumping the gate valve handle can inadvertently charge the hose bundle before it's laid out, creating a pile of kinks and spaghetti. Others maintain that splitting the flow between two lines isn't as sufficient as supplying a single 2½-inch line or that a standpipe cannot supply multiple lines coming off a single discharge valve.

On the surface, the hydraulic rationale against using wyes may sound convincing, but in reality, unless you have the staffing readily available, it will probably create more problems than it will solve. Properly supplied, the gpm flow from a 4- or 6-inch standpipe can easily flow 600 gpm or more of water, and this is a reasonable expected flow. Two-and-a-half-inch wyes or 1½-inch wyes can allow for a combination of hose lines and nozzles that are within those 600 gpm. The friction loss through wyes is negligible. What you need to remember is that the lines are not all flowing water simultaneously. The hose line flowing the most water is the attack line. A backup line or an interior exposure line should not be flowing water unless needed, and then those streams will be intermittent and operated for a short period of time. The wyes provide options. The concern for accidentally charging the line by bumping the gate valve handle can be eliminated by the department purchasing wye models with valve-handle locks to prevent such occurrences. Wyes can bleed the air out of a standpipe faster than a nozzle can. The wyes allow for prebundled, predetermined hose loads of the same length to be connected at the same location, so a 200-foot backup line will be able to cover a 200-foot attack line.

FIGURE 10-49 The pig tail connects to the standpipe outlet. The wye is connected to the pig tail and allows hose connections to be made on the floor, taking the weight off the standpipe valve.

Photo by Capt. Steve Baer.

Single Hose Method (2½-Inch)

The method for taking a single 2½-inch hose line per floor can complicate the attack or cause it to fail without significant staffing. Very few fire departments, except for New York City, Chicago, and Los Angeles, have the immediate resources available. Consider the hose lines that need to be laid up the stairs and allow 50 feet of hose per floor; the 200-foot attack line is connected on the floor below the fire. If the exposure line is needed on the floor above, before the backup line, that hose can be connected two floors below the fire and should lay up to the floor above the fire; that is 150 feet of hose just for the stairs. The backup line must connect to the standpipe three floors below the fire and also requires 150 feet of hose for the stairs, plus an additional 150 feet to be able to reach the attack nozzle: that's 300 feet of charge 2½-inch hose, which is not an easy task. Where will the rapid intervention team line be connected? Four floors below the fire? Or from another standpipe? You can see how this gets complicated. Plus, you need a sufficient number of companies to manage this much hose. New standpipe designs are now putting two 2½-inch discharge ports on each landing; that's the same as a 2½-inch wye.

Hose Layering

Hose layering also needs to be considered. We have all seen hose lays forcefully expand when the line is charged. The first line laid is the attack line, and thus that hose is on the bottom. When the 2½-inch backup

FIGURE 10-50 Three charged 2½-inch lines in a single stairwell are unmanageable. The attack line is buried under the backup line and exposure lines, making it extremely difficult, if not impossible, to advance the attack line without an army of firefighters.

Courtesy of John Odegard.

line and the 2½-inch exposure lines are charged, the weight of the hoses and friction against the jackets make it impossible for the attack line to advance **FIGURE 10-50**. That can be disastrous. Without an army of firefighters, three charged 2½-inch lines in a single stairwell are unmanageable. If this fire cannot be extinguished with the firefighters available to position and maneuver two 2½-inch hoses effectively, a direct attack will most likely fail. You need the options and flexibility, as well as the speed and maneuverability, of 1¾-inch hose lines. Therefore, you need the wyes.

Dangers in the Firefighting Stairway

There are a lot of variables to consider in every fire based on the building's fire protection system; heating, ventilation, and air-conditioning (HVAC) capabilities; the vertical ventilation plan; and if the fire is contained to the unit or has spread into the hallway. However, unless the entire fire attack stairwell can remain pressurized, connecting hose lines above the fire floor landing should be avoided. Fire and smoke entering the hallway means that, as soon as the offensive attack is initiated on the fire floor, a low-pressure atmosphere is created at the landing; flames, heat, and smoke will immediately travel out the door and convection currents will rise up the stairway to the top floors, trapping any civilian occupants or firefighters who may be in the stairwell above the fire floor **FIGURE 10-51**.

If the top of the stairwell is vented, a high-low flow path *chimney effect* can be created, accelerating smoke and heat conditions in the stairwell. If a window breaks in the fire compartment, a wind-driven fire scenario can develop, and the chimney flow path can convert to extreme temperatures and flames, traveling through the stairwell with blowtorch force. This makes the stairway above the fire floor untenable and deadly, killing anyone caught in the flow path **FIGURE 10-52**.

This is why it's important to control the entry door in addition to establishing a firefighting stairwell and an evacuation stairwell. New York City Firefighter John King of Engine Company 23 was killed in such a situation on December 27, 1961. As the crew was ready to make entry, he was flaking the excess hose on the stairs above, a practice we all use today to make it easier to advance the hose line onto the fire floor. When the hallway door was opened, smoke and flames trapped Firefighter King on the stairs above the fire floor, killing him. Four civilians died in a superheated, smoke-filled stairway on December 24, 1998, in New York City. The fire started on floor 19, and the victims were found in the firefighting stairwell on floors 27 and 29. On January 6, 2014, again in New York City, one man was killed and another critically injured when they tried to escape a fire in their high-rise apartment. The fire started on floor 20. The two men unknowingly entered the firefighting stairwell on floor 38. As the firefighters opened the hallway door to advance to the fire floor, heat and smoke filled the upper stairwell. The men only made it down to floor 31.

These examples illustrate important actions to consider during standpipe operations before initiating the offensive attack from the stairwell:

- Building occupants don't know the difference between a firefighting stairway and an evacuation stairway. If there is a building public address system at the fire alarm panel, the IC should announce that occupants should use a specific stairway or remain in their units and shelter in place.
- If forcible entry is required to the fire floor, the door must be controlled until the upper stairs are cleared.

FIGURE 10-51 Initiating a fire attack from the stairwell landing introduces smoke and heat into the stairway above the fire floor, trapping anyone caught on the upper floors.

- Before the fire floor door is opened, a search team should check the upper levels in the firefighting stairwell for evacuees and quickly move them below the fire floor or move them inside an upper hallway and direct them to the evacuation stairway, but they must be moved out of the firefighting stairway before the attack begins **FIGURE 10-53**.
- If time allows, the exposure line to the floor above the fire should be in place, or it should at least be out of the stairway and into the hallway before the fire floor door is opened.
- The exposure floor door should be opened just enough to allow for the hose to pass through the threshold to limit smoke from the stairway entering the hallway.
- If the excess hose of the attack line is stretched to the upper stairs or landing to assist in advancing the hose, this should be done *before* the fire floor hallway door is opened. The officer should make sure the upper stairs are clear and that the entire crew is back on the hose line before opening the door to the fire floor.
- If the building has more than one stairway with a standpipe, the exposure lines to the upper floors can be taken off a different standpipe to avoid the heat and smoke from the upper firefighting stairway. This option may not be possible if it compromises the evacuation stairway, but it is another option to think about.
- In buildings with more than one stairway, the interior exposure line can be connected to the

FIGURE 10-52 A wind-driven fire scenario can convert a chimney flow path to one with extreme temperatures and flames.

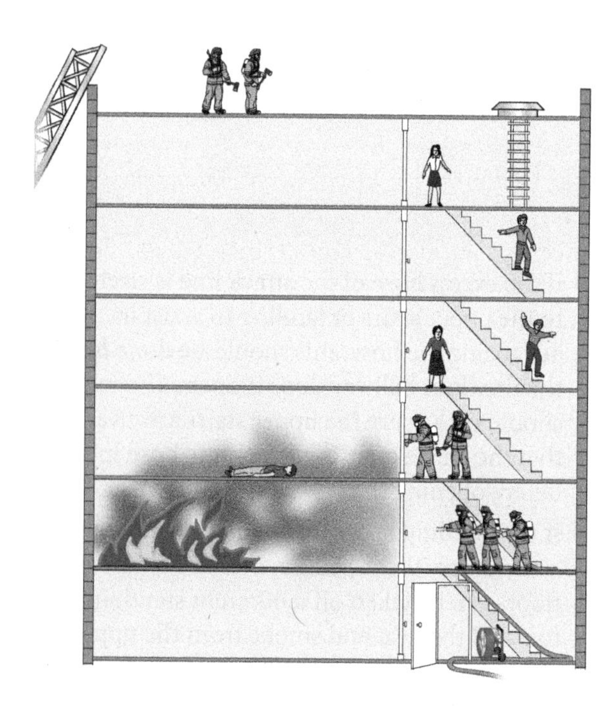

FIGURE 10-53 Before the fire floor door is opened, a search team should check the upper levels in the firefighting stairwell for possible evacuees; they must be moved out of the firefighting stairway before the attack begins.

standpipe in the firefighting stairwell, then laid through the hallway below the fire floor and up to the floor above the fire, using a different stairwell to be clear of the smoke and heat rising above the fire floor in the firefighting stairway.

- The search teams should use the evacuation stairway instead of the firefighting stairway so they can move freely between upper floors without being in the heat and smoke that contaminates the firefighting stairway. If the firefighting stairway must be used, the search crew should carry a set of irons in case they have to force a hallway door or breach a wall to escape the firefighting stairway should conditions become untenable.
- While the crew hooks up to the standpipe and lays out the hose, the company officer should look into the floor below the fire to get a feel for the layout of the hallway. He or she should note how many doors are on the left and how many doors are on the right. He or she should also note the ceiling height and the length of the hallway, the location of other exits within the hallway, and the construction material of the doors to the individual units. Wooden doors are easier to force than metal doors. Metal doors hold back the fire better than wooden doors. Are the walls made of gypsum board or concrete? The construction makes a difference if a wall has to be breached for any reason. Having the time and opportunity to look inside a unit would be even better. The layout in the fire unit will most likely be identical and knowing the layout aids in keeping you and your crew oriented. This type of survey is quick and can be completed in about 60 seconds.
- The fire attack engine officer should announce on the radio that firefighters are commencing the attack into the hallway, which is a verbal warning to everyone else that conditions may quickly change for the worse in the firefighting stairwell.

Making Entry into the Fire Room for Initial Attack

Once the door to the fire unit is reached, spray the surface of the door; water quickly turning to steam indicates high heat inside the compartment. Open or force the door. Maintain door control in case it needs to be closed immediately due to extreme fire conditions **FIGURE 10-54**. The nozzle and hose line should be worked to use the reach of the straight stream. The

FIGURE 10-54 Maintain door control in case the door needs to be closed immediately due to extreme fire conditions.

FIGURE 10-55 The nozzle and hose line should be worked to use the reach of the straight stream.

officer should use a TIC to check for victims close to the door and to obtain thermal readings **FIGURE 10-55**. Maintaining thermal balance keeps the area clear enough to search for any victims or locate the seat of the fire; however, if the intensity of the flames are growing and temperatures are climbing, the ceiling area must be cooled by deflecting the stream off the ceiling to prevent flashover. Ideally, the seat of the fire should be hit with a straight stream, but if the room is involved, Z-, T-, or clockwise O-patterns should be used to cover all areas of the room, including the floor. The majority of the fire load is in the lower half of the room. A straight stream at mid-level between the floor and ceiling will most likely blow out any windows, aiding in horizontal and hydraulic ventilation. Once this exhaust portal is created, PPV fans can be used to bring in fresh air and accelerate the ventilation of smoke and fire gases.

As conditions improve, all shards of glass should be removed so the window opening is totally clear. Keep broken glass inside the fire unit unless it is safe to toss glass outside without injuring firefighters working on the perimeter. After knockdown, shut the line down and allow the smoke and steam to vent. Perform a primary search of the area and then follow with a secondary search by a different crew. The officer can reevaluate to find hot spots. After final extinguishment, keep one crew to check the fire. Rotate crews. Try to determine the cause of the fire. Perform a new size-up, safety survey, and risk-benefit analysis before salvage and overhaul begins.

As with any interior fire, if a 1¾-inch hand line isn't knocking down the fire after 30 to 60 seconds of water application, bring up the 2½-inch backup line, and bring in another 2½-inch backup line or portable master stream. If the initial attack strategy hasn't extinguished the fire within the first 20 minutes, withdraw all interior crews and prepare for defensive operations.

Ventilation

Ventilation is the systematic removal of heat, smoke, steam, and other flammable and toxic fire gases from the building and replacing them with fresh air. Proper ventilation can enhance a successful outcome of a fire, while uncoordinated and improper ventilation can have disastrous results by creating a flow path that greatly increases the intensity of the fire. Ventilation without the immediate application of water intensifies the fire by giving it more oxygen. Ventilation is normally assigned to ladder companies, but it must be accomplished at every fire regardless of who is assigned the task. Many fire departments do not have ladder or truck companies, so the responsibility falls to the engine company.

The operations discussed in this section are enhanced by proper ventilation, which allows firefighters on the hose lines to move more easily and quickly to the proper firefighting positions. Ventilation of stairways and hallways is of utmost importance, but venting the stairs can also result in unintended consequences because fire and smoke can be pulled into the stairway from fire on lower floors. It is essential that firefighters understand what happens when a high-low flow path (low air intake–high vent exhaust) is created (Figure 10-53). They must also be able to predict where pressurized atmospheres seek the path of least resistance in order to equalize. Engine company members should position attack lines in key positions to protect firefighters advancing to upper floors of the building for search and rescue and ventilation **FIGURE 10-56**. Firefighters attempting to advance attack and backup lines to the fire should vent as they go if that action helps them to advance. For growth stage and fully developed fires, the general procedure is to use a coordinated fire attack, which

FIGURE 10-56 Engine company members should position attack lines in key positions to protect firefighters advancing to upper floors of the building for search and rescue and ventilation.

ideally provides ventilation and fire attack simultaneously. **FIGURE 10-57** shows a coordinated fire attack in which one team of firefighters attacks the fire directly, while another team advances to the roof to ventilate the area above the fire. Remember, ventilation without hose lines in place accelerates and spreads the fire. If hose teams are in place, they should not hesitate to apply fast water immediately. Water first, then vent.

There are two types of ventilation: vertical and horizontal. There are various methods to carry out vertical and horizontal ventilation: ventilating for fire control, ventilating for life, mechanical ventilation, hydraulic ventilation, and natural ventilation.

Vertical Ventilation

As the name implies, vertical ventilation allows heat and smoke to rise and escape from the highest points of the structure: skylights, atriums, roof access doors, bulkheads, shafts, vents, scuttles and the like, and also by firefighters cutting holes into the roof **FIGURE 10-58**. Vertical ventilation prevents smoke, heat, and fire gases from accumulating and banking down (mushrooming) onto firefighters and occupants, thus improving their survivability profile. When smoke and heat are allowed to escape and fresh air replaces the smoke, interior temperatures drop, backdrafts and flashovers are averted, horizontal fire spread decreases, visibility improves, and oxygen content increases.

Allowing the smoke and heat to *lift* enables the attack crew to locate the seat of the fire and extinguish it more quickly and easily. Improved visibility provides the ability to conduct a fast and thorough search for occupants. Ventilation is a calculated balance of cause-and-effect actions. While smoke and heat are exhausted from the compartment, fresh air enters the atmosphere, thus increasing the oxygen percentage. This is good for possible trapped occupants, but it also increases the

FIGURE 10-57 Ideally, a coordinated fire attack includes ventilation happening just before or simultaneously with extinguishment.

FIGURE 10-58 Vertical ventilation allows heat and smoke to rise and escape from the highest points of the structure through designed ventilation openings or by firefighters cutting holes into the roof.

burn rate and intensity of the fire unless hose lines are used to intervene with fast-water application.

Vertical ventilation can also change the direction of the fire. For example, it stops the horizontal spread of smoke and fire in a long attic space. The ventilation hole created by firefighters creates a new path of least resistance for the pressurized smoke to escape FIGURE 10-59. Remember that wherever you cut the vertical hole, that becomes the path of least resistance, and smoke and flames will be drawn toward that hole. FIGURE 10-60 illustrates what happens when you ventilate in the wrong location. These firefighters inadvertently cut off their escape route and jeopardized the protection provided by the hose line—another reason to provide a secondary means of escape during rooftop ventilation evolutions.

Vertical ventilation, especially trench cuts (which are made across the entire width of the roof), can be labor-intensive and time-consuming depending on the resources available and the construction of the roof. Vertical ventilation is best suited for the following:

- Fires on the top floor of a multistory building.
- Fires in attic spaces and cocklofts.
- Fires in balloon-frame construction.
- Fires in single-story commercial and residential structures.
- When trying to prevent a backdraft from occurring in a superheated, over-pressurized structure.
- Ventilating stairways, shafts, atriums, and other vertical spaces that terminate at the roof.

Except for these situations, horizontal ventilation should be used.

The Roof Package

The tools for vertical rooftop cutting operations can include the following:

- All members wearing a pickhead ax and scabbard

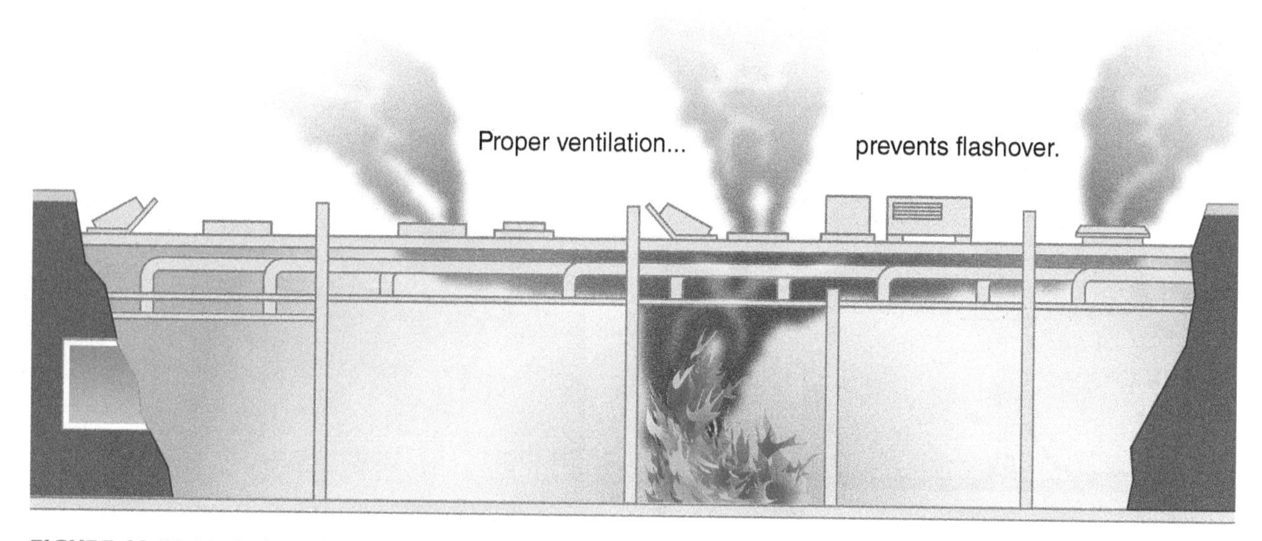

FIGURE 10-59 Vertical ventilation can change the direction of horizontal smoke and fire spread in a long attic space.

FIGURE 10-60 Cutting a vertical hole creates the path of least resistance, and smoke and flames will be drawn toward that hole. If the vertical hole is cut in the wrong location, it can trap the firefighters.

- A ground extension ladder large enough so that the tip extends at least five rungs above the roof line
- A roof ladder
- Two chainsaws
- One rescue saw (circular blade saw)
- Two D-handle pike poles (length based on attic pitch)
- Two roof hooks
- One Halligan tool (if used for a foothold)
- Sledgehammer or maul
- TIC

Cutting the Roof

Ideally, roof-cutting operations should start at the highest point of the structure that is closest to, or directly over, the fire, as safety allows. An experienced fire officer should be in charge of this operation. Some signs that help the officer or lead firefighter determine the location of the fire from the roof include:

- Steam coming off a wet roof
- A dry area over a wet roof
- Melted snow
- Bubbling tar
- Increased smoke coming through the roof decking and shingles
- Looking over the edge of the roof to check for the location of smoke and fire
- TIC images indicating the presence of fire

A TIC can also help identify the direction of the rafters and other solid structural members of the roof assembly. The strongest parts of the roof are the corners, exterior walls, ridges, valleys, rafters, and hips. To avoid injury, the first firefighter up the ladder should sound the roof decking aggressively with the roof hook to check its integrity before stepping on it **FIGURE 10-61**. The lead firefighter continues to sound the roof forcefully to check for its integrity while remaining on the strong portions already found and/or those mentioned above. Small inspection cuts should be made periodically with the chainsaw to check the smoke conditions underneath. Inspection cuts also allow you to verify physically the direction of the roof rafters. An increase in smoke production from the inspection holes tells the lead firefighter that she or he is getting closer to the fire. Confirm with the TIC.

While this inspection takes place, the officer should request that another ground ladder be placed for a

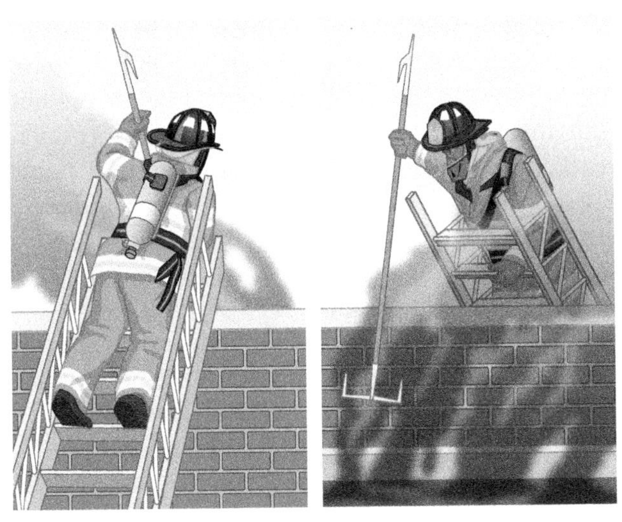

FIGURE 10-61 To avoid injury, the first firefighter up the ladder should sound the roof decking aggressively with the roof hook to check its integrity before stepping on it.

FIGURE 10-62 The louver cut.

secondary means of egress. Ladders should not be placed over windows or any opening where fire can break out and engulf the ladder, cutting off the escape route.

The Louver Cut

Note: Firefighters should not reach down or put their faces into the ventilation hole to punch the ceiling, and all members should be covered and on air, with their SCBA during the entire operation described here.

Several types of vertical cuts, including rectangular, square, triangle, and trench cuts, can be made to ventilate different styles of roofs, along with various cut sequences. The louver cut is one of the most popular vertical ventilation cuts for residential structures. It is particularly suitable for flat or pitched roofs with plywood decking. It is quick, efficient, and easy to extend if a larger ventilation hole is needed FIGURE 10-62. With any vertical ventilation evolution, keep the wind at your back for safety and visibility.

- **The first cut:** When the roof team reaches the ventilation point, the first cut of the ventilation hole is a horizontal top cut, or "ID cut." This cut is made parallel to the ridge rafter toward the fire, and away from the egress route, stopping at the first rafter encountered. This cut is made about a foot below the ridge rafter to avoid hitting any metal flashing or rafter brackets. The ID cut is made to identify the starting or outside rafter.
- **The second cut:** The second cut is the top cut, or "head cut." It should connect with the ID cut, and it is a horizontal cut away from the fire and toward the egress route. The head cut should roll one rafter and stop at the next one.
- **The third cut:** The third cut is a vertical "down cut" starting just inside the outside rafter closest to the fire and farthest from the egress route. The down cut should extend down about 4 feet.
- **The fourth cut:** The fourth cut is a horizontal "bottom cut" connecting with the first down cut. It is made toward the egress route and away from the fire. The bottom cut should roll one rafter and stop at the next rafter.
- **The fifth cut:** The fifth cut is a vertical "down cut" completing the first louver. It is made on the fire side of the rafter, but on the egress side of the ventilation hole where the head and bottom cuts stopped. Start at the top by the ridge and cut down to complete the louver.

Using the rubbish hook, punch down the louver so it opens toward the fire, which helps protect the roof team from the flames once the ceiling is punched through. Then punch through as much of the ceiling as possible to vent the smoke. Prepare for flames to erupt when the hot smoke and gases get the proper air mixture. Firefighters should not reach down or put their faces into the ventilation hole to punch the ceiling, and all members should be covered and on air, with their SCBA during the entire operation.

The louver can be extended in a similar cut sequence if the hole needs to be enlarged. Extending the louver is preferable to starting a new ventilation hole. Once sufficient vertical ventilation has been established, the entire crew should get off the roof. If additional ventilation operations are required, a charged hose line should be brought up to the roof to protect the firefighters and to wet down the roof and thus protect it from the radiant heat of the flames.

Horizontal Ventilation

As the name implies, horizontal ventilation uses windows and doors on the fire floor and the floors above as entry and exhaust portals to accomplish the same objectives and benefits that vertical ventilation does. Horizontal ventilation pushes smoke, heat, and fire gases away from firefighters and occupants, thus improving their survivability profile. When smoke and heat is forced out of an exhaust portal, interior temperatures drop, flashover is averted, visibility improves, and oxygen content increases as fresh air replaces smoke. All of this allows the attack crew to locate the seat of the fire and extinguish it faster and easier. Improved visibility provides the ability to conduct fast and thorough primary and secondary searches for occupants.

Horizontal ventilation is a calculated balance of cause-and-effect actions. While smoke and heat are being exhausted from the fire room, fresh air is entering the atmosphere, increasing the oxygen levels. This increases the survivability profile for possible trapped occupants, but it also increases the intensity and heat release rate of the fire unless hose lines apply effective water application without delay.

Horizontal ventilation needs to be coordinated by the ladder company or an outside team. Breaking out windows indiscriminately, before the hose lines are in position and ready to apply water, accelerates the fire and draws it right to the broken windows. Accelerating the fire increases heat temperature and smoke production, decreasing the survivability profile in the room and accelerating the time to flashover.

If the fire room can be determined from the outside by looking for black-stained, crazed glass, the outside team should prepare to break out the window with an axe or a pike pole. The darker the window is stained, the likelier the room to which it leads is the room of origin. A black-stained, cracked window is closer to the seat of the fire than a window that has water droplets running down the glass.

When the hose team is ready to make entry, the outside team breaks out the entire window and allows the fire to vent. This relieves the interior pressure buildup. The fresh air from the entry door creates a horizontal flow path with the broken-out window, and the attack team can extinguish the fire while pushing the products of combustion out the window. Once the horizontal flow path is established with the hose line in place, a PPV fan can be set up at the entry door to enhance and speed up horizontal ventilation **FIGURE 10-63**.

If an outside team isn't available, the nozzle firefighter can create the same conditions by spraying the straight stream across the fire room in a horizontal sweep, halfway between the ceiling and the floor. Any window in the fire room should be broken out with the force of the stream, initiating horizontal ventilation. As soon as possible, the entire window should be broken out to maximize the size of the exhaust portal.

Both these scenarios can be backed up with PPV and hydraulic ventilation.

Nearly any structure fire we face, from high-rise fires to one-story, single-family residence fires, can be ventilated effectively with horizontal ventilation. With career-ending incidents like the one on March 29, 2015, in Fresno, California, when Fire Captain Pete Dern fell through the roof of a burning garage involving lightweight trusses, many fire departments are no longer willing to accept the risks associated with vertical ventilation operations when horizontal ventilation can accomplish the job. The only time when horizontal ventilation should *not* occur is when a building is over-pressurized, with high heat and smoke presenting classic conditions for a backdraft. A firefighter opening a door or lower window for any reason introduces the needed oxygen for the explosion to occur with deadly force **FIGURE 10-64**.

Ventilating for Fire Control

Ventilating for fire includes all of the benefits and points mentioned in the previous sections, and there is no need to repeat them here. But to clarify, ventilation helps control the fire. Ventilation is not necessarily an isolated tactic. It is actually part of the initial fire attack and goes hand in hand with extinguishing the fire. The smoke and other products of combustion have to be removed from the building.

Ventilating for Life: Firefighter Safety

Ventilating for firefighter safety occurs when the ventilation plan, originally designed for fire control, is not working and interior crews are taking a beating. Smoke conditions may be so thick that there is a severe risk of firefighters becoming disoriented or lost. Heat temperatures may be increasing, and interior conditions may become untenable before firefighters will be able to escape safely from a flashover or succumb to the effects of high heat.

Another scenario may be a rapid intervention rescue in progress. In such conditions, smoke and heat need to be released and exhausted by the quickest means possible. It may even require a combination of ventilation methods. At this point, the concern for the safety of the firefighters outweighs the effects that ventilation will have on the fire.

FIGURE 10-63 When the hose team is ready to make entry, the outside team breaks out the entire window of the fire room and allows the fire to vent horizontally.

Ventilating for Life: Civilian Life Safety

The most routine example of ventilating for life is vent-enter-isolate-search (VEIS). When a room is selected for VEIS, the window is opened to allow the room to begin ventilating. Before a flow path can develop and draw the fire into the selected room, the door to the room is closed, thus isolating the room from the products of combustion. Then the search of the room begins. Although the improvements may last only a few seconds, atmospheric conditions are improving, and the room temperature is decreasing, which increases the survivability profile.

In fires involving overcrowded, unsprinklered public assemblies, the company officer needs to recognize that time is of the essence. A fire like this can spread faster than your crew can get hose lines in service and water on the fire. Perhaps the only action you can take to save as many lives as possible is to vent for life by breaking all available windows and forcing as many doors as possible to create exits that permit the escape and removal of viable victims. This procedure is called venting for life. The numerous ventilation portals accelerate the fire, which will most likely become a defensive fire.

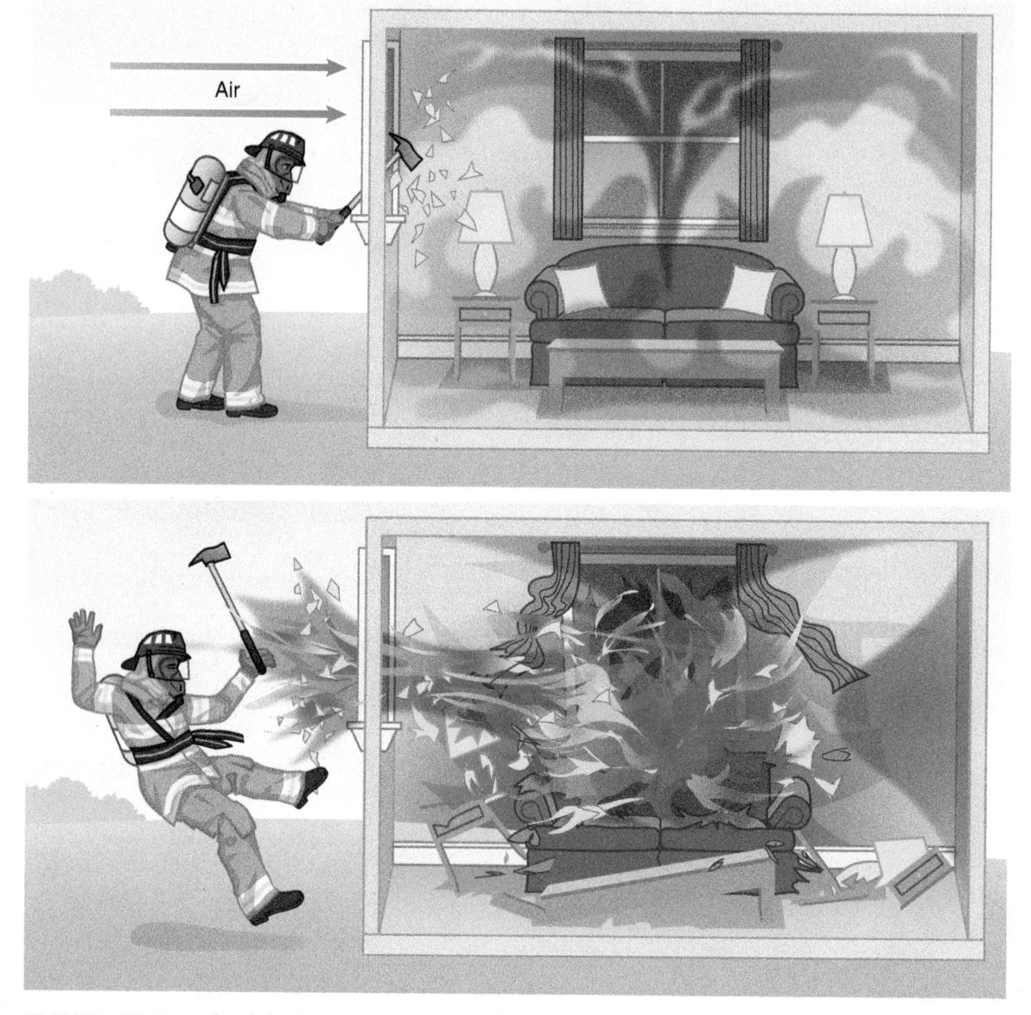

FIGURE 10-64 A backdraft occurs when a firefighter, for any reason, opens a door or lower window of a compartment with a high-heat, oxygen-deficient atmosphere.

Mechanical Ventilation

Mechanical ventilation uses electric smoke ejectors; gas and electric PPV fans or blowers; and building exhaust fans and heating, ventilation, and air-conditioning (HVAC) systems. The most popular piece of equipment is the PPV fan. These units range between 18 inches and 27 inches (between 460 mm and 690 mm) in diameter and can deliver 7,000 to 24,000 cfm (11,900 m^3/hr. to 40,600 m^3/hr.) of air. Mechanical ventilation using PPV enhances and speeds up the process of horizontal ventilation by pressurizing the structure and creating a flow path that can be directed by the ventilation team by establishing an exhaust portal as close as possible to the seat of the fire. The exhaust portal is typically a window in the fire room. The entry portal is usually the door used for access by the initial attack team **FIGURE 10-65**. If the inside pressure is higher than the outside atmospheric pressure, the positive pressure seeks the path of least resistance in an effort to equalize, carrying all the smoke and other products of combustion with it. Usually a single PPV fan is sufficient for a small residential fire when it is set 8 to 10 feet from the front door. But additional fans can be used in series, parallel, or in a V-pattern depending on the size of the opening and

FIGURE 10-65 With proper use of PPV, the exhaust portal is typically a window in the fire room. The entry portal is usually the door used for access by the initial attack team.

the size of the structure. The PPV currents entering the structure should always be at the backs of firefighters working inside the structure. Firefighters should always be between the fan and the fire, never between the fire and the exhaust portal.

The use of PPV and positive pressure attack requires extensive department-wide instruction in theory and hands-on training, and this topic is beyond the scope of this book. When used correctly, PPV is safe and efficient in ventilating a structure quickly, replacing heat and smoke with fresh air to cool the atmosphere and improve survivability profiles, improving visibility to find the seat of the fire quickly for extinguishment, and reducing the exposure of carcinogenic by-products to firefighters.

Hydraulic Ventilation

Hydraulic ventilation is the first action the nozzle firefighter should take after knocking down and extinguishing the fire to remove smoke and heat from the fire room. It is best accomplished with a combination spray nozzle using a 60° fog pattern placed about 2 to 4 feet from the windowsill. The fog pattern should cover about 85% to 90% of the window opening. The fog entrains the smoke by creating a positive-pressure and negative-pressure venturi, forcing the smoke out the window. It is an effective initial action that can move close to 7,000 cfm of smoke **FIGURE 10-66**.

A

B

FIGURE 10-66 **A**. For hydraulic ventilation, the combination spray nozzle should use a 60° fog pattern placed about 2 to 4 feet from the windowsill. **B**. The fog pattern should cover about 85% to 90% of the window opening.

Natural Ventilation

Natural ventilation uses doors, windows, and other structural openings to allow the fire to vent without the use of water streams, PPV, smoke ejectors, or building ventilation systems. This is especially effective when there is a steady 15 to 20 mph breeze. The leeward side window can be opened first, then the windward side window can be opened for effective horizontal ventilation of a fire room **FIGURE 10-67**.

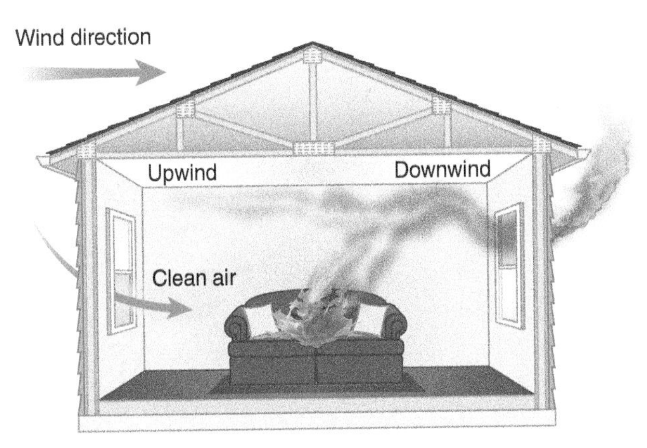

FIGURE 10-67 With natural ventilation, the leeward side window is opened first, then the windward side window can be opened for effective horizontal ventilation of a fire room.

Ventilation-Limited, Smoldering, and Decay-Stage Fires

Ventilation-limited, *smoldering*, and *decay-stage* are terms that are often used interchangeably, but they are not exactly the same thing. However, all three types of fires occur on the downward side of the standard time-temperature curve. Ventilation-limited fires need only fresh air to continue the combustion process and can reignite with explosive force. A smoldering fire can be a ventilation-limited or a fuel-limited fire. Decay-stage fires are usually fuel-limited, meaning that there may be unlimited air, but the combustible fuel has already been consumed by the fire and no more thermal energy is being produced. A fire that has "burned itself out" with ample oxygen is a decay-stage fire. Ventilation-limited is a relatively new term for the fire service, and the common term used to describe such fires is still smoldering.

Firefighters responding to an alarm may encounter a fire that is not free burning or at least does not appear to be. There may be enough smoke to indicate that there was a fire, but there are no flames. In this case, and lacking information about the exact situation, the firefighters must assume that they have come upon a ventilation-limited fire. If it is not handled properly, this type of fire can be disastrous. The first-in company officer must perform a 360-degree size-up and risk-benefit analysis, and gather as much information about how the alarm was reported, who reported the alarm, and whether the alarm came from a monitored home detection system. Critical information from face-to-face contact with the owner, neighbors, or witnesses can help the officer determine if this is a ventilation-limited fire or a smoldering fire in the decay stage. Nothing can be taken for granted, especially when dealing with modern construction. Firefighters must prepare for fire attack but resist the urge to rush in before a few vital factors are confirmed.

There are two types of smoldering fires: one that can occur on the incline side of the standard time-temperature curve, when temperatures are high and there is still plenty of fuel available to burn, and one on the decline side of the curve, when temperatures are cooling and the majority of the fuel has been consumed (fuel-limited). Both involve ventilation-limited or oxygen-deficient atmospheres, and they are often found in well-sealed, well-insulated homes and structures that contained the fire until it ran out of oxygen. Modern homes often have double- and triple-pane windows that can withstand the heat from room and content fires, versus single-pane glass windows that can fail easily. When a window fails in a ventilation-limited compartment, an unlimited supply of oxygen is available for the fire to grow, initiating a secondary growth stage, and flashover occurs usually before the fire department arrives or around the time of its arrival.

The standard time-temperature curve, a bell curve, of a fire is a graph that firefighters are familiar with. The temperature differential between a flashover and a backdraft is very small. Both are high-temperature events on the same side of the curve: the upward incline of the graph where the temperatures are high. When a fire starts, the temperature is just above ambient, and the pressure inside the room is low. As the temperature rises, so does the pressure. The upside incline of the graph indicates that the pressure inside the structure is building, but it is not yet pressurized **FIGURE 10-68**. Firefighters must train themselves to look for the signs of a potential backdraft situation when they are on scene of fires in closed-up structures and resist the urgency to rush in.

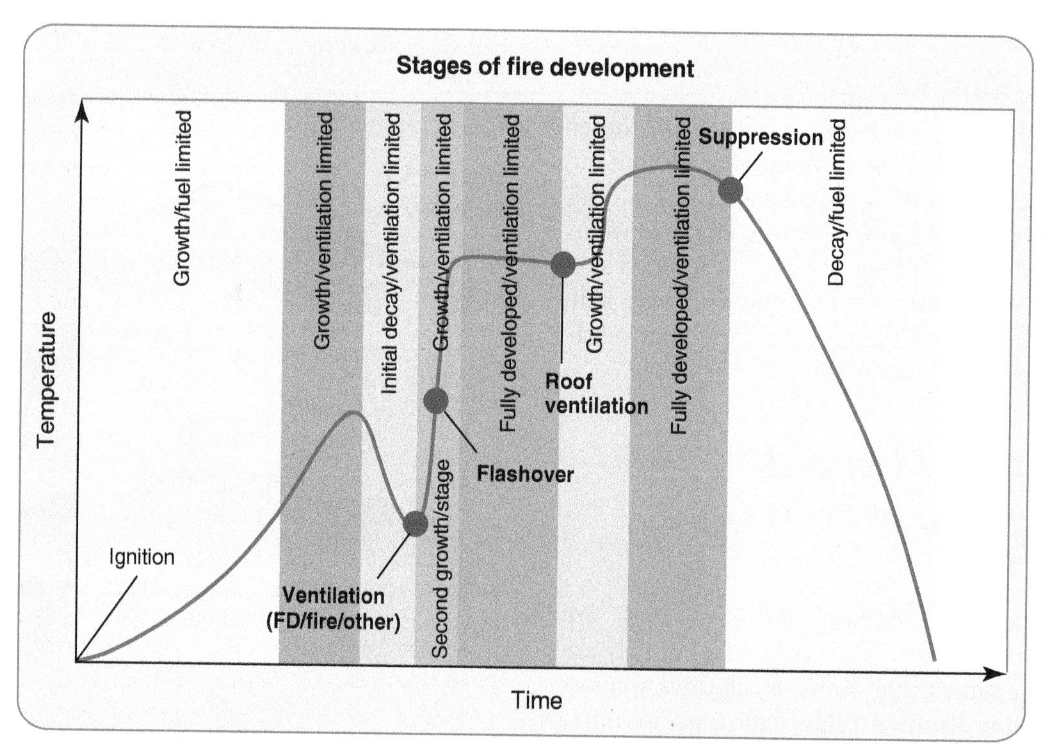

FIGURE 10-68 Fire growth in a single-family residence is often ventilation-limited when the fire department arrives. Ventilation without immediate water application results in an explosive secondary growth stage that can lead to flashover.

Courtesy of UL.

Indications of a Ventilation-Limited Smoldering Fire

A ventilation-limited smoldering fire can be indicated by one or more of the following conditions:

- Smoke is visible, but little or no fire is visible from the outside.
- The smoke rises rapidly as it comes from the building, indicating that it is hot (however, humid weather may hold down the smoke).
- Usually the building is tightly sealed with windows and doors and other openings to the outside tightly secured.
- Smoke leaves the building under pressure from around windows, doors, eaves, and other openings.
- The smoke may be yellow or dirty brown in color.
- Although no flames are showing, the window glass is brown or stained from heavy carbon deposits created by the smoke.
- There are signs of extreme heat.
- All windows are darkened, with linear crazing or cracks.
- Smoke exits the building and appears to be sucked back into the building (in other words, the building appears to be breathing).

This is a list of pre-backdraft conditions. Whenever any of these conditions are present or are suspected, the fire must be handled as a ventilation-limited fire for the safety of firefighters and for proper firefighting operations.

A ventilation-limited smoldering fire has sufficient heat and fuel to become free burning. The heat comes from the fire, which was probably burning freely at one time. The fuel is mainly carbon monoxide gas from the original fire, the contents of the building, and perhaps the building itself. The carbon monoxide has filled the building and surrounded the smoldering fire, cutting it off from an unlimited supply of oxygen. Thus, a smoldering fire needs only oxygen to burst into flames. A fire may be smoldering in a building of any size or type or, in some cases, in only one area of a large structure such as an attic, storage area, or other concealed space.

Backdraft

Not every ventilation-limited fire is a backdraft condition, but every backdraft situation is a ventilation-limited smoldering fire, and it must be ventilated before it is attacked—that is, the carbon monoxide must be cleared from the building before air is allowed to enter it. The addition of any oxygen to the heat and fuel, even as little oxygen as might enter the building when an outside door is quickly opened and closed, can lead to immediate ignition of the fire.

A backdraft is a dangerous situation. The sudden ignition can take any of several forms. A backdraft can occur naturally when a window fails due to excess heat and pressurized smoke, or by accident when a firefighter opens a closed door or breaks a window for ventilation. The gases and preheated combustibles might simply burst into flames, engulfing the building in fire. Or the force of the ignition can exert enough force to blow windows, doors, and firefighters out of the building. There could also be an explosion strong enough to cause structural damage to the building and grave injuries to firefighters and any other occupants. Just what will happen cannot be determined beforehand, but it is certain that the addition of oxygen will cause some sort of ignition. That ignition is referred to as backdraft.

A backdraft is a high-temperature event caused by the sudden introduction of oxygen into an enclosure where superheated smoke, gases, and contents are at or above their ignition temperature (+1,200°F [+649°C]) but previously did not have sufficient oxygen to burn, or the gases were too rich to burn. The rush of fresh air causes the sudden explosive ignition of fire gases.

A fire burning in a closed box, meaning a room or structure that is ventilation-limited, can consume all the available oxygen within the compartment, slowing the combustion process and the intensity of the fire; however, the room is filled with superheated fuel in the form of smoke. At this stage, the fire is starving for oxygen **FIGURE 10-69**. The sudden opening of a door

FIGURE 10-69 This building is charged with superheated fuel in the form of smoke. At this stage, the fire is starving for oxygen. The sudden introduction of air from a door or a window being opened will most likely cause a violent backdraft.

Courtesy of Keith Muratori.

or a window (often by firefighters making the initial entry) introduces a fresh supply of oxygen to a superheated, fuel-rich environment, and the four elements of the fire tetrahedron quickly come together, causing the room to burst violently into flames.

A backdraft is sometimes confused with a flashover. They are not the same, even though both occur in very high temperatures on the upward side of the standard time-temperature curve. They usually can occur between the growth stage and fully developed stage of a fire. These events are also sometimes called smoke explosions, but that term is also incorrect. Again, backdrafts and flashovers are not the same thing. A backdraft involves superheated temperatures; a smoke explosion typically occurs during the decay stage of a fire, with cooler atmospheric temperatures for the smoke. Signs of a potential backdraft include:

- Extreme heat (more than 1,200°F).
- Signs of extreme heat, including melted window frames and melted siding.
- No smoke showing.
- Any confined fire with a large heat buildup in a tightly sealed building.
- No visible flame from the exterior of the building.
- Dull red flame in the thick of the smoke.
- Smoke puffing from seams and cracks around windows and doors and then being sucked back into the building.
- Heavy black smoke pushing out under pressure from cracks and seams.
- Smoke-stained windows that are hot to the touch (an indication of a significant fire), windows may be bowed, sooty, and/or crazed from heavy carbon deposits.
- Turbulent smoke.
- Thick yellow-brown or yellowish-gray smoke (indicating that it contains sulfur compounds).

When smoke has filled large open-area structures, like supermarkets, bowling alleys, department stores, as well as large open-void spaces, that is, atriums, giant attic spaces, and cocklofts between hanging ceilings, the potential for backdraft is great because they can accumulate a tremendous volume of volatile smoke. Remember, smoke is fuel. Such volumes can create a tremendous explosion sufficient to destabilize the building.

Ventilation for Preventing a Backdraft

When backdraft conditions exist, fire departments have no choice but to ventilate the structure. There are basically two strategies to prevent a backdraft from developing, both of which need to release as much heat and unburned fuel gases as possible. The first is to ventilate vertically at the highest point of the structure, preferably from an aerial ladder or elevating platform. This will draw the pressurized hot gases and smoke and allow them to escape into the atmosphere, but only if ventilation is directly above the smoke-charged compartment. Vertical ventilation isn't helpful if the potential backdraft exists on the third floor of a six-story apartment.

The second strategy, which is much more dangerous, is to ventilate the structure horizontally from a safe distance with overwhelming water streams from 2½-inch hose lines, a deluge monitor, or a deck gun aimed at the highest windows of the structures **FIGURE 10-70**. This tactic can be used when no aerial apparatus is available. Keep personnel clear of all building openings. This defensive tactic either ventilates the structure successfully or triggers a backdraft, but with the immediate application of

FIGURE 10-70 If an aerial ladder isn't available, a straight stream from a hose line or master stream can be used from a safe distance to break out the highest windows in the structure. This action either ventilates the structure successfully or triggers the backdraft.

Courtesy of John Odegard.

overwhelming water streams, the interior temperatures of the gases and fuel are quickly reduced, lessening the effects. Remember, there are no viable lives inside a pre-backdraft compartment. Because this is a deliberate tactic, the backdraft won't catch firefighters off guard, but it can still produce a shock wave that can destabilize the building. There are no other options other than standing back and letting events play out naturally.

Smoke Explosions

Flashovers do not occur at every fire. In fires that remain ventilation-limited or oxygen-deficient, a flashover may not occur. A ventilation-limited fire produces a limited amount of heat energy. If the oxygen supply is not increased, the fire starts to enter the decay stage. During the decay stage, heat continues to pyrolyze fuels in the room. The pyrolysis produces additional vaporized fuel in the form of smoke. A fire in the decay stage can flash quickly if oxygen is introduced to the smoldering fire. The fire gases in smoke are already heated to their ignition temperatures (or just below) in the decay stage. This is where smoke explosions can occur.

Typically, a smoke explosion occurs during the decay stage of a fire that is on the decline side of the time-temperature curve. The fire has already used up the available oxygen, and incomplete combustion is occurring or has occurred. As the oxygen levels decrease in the structure, the production of carbon monoxide (CO) increases. Temperatures are cooling, but the building is now pressurized **FIGURE 10-71**. A smoke explosion occurs when a mixture of carbon monoxide, flammable gases, and oxygen is present, usually in a void space or other pocketed area separate from the fire room. The building is as pressurized as it can be, so smoke can travel quite a distance away from the fire. The smoke can be right at its flash point or cooled just below its ignition temperature. Many of the flammable gases are already within their flammable range; these are sometimes called ladder fuels. Remember, carbon monoxide has a very wide flammability range (12% to 75%). When it comes in contact with an ignition source, like a smoldering ember, the flammable mixtures can quickly ignite, often violently. The conditions needed to produce a smoke explosion often include the presence of void spaces, combustible building materials, a ventilation-controlled fire that produces unburned fuel (primarily carbon monoxide), relatively cool temperatures, and an ignition source.

Although smoke explosions usually occur during the decay stage of a fire, they are not limited to this stage. Smoke explosions result in a violent explosion during final extinguishment or overhaul when smoldering embers may be uncovered, which provide the ignition source. In a smoke explosion, there is no change to the ventilation profile, such as opening a door or window; rather, it occurs from pressurized smoke traveling within the structure to an ignition source.

Ventilation of a Smoldering Fire

Not every smoldering fire is a backdraft condition. Although smoldering fires are not spectacular, as are free-burning fires, in some ways, they are even more dangerous because the fire gases are already in the

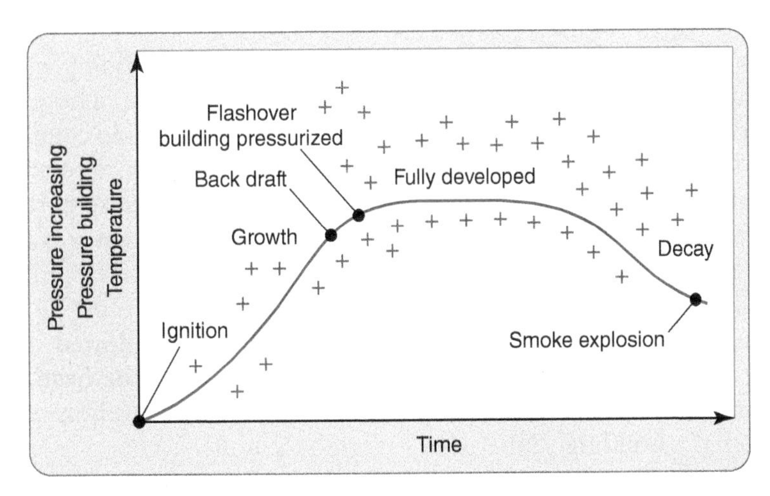

FIGURE 10-71 A smoke explosion typically occurs during the decay stage of a fire.

From the National Institute of Standard and Technology (NIST).

FIGURE 10-72 Firefighters should ventilate a smoldering fire at the highest point possible before making entry for fire attack.

proper physical state for rapid combustion, and the situation can change unexpectedly with lightning speed. Carbon monoxide, other heated gases, and smoke convect upward and collect at the top of the structure. An opening must be made on the building as high as possible to release these gases and allow them to move out of the fire area **FIGURE 10-72**. This relieves the explosive situation and greatly reduces the chance of creating a backdraft or a sudden reignition of the fire.

Again, this ventilation must take place before any attempt is made to enter the building. Otherwise, air entering the building with firefighters will cause immediate ignition. These tactics must be coordinated between engine companies who are preparing to enter the structure to attack the fire and ladder companies who must perform proper ventilation procedures before entry is made. A smoldering fire must be ventilated before it is attacked; that is, the carbon monoxide must be cleared from the building before air is allowed to enter it. The addition of any oxygen to the heat and fuel, even a minute amount, will lead to immediate ignition of the fire, or backdraft.

Initiating the Coordinated Attack

A coordinated fire attack is not always possible when waiting for a ladder company to arrive on scene and set up ventilation, but when all units arrive in a timely manner, attack lines should be charged and ready for use during the ventilation process. Firefighters on the attack lines should be in safe positions, protected from flying glass, and ready to enter the building as soon as ventilation is completed. Apparatus and drivers in a direct line with the building should remain alert and take cover behind the apparatus until the structure is vented and the smoke pressure is released. After the gas, hot air, and smoke are released from the building, outside air will enter and cause the previously smoldering fire to burst into flames. This is a sign that ventilating is complete. The danger of explosion is minimal, but firefighters have only a limited idea of the size of the fire at this point.

The initial direct attack should begin after ventilation is complete. Firefighters making entry allow additional air to reach the fire, and it may be burning freely by the time the seat of the fire is reached. Additional attack and exposure lines should be stretched above and around the fire to contain it, and search and rescue operations should be initiated. More detailed aspects of ventilation will be covered in the forthcoming book, *Ladder Company Fireground Operations*, 4th edition.

After-Action REVIEW

IN SUMMARY

- At a working structure fire, the placement and operation of initial attack lines protect occupants and firefighters, and provide the first water on the fire for extinguishment.
- The initial IC should conduct a 360-degree size-up and perform a risk-benefit analysis before the decision is made to allow firefighters to enter a building to conduct an aggressive interior attack.
- This decision making begins when the first company arrives on the fireground and the company officer assumes command.
- A decision needs to be made about the strategy of operation used: offensive, transitional, or defensive.
- A decision also has to be made regarding the size of hose line to be deployed during the initial attack, how many hose lines are needed, and where they should be positioned.
- The type of nozzle must also be considered because it determines the flow and shape of the stream.
- Spray nozzles operate with a stream pattern from straight stream to wide-angle fog, and smooth-bore nozzles operate with a solid-stream pattern.
- During an offensive fire attack, an engine company must also consider whether to use a direct or indirect attack.
- A direct attack is one in which a solid or straight hose stream is used to deliver water directly onto the base of the fire.
- An indirect attack is one in which a solid, straight, or narrow fog stream is used to direct water at the ceiling or to deflect it off the top window header to cool superheated gases in the upper levels of the room, with the objective of preventing flashover. This technique converts the water to steam, absorbs heat, and extinguishes the fire.
- Ventilation must be managed with suppression efforts, and a coordinated fire attack, supervised by the IC, ensures that both tasks take place simultaneously.
- Ventilation is the systematic removal of heat, smoke, steam, and other flammable and toxic fire gases from the building and replacing them with fresh air.

- Proper ventilation can enhance the successful outcome of a fire; uncoordinated and improper ventilation can have disastrous results by creating a flow path that greatly increases the intensity and spread of the fire.
- Ventilation is normally assigned to ladder companies, but it must be accomplished at every fire regardless of who is assigned the task. Many fire departments do not have ladder or truck companies, so the responsibility falls to the engine company.
- There are two types of ventilation: vertical and horizontal. There are various methods to carry out vertical and horizontal ventilation: ventilating for fire, ventilating for life, mechanical ventilation, hydraulic ventilation, and natural ventilation.
- Vertical ventilation can change the direction of the fire by stopping the horizontal spread of smoke and fire.
- The ventilation hole created by firefighters creates a new path of least resistance for the pressurized smoke to escape.
- A vertical hole, whether created naturally or intentionally, becomes the path of least resistance, and smoke and flames are drawn toward that hole.
- The louver cut is one of the most popular vertical ventilation cuts for residential structures. It is particularly suitable for flat or pitched roofs with plywood decking. It is quick, efficient, and easy to extend if a larger ventilation hole is needed.
- For ventilation-limited fires or pre-backdraft conditions, firefighters should ventilate a smoldering fire at the highest point possible before making entry for fire attack.
- To prevent a backdraft explosion, a smoldering fire should be ventilated before any attempt is made to enter the fire building.
- After ventilation, a smoldering fire burns freely and may be attacked using fire department standard operating guidelines.

KEY TERMS

auto-exposure Vertical fire extension from floor to floor through the failure of exterior windows.

combination attack A type of fire attack employing both the direct and indirect attack methods.

direct attack Firefighting operation involving the application of extinguishing agents directly onto the burning fuel.

fast water The application of water onto the fire by the fastest means possible.

indirect attack Firefighting operation involving the application of extinguishing agents to reduce the buildup of heat released from a fire without applying the agent directly onto the burning fuel. (NFPA 1145)

opposing lines The inadvertent positioning of two attack teams approaching each other on the same level from opposite sides of the structure.

transitional attack An offensive fire attack initiated by an exterior, indirect hand-line operation into the fire compartment to initiate cooling while transitioning to interior direct fire attack in coordination with ventilation operations. Also known as *blitz attack, defensive-to-offensive exterior attack.*

REFERENCES

A Career Lieutenant and Firefighter/Paramedic Die in a Hillside Residential House Fire—California. Executive Summary, National Institute for Occupational Safety and Health (NIOSH) F2011-13, March 1, 2012, Cincinnati, OH.

Angle, James, Michael Gala, David Harlow, William Lombardo, and Craig Maciuba, *Firefighting Strategies and Tactics Enhanced,* Third Edition. Burlington, MA: Jones & Bartlett Learning.

Evidence-Based Practices for Strategic and Tactical Firefighting. Burlington, MA: UL/NIST/NFPA/Jones & Bartlett Learning.

Fundamentals of Fire Fighter Skills and Hazardous Materials Response, Fourth Edition. Burlington, MA: Jones & Bartlett Learning.

Klaene, Bernard. *Structural Firefighting Strategy and Tactics,* Third Edition. Burlington, MA: NFPA/Jones & Bartlett Learning.

Nine Career Fire Fighters Die in a Rapid Fire Progression at Commercial Furniture Showroom, South Carolina. NIOSH F-2007-18, February 11, 2009.

Norman, John. *Fire Officer's Handbook of Tactics*, Fourth Edition. Fire Engineering.

Richman, Harold. *Ladder Company Fireground Operations*, Third Edition. Burlington, MA: NFPA/Jones & Bartlett Learning.

Rose, Stewart E. *Strategy & Tactics MCTO*. InSource. June 1995.

Routley, J. Gordon. *Firefighter Fatality Investigative Report: Sofa Super Store Fire*, 1807 Savannah Highway, Charleston, South Carolina, June 18, 2007. Also by: Michael D. Chiaramote, Brian A. Crawford, Peter A. Piringer, Kevin M. Roche, and Timothy E. Sendelbach.

Three Firefighters Die in Pittsburgh House Fire. U.S. Fire Administration Technical Report Series 078/February 1995, by J. Gordon Routley, Emmitsburg, MD.

CHAPTER 11

Backup Lines

LEARNING OBJECTIVES

- Explain the reasons that a backup line should always be laid after the initial attack line and why it should follow the path of the initial attack line.
- Realize that backup lines must be stretched whenever it is obvious that the fire will not be extinguished quickly with initial attack lines.
- Recognize that smaller initial attack lines should be shut down if they are ineffective and larger backup lines should be placed in operation.
- Examine the need to have adequate personnel on scene to operate initial attack lines and also to stretch and operate a larger-diameter backup line.
- Realize that the implementation of the two in/two out rule and the advent of RITs have settled the issue about whether backup lines are needed.

Introduction

This chapter discusses: the definition of the term *backup line*; the sizes, purposes, and uses of backup lines; and the personnel needed to operate them. It addresses the positioning, sizing, and use of backup lines. It also describes personnel considerations. Backup lines should be deployed on all structural fires that start with lines laid for an offensive attack.

Backup lines are the engine company's first line of defense. A **backup line** is an additional hose line used to reinforce the work of and to protect personnel when the initial attack proves inadequate or when a problem occurs with the initial hand line. Problems include the following:

- A split or tear in the hose jacket
- A ruptured attack hose
- A kink that stops the flow of water
- A hose line getting snagged or buried under falling debris
- A clogged nozzle
- A problem with a crew member, for example, a medical issue or self-contained breathing apparatus (SCBA) failure
- A floor or ceiling collapse
- A flashover
- An explosion
- A wind-driven fire
- Extreme fire behavior

Sizes and Purposes of Backup Lines

Backup lines are to be used when initial attack lines cannot control the fire quickly enough. Backup lines should be the same size diameter as or larger than the diameter of the initial attack line. They are not used for dedicated exposure coverage or to attack the fire at other positions. Rather, they are held in a position of readiness to back up the attack lines, in the same general area, when and if they are needed **FIGURE 11-1**. This is not to say the backup line can never be opened for interior exposure control—it can. Protecting the immediate interior exposures surrounding the initial attack team is part of protecting and backing up that attack team. If interior exposure control is required, however, another hose line dedicated to that task should be called for. Depending on what's burning and the speed of extension, an exposure line may sometimes need to be laid first to stop the fire spread.

FIGURE 11-1 Backup lines should be the same size diameter as or larger than the diameter of the initial attack line.

When discussing initial fire attack on a burning structure, the question of selecting the proper hose size and nozzle is often framed as a dilemma for the company officer. It is not. Although some fire departments still use 1½-inch hose lines, and others also use 2-inch hand lines, the majority of fire departments use 1¾-inch and 2½-inch hoses for fire attack. The reality is that there are usually two choices: a 1¾-inch hose line or a 2½-inch hose line. It's the same for the nozzle; the company officer has basically two choices: a combination nozzle or a smooth-bore tip nozzle. For the most part, that decision is already made by your fire department. Department policy and standard operating procedures (SOPs) must establish the standardization of apparatus and equipment within the department. You cannot have oncoming crews show up for each shift to new hose loads and equipment caches based on the preference of the previous shift. Not every apparatus is identical, so there are variations, but hose loads are usually standardized, including what type of nozzle is attached to the respective loads. Thus, the choice comes down to pulling the 1¾-inch hand line or the 2½-inch hand line for offensive structure fires.

In Chapter 8, we discussed value-time-size (VTS). The S represents the "size" of the fire and how many gallons per minute (gpm) of water are needed to overwhelm the British thermal units (BTUs). This is determined by using the National Fire Academy (NFA) fire flow formula or the Royer-Nelson formula. Knowing the calculated fire flow requirement is a better indicator of the size of the hose that should be used for a backup line rather than simply making the general statement that the backup line should be a larger-diameter hose.

Determining the size of the backup line includes the consideration of other criteria. An experienced engine officer can ask if this fire can be extinguished with 500 gallons or less. This is the standard capacity of the water tank on most engine apparatus. One or two room and contents fires can usually be extinguished with the initial hand line using less than 500 gallons. The last factor is determining if you have enough firefighters to handle a 2½-inch backup line, a task that takes at least three. If not, that pretty much settles the issue. Taking a 2½-inch hose line into a residential house fire is not easy. With the weight of the charged 2½-inch hose (approximately 106 pounds [48 kg] per 50-foot section), and accounting for all the corners and bends that will be encountered, it could take a crew of four or five to overcome the friction to advance this hose line effectively. For quick mobility in residential house fires, a 1½-inch backup line is not an advantageous choice. Besides, the new modern 1¾-inch nozzles attached to a 1¾-inch hose line can flow upward of 250 gpm. That's a lot of water for a house fire with standard household furnishings. A 1¾-inch hose also has the reach to penetrate through the majority of single-family apartments and residences. Remember, "Little fire, little water; big fire, big water."

Still, backup lines should at least be the same size as the initial attack line, if not bigger. Again, that leaves you with only two choices. Don't misinterpret the simplicity of this issue. Depending on the threatened exposures and the speed of fire spread, the company officer has the discretion to select any size hose or type of nozzle from the apparatus. But remember that experience and past successes have already entered into the development of department SOPs, which takes a lot of the guesswork out of the initial decision making. There's not a wrong choice, just a better one. That being said, firefighters stand a better chance of controlling and extinguishing the fire in a shorter period of time, preventing further property damage, if they select a hose size the crew can effectively manage.

Nevertheless, there is always the possibility of a sudden increase in fire activity after the initial attack has begun. Flashover, particularly with modern fuels; the spread of fire to a new fuel source; interior construction assemblies that do not limit fire spread; or the creation of a wind-driven scenario can quickly accelerate the fire intensity to the point where the initial attack line is incapable of extinguishing the fire. Other such extreme fire behavior events could occur on the fireground, including backdraft, a smoke explosion, the sudden involvement of hazardous materials or exotic fuels. The chance of any of these events occurring is great enough that the engine company equipped with a 2½-inch backup line should be ready to intervene.

The mind-set in attacking residential fires shouldn't be applied to commercial structure fires. Depending on what is burning and the burning stage of the fire, attacking a commercial fire with a 1¾-inch hand line may be sufficient for the initial attack line, but due to the substantially larger footprint of the building and substantially larger fire load, a 2½-inch hand line is usually the better choice for fire attack. A 2½-inch hand line for the backup is the best choice for commercial structure firefighting **FIGURE 11-2**.

The backup line should stand ready; the backup firefighters should not fight the fire along with the initial attack team. If another hose line is needed for fire attack, a second line should be laid. The exception is fighting Type II noncombustible building fires, where nonprotected, open web steel bar joists are used for roof truss and floor joist assemblies **FIGURE 11-3**. **Nonprotected steel** means that the steel is not covered or coated with fire-resistant material like it is in Type I fire-resistant buildings. The bare steel is exposed to the heat and flames, where ceiling temperatures can easily reach over 1,000° F (538°C). At temperatures between 1,000° and 1,100°F (538°C and 593°C), steel expands, warps, twists, and loses its tensile and compression strength; it can collapse under its own weight within 10 or more minutes of being subjected to this high-temperature thermal assault. In fighting Type II noncombustible fires, the 2½-inch backup line should be used to cool the open web steel bar trusses at the ceiling and the steel columns that support them, while the attack team focuses on the seat of the fire.

FIGURE 11-2 Depending on what is burning and the burning stage of the fire, attacking a commercial fire with a 1¾-inch hand line may be sufficient for the initial attack line, but a 2½-inch hand line is usually the better choice for fire attack. Note the hose layering. The 2½-inch hose line is on top of the 1¾-inch line. The 1¾-inch line was laid first.

Courtesy of John Odegard.

FIGURE 11-3 The 2½-inch backup line should be used to cool the open web steel bar trusses at the ceiling and the steel columns that support them, while the attack team focuses on the seat of the fire.

Positioning of Backup Lines

Backup lines should be stretched whenever it is obvious that the fire will not be extinguished quickly with initial attack lines. This means that backup lines should be taken into the fire building right after the initial attack lines enter the structure. They should be charged, follow the initial hose lay, and positioned to cover the same area as the initial attack line. The team-assigned backup should not enter the building from a different access point. There are some who believe that as long as the backup hose ends up in the same area as the attack team, the protection level is the same. Not so; if conditions suddenly change for the worse, the attack team is trained to follow their hose line back out of the building. If the backup line is laid in from a separate entrance, the backup team is no longer in a position to back out with the attack team or protect the attack team from thermal assault. The backup line serves as the initial RIT line until a dedicated RIT line is laid.

Backup lines are of little use if they remain on the pumper. In the time it takes to get a backup line off the pumper, stretched into position, charged, and placed into operation, a fire can become completely out of control. Backup lines must be where they are needed, at the time they are needed, to protect and support the attack team if they are to establish control of the fire.

Hose Order and Hose Layering

Hose Order

In a residential structure fire, a minimum of four hose lines should be pulled: an attack line, a backup line, an exposure line, and a rapid intervention team (RIT) line. We can all agree that the attack line is the first hose line pulled, but which ones are second, third, and fourth? Most argue that a backup line should be laid immediately after the attack line. Others say that an exposure line needs to be laid first to prevent fire spread. The RIT line should be fourth because a backup line can act as an RIT line should an incident occur before a dedicated hose line can be set up for the RIT team. Remember, the two in/two out initial RIT needs to be established when the attack team enters. Though highly advisable, a dedicated RIT hose line may be the second line laid, but it is not required at this time.

So which line is second and which one is third? It depends. Rather than having a hard-and-fast rule, some factors must be considered and depend on the situation and the current fire conditions. Let's say that you have a one-story wood-frame house with fire blowing out the window on the B side, and the flames are impinging on the D side of the exposure house next door. In this scenario, the first hose line would be the attack line and the second hose line would be the exterior exposure line. The third line would be a backup line to the attack line **FIGURE 11-4**.

Consider this example: a two-story, 40-foot × 60-foot wood-frame house is on fire, with the fire showing from the kitchen on floor 1 at the C-D corner of the house and extending to floor 2. There are interior exposure bedrooms on floor 2, and flames are starting

FIGURE 11-4 If a wood-frame house has fire blowing out the B-side window, and the flames are impinging on the D side of the exposure house next door, the second hose line is the exterior exposure line. The third line is a backup line to the attack line.

to show from the bedroom above the fire. The 1¾-inch attack line would be the first line stretched into floor 1. A decision to take the second 1¾-inch hose to floor 2 as an exposure line is a sound tactic and the correct decision **FIGURE 11-5**. A 2½-inch backup line for the attack line is the third line laid. An RIT line would be the fourth.

Hose Layering

At some point in the majority of hose lays, three lines will cross, or be layered over each other. The 2½-inch line, if used, is the heaviest of the three hoses, and it will be very difficult for the attack line or the exposure line to advance if either one is weighted underneath subsequent hose lines. Thus, **hose layering**, or spaghetti, as it is sometimes called, must be considered. In Figure 11-5, the 2½-inch backup line might be laying on top of the attack line on floor 1 and the exposure line to floor 2, making it difficult for them to advance. If the 2½-inch backup line is laid second, and the 1¾-inch exposure line to floor 2 is laid third, the exposure line could advance or reposition with very little effort because this hose is on top of the other two. Because the 2½-inch backup line is on the same floor as the attack line—and is not necessarily flowing water—the backup team can assist the attack hose team in advancing or repositioning the attack line.

When hoses are layered over each other, there is significant weight and friction to deal with, especially when hose layering happens around corners or up stairways. Hose layering is also difficult to deal with in multistory commercial complexes and in the stairwells of high-rise buildings. In those cases, no more than three hose lines should be laid in the same stairwell in the same general area. It becomes too difficult, if not impossible, to advance any one hose line, especially the one on the bottom, which is the first line laid **FIGURE 11-6**. The ability to operate multiple interior 2½-inch lines depends on the available staffing responding to the fire. Realistically, only two 2½-inch lines can be effectively advanced in stairwells and hallways.

In most scenarios, the backup hose line should be the second line laid. But if immediate exposure problems threaten the lives of trapped occupants, or flames impinge on an exposure and thus greatly increase the scale of the emergency, the exposure line should be second and the backup line third. If the second and third engines arrive in short sequence or at the same time, the decision isn't as critical because both tactical positions will be covered within a few minutes of their arrival.

FIGURE 11-5 A fire in the growth stage or the fully developed stage on floor 1 is extending and endangering interior exposures on floor 2, with a possible life hazard. In this situation, the second line stretched should be an exposure line to floor 2. The backup line should be the third line stretched, and the RIT line the fourth.

FIGURE 11-6 Hose layering. Three 2½-inch hose lines will be extremely difficult, if not impossible, to advance up a stairway. Realistically, two 2½-inch hose lines should be the limit.

Courtesy of John Odegard.

FIGURE 11-7 Adequate personnel must be available to advance and operate the charged 2½-inch hose lines effectively around corners and down hallways.

Courtesy of John Odegard.

Sizes of Backup Lines

If the initial 1¾-inch attack lines are not gaining control of a fire within 30 to 60 seconds, it is probable that the streams are not reaching into the seat of the fire or that these hose lines are not delivering enough water to cool the area. In this scenario, the use of additional hose lines of the same size will not solve the problem. They will certainly not penetrate any farther into the fire than the original attack lines, and they will do little toward increasing the cooling effect.

For these reasons, the backup lines should have greater reach and deliver more water than the initial attack lines. Thus, in general, backup lines should be at least one size larger than the initial attack lines. Again, you must have the personnel available to advance and operate the charged 2½-inch hose lines effectively around corners and down hallways; otherwise, it will be extremely difficult to launch an effective attack because of the weight of the hose lines **FIGURE 11-7**.

Backing Up 1¾-Inch Lines

Some fire departments still insist on using 1½-inch hose lines for attack lines on certain fires. Their rationale is that using fewer gpm reduces water damage, thus conserving property, and it extends the operational time for preconnected lines running off the 500-gallon tank supply. But modern-fuel fires require more water to extinguish, and the majority of fire departments have already made the switch to 1¾-inch hose lines, which is now the standard initial attack hose line. Some fire departments use 2-inch hose lines for fire attack, which is also very beneficial and easier to manage than a 2½-inch, but they are still connected with 1½-inch couplings. (According to information provided by national fire hose manufacturers, 1¾-inch and 2½-inch account for the majority of their business sales of attack hose across the United States.) Based on the increased energy and heat-release rates emitted from modern synthetic fuels produced in fires today, combined with lightweight construction, more gpm are required to overwhelm the generation of the BTUs emitted. A 1¾-inch hose delivers more gallonage that a 1½-inch hose, and it can be handled easily by a single firefighter. If a fire department hasn't switched over to 1¾-inch hoses, they should—now.

When two 1¾-inch hose lines are used for initial attack, the backup line should be a 2½-inch hose line **FIGURE 11-8(A)**. A 2½-inch hose line delivers almost as much water, with far greater reach, than two 1¾-inch hand lines. It allows deeper penetration into the fire and a lot more cooling if a solid stream with a 1⅛-inch or 1¼-inch smooth-bore tip is used. A 2½-inch attack line should always be backed up by another 2½-inch hose line **FIGURE 11-8(B)**.

Adding another 1¾-inch attack line would do little except add to the water load in the fire building. If two 1¾-inch lines are used and they are ineffective, three

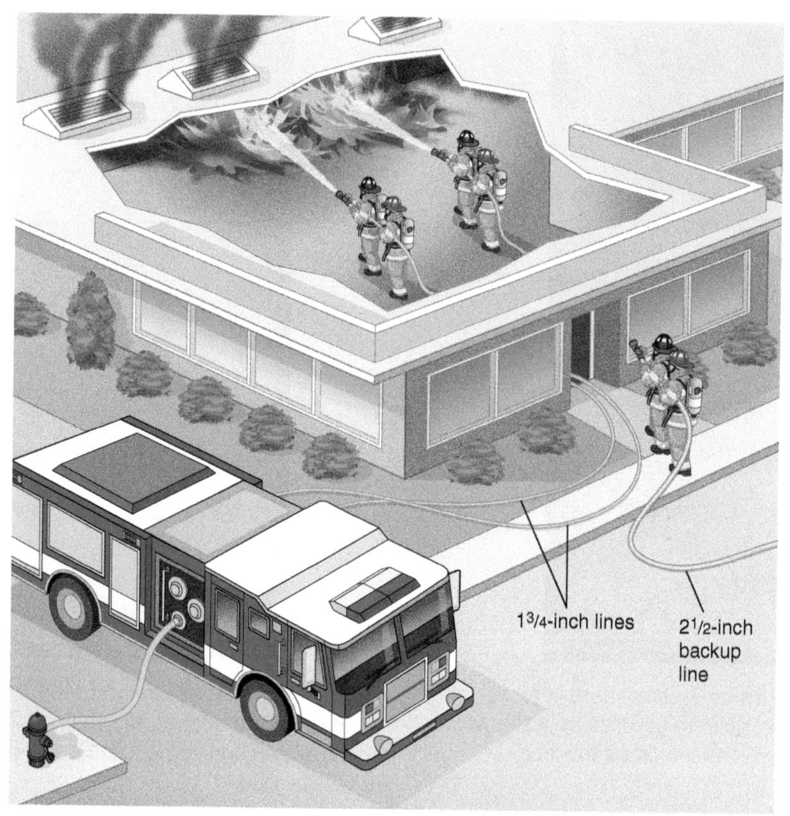

A

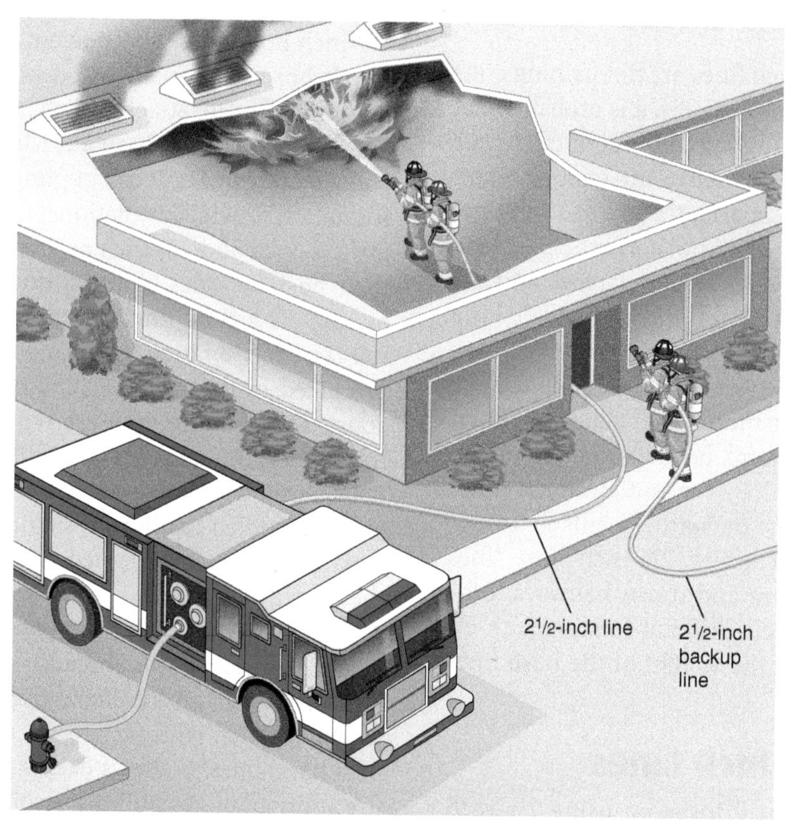

B

FIGURE 11-8 **A**. The 1¾-inch attack lines are reaching but not penetrating the fire, and the water volume is not sufficient to knock down or extinguish the fire. **B**. The 1¾-inch hand lines are backed out while the 2½-inch attack line is brought in. The 2½-inch attack line is reaching and penetrating the fire with sufficient water volume and increased gpm to begin control of the fire.

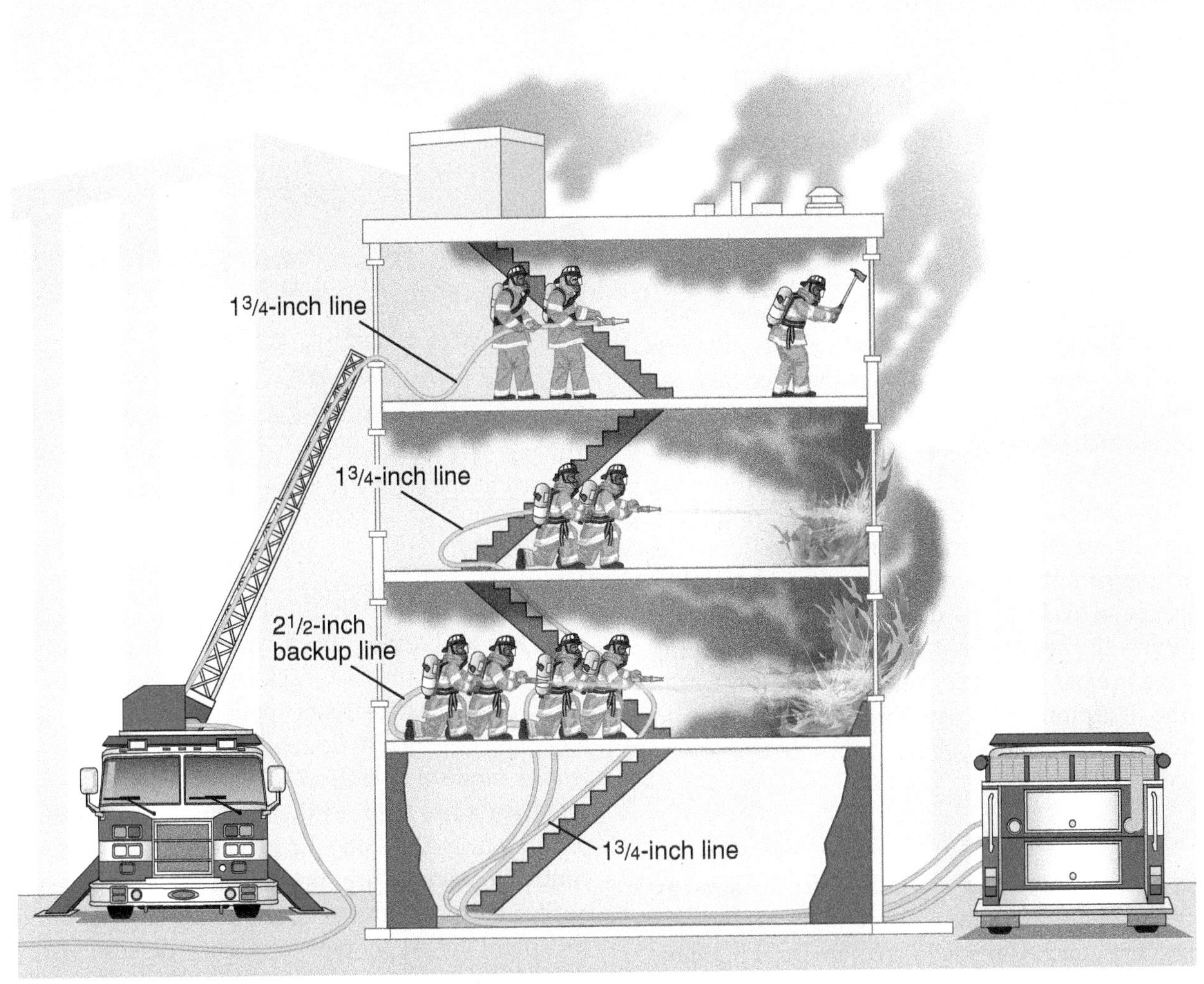

FIGURE 11-9 One 2½-inch hose line may be sufficient to back up two or more 1¾-inch attack lines when they are operating within one or two floors of each other.

or four of them would be just as ineffective. The result would be more water damage, not more fire control. The penetration and cooling effect of larger hose lines are what is needed.

The fire situation and the location and number of 1¾-inch lines determine how many 2½-inch backup lines are required. For example, one 2½-inch line may be sufficient to back up two or more 1¾-inch lines on the same floor if the floor is divided into apartments, work areas, or offices. When a large open area is involved, the number of backup lines might have to be increased. A 2½-inch backup line may be advanced to a particular floor to back up 1¾-inch lines on that floor and on the floor above FIGURE 11-9.

Backing Up 2½-Inch Hose Lines

If 2½-inch hose lines are used for the initial attack, backup lines should be 2½-inch hose lines FIGURE 11-10. If the 2½-inch initial attack lines are

FIGURE 11-10 If 2½-inch hose lines with combination nozzles are used for the initial attack, backup lines should be 2½-inch hose lines with smooth-bore tips.

Courtesy of John Odegard.

using combination nozzles, the backup lines should be equipped with smooth-bore tips. 2½-inch combination nozzles deliver up to 300 gpm. They can be adjusted from a wide-angle spray of almost 90° to a straight stream; however, the water delivery rate cannot be increased over their maximum in any nozzle position, and requires a tip pressure of about 100 pounds per square inch (psi).

Many 2½-inch smooth-bore nozzles carry stacked tips. Depending on the size of the tip, a smooth-bore 2½-inch backup line delivers more water with less pressure, and the stream penetrates farther into the fire. For example, at 50 psi, a 1-inch tip can deliver 225 gpm, a 1⅛-inch tip can deliver 265 gpm, a 1³⁄₁₆-inch tip can deliver 300 gpm, and a 1¼-inch tip can deliver 325 gpm. This stream has a better chance of controlling the fire than a straight stream from a spray nozzle; in the long run, it can gain effective control with less water expended. In many departments, the 2½-inch hose line is the largest weapon available for attacking a fire. If a fire has gained considerable headway by the time the first pumpers arrive, these hose lines should be placed in service without delay.

Master Streams for Backup

Master streams are typically used for defensive strategy fires. When master streams are used for backup, it will be when the attack is made in a commercial building where there is still substantial savable property, or in commercial open-space high-rise buildings. When a fire is so serious that the heaviest hose lines (2½-inch hose lines with 1⅛-inch and 1¼-inch tips) must be used for the initial attack, the chances of fire spread and increased intensity are great. It is imperative that master stream appliances be used for backup **FIGURE 11-11**. Portable deluge sets, monitors, deck guns, ladder pipes, and platform nozzles should be set up and ready for use if an adequate water supply is available. If the responding department does not have such equipment but a neighboring department does, then that department should be called to the scene as soon as the potential of possible involvement is determined. If a master stream is needed for backup, the fire is well-involved, and the strategy will most likely shift to defensive.

The master stream water supply should be charged to a forward control point like a tri-gate or manifold, or to the device itself if it is equipped with a shut-off so the appliance can be placed in service with no loss of time. In many cases, master stream appliances have been necessary to cover the escape of firefighters from particularly dangerous positions because such appliances can keep away heat and fire as the firefighters make their way to safety.

FIGURE 11-11 When a fire is so serious that 2½-inch hoses are used for the initial attack, the master stream appliances should be used for backup.
Courtesy of Task Force Tips.

When an interior attack using 2½-inch hose lines fails to control the fire, back out all firefighters, and attack using master stream appliances. This tactic calls for changing strategy from an offensive mode to a defensive mode of operation.

When a fire requires the use of master stream appliances from the start, the problem is usually that of controlling overall fire spread. In this case, the need is not for backup lines but rather for additional heavier streams of the proper type for general defensive fire attack. Collapse zones must be established. Operations requiring master stream appliances from the start are covered in detail in Chapter 13.

Use of Backup Lines

If all goes well, the initial attack lines extinguish the fire, and the backup lines can be repositioned to check for extension, used for overhaul, or they can be withdrawn; however, if the backup lines have to be placed in service, the initial attack lines should be shut down if they are ineffective against the fire. There is little that they can do in support of the heavier streams, and keeping them in operation robs water from the water main supplying appliances that require more volume. An incident commander should not allow firefighters to operate ineffective attack lines while the building burns down around them. Ineffective hand lines should be shut down, and a concerted effort should be made toward getting additional heavier streams into operation. After the backup lines have gained control of the fire and an advance can be made, the smaller lines can be used to perform final fire extinguishment and overhaul (mop-up) operations. You may redeploy firefighters from the attack line to the backup line.

FIGURE 11-12 For final extinguishment and overhaul, 1¾-inch hose lines utilizing foam or water are efficient and effective.

If a 2½-inch hose has been used for both initial attack and backup lines, then a 1¾-inch hose line can be attached to the 2½-inch nozzles or wyes for extinguishment and overhaul operations. Again, the purpose is to minimize the amount of water delivered into the building, consistent with complete and efficient extinguishment of the fire **FIGURE 11-12**.

Personnel

Hose lines are needed to attack the fire, to get above and around the fire, and to back up initial attack lines. These operations take an adequate number of personnel. The backup assignment must not be overlooked, but avoid giving this assignment to firefighters who have already been engaged in initial attack operations. These firefighters are now tired and beaten down by the physical activity, heat stress, and the weight of personal protective equipment (PPE) and SCBA used during the initial fire attack. They need rehabilitation before being reassigned to the fireground. If they are pulled off the initial attack lines and assigned to backup lines, the backup operation will be understandably limited; avoid using them if possible **FIGURE 11-13**. Additional (fresh) firefighters must be assigned to backup duties. Backup lines should be stretched while the initial attack is in progress. The incident commander is responsible for maintaining an adequate number of personnel on scene to perform the necessary tasks. Remember, this fire may come at the beginning of the shift; there may be more fires to fight by these same firefighters before the end of their shift.

FIGURE 11-13 Firefighters rotating out of the initial fire attack have taken a beating and need a break. Avoid reassigning firefighters who were engaged in the initial fire attack to a backup line.

Courtesy of Lt. Mike Heaton.

If there are not enough personnel at the fire scene, additional firefighters should be summoned promptly. In volunteer departments, backup duties can be assigned to firefighters who arrive after the first companies. If this is impractical, neighboring departments can be called to handle this assignment. In paid career departments with additional on-duty companies, there should be no hesitation in getting the necessary engine companies on the scene; call for an additional alarm. Smaller paid departments should also plan for backup operations, even if arrangements must be made ahead of time with neighboring departments through a mutual aid agreement.

By the same token, master stream appliances needed for backup should be set up by incoming engine companies laying the proper supply lines to operate the appliances. This eliminates much work on the part of firefighters. Hand-laying large-diameter hose or multiple supply lines, called cross-lots, is sometimes necessary. This action can be slow and labor-intensive, and it can consume a lot of energy. The master stream appliances should be charged and operated by fresh crews, again for speed and efficiency. If firefighters assigned to the initial attack are forced to withdraw from the fire building, they can hook up their 2½-inch hose lines into the master stream appliances for increased water supply, if this has not been done already.

In addition to backup lines, an RIT of at least four firefighters equipped with full PPE, SCBA, and necessary tools and equipment should be standing by, ready to enter the building. RIT teams have their own dedicated 1¾-inch hose to lay over existing hose to advance the line without the problems encountered with hose layering. This crew consists of personnel who may need to respond immediately to assist in locating or removing a lost or trapped firefighter or to respond to any incident involving a firefighter in distress.

After-Action REVIEW

IN SUMMARY

- A backup line should be laid at any fire where an initial attack line is deployed.
- Typically, the backup line is the second hand line laid.
- Under certain conditions, the exposure line should be the second line laid and the backup line is the third.
- Backup lines should be larger than the initial attack lines, measured in gpm flow, and they should be laid out, charged, and assigned a crew to operate them when it becomes necessary.
- Two-and-a-half-inch hand lines should be the backup lines used at commercial structure fires.
- Fighting fires in Type II noncombustible buildings requires that a 2½-inch backup line operates during the fire attack to cool unprotected steel bar open web trusses and the unprotected steel columns that support those trusses.
- Fully developed fires in Type II buildings should be fought from the outside using the reach of the hose streams.
- Use a smooth-bore nozzle on backup lines. Advance backup lines into the building in the same general area where initial attack lines were advanced.
- The crew assigned to the backup line must be in full PPE and SCBA so that they can enter the building quickly to assist other firefighters if necessary.
- At large fires where 2½-inch hand lines are used in the initial fire attack, the 2½-inch lines should be backed up by master stream appliances ready to operate.

KEY TERMS

backup line An additional hose line used to reinforce and protect personnel in the event that the initial attack proves inadequate. (NFPA 1410)

hose layering When two or more hand lines are laid over each other in the same hallway or stairway.

nonprotected steel Bare steel used in Type II noncombustible buildings for trusses, joists, and columns, without any fire-resistive material covering or coating the steel.

CHAPTER 12

Exposure Protection

LEARNING OBJECTIVES

- Define the term *exterior exposure*.
- List the ways that fire can spread to exterior exposures.
- List the actions that can be taken to prevent fire spread to an exterior exposure.
- Define the term *interior exposure*.
- List the actions that can be taken to prevent fire spread to interior exposures.
- Define a movable exposure.
- Explain initial response considerations affecting the severity of external exposure problems.
- Explain when and how master streams are used to protect exposures.
- Examine and compare the tactics used to control basement fires.
- Describe how basement fires spread vertically and horizontally.

Introduction

In this chapter, both exterior and interior exposures are discussed. The latter are the less spectacular, but interior exposures may demand as much effort from firefighters to control. Basement fires, which can lead to serious interior exposure problems, are discussed in the final section of the chapter.

Definition of Terms in Exposure Protection

Firefighting operations can be difficult and dangerous for fireground personnel attempting to protect exposures. Exposure protection is needed to shield a building or part of a building that has been subjected to radiant and convected heat, as well as direct flame impingement from the main body of fire **FIGURE 12-1**. An adequate water supply must be initiated, and firefighters must place hose lines or master streams in strategic locations to cover the maximum amount of exposed area. These positions should afford them protection from the effects of the fire while providing a vantage point to protect exposures and stop the spread of fire into uninvolved areas of a building.

As in any fireground activity, the incident commander (IC) must consider the size of the fire and the risks to firefighters performing exposure protection operations. The IC must continually evaluate the current conditions to ensure the safety of firefighters working in designated positions. This can be accomplished by performing an ongoing size-up and conducting a risk-benefit analysis. Firefighters must work in teams, follow directions, and refrain from freelancing on the fireground. They must pay attention to their surroundings and remain continually aware that they are working in an uncontrolled, dangerous environment.

Exposure coverage is second only to rescue on any list of the basic objectives of a firefighting operation. Structures near a fire building (the **exterior exposures**) and parts of the fire building not yet involved (the **interior exposures**) must be protected to minimize the danger to their occupants as well as to contain the fire. **Movable exposures** are exposures that can be protected by physically moving them away from the reach of the fire. Examples are trucks, tractors, automobiles, boats, yachts, and any other motorized equipment that can be driven or towed away from the fire **FIGURE 12-2**. At many fires, the efforts of engine companies to obtain an adequate water supply quickly, get initial attack lines in service, and back up those attack lines with other hose lines are necessitated by exposure problems.

As in every phase of firefighting, area company inspections and pre-incident planning are important parts of exposure protection. A **pre-incident plan** helps to locate **exposure hazards** (conditions or situations in or around a structure that can promote the spread of fire), as well as areas or neighborhoods in which fire spread is especially likely. Pre-incident planning should ensure that sufficient equipment and personnel are dispatched on the first alarm to cover these exposures. If the incident is quickly developing, the first alarm response should be increased by calling a second alarm or mutual aid.

An adequate number of firefighters is the key to full exposure protection. Implicit throughout this chapter is the need for firefighters and equipment over and above those engaged in fighting the fire itself. Firefighters are needed to check for secondary

FIGURE 12-1 Exposure protection is needed to shield a building or part of a building that has been subjected to radiant and convected heat, as well as direct flame impingement from the main body of water.
Courtesy of Dave Blair.

FIGURE 12-2 Movable exposures can be protected by physically moving them away from the reach of the fire.
Courtesy of John Odegard.

fires and control them, direct hose lines on exposed structures, set up additional water supply lines, open interior channels in which fire may be spreading, and carry out other tasks that may need attention. Personnel and the necessary equipment must be available.

Responding ladder companies can perform some of these jobs; if they are not available, engine company personnel must do this work. Additional alarms may be needed to summon the desired resources to combat the fire. Prearranged mutual-aid agreements between neighboring communities, agencies, and/or counties ensure the availability of additional resources.

Any special equipment that aids in firefighting should be dispatched to the fire scene on the first alarm. If fire departments do not call for this equipment until they reach the fire scene and find out it is needed, the loss of time may have serious consequences. For example, if a fire department does not have an aerial ladder or additional mobile water supply apparatus available and they will be needed at a fire event, they should be requested as soon as possible.

Another example is requesting the water, gas, and electric utilities. The water department should at least be notified that a major fire is taking place in case it needs to redirect or increase volume and pressure to the affected grid. Unless the incident involves a major gas leak, the gas company is usually not needed at structure fires unless gas shutoffs are underground. Most exterior gas meters, residential or commercial, can be shut off by the fire department. Electricity is always a hazard at any structure fire, and the fire department is ill-equipped or is not trained to deal with it. Other than throwing the main circuit breaker switch at the electrical panel, fire department personnel should not pull electrical meters or tinker with any other vaulted electrical equipment. The electric power utility should automatically be dispatched at any confirmed fire with the rest of the initial alarm assignment instead of waiting for the IC to request its presence **FIGURE 12-3**.

Requesting the power company's presence is sometimes an afterthought, much like requesting the police for traffic control. It's a hazard that threatens and has killed many firefighters and should be mitigated immediately by trained electrical employees of the utility who can drop the main line coming into the weatherhead **FIGURE 12-4**. This is the surest way to ensure that all power has been cut to the fire building, so call for the electric utility early—even while en route to the incident. Electricity is invisible, and firefighters have enough hazards to deal with. Knowing the power has been cut to the fire building can bring a sense of relief to firefighters who are pulling ceilings and opening void spaces where electrical conduits run throughout the building.

FIGURE 12-3 The electric power utility should automatically be dispatched at any confirmed fire with the rest of the initial alarm assignment.

Courtesy of David Dodson.

FIGURE 12-4 Electricity is a hazard that threatens and has killed many firefighters, and it should be mitigated immediately by trained electrical employees of the utility who can drop the main line coming into the weatherhead.

Exterior Exposures

The term *exposure fire* applies to the outside exposure; such a fire is regarded as one that spreads from one structure to another or from one independent part of a building to another, for example, across a court or between the wings of a building **FIGURE 12-5(A)**. An exposure hazard is an area that promotes the spread of fire if a fire should start in or reach that area **FIGURE 12-5(B)**.

Unpierced, intact fire walls and spaces that check the fire from spreading between buildings and stacked materials are the greatest deterrents to exposure fires

A

B

FIGURE 12-5 **A**. Firefighters should protect buildings from exposure fires, which can occur when radiant heat spreads a fire from one building to another. **B**. Firefighters here are protecting an exposure house from radiant heat.

and are of great assistance to firefighters when severe outside fires are burning. Outside sprinklers and spray systems are also a great help, but unfortunately, they are rare items in exterior fire protection equipment except in special installations.

Initial Response Considerations

Factors affecting the severity of an exterior exposure problem include the following:

- Recent weather
- Present weather, especially wind conditions
- Delay in reporting the fire to the fire department
- Traffic conditions
- Proximity of the fire to the exposures
- Building construction design and materials
- Intensity and size of the fire
- Location of the fire
- Availability and combustibility of fuel
- Size of the firefighting force
- Firefighting equipment and apparatus on hand
- Available water

The worst combination of factors might be recent dry weather, strong winds blowing toward the exposures, an area of closely spaced frame buildings, a severe fire that is difficult to reach, plenty of easy-to-ignite materials located between the fire building and exposures, limited personnel and apparatus response to the first alarm, and response time.

Of these factors, the fire department nominally has control of only the fire force and equipment responding to the first and additional alarms. For this reason, pre-incident planning is crucial. The fact that the department has control over only two of the 12 factors affecting exposure problems gives those two factors special importance. Equipment must be available when it is needed, firefighters must be trained in exposure coverage, and sufficient personnel must be available to perform all fireground tasks.

If pre-incident planning has been done properly and firefighters have been trained, responding companies will be aware of the construction and spacing of buildings and the availability of fuel in and around the fire area. Sufficient personnel and equipment should be available and ready to respond. What happens in the first few minutes on the fireground dictates the results over the course of the incident.

Recent weather is a matter of record, and present weather must be observed.

What may not be known at the time of the alarm are the size and intensity of the fire and its location, either in the building or the area. These factors have a direct effect on the amount of heat radiated from the fire building. The size and intensity of the fire, combined with the extent of structural involvement, determines the amount of radiated heat. Radiant heat not only keeps firefighters away from the fire building, but it can also severely damage parked fire apparatus and add to exposure problems **FIGURE 12-6**. **FIGURE 12-7** shows some of the characteristics of radiant heat.

Other factors that cannot be evaluated until the company arrives at the fireground are wind direction and wind speed. Hot air, smoke, gases, and embers rise in convection currents and are carried downwind. The result could be a chain of exposure fires. This was the case in the Santana Row fire on August 19, 2002, in San Jose, California.

Santana Row Case Study

Santana Row is an upscale, residential and commercial urban village in San Jose, California. It is one of Silicon Valley's most affluent shopping districts and one of the most successful examples of mixed-use urban developments in the United States.

On Monday, August 19, 2002, during the construction phase, the largest building at Santana Row, Building 7,

FIGURE 12-6 Radiant heat not only keeps firefighters away from the fire building, but it can also severely damage parked fire apparatus.

Photo by Rick Mcclure.

caught fire. Building 7 was a six-acre wooden structure in the framing stage. The fire quickly spread, and soon the entire construction complex became fully involved. The San Jose Fire Department (SJFD) routinely inspected the construction site, and it was aware of the many hazards present, including the fact that this was the largest wood-frame building in the city. Before the entire incident was over, it would grow to a total of 11 alarms: the main body of fire required five alarms to control, and flying embers, which ignited several buildings downwind, ultimately developed into a separate six-alarm blaze. Extinguishment required the combined efforts of 221 firefighters and 65 pieces of apparatus. It was the largest structure fire in the history of San Jose and caused over $130 million in damage. In the end, six buildings and 40 apartment units were destroyed, three buildings were severely damaged, and 130 residents lost their homes.

The incident occurred just as the afternoon rush hour was beginning. Heavy traffic would delay the arrival of all multiple-alarm companies. The ambient temperature was 75°F (24°C). Skies were clear, and the winds were mild. The fire would create its own weather, however, principally high winds. These winds would carry burning embers into the air and ignite exposures south of Santana Row.

Convection Exposures. Flying firebrands and embers carried by convection and wind currents can cause serious fire containment problems. They are especially dangerous in large structures under construction in the wood-framing stage and in lumber yards and other open storage areas. A severe building fire, especially a fire that is burning through the roof

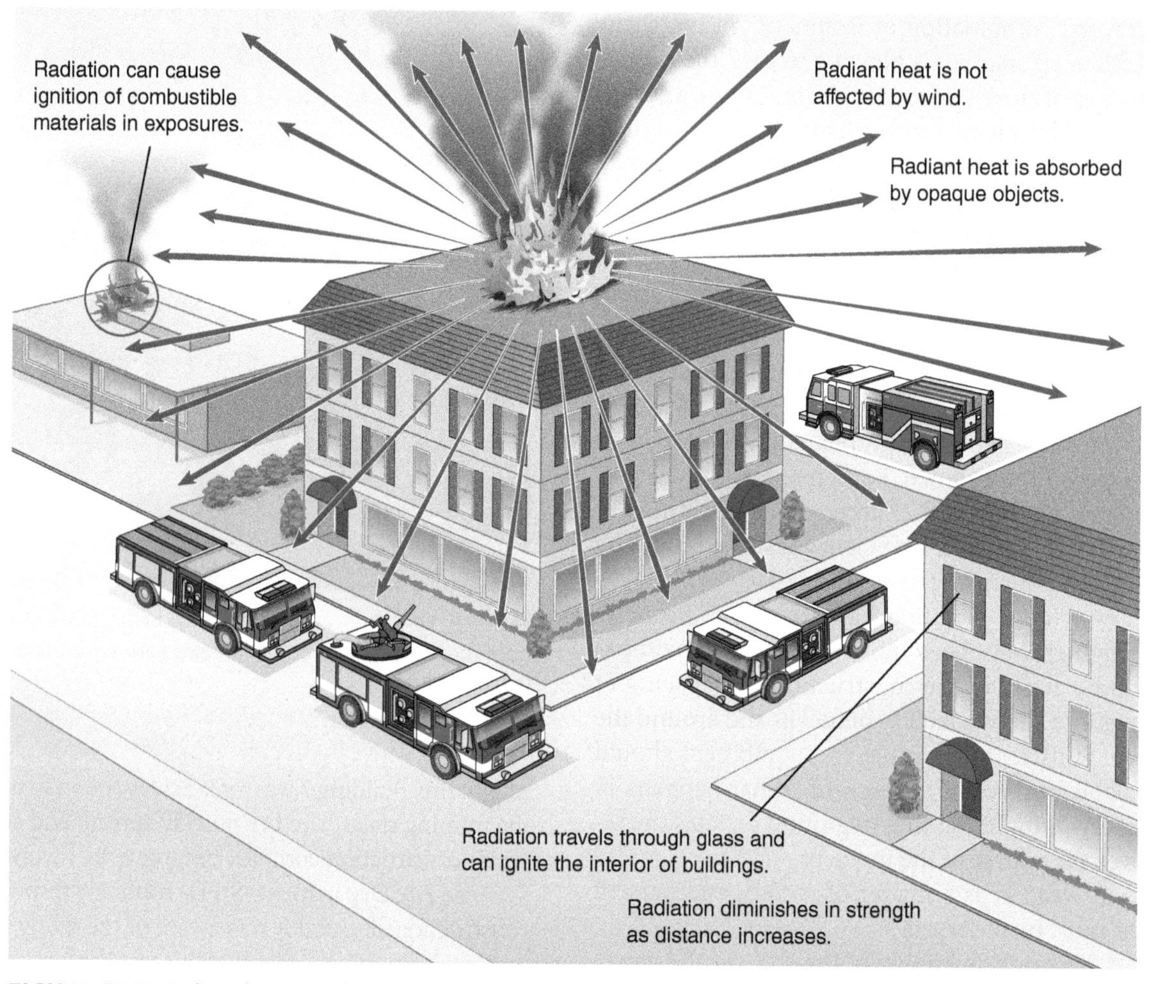

FIGURE 12-7 Radiant heat travels in straight lines in all directions.

or a fire in a building in the wood-framing stage of construction, can create strong convection currents above the building. In such situations, many fire departments have experienced exposure fires at great distances from the original fire, sometimes creating a problem worse than the initial fire **FIGURE 12-8**. Wood shingle roofs can also contribute to the firebrand problem; there are still areas of the country where wood shingle roofs exist.

The Main Fire. At 15:36, someone in a nearby high-rise building called 911 and stated that he could see flames and smoke billowing from the Santana Row construction site. A full response was dispatched. While en route, Engine 10 could see a heavy column of black smoke rising from the vicinity of the reported fire and called for a second alarm. Engine 10 arrived at 15:41 and reported a working fire on the upper level of Building 7. Building 7 was in the wood-framing stage, so no sprinklers were in place at the time. It was primarily wood, light-gauge steel, and Sheetrock—essentially a giant lumberyard fire producing massive amounts of flying fire brands and embers into the air and generating a tremendous amount of radiant heat. Construction workers who were preparing to leave for the day were scrambling down the scaffolding to escape the intense flames. Engine 10 attempted to access the vehicle ramp that led to the interior of the complex, but the size of the fire and the intensity of the radiant heat made it unsafe and untenable for the engine and crew to enter the area.

While en route, Battalion 10 requested a third alarm at 15:40. Assuming command at 15:42, he declared a defensive strategy on the fire and ordered master streams be placed in service to protect exposures and knock down the flames. The chief requested a fourth alarm at 15:49 and a fifth alarm at 15:52. The fourth alarm companies were all mutual-aid companies because the fire was located in the western edge of the city, and these companies were much closer to the fire than the next-due SJFD units.

Flying embers and radiant heat ignited vehicles, forklifts, portable toilets, and dumpsters. The water

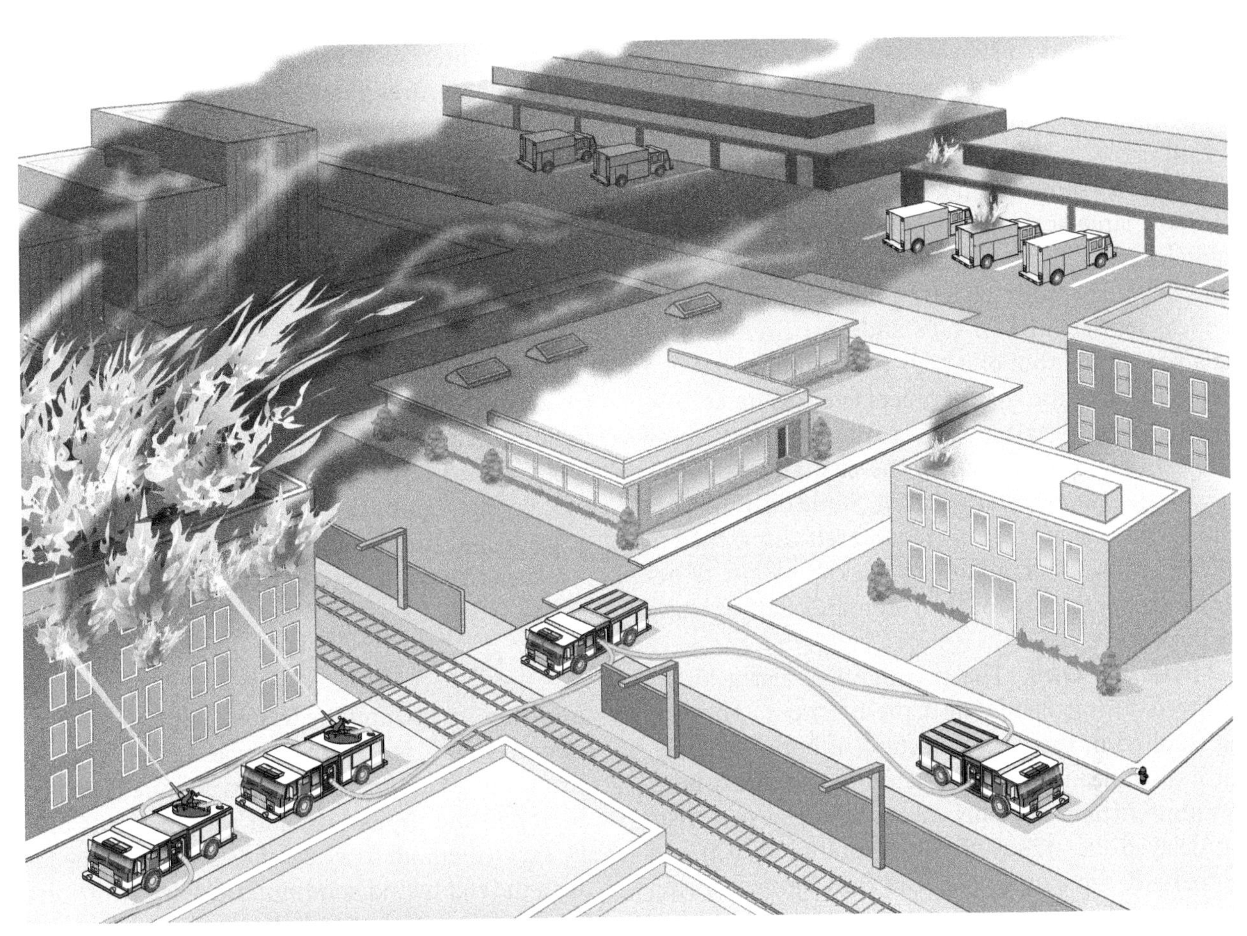

FIGURE 12-8 Convection can spread firebrands and embers to exposures some distance away.

department boosted system pressure in the area to maximum capacity. So much water was pumped onto the fire that the runoff flooded the underground parking garage, damaging approximately 160 vehicles. The fire at the construction site was held to five alarms and required the efforts of 119 personnel (11 chief officers, 103 firefighters, 5 dispatchers) and 31 pieces of apparatus to bring it under control.

Secondary Exposure Fires. At 15:53, one minute after the fifth-alarm companies were dispatched to Santana Row, a 911 call reported roof fires approximately ½ mile south of the fire. Flying embers, some as large as two-by-fours, were being carried off by the winds generated from the convective currents of the massive fire, and they ignited a number of fires downwind from Building 7. Exposure fires were started at the 68-unit Moorpark Garden Apartments and several townhouses at the Moorpark Village complex. The flying embers spread to 13 apartment buildings, some with wooden shingles. The fires also spread to the Huff/Moorpark area by the wind-carried embers. They ignited numerous wooden roofs in the single-family residential neighborhood, which was also a half-mile downwind. Communications advised the Santana Row IC that numerous calls of possible structure fires on Moorpark Avenue were received and a full response was dispatched at 15:59. While en route, Battalion 13 requested a second alarm at 16:06. Anticipating the request, Communications had already dispatched a second alarm at 16:01.

Engine 9 was the first company on-scene and reported a two-story apartment building with flames through the roof. The firefighters set up a master stream to protect exposures and deployed hand lines to attack the fire.

Battalion 13 arrived at 16:11 hours, assumed command, and declared the incident to be a defensive operation. Flying embers continued to ignite buildings in the area, including the several townhouses at the Moorpark Village complex. All the buildings in the complex that sustained fire damage had wood-shake roofs. Buildings with composition roofs largely escaped the conflagration.

The IC requested two Strike Teams from the county (third and fourth alarms), and at 16:17, an out-of-county Strike Team was ordered (the fifth alarm). A sixth alarm was requested at 16:56. The incident was held to six alarms and required the efforts of 102 personnel (93 firefighters and 9 chiefs) and 34 pieces of apparatus to bring these fires under control.

Within a 2-hour period, the SJFD was confronted with two large-scale events that would overtake the capabilities of all but the largest of fire departments. Still, the SJFD continued to respond to 12 emergency medical services (EMS) calls and four fires, including a fire on the roof of a high-rise building that had been ignited by the flying embers from Santana Row. When the fire was reported at the high-rise, there were no chief officers left in the city, and by 17:00, only 14 SJFD companies were left. Some off-duty firefighters staffed reserve engines and self-dispatched to the second incident, adding to the complexity of accurate accountability for the command staff working the incident.

Time of Day. The two multiple-alarm blazes occurred during the middle of the afternoon. The timing resulted in the early detection of both incidents, but it also meant heavy traffic during the rush-hour commute, which not only inconvenienced motorists but delayed the arrival of all the multiple-alarm and mutual-aid companies. Had the fires ignited during the night, when the residents of the Huff and Moorpark neighborhoods were asleep, it could have resulted in multiple fatalities. Miraculously, there were no deaths or serious injuries to firefighters or civilians. The cause of the fire was never determined.

Ineffective Master Streams. At the Santana Row and Huff/Moorpark incidents, many of the master stream appliances were equipped with fog nozzles. Due to the intense radiant heat and the potential for structural collapse, numerous appliances were positioned beyond the effective reach of a fog stream. Smooth-bore tips would have provided a longer effective reach under such conditions. During defensive operations, engine officers must be alert for the need to change over to straight tips in order to ensure the effectiveness of their efforts. Likewise, pre-plumbed deck guns are quicker to place into service and have a higher vertical reach than portable monitors that are normally positioned on the ground.

When fires occur in lumberyards or significant new construction projects in the wood-framing stage, a fire spreads rapidly in open air. There is plenty of fuel and plenty of oxygen and, once the fire gets going, plenty of heat—the perfect fire triangle. Construction projects in the wood-framing stage have few if any fire prevention features that would make any difference. Everything is primarily wood; vertical extension is swift in such circumstances, and a fire releases tremendous radiant heat, making it untenable for firefighters to have a good vantage point for hand lines. These fires require an immediate transitional attack, defensive to offensive; if the fire is gaining momentum, a defensive attack should immediately be initiated. Collapse is quick in the wood-framing stage, and a collapse zone must be established and maintained.

Because the wall studs are exposed and wide open in buildings in the framing stage, a visual search for construction workers or unauthorized occupants can be made from the exterior perimeter. Because of the quick and unpredictable fire spread potential along with the potential for early collapse, interior fire attacks should not be made until the fire is knocked down from the outside and the structural integrity of the building is evaluated by the IC and the Safety Officer.

Fire Patrols

Most important, and the lesson learned from the Santana Row fire, is that these types of *lumberyard fires* throw out giant burning embers and fire brands numbering in the thousands, and wind currents can easily carry them up to a half-mile away—maybe farther with the right wind conditions. Company officers and chiefs must anticipate the development of this type of scenario, and the strategy of exposure protection must be expanded to include fire patrols.

Under a Unified Command, police officers and other authorized, radio-equipped personnel, such as public works employees or employees of local utilities, can be assigned to the **fire patrol**. Cell phones can be used to communicate with the fire department. If personnel are to be efficient and if the fire patrols are to be effective, training sessions between agencies should be held to ensure smooth operations when fire patrols are required. Members of fire patrols must be taught how to conduct a systematic search grid around the neighborhood, starting with streets and alleys closest to the fire and then fanning out. They must be trained on what to look for, and they must learn how to report accurate locations and information to the fire alarm center.

Patrols are extremely important when wind conditions are severe. A secondary fire that is started from wind-carried firebrands can be quickly extinguished if it is caught early in the incipient stage. If it is permitted to burn unchecked, such a fire could become as serious as the original fire in a very short time. Many companies may already be committed at the original fire and using large volumes of water. If a second serious fire develops close by, there may be insufficient apparatus, personnel, or water to control both fires. The potential for disaster is great. It is highly recommended to read and study the entire Santana Row fire report online; it was an exposure worst-case scenario.

The 911 fire alarm center must be diligent in screening calls and gathering specific location information during a large fire **FIGURE 12-9**. Large fires can be seen for miles, and well-meaning citizens will call from different vantage points. They may report a fire that they can see, but they may be unsure of the exact location. The dispatcher may think the callers are reporting the original fire—where the fire department is already on scene—when they actually may be reporting an exposure fire very close to the main fire. The worst thing a dispatcher can say is: "Thank you. We've got that. Fire units have responded and we're on scene." If there are no fire units available to investigate, the fire patrols should check these calls for accuracy. If any call results in the fire patrol finding an exposure fire caused by flying embers and firebrands from the main fire, the patrol can notify the alarm center that this is a separate fire incident.

FIGURE 12-9 Dispatchers must be diligent in screening calls for specific location information during a large fire. A caller may actually be reporting a separate fire close to the ongoing incident.

Radiant Heat Exposures

Radiant heat moves away from the fire building in all directions; winds do not affect it. Thus, fire may spread by radiation to any building near enough to the fire building to absorb sufficient heat. The only way to protect exposures from radiant heat is to cool them by the application of water. If exposures are wet, they won't burn.

A water curtain is not an exposure stream. Water itself is transparent, and radiant heat passes through it **FIGURE 12-10**. Water curtains or operating a stream of water between the fire and the exposure does not protect the exposure. Radiant heat moves through the stream and heats the surface of the exposure to its ignition point. Instead, the stream must be directed onto the surface of the exposure so that the water washes down its walls. The water absorbs heat from the exposure and thus keeps it from igniting. Again, if the exposure is kept wet, it won't burn. Water curtains are effective in extinguishing flying embers and firebrands rising in the thermal column—in this sense, a water curtain aimed into

FIGURE 12-10 Radiant heat can pass through a transparent water curtain stream and ignite an exposure; however, the cooling effect of water on the surface area of the exposure helps prevent ignition.

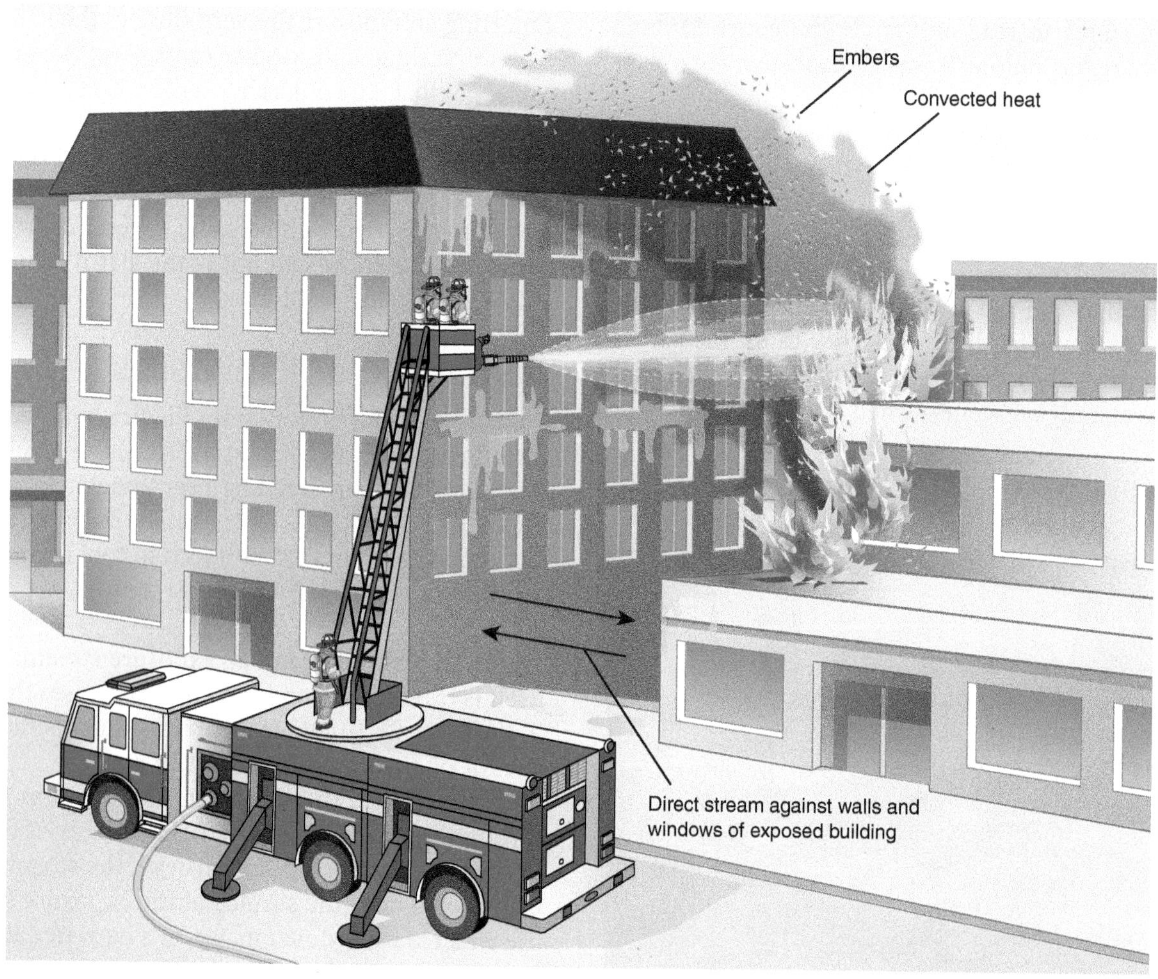

FIGURE 12-11 A water curtain is not an exposure stream, but a water curtain is effective in extinguishing flying embers and firebrands rising in the thermal column, which in turn helps protect exposures downwind.

the thermal column does help protect exposures downwind **FIGURE 12-11**.

Radiant heat also passes through transparent glass and ignites materials within a building. If the outside surface of a building is in danger of ignition from radiant heat, the areas within its windows constitute an equal hazard. In this case, the building should be entered, and each floor checked. It may be necessary to use forcible entry into an exposure building in order to protect it. Don't hesitate to make this call. Hose lines should be taken into the exposure building so that they can be placed into operation if needed. Curtains, shades, blinds, and other window coverings should be removed. All furniture, wall hangings, pictures, and other combustible material should be moved away from the walls exposed to the fire (facing the fire) to the opposite side of the room. If the building has a standpipe system, it should be charged. If the exposure has a sprinkler system, a pumper should be connected and readied to charge the system if necessary. (The use of standpipe and sprinkler systems is discussed in more detail in Chapter 15.)

Exposure Coverage

The size and location of the fire, and the wind conditions—factors not known until the company arrives at the fireground—determine what actions are needed to cover exposures. The structures or the materials nearest the fire must be covered, beginning first on the leeward or downwind side. The combination of convected and radiant heat makes the leeward side most vulnerable **FIGURE 12-12**. After the leeward side is covered, other areas must be protected because of the spread of fire by radiant heat alone.

When the exposures are close to the fire building, the most vulnerable areas are the parts of the exposed buildings just above the fire **FIGURE 12-13**. The radiant heat and the hot air, gases, and smoke tend to concentrate there. To be effective, streams should be

FIGURE 12-12 The structures nearest the fire must be covered, beginning first on the leeward or downwind side. The combination of convected and radiant heat makes the leeward side most vulnerable.

directed onto the exposure at a level somewhat above that of the fire.

Liquified petroleum gas (LPG) tanks, which can serve as the primary heating source for many structures, can sometimes be placed right up against the building. When the building is on fire, radiant heat or direct flame impingement can make this type of exposure extremely volatile and must be protected. The officer must assign a firefighter to shut off the main valve, then order an exposure line to cover the tank specifically. The greatest hazard potential is a boiling liquid expanding vapor explosion (BLEVE) if the tank isn't kept cool with a fog stream **FIGURE 12-14**.

Hose Lines and Nozzles

Water Streams

The hose lines and nozzles must be of the proper size to deliver the required volume of water in order to control the spread of fire and extinguish it. You might think that covering the exposures is not part of the fire attack—on the contrary, it is an integral part of fire attack. Streams developed by nozzles should deliver enough water to cool the exposure and should be large enough to reach the exposure, penetrating the fire if necessary, without being dissipated by the heat and draft of the fire. The streams must carry through the fire area to the exposure with enough water to prevent ignition. In some instances, one firefighter with a 1¾-inch line can protect an exposure; in others, 2½-inch hose lines or master stream appliances may be required.

Although handheld hose lines can be effective, they require more staffing, place firefighters closer to the building, and provide less water than master streams. Unattended 2½-inch hand lines can be set up quickly by securing them to ground ladders or other stationary objects and aimed at the fire, freeing up firefighters to perform other critical tasks **FIGURE 12-15**. Sometimes a handheld hose line is the only way to protect an exposure, but master streams provide safer and more effective streams for most situations requiring exposure protection.

Streams must also be large enough so that they are not affected by winds. A strong wind may necessitate the use of streams that are larger than usual, especially on the windward side of the fire. Streams developed by combination spray nozzles, although effective in cooling exposures, are very susceptible to breakup and reduction of efficiency by winds.

A

B

FIGURE 12-13 **A**. The red arrow points to the area of greatest danger to this exposure. Cover the exposure with water. **B**. The most vulnerable area on the exposure building is the area directly above the level of the fire.

A

B

FIGURE 12-14 A. LPG tanks are sometimes placed right against the building. The tank presents a potential BLEVE if it is exposed to radiant heat or direct flame impingement, and left unprotected by an exposure line. B. A firefighter must shut the main valve off, which is often at the top of the tank, then man an exposure line to cover the tank specifically.

A: Photo by Raul Angulo; B: Courtesy of Amanda Mitchell.

A

B

C

D

FIGURE 12-15 The four photos show how straight streams and fog patterns can be used with roof ladders as unattended exposure lines.

A-D: Photos by Raul Angulo.

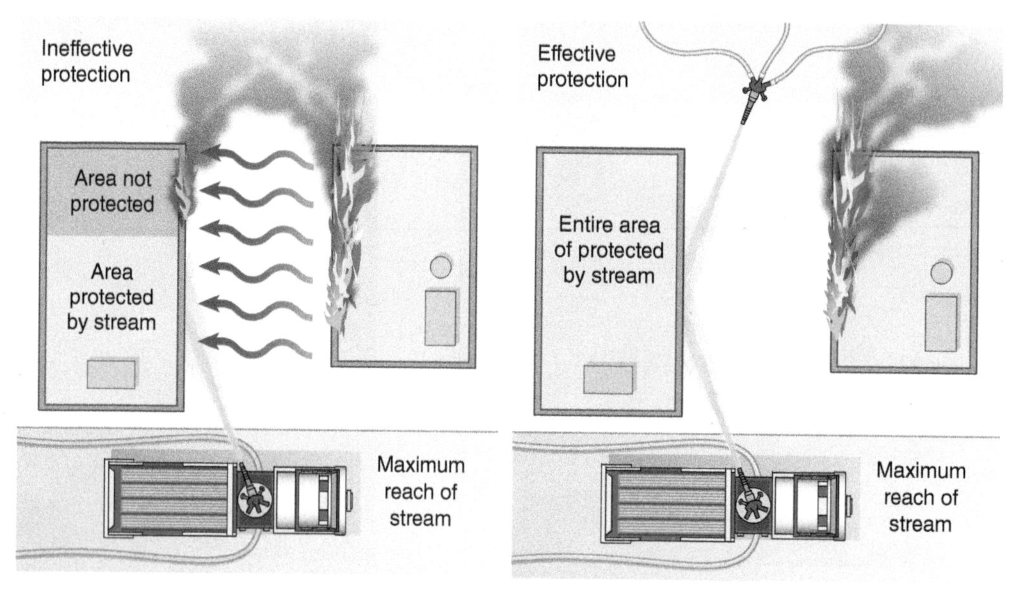

FIGURE 12-16 When one stream position does not cover the entire exposure area, others should be initiated to ensure complete coverage.

Streams developed by smooth-bore nozzles hold together better and therefore penetrate winds better than spray-nozzle streams. If the winds are too strong for spray-nozzle streams, the heaviest smooth-bore streams, including those produced by master stream appliances, might have to be used for exposure protection. Streams produced by smooth-bore nozzles can be powerful and could cause damage to buildings and their appendages like cornices, marquees, signs, and light fixtures. Straight streams can break glass on the exposure, making it more vulnerable for fire extension within the building, so use caution. Combination fog nozzles set to straight stream where strong winds are not a factor are probably the better choice for exposure protection.

Hose-Line Positions

The positioning of exposure lines is especially important. Exposure lines must be placed where they will cover the maximum amount of exposed area. If one stream does not cover the exposure completely, then additional streams should be used **FIGURE 12-16**. Positioning a stream so that the exposure is protected while the stream is brought to bear on the part of the fire nearest the exposure is an effective tactic; however, concern for hitting the fire should not be allowed to reduce the effectiveness of exposure coverage.

The ideal position is one that maximizes the reach and effectiveness of the stream and at the same time provides firefighters protection from radiant heat and keeps them out of the collapse zone. Adjoining roofs; secure and safe stacks of noncombustible stored materials; and buildings across alleys, courts, or narrow streets often make good positions for exposure lines. Companies should be aware of such locations; onsite training drills are a great help in developing a plan of attack prior to an actual incident. **FIGURE 12-17** shows various vantage points in protecting exposures. Hose lines should be brought into exposed buildings, and protective systems in both fire and exposed buildings should be supplied.

When the affected side of an exposure building is beyond the reach of exterior streams, water from hose lines positioned on the roof or out of upper floor windows can protect it by flowing water down the sides of the exposure building to keep it wet.

Interior Exposures

The objective of interior exposure coverage is the same as that of exterior exposure coverage: to keep the fire from spreading to uninvolved areas. Interior exposure fires could require the use of hose lines in many positions inside the building to stop the spread of fire **FIGURE 12-18**. The mobility of smaller hose lines, such as the 1¾ inch, makes them desirable for interior exposure coverage. On occasion, however, the severity of interior fire spread may necessitate the use of larger hose lines, such as the 2½ inch, especially in commercial structure fires.

In addition to getting an exposure line to the room and floor above the fire, the IC should consider exposure lines in adjacent rooms to the right side of the

FIGURE 12-17 Aerial platforms, adjacent windows, and adjacent roofs provide several vantage points in protecting exposures.

FIGURE 12-18 The objective of interior exposure coverage is the same as that of exterior exposure coverage: to keep the fire from spreading to upper floors and other uninvolved areas.

fire, the left side of the fire, the room behind the fire, and even the room or floor below the fire. Fire can spread horizontally through concealed channels and spaces above a suspended ceiling that contains various building systems. As the material burns, it can fall onto the suspended ceiling tiles, knocking them out of place or collapsing the ceiling. As burning debris falls to the floor below, it allows for downward vertical extension.

Such was the case in the Club Marena fire on February 12, 2003, in Rosarito, Baja California, Mexico. A fire started in a unit on the sixth floor of a seven-story oceanfront condominium. Falling embers and burning debris from the balcony was blown by strong winds onto a fifth-floor balcony, igniting the patio furniture. When the sliding door failed, the wind blew the fire into that fifth-floor unit. In addition, the concrete floor of the fire unit superheated the horizontal space above the ceiling of the unit below. Insulation foam panels dislodged from the concrete and began to burn. The ceiling collapsed, and fire spread to the unit below. These unusual but predictable forms of fire spread caused the fire to extend down (as well as up to the top floor). This process continued, and the fire spread all the way down to the first floor. The entire building was destroyed by downward vertical exposure.

Whereas outside exposure fires are obvious and easy to see, many interior exposure fires are not at all obvious; they must be sought out and located by firefighters on the scene. If fire is spreading up stairways, through halls and corridors, or up elevator shafts, there is a very good chance that it is also spreading vertically and horizontally through walls, ceilings, and other concealed spaces **FIGURE 12-19**. Using thermal-imaging cameras (TICs) is the best way to identify the location of these hidden fires. Performing these tasks properly and safely requires adequate personnel. Many structure fires have resulted in the complete and total destruction of the buildings on fire because inadequate staffing meant that there weren't enough firefighters to get ahead of the fire in concealed spaces. This is why "exposures" is the second strategic objective in rescue, exposures, confinement,

FIGURE 12-19 Using a TIC is the best way to identify the location of these hidden fires.

extinguishment, overhaul/ventilation and salvage (RECEO/VS). The fire does not take a timeout while the IC musters adequate staffing to combat the fire.

Fire in Concealed Spaces

There may or may not be signs that indicate to firefighters that fire is spreading within a concealed space. Alterations to the building might have resulted in two or more suspended ceilings, which may hide the exposures to fire **FIGURE 12-20**. If there is any possibility of fire in a horizontal or vertical space or shaft, that space must be opened and inspected visually. A TIC helps with accurate verification of a fire spreading in these spaces. If necessary, streams must be directed into the concealed space, and then it must be ventilated.

These actions can cause damage to the building, but there is little choice in the matter: either open up shafts, walls, partitions, ceilings, floors, and so on, or let the fire destroy the building completely. Fire officers shouldn't hesitate with this decision: the building is already damaged by smoke and water, and some buildings—disposable buildings—will be torn down anyway because it is financially cheaper to build a new one rather than repair a building that has been severely damaged and weakened by fire. That decision lies with the property owner. Our job is to extinguish the fire. Nevertheless, as part of our mission statement and fourth incident priority, every effort should still be made to minimize damage to the building and its contents.

Openings must be large enough for inspection, hose-line manipulation, and ventilation activities. The openings must also be able to take in enough water to extinguish the fire. Fires in concealed spaces are the exact conditions for which piercing nozzles and round

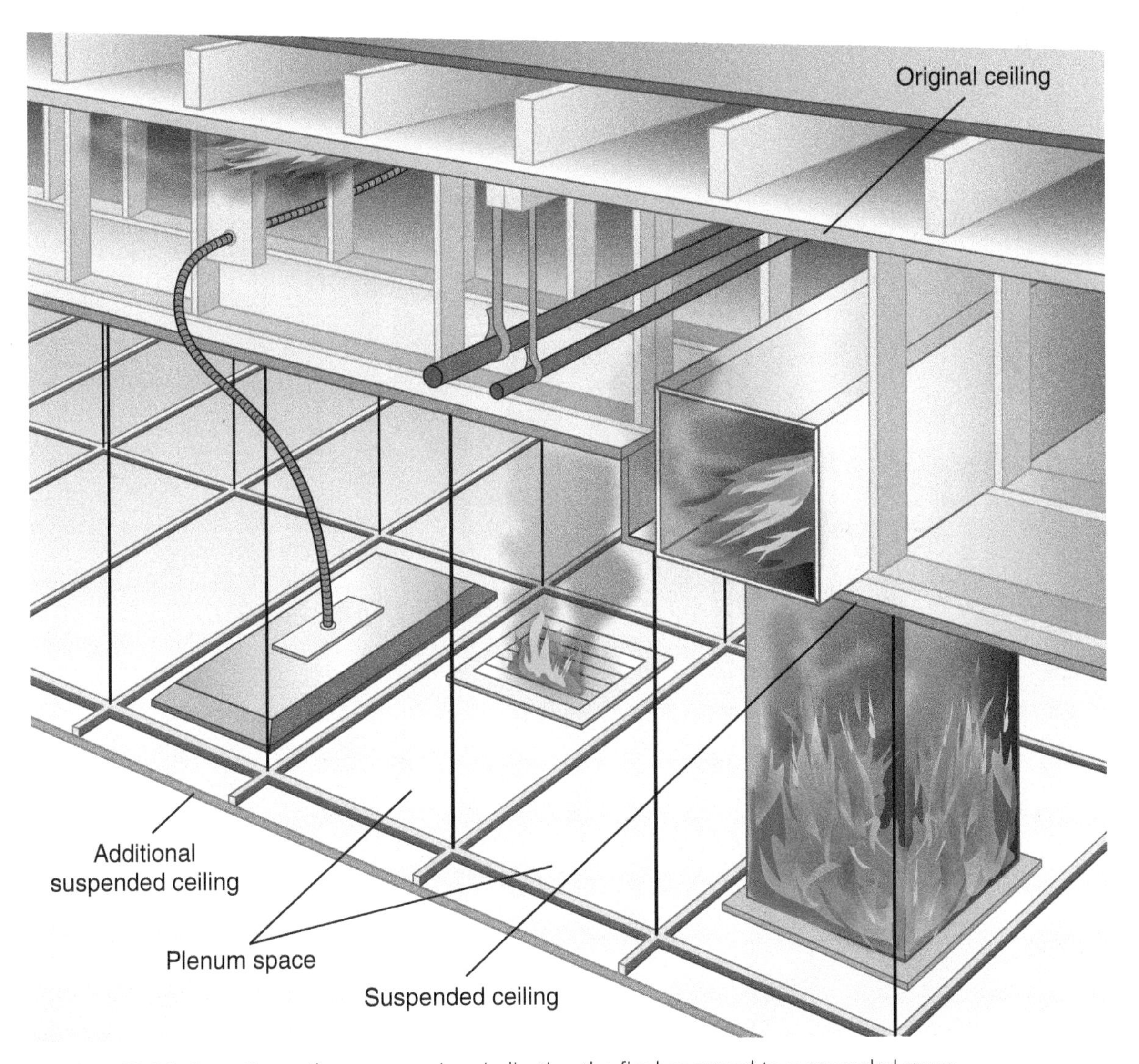

FIGURE 12-20 Sometimes, there are no signs indicating the fire has spread to a concealed space.

nozzles were designed. These highly effective extinguishing tools are underutilized.

Opening concealed spaces and creating ventilation outlets is ladder company work and should be assigned as such; however, if there are no ladder companies at the fire scene or if they are not part of the available fire force, engine company personnel must do the job. Ladder company operations are performed at every fire, regardless of who actually does them. Fire departments should place value in cross-training crews between ladder company and engine company tasks and assignments.

Vertical Fire Spread

Fire can travel vertically inside walls and partitions and through pipe shafts, dumbwaiters, air shafts, and similar pathways in a building **FIGURE 12-21**. Many structures contain concealed vertical shafts that house building utilities such as water, gas, and electric lines or sewer system vent pipes. Many single-family dwelling and apartment houses have central heating system vents that extend through the building from the basement to a chimney fixture on the roof. If walls need to be opened, start at waist height to check the space. This allows the nozzle to be aimed up the stud channel toward the top plate, or down the stud channel toward the bottom plate **FIGURE 12-22**.

Balloon-frame construction homes still exist by the hundreds of thousands throughout the United States and Canada. This construction started in 1833 and became the standard for multistory wooden buildings until the 1950s. In balloon-frame construction, the studs run two or more stories in height, from the foundation to the eaves. They are notorious for not having vertical fire stops. A horizontal ribbon board is nailed to the studs at the floor line. The joists rest on the ribbon board. The channels between the studs may be open from the basement to the attic, and the horizontal space between the joists are open to the vertical stud channels. As you can imagine, fire can spread in every interconnected space—from the cellar or basement all the way up into the attic **FIGURE 12-23**.

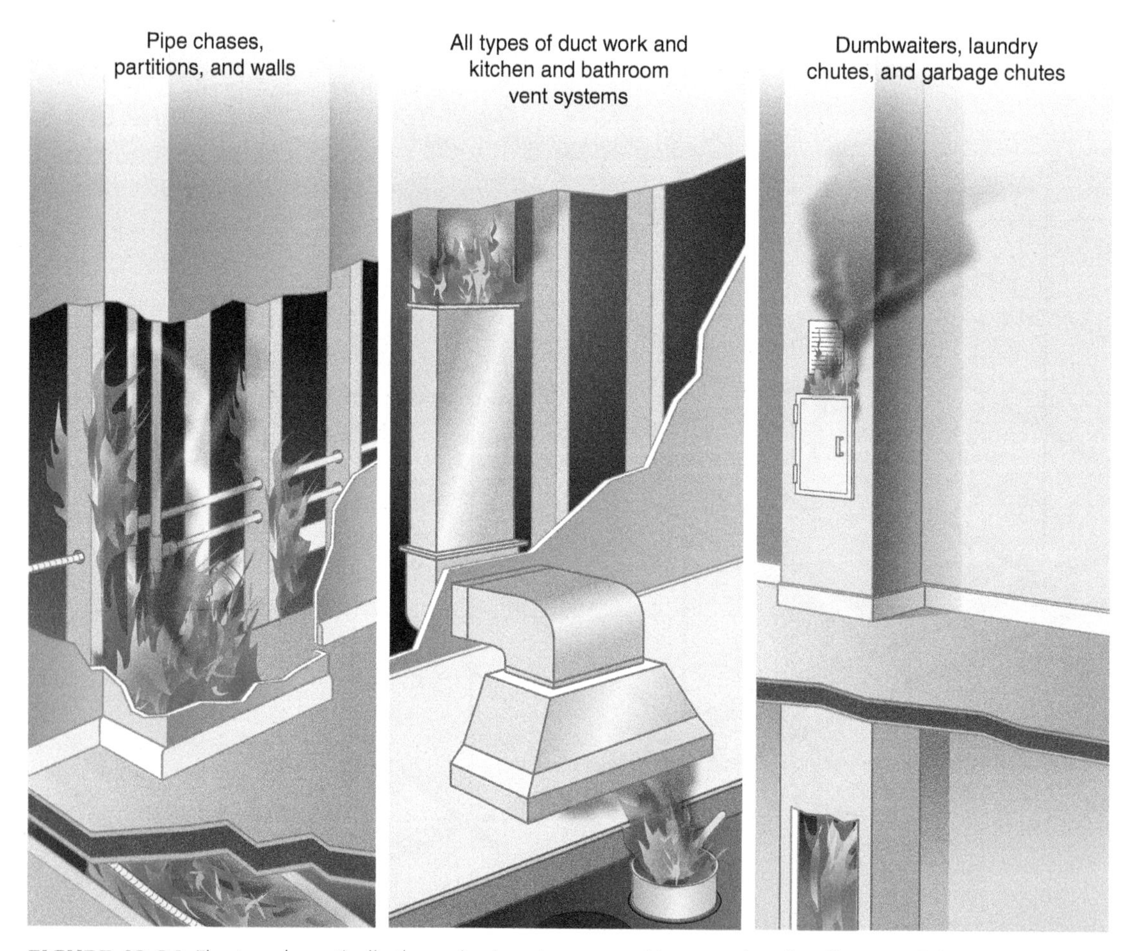

FIGURE 12-21 Fire travels vertically through pipe chases, partitions, and walls; all types of duct work and kitchen and bathroom vent systems; and dumbwaiters, laundry chutes, and garbage chutes.

FIGURE 12-22 When opening walls, start at waist height to check the space.

FIGURE 12-23 In balloon-frame construction, the channels between the studs are open from the basement to the attic, and the horizontal space between the joists are open to the vertical stud channels. In the event of fire, an exposure line must be deployed immediately to the attic space.

It can also spread horizontally between floor and ceiling spaces.

If a fire occurs in a balloon-frame building, an exposure line must be deployed immediately to the attic space, even if this space is accessed by a ground ladder from the outside. The quickest and most efficient way to check for vertical fire spread in balloon-frame buildings is to open the siding panels from the outside of the fire area. This tactic exposes all the vertical stud channels for extinguishment without smoke obscuring the view.

In commercial structures, stores, and shopping centers, these vertical vent channels are normally placed toward the rear of the building. In apartment buildings, the vertical shafts follow the pattern of the apartment layout. They are most often found near the kitchens and bathrooms, and each shaft is usually placed so that it serves two, four, or more apartments. The great variety in the design of single-family dwellings means that vertical shafts could be located almost anywhere in these structures; however, the locations of vent pipes and kitchen vents on the roof are good indications of where these shafts can be found **FIGURE 12-24**.

Signs of Vertical Spread

If there is a working fire inside a building, firefighters should assume that flames have entered concealed spaces until they have determined otherwise. As they arrive at the fireground, engine company personnel should be looking for signs that fire has extended into vertical channels within the building. Use a TIC and listen for the audible sound of smoke detectors, which provide a clue to how far the smoke has traveled.

Smoke or fire showing around roof features such as vent pipes is an obvious sign that fire has spread to the shafts leading to these outlets. Close examination may show that the tar on some parts of the roof is soft and shiny, which means that the tar has been melted slightly by heat in a shaft below it. In rain or snow, a clear or dry spot around a roof feature over a shaft may indicate that heat from the fire is in the shaft or that fire itself has entered the shaft. Again, a TIC clearly registers heat and temperature differentials on the roof. It even reveals the direction of the roof joists or rafters for vertical ventilation.

Inside the building, signs such as smoke and flames issuing from walls, or blistering or discoloration of paint or other wall coverings, indicate the presence of fire within concealed shafts, walls, or partitions. A wall that is hot to the touch is probably concealing fire. Whenever you check a door or wall for heat, it must be done with an ungloved hand. Use the back of your hand so you don't burn your palm or your fingers—a more debilitating injury. It is best to use a TIC to verify temperature variations with accuracy. If you find any signs of heat or charring, the wall must be opened **FIGURE 12-25**.

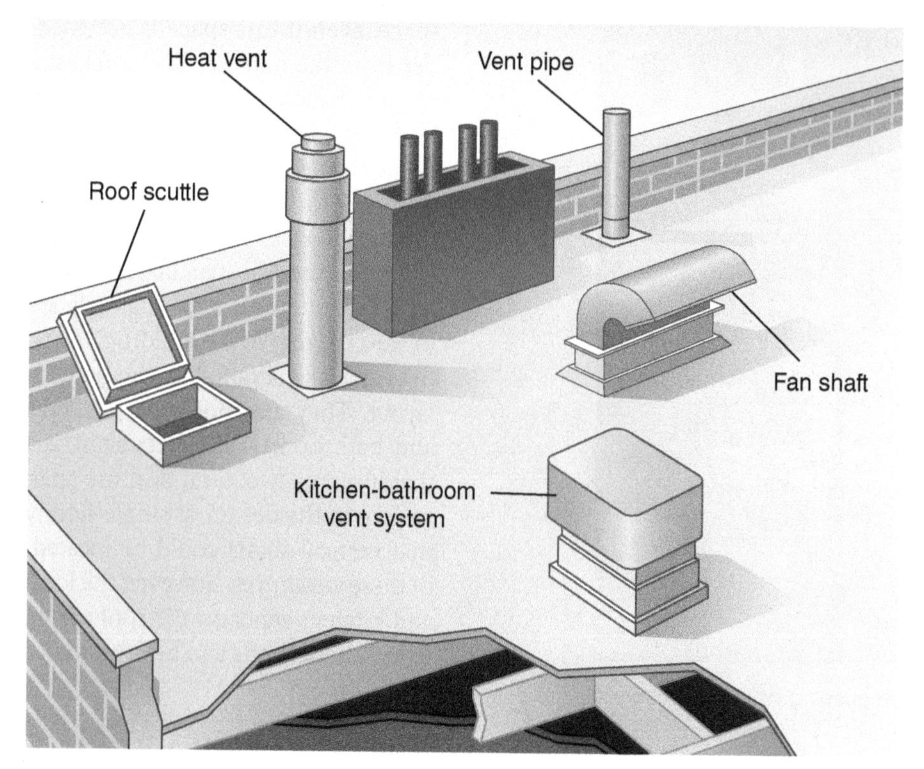

FIGURE 12-24 Rooftop fixtures indicate the presence and location of vertical shafts in a building.

FIGURE 12-25 When there are signs of fire in walls, partitions, and vertical shafts, the wall should be opened. The fire can then be attacked with an attack line of the proper size.

Control of Vertical Spread

If it is known or suspected that a fire has entered a vertical shaft, hose lines should be directed into the shaft, and it should be opened and inspected from the roof. Although intense, fire in these channels is confined to a comparatively small area; thus, 1¾-inch hose lines are most useful for this operation. In addition, the mobility of these hose lines permits them to be moved rapidly from one location to another with a minimum number of personnel.

FIGURE 12-26 Vertical ventilation is essential when a fire is burning in an attic or cockloft to keep the fire contained and limit horizontal movement of the fire.

FIGURE 12-27 Fire and moisture cannot exist in the same molecular space. Piercing nozzles and round nozzles create a sprinkler system in the attic space and need to be inserted *before* any vertical ventilation occurs. The attic space is filled with smoke and hot gases, but not fire. The atmosphere is oxygen-deficient and cannot support combustion.

Fire travels vertically as long as it is allowed to do so. Opening the roof encourages vertical travel and slows the horizontal movement of the fire **FIGURE 12-26**. In a sense, when you create a path of least resistance, the fire is led out of the structure and away from horizontal spread through cocklofts and attics, spaces between floors and ceilings, and other horizontal channels. Opening a skylight in an attic may not ventilate the cockloft area if it is enclosed. A vent hole that accesses the attic space must be cut to vertically ventilate it. If a vertical shaft does not terminate at the roof in the best location to ventilate it, the roof opening should be made directly above the shaft or as close to that point as possible. A hose line should be advanced to the attic to attack any fire that has spread there.

Whenever you use vertical ventilation, a hose line should already be in place, either at roof level or the ceiling below. The vent hole releases pressurized gases and flames will quickly follow. This brings the flames right into the attic and through the trusses. A flow path is created, and vertically vented fires often produce spectacular flames through the roof, accompanied by a tremendous amount of radiant heat. These flames need to be extinguished before the trusses become fully involved, preferably from below. If they are not extinguished, the direct flame contact will cause the trusses to collapse.

Another safe and effective method for extinguishing fires that have spread vertically into the attic space is using piercing nozzles or round nozzles. Round nozzles resemble fog applicators, and they work the same way by throwing a 360° sphere of water. If this indirect attack is attempted, it needs to be done *before* any vertical ventilation occurs **FIGURE 12-27**. Before ventilation, the attic space is filled with smoke and hot gases, but not fire. In other words, there is no fire in the trusses. The atmosphere is oxygen-deficient, with 16% or less oxygen concentration, and cannot support combustion. The ventilation process is what brings in the fresh air that raises the oxygen concentration above 19% and creates the flames that can be seen through the roof. Therefore, cutting a hole into the roof just large enough to insert the round nozzle or to insert the piercing nozzle into this confined heated space causes the water to turn to steam. The water expands to 1,700 times its volume and absorbs the heat, thus cooling the fire gases and smoke below their ignition temperature and preventing the fire from entering the attic. Fire and moisture cannot exist in the same molecular space. After water application, vertical ventilation can proceed. However, it may not be necessary to open the roof if the fire is extinguished from below. The attic can be horizontally ventilated through the attic vents, thus preserving the integrity of the roof. This is a much safer evolution than vertical ventilation, especially on lightweight trussed roofs. The reason this indirect attack is underutilized is because the modern fire service hasn't taught it. Most officers and firefighters either don't understand the physics, don't trust the science, or just simply ignore it and opt for rooftop ventilation because it's more dangerous and the results are spectacular.

Consider the design and function of a sprinkler system. If an attic on fire is sprinklered, the affected heads would fuse when the temperature reaches 165°F (74°C). The 8 to 24 gallons per minute (gpm) of water converts to steam and extinguishes the fire without the help of engine or truck companies. The famous One Meridian Plaza high-rise fire in Philadelphia, Pennsylvania, killed three firefighters and injured 24. The fire started on floor 22 and consumed eight floors before it was extinguished on floor 30 by 10 sprinklers. Using an indirect attack with a piercing nozzle or round nozzle placed into a confined heated attic space before ventilation is basically introducing a sprinkler system into the attic. The results are the same.

To attack the fire from below, an opening should be made in the wall or vertical space that shows signs of fire. An opening that is about waist high allows a firefighter on one knee to handle a nozzle easily and to direct the stream up into the opening. An opening at this level also allows the nozzle to be directed down if the fire is coming from below.

It is important that at least one stream be directed upward toward the fire to control embers, hot gases, and smoke as well as the fire itself. If using a spray nozzle, adjust the stream pattern to provide maximum reach and coverage within the shaft or partition. If it is evident, because of the height of the building, that this one hose line will not suffice, more hose lines must be stretched above or below to ensure full coverage of the exposure area. If the fire is traveling in several separate but parallel shafts or between the studs in walls and partitions, the area should be opened completely, again to ensure full coverage.

In a large vertical opening such as an elevator shaft or a stairway, it may be necessary to use 2½-inch attack lines to control the fire because of the volume of fire that such openings can support **FIGURE 12-28**. This is especially true when several elevators are arranged in a bank in one shaft, or where stairways are unusually wide. The vertical spread continues until it reaches a horizontal barrier, then the fire spreads horizontally. Once the fire reaches a vertical barrier, it starts to mushroom down. An open cockloft can allow the fire to spread beneath the entire roof structure.

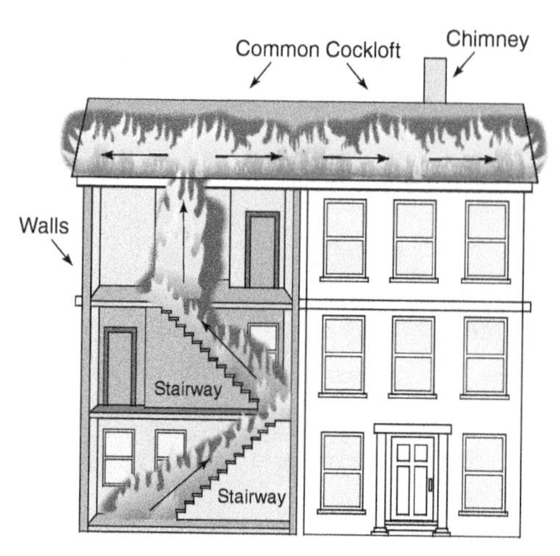

FIGURE 12-28 Vertical fire spreads until it reaches a horizontal barrier, then the fire spreads horizontally.

Horizontal Fire Spread

Although fire tends to travel vertically, it also travels horizontally through any available paths. Fire may travel horizontally through the spaces between ceilings and floors, over false or hanging ceilings, along ductwork and utility conduits, through conveyor tunnels in industrial buildings and warehouses, and through similar channels in other buildings. In addition, construction features may cause concealed horizontal channels to be formed within walls, floors, and ceilings. Such channels can permit fire to spread horizontally through the building.

Fire may also move horizontally from one building or occupancy to an adjoining one through ducts, ceiling spaces, and walls **FIGURE 12-29**. An example is the spread of fire in a row of stores or from one apartment to another on the same floor.

Exposure lines must be positioned inside the fire building as well as in adjoining units. Ceilings, floors, and other horizontal paths suspected of or showing signs of fire travel must be opened for inspection and, if necessary, fire attack operations must be initiated. This is why the engine officer, in addition to carrying a TIC, must carry a 6-foot pike pole. The officer needs the reach in order to pop ceiling tiles or to pull ceilings to check for fire above the exposure team **FIGURE 12-30**. An ax, Halligan bar, or an A-tool does not provide the reach necessary to check above 8- or 10-foot ceilings. If necessary, call for a baby ladder, an attic ladder, or a folding ladder. You must have the ability to check the horizontal spaces above you thoroughly.

For the most part, fire spread through horizontal channels exhibits few exterior signs, except when the flames reach and involve exterior walls. The interior signs are the same as the signs of vertical fire spread listed in the previous section. Horizontal exposures are controlled in the same way as vertical exposures.

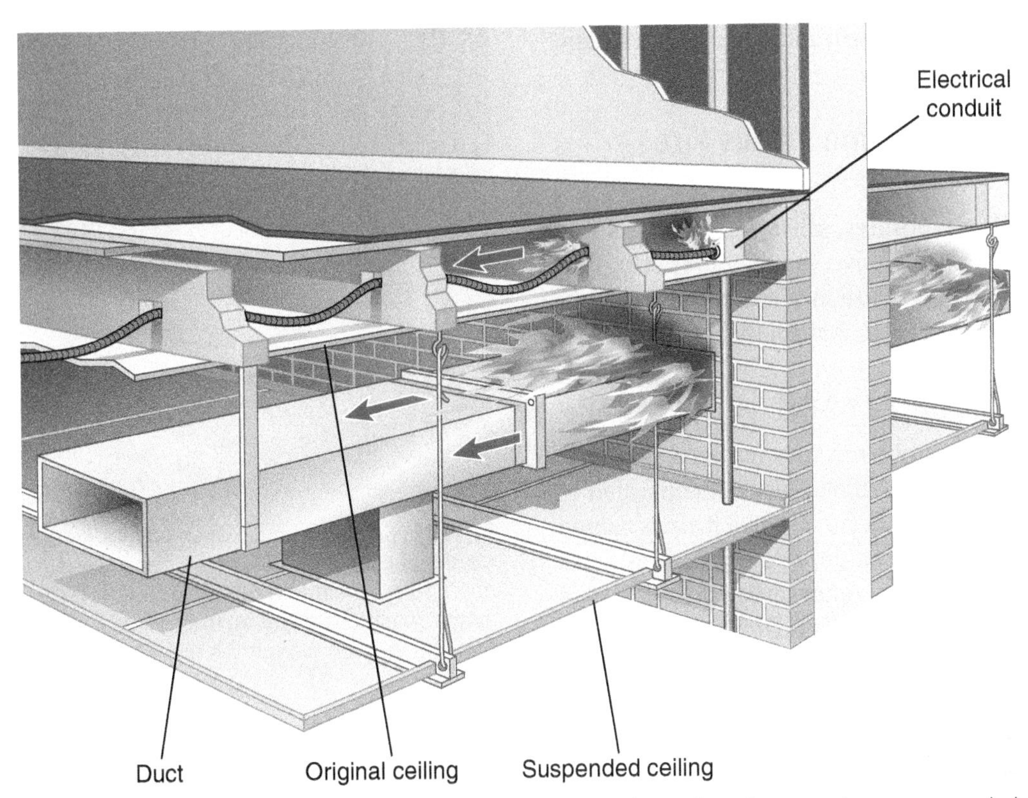

FIGURE 12-29 Fire travels horizontally through concealed channels and spaces above a suspended ceiling. Burning material can fall onto the suspended ceiling tiles, knocking them out of place or collapsing the false ceiling and allowing the fire to spread to the floor below.

FIGURE 12-30 A check for horizontal fire spread should be made on each side of the main body of fire.

The channel must be opened, and an adequate stream must be directed at the fire.

Property Conservation Versus Fire Damage

While protecting exposures, salvage crews should be assigned to cover or remove valuable property, thus keeping damage and loss to a minimum. The most important goal, however, is to control the fire.

Open Interior Spread

One other interior problem can often be serious, although the exposure is not concealed. Unconcealed or open interior fire spread occurs when a fire has gained control of a large enclosed upper area, such as a supermarket, warehouse, or other large enclosed space that includes little or no construction features to deter the horizontal spread of fire. The problem is essentially that of a large outside exposure; however, the fire is constrained by the walls and roof of a very large building. Although the fire may not be endangering adjoining structures, the intense heat and the smoke and gases contained by the building pose severe problems for firefighters trying to make entry.

The IC must perform a 360-degree size-up and conduct a risk-benefit analysis before committing firefighters to interior structural firefighting operations. In this situation, fire attack operations might be more closely related to exterior firefighting than to interior, even though the fire is inside. Two-and-a-half-inch hand lines and even master stream appliances might be needed inside the structure to cover exposed areas and to knock down the fire **FIGURE 12-31**. In such a building, the roof can become a problem in terms of both firefighting and safety. If interior operations cannot be conducted safely, firefighters should be withdrawn from those positions, and the operation should be changed from an offensive to a defensive strategy.

The unprotected steel roof of a typical supermarket is a good example. The rising heat and hot combustion products of a major fire could cause the roof supports to buckle and the roof to collapse. A large fire often has to be controlled before interior exposures can be protected. Until the fire is extinguished, the possibility of roof collapse makes the interior of the store extremely hazardous to firefighters. Collapse may occur within 10 or more minutes with steel bar joist roofs. Bowstring truss roofs, which are often found in supermarket and bowling alley construction, are also notorious for early collapse. The unprotected steel tie rods and connectors that hold the trusses together, and the unprotected steel girders that support them, lose their strength and can fail when the steel is subjected to fire temperatures over 1,000°F (538°C) for 10 or more minutes.

FIGURE 12-31 When an open interior fire has gained control of a large enclosed area, like a warehouse that doesn't have walls to deter the horizontal spread of fire, fire attack operations might be more closely related to exterior firefighting, even though the fire is inside.

Courtesy of John Odegard.

FIGURE 12-32 Unprotected steel tie rods and connectors that hold the trusses together, and the unprotected steel girders that support them, lose their strength and can fail when the steel is subjected to fire temperatures over 1,000°F (538°C) for 10 or more minutes.

It is no longer safe to vertically ventilate the roofs of these structures until the fire has been knocked down or extinguished with 2½-inch lines or master streams from a safe location close to the entrance **FIGURE 12-32**. This tactic can have a great effect on slowing the tendency of the fire to spread through the enclosed area. Once the fire is knocked down, the thermal assault on and the deterioration of the steel diminishes. Then it might be permissible to attempt vertical ventilation. Mobile ventilation units (MVUs), which are giant positive-pressure ventilation (PPV) fans mounted on a truck or trailer, are safer for horizontal ventilation on a grand scale. Their use is much safer than risking the lives of firefighters on a roof when there is no way to gauge the strength of the steel within the trusses.

If the involved occupancy adjoins other buildings, the adjoining buildings should be checked for fire spreading through ceiling spaces and other channels. If the fire is large and numerous exposures exist, it is necessary to handle two kinds of exposure problems at the same time: one requiring big hose lines and larger flows of water; the other requiring smaller, fast-moving hose lines.

Basement Fires

Basement fires are one of the most difficult and dangerous types of incidents that firefighters encounter. There may be only one or two entry points into the basement area from either the inside or outside of the building. In addition to limited access and limited egress, firefighters must contend with:

- Narrow stairs
- Limited or no windows
- Small or narrow windows
- The nature of stock, product, and other items stored in basements
- Low ceilings
- Trapped heat and smoke

If a basement does exist as a bona-fide living area, there are many potential scenarios where human factors can lead to a fire that will ultimately extend upwards.

Fires in unfinished basements pose another dangerous hazard. Unfinished basements allow the fire to attack the floor above because the floor joists, which are often constructed with TJI (TGI) lightweight prefabricated wooden I-beams, and the lightweight galvanized steel hangers and gusset plates that support the floor assembly are exposed to the heat of the fire. These assemblies can collapse in as little as 2 to 5 minutes after direct flame contact. Such a collapse will contribute to massive interior fire spread to the upper floors.

When attacking or covering exposures during a basement fire, every effort must be made to create a sprinkler system in the basement by using hose lines, fog applicators, round nozzles, Bresnan distributor nozzles, or cellar nozzles. (There's a reason they are called cellar nozzles; to use them on cellars!) Some of these tools can be inserted into the basement area from the outside. That means firefighters are outside the immediately dangerous to life or health (IDLH) area. That means firefighters are not taking a beating when the only life hazard is themselves, and they are not going down stairs when there is no way to determine if the stairs are strong enough to support the weight of an entire crew or even the weight of a single firefighter. Once the fire is knocked down, firefighters can make an interior entry to the basement for search and final extinguishment with a much wider margin for safety. Ventilation openings in basements may be scarce, making ventilation challenging, but not impossible. Water first, then ventilation.

The higher a fire is in a building, the better off firefighters and building occupants are in terms of fire spread. A fire in the basement of an apartment building is much more dangerous to occupants throughout the building than is a fire on an upper floor **FIGURE 12-33**. Heat, smoke, gases (especially carbon monoxide), and hot embers traveling vertically from the basement through the building can ignite secondary fires and overcome occupants on all floors. This is why it is important to get exposure lines to the floors above when dealing with a basement fire.

Controlling Basement Fires

The first-arriving company officer is the initial IC and must perform a thorough size-up to identify problems and determine the location of the fire within the basement, the stage of the basement fire, and where it is likely to spread based on the construction features of the building.

Incipient Stage Basement Fires

Part of size-up and risk management is gathering information. If tenants or employees give you accurate information about what's burning, where it's burning, how long it has been burning, and how to access the basement, and smoke and heat conditions indicate the fire is in the incipient stage, taking a preconnected attack line down the interior stairs and into the basement to extinguish the fire may be the right tactic and the right decision to make. Vertical extension from a fire in the incipient stage is negligible. The urgency to get exposure lines in place above the fire is lessened, but checking for extension will still be necessary.

Growth Stage and Fully Developed Stage Basement Fires

When fire is blowing out basement windows in residential or commercial buildings that have basement doors to the outside, it is best to use these doors for an aggressive interior attack, which should be made

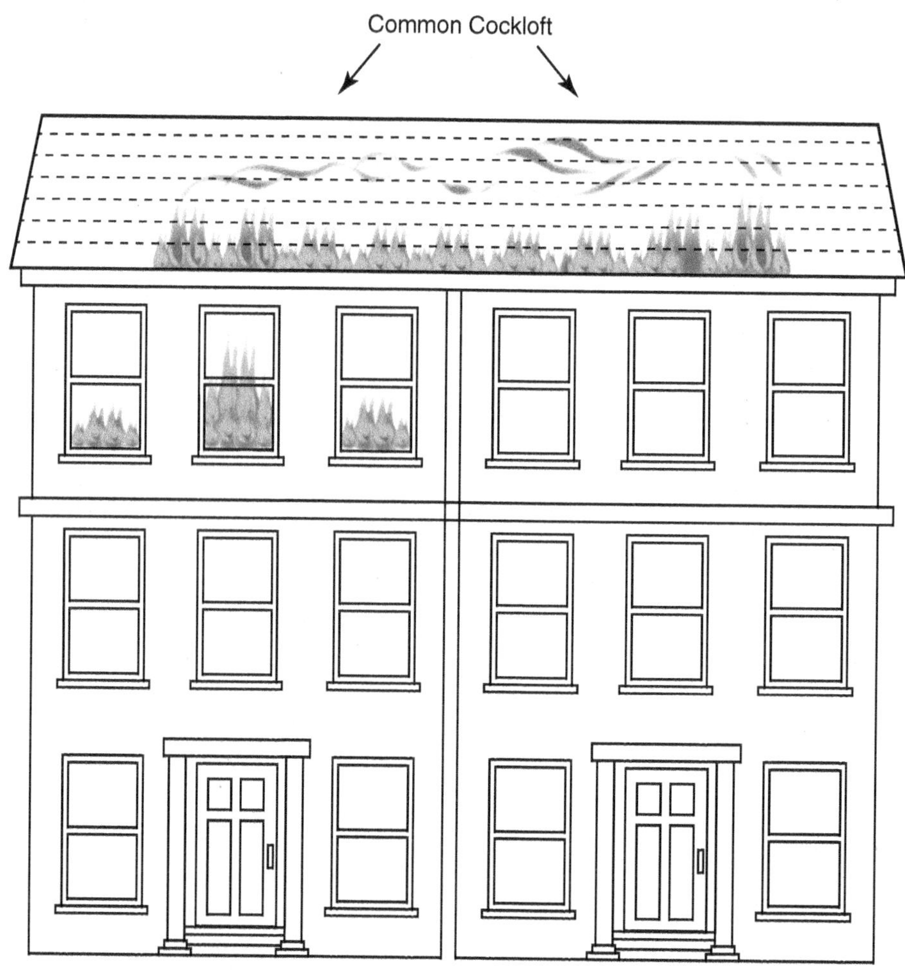

FIGURE 12-33 The higher a fire is in a building, the better off firefighters and building occupants are in terms of fire spread.

from the unburned side toward the burning side. Because the fire is already taking the path of least resistance and venting from the window closest to the fire, utilize the reach of the hose stream and continue to advance in the direction toward the fire. A PPV fan can be set up at the entry door, keeping cool fresh air at your back and moving the smoke toward the ventilation window **FIGURE 12-34**. If other windows are close to the ventilation window on the fire side of the structure, they should be broken out to ventilate the basement more quickly and effectively. *Note:* The PPV fan may accelerate the flame out the ventilation window with blowtorch force, so an exterior exposure line should be positioned on that side of the structure. If the air pressure entering the basement is too strong for the volume of smoke venting out the window, the smoke will look for another path of least resistance and start to exit the entry door. Lower the engine revolutions per minute (rpm), move the fan farther away from the door, or shut off the fan until the air currents reverse. Ventilation is a key element in controlling the fire by permitting the heat and smoke to escape while allowing firefighters to advance the interior attack into the basement.

Firefighters must keep ahead of the fire. If fire is burning up stairways or other vertical openings into the first floor, an exposure line must be positioned to protect the integrity of the stairway door. Utilize the reach of the stream and operate from a safe position, and avoid the floor area directly above the fire. Don't hesitate to flow water on the floor to cool it, like you would do for any exposure. Water seeping through the floor may also help extinguish any fire moving up the vertical cracks. Sufficient exposure lines must be provided with an adequate number of firefighters to accomplish the objective of protecting the first floor and the floor above. Once exposure lines are in place and adequate hose lines are attacking the fire, another PPV fan can be set up to pressurize the first floor. The pressurization will be stronger than the smoke and flames trying to extend vertically and should reverse their direction back toward the established ventilation portal.

If the IC suspects a basement fire in a balloon-frame construction building, the other objective, in addition

FIGURE 12-34 When fire is blowing out a basement window, it is already taking the path of least resistance.

to keeping ahead of the fire, is to keep the fire from extending to upper floors, including the attic and cockloft areas. *No PPV fans should be used in balloon-frame construction during the fire attack.* The vertical and horizontal open channels will spread the fire everywhere. PPV is very effective when you can control the exhaust portal, and it is disastrous if you can't. Firefighters need to be assigned above the fire to open up vertical or horizontal spaces suspected of containing fire. Hose lines must be positioned and operated by engine companies if these conditions are encountered **FIGURE 12-35**.

When the fire is raging at the very bottom of the structure and forcing its way up into the building, but there are no exterior doors from which to make access, it is important to control the main body of fire as quickly as possible. Windows, access chutes, trapdoors, and any other openings should be used for positioning hose lines to attack the fire. The best option is the window(s) **FIGURE 12-36**. A 2½-inch hose line with a Bresnan distributor or Bulldozer nozzle can be secured with webbing on top of the rungs of a 12-foot roof ladder. The hooks support the coupling, and the round nozzle can spin 360° freely **FIGURE 12-37**. The ladder, with the distributor or nozzle, can now be inserted into the window for deeper penetration into the basement to create a 290-gpm sprinkler head

FIGURE 12-35 Firefighters need to be assigned above the fire to open any vertical or horizontal space suspected of containing fire.

FIGURE 12-38. The ladder can be moved up and down, left and right, while the firefighter remains outside the IDLH atmosphere. One firefighter can handle the positioning of the roof ladder with the charged Bresnan

FIGURE 12-36 All openings into the basement should be located and considered as points of attack, but windows provide the best access for applying fast water.
Photo by Raul Angulo.

FIGURE 12-37 In this demonstration, a 2½-inch hose line with a Bresnan distributor or Bulldozer nozzle can be secured with webbing on top of the rungs of a 12-foot roof ladder.
Photo by Raul Angulo.

distributer. Cellar pipes and Bresnan distributor nozzles can be placed in service from the floor above into positions that permit them to knock down or assist in extinguishing the fire. A backup line should be laid to protect the firefighters carrying out this tactic.

Once the fire is knocked down, engine crews can enter the basement for search and final extinguishment. Obviously, outside attack lines are not to be used from openings into the basement, such as windows and doors, if firefighters have already made entry and are working inside the basement.

The faster the fire is extinguished, the less there is a need for exposure lines. Conversely, if the basement fire attack is stalled or meeting with resistance, exposure lines should be deployed and in place to get ahead of the fire to cut off vertical extension. Along with the application of sufficient streams, a basement

FIGURE 12-38 In this demonstration, the ladder, with a Bresnan distributor or Bulldozer nozzle, can now be inserted into the window for deeper penetration into the basement to create a 290-gpm sprinkler head.
Photo by Raul Angulo.

fire requires full ventilation of both the basement and the first floor. Basement windows, preferably opposite those being used for fire attack, should be used for ventilation. The first floor should be ventilated completely to allow efficient and safe operations. Forcing the products of combustion out of the building through the basement and first floor reduces the chance that they will create exposure problems on upper floors **FIGURE 12-39**.

Basement Fires in Mercantile and Commercial Stores

If outside entrances are available but there are no windows to the basement, the situation becomes more difficult. Solid streams from 2½-inch hand lines or portable monitors are needed. The fire may travel vertically more quickly, and the first floor is in greater peril. If the attack appears to have a quick effect on the fire, ventilation of the first floor may suffice. If the attack does not have a quick effect, openings should be made in the first floor just inside windows; then the windows should be removed or opened to provide ventilation of the basement, improve fire attack effectiveness, and lessen interior exposure problems by providing the path of least resistance for smoke and fire.

In a basement fire in a store, the storefront and display window area can be used for ventilation depending on the fire situation and the presence of other basement openings. If there is a rear door but no other openings to the basement, the attack can begin from that position, and ventilation efforts should be conducted through openings at or near areas already burned **FIGURE 12-40**. If there are other openings

FIGURE 12-39 Available openings to the basement can be used for attack from one direction while ventilation is conducted from another.

FIGURE 12-40 Smoke and flames pushing up from cracks in the sidewalk at the base of the building indicate a basement fire and the need for exposure lines to the floors above.

such as windows or sidewalk doors, opening of the storefront might not be necessary. If the basement has no outside opening but only an interior stairway or a trapdoor entrance, the display window openings can be used for ventilation, and the fire can be attacked with hose lines through interior openings or with Bresnan distributor or cellar nozzles from the floor above.

Protecting Exposures

Stairways and other openings from the basement to the first floor are major channels for the vertical spread of a basement fire. These exposures must be covered as soon as possible. Attack lines should be brought in to first-floor landings to knock down any fire that may have spread from the basement; walls and partitions should be checked with a TIC for vertical fire spread, and baseboards should be removed in suspect areas.

These first-floor operations are extremely important, but the safety of firefighters and occupants is even more essential. When a basement fire has gained considerable headway or fully involves the basement, the IC must assess the risks versus the benefits and decide whether firefighters should enter the building. If a decision has been made to allow entry into the building, firefighters must pay particular attention to conditions within the structure. They should operate exposure lines from a safe location on solid flooring. The floor should be checked periodically for a soft or spongy feeling. Smoke coming through the floor should not be ignored, especially at outside entrances, along sidewalks and walls, and at other places that may involve main structural members. These signs confirm a significant basement fire because the smoke is pressurized and thus pushing through the cracks of the foundation or the sidewalk. Pressurized smoke also indicates the deterioration and weakening of the floor, so firefighters should be withdrawn from the building immediately.

Due to substantial live-load and dead-load weight combinations, first-floor collapses can be sudden, giving little or no warning.

Do not proceed any farther over floors showing the signs discussed in this section of the chapter. If occupants still need to be evacuated or firefighters are working on upper floors, ground ladders should be placed to the second story for emergency egress. Once firefighting efforts on the basement level and first floor are abandoned, a structural building collapse will follow, and the decision should be made to change to a defensive strategy.

Vertical Spread

Many of the vertical openings in a building originate at the basement level, and thus there is nothing to keep fire from entering the openings and spreading up through the building **FIGURE 12-41**. Vertical openings should be opened at the first floor if there are any signs of fire spread there. Confirm any fire spread using a TIC. In any case, these openings should be opened at roof level to allow the heat, smoke, and gases to leave the building *after* water has been applied to the spaces. Otherwise a high-low flow path is created, and you may end up with

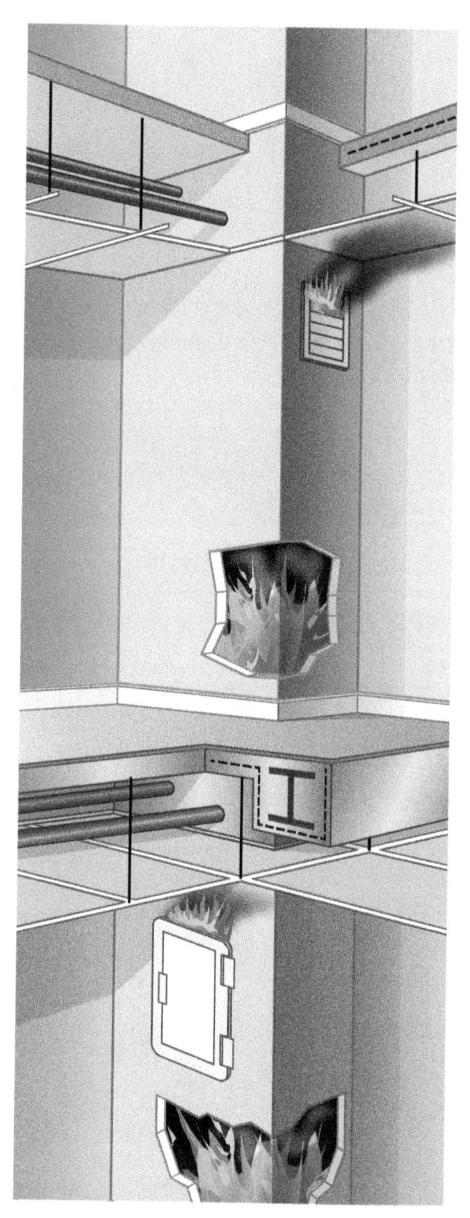

FIGURE 12-41 Many of the vertical openings in a building originate at the basement level, and thus there is nothing to keep fire from entering the openings and spreading up through the building.

FIGURE 12-42 In older structures, where deterioration, remodeling, and additions might have destroyed the integrity of the walls between buildings, horizontal breaches can allow the spread of fire.

a fire in the basement and one at the top of the building, with fire in between. Water first, then vent.

Horizontal Spread

Adjoining basements must be checked for horizontal fire spread. Hose lines must be taken to those locations to extinguish extending fire. This is particularly important in older structures, where deterioration, remodeling, and additions might have destroyed the integrity of the walls between buildings. Party walls, those that support beams or joists from two buildings, often have openings where the beams rest on the walls. Fire can easily spread from basement to basement through these openings **FIGURE 12-42**. Regardless of their construction, adjoining basements must be checked.

The same is true for adjoining attics, especially if the fire building was not vented quickly. Whether the fire originated in the basement or in some other area of the building, the tops of adjoining buildings must be checked and, if necessary, vented.

Basement fires are typically viewed as one of the most dangerous and difficult fires we face—but they don't have to be. In fact, they can be one of the safest if we start our fire attack by first knocking down the fire from the outside.

Fast water takes the energy away from the fire. Everything gets better when the fire is extinguished. Quick extinguishment is not only better for trapped occupants, it also reduces or eliminates the extent at which the fire will spread, lessening the urgency for exposure lines.

After-Action REVIEW

IN SUMMARY

- An exterior exposure is a structure or other object to which the main body of fire can extend. It may be a separate structure or an independent part of the main fire building, such as a separate wing or a building separated by a courtyard.
- Exposure protection is initiated to protect a building or part of a building that has been subjected to radiant and convected heat as well as direct flame impingement from the main body of fire.
- On the fireground, it is imperative that sufficient numbers of personnel and equipment are available to cover exposures while the main body of fire is attacked.

- An adequate water supply must be obtained, and sufficient hose lines must be placed in strategic locations to cover the maximum amount of exposed areas.
- Large fires present a difficult situation because of the amount of radiant and convected heat created by the fire.
- Exposure lines should apply water directly onto the surfaces of the exposures. Direct water application and water running down the sides of the exposures keeps combustible materials from reaching their ignition temperature.
- During large fires, the use of master stream devices can deliver a significant amount of water onto an exposure from a safe distance.
- Radiant and convected heat, embers, and wind currents must be considered when protecting exterior exposures.
- At a large fire, strong winds can carry burning embers and firebrands a considerable distance downwind and start exposure fires. Fire patrols need to be assembled when engine companies are not available to check the impact of falling embers in a neighborhood.
- Interior exposures are objects or rooms in the immediate vicinity of the original interior fire that are not yet burning. These rooms and objects exposed to radiant heat, conduction, convection, and direct flame contact cause the interior fire to spread to uninvolved areas unless exposure lines are in place.
- The most threatened interior exposure is the room and floor above the fire. Exposure lines should be laid after the backup line to protect these areas.
- Interior exposures are protected from radiant heat, convected heat, and direct flame impingement with hand lines deployed after the backup lines.
- Vertical and horizontal channels or voids such as walls, ceilings and subfloors must be opened to expose and extinguish any fire that may be traveling in these areas. Confirmation must be made using a TIC.
- A basement fire presents a difficult and dangerous situation to firefighters because the fire might be hard to reach, and the entire building may be exposed to the spread of fire.
- Engine companies must use the safest entry point to attack a basement fire, and sufficient hose lines must be available to accomplish the task of extinguishing the fire.
- Initial attacks for basement fires should start with firefighters spraying water from the outside and through a window to knock down the fire.
- After a basement fire is knocked down from the outside, crews can advance down the interior stairway for search and final extinguishment.
- Bresnan distributor nozzles, round nozzles, and cellar pipes are very effective extinguishing tools that are highly underutilized. Using these nozzles basically creates a sprinkler system into the basement area to knock down and control the fire.
- Ventilation must be conducted simultaneously or immediately after the fire is knocked down. Without ventilation, low ceilings and limited access areas retain the heat and smoke, making it difficult for the engine crew to finish extinguishing the fire and search for victims.
- The type and quantity of stock and product stored in mercantile and commercial basements makes fires in these basements extremely dangerous, especially if they can't be brought under control quickly. Firefighters must pay attention to their surroundings, maintain orientation within the area, and have a secure means of egress.
- The first floor should be checked for a soft or spongy feeling. Pressurized smoke coming through the floors, foundation, sidewalks, walls, and other main structural members indicate a significant basement fire. Pressurized smoke also indicates the deterioration and weakening of the first floor. Firefighters should be withdrawn from the building immediately.
- Quick extinguishment using fast water by any means possible reduces the threat of fire spread the urgency to get exposure lines in place.

KEY TERMS

exterior exposures Outside structures that can catch fire by radiant heat or direct flame contact from the original fire.

exposure hazard A condition that promotes the spread of fire if a fire starts in or reaches that area.

fire patrol Other non-fire municipal employees equipped with a radio or cell phone assigned to driving neighborhood streets downwind from a large fire to look for falling embers and firebrands that could start new fires.

interior exposures Objects or rooms in the immediate vicinity of the original interior fire that are not yet burning. These rooms and objects exposed to radiant heat, conduction, convection, and direct flame contact cause the interior fire to spread to uninvolved areas.

movable exposures Exposures that can be protected by physically moving them away from the reach of the fire. Examples are trucks, tractors, automobiles, boats, yachts, and any other motorized equipment that can be driven or towed away from the fire.

pre-incident plan A document developed by gathering general and detailed data that is used by responding personnel in effectively managing emergencies for the protection of occupants, responding personnel, property, and the environment. (NFPA 1620)

REFERENCES

Cook Jr, J. Lee. *Santana Row Development Fire, San Jose, California*. FEMA/U.S. Fire Administration/Technical Report Series 153, August 2002. Emittsburg, MD.

Routley, J. Gordon. *Four Firefighters Die in Seattle Warehouse Fire, Seattle, Washington*. FEMA/U.S Fire Administration, Technical Report Series 077, January 1995, Emittsburg, MD.

CHAPTER 13

Master Stream Appliances

LEARNING OBJECTIVES

- Compare the two types of transitional attack.
- Describe how a blitz attack works.
- Identify the three types of master stream appliances and their application on the fireground, including the advantages and disadvantages of each.
- Recognize the various types and sizes of nozzles used to deliver the proper flow rates and stream patterns when using a master stream appliance.
- Explain how a master stream appliance can best be used to control and extinguish a fire.
- Describe the proper positioning of a master stream appliance.
- Compare the three angles of a straight stream into a window and understand the results that these angles produce in fighting the fire.
- List the signs of a probable structural collapse.
- Explain how to set up a proper collapse zone.

Introduction

The most important question an incident commander (IC) must ask is, "At what point does this building no longer have any value?" When the life safety priority has been addressed and there is no more savable property to protect, the primary mission is over. It's time to put this fire out with master streams or let the fire burn itself out. Once the decision is made to go defensive, no firefighter lives should be placed in danger, and no more risks should be taken inside the fire building. Master stream appliances are large-caliber devices used primarily during defensive strategy operations; they are also known as heavy-stream and large-caliber stream devices. This chapter discusses the types of appliances and nozzles that are available, water supply for master streams, and the use of master stream appliances on the fireground.

Transitional Attack

Master stream appliances are also used in a transitional attack. There are two types of master stream transitional attacks: offensive to defensive, and defensive to offensive. An offensive to defensive attack is the riskier of the two **FIGURE 13-1**. The IC can select this strategy when the fire requires master streams to extinguish it, but there is still a probable or known life hazard to resolve. For the time it takes to perform the search when the survivability profile is still good, or the time it takes to effect an actual rescue, the strategy should be offensive and use master streams. Once the life hazard is mitigated, the IC withdraws all units from the interior and switches to a defensive strategy.

A transitional attack, defensive to offensive, is used when there is more fire than 1¾-inch and 2½-inch hand lines are able to handle, but the building still has value and is savable **FIGURE 13-2**. The blitz attack using a pre-plumbed monitor is an example of a defensive to offensive transitional attack **FIGURE 13-3**. A 2½-inch line flows 300 gallons per minute (gpm). If it doesn't have the reach and penetration to overwhelm the fire, adding a second 2½-inch line, even though you're now flowing 600 gpm, won't penetrate the fire any farther than the first 2½-inch hand line. In addition to more gpm, velocity is needed to reach and penetrate the fire to knock it down to a size where hand lines can be effective again. The concept is the same as fast water on an offensive attack, only on a grander scale. The goal is to knock down and reset the fire back to hand-line size. The additional velocity and gpm can be attained only by putting master streams in service for a short period of time. Once the fire is knocked down, the strategy changes to offensive.

FIGURE 13-1 An offensive to defensive attack is used when the fire requires master streams to extinguish but there is still a probable or known life hazard.

Courtesy of Michael Gala.

Defensive Operations

In defensive strategy, conditions on the fireground are deteriorating or have already reached a point when master stream appliances are placed in service. During this time, firefighters must be completely aware of their surroundings on the fireground. In a defensive mode when master streams are placed in service, the IC has already made the decision that this building no longer has any value. The building is written off when it's time to "surround and drown," and no personal risks of injury or death should be taken **FIGURE 13-4**. A defensive fire *should* be one of the safest fires we fight.

Master stream appliances mean that a large volume of water is being applied on the fireground. Firefighters must use extreme caution working around hose lines, pumpers, and the appliance itself. Precautions must be observed around buildings or other structures at which the appliance is being directed. At this stage, the building is severely damaged by fire, and this is the time when roof and wall collapses are likely to happen **FIGURE 13-5**. If the potential for wall collapse exists, a collapse zone must be established. A collapse zone is the area endangered by the potential wall collapse of a building, and it is calculated as the distance out from the wall equal to 1½ times the height of the involved building **FIGURE 13-6**.

Both pumpers and aerial fire apparatus should be positioned so that they are not subject to the damage caused by radiant heat, convective heat, burning embers, or structural collapse. Master stream appliances should be as far away from the fire as possible without compromising their effectiveness. If there is any indication that the structural integrity is in question,

FIGURE 13-2 A transitional defensive to offensive attack.The attack team is in a safe and ready position waiting for the fire to get knocked down.

Ladder Company Fireground Operations, Third Edition, Harold Richman, Steve Persson, NFPA © 2008 Jones & Bartlett Publishers.

FIGURE 13-3 The blitz attack using a pre-plumbed monitor—an example of a defensive to offensive transitional attack.

Courtesy of John Odegard.

FIGURE 13-4 In a defensive fire, the IC has made the decision that the building no longer has any value and it is written off.

Photo by Rick McClure.

a perimeter should be set up a safe distance from the collapse zone and all personnel and equipment should be kept out of this area **FIGURE 13-7**.

Command must manage the use of master stream appliances on the fireground. Master stream appliances should not be placed in service in a defensive mode while an offensive interior attack is in operation. Serious injury to firefighters could result as fire, heat, and products of combustion are forced onto them, in addition to the assault by the blunt force of the water stream **FIGURE 13-8**.

FIGURE 13-5 When the building is severely damaged by fire, roof and wall collapses are likely. These firefighters are working within the collapse zone with power lines overhead.
Courtesy of Mr. David J. Jones, Cincinnati, Ohio.

Master stream appliances deliver more water and can reach farther than the largest handheld hose lines. They are placed into operation when handheld hose lines are ineffective in fire attack, when exposure protection is needed, as backup lines, and to knock over freestanding walls and chimneys. Several types of appliances are available, with several nozzle sizes, including both spray nozzles and smooth-bore tips.

Master stream appliances are not special equipment. They should be considered standard firefighting tools and thus should be part of pre-incident planning and training evolutions. To be effective, these appliances require water flows from 350 to 2,000 gpm, depending on the size of the appliance, the type of nozzle used, and the volume of water available. (Large ocean-tug fireboats can pump up to 20,000 gpm.) Engine company personnel should be fully trained in the operation of these appliances, from the laying and charging of supply lines to the use of the appliances in fire attack and exposure protection.

Types of Master Stream Appliances

There are essentially three types of master stream appliances. The first two types are portable and fixed appliances, which are carried on the fire apparatus and operated from that apparatus. In addition, portable

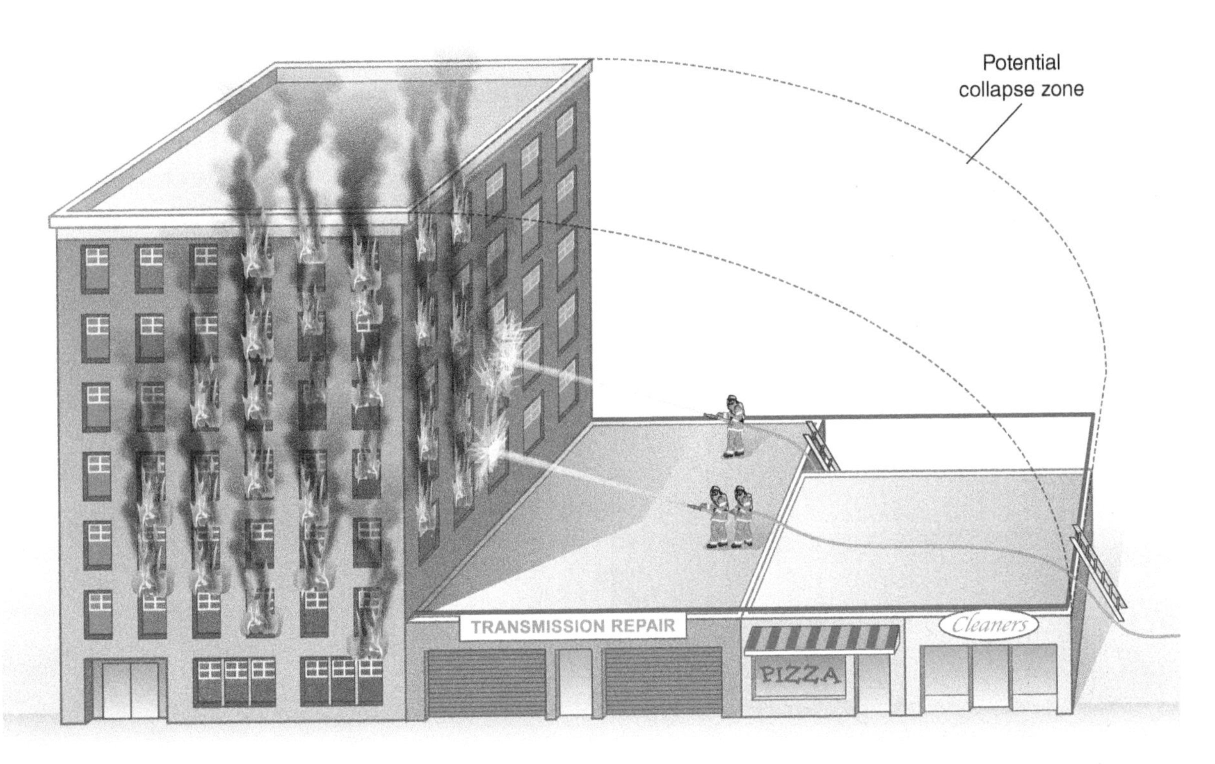

FIGURE 13-6 If the potential for wall collapse exists, establish a collapse zone of at least 1½ times the height of the involved building.

appliances can be operated on the ground or from other remote positions. The third type, the elevated master stream appliance, is operated from aerial fire apparatus ladders or tower ladder platforms. Each type has its advantages and disadvantages in a particular fire situation.

Portable Master Stream Appliance

Portable master stream appliances are often referred to as deluges, deck guns, monitors, or ladder pipes. They are carried on the apparatus and generally are operated from there, although they are designed to permit operation from either the apparatus or the ground. A number of varieties are available, but all are operated similarly.

FIGURE 13-7 When the structural integrity of a building is in question, establish a safety perimeter.

Courtesy of Martin Grube/FireRescue TV.

Many portable master stream appliances are pre-plumbed and can be operated from a fixed position on the pumper and placed into service quickly. If the pumper can be placed in a strategic location so that the appliance can be operated on the fire or operated to protect an exposure, then it may be advantageous to work from that position **FIGURE 13-9**. If the pumper cannot be located near the fire, the appliance can be removed from the pumper and operated separately from the apparatus.

Portable master stream appliances are provided with spikes on the bottom of the stabilizing legs, as well as a chain or straps. The spikes dig into the ground, while the chain and straps are attached to a stationary object to prevent the appliance from shifting during operation **FIGURE 13-10**.

When water flows through the barrel at too low an angle, the nozzle reaction force created can cause the device to move, shift from its position, or whip around. A safety lock is provided so that the lock must be manually released before the barrel can be lowered below a 35° angle. Some portable appliances employ a safety shutoff valve that automatically shuts down the flow of water if the appliance moves. This feature reduces the risk of injury to firefighters from an

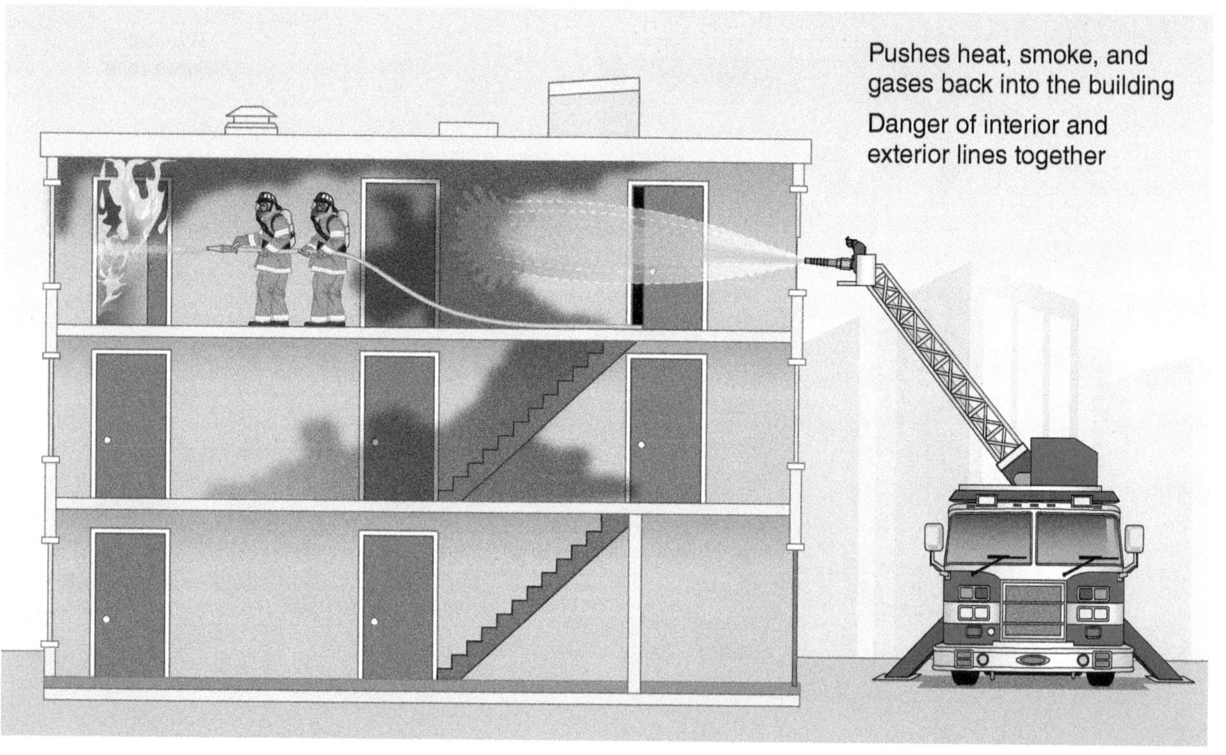

FIGURE 13-8 Master stream appliances should not be used while an offensive interior attack is in operation. The risk of injuring firefighters is too great.

Ladder Company Fireground Operations, Third Edition, Harold Richman, Steve Persson, NFPA © 2008 Jones & Bartlett Publishers.

FIGURE 13-9 Many portable master stream appliances are pre-plumbed and can be placed in service quickly to attack a fire or to protect exposures.

Courtesy of John Odegard.

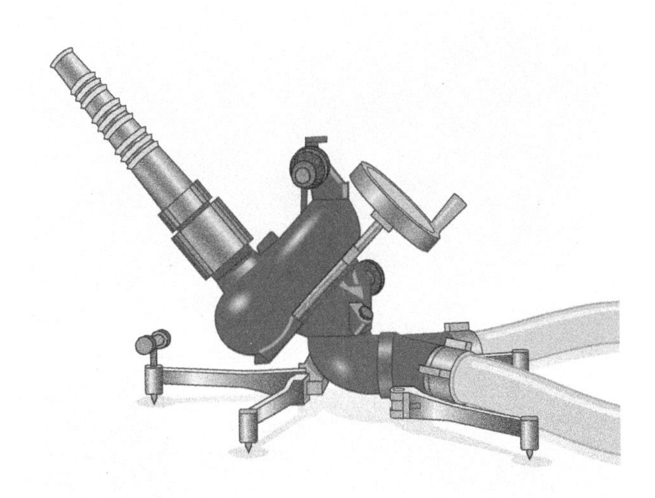

A

FIGURE 13-10 Portable master stream appliances are provided with spikes on the bottom of the stabilizing legs.

© Jones & Bartlett Learning; Photograph by Glen E. Ellman.

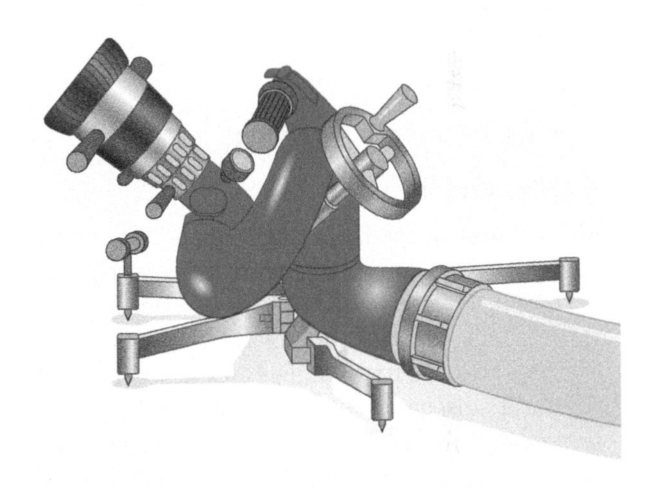

B

FIGURE 13-11 Portable master stream appliances can have A. two inlets or B. one inlet.

A-B: *Engine Company Fireground Operations*, 3rd Edition, Harold Richman, Steve Persson, NFPA © 2008 Jones & Bartlett Publishers.

out-of-control appliance. Pump operators should be alerted to shut down the discharge port at the panel quickly should a monitor get out of control. Some portable master stream appliances are made to take advantage of a hose loop to prevent movement. This feature should be used if the master stream appliance is so equipped.

FIGURE 13-11 shows two types of portable master stream appliances with one and two inlets. These inlets are designed to be supplied by a large-diameter hose (LDH) or 2½-inch supply lines. Both types of portable master stream appliances are capable of operating with smooth-bore tips or combination nozzles.

Today's modern fire apparatus are usually equipped with a pre-plumbed deck gun that can be quickly supplied with water from the tank using a dedicated discharge gate on the pump panel, or they can be taken off the apparatus, placed into a mounting bracket, and operated on the ground or other remote location FIGURE 13-12. They may be equipped with a telescoping feature, which lowers the device for storage and raises it when operating to give greater clearance from other equipment, including raised cab roofs and ladder brackets above the hose bed FIGURE 13-13(A). In addition, an appliance may also be remotely controlled from its position on the pumper. Engine company members must routinely

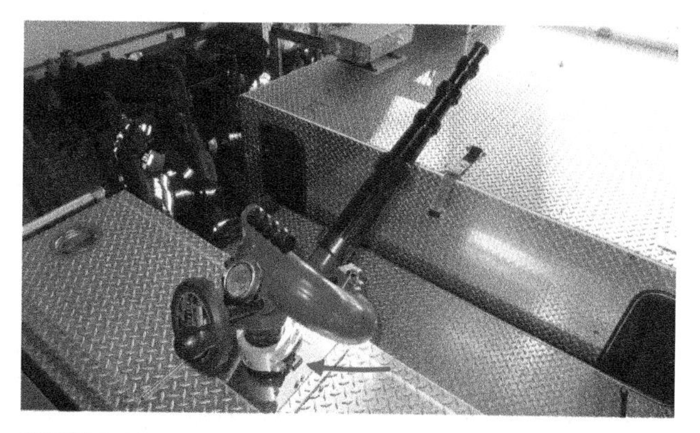

FIGURE 13-12 Some pre-plumbed deck guns can be taken off the apparatus, placed into a mounting bracket, and operated on the ground as a portable unit.
Courtesy of Raul Angulo.

A

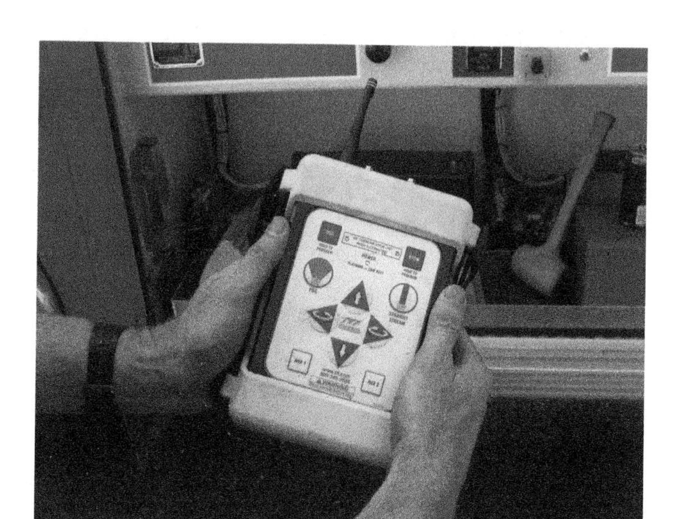

B

FIGURE 13-13 **A**. Some pre-plumbed monitors have a telescoping feature. **B**. New pre-plumbed monitors can be operated by remote control.
Courtesy of Raul Angulo.

check the batteries in the remote control unit and be familiar with the manufacturer's operating instructions for use from either the apparatus or the ground **FIGURE 13-13(B)**.

Fixed Master Stream Appliances

Master stream appliances can be mounted permanently or fixed to the engine, or other specialty apparatus and vessels **FIGURE 13-14**. Water is supplied to fixed appliances in one of two ways. In the first method, water is pre-plumbed to the appliance from a separate discharge gate on the pump. If the engine is parked in front of the fire building, water from the tank can be used for a blitz attack to knock down the fire. The sequence of this evolution needs to be practiced ahead of time. Depending on the size of the tip carried on the monitor, it can flow between 500 and 1,000 gpm, which gives only between 30 to 60 seconds of water. The monitor has to be aimed just right before the gate valve is opened. Valuable water can't be wasted with major adjustments to the stream **FIGURE 13-15**. It is always best to hook up to a hydrant water supply by connecting LDH lines to the intake ports of the engine before the monitor is placed in service.

In the second method, the appliance is supplied by hose lines with one or more connections to the pumper's discharge outlets. Devices supplied by the second method can also be mounted on ladder trucks (called ladder pipes), special service vehicles designed to carry such appliances, and other vehicles without pumps. The apparatus on which a fixed master stream appliance is mounted must be positioned carefully and correctly at the fire scene.

Elevated Master Stream Appliances

Elevated master stream appliances, or ladder pipes, are found on aerial ladders, elevating platforms, and water towers. Unless the aerial fire apparatus is equipped with a fire pump, an engine company must supply water to the appliance.

When a pre-plumbed waterway is provided on an aerial ladder, the waterway system must be capable of flowing 1,000 gpm at 100 pounds per square inch (psi) nozzle pressure at full elevation and extension **FIGURE 13-16**. A permanently attached monitor must be provided with a 1,000-gpm nozzle **FIGURE 13-17**. Where a pre-plumbed waterway is not provided, a ladder pipe with clamps to secure the appliance to the aerial ladder should be provided **FIGURE 13-18**. In addition,

A

B

C

FIGURE 13-14 Fixed master stream appliances can be operated **A**. from a fire boat, **B**. via remote control modern fireboat monitors, **C**. and from a pumper.

Courtesy of Raul Angulo.

appropriate tips, hose, hose straps, and halyards should be provided to operate the ladder pipe properly.

Elevating platforms of 110 feet or less rated vertical height must have a permanent water delivery system installed capable of delivering 1,000 gpm at 100 psi nozzle pressure with the elevating platform at its rated vertical height **FIGURE 13-19**. One or more permanently installed monitors with nozzles capable

FIGURE 13-15 Using a deck gun for a blitz attack using tank water gives 30 to 60 seconds of water. The monitor must be aimed carefully before the gate valve is opened.

Courtesy of John Odegard.

of discharging 1,000 gpm must be provided on the platform **FIGURE 13-20**. The permanent water system must supply the monitor. Permanent waterways on both an aerial ladder and elevating platform must be arranged so that they can be supplied at ground level through an external intake that is a minimum of 4 inches.

Nozzles for Master Streams

Various sizes of smooth-bore tips are available for use with master stream appliances. The most common are the 1⅜-, 1½-, 1¾-, and 2-inch tips. These tips are all normally operated at a nozzle pressure of 80 psi. The 1⅜-inch tip discharges 500 gpm; the 1½-inch tip discharges 600 gpm; the 1¾-inch tip, 800 gpm; and the 2-inch tip, 1,000 gpm at 80 psi nozzle pressure. When water is delivered to any of these nozzles at an insufficient rate, the stream tends to break up. This decreases the reach of the stream. If water supply is a problem, a smaller tip size should be used.

A 1¼-inch tip can be used on a master stream appliance when there is not enough water volume to feed the tip sizes listed in the paragraph above. Although generally considered an attack line tip (325 gpm at 50 psi), it can develop a fairly heavy stream with a flow of 400 gpm at a nozzle pressure of 80 psi. A 1¼-inch tip is often furnished as part of a ladder pipe's assortment of smooth-bore tips. Tips of this size could also be used on other master stream appliances when necessary.

Combination master stream nozzles are available in many sizes. They are designed to operate at 100 psi, with water delivery rates generally from about 300

A

B

FIGURE 13-16 **A**. Most aerial ladders have a pre-plumbed waterway for the elevated master stream appliance. **B**. The waterway system must be capable of flowing 1,000 gpm at 100 psi nozzle pressure at full elevation and extension.

B: Courtesy of Los Angeles County Fire Department.

FIGURE 13-17 A permanently attached monitor must be provided with a 1,000-gpm nozzle.

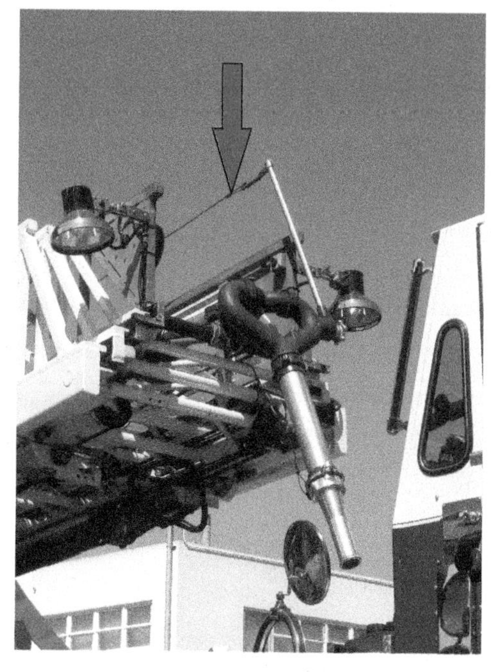

FIGURE 13-18 When a pre-plumbed waterway is not provided, a ladder pipe with clamps to secure the appliance to the aerial ladder should be provided.

Courtesy of Al Hom.

to 1,250 gpm. Some combination spray nozzles have greater flow rates available. Again, these nozzles must receive a sufficient water supply and operate at the proper nozzle pressure in order to be effective. Some combination spray nozzles are constructed so that their flow rate can be varied (the flow rate is selectable), generally in increments of 250, 350, 500, 750, and 1,000 gpm. They allow the volume of water to be matched to the capability of the supply system, and they are recommended for use in areas where water supply delivery rates vary from location to location. Some are available with a factory-ordered fixed orifice, whereas others may have an automatic pressure control.

A master stream appliance is only as good as its water supply; no device can be effective with an inadequate flow of water. A water flow rate that is inadequate for a particular size of smooth-bore tip may be just right for a smaller diameter smooth-bore tip. A 500-gpm stream from a 500-gpm tip is much more effective in controlling a fire than an inadequate stream from a 1,000-gpm tip. For this reason, it is important that smooth-bore tips of several sizes be carried on the apparatus for use as required.

FIGURE 13-19 Elevating platforms of 110 feet or less rated vertical height must have a permanent water delivery system installed capable of delivering 1,000 gpm at 100 psi nozzle pressure with the elevating platform at its rated vertical height.

FIGURE 13-20 One or more permanently installed monitors with nozzles capable of discharging 1,000 gpm must be provided on the platform.

Water Supply for Master Stream Appliances

Master stream appliances operate at high flow rates; such operation increases friction loss in supply lines and thus requires higher pressures and increased water flow from pumpers. As explained earlier, the effectiveness of a master stream appliance depends almost completely on an adequate supply of water; therefore, it is important that friction losses be minimized and that pumpers be used most efficiently. Recall the following points when supplying a master stream appliance:

- Locate an engine at a hydrant or other water source capable of flowing a volume of water sufficient to supply a master stream appliance on the fireground. If you want to supply a ladder pipe, this can be accomplished by laying reverse from the aerial ladder to the hydrant. Never lay forward from the hydrant to the aerial ladder for ladder pipe operations unless another engine is hooked up to the hydrant.
- Use an LDH or an adequate number of 2½-inch supply lines laid between pumpers if you are engaged in relay pumping operations.
- If the portable master stream appliance is to be operated away from the pumper, use an LDH or an adequate number of 2½-inch supply lines to maintain adequate volume to the appliance.
- To prevent excess friction loss in the supply line, maintain a minimum distance between the pumper and master stream appliance if 2½-inch supply lines are being used. The use of an LDH usually allows distances between the pumper and master stream appliance to be increased significantly.

Before discussing these recommendations individually, consider the following three scenarios. In the first two scenarios, 2½-inch hose lines are used to supply the master stream appliance from a pumper; in the third scenario, a 4-inch supply line is used. For these scenarios, the pumper and master stream appliance are 100 feet (30.5 m) apart **FIGURE 13-21**.

Scenario 1

A pumper supplies a 1¼-inch smooth-bore tip, the smallest used on master stream appliances. This tip requires a nozzle pressure of 80 psi, which will develop a 400-gpm stream. This flow can be carried by a single 2½-inch hose line. Friction loss is approximately 35 psi per 100 feet; therefore, the pump pressure required for the 100-foot lay is 115 psi:

- Nozzle pressure: 80 psi
- Friction loss: 35 psi
- Add them together and the pump pressure = 115 psi

Scenario 2

A pumper supplies a 750-gpm combination spray nozzle on the master stream appliance. The combination spray nozzle requires 100 psi. Two 2½-inch supply lines, with 375 gpm flowing in each, are used. Friction loss is approximately 30 psi per 100 feet; therefore, the pump pressure required for the 100-foot lay is 130 psi.

- Nozzle pressure: 100 psi
- Friction loss: 30 psi
- Add them together and the pump pressure = 130 psi

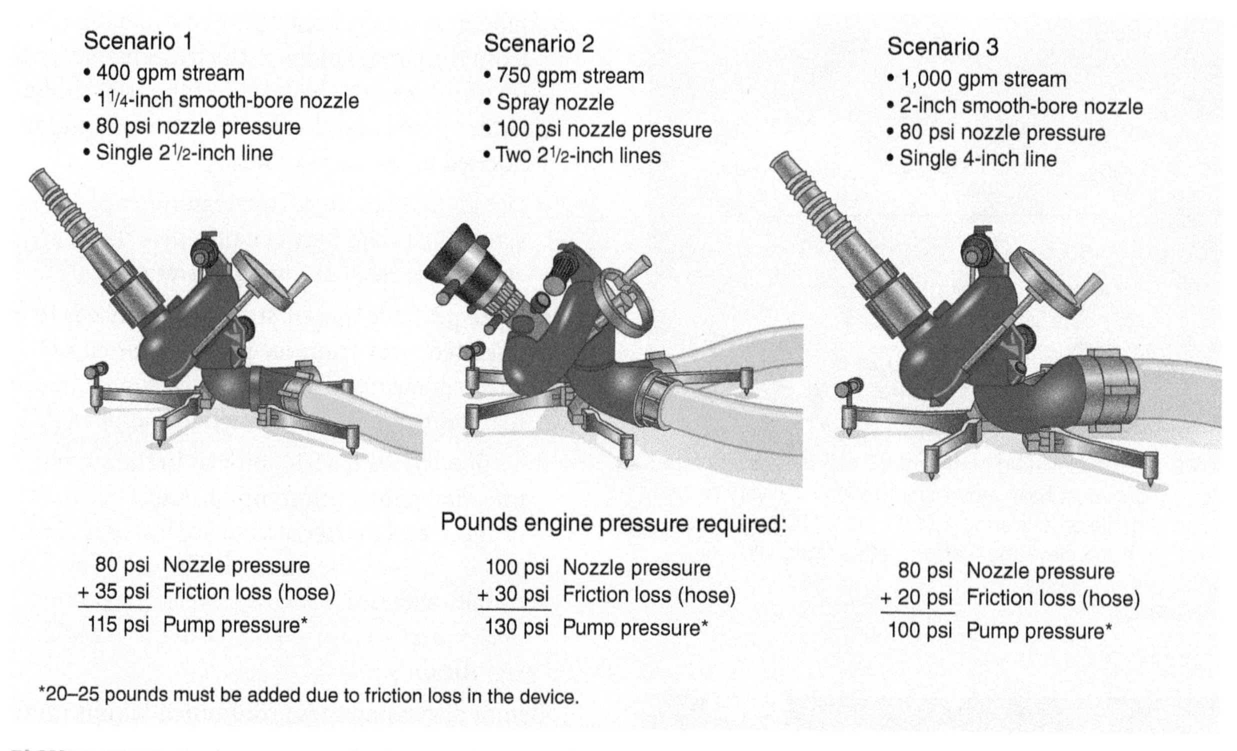

FIGURE 13-21 Engine pressures for increased water volume to a master stream appliance vary little if specified recommendations are followed.

Engine Company Fireground Operations, 3rd Edition, Harold Richman, Steve Persson, NFPA © 2008 Jones & Bartlett Publishers.

Scenario 3

A pumper supplies a 2-inch smooth-bore tip on a master stream appliance. This tip requires a nozzle pressure of 80 psi, which develops a 1,000-gpm stream that is carried through a 4-inch supply line. Friction loss is approximately 20 psi per 100 feet; therefore, the pump pressure required for a 100-foot lay is 100 psi.

- Nozzle pressure: 80 psi
- Friction loss: 20 psi
- Add them together and the pump pressure = 100 psi

Very little difference exists among the three calculated pump pressures, although different gpm flows were achieved using different size hose lines, nozzles, and nozzle pressures. If a single 4-inch supply line had been used for all three scenarios, it would have been able to supply the master stream appliances over longer distances with negligible friction loss.

Pumper-to-Pumper Operation

The most effective operation for delivering a large volume of water to a fire is to have pumpers at the hydrants or other adequate water sources to discharge their water into pumpers at the fireground. This allows the use of a short lay of hose from the pumper at the fire to the master stream appliance, with the friction loss minimized **FIGURE 13-22**. If an engine company operates a master stream appliance from the apparatus, it too should be supplied by a pumper at a hydrant or an adequate water source, whether or not the appliance is permanently connected to its pump. Supplying an engine from another pumper may not be necessary if an LDH is used from an adequate water source. Many fire departments with adequate water delivery systems use 5-inch LDH to reduce friction loss further. A disadvantage of relay pumping is that two engines are required to supply water to a single master stream appliance.

Adequate Number of Supply Lines

In the three scenarios described above, the number of hose lines or their size between the pumper at the fire and the master stream appliance was increased when the required flow rate increased. This actually lowered the friction loss and kept the engine pressure fairly constant. If hose lines with less carrying capacity had been used in the last two examples, the friction loss would have increased to make the desired performance impossible.

The intake of a master stream appliance is actually a large Siamese, a device that collects water from two

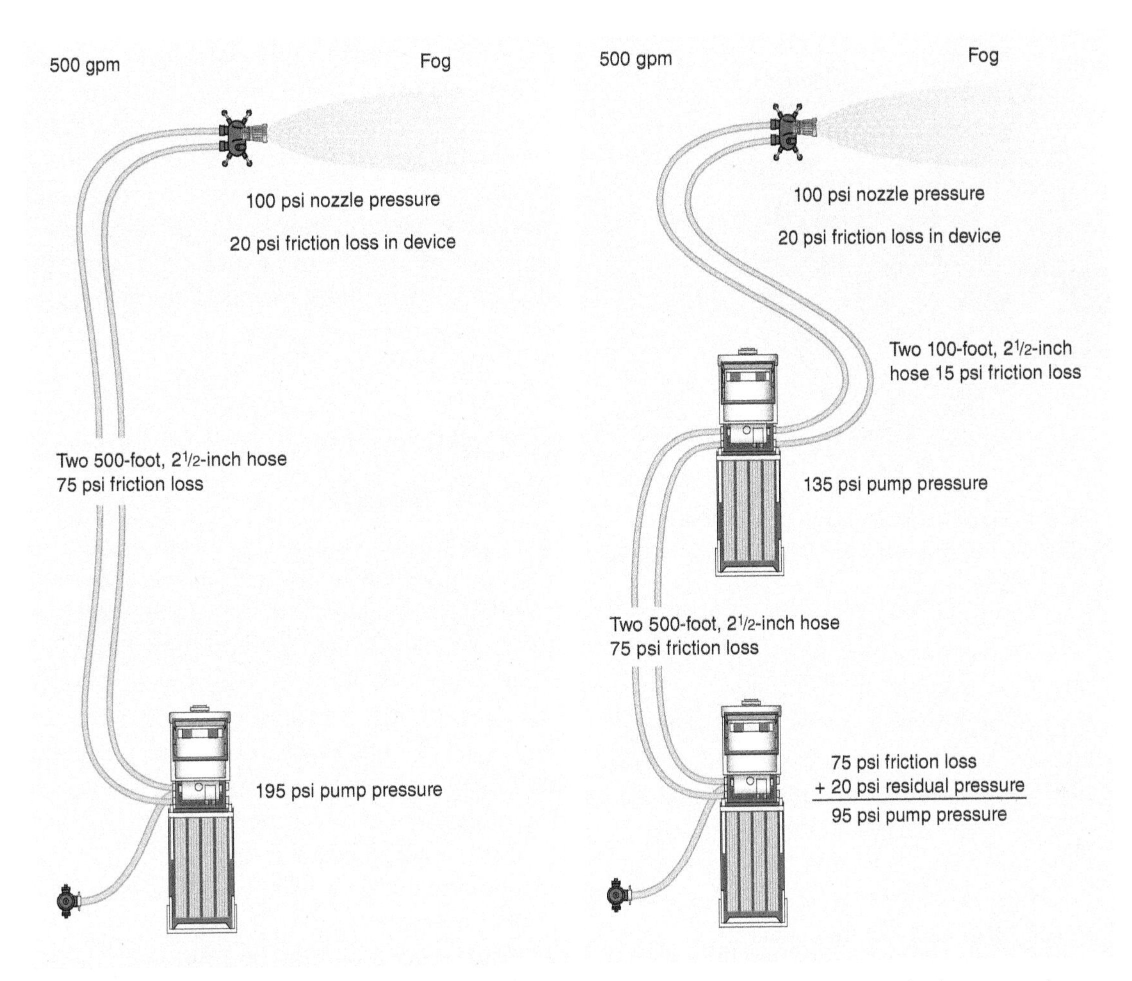

FIGURE 13-22 The most effective delivery of a large volume of water is a pumper at a hydrant pumping to a pumper at the fire.

Engine Company Fireground Operations, 3rd Edition, Harold Richman, Steve Persson, NFPA © 2008 Jones & Bartlett Publishers.

or more hose lines and delivers it to a single line, which in this case is the nozzle **FIGURE 13-23**. The intake can also be a single LDH connection. More water meeting less resistance flows through the multiple lines or a single LDH line to reach the appliance. Thus, less pump pressure is required, and a more efficient operation is carried out.

The 2½-inch hose was used in the first two examples only because it is the smallest supply line. A larger-diameter hose would give even better performance because of greater carrying capacities and lower friction loss. Some master stream appliances are manufactured with a single intake, which is designed for use with LDH.

It is suggested that a maximum length of 100 feet of hose be used between the pumper and a master stream appliance when using smaller supply lines. This recommended maximum is intended to limit the friction loss in the hose lines between the pumper at the fire and the appliance. Occasionally, there will be situations in which the master stream appliance must be more than 100 feet from the pumper at the fire because the pumper cannot be brought to the position necessary for proper operation. In such cases, it is important to remember that longer-length hose lines require greater pressures. Again, the use of LDH allows greater carrying capacities with less friction loss than smaller supply lines.

Standard Operating Guidelines

Standard operating guidelines (SOGs) differ from jurisdiction to jurisdiction. A fire department must establish an SOG that reflects its specific needs predicated on the water supply available; pumping capacities; and the water delivery system, including supply hose, appliances, and the style of hydrant ports.

An additional benefit can be derived from following the four recommendations given earlier: They are the basis of an SOG for placing master stream appliances into service. The recommendations themselves specify the size and length of hose to be used and the operation of pumping from one pumper to another. This combination is effective because it minimizes pump pressures and thus maximizes water delivery capability.

FIGURE 13-23 If not part of the appliance, a Siamese device can be used to increase the water flow into a portable master stream appliance.

Engine Company Fireground Operations, 3rd Edition, Harold Richman, Steve Persson, NFPA © 2008 Jones & Bartlett Publishers.

The recommendations determine the discharge pressure at which the pump at the fire should be set, at least initially. The required pump pressure was close to the same in all three scenarios: about 115 to 130 psi. The friction loss in most master stream devices is 20 to 25 psi, which must be added to the engine pressures calculated in the examples. Each department must initiate a set of guidelines predicated on their current equipment and capabilities.

In most cases, a bit too much pressure isn't a problem because there are no firefighters trying to fight the nozzle reaction while holding the master stream appliance steady. However, a great deal of extra pressure causes a solid stream to break up as soon as it leaves the nozzle, thereby making it ineffective. Its reach will be greatly reduced, and it will be as vulnerable to winds as a fog stream. In effect, too much pressure turns a solid stream into a poor fog stream, a heavy spray of water accomplishing little if anything in extinguishing a fire or protecting an exposure. Fortunately, the results of a pressure that is too high are obvious to a trained firefighter observing the stream and a call for a reduction in pump pressure solves the problem quickly.

Use of Master Stream Appliances

As noted earlier, several sizes of smooth-bore tips and spray nozzles should be carried on the engine so that firefighters can choose the correct nozzle for the fire situation and the water supply. It is just as important that both spray nozzles and smooth-bore tips are available for use as the situation demands.

A master stream appliance may be used for fire attack or exposure protection, or to back up an existing hose stream. Most often, the appliance is positioned on the outside to deliver water into the fire building through windows or doorways or to protect an exposure. Wind conditions and the distance to the fire or exposure determines which type of nozzle is used and how it should be operated. Spray nozzles should be used to cover exposure buildings; smooth-bore tips are better suited for firefighting.

Solid-Stream Nozzles Versus Spray Nozzles

Because wind conditions vary, and some fires create their own wind, the strong draft created by a large fire can destroy the effectiveness of the water stream between the nozzle and the building. The solid stream from a smooth-bore tip has proven to be the most effective for fire attack when using a master stream appliance **FIGURE 13-24**. It penetrates farther into the building, covering more area of the fire while delivering more gpm, and is the least affected by the wind.

For exposure protection, a spray pattern from a combination nozzle using a fog stream may be superior to a solid stream if it is not affected by the wind or the distance from the fire building. The fog stream covers a wider area than a solid stream and requires less movement.

A combination spray nozzle using a straight stream, even from a master stream appliance, loses its effectiveness if it must be applied over any distance. If there is any question about the ability of a straight stream to reach a fire or exposure, a solid stream from a smooth bore should be used. For example, the intensity of a fire and/or the structural integrity of the building might prevent the positioning of master

FIGURE 13-24 The smooth-bore solid stream has proven to be most effective for fire attack when using a master stream appliance.

Photo by Mike Heaton.

FIGURE 13-25 The straight stream from the master stream appliance (left) is breaking up from the effects of wind. This adverse effect may be avoided by moving the appliance closer to the fire building if conditions allow. Another alternative is to switch to a large smooth-bore tip (right).

Engine Company Fireground Operations, 3rd Edition, Harold Richman, Steve Persson, NFPA © 2008 Jones & Bartlett Publishers.

stream appliances near enough for straight streams to be effective. In such cases, solid-stream nozzles with appropriate size tips would have to be used.

Strong crosswinds adversely affect both fog and solid streams, but they can render a fog stream completely ineffective exposures **FIGURE 13-25**. Again, in a strong wind, a solid stream should be used. At the same time, company officers must be aware that using a large smooth-bore tip to cover an exposure building can cause structural damage **FIGURE 13-26**. The heavy straight streams are powerful and can damage exposure buildings by tearing off siding and decorative cornices, marquees, logo signs, neon signs and other exterior lighting, cell phone towers, water tanks, antennas, and the like. It can tear off awnings, weaken mortar, dislodge bricks and concrete, and even topple weakened walls. Glass can be broken, making the building more vulnerable for fire extension within the building.

FIGURE 13-26 Large smooth-bore tips produce heavy, powerful straight streams that can cause structural damage to exposure buildings.

Positioning the Master Stream Appliance

To be effective, a spray nozzle using a straight stream must be positioned closer to the fire structure. Solid-stream nozzles, on the other hand, perform as well or better if they are positioned some distance from the building. For example, a solid stream may be used to attack a fire that is three or four floors above the level of the master stream appliance. For this operation, the appliance must be placed some distance from the building to achieve the proper angle of entry.

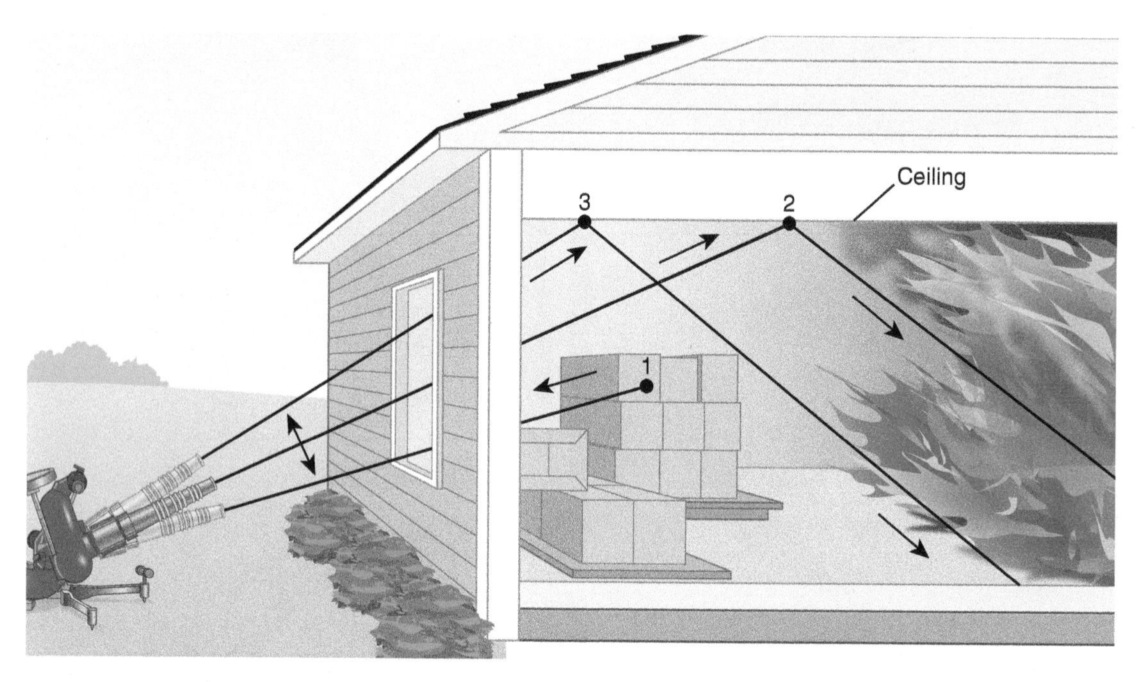

FIGURE 13-27 A master stream appliance must be positioned so that it hits the fire at an effective angle. In this figure, angle 2 is the best angle.

Engine Company Fireground Operations, 3rd Edition, Harold Richman, Steve Persson, NFPA © 2008 Jones & Bartlett Publishers.

If the appliance is too close to the building, the angle will be too steep. The stream then might not reach the fire area but rather flow water needlessly into an uninvolved area.

A properly positioned solid stream from a ground monitor enters the building at an angle that causes the stream to be deflected over a wide area when it strikes the ceiling or other overhead **FIGURE 13-27**. The low angle of stream 1 shown in the figure, just over the windowsill into the building, is best for maximum penetration but may be affected by stationary obstructions. Nonstationary obstructions will most likely be toppled over by the powerful straight stream. The angle of stream 2 is also effective, but it may not penetrate as far into the building as stream 1. Stream 3 has the least penetration; therefore, the monitor should be moved farther back from the window or repositioned to a doorway or other opening, unless the desired effect is to deflect the stream off the ceiling. The stream should be moved up and down to achieve the best combination of effective results.

For aerial ladders, tower ladders, and elevating platforms, the angle of streams 1, 2, and 3 in Figure 13-27 should be the goal no matter what floor the fire is on within the reach of the aerial and ladder pipe **FIGURE 13-28**. In upper floors, if these angles cannot be achieved, you've reached the limits of your aerial device or elevating platform for the delivery of effective master streams. On average, aerial ladder pipes and tower ladder master streams are effective up to the twelfth floor. Fires above the twelfth floor are out of reach for most aerial master streams operating from the exterior of the building, although they may help to prevent the fire from lapping (autoexposure) up to the sixteenth floor **FIGURE 13-29**.

Directing a Master Stream

To be most effective in fire control, a master stream should be moved horizontally back and forth across the fire area it is covering. The stream also should be moved up and down so that it reaches the full depth of the fire area. The amount of movement depends on the extent of the fire and existing conditions. Although nothing may burn directly under a stationary stream, the fire could spread away from it. Movement of the stream ensures coverage of a large enough area in and around the fire **FIGURE 13-30**. Less movement might be required for exposure coverage, especially when fog streams are used. The water should be allowed to rain down on the exposure to keep it wet and thus prevent it from burning.

Sometimes a master stream appliance is set up and then left unattended. This may occur because of a shortage of personnel, or perhaps a situation has become unsafe for firefighters, such as the possibility of a building collapse or a hazardous materials incident **FIGURE 13-31**. However, a master stream appliance should never be set up and then left unattended indefinitely. The stream and its effect on the fire or the exposure should be monitored carefully. A poorly aimed stream or one that goes off target is a tremendous waste of gallonage. Command needs to be kept aware of the current conditions because the appliance may need to be relocated to perform properly. If fire breaks out on an exposure, the exposure stream may have to be moved immediately to control the fire.

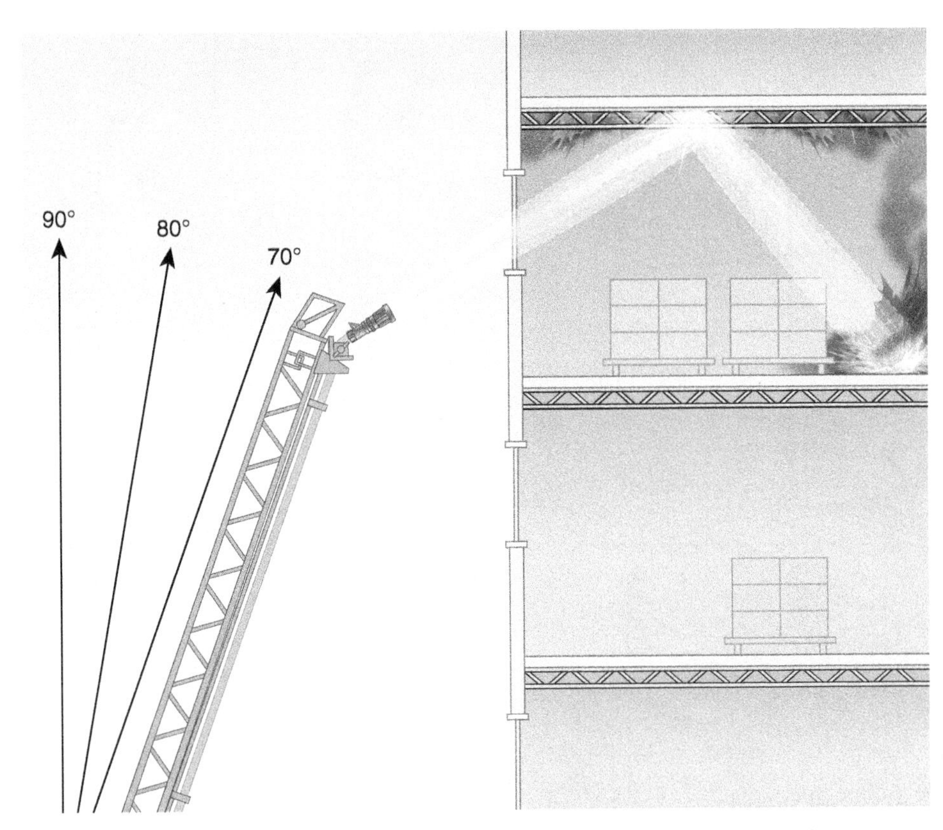

FIGURE 13-28 Aerial ladders, tower ladders, and elevating platforms must achieve an effective angle.

Ladder Company Fireground Operations, Third Edition, Harold Richman, Steve Persson, NFPA © 2008 Jones & Bartlett Publishers.

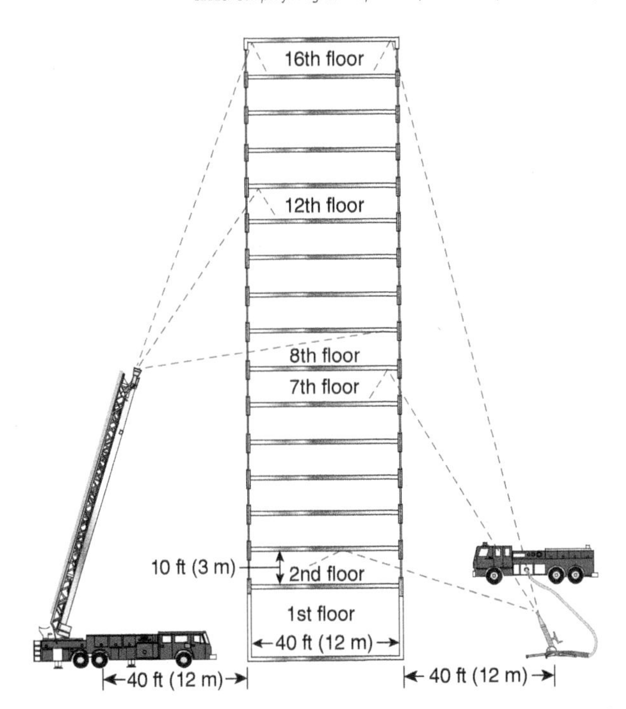

FIGURE 13-29 Fires above the twelfth floor are out of reach for most aerial master streams, although these streams may help to prevent the fire from lapping up to the sixteenth floor.

© Jones & Bartlett Learning.

If a stream does not seem to be having any effect on a fire, it might be positioned improperly. The water must reach the base of the fire, and the master stream appliance must be positioned accordingly. Sometimes the effectiveness of the stream can be increased by moving the appliance forward or back, so that the stream hits the fire from a better vantage point. If repositioning is not the problem, the size of the stream may need to be increased, or additional streams may need to be directed onto the fire.

In heavy smoke, it can be difficult to determine whether a stream is entering the building. Firefighters directing the stream might not be able to see the windows. An officer should visually check the building using a thermal-imaging camera (TIC) if it is possible to do so safely. If this is not feasible, the officer should listen for the loud sound of the stream hitting the building and should look for heavy water runoff. Both signs indicate that the stream is not entering the building. The stream then should be adjusted until both signs have disappeared. If the water runoff stops, and the noise of the powerful stream hitting a wall ceases, the stream is aimed back into the window. At that point, the stream should be operating effectively.

Shutdown

A stream from an appliance should be used only as long as fire is visible in the area covered by the stream. A check should be made for visible fire. Steam and/or white smoke is an indication that the main body of fire in that area has been knocked down

FIGURE 13-30 Master streams must be moved both vertically and horizontally to cover the entire area involved.

Engine Company Fireground Operations, 3rd Edition, Harold Richman, Steve Persson, NFPA © 2008 Jones & Bartlett Publishers.

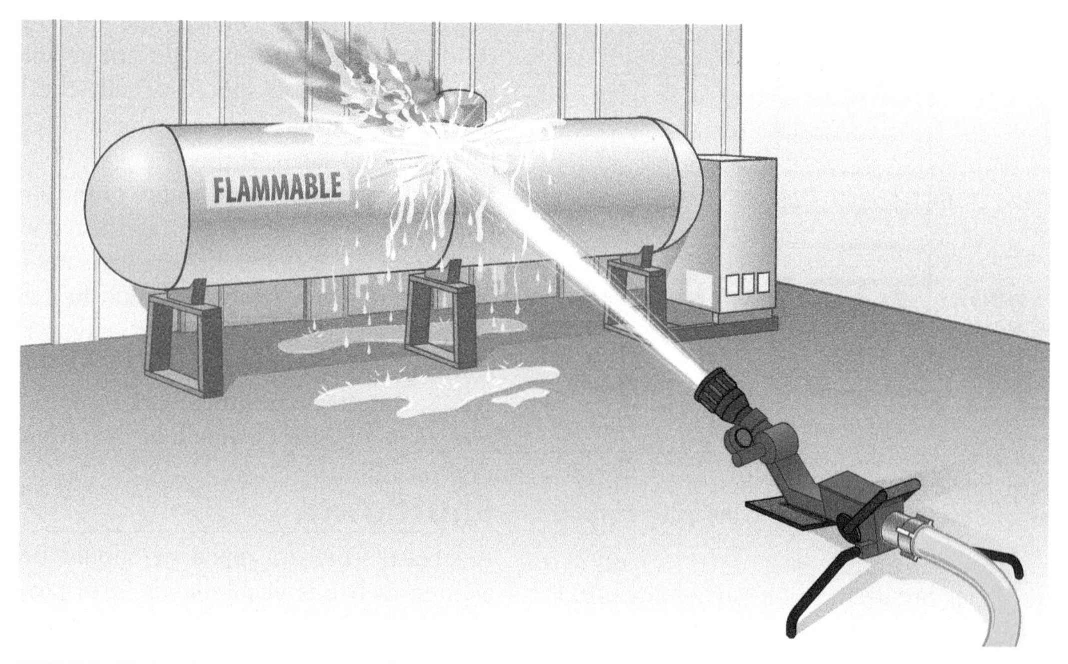

FIGURE 13-31 A master stream appliance may sometimes need to be left unattended.

FIGURE 13-32 Steam and white smoke indicate that the water stream is having a positive effect on the fire.

Engine Company Fireground Operations, 3rd Edition, Harold Richman, Steve Persson, NFPA © 2008 Jones & Bartlett Publishers.

FIGURE 13-32. When steam and/or white smoke are no longer visible, the fire has apparently been put out. The master stream should then be shut down or moved to cover another area; its work here has been completed. Continued operation at this point would only add to the water load on the fire building and the strain on the water supply system. If necessary, hand-held hose lines can be used for final extinguishment as long as the building is deemed safe to enter by the safety officer and the IC.

Elevated Master Streams

The same rules of operation discussed for portable and fixed master stream appliances apply for elevated master stream appliances. Because they are elevated above ground level, they can be used for fires on upper stories of buildings. They can be directed through windows or other openings by placing the nozzle at or near the window during defensive firefighting operations. When operating in a defensive mode, the chance of a building collapse must be considered. As long as their streams are effective, it is best to position the aerial apparatus at the corners of the building or at a distance away from the building that will afford protection for firefighters and the apparatus **FIGURE 13-33**. This should safeguard them from radiant and convective heat, burning embers, and building collapse. Elevated master streams are also useful in the protection of exposures, especially on upper floors. Directing water on the exposure is more effective than directing a stream between the fire and the exposure, unless a water curtain is specifically set up to extinguish flying embers, as in a large lumberyard fire.

A solid stream is most effective for fire attack when using an elevated master stream appliance. The stream will penetrate farther into the building and thus cover more area of the fire. For exposure protection, a combination spray nozzle using a fog stream may be superior to a solid stream if it is not affected by the wind or the distance from the fire building. The fog stream covers a wider area than a solid stream and requires less movement. Several sizes of smooth-bore tips and spray nozzles should be carried so that firefighters can choose the correct nozzle for the fire situation and the available water supply.

Elevated master streams, especially fog streams, must never be directed into natural exhaust openings, such as skylights, scuttles and hatches, and holes made in the roof for venting the smoke, or where fire has already burned through the roof. Firefighters vent roofs to allow smoke, heat, gases, and fire to escape the building so that an interior attack can be carried out. Vertical ventilation allows firefighters to perform a primary search, get to the seat of the fire, and apply water to extinguish the fire. The mistake of directing elevated master streams into an opening or vent is essentially placing a lid over the ventilation exhaust hole and thus prevents the fire, heat, and gases from escaping the building. It causes the ventilation flow path to reverse itself and seek another path of least resistance in order to vent. The air entrained from a master stream fog nozzle causes this reversal of smoke and fire gases to occur with tremendous force and speed and only spreads the smoke, heat, gases, and fire downward throughout the building, making it untenable for any victims as well as firefighters who may still be engaged in interior firefighting operations **FIGURE 13-34**.

If it is necessary to protect the roof from fire and radiant heat coming through these exhaust heat holes, a stream of water should be directed onto the roof adjacent to the exhaust hole so that the stream will not flow into the opening. During a defensive operation when a major portion of the roof or large area is destroyed by fire, elevated master streams can be directed at visible fire inside the building without the danger of reversing ventilation.

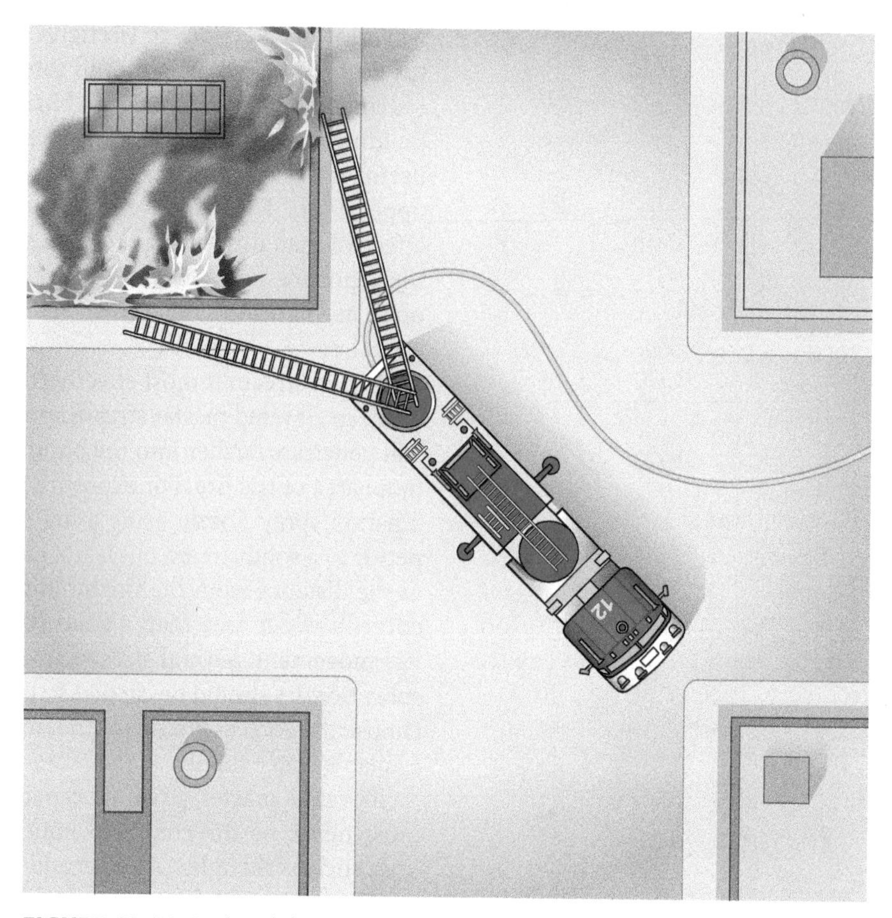

FIGURE 13-33 During defensive operations, the chance of a building collapse must be considered. If their streams are effective, it is best to position aerial apparatus at the corners of the building.

Ladder Company Fireground Operations, Third Edition, Harold Richman, Steve Persson, NFPA © 2008 Jones & Bartlett Publishers.

FIGURE 13-34 A fog master stream should never be directed into a ventilation exhaust roof opening. Doing so is the equivalent of putting a lid on the fire and forces the smoke and flames to spread throughout the building, making it untenable for anyone left inside.

Ladder Company Fireground Operations, Third Edition, Harold Richman, Steve Persson, NFPA © 2008 Jones & Bartlett Publishers.

Structural Collapse

By the time a building fire is declared defensive, the IC has written off the building. There's nothing left to save. The goal is to keep the fire from spreading to other exposed structures. The building, or what's left of it, has suffered extensive fire damage, and conditions are ripe for structural collapse **FIGURE 13-35**. By this time, the fire has burned away wooden structural members, and/or the intense heat has weakened unprotected steel components. The building may have sustained an explosion from natural gas or a backdraft.

Hundreds of thousands of gallons of water have been dumped into the building, adding tremendous weight to the roof and floor assemblies. Consider that one gallon of water weighs 8.34 pounds. (One liter of water weighs 1 kilogram.) For every 1,000 gpm from a ladder pipe, 8,340 pounds per minute is being added to the structure. A cubic foot of water weighs 62.4 pounds. One foot of standing water on a 20 foot × 20 foot roof weighs 25,000 pounds. One foot of standing water on a 40 foot × 40 foot roof weighs about 100,000 pounds. These are the effects on the building from the firefight. If preexisting conditions, for example, design flaws, inferior building construction methods and materials, and illegal modifications to the structure are added, you can begin to appreciate the extreme danger that exists with a defensive fire building. However, firefighter safety can still be maintained.

Signs of Collapse

The next time you and your crew are fighting a defensive fire, consider it a live-fire laboratory. All the signs that precede a building collapse are happening right in front of you. Sometimes sudden collapses occur without warning, but the majority of buildings deteriorate at a steady pace and you can note the interior and exterior warning signs. The interior warning signs include the following:

- Spongy roof and sagging roof decking
- Sagging floors or fire burning through the floors
- Ceilings dropping
- Structural movement or shaking of floors, bearing walls, beams, and columns
- Loud noises accompanying all of the above

Examples of exterior warning signs are the following:

- Fires burning out of control longer than 20 minutes without progress
- No water runoff from stairways and doors
- Cracks in the walls
- Walls bulging out or out of plumb
- Water and smoke pushing through cracks in the walls
- Separation of the roof from the walls
- Fire visible on every floor from every window

FIGURE 13-35 By the time a building fire is declared defensive, the IC has written off the building.

Collapse Zones

When signs indicate that a wall collapse is eminent, the IC must set up a collapse zone on the affected side of the fire building. The collapse zone is the area endangered by the potential wall collapse of a building and is figured at a distance out from the wall at 1½ times the height of the involved building. This allows for scattering debris, which can travel with deadly force when the wall hits the ground. When the building is taller than the street, the street should be cordoned off on the affected side. For example, if the D side of the building is in danger of collapsing, the entire street on the D side needs to be shut down and cleared of all firefighters, hose lines, master stream appliances, and apparatus. Aerial ladders, tower ladders, or elevated platforms operating master streams should be repositioned at the corners of the building outside the collapse zone **FIGURE 13-36**.

If any exposure buildings or homes are inside the collapse zone, they must be evacuated immediately. Fire scene tape can be used to mark off the collapse zone, but it has its limitations. Having sufficient rolls of scene tape readily available and finding a stationary object to tie it to can be a problem. An excellent alternative is utilizing uncharged fire hose. Using

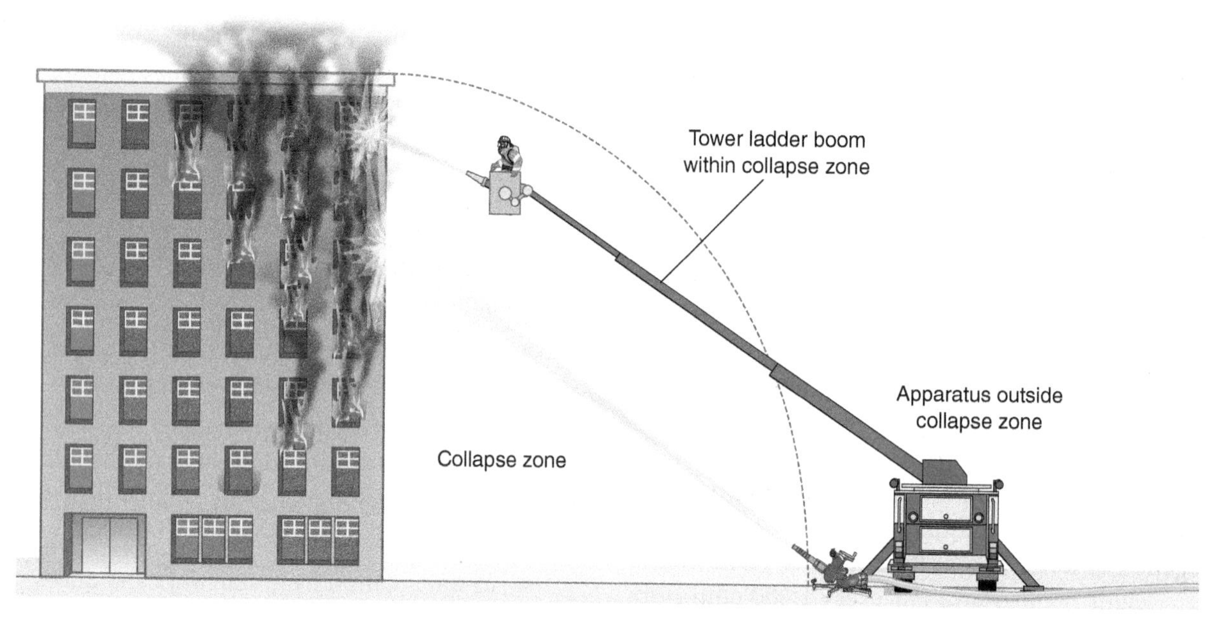

FIGURE 13-36 Aerial ladders, tower ladders, or elevated platforms operating master streams should be repositioned at the corners of the building outside the collapse zone.

uncharged fire hose to mark off a collapse zone clearly delineates the danger area. The hose is not subject to wind, and stationary objects are not needed for tying. It can be supplemented with traffic cones, and there is an ample supply of hose on any fireground. Firefighter safety is maintained by creating distance between firefighters and the building that is about to collapse.

If there is insufficient water pressure to put out a defensive fire, cover the exposures and let the fire burn itself out. There is no rush. Do not risk any firefighter lives at this stage of the game. Remember: There are no fires still burning today because we didn't have enough water to put them out; sooner or later, every fire goes out.

After-Action REVIEW

IN SUMMARY

- Master stream appliances are large-caliber devices that are used primarily during defensive strategy operations.
- Master stream appliances are placed in service when the streams from handheld hose lines are not effective in fire control or in exposure protection or as backup lines. They are used primarily during defensive operations.
- Master streams are also used in a transitional attack. There are two types of master stream transitional attacks: offensive to defensive, and defensive to offensive.
- The blitz attack using a pre-plumbed monitor is an example of a defensive to offensive transitional attack. Velocity and more gallons per minute are needed to reach and penetrate the fire and knock it down. The concept is the same as fast water on an offensive attack, only on a grander scale. The additional velocity and gallons of water can only be attained using master streams for a short period of time. Once the fire is knocked down, the strategy changes to offensive.
- When master streams are in service in a defensive mode, the IC has already made the decision that the building no longer has any value. No risk of injury or death should be taken. A defensive fire *should* be one of the safest fires we fight.
- Engine company personnel should be well trained in the operation of these appliances. This includes ensuring that a proper water supply is available as well as selecting the proper supply line or lines to provide a sufficient water supply to operate the appliance safely and effectively.

- The most effective method for supplying a master stream appliance involves the use of two pumpers: one at the water supply and one at the fire.
- Both spray nozzles and smooth-bore tips, in a range of sizes and styles, are available for use with these appliances. The size and style of the nozzle depends on the fire situation and the available water supply.
- Effective operation of master stream appliances is mainly a matter of observing stream performance. The appliance should be positioned so the stream reaches into the fire and should be operated only as long as the stream is doing its job.
- When the fire is extinguished, the appliance should be shut down. If the stream has little effect on the fire, it should be replaced either with a larger stream or supplemented with additional streams. The basic objective is maximum use of the available water supply.
- When a fire is declared defensive, the IC has written off the building and there is nothing left to save. The building has suffered extensive fire damage, and conditions are ripe for structural collapse. The fire has burned away wooden structural members and/or the intense heat has weakened unprotected steel.
- In a defensive strategy, hundreds of thousands of gallons of water are dumped into a building, adding tremendous weight to the roof and floor assemblies. Preexisting conditions, for example, design flaws, inferior building construction, and illegal modifications, make these structures extremely dangerous. Firefighter safety must be maintained by creating distance between firefighters and the building.
- There are interior and exterior warning signs preceding the majority of structural collapses.
- Interior warning signs include structural movement and shaking of floors, bearing walls, beams, and columns accompanied by loud structural noises. Exterior warning signs include fires burning out of control longer than 20 minutes without progress, no water runoff from stairways and doors, cracks in the walls, walls bulging out or out of plumb, water and smoke pushing through cracks in the walls, separation of the roof from the walls, and fire visible on every floor from every window.
- When the signs of eminent collapse exist, the IC must set up a collapse zone. The collapse zone is the area endangered by the potential wall collapse of a building. It is calculated as 1½ times the height of the involved building out from the wall.

KEY TERMS

blitz attack An aggressive fire attack that often utilizes a 2½-inch hand line or deck gun and occurs just prior to entry, search, and tactical ventilation. Also known as *defensive to offensive exterior attack*, *softening the target*, and *transitional attack*.

collapse zone The area endangered by a potential building collapse; it is calculated as 1½ times the height of the involved building.

master stream appliance A large-capacity nozzle that can be supplied by two or more hose lines or fixed piping and can flow in excess of 350 gpm. It includes ladder pipes, deluges, deck guns, and portable ground monitors. Also known as *heavy-stream devices*, *large-caliber stream devices*.

transitional attack An offensive fire attack initiated by an exterior, indirect hand-line operation into the fire compartment to initiate cooling while transitioning to interior direct fire attack in coordination with ventilation operations.

CHAPTER 14

Fire Protection Systems

LEARNING OBJECTIVES

- Examine the types and classes of standpipe systems and their purpose of providing water throughout a building for firefighting operations.
- Examine the types of sprinkler systems and the impact a system has on protecting a building and its contents from fire.
- Describe the types of sprinkler heads.
- Describe the difference in operation between how a fusible link sprinkler head operates compared to a glass bulb sprinkler head.
- Become familiar with the different temperature ratings for the various sprinkler heads and where they are used.
- List the reasons why sprinkler systems may be out of service or can fail altogether.
- Consider other protective systems that use various extinguishing agents in buildings or other special processes or uses.

Introduction

The safety of occupants is the first priority of firefighters when confronted with an incident of the scope and magnitude of a high-rise fire. To assist firefighters in combating fires in skyscrapers and other large occupancies, built-in fire protection systems are often required. A standpipe system and a sprinkler system—when properly designed, installed, tested, and maintained—can assist fire department members if they are properly trained in the use of each system. Firefighting operations should be based on a thorough knowledge of the building and the installed system. This knowledge is based on training, company inspections and pre-incident planning. In addition, a fire department must have written standard operating guidelines (SOGs) in place and an incident management system that provides for the safety of firefighters working in these hostile environments.

For firefighters to work safely and effectively under fire conditions in buildings provided with a built-in fire protection system, they must know the type of system installed and how it functions. In addition, members must ensure an adequate water supply to the system so that a high-rise fire can be suppressed and controlled. If a system is not functioning properly, firefighters must be able to address the problem and provide alternative solutions, which is why fire protection systems are covered first in this chapter before proceeding to high-rise firefighting strategy and tactics in Chapter 15.

Standpipe systems and sprinkler systems are both built-in fire protection systems, but they are very different in operation. Standpipe systems are usually installed in large structures of four stories or more, including high-rise apartments, office buildings, and warehouses, to provide a water source in upper elevations or extended distances. A standpipe replaces hose and therefore reduces hose friction loss, and eliminates the need for long stretches of hose lines from the street to the fire floors. They are designed to provide firefighters with water supply outlets close enough to any fire to allow quick fire attack and extinguishment. Standpipe systems are installed in buildings for use by firefighters and in some cases by building occupants. These vertical risers are most often found inside the interior stairway of modern buildings, but in older buildings, many Class I standpipes are found on the outside of the building, usually alongside an exterior fire escape **FIGURE 14-1**. Firefighters may have to make the hose connection by standing on the exterior fire escape landing or by reaching through a window at the end of the hallway. A roof ladder or pike pole may be needed to access the fire escape from the ground and, depending on its structural integrity, the fire escape may be a little precarious.

FIGURE 14-1 In older buildings, many Class I standpipes are found on the outside of the building alongside an exterior fire escape.

Engine Company Fireground Operations, 3rd Edition, Harold Richman, Steve Persson, NFPA © 2008 Jones & Bartlett Publishers.

An automatic sprinkler system, on the other hand, is designed to detect a fire in its earliest stages and apply fast water. As the name implies, it does so automatically without the help of firefighters or building occupants, and it is not deterred from operating due to problems firefighters face with heavy smoke, high heat, disorientation, or running out of air. Sprinkler heads are like nozzles unaffected by human limitations, ready to spray water effectively on the fire at the right time. Firefighters ensure that the water supply is maintained, and that the system is functioning properly while it is in operation at a fire. Both types of systems are described in this chapter, with the use of standpipe systems discussed in some detail.

Automatic fire sprinklers are designed to suppress (rapidly reduce fire growth rate), or control (stop fire growth rate from increasing) fires. Most automatic fire sprinklers will contain the fire but there are very few fire sprinkler systems which will completely extinguish a fire; therefore, firefighters should expect to enter the fire area with hose lines for final extinguishment. There are occasions, depending on the arrangement of the fire load, where the fire sprinklers might have extinguished the fire.

Automatic fire sprinklers use a *design density*: a specific area with a calculated flow to reach the design intent of suppression or control. The area is large enough to ensure a fire in that area will be suppressed or controlled. However, if the combustible fuels burn hotter or faster than the system was designed for, the fire could burn beyond the design area. Potentially, this could require more water than is available and the fire could continue to spread, opening additional fire sprinklers. As more sprinkler heads open, the flow from each sprinkler head will decrease, resulting in an uncontrolled fire. The IC should ensure the fire sprinkler fire department connection (FDC) is supplied as soon as possible. This simple action might increase flow sufficiently to control the fire before firefighters reach the fire area with hose lines.

Standpipe Systems

A **standpipe system** is a piping arrangement that carries water vertically from floor to floor, and sometimes horizontally, for two or more hose connections through a building for firefighting operations. Horizontal standpipes are found inside large, one-story shopping malls and on long waterfront piers. The designed purpose of a vertical and horizontal standpipe are the same; to provide a means of getting water to a fire without long, time-consuming hose lays and the friction loss associated with with elevation and long lays. For the most part, standpipe systems came into being as the result of ordinances passed when increasing heights and floor areas of new buildings began to pose severe firefighting problems. The time consumed in advancing hose lines several stories up to a fire floor by stairway, ladder, or hoisting with rope often meant that a fire had enough time to get out of control easily. Some buildings were so tall that they did not allow any hose-line operations at all due to pressure needed to overcome the elevation losses. The fire service urged that large structures be required by law to contain standpipe systems to eliminate the loss of time between the arrival of firefighters at the fireground and initiation of the fire attack.

The first laws were passed in major cities and required standpipes in all buildings more than 75 feet high. This is probably because, at the time these laws were passed, the 75-foot wooden aerial ladder was normally the longest ladder available; however, many thousands of buildings constructed under the old laws lack standpipes if they are more than three stories but fewer than 75 feet high. Fortunately, many cities passed or modernized standpipe ordinances, which compelled the owners to install standpipe systems in new buildings and retrofit older ones.

National Fire Protection Association (NFPA) 13, *Standard for the Installation of Sprinkler Systems* and 14, *Standard for the Installation of Standpipe and Hose Systems*, are generally referenced by local, state, and model building codes and insurance standards when designing a system. Other codes and standards adopted by the authority having jurisdiction also may influence system design.

Classes of Standpipe Systems

Standpipe systems are designated Class I, Class II, and Class III according to their intended use. The system could be designed for fire department use, fire brigade or **first-aid firefighting** (trained personnel firefighting), or both. Standpipes also have different water status types, as described here.

Class I

Class I systems are vertical dry pipe risers with 2½-inch (65-mm) hose discharge connections at designated locations, usually on every floor or stairway landing inside the building for full-scale firefighting. These systems are intended for use by the fire department rather than fire brigades and building occupants. A Class I system is required to have a minimum flow rate of 500 gpm.

Class II

Class II systems provide 1½-inch (40-mm) hose connections at designated locations in the building. These systems are generally intended for use by trained personnel to attack an incipient fire before the fire department arrives and therefore have lower water pressure and flows. This type of system, installed at various interior locations, is usually supplied with a nozzle attached to 100 feet (30.5 m) of 1½-inch hose flaked onto a hose rack, and stored in a wall-mounted hose cabinet. Often a fire extinguisher is included in the hose cabinet **FIGURE 14-2**.

Class III

Class III systems are a combination of Class I and Class II systems. They can have a 1½-inch hose valve and a 2½-inch hose valve, so they can be used for full-scale fire department and first-aid firefighting. Class I and Class II hose connections are provided with this system.

FIGURE 14-2 A hose cabinet is part of the Class II standpipe system for trained personnel use in first-aid firefighting on incipient fires. It is equipped with a nozzle attached to 100 feet of 1½-inch hose flaked onto a hose rack.

Courtesy of Ralph G. Johnson (nyail.com/fsd).

FIGURE 14-3 A combined standpipe system supplies both riser hose connections and automatic sprinklers.

Courtesy of Raul Angulo.

Per NFPA 14, 7.10.1, Class III standpipes have a minimum required flow rate of 500 gpm (1,893 L/min) and a maximum flow rate of 1,000 gpm (3,785 L/min) for buildings that are sprinklered throughout in accordance with NFPA 13 and 1,250 gpm (4,731 L/min) for buildings that are not sprinklered throughout in accordance with NFPA 13.

Types of Standpipe Systems

Standpipe systems are also classified by "type," depending on whether the piping is filled with water or whether the piping is dry. This classification also depends on whether the water supply for firefighting is automatically, semi-automatically, or manually available. For the firefighter on the fireground, the system is either wet or dry. But as a professional, it is still important to build your knowledge base on the various systems you rely on to extinguish fires. NFPA 14 lists seven types of standpipe systems:

- **Automatic Dry Standpipe.** Automatic dry systems have piping that is normally filled with pressurized air. These systems are arranged, through the use of a device such as a dry-pipe valve, to admit water automatically into system piping when a hose valve is opened. They are connected to an automatically available water supply capable of meeting the water demand necessary for firefighting.
- **Automatic Wet Standpipe.** Automatic wet systems have piping that is filled with water at all times. The automatically available supply of water is capable of supplying the water demand pressure and flow necessary for firefighting.
- **Combination (or Combined) Standpipe.** A combination standpipe system supplies both hose connections and automatic sprinklers FIGURE 14-3.
- **Manual Dry Standpipe.** Manual dry systems have piping that is normally filled with nonpressurized air; these systems do not have a preconnected water supply. A fire department connection (FDC), usually a 2½-inch connection, must be used to supply and maintain water for firefighting.
- **Manual Wet Standpipe.** Manual wet systems have piping that is normally filled with water for the purpose of allowing leaks to be detected and minimize time to fill the pipe with water. The water supply for these systems is typically provided by a small connection to domestic water piping. This water supply is not capable of supplying firefighting water demands without a fire department apparatus providing flow and pressure through the FDC.
- **Semiautomatic Dry Standpipe.** Semiautomatic dry systems have piping that is normally filled with air that may or may not be pressurized. These systems are arranged through the use of devices such as a deluge valve to admit water into the system piping when a remote activation device located at a hose station, such as a pull station, is operated. They also have a preconnected water supply capable of supplying the water demand necessary for firefighting.

- **Wet Standpipe.** A wet standpipe is a piping system that has water in it at all times and is further clarified as a manual or automatic wet system (see above).

To summarize, wet systems have water-filled piping, and dry systems do not. Automatic systems provide water supply for firefighting by simply opening a hose valve. Semiautomatic systems are connected to a water supply for firefighting but require activation of a device at a hose valve in addition to opening the valve to get water. Manual systems do not have a preconnected water supply for firefighting, and these systems must be supplied manually by connecting hoses from a fire department pumper to an FDC.

Dry System Features and Uses

A dry system is simply a vertical or horizontal pipe, or riser, running through or along the outside of the building. Dry standpipe systems may be found in tall or wide-area structures and can be located inside or outside the building. They are used in unheated buildings and areas subject to freezing temperatures. An interior dry system usually has at least one discharge valve outlet on each floor inside the building; the fire department intake is located outside the building. An exterior dry system usually runs adjacent to the outside fire escape (Figure 14-1), on the side of the building, or at the C side, rear in an alley **FIGURE 14-4**. There is one discharge hose outlet at each landing, with a Siamese inlet appliance placed at the bottom of the pipe, a few feet above the ground. Such an exterior setup is usually the result of laws that are newer than the building. Only fire department pumpers can supply dry systems; thus, dry systems are not equipped with interior lines for use by occupants.

FIGURE 14-4 Dry standpipe systems may be found in tall or wide-area structures and can be located inside the building, outside the building on the side, or at the C side rear in an alley. They are used in unheated buildings and areas subject to freezing, and can only be supplied by the fire department.
Courtesy of Raul Angulo.

Dry systems may be found in unheated buildings in areas subject to freezing temperatures, in buildings in cities that do not require wet systems, and in buildings built before retroactive standpipe laws were passed. Where laws permit, dry systems are preferred in some heated occupancies to prevent expensive water damage due to tampering by tenants.

A building that is spread over a wide area may be equipped with two or more standpipe systems, each protecting a portion of the building. These systems might be completely separate, with each riser being supplied only by its own FDC intake. It is important that engine company personnel become familiar with such separate systems during pre-incident planning or building inspections. Particular attention should be given to the location of the intake that supplies each riser.

Obviously, the effect of pumping water into the wrong intake of a separate multiple-riser system would be disastrous by delaying time to provide water. The International Fire Code and NFPA 14 requires such intakes to be placarded or labeled, indicating what portions of the building are supplied, to prevent errors or confusion by connecting to the wrong system **FIGURE 14-5**. Newer fire codes require multiple-riser dry systems to be interconnected so that water pumped into any intake supplies all the risers.

One problem encountered with multiple-riser dry systems is the time that it takes to drive the air out of a large system. This leads to a lag between the time when water is first pumped into the intake and the time when effective hose streams can be developed. Many departments have been caught off guard by this time lag, and the results can be serious losses and unnecessary injuries to personnel. This is one reason why 2½-inch and 1½-inch gated wyes should be included in dry standpipe hose lays. While the hose is being connected to one discharge port of the wye, the other discharge port can bleed the air. Bleeding the air through an open gated wye is much quicker than bleeding it through the nozzle at the end of a hose line. Nevertheless, these systems should be time-checked during a pre-incident planning session with the building engineer, or when the system is service tested so that the local fire companies know what to expect when using them.

Wet System Features and Uses

An automatic wet standpipe system is connected to a water source and contains water at all times **FIGURE 14-6**.

A

B

C

D

FIGURE 14-5 The International Fire Code requires standpipe and sprinkler intakes to be labeled with a plate **A**. or placarded **B**. to prevent errors or confusion by connecting to the wrong system. The wording can also be stamped on top of the FDC. **C**. Note the deterrent **D**. to prevent people from using the FDC as a sitting stool.

Courtesy of Raul Angulo.

In this type of system, the water is under enough pressure to allow fire attack without the support from fire department pumpers. This can be the "street pressure" of the water mains, or pressure provided from fire pumps. Meeting these residual pressure requirements ensures sufficient water flow for normal initial attack operations.

NFPA 14 requires that a water supply for Class I and Class III systems be able to deliver a residual pressure of 100 pounds per square inch (psi) at the outlet of the topmost hose connection on each standpipe. This pressure must be available while flowing 250 gallons per minute (gpm) from each of the two topmost hose connections of the hydraulically most remote standpipe, plus 250 gpm from each additional standpipe, up to the required maximum (500 gpm [1893 L/min]). NFPA 14 requires that the water supply for a Class II system be capable of delivering 100 gpm for 30 minutes. It also must be strong enough to maintain a residual pressure of 65 psi at the outlet for the hydraulically most remote hose connection with 100 gpm flowing. (NFPA 14 was modified in the 1993 edition to increase the minimum outlet pressure for Class I and Class III outlets from 65 psi to 100 psi because of questions raised regarding the adequacy of a 65-psi minimum design pressure on automatic and semiautomatic standpipes.)

Although these are current requirements, many buildings were built under earlier fire code regulations with lower flow and pressure requirements for all standpipe classes. These buildings are typically grandfathered in and are rarely required to retrofit their systems to meet modern pressures and flows. Firefighters should plan for this potential.

Water Supply to Wet Standpipe Systems

A fire department with a reliable and adequate source of water that is accessible nearby may supply

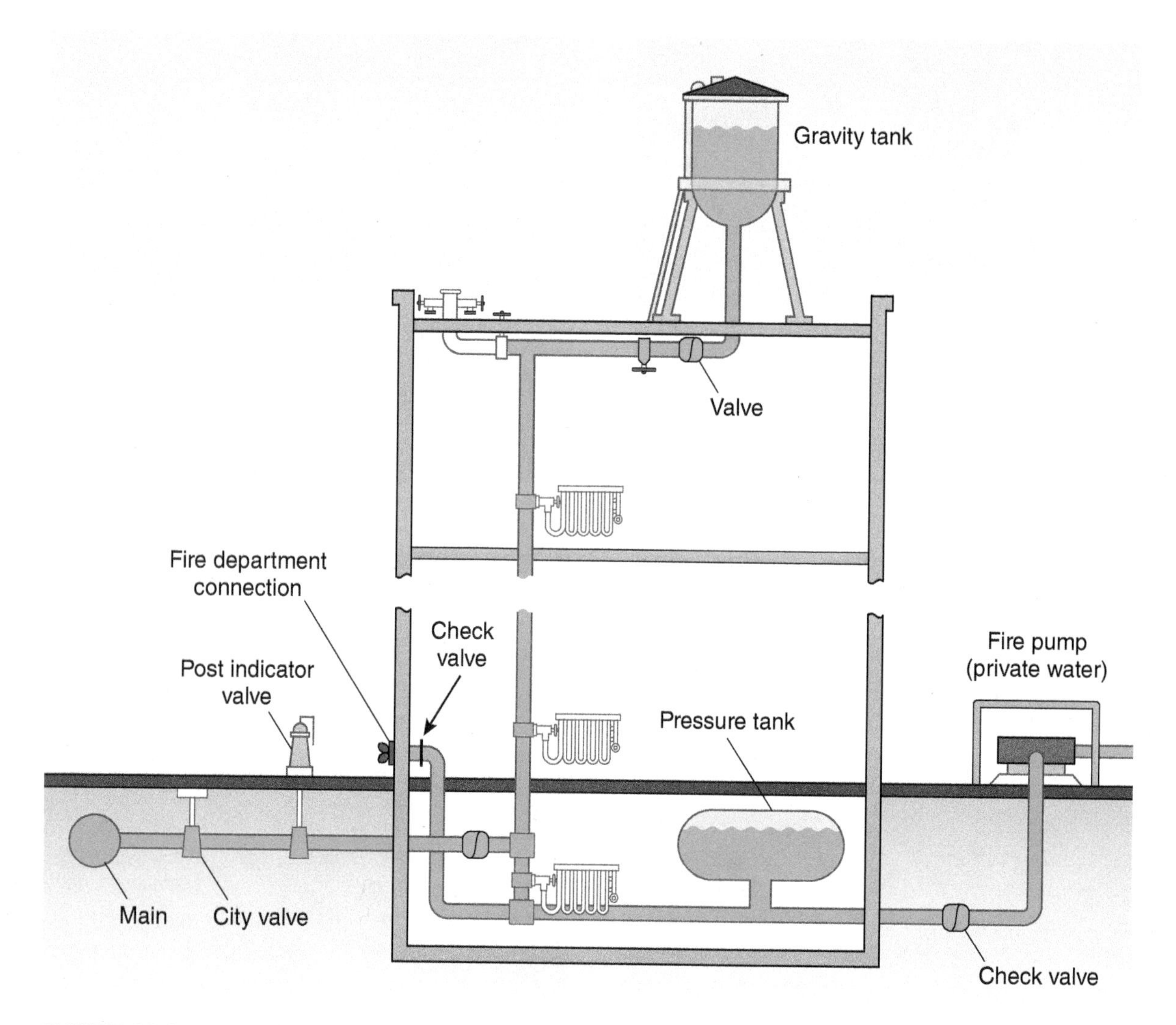

FIGURE 14-6 Wet standpipe systems may be supplied by one or more methods: public waterworks systems, gravity-fed tanks, pressure tanks, and private water supplies with pumps.

Engine Company Fireground Operations, 3rd Edition, Harold Richman, Steve Persson, NFPA © 2008 Jones & Bartlett Publishers.

manual standpipe systems. Automatic and semiautomatic standpipe systems require a minimum of one preconnected water supply that can supply the standpipe system's hydraulic demand for the minimum required duration. This will likely only supply two to three handlines depending on hose size. In addition to the primary water supply on an automatic or semiautomatic standpipe system, one or more FDCs with a reliable water source accessible nearby are required.

For high-rise buildings, two remotely located FDCs are required for each zone within the pumping range of fire apparatus in addition to the automatic water supply. Two FDCs reduce the possibility of all standpipe and sprinkler supply hose being cut by falling glass and thus interrupting the secondary water supply during a fire **FIGURE 14-7**. Departments rig protective horizontal barriers using ground ladders and plywood to protect the hose connections from falling glass. The sources of water most often used include hydrants, public waterworks systems, gravity-fed tanks, pressure tanks, and fire pumps connected to a reliable fixed water source such as a reservoir.

When two or more sources feed a wet standpipe system, the source providing the highest pressure is the one that provides the water. If the supply in use fails, such as the building's fire pump or a fire department pumper, the remaining source with the most pressure kicks in and supplies the system. A double check valve arrangement keeps possibly contaminated water from entering the potable water system.

Public Waterworks and Fire Hydrants

Fire hydrants, the most common public waterworks system for firefighting, are almost always used as the primary source of water. Other sources are used only

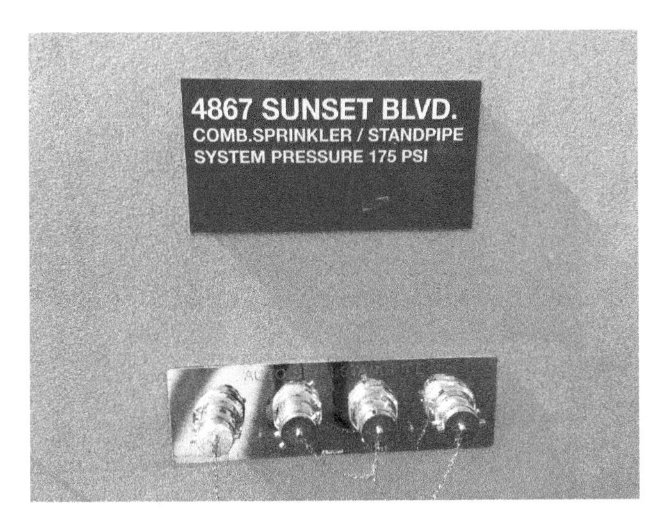

FIGURE 14-7 For high-rise buildings, two remotely located FDCs reduce the possibility of all standpipe and sprinkler supply hose being clogged by debris or cut by falling glass and thus interrupting the secondary water supply during a fire.

Courtesy of Raul Angulo.

because of extenuating circumstances, such as a broken or frozen hydrant or water main. If the public waterworks water main system supplies enough pressure to satisfy code requirements, no other source will need to be used. The exceptions are commonly industrial, private, warehousing, and similar occupancies, where small water mains or low-pressure systems exist, and additional water sources may be required. A wet standpipe system, supplied by public water mains, has become the most common arrangement in medium- and high-rise apartment and office buildings.

Firefighters must be aware of the dependability of public waterworks supplies in their areas of responsibility and especially of changes in pressure that vary with the time of year, time of day, and so on. For example, since public waterworks systems are used for watering grass and landscape foliage, will the higher domestic water demand at 16:00 hours on August 15 be the same at 16:00 hours on December 15? Will the demand on the water system during the day for domestic and industrial use have the same effect as in the middle of the night when these demands are very low? Will drought conditions affect the system? It is vital that the answers to such questions be determined during pre-incident planning for reference during a fire situation. Many cities have at least one crew from the water department available around the clock for emergencies. A water department representative should be contacted during multiple alarm fires or known low water availability times to ensure that water pressure and flows are increased in that area of the water grid to support fire operations.

Other Sources

If the public waterworks system is unavailable or lacks flow capacity for a wet standpipe system, other water sources must be used. Where waterworks systems do not provide enough pressure, building pressure pumps are used to boost the pressure in the primary supply, which is usually the public waterworks system but could be a water tank, such as a gravity-fed tank system. Water tanks are typically mounted on the ground when supplying one- and two-story buildings, or aboveground when supplying taller structures. They may hold as much as 100,000 gallons of water. Gravity supplies the pressure required in the standpipe and sprinkler systems. Don't confuse a water tank pump with a fire pump. A water tank pump is used to fill the tank but not to aid in developing pressure.

Generally, pressure tanks are found on smaller systems where they supply house lines with water for initial attack by the occupants (first-aid firefighting). They may also be used to augment another source in case of failure in the main supply. Pressure tanks rarely exceed 3,000 gallons in capacity and can be located anywhere in the water supply system. The pressure tank differs from the gravity tank in that typically the top third of the tank contains air under pressure, which ensures sufficient water pressure while the tank is supplying water to the standpipe and sprinkler systems.

Remember that a water tank has a fixed capacity. After it is emptied of water, it is useless as a source until recharged by gravity or building pumps. For this reason, pumpers should be used to supply any standpipe system, wet or dry, when firefighting operations begin.

Some older buildings included large supply tanks as the public water supply lacked enough flow for large-scale fire operations. This was a critical factor in the 1988 First Interstate fire in Los Angeles, California, whose designers included an 85,000-gallon tank reservoir for firefighting.

Newer high-rise buildings may be required to have "in building" tanks to supply water for firefighting through *in-building pumps* in case public water mains are damaged from earthquakes. These water tanks can exceed 30,000 gallons or be considerably smaller.

Multiple Wet Systems

As with dry systems, wet systems may consist of two or more risers that may be separate or interconnected. In any case, the risers are all supplied by the same source, although often from different points in the public water supply system. When the risers are separate, there is one fire department intake for each riser; each intake serves one riser and one riser only. When the risers are interconnected, the system must have two or more intakes **FIGURE 14-8**. Again, it is important that

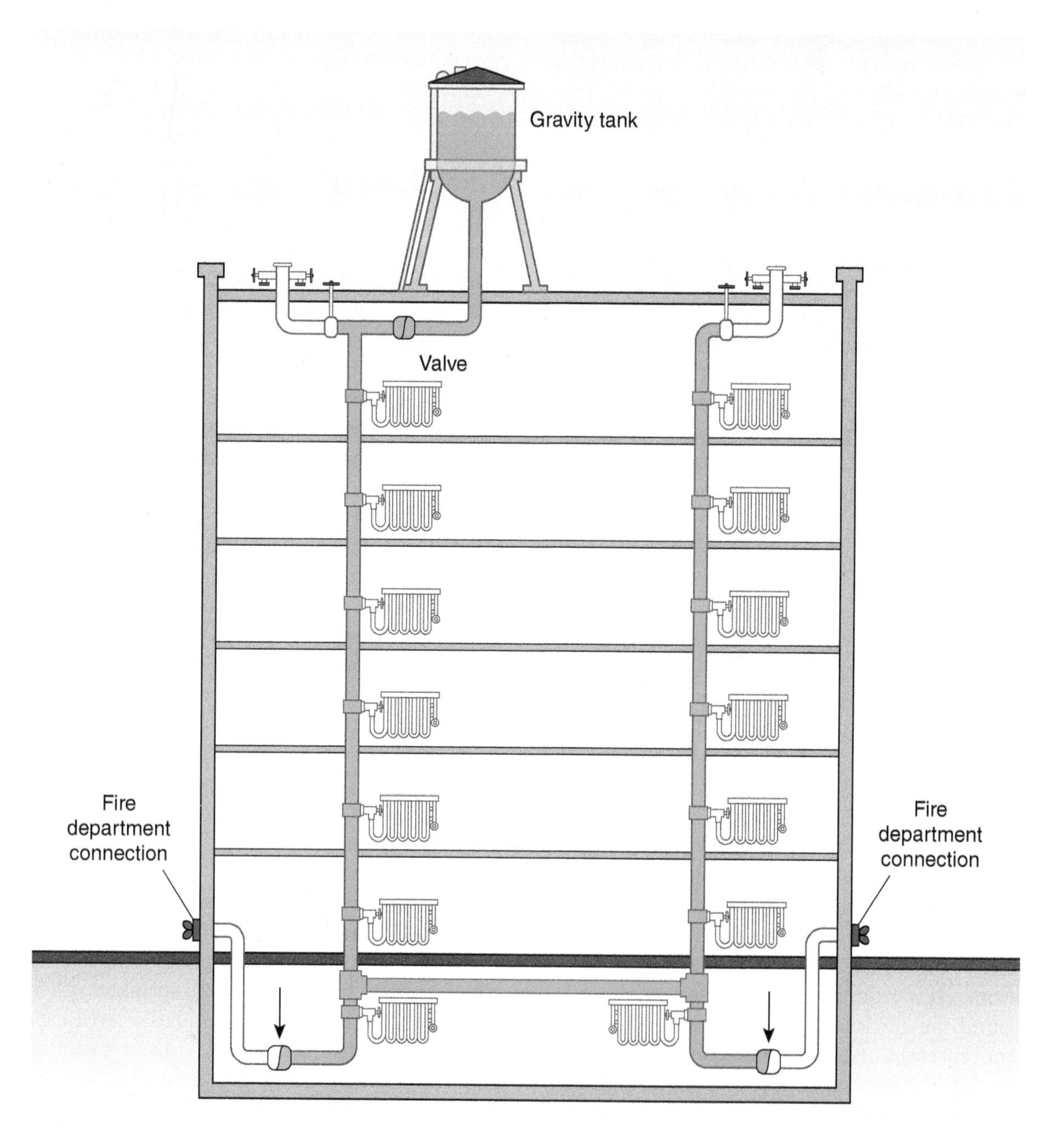

FIGURE 14-8 Two standpipes may be interconnected and have one water supply and two intakes.

firefighters be aware of the type of system with which they are working.

In separate-riser systems, the multiple intakes allow water flows to various areas of a building without the need for long hose-stretching operations. In addition, if one riser is out of service, the second can be used for fire attack, although a long stretch of hose may be required on the fire floor. If a 1¾-inch hose line is used for the firefighting operation, a 2½-inch hose line should be used to bring the water close to the fire area. It then could be reduced to a 1¾-inch hose line for easier maneuverability, as a long stretch of 1¾-inch hose would create excessive friction loss and reduce the water delivery. It may be advantageous to use the 2½-inch hand line for fire attack and forgo the use of the smaller hose line, but this depends on whether there are enough firefighters to advance and operate a charged 2½-inch hose line down a long hallway—not an easy task for two firefighters. Advancing the uncharged 2½-inch hose line down the hall should be done as far as safety allows. The increased water delivery of a 2½-inch hose line is important but won't do any good if you can't move it to the seat of the fire. The speed and mobility provided by 1¾-inch hand lines, when they are properly supplied, shouldn't be dismissed. Also, the building water supply system or limits on standpipe size (4 inches) can severely limit use of 2½ lines as the available flow is simply inadequate for more than three or four lines.

Valves

A wet standpipe system must have some type of sectional cutoff or control valve between the system and the water main or other water source in order for the system to be shut down for repairs and maintenance. Post-indicator valves (PIVs) are usually found on industrial and warehouse properties; gate valves or outside stem-and-yoke (OS&Y) valves are usually installed on the exterior walls of apartment buildings, office buildings, and stores **FIGURE 14-9**.

FIGURE 14-9 A wall-mounted PIV is often found on the exterior of industrial and warehouse properties.

Courtesy of Bob Markford, EMS Safety Chief, Palm Harbor Fire Rescue.

Larger systems can be divided into control zones or areas, permitting only part of the protection to be cut off to allow repairs. In such cases, there may be several sectional valves. While approaching the building before committing to a standpipe system, firefighters should check, if it is convenient, to make sure the PIVs are in the "open" position. The fire code requires these valves to be chained in the open position or to have an electronic monitoring device that alerts a supervised security company if the valve is shut off **FIGURE 14-10**. PIVs may be found on wet standpipe systems. Two popular types are the window type and the butterfly type **FIGURE 14-11**. You are able to read the words "open, shut, or closed" on a window-type PIV.

FIGURE 14-10 Wet standpipe and sprinkler system main valves need to be chained in the open position or have an electronic monitoring device that alerts a supervised security company if the valve is shut off. These OS&Y valves have both.

Courtesy of Raul Angulo.

When a PIV is not placed on the outside of the building, a gate valve or OS&Y valve might be located on the intake pipe in the basement, or just inside the first floor if the building has no basement **FIGURE 14-12**. These valves are normally placed close to the outside wall of the building, with many variations in their exact locations. Fire department members should know the locations of these valves during an incident, and they should also be noted on pre-incident maps and plans. If water flows to fire sprinklers or hose lines is nonexistent or limited, this may be a result of one or more of these valves being fully or partially closed. When the stem of an OS&Y valve protrudes beyond the shutoff wheel, it indicates the valve is open. If the stem is not visible, the valve is closed.

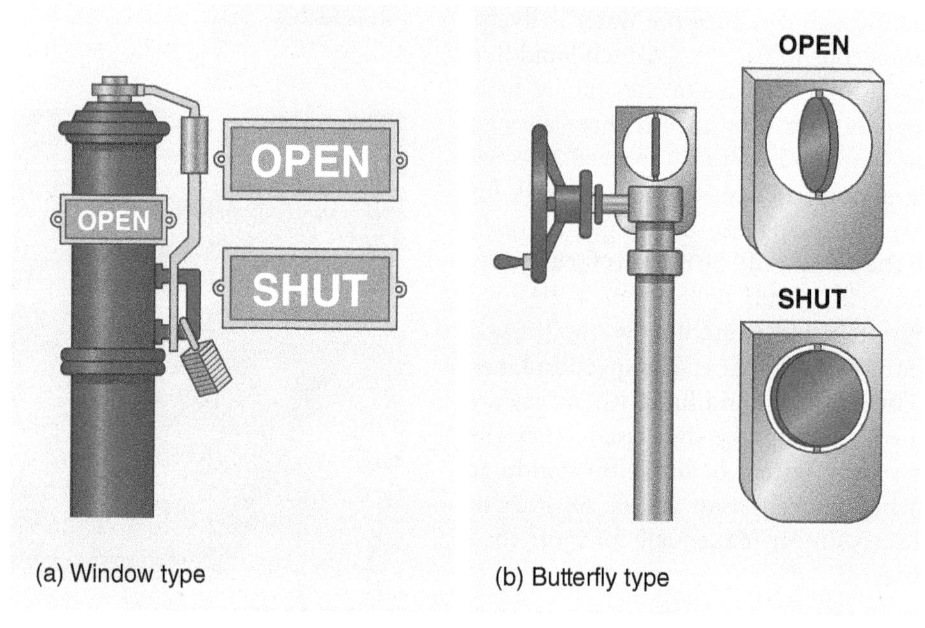

FIGURE 14-11 The window type **A**. and butterfly type **B**. are two commonly used PIVs.

A: Courtesy of Craig Macluba; B: *Engine Company Fireground Operations*, 3rd Edition, Harold Richman, Steve Persson, NFPA © 2008 Jones & Bartlett Publishers.

FIGURE 14-12 The OS&Y main valve to a wet standpipe or sprinkler system is often in the basement sprinkler room.

© Jones & Bartlett Learning. Photographed by Glen E. Ellman.

Automatic Sprinkler Systems

Sprinklers are highly effective and very reliable parts of a building's fire protection system. The value and tactical advantage of an automatic sprinkler system quickly controlling a fire in a high-rise building cannot be overstated. The superior protection and reliability in protecting lives and property has been demonstrated time and again. A sprinkler system is like having a battalion of firefighters pre-staged in a building with nozzles in hand 24 hours a day. They never tire, never fall asleep, never take a day off. They are immune to the effects of smoke; they don't become disoriented, or lost, or run out of air; and, when needed, they are at the seat of the fire. When properly designed, installed, maintained, and supported by the fire department, a sprinkler system can apply water directly to the fire much more effectively than can the

FIGURE 14-13 A sprinkler system can apply water directly to the fire much more effectively than can the fire department using manual fire-suppression methods.

fire department using manual fire-suppression methods when fire conditions prevent close approach by firefighters FIGURE 14-13.

According to the 2017 NFPA report on *U.S. Experience with Sprinklers*, between 2010 and 2014, sprinklers were present in 10% of reported U.S. fires. The death rate per 1,000 reported fires was 87% lower in properties with sprinklers than in properties with no automatic extinguishing systems (AESs). The civilian injury rate was 27% lower, and the firefighter fireground injury rate per 1,000 fires was 67% lower in sprinklered properties than in fires in properties without an AES. In fires considered large enough to activate the sprinklers, sprinklers operated 92% of the time. Sprinklers were effective in controlling fire in 96% of the fires in which they operated. Aggregating these statistics means that sprinklers both operated and were effective in 88% of the fires large enough to operate them. In three-fifths of the fires in which the sprinklers failed to operate, the system had been shut off—sometimes deliberately, often a direct result of human error.

At the First Interstate High-Rise Fire on May 4, 1988, in Los Angeles, California, the entire building was being retrofitted with sprinklers. The work was 90% complete at the time of the fire and the floors that burned already had the piping and sprinkler heads installed. However, the decision was made to activate the system only on completion of the project, after the water flow alarm could be tied into the building's automatic fire alarm panel, so the valves controlling the sprinklers on the completed floors were shut off. Five floors were destroyed by a fire that took 3 hours and 39 minutes to extinguish.

At the One Meridian Plaza Fire on February 23, 1991, in Philadelphia, Pennsylvania, firefighters were prevented from fighting the fire by pressure-reducing valves that were improperly set too low to produce effective hose streams. Tools and trained personnel with the required expertise to adjust the valve settings did not become available until it was too late. The fire started on floor 22 and, after 11 hours of uncontrolled fire growth and spread by autoexposure, eight floors were consumed by the fire. However, the thirtieth floor was sprinklered (by the request of tenants on that floor), and the vertical spread was stopped solely by the activation of 10 sprinkler heads. These sprinklers held the fire. Due to the concern for building collapse, the fire was allowed to burn itself out, which finally occurred more than 19 hours after it started.

Sprinkler System Theory

In an automatic sprinkler system, water is piped from a source to the protected area, where the piping is branched, usually at ceiling level, to cover the entire area. Small discharge appliances, previously known as sprinkler heads, are placed at specific intervals in the piping. The various heads are sensitive to different heat temperatures; even a fairly small fire causes a traditional head to open and apply water to the burning area below. As long as enough pressure is available, each head distributes water over an area of 100 or more square feet in area sufficient to control or extinguish most fires. System piping and heads vary, but water flows range between 15 and 110 gpm FIGURE 14-14.

Sprinkler Heads

There are three types of sprinkler heads commonly used inside of buildings FIGURE 14-15:

- Pendent
- Upright
- Sidewall

The deluge system will be covered separately later in the chapter. A pendent sprinkler head hangs down from the ceiling and has a flat deflector plate. When the glass bulb operates, water flows from the sprinkler orifice, hits the deflector, and throws a circular spray pattern. An upright head projects up from the sprinkler orifice. It is often used in mechanical rooms or in inaccessible areas to provide better coverage between obstructions, like beams and vent ducts. It has an umbrella-type deflector that also sprays water in a circular pattern. The spray pattern from either pendent or upright heads can be obstructed by material stored within

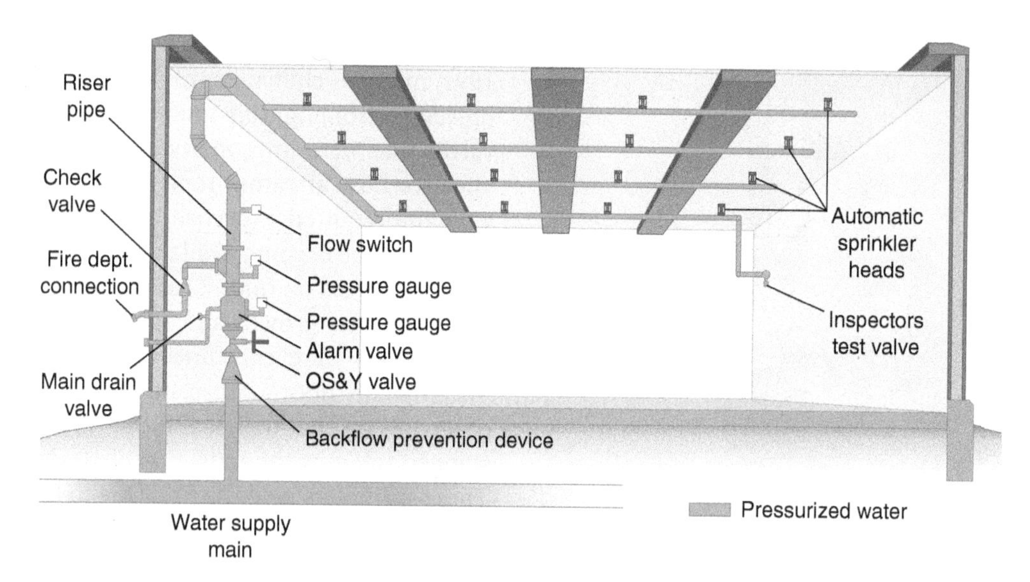

FIGURE 14-14 The basic components of an automatic sprinkler system include sprinkler heads, piping, risers, control valves, and a water supply.

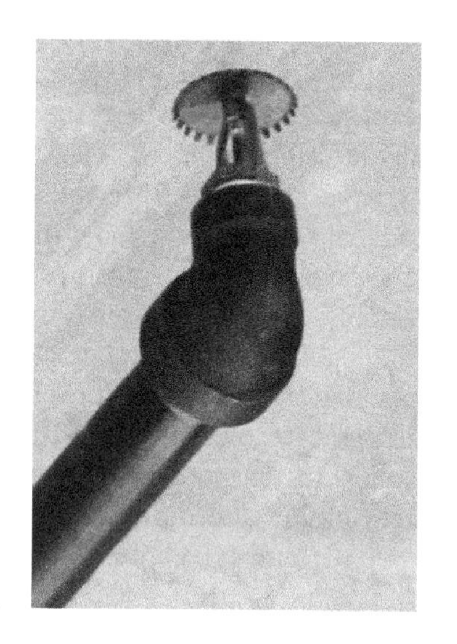

FIGURE 14-15 Types of sprinkler heads: **A**. upright, **B**. pendent, and **C**. sidewall.

FIGURE 14-16 The spray pattern from either pendent or upright heads can be obstructed when combustible material is stored within a rack system.

FIGURE 14-17 The sprinkler head components consist of a frame, a heat-sensitive element, an orifice, a pip or an orifice plug, and a deflector.

Courtesy of A. Maurice Jones, Jr.

FIGURE 14-18 A fusible link sprinkler head has a two-part metal element that is fused by a heat-sensitive alloy. The link holds the pip cap, or plug, in place.

a rack system FIGURE 14-16. Big-box warehouse stores are often required to have a sprinkler system inside the racks for high-piled stock. A sidewall sprinkler head comes out from the upper level of a wall. A tilted deflector sprays the water away from the wall, and a small second deflector sprays water back at the wall for protection; thus, this is not a circular spray pattern. A sidewall sprinkler is popular in motels and hotels, and they are used when a pendent head cannot be mounted to the ceiling or when low structural heights prevent sprinkler piping.

Fusible Link and Glass Bulb Sprinkler Heads

The sprinkler head components consist of a frame, a heat-sensitive element, an orifice, a pip or an orifice plug, and a deflector FIGURE 14-17. All wet-pipe sprinkler heads are held closed by either a fusible link or a glass bulb that contains a heat-sensitive liquid.

A fusible link sprinkler head has a two-part metal element that is fused by a heat-sensitive alloy with a specific melting temperature. The link holds the pip cap, or plug, in place. Once the ambient temperature around the sprinkler head reaches a specified temperature, the alloy releases and the metal elements separate, which causes the pip cap to fall away. Water is then released. Note that water is only released by sprinkler heads where the ambient temperature reaches a specified level; therefore, water is only released in the area of a fire, which helps limit water damage and limits flow to that area FIGURE 14-18.

Glass bulb sprinkler heads have a small reservoir that holds a heat-sensitive liquid. This glass bulb holds the pip cap in place. When the ambient temperature of the liquid reaches a certain level, the liquid expands and causes the glass bulb to break, which allows the pip cap to fall away, releasing the water. As with the fusible link heads, water is only released where the ambient temperature reaches a certain level, which

FIGURE 14-19 A glass bulb sprinkler head has a small reservoir that holds a heat-sensitive liquid. When the ambient temperature of the liquid reaches a certain level, the liquid expands and causes the glass bulb to break. The plug falls away and releases water.

helps limit water damage. Other specialized types of heads operate in the same manner **FIGURE 14-19**.

Sprinkler Response Temperatures

Because ambient air temperatures can vary, for example boiler rooms may have high ambient temperatures, sprinklers come rated for different activation temperatures depending on the use of a space or surrounding environment, and the amount of heat energy the burning fuel will produce **FIGURE 14-20**. The data in **TABLE 14-1** is based on NFPA 13, Table 7.2.4.1; however, manufacturers may have sprinklers with different temperature ratings and colors.

Temperature		Color of liquid alcohol inside bulb
°C	°F	
57	135	Orange
68	155	Red
79	174	Yellow
93	200	Green
141	286	Blue
182	360	Purple
227 260	440 500	Black

FIGURE 14-20 Temperature color code rating chart for temperature-sensitive liquids for glass bulb sprinkler heads.

Microbiologically Influenced Corrosion (MIC)

Microbiologically influenced corrosion (MIC) is the result of microbiologically induced corrosion of pipe interior, which leads to both buildup of corrosion within the sprinkler pipes with heads that can limit or prevent water flow and the weakening of pipes. It is also found in standpipes and discharge valves. Most new sprinkler systems use steel pipe. MIC develops only in systems using black steel or galvanized or stainless-steel pipes. MIC buildup is more critical in smaller-diameter pipes because it takes less debris to occlude. Similar to clogged arteries restricting blood flow in the body, MIC has the ability to cripple or disable a sprinkler system by reducing its flow or clogging the line completely so there is no water flow.

Many factors determine the corrosion rate for pipes. It can take a few years or many years, and is more prevalent inside dry systems where water and moisture can accumulate in the bottom surface of the piping left over from testing. The moisture mixes with air and oxidizes the pipe. Wet systems that displace the air, or dry systems that use nitrogen gas instead of air, have a much slower rate of corrosion due to lack of oxygen. The only way to determine if there is serious MIC buildup is to have the water tested in a lab. Since buildings are required to have their sprinkler system piping examined at least every 5 years, the test would indicate the presence of corrosion or MIC and should be addressed by the building owner. Testing for MIC serves no purpose for a firefighter on scene, but company officers need to conclude quickly that any problem with flow rates, reduced operating flow rates, or no flow rates resulting in the complete failure of a sprinkler system is probably due to MIC corrosion buildup and not the fault of the pump or pump operator. Tactics have to be adjusted as if there were no sprinkler system in the occupancy **FIGURE 14-21**.

Types of Sprinkler Systems

There are four basic types of sprinkler systems: wet pipe, pre-action, dry pipe, and deluge.

Wet Pipe System

In the wet pipe system, the piping is always filled completely with water from the source to the heads. When a sufficient amount of heat is generated from a fire, water is discharged immediately from an opened sprinkler head or sprinklers. This is the simplest and one of the least expensive types of system, and it requires little in the way of special equipment; however, it cannot be used in unheated buildings or for exterior protection in areas where the temperature might go below freezing,

TABLE 14-1 Temperature Ratings, Classifications, and Color Codes

Classification	Maximum Ceiling Temperature	Sprinkler Activation Temperature	Glass Bulb Color	Fusible Link Color
Ordinary	100°F (38°C)	135°F–170°F (57°C–77°C)	Orange 135°F (57°C) Red 155°F (68°C)	Black; no color
Intermediate	150°F (66°C)	175°F–225°F (79°C– 107°C)	Yellow 175°F (79°C) Green 200°F (93°C)	White
High	225°F (107°C)	250°F–300°F (121°C–149°C)	Blue	Blue
Extra high	300°F (149°C)	325°F–375°F (163°C–191°C)	Purple	Red
Very extra high	375°F (191°C)	400°F–475°F (204°C–246°C)	Black	Green
Ultra-high	475°F (246°C)	500°F–575°F (260°C–302°C)	Black	Orange
Ultra-high	625°F (329°C)	650°F (343°C)	Black	Orange

Reproduced from NFPA 13, *Standard for the Installation of Sprinkler Systems*, 2019 Edition, Table 7.2.4.1.

FIGURE 14-21 Although not visible here, MIC is the buildup of corrosion within the sprinkler pipes and heads. Such buildup can limit or prevent water flow, creating complete failure of the sprinkler system.

where other forms of sprinklers are used. Nevertheless, it is the most common type of sprinkler system.

Dry Pipe System

The dry pipe system was developed for use in unheated areas where the piping in a wet system might freeze. The system contains water only from the source to a control valve, known as the dry pipe valve **FIGURE 14-22**. The piping from the valve to the heads contains compressed air. The air pressure, maintained by a compressor, holds the dry pipe valve closed. Most dry pipe valves act on a pressure differential principle, in which the surface area of the valve face on the air side is greater than the surface area on the water side. When a sprinkler head opens, air escapes from the lines. This reduces the pressure and trips the dry pipe valve, which opens and allows water to flow into the sprinkler piping and then to the open head or heads. In unheated areas, the dry pipe valve and wet section of pipe is enclosed in an insulated and heated closet or cabinet to protect it from freezing. The enclosure usually is small enough to allow protection with minimum heating and large enough to permit entry for maintenance.

Pre-Action System

Pre-action systems are similar to dry pipe systems, with/or without pressurized air, designed to prevent accidental activation. For water to discharge, two separate events are required. A supplemental detection system activated by a manual pull station, smoke, or special heat detection such as rate of rise, or a lower level heat level detector opens a valve allowing water to enter the sprinkler pipe. If heat is high enough, the sprinkler head will fuse, and water

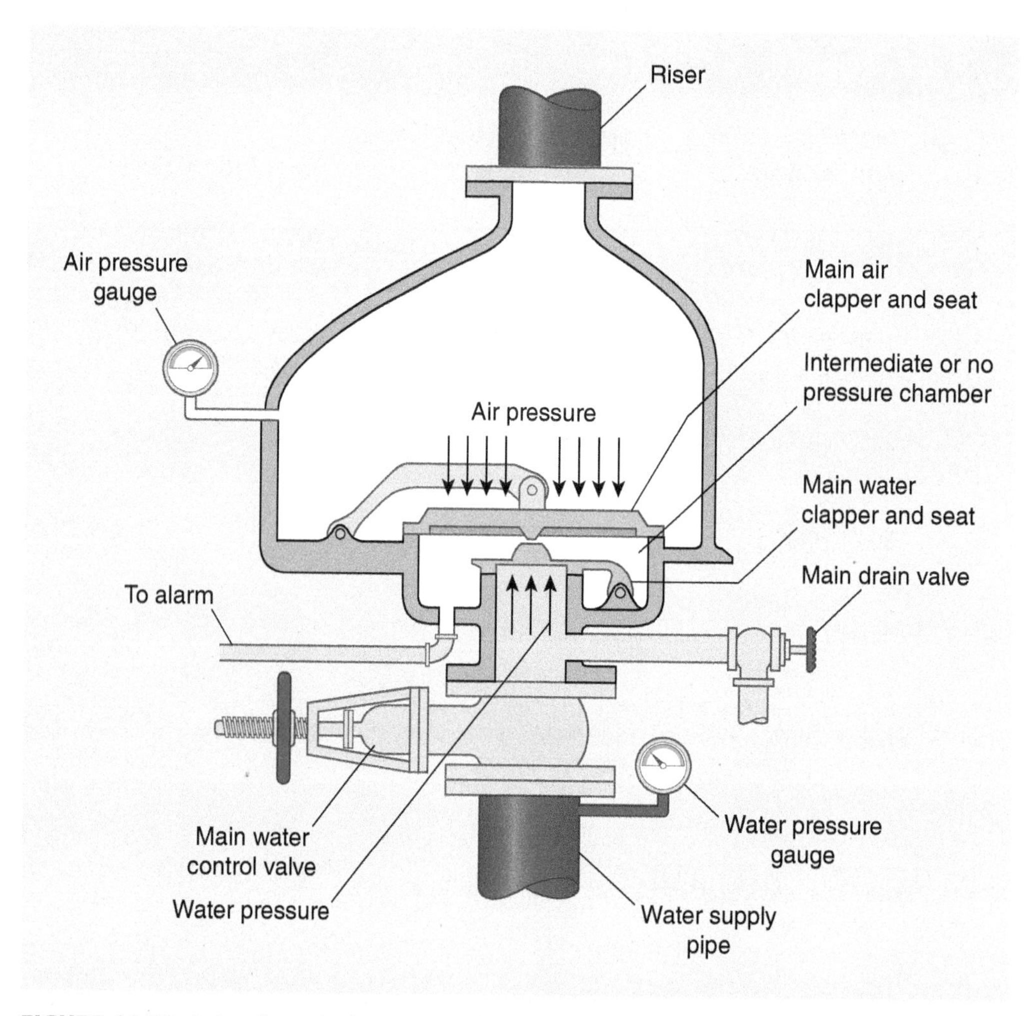

FIGURE 14-22 A dry pipe valve keeps water from entering the pipes until the air pressure is released from the valve.

Engine Company Fireground Operations, 3rd Edition, Harold Richman, Steve Persson, NFPA © 2008 Jones & Bartlett Publishers.

will flow. The supplemental detection allows early notification of possible fire prior to water flowing. These systems are used to protect high-value properties which may be damaged by accidental activation, or special occupancies such as cold storage where accidental water flow could create significant freezing problems.

Deluge System

The deluge system differs from the systems discussed previously; it is designed to deliver large volumes of water quickly. A deluge system only contains water from the water source to a control valve. The sprinkler piping from the valve to the heads are open at all times **FIGURE 14-23(A)**. The orifice on a deluge sprinkler head is always open. There is no pip or plug holding the water back. If there is a fire, the valve is activated by a supplemental fire detection system. The valve then opens and allows water to flow through the piping and out all of the heads simultaneously, which floods the designed entire area quickly **FIGURE 14-23(B)**. Deluge systems are usually installed in high-hazard locations such as flammable liquid storage facilities, chemical plants, laboratories, transformer rooms, aircraft hangars, or other areas where fire might spread rapidly. These systems must have good water supplies in order to function effectively. Low water flow to a building or other water flow restrictions may result in inadequate water flow to suppress a fire.

Water Supply for Sprinkler Systems

According to NFPA 13E, *Recommended Practice for Fire Department Operations in Properties Protected by Sprinkler and Standpipe Systems*, when arriving at a property protected by an automatic sprinkler system, fire companies should take prompt action to supply the system as the fire might have resulted in more flow than the system was designed for, effectively making the available flow less effective. Any of these systems can be supplied by a public waterworks system, gravity-fed tanks, pressure tanks, or fire pumps connected to a reliable fixed water source.

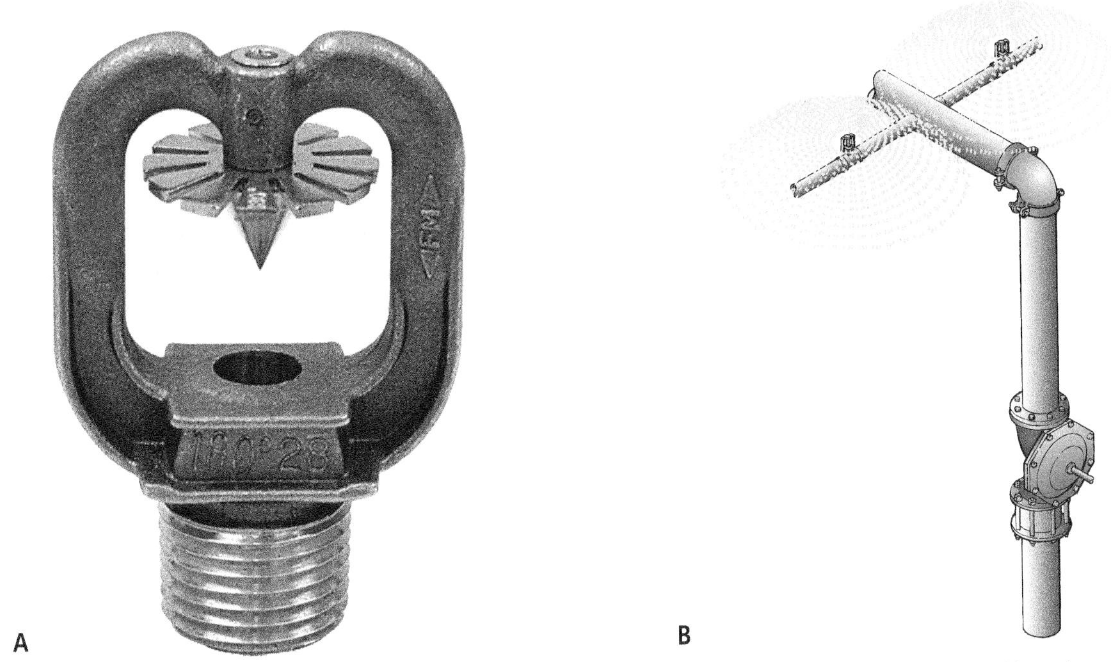

FIGURE 14-23 **A**. The orifice on a deluge sprinkler head is always open. There is no pip or plug holding the water back. **B**. Water flows simultaneously from all the heads on a deluge system as soon as the system is activated.

A: Courtesy of Tyco Fire & Building Products; **B:** © Jones & Bartlett Learning.

Every sprinkler system is required to be equipped with a fire department intake. Following best practices and NFPA 13E, at least one fire engine, on arrival at a fire in a sprinkler-protected structure, should immediately hook up to the FDC labeled "Sprinkler" or "Combined" (Combination), and prepare to pump into the system.

In older buildings built between approximately 1900 and 1960, that are less than four stories, you may encounter a single intake FDC with a female swivel. This intake is typically for a manual dry sprinkler system, in the basement or a small underground garage **FIGURE 14-24**.

Some buildings have multiple sets of FDC connections, each with either two or four intakes. Each individual FDC supplies a different area. Often this is a combined standpipe/sprinkler for grade - to upper floors, and a separate FDC, often with two intakes for a dry sprinkler system located below-grade or in the basement **FIGURE 14-25(A)**. These FDCs should be placarded for the areas served, and usually—but not always—the two intakes for the dry system are on the bottom **FIGURE 14-25(B)**.

Firefighters should check the sprinkler system, including the main valve(s), the sprinkler riser, and the fire pump, if so equipped, to ensure that all components are functioning properly. An actual check of the fire pump is needed; the incident commander (IC) needs to know whether the fire pump is functioning. The fire pump often has an indicator light at the fire alarm panel to show it is operating. Another quick way to tell is by putting your ear down to the sprinkler Siamese or on one of the sprinkler risers. If the fire pump is operating, there will be a detectable vibration through the pipes. If the pump is not functioning, it may be possible to have the pump manually started by a building engineer. During an actual fire incident and if staffing allows, two firefighters may be assigned to the main valve and fire pump to ensure continuous operation of the system. If there is a confirmed fire, a pumper should be hooked up to the sprinkler system FDC to ensure a reliable water source in case the building's system is ineffective or fails. Given a continuous supply of water, the sprinkler system will fight the fire effectively; the job of the fire engine pump operator is to ensure that the water supply is adequate.

The fire engine pump operator should wait for orders to begin pumping. Department SOGs should dictate the hose layout and connection to the sprinkler Siamese, as well as the initial pump pressure. Publications on supplying sprinkler systems recommend a pumping pressure of 150 psi, but department pre-fire planning and guidelines may modify this number. After pumping begins, the operator must monitor the discharge pressure carefully. The pressure tends to decrease as more heads open and may require readjustment from time to time.

If necessary, additional supply lines should be laid to the sprinkler Siamese connection. Firefighters should be prepared for an offensive attack, and hose

FIGURE 14-24 This single intake is for a dry sprinkler system in the basement or a small underground garage.
Courtesy of Raul Angulo.

A

B

FIGURE 14-25 **A**. Some buildings have FDC connections grouped with four intakes, and a separate group of two intakes, for a total of six intakes with female swivels. **B**. The two separate intakes are most likely for a dry sprinkler system located below-grade or in the basement.
Courtesy of Raul Angulo.

lines should be in place to support this effort. These hose lines may be needed to conduct a search and rescue operation or to protect exposures, or they may be used if the building's sprinkler system is not working or is inadequate. In addition, hose lines may be needed during overhaul after the sprinkler system has been shut down.

During the fire, a check must be made to ensure that an adequate water supply is available to support the sprinkler system. An additional source of water may need to be acquired to sustain the operation. If hose lines are being used in the operation, they should be supplied from a different water system if possible, other than the one supplying the sprinklers.

Exposure Protection

Sprinkler systems are as effective for exposure protection as they are for fire control, provided that they are supplied with enough water. For this reason, pumpers should be hooked into the sprinkler systems of exposed buildings as well as into the system in the fire building. Pumping into an exposure should begin when the building's sprinkler system is activated. This is especially important when several sprinkler systems, or sprinkler systems and pumpers, are operating from the same water source. The one deterrent to effective sprinkler system operation, other than human error, is inadequate water supply. Pumpers should be used to maintain the supply to sprinklers, not to rob them of their water.

Out-of-Service Sprinkler Systems

A sprinkler system can be out of service or inoperable for a variety of reasons, including the following:

- Shut off for repairs
- Shut down by human error or lack of system knowledge
- Intentionally shut down, which comes under the category of arson
- System component damage; for example, valve damage by a forklift
- An earthquake damaging the water supply to the building
- Frozen pipes
- Broken water pipes from freezing, then thawing
- Building collapse

Sprinkler System Failure

According to the Ahrens' *U.S. Experience with Sprinklers* NFPA research paper report, in 59% of incidents in which sprinklers failed to operate, the system had been shut off. This is the number one cause of system failure. Manual intervention defeated the system in 17% of the incidents. In some cases, someone turned off the system prematurely. In 7% of the incidents in which sprinklers failed to operate, the system was inappropriate for the type of fire. The purchase and sale of buildings often result in a change of use and occupancy type. A system that was designed for the original occupancy purpose may not be sufficient to meet the requirements of the new change of occupancy use, for example, when manufacturing hazards are increased. In another 7% of sprinkler failures, system components were damaged. This can result when piping or heads are broken by forklifts, trucks, earthquakes, or a partial building collapse.

Shutting Down the System

After the fire is extinguished, the sprinkler system should be shut down by turning off the power to the fire pump or closing the valves when no fire pumps are present (closing the valve supplying water to a fire pump without turning off the pump power first will destroy the pump); however, two firefighters should remain at the control valves (or fire pump control panel) to turn on the system quickly if necessary. Shutting down a system prematurely is a common mistake made when automatic sprinklers are in operation. Premature shutdown is usually done to decrease water damage inside the structure or to get a better look at the area involved. The sprinkler system may have been effectively controlling the fire, though not completely extinguishing it. Premature shutdown of the sprinklers can allow the fire to regain intensity, causing further property damage and financial loss.

Depending on the occupancy, location of the fire, the products in the structure, or stock warehoused in the building, and the method in which these items are stored, the fire may be contained but not completely extinguished. Firefighters must be ready to advance hose lines to extinguish any remaining fire. The emphasis should be on complete extinguishment before worrying about water damage. The building is already damaged by the fire, heat, water, and smoke.

Placing the Sprinkler System Back in Service

After the sprinkler system is shut down, the system should be restored to operational level by the fire department if possible. Once flow pressure has stopped

FIGURE 14-26 Spare sprinkler heads should be kept in a special box, along with a sprinkler wrench and Teflon tape, in the sprinkler room near the main valve.

FIGURE 14-27 A water motor gong.

Courtesy of Raul Angulo.

from the fused head or heads, they should be unscrewed and replaced. There are usually spare sprinkler heads kept in a special box, along with a sprinkler wrench and Teflon tape, in the sprinkler room near the main valve **FIGURE 14-26**. Screw the new heads into the piping and open the main valve. Dry systems and pre-action systems need to be serviced by a sprinkler technician, but unless temperatures are below freezing, the fire department must leave it charged and wet for the interim to provide fire protection. If many sprinklers have opened during a fire, there might not be enough replacement heads available, and/or the system may be too complex for the fire department to restore. The IC should inform the building's owner or the management company that the sprinkler heads temporarily installed by the fire department should be replaced, and the entire system checked by qualified contractors responsible for the maintenance and restoration of the sprinkler system. This is a standard procedure for many fire departments because of possible legal repercussions. The building's owner should take the proper steps to ensure that the system is operational as soon as possible.

When it is not possible to place the sprinkler system in service soon after a fire, the premises should be guarded by building security, or the owner should hire a security guard for a fire watch with the sole purpose of patrolling the premises for fires and reporting them to the fire department in the interim. As a last resort, the fire department may have to establish a fire watch detail. The department Fire Marshal can clarify these requirements.

Many sprinkler systems are equipped with a monitoring alarm system that notifies a 24-hour guard, a private supervisory system, or the fire department alarm center when a sprinkler head is activated. All alarms should be treated as actual alarms by the fire department, but water flow alarms, particularly when the water is flowing from the drain valve, indicates that a fire is occurring or has occurred. When the fire department shows up on scene and water is flowing from the drain valve of the water gong, especially after hours when an occupancy is closed, a fire should be anticipated. The alarm system should also be returned to service before firefighters leave the scene **FIGURE 14-27**.

Other Fire Protection Systems

Other kinds of fire protection systems, usually operating similar to automatic sprinkler systems, deliver water fog spray to enclosed rooms such as ship engine rooms or direct water to specific target areas. Other systems may discharge foam or special extinguishing agents for spaces with potential flammable liquid spills, or some other exotic chemical processes **FIGURE 14-28**. The systems and extinguishing agents vary with the type of building and/or its use and are more common in high-hazard facilities. Firefighters should be aware of any such special systems in their area of responsibility and be thoroughly familiar with their operation.

Some fire protection systems require a manual procedure for starting. A pump might have to be activated, a valve might have to be opened, or a water supply might have to be hooked up. Such operations are normally assigned to building personnel, but firefighters must check to see that the operations have been performed properly and that fire protection

FIGURE 14-28 A CO_2 AES onboard a marine vessel is an example of a specialized system.

systems are operating. As with sprinkler systems, fire department personnel must be sure to leave any fire protection system in operating condition or at least in the charge of a responsible party for its proper operation.

The newest type of automatic fire protection system is the relatively low-cost, residential sprinkler system developed in the early 1980s primarily for residential use. Residential sprinkler systems are an effective means of controlling fires in the home, allowing occupants the time to escape or be rescued. Several different systems exist, and research is ongoing to develop systems that are increasingly efficient and cost effective. More and more cities are requiring residential sprinklers for new home construction. Residential sprinklers provide fast water on the fire, keeping temperatures and smoke production down, while limiting, if not eliminating, the chance for flashover. In the long run, residential sprinklers will be better for occupants and firefighters alike.

After-Action REVIEW

IN SUMMARY

- An IC should consider a fire-suppression system inside a building as the best resource available for use during a fire incident.
- A standpipe system is simply a vertical piping system in which water flows to various discharge valves located on each floor within a building.
- Standpipe systems are designated Class I, Class II, and Class III according to their intended use.
- Class II systems provide 1½-inch hose connections at designated locations in the building. These systems are generally intended for use by trained personnel before the fire department arrives. This type of system is usually supplied with a nozzle attached to 100 feet of 1½-inch hose flaked onto a hose rack, and stored in a wall-mounted hose cabinet.
- A sprinkler system is a piping system distributed throughout a building, with water applied to the fire through sprinkler heads.
- A fusible link sprinkler head has a two-part metal element that is fused by a heat-sensitive alloy. The link holds the pip cap, or plug, in place. Once the specified temperature around the sprinkler head is reached, the alloy releases and the metal elements separate. The pip cap falls away and water is released.
- A glass bulb sprinkler head has a small glass reservoir that contains a heat-sensitive liquid and holds the pip cap in place. When the temperature of the liquid reaches a certain level, it expands, and the glass bulb breaks. The pip cap falls away and releases the water.
- There are four basic types of sprinkler systems: wet pipe system, pre-action system, dry pipe system, and deluge system.
- When an automatic fire protection system is present, the fire department should support the system and let it do its job.
- MIC is the buildup of corrosion within the sprinkler pipes and heads; it can limit or prevent water flow.
- Properly supplied with water, an automatic sprinkler system usually suppresses or at least controls the fire by itself. Sprinkler systems may sometimes extinguish the fire.

- A sprinkler system can be out of service for a variety of reasons, including needing repair, human error, arson, damage by a forklift or an earthquake, frozen pipes, thawing, and building collapse.
- The premature shutdown of the sprinklers, a common mistake, can allow the fire to regain intensity, thus causing further property damage and financial loss.
- A water flow alarm, particularly when the water is flowing from the drain valve, indicates that a fire is occurring or has occurred. When the fire department arrives on scene and water is flowing from the drain valve of the water gong, especially after hours when an occupancy is closed, a fire should be anticipated.
- The newest type of automatic fire protection system is the low-cost, lightweight, quick-response sprinkler system developed primarily for residential use. Residential sprinkler systems are an effective means of controlling fires in the home, allowing occupants the time to escape or be rescued.
- To be able to use fire protection systems fully, firefighters should be aware of the location, operation, and proper use of every system in their area of responsibility.

KEY TERMS

automatic sprinkler system A system of pipes filled with water that is under pressure; it discharges water immediately when a sprinkler head opens.

first-aid firefighting Part of the Class II standpipe system. The hose cabinet is equipped with a 2½-inch discharge valve, a reducer coupling to a 1½-inch male threads, 100 feet of 1½-inch hose line, and an attached spray nozzle designed for trained personnel.

microbiologically influenced corrosion (MIC) The buildup of corrosion within the sprinkler pipes and heads that can limit or prevent water flow. MIC develops only in steel metal piping. Known in the firefighting industry as "MIC."

standpipe system An arrangement of piping, valves, and hose connections installed in a structure to deliver water for fire hose.

REFERENCES

Fundamentals of Firefighting Skills and Hazardous Materials Response, Fourth Edition, Jones & Bartlett Learning, Burlington, MA.

Structural Firefighting Strategy and Tactics, Third Edition, Bernard "Ben" J. Klaene, Jones & Bartlett Learning, Burlington, MA.

U.S. Experience with Sprinklers, NFPA Research Paper, Marty Ahrens, July 2017.

© Rick McClure, Los Angeles Fire Department, Retired.

CHAPTER 15

High-Rise Firefighting

LEARNING OBJECTIVES

- Define a high-rise building.
- Describe center-core and side-core construction.
- List the specialized problems and hazards encountered in high-rise firefighting.
- Identify the keys to success for fighting fires in high-rise buildings.
- List the various high-rise building systems.
- List the responsibilities of lobby control.
- Examine the sequences of engine apparatus positioning.
- List and describe the dangers that can occur in the firefighting stairway.
- Explain the multiple strategies and tactics that can be used in high-rise firefighting.
- Explore the various fire attack methods using 1¾-inch hose, 2-inch hose, 2½-inch hose, and master streams.
- Understand the change in emphasis from search and rescue (SAR) to search-evacuate-rescue (SER) in high-rise firefighting.
- Explain the defend-in-place strategy.
- Explain the reasons ventilation should be withheld until after the fire is extinguished.

Introduction

This chapter covers the basic strategy, tactics, and responsibilities that may be assigned to the engine company in moving and delivering water to a high-rise fire. It also covers other support tactics involved in high-rise firefighting. The official definition of a high-rise building is any building where the floor of an occupied story is greater than 75 feet (23 m) above the lowest level of fire department vehicle access. That means you don't have to be a major city fire department to deal with all the problems encountered in high-rise firefighting.

Between 1977 and 1996, 16 firefighters were killed from traumatic injuries while fighting fires in high-rise buildings—a relatively high number compared to the actual confirmed high-rise fires we respond to in the United States. A high-rise apartment or office building that is on fire is a complex environment that presents several difficult challenges for firefighters to overcome. The primary goals are still to save lives and quickly put out the fire. Most firefighters, even those in major cities, may go their entire career without battling a true significant high-rise fire. Those who have actual experience are a select few. Almost all of us must rely on study, theory, fire science, high-rise case studies, computer modeling, and fire simulators, as well as diligent new and old construction inspections; pre-incident planning (prefires); and on-site, hands-on training.

Findings from the NFPA Research Report and Others

The economic opportunities in urban cities increase the population, and new construction projects can accommodate more people with a greater profit margin if construction is shifted from horizontal to vertical development. Because of the tremendous fire load and occupancy load potentials, the loss of life can be great. According to the National Fire Protection Association (NFPA) 2016 research report *High-Rise Building Fires,* there were, on average, 40 civilian deaths, 520 civilian injuries, and $154 million in direct property loss occurring annually between 2009 and 2013. Yet because of Type I construction and built-in fire protection systems in modern high-rises, fires in high-rise buildings are less likely to spread beyond the room and floor of origin than are fires in shorter buildings, making high-rise structures actually safer than all the other types of buildings we respond to. The NFPA report on *High-Rise Building Fires* breaks down high-rise buildings into six property classes:

- Apartments (including multifamily housing)
- Hotels
- Dormitories
- Offices (business and commercial)
- Facilities that care for the sick (hospitals, retirement nursing homes)
- Other (multi-use occupancies)

During the 4-year study, U.S. fire departments responded to an average of 14,500 high-rise structure fires per year. Sixty-two percent of all high-rise fires occurred in apartments, and 64% of all the high-rise civilian fatalities also occurred in apartments. Of all the property classes, residential buildings are least likely to be sprinklered. Regardless of the building height, the leading cause of fires in all the property classes involved fires in the kitchen, in cooking areas, and with cooking equipment—in other words, *food on the stove.*

The primary factor that makes fighting fires in skyscrapers so challenging is their height. Fires are beyond the reach of ground-based operations and deprive the fire department of exterior rescues and firefighting, even with aerial apparatus. Ladder pipe master streams may reach the sixteenth floor to fight against auto exposure, but it is unlikely these streams will be effective in penetrating fires above the eighth floor **FIGURE 15-1.**

On the upside, most of the high-rise building fires in the NFPA 2016 report began on floors no higher

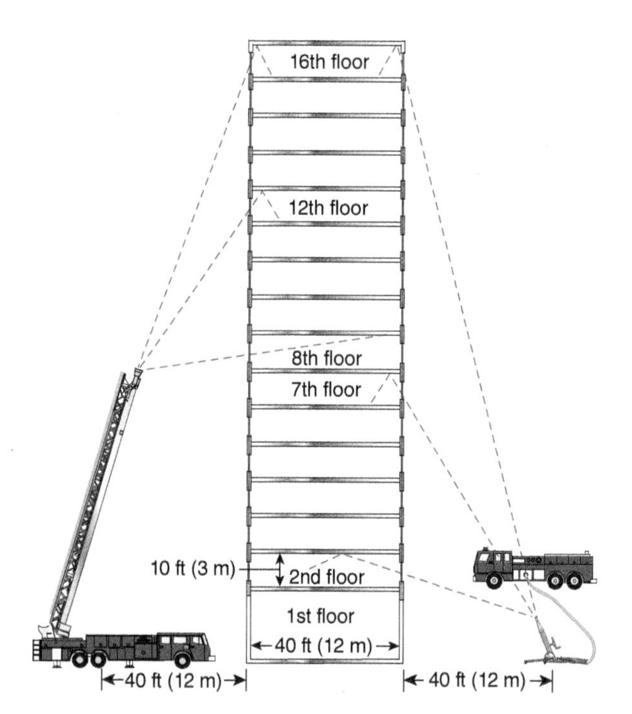

FIGURE 15-1 In high-rise buildings, it is unlikely that elevated master streams will be effective in penetrating fires above the eighth floor.

than the sixth floor. This report contains other valuable statistical information, and the findings read like a pre-incident plan on what we can expect. Not all high-rise fires will be as spectacular as the First Interstate Fire, but if we're going to respond to one, more than likely it will be a high-rise apartment building fire.

Ladder companies are traditionally assigned reconnaissance (recon), lobby control, elevators, forcible entry, search-evacuate-rescue (SER), ventilation, occupant management, and many other support activities. Although the engine company's primary responsibility is fire suppression, many supporting tactics, tasks, and responsibilities under the incident management system/incident command system (IMS/ICS) fall to an engine company during a high-rise fire. It is assumed that all engine and ladder companies will work together within the IMS, but this topic goes beyond the scope of this book.

The One Meridian Plaza Fire and the First Interstate Fire are referred to extensively throughout this chapter for the lessons learned. These are not just old stories of historical fires; these are examples of the realities we may face before the end of shift. In these two spectacular fires combined, everything that could go wrong did go wrong—and both fires can play out exactly the same way today. Standard operating procedures (SOPs) were drastically changed within their respective departments, and they should serve as blueprints about what to expect and how every fire department should approach fire attack in high-rise buildings. The One Meridian Plaza Fire claimed the lives of Captain David Holcombe, Firefighter Phyllis McAllister, and Firefighter James Chappell, all from Engine Company 11. We owe it to them to learn and apply the lessons learned so their sacrifices weren't in vain. They died, so that you could learn and live. Engraved on the One Meridian Plaza Firefighter Memorial are the following words:

> *To Sacrifice One's Own Safety In The Service To Others Requires A Courage That Is Rare. Those Among Us Who Do Are True Heroes.*

Case studies and National Institute of Standards and Technology (NIST) experiments have shown that fire growth and heat release rates have increased with modern synthetic fuels used in residential and commercial furnishings. In compartments with office furnishings and computer equipment, total room involvement has occurred within 7 minutes of ignition; therefore, the speed at which firefighters can access, confine, and extinguish the fire greatly increases the safety of building occupants and the firefighters themselves. The increased heat release rate coupled with the toxicity of the deadly smoke makes it essential that the initial attack effort be planned to overwhelm or at least contain the fire effectively.

Specialized Problems and Hazards

The major issues that firefighters face is heat, elevation, water, breathable air, and wind. Due to the elimination of ground-level access and the inability to ventilate a high-rise fire quickly, intense sustained heat can quickly fatigue firefighters during interior operations **FIGURE 15-2**. Frequent rotation of crews helps prevent heat exhaustion. Elevation affects the physical stamina of firefighters and creates possible water pressure and supply problems. Spare self-contained breathing apparatus (SCBA) cylinders may not be readily available in the initial stages of the fire attack. Unlike house fires, gaining quick access to a window or door for fresh air is not possible while fighting a fire inside a high-rise building. Wind, along with stack effect (which will be covered later), affect the movement of smoke within the building; often the wind can create unpredictable air movement and complicate the fire problem by spreading heat and smoke.

FIGURE 15-2 Working in sustained intense heat along with elevation can quickly fatigue firefighters. Frequent rotation of crews allows them to recover quickly and prevents heat exhaustion.

Courtesy of Mike Handoga.

Other problems and hazards include, but are not limited to:

- Large open spaces allowing for rapid fire spread
- Large fire loads
- Low water pressure
- High occupancy loads
- Numerous forcible entry problems
- Cubicle configurations, maze-like spaces, obstructions, and barriers
- Dead-end corridors
- Plenums, which contribute to horizontal fire spread
- Poor maintenance of building systems, especially in older buildings
- Lack of training for firefighters in high-rise hydraulics
- Lack of training for firefighters in dealing with all the high-rise building systems
- Inadequate prefire planning
- Lack of continual review of existing prefire plans at the company level
- Unfamiliarity with new security technology
- Inadequate training and education on high-rise fire and emergency procedures for building occupants

A high-rise acronym that existed in the early 1970s is worth reviving for the newest generation of firefighters:

Hazards
In
Greater
Heights

Reach is beyond the capabilities of aerial apparatus and ground-based operations.
Interior fire attack is essential on upper floors.
Significant smoke and stack effect potential is present.
Evacuation time required to remove all building occupants is unreasonable and impractical.

Variables in High-Rise Firefighting

This chapter introduces you to the basics of high-rise firefighting, along with the many engine company responsibilities and problems that may (or will) be encountered in this low-frequency, high-risk event. The many variables include the size of the fire department; the number and types of apparatus available, including specialized apparatus; the number of on-duty personnel; their level of training in high-rise operations; the availability of well-trained mutual aid resources; water supply within the building; the existing public water supply for the incident; water delivery capability within the building; type of occupancy; building size, height, and design; and building systems.

Automatic fire alarm systems are not all the same; some are basic, some are very sophisticated and divided into zones. Elevator recall and emergency control systems are different. Heating, ventilation, and air-conditioning (HVAC) systems are also unique, with some having the ability to ventilate smoke while keeping the stairways pressurized—even with two or three doors open on different floors. In other words, all these variables combined make each high-rise building unique.

The fire attack is based on the building design, the number of occupants, the number of stairways, their configuration, and their proximity to the seat of the fire. There is no simple solution or single way to fight fires in high-rise buildings. Fire behavior, fire load, the location and size of the fire within the building, the heat, and the speed of smoke spread, as well as the extra time and physical exertion to initiate the fire attack and conduct every phase of firefighting, increase the complexity of high-rise operations.

Successful High-Rise Firefighting

The secrets to success in fighting fires in high-rise buildings are pre-incident planning, consulting with the building or maintenance engineer, developing a prefire plan, and following the IC high-rise fire checklist. All the possible scenarios, foreseeable problems, and questions regarding water pressure, sprinklers, fire pumps, standpipes, pressure-reducing valves (PRVs), elevators, the capability of the HVAC system to pressurize and ventilate the structure, electrical system for the building, and emergency generators need to be identified and discussed with the building engineer or maintenance supervisor and entered into the prefire plans, which are now accessible on in-cab mobile data computers. A department incident commander (IC) high-rise fire checklist should be developed and carried on every fire apparatus. This checklist ensures that important considerations are not missed or overlooked; however, following a high-rise checklist is not an initial action plan. As for any structure fire size-up, problems still need to be identified before strategies and tactics can be selected to solve them. Once the incident action plan is developed, follow it.

High-rise buildings are rarely left unattended. There is usually some responsible person on duty 24 hours a day who has knowledge of the building systems or can quickly contact a knowledgeable person by phone. If the fire safety director or the building engineer is on site, that person needs to have building and system plans at the ready and stay glued to the IC at the command post. Once the prefire plan is in place, stick to it. It is impossible for a company officer to be knowledgeable of all the buildings and systems within her or his jurisdiction. The importance of building surveys; on-site, hands-on training; and prefire planning cannot be overemphasized.

General Information on High-Rise Building Construction

High-rise building construction falls into three groups:

- High-rise structures built before 1945
  - Heavyweight building: 20–30 pounds per cubic foot
  - Structural steel component encased in concrete
  - Exterior walls are masonry
  - Floors are reinforced concrete
  - Lack of plenum spaces
  - Exterior walls are tied to the floors
  - Steam heated; lack of HVAC systems
  - Core construction was not used
  - Most exterior windows can be opened
- High-rise structures built between 1945 and 1968, when the post–World War II economy forced changes in the building industry
  - Medium weight: 10–20 pounds per cubic foot
  - Fire or smoke towers
  - Many exterior windows can be opened
- High-rise structures built after 1968
  - Lightweight buildings: 8–10 pounds per cubic foot
  - Lack of compartmentation
  - Wide open, unobstructed floor space
  - Spray-on fireproofing for steel structural members
  - Exterior and curtain walls are a combination of glass and steel
  - Exterior windows do not open
  - Open plenums with lack of fire stops
  - Q decking used for floors
  - Center-core construction used extensively

Center-Core and Side-Core Design

Center-core construction is used in the majority of super-high-rise buildings. It is also known as central-core design. Constructed of reinforced concrete, it is a giant vertical shaft that runs the entire height of the skyscraper and contains the elevators, enclosed stairways, standpipes, sprinkler system piping, power, water and plumbing, restrooms, air supply, return air shafts, and other building systems **FIGURE 15-3**. Central-core design allows for 360° views with unobstructed floor space. The potential problem with central-core design is that the fire can also whip around 360° and flank the fire attack team. The fire attack strategy may have to be a pincer or double-flanking attack using a second attack team accessing the fire floor from the same or a second stairway to confine the fire and prevent it from circling around.

Side-core design puts the giant vertical shaft with all the major building systems to one side of the building **FIGURE 15-4**. There may still be additional stairways with standpipes (if required) on the opposite side of the structure for access and egress. Fire attack strategy in side-core construction should most likely be a direct frontal attack.

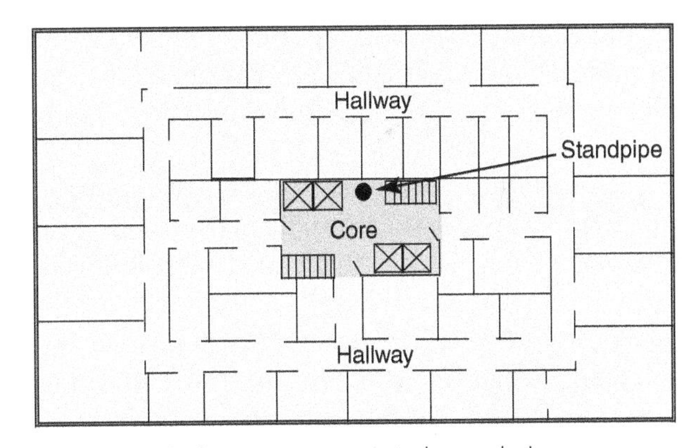

FIGURE 15-3 Center-core or central-core design.

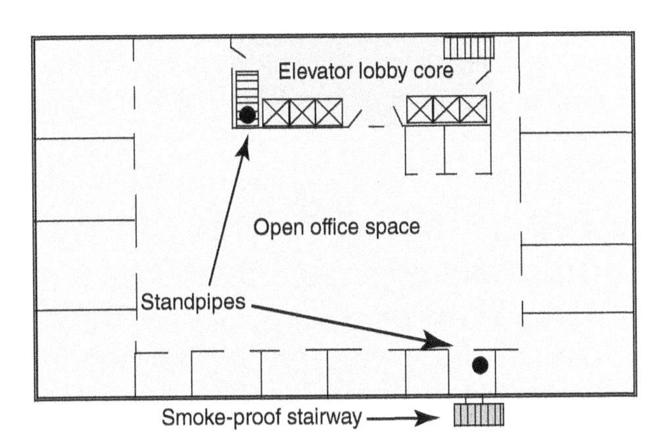

FIGURE 15-4 Side-core design.

Non-core design is found in low-rise buildings and older high-rise buildings, and it will require standpipe operations above the third floor. Access and egress stairways that would be used for fire attack and evacuation are found on either end or side of the building, with elevators and an additional stairway in the center or elsewhere.

Floor Area

Floor area or square footage in a high-rise building can range between 10,000 square feet and well over 300,000 square feet, as found in city-block buildings, Las Vegas hotels and casinos, as well as other exotic resorts; 80,000 square feet per floor in a skyscraper is not unusual. Total floor fire involvement of such structures requires 6,000 to 10,000 gallons per minute (gpm) to control, which is an unrealistic number obtainable from manual firefighting efforts.

Large fires in such structures may end up being a "controlled burn" that relies on the fire resistance integrity of Type I construction until the fuel load is consumed. In the controlled burn scenario, the sprinkler valves (often two) might need to be turned off because all sprinklers on the floor will fuse open from the heat. Significant water flow onto the fire floor could occur depending on the available water pressure and sprinkler pipe size. This flow could be high enough to limit water available for hose lines from the same standpipe on other floors.

Downtown high-rise buildings can have up to six stairways, but super-structures, like resort hotels, can have many more. Elevator shafts vary. Some buildings have low-rise banks, midlevel banks, and high-rise and express banks. High-rise buildings also fall into three use categories: commercial, residential, and mixed use. Mixed use can have residential units on the upper floors, while the main street level can be offices, stores, banks, showrooms, restaurants, and other public assemblies. All vary by design and are uniquely complex among themselves. Again, the keys to success are prefire planning specific to the high-rise building, identifying the problems, then following the incident action and prefire plans.

Heating, Ventilation, and Air Conditioning (HVAC) Systems

Many types of HVAC systems are used in high-rise buildings. The **plenum** is the space between the structural ceiling and the dropped ceiling of a floor/ceiling assembly—the underside of the floor above. It is used to house and conceal telecommunications cables for telephone and computer networks. It is also used to support and conceal a labyrinth of HVAC ductwork containing forced-air supply shafts; forced-air return shafts, which are usually the plenum spaces themselves; supply air fans; return air fans; outside air intake dampers; exhaust dampers; mixing dampers; fire dampers; air diffusers; air filters; humidity control equipment; and wiring for smoke and heat detectors.

To keep them heated or cooled, high-rise buildings are well-sealed to control interior temperatures. The lack of functional windows makes ventilation during the fire attack nearly impossible, and because wind currents can make fire behavior unpredictable, opening windows is not recommended during fire attack. Ventilation tactics often take place after the fire has been controlled or extinguished. The lack of ventilation creates the greatest physical stressor—trapped heat—that firefighters must deal with in high-rise firefighting.

There are system variables, but the strategic objective for smoke control is to rely on the HVAC system to limit the spread of fire and control the movement of smoke within the building during the fire by pressurizing the stairways. In many high-rise buildings built during the 1960s, 1970s, and 1980s, only the stairways were pressurized. Usually the elevator shafts were not. Unfortunately, many of these systems are not tested on a regular basis. In older systems, some stairways may be underpressurized, allowing for smoke to enter the stairway, especially when many occupants are self-evacuating from numerous floors. In other cases, the stairways may be overpressurized on floors in close proximity to the pressurization fans, making some doors very difficult to open from the inside of the hallway as they open against the stairwell pressure. This can cause problems and panic for elderly occupants, children, or those with disabilities who may not have the strength to push the door against the pressure in the stairway. Therefore, annual testing of stair pressurization is important to prevent over- or underpressurization of the system when operating in alarm.

Many practices still recommend completely shutting down the HVAC system during a fire alarm, so the fresh air intakes don't provide additional oxygen to the fire. Smoke and heat detectors within the air ducts should automatically close the dampers to isolate the fire area, but if they don't, an operating HVAC system can spread smoke throughout the entire building. This was the case on November 21, 1980, at the MGM Grand Hotel Fire in Las Vegas, Nevada. Workers were tired of having the air conditioning shut down in the casino and hotel after years of false alarms and having to reset the system and smoke dampers, so they decided to solve the problem by screwing the smoke dampers open. This action prevented the HVAC system from automatically shutting down during the fire, and 85 people were killed.

When the movement of air is intensifying the fire, shutting down the HVAC system is the correct action to take. On the other hand, shutting down the system may also eliminate the ability of the system to pressurize the elevator shafts, stairwells, and floors above and below the fire—something that is essential for high-rise firefighting because it prevents smoke from entering these areas. In most high-rise modern construction, the elevator shafts and floors above and below the fire are now pressurized, sandwiching the fire floor to prevent its spread. Newer HVAC systems are extremely powerful and efficient in pressurization and removing smoke from the affected areas. They are designed to remain in operation during a fire, and they protect firefighters from the development of a flow path.

It is impossible for a fire officer to know every working detail about the various HVAC systems. The only way to be sure if the HVAC system should be shut down or left running as designed in order to utilize the system to its maximum potential during a fire is for the IC to rely on the expertise and judgment of the building engineer or the building maintenance supervisor, or to refer to the recommended procedures that have already been determined in the prefire plan. All units should report to the IC any adverse effects of smoke movement or of the fire intensifying from the HVAC system so it can be shut down.

Automatic Sprinkler Systems

An automatic sprinkler system is the most effective way to prevent, control, or suppress a major high-rise fire. Firefighters must rely on the built-in fire protection systems to help them protect occupants and property. For these systems to work as designed, it is essential that they are installed, tested, inspected, and maintained according to NFPA 13, *Standard for the Installation of Sprinkler Systems* and NFPA 25, *Standard for Inspection, Testing, and Maintenance of Water-Based Fire Protection Systems.* Major fires have occurred in high-rise buildings where the fire protection systems failed to work properly, creating tactical and logistical nightmares where even the most experienced and well-equipped fire departments could not control the fires.

At the One Meridian Plaza Fire, only the below-grade levels were sprinklered. When the building systems were upgraded, the dry standpipe system was converted to a wet system with two fire pumps, one serving the lower floors and a mid-rise pump serving the upper floors. This conversion facilitated the installation of sprinklers throughout the building, but it still wasn't required. At the request of the tenants, automatic sprinklers were installed on floors 30, 31, 34, and 35. After all the Philadelphia Fire Department efforts and sacrifices, the fusing of 10 sprinkler heads on floor 30 finally extinguished the fire, but not before it consumed eight floors of the building.

At the First Interstate Fire in Los Angeles, California, a major sprinkler renovation was nearly complete: 90% of the building was fully sprinklered, including all the fire floors. But the building manager and sprinkler contractor decided to wait until the water flow alarms were installed before charging the system. Sectional valves were closed between the standpipe risers and the sprinkler system on each floor. The fire pumps were also shut down. Los Angeles Fire Department (LAFD) firefighters were left with static head pressure in the standpipe to fight the fire initially. Fortunately, the sprinkler installation supervisor was still on site, but he was trapped on the roof. He was rescued by a police helicopter and eventually returned to the command post. He was taken to the basement where he manually started the two fire pumps. Then he coached firefighters to open the sprinkler sectional valves on floors 17 and 18 in anticipation of the fire lapping past floor 16; however, the fire was stopped on floor 16, with the *flow the floor* strategy, which will be covered later in this chapter.

Nozzle Water Flow and Pressure Problems in Fully Sprinklered High-Rise Buildings

This section on nozzle water flow is offered by Deputy Chief Gary English, Assistant Fire Marshal (Retired) of the Seattle Fire Department:

Water flow and pressure problems at the nozzle are caused by many reasons, such as kinked hose, multiple hand lines flowing, relying solely on the building's fire pump, the engine pumping at too low a pressure into the fire department connection (FDC), supply lines from the engine to the FDC too small, or too few, standpipe friction loss (common with 4-inch standpipes), in-line valves partially closed, or the simple lack of water in the municipal water system from the hydrants. But there is another, albeit not altogether common, problem that might occur in buildings with fire sprinklers. Simply, the fire sprinkler flow is so high that the available water flow for hose lines from a combination (sprinkler and fire hose) standpipe is reduced. How? Sprinklers are expected to flow a set amount of water in a finite "design area," that is, where a handful of sprinklers are designed to prevent fire spread, or control the fire to a limited area. This is based on the hazard class and area size, along with other variables. For example, the flow for a designed area might be 90 gpm, delivered by 6 sprinkler heads, flowing at 15 gpm each. A flow requirement is based on the hazard and fuel load of what is "expected" to be in the design fire area.

We know from hard-earned experience and the UL/NIST tests that a fire consuming legacy fuels (office furniture made up of organic material like cotton, paper, wool, and wood) grows slower and produces less heat than similar furniture made up of modern fuels of hydrocarbons (synthetics and plastics). When sprinklers designed for a slower, cooler fire (organics) are used to cover new materials that create much hotter, faster fires (hydrocarbons), this can result in inadequate flow from the "unexpected" fire load. In most cases, the difference is not extreme enough to create problems unless there is much more fuel in one area, that is, an area with high concentrations of hydrocarbons such as a large computer server area.

Because the hydrocarbons burn hotter and faster, the hazard is greater, but the volume of water produced by the sprinkler stays the same for the legacy fuel-designed area. As this fire spreads beyond the designed area, more and more sprinklers open, using more and more water from the combination standpipe. More flow for sprinklers equals less flow for hose lines and potentially less pressure available.

This phenomenon is not limited to changes in office furniture construction and adding computers; it can happen any time the hazard is greater than the sprinkler design. A potent example is using a large area to store new furniture and new computers while the office is in transition, and piles of empty cardboard boxes are created when new computers are set up. Piling hundred-plus sets of boxes, Styrofoam, plastic wrap, computers, furniture, and so on, in a concentrated area within an open floor plan space creates a fire hazard greater than the sprinklers were designed to control. A fire in this pile could rapidly overwhelm the fire sprinklers in the design area and continue to spread. Although some may think the sprinklers activating ahead of the fire will prevent fire spread, this is not necessarily true because the lower levels of the pile are protected from the sprinkler water by the layers in the pile itself.

Fire sprinklers do not shut off once the fire has burned out in their design area or in the area beyond the design area; rather they simply just keep flowing until they're manually shut off at the valve. An uncontrolled fire might send heat well beyond the combustion area, opening fire sprinkler heads across a ceiling area where nothing is burning. In this example, rather than a 90-gpm flow, the additional fused sprinkler heads operating in an area five times larger could result in a flow of 450 gpm. As a practical matter, the pipe sizes supplying the sprinkler might not allow this full flow, but even a sprinkler flow of 350 gpm is greater than a 2½-inch hand line.

How would we recognize this reality and what can we do to prevent this occurrence? Exterior size-up might reveal visible flames that, in a building with sprinklers throughout, would indicate either sprinkler system failure or excessive fire loads not being controlled by sprinklers. When firefighters arrived at the First Interstate Fire, flames were visible from every window of the twelfth floor. This would be the first onsite clue that they were dealing with a nonsprinklered building or that the sprinkler system was not working. Eyewitnesses or floor cameras might have revealed dramatic fire buildup before cameras were obscured. First-arriving units at the fire floor might find heat levels greater than what would be expected if sprinklers were controlling the fire.

In these types of scenarios, command should consider the possibility that sprinklers are not controlling the fire; turning off sprinklers to the floor helps preserve adequate flows and pressures for the operation of hand lines. This was the concern at the First Interstate Fire. The fire floors were sprinklered, but the system was shut down for renovation. Because the fire was so advanced on the involved floors, the sprinkler technician felt the sprinkler demand would only rob water from the fire attack hose lines, so the decision was made to leave the sprinkler valves closed.

Turning off sprinklers may be more complicated than you might expect, especially when the fire sprinkler valve is located in the same area as the fire. Even when sprinkler valves are located in stairwells (and these valves are not always located there), the correct valve must be located for the area to stop sprinkler flow. For looped sprinkler systems, closing more than one valve might be necessary. Examining sprinkler layout on plans is not always possible, but a quick survey of a lower floor might indicate valve locations for the fire floor (sometimes hidden in drop ceilings). It is most effective to confer with building engineers because they will have information on sprinkler zones and valves.

Using standpipes that are *not* interconnected with a sprinkler system (Class 1 standpipes) might be more effective. For interconnected combination standpipes, it may be necessary to turn off some or all of the sprinkler flow before an attack can be initiated to allow adequate hand-line flow with the required minimum tip pressure. The potential for low-pressure flows makes the use of 2½-inch hose lines with 1⅛-inch smooth-bore tips for attack lines a safe bet for maximum efficiency in the worst-case scenario.

Keep in mind that the sprinkler flow from a standpipe may significantly affect all hand lines on all floors from that particular standpipe. Thus, flows from exposure lines on upper floors must be considered. Low-flow pressures from interconnected standpipes also results from the activation of a combined sprinkler system. Once sprinkler valves are located for the fire area, a firefighter should be assigned to close and control the valves slowly and be ready to reopen them quickly if sprinklers are needed to support the fire attack. Closing the valves too early, before the attack team gains control of the fire, could allow for rapid fire growth.

Simply being aware of this potential problem is not enough. Building inspections that identify unusually high fire loads in any sprinklered building should result in having the owner or tenant remove or reduce the fire load or verifying that the fire sprinklers are adequate for the load through the department fire marshal.

NFPA 14

NFPA 14, *Standard for the Installation of Standpipe and Hose Systems*, is a must-read reference document for any firefighter who will be involved in standpipe operations. NFPA 14 has evolved over time; thus, a standpipe system built according to the 1970s version of NFPA 14 could be very different from a new building built to the latest standards. Understanding the standpipe requirement differences by year is important to understanding the capacity and pressure availability and system limits. NFPA 14 provides minimum requirements for the design of the standpipe system, which is determined by the height of the building, the floor area of each story, the occupancy classification, egress system design, required flow rate and residual pressure, and the distance of the hose connections from the sources of the water supply.

Early high-rise buildings used 4-inch standpipes. Newer buildings (since the late 1980s) use 6-inch standpipes. Standpipes and hose are classified by the intended user. Class I is for firefighter use, Class II is for tenant use, and Class III has both Class I and Class II features that can be used by both tenants and firefighters. Both Class II and Class III also have 1-inch-diameter hose lines with a nozzle installed in building hose cabinets for ready use by tenants. Because tenant hose is designed for tenant use, the pressure and flows are much less than those for firefighter hose. When properly supplied by an adequate municipal water supply and engines pumping in tandem,

FIGURE 15-5 The fire code requires standpipe and sprinkler FDCs to be labeled. If there is signage for pressure requirements, pump to the posted pressure.

Courtesy of Raul Angulo.

you can attain approximately 1,000 gpm from a 4-inch pipe and approximately 2,500 gpm from a 6-inch pipe. Because water sources and apparatus vary, attaining 1,000 gpm from a standpipe is the most realistic expectation.

The following list highlights some key points in NFPA 14 that can be helpful in identifying potential dilemmas and takes the guesswork out of possible solutions. It also provides clues for quick decision making by the company officer and the engine apparatus driver. Although codes and standards have requirements, not all buildings comply. Many older buildings may be in compliance with older requirements, and these buildings are not required to meet newer codes and/or standards. Therefore, the following may not always be true for older buildings.

- The fire code requires standpipe at sprinkler FDCs to be labeled. If there is signage for pressure requirements, pump to the FDC at the posted pressure **FIGURE 15-5**.
- Where the standpipe connections (FDCs) serve more than one building or location, signage should be provided by the FDC for the buildings and locations served **FIGURE 15-6**.
- Where two or more standpipes are installed in the same building or section of the building, the standpipes will be interconnected (with some exceptions).
- There should be a fire hydrant within 100 feet of an FDC.
- If the required inlet pressure at the Siamese is 150 pounds per square inch (psi) (10.3 bar) or less, there will be no signage.
- Maximum pressure at any point in the standpipe system shall not exceed 400 psi (28 bar) (with some exceptions).

FIGURE 15-6 When standpipe FDCs serve more than one building or location, signage should be provided for the buildings and locations served.
Courtesy of Raul Angulo.

- Buildings constructed before 1993 have a minimum flow rating of 500 gpm at 65 psi. These are low-pressure standpipes. Multiple 2½-inch hand lines can exceed the flow capability depending on nozzle selection and use, and you will run out of water. (Some cities, such as Seattle, Washington, required retrofits of older buildings and all newer construction to have a minimum pressure of 125 psi but reduced gpm to 300 gpm at outlets. This allowed smaller, more maneuverable hand lines in the sprinklered buildings).
- Where the static pressure of a 2½-inch (65-mm) hose connection exceeds 175 psi (12.1 bar), a listed pressure regulating device must be installed to limit the static and residual pressures at the hose connections to no more than 175 psi. (Expect a pressure-reducing valve [PRV].)
- A PRV is designed to reduce the downstream water pressure under *both* flowing (residual) and nonflowing (static) conditions.
- A pressure-restricting device (PRD) is designed to reduce the downstream water pressure under flowing (residual) conditions *only*, which allows higher static pressure in the line.
- For Class I, the minimum flow rate of the most remote standpipe is 500 gpm (1,893 L/min) through the two most remote 2½-inch (65-mm) hose connections.
- The minimum flow rate for additional standpipes is 250 gpm (946 L/min) per standpipe for buildings with floor areas that do not exceed 80,000 square feet (7,432 m^2) per floor.
- For buildings that exceed 80,000 square feet per floor, the minimum flow rate for additional standpipes is 500 gpm (1,893 L/min) for the second standpipe and 250 gpm (946 L/min) for the third standpipe if the additional flow is required for an unsprinklered building.
- The maximum standpipe flow rates are 1,000 gpm (3,785 L/min) for buildings that are fully sprinklered, and 1,250 gpm (4,731 L/min) for buildings that are not sprinklered.
- For Class II systems, the maximum flow required from a 2 ½-inch hose connection is 250 gpm.
- Based on the travel distances required to cover all portions of each floor level of the building and starting from the hose connection, the travel distance shall be 200 feet (61 m) for sprinklered buildings built after 2005 (older buildings have shorter travel distances, e.g., 150 feet). These distances are measured in a straight line; therefore, longer hose is needed.
- The travel distance shall be 130 feet (39.7 m) for nonsprinklered buildings. Two hundred feet of hose should be the minimum to start any fire attack.
- A hose connection should be provided at the highest landing of the stairway with roof access, or it should be provided on the roof.
- In stairways without roof access, there may, be a hose connection at the top of the standpipe on the roof.

Solid strategy and tactics are rooted in the engine officer's understanding of NFPA 14.

Strategic and Tactical Limitations

Tactical Clarifications

Consideration must be given to the size of a fire department and its established SOPs. The high-rise tasks and responsibilities in this chapter are written in a logical sequence, but the sequences may vary and even take place simultaneously depending on the practices of your department and the size of the initial response. Because the number of firefighters and apparatus available also vary, any support assignment or function that is usually assigned to the ladder company is most likely assigned to an engine company. Staffing for specific assignments also varies; it is not specific to any unit. Members of a ladder company may easily find themselves operating hose lines on a large fire because the ability to complete their assigned tactical objectives are prevented until the fire is knocked down. For example, if 2½-inch hose lines are utilized, it will obviously require more firefighters and maybe even two companies to advance and operate a single line than if 1¾-inch hand lines are used. This chapter does not address how many firefighters are needed to advance a 2½-inch hose line or how far the firefighters should be spaced out. These specifics depend on the experience and physical ability of every firefighter. Although some hose and nozzle techniques are covered, 2½-inch hose-handling techniques are part of Firefighter I and II training, and efficient hose-handling techniques should have already been developed through on-site and hands-on training. Nevertheless, certain tactics and responsibilities need to be carried out regardless of who is assigned the task. If more firefighters are needed to accomplish essential tactics, call for them; if they are not available, the IC has to manage the available resources as best as can be expected.

Firefighter Safety Considerations

High-rise building fires and standpipe operations require the efforts of many firefighters. Every person engaged in firefighting or rescue operations must be in full personal protective equipment (PPE), including SCBA, and no one is to be assigned to work in the building alone. All assignments should be given to teams of at least two firefighters. The only exception is the firefighter assigned to run the elevator car in Phase II. Because this firefighter is shuffling crews between the ground floor and staging, he or she is not alone, even though technically it is a stand-alone assignment.

Smoke cannot usually be removed or ventilated quickly in high-rise buildings; therefore, visibility can be limited. Firefighters working together can keep track of each other by sight, touch, voice, radio communications, or they can remain in contact with the hose line or search rope stretched for that purpose. A firefighter working alone is in great danger of being injured or lost no matter how menial the task, so maintain the buddy system. Numerous firefighters have died in high-rise fires as a result of becoming disoriented and then lost in the thick smoke.

The decisions on what fire attack actions to take may ultimately be made by the fire. The amount of radiant heat energy is tremendous, and it is hard to convey its power with words. Keep in mind that your PPE has a threshold rating of 572°F (300°C) for 15 to 30 seconds. Like everything else in the room, your bunker gear absorbs heat, and once it's heat-saturated, you will get burned. No PPE has been invented to counter the effects of the heat that a firefighter endures during prolonged operational periods in a high-rise fire. Basic maneuvers, like the ability to advance a charged hose line, take longer and become more difficult, and crews may succumb to heat exhaustion much faster in a high-rise fire than in other types. Flashover and post-flashover ceiling temperatures can be up to 2,000°F (1,093°C), melting everything in the overhead and all office furnishings throughout. These drips of burning plastic land on the carpeting, igniting the floor and the adhesive mastic underneath. Floor temperatures when the floor is on fire can be around 500°F (260°C).

At the DeWitt-Chestnut Fire in Chicago, Illinois, firefighters were pushed back numerous times attempting to attack the fire on floor 36. Ceiling temperatures in the hallway were well over 1,500°F (816°C). Every time the firefighters attempted entry, these superhot temperatures rushed up the firefighting stairway, so much so that they melted the floor 37 placard on the landing above **FIGURE 15-7**.

The lens and the facepiece of SCBA is often considered the weakest component of a firefighter's protective gear ensemble in high-heat conditions. The literature from different manufacturers varies, but the glass transitioning or the *softening* of the polycarbonate lens occurs between 293°F and 302°F (145°C and 150°C), and melting of the lens can occur in varying temperatures between 419°F and 640°F (215°C to 338°C). In tests conducted with NIST, U.S. Fire Administration (USFA), and the Chicago Fire Department, fire experiments performed in furnished townhouses demonstrated the effects of a range of realistic firefighting environments on eight different makes of SCBA facepiece lenses. The maximum exterior lens temperatures were as high as 572°F (300°C). Thermal degradation of SCBA facepiece lenses was observed in all cases when the facepiece lens temperature exceeded

FIGURE 15-7 The floor 37 placard melted from the heat on floor 36 at the DeWitt-Chestnut Fire in Chicago, Illinois.
Photo by Battalion Chief Michael Wielgat, Chicago Fire Department.

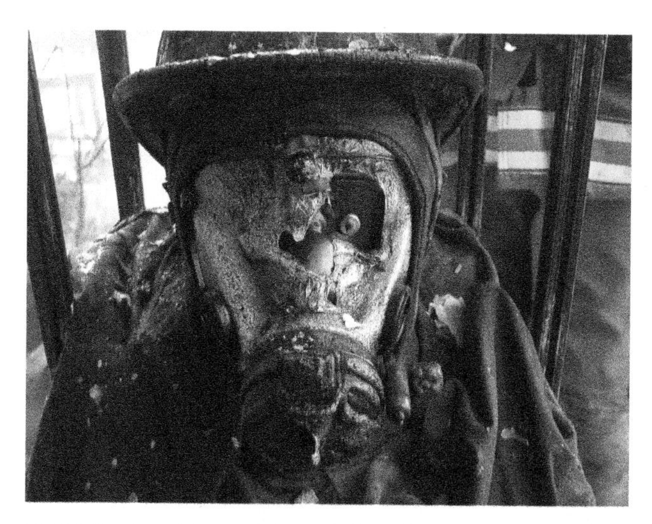

FIGURE 15-8 Thermal degradation of an SCBA facepiece.
Courtesy of Raul Angulo.

the lower end of the melting temperature range for polycarbonate, and the integrated heat flux exceeded 3.1 MJ/m2. These lenses exhibited bubbling and loss of visual acuity, as well as severe deformation; in one case, a hole was evident **FIGURE 15-8**.

The Utilization and Preservation of Firefighters

Considering the information provided so far in this chapter, the firefighter's greatest adversary in a high-rise building fire is the heat. One of the first safety considerations that must be anticipated is the frequent rotation of crews. In a fast-moving fire that can double in size every minute, operational times and the amount of interior firefighting accomplished by members may be shortened due to fatigue. This is the primary reason that high-rise fires require multiple alarms and mutual aid. NFPA 1500, 8.9.1.1 states that crews shall not be permitted to use more than two SCBA cylinders before being rotated through rehab. To sustain a prolonged attack strategy, crews must be relieved frequently on a regular rotation. Keep in mind that every 50-foot (15.24-m) section of charged 2½-inch hose weighs approximately 105 pounds (48 kg). Five to 10 minutes of wrestling with a charged 2½-inch hose line on air, in full PPE, in a hot smoky environment is about the limit for the average firefighter. After that, firefighters begin to lose their strength and may be too weak to advance the heavy hoses effectively. Frequent rotations allow them to recover quickly and reenter the battle. Over the long haul, this strategy extends the operational period of assigned crews for the duration of the incident.

Company officers may meet with resistance because firefighters naturally feel a sense of failure and defeat when they're pulled off the fire before they're ready. Pre-incident coaching should convey that every attempt should be made to attack the fire and search for occupants with speed but that a high-rise fire attack should be viewed as running a marathon, not running a sprint. Frequent rotation of crews actually prevents injuries from heat and fatigue. Firefighters and company officers who push themselves beyond a self-limiting rotation may actually succumb to the effects of heat exhaustion and be taken off the fire incident completely. Thus, the IC loses valuable resources and must replace them. Also, pushing past a self-limiting point may lead to a firefighter collapsing and triggering a rapid intervention rescue, thus changing the incident priority and shifting resources from extinguishing the fire to rescuing a downed firefighter.

Conservative Approach for Smaller Fire Departments

Not every fire department has the resources of a large city fire department, but small cities and even towns still have to deal with high-rise structures due to urban growth. The heavy work is getting sufficient hose lines in place and in operation on upper floors. Once set, conserve your resources by allowing only the number of firefighters required to operate the lines. Pull the excess crews back and hold them in reserve for rotation. There is no need to expose all your limited resources at once to the effects of heat. Smaller departments may have to take a more conservative approach by attacking a large high-rise fire defensively with the use of portable master streams or by taking up a defensive position from a place of shelter, thus giving up additional property loss to the fire. This may mean relying on the integrity of Class I fire-resistive construction to contain the fire to the floor of origin and taking up exposure positions on the floor above after the life hazard on the fire floor has been mitigated. Uncontrolled high-rise fires cannot be contained without a large contingent of firefighters. Preserving the effectiveness of crews with frequent rotations actually extinguishes the fire faster by extending the working capability of limited resources. This safety strategy inevitably concludes the incident in a shorter time with the fewest amount of injuries.

Firefighter Accountability

The NFPA 1500 safety standard states, "[A]n accountability system shall be used on all emergency incidents, and the initial IC shall maintain an awareness of the location and function of all companies or crews at the scene and must be maintained throughout the incident." Officers that are assigned to divisions or

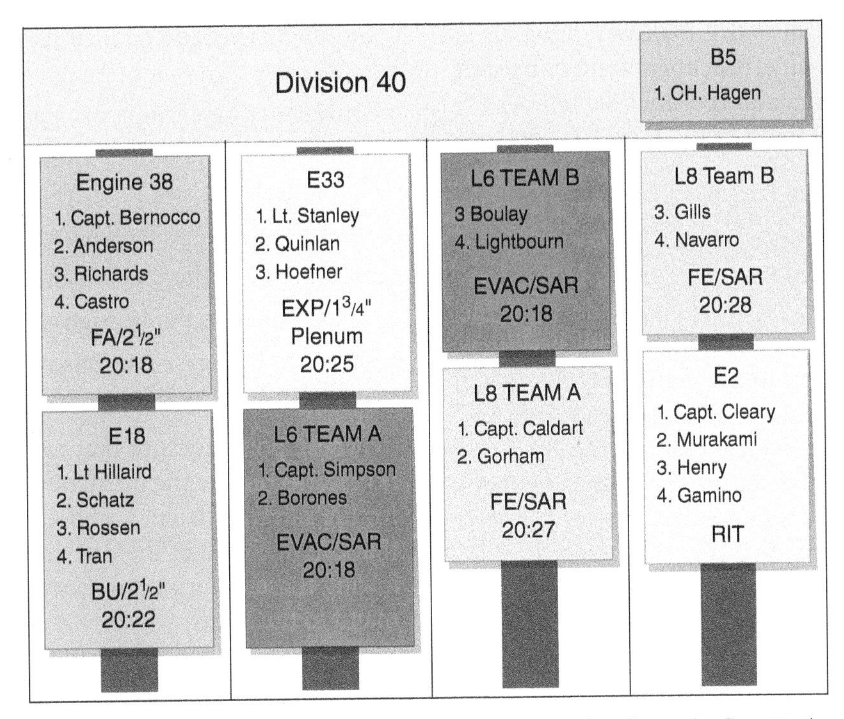

FIGURE 15-9 This accountability status board accurately reflects the fire attack on division 40 (floor 40).

Courtesy of Raul Angulo.

special tactical groups have the responsibility to supervise and account directly for all companies operating in their specific area of responsibility. Company officers shall maintain on ongoing awareness of the location and condition of all crew members. A significant number of firefighters are required to accomplish assigned tasks with the proper tools and equipment, making accountability at a high-rise extremely challenging—but not impossible. The key to keeping accurate accountability is to work the status board as the assignments are being made.

The accountability status board in **FIGURE 15-9** depicts a fire attack on the fortieth floor. The companies, the names of crew members, their assignments, and the times they made entry onto floor 40 are accurately accounted for and are listed below:

- E38 started the fire attack with a 2½-inch hose line at 20:18.
- L6 split their crew and also entered at 20:18. Both teams are ensuring complete evacuation on floor 40, then commencing the primary search.
- E18 is using a 2½-inch backup line and entered the fire floor at 20:22.
- E33 took a 1¾-inch exposure line and entered at 20:25. They protect the search teams and are also responsible for checking for horizontal fire spread in the plenum.
- L8 also split their crew to cover more ground. They are forcing any doors and assisting with search and rescue (SAR).
- E22 is the rapid intervention team (RIT) for floor 40.
- B5 is division 40.

If any member on the hose line calls a Mayday, division 40 and the RIT know which company and hose line she or he is on. By checking the stencil on the hose jacket, the RIT can follow the correct hose line to the nozzle. If a member on the ladder company calls a Mayday, division 40 can notify the rest of the company. The ladder officer or crew can direct the RIT to her or his location. If a member panics and yells that he or she is out of air, more than likely it's a member of E38 because they have been on air the longest.

Command Priorities

Command must have a plan for fire attack. The four incident priorities remain the same for high-rise building fires, except on a much grander scale:

- Life safety
- Incident stabilization
- Extinguishment
- Property conservation

For example, life safety by itself may be ensuring the safe evacuation of several thousands of people.

Although evacuating an entire high-rise building is not desirable nor advisable, the cooperation or trust of the public may not be guaranteed since September 11, 2001, when the North Tower of the World Trade Center in New York City was struck by the jet airliner. It was struck before the South Tower was struck by the second jet airliner, yet the South Tower collapsed first: 29 minutes before the North Tower. In the confusion, many announcements were made over the loudspeaker and by floor wardens in both buildings instructing the tenants of uninvolved floors to remain at their workstations and that evacuating was not necessary because everything was under control.

A first-in company officer may have to assign personnel immediately to coordinate and direct a massive evacuation that is already in progress. This aligns with life safety as the first incident objective; however, large evacuations crowd the stairwells and make firefighter ascent into those stairwells very difficult.

Incident stabilization and extinguishment also take place on a grander scale. For example, it might take the first company 10 or more minutes before the initial hose line can apply water on a fire on upper floors. Controlling the fire and extinguishment takes even longer. At the First Interstate Fire in Los Angeles, California, it took the fire department approximately 27 minutes after arrival before the initial attack was commenced on the twelfth floor.

Rescue, exposures, confinement, extinguishment, overhaul/ventilation and salvage (RECEO/VS) also takes more time to initiate and accomplish. For example, the search, evacuation, and rescue of possible occupants may require a search of stairwells and access to numerous floors before the primary search is complete. It may even require that the fire is first knocked down and contained. Due to heavy smoke and heat in all the four stairways at the First Interstate Fire, the trapped occupants on floor 37 and floor 50 were not able to be rescued until after the fire was knocked down: over three hours from the start of the initial attack. The seven major problems that present themselves at any fire also present themselves on the fire floor of a high-rise building: possible trapped occupants, visible trapped occupants, gaining access, exposures, fire, smoke, and the presence of hazardous materials. The initial IC should remain calm and approach a high-rise emergency systematically by:

- Performing a complete size-up and risk-benefit analysis
- Remaining guided by the four incident priorities
- Remaining guided by the strategic and tactical objectives of RECEO/VS
- Identifying and confirming the location of the fire
- Identifying existing problems
- Assigning crews to address the problems tactically
- Anticipating potential problems
- Sticking to the pre-fire and incident action plans
- Referring to the IC high-rise fire checklist
- Maintaining a close liaison with the building engineer or maintenance supervisor

These points ensure that every professional effort is made to resolve the incident with the highest regard for life safety, efficiency, and effectiveness. That is all that can be asked of the first-in company officer. Following these points allows a smooth transfer of command to the chief.

Initial Size-Up for High-Rise Buildings

All the size-up points covered in Chapter 8 also apply to fires in high-rise buildings but, again, on a grander level. The gathering of significant information happens when the first-in company officer actually enters the lobby to access the fire alarm control panel (FACP). It may also take a considerable amount of time to confirm exactly what's happening within the building.

Time of day, day of the week, and weather can contribute to stack effect. Holidays, ongoing road construction, and the schedule of large public assembly events are all known factors before the alarm is even received. The movement of apparatus, adjacent companies that are out of service on alarms or out of position for in-service department training should be monitored throughout the shift, as should the beginning and end of commuter traffic. The address, building name and occupancy, and the anticipated occupancy load is learned at dispatch before the company even leaves the station. The size of the initial response is also an indication of the caller information that was given to the dispatcher. A reduced response, for example, two engines and a ladder company, may mean an automatic alarm has been activated with no other information. An unusually large initial full response with extra units dispatched indicates additional information was received by the dispatchers that confirm an actual fire.

While en route, the company officer can notice the presence or lack of wind conditions. Taking note of flags flying from buildings and cranes, smoke from chimneys, steam plumes from industrial facilities, and branches blown about by breezes gives clues to wind direction and wind speed—factors that may lead to unpredictable or extreme fire behavior, including stack effect. The radio report from the dispatcher provides all the currently available information that was reported by callers, the building staff, and the monitoring alarm company.

As the first-due engine approaches the address, a 360-degree approach assessment should be made to view exterior conditions and as many sides of the building as possible. There may be nothing visible from the outside of the skyscraper. Fog and cloud cover during inclement weather may also shroud the upper stories of the building from visual confirmation of smoke or fire. If flames are visible with the magnitude of the One Meridian Plaza Fire or the First Interstate Fire, where the entire twelfth floor was involved when the fire department arrived, it is the first on-scene indication that you're dealing with a nonsprinklered building or that, for a variety of reasons, the sprinkler system is not operating.

Although driving around the block may seem impractical, the extra time taken may pay big dividends in determining the safest and most advantageous location to connect to the FDC and access the building. For example, if flames are visible and it appears that autoexposure is imminent, connecting to an FDC on the side below the fire may prove to be extremely dangerous for the pump operator and the exposed FDC supply hoses, which could be subject to falling burning debris and large shards of glass from broken windows. A reverse lay to the next hydrant is warranted to increase the safety margin for the pump operator and apparatus. Supply lines have to be protected. If there is another FDC on the opposite side of the building and the standpipes are interconnected, the second-due engine can connect to that FDC and the hazards can become non-issues unless the fire spreads to the other side of the building. These initial actions with the first two engines need to be deliberate to set up the fireground properly.

The urgency, adrenaline, and pressure of all the incoming units quickly descending on the closest intersection can be controlled by the first-in officer announcing that all units shall stand by two blocks from the building except for the first two engines and first-due ladder company. Announcing the numbers of those companies reduces confusion, for example, "All units stand by at 1st Avenue and Denny Way except for Engine 2, Engine 8, and Ladder 6." This single action sets up the command presence of the first-in company officer and allows time for gathering information, identifying problems, and making rational well-thought-out decisions instead of making random assignments because units are anxiously waiting for orders.

The initial radio report should be given and should include the arrival of the first-in unit; confirmation of the address and the occupancy; the building description, including dimensions; any visible smoke and fire conditions; and current wind direction. The officer should announce if this is a high-rise office building, a high-rise residential building, or a high-rise mixed-use building, and whether the building is fully sprinklered or nonsprinklered. This can easily be determined by looking at the FDC signage, asking the pump operator at the FDC, or looking for sprinkler heads in the lobby. If the building is known to have a high or large interior atrium, this information also needs to be announced to incoming units because this building feature poses a significant smoke accumulation hazard to everyone inside, including firefighters. The first-in company officer needs to establish command and announce the initiation of lobby control. The initial command post is usually in the lobby, close to the FACP or inside the building's fire control center (FCC), where additional information is available along with the controls for the HVAC system, interior communications, and the elevators.

Lobby Control

Lobby control consists of the following tasks:

- Verifying the alarm location
- Verifying the location of any sprinkler activation
- Establishing contact with the building engineer or safety supervisor
- Acquiring emergency procedures documents, including building plans
- Selecting the firefighting and evacuation stairways
- Managing the evacuation of occupants to a safe location
- Gaining control of the building systems, including the elevators, the HVAC system, access keys, communications, electrical generators, and fire pumps
- Directing incoming units to the up elevator or the firefighting stairwell

Verifying the Fire Location

The first detail that the company officer needs to determine when arriving at a high-rise fire is to verify the initial location of the fire or the location of the fire alarm activation on the FACP. (There may be more than one fire.) This is done by pushing the acknowledgment (ACK) button, which notifies the monitoring company that the fire department is on scene at the control panel. Pushing the ACK button turns off the alarm buzzer in the FACP but leaves the alarm status lights and building alarm on. For newer FACPs, this also starts scrolling specific information in the viewing window. Automatic fire alarm systems vary; some provide general information, while newer systems provide very specific information **FIGURE 15-10**. For example, basic systems may have a red light on the panel indicating a fire alarm on a certain floor, while other sophisticated systems give you a specific zone; location on the fire floor; location in front of a room number or inside a room number; and whether the alarm is a pull station, a smoke detector, a heat detector, or a water flow from a sprinkler. Often there is a remote annunciator panel right inside the front entrance to the building. This mini-panel reflects the information on the main panel with select buttons, including the ACK button. This information identifies the problem and allows the initial IC to formulate the plan of attack as well as gives firefighters notice about what to expect. Firsthand information should also be gathered from occupants evacuating the building; this information may include visible signs of flame or smoke; elevated heat; difficulty breathing with or without smoke; status of evacuation stair door locks (are they all unlocked?); status of other occupants, for example, the specific location of people trapped or having trouble evacuating, known occupants with limited mobility, suspicious behaviors, and so on.

FIGURE 15-10 Automatic fire alarm control panel.

Courtesy of Raul Angulo.

Other building system control panels for elevators and the HVAC system may also be mounted on the wall next to the automatic fire alarm panel in the FCC. **FIGURE 15-11** shows a smoke control panel that illuminates which floor is in alarm, along with a smoke detector alarm for the elevator mechanical room. It also has automatic mode and manual switches for the stairwell pressurization fans and elevator shaft pressurization, damper and pressure relief switches for the shafts, and a smart-lock switch that unlocks all the doors in the stairwells for exit and reentry to any floor.

Determine the fire floor and other floors that show in alarm. The fire floor must be physically verified as soon as possible. *Everything hinges on this information.* This may be accomplished by the maintenance supervisor or security personnel, but unless you're willing to bet on the veracity of their conclusions, it's best to send two firefighters from the first-in ladder company to confirm this information and find out exactly what's on fire. Pressurized smoke rises, and, depending on the status of the HVAC system and stack effect, the smoke may have spread and activated one or multiple smoke detectors several floors above the actual fire. The activated smoke detectors also register on the FACP, and it may seem as though there are numerous fires on upper levels when the actual seat of the fire is several floors below. In an example from Seattle,

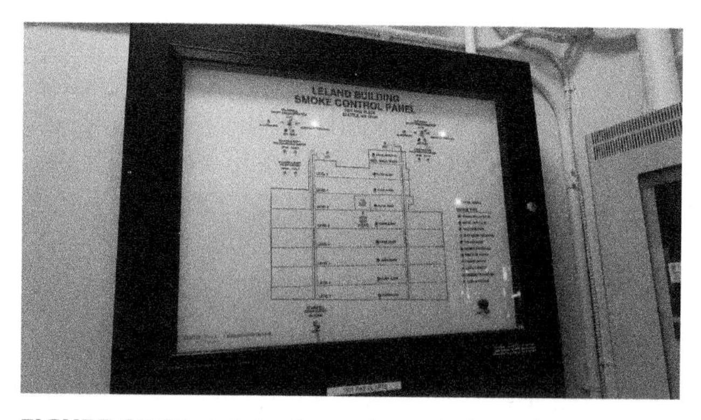

FIGURE 15-11 Automatic smoke control panel.

Courtesy of Raul Angulo.

Washington, a fire occurred on the fourth floor of an eight-story building. When fire units arrived on scene, heavy smoke was showing from the eighth floor. The door to the fire apartment was left open, and the hallway doors to the exit stairs were propped open throughout the building—a common fire code violation. The smoke from the hallway entered the stairway and rose to the eighth floor. The recon team entered the stairway and discovered the source of the smoke was actually from floor 4. Only the smoke from the eighth floor was visible from the exterior, and everyone was convinced the fire was on floor 8. The standpipe was in the opposite stairway. Had a recon team not been assigned to confirm the location, the hose teams would have hiked to floor 7.

It is best to use ladder company members for this initial recon assignment because the initial commitments to high-rise firefighting are to utilize the engine companies to establish a water supply and move crews, hose, and equipment into position within the building to initiate a direct frontal attack and thus confine and extinguish the fire. Along with determining the location of the fire, the recon crew should report heat and smoke conditions in the stairway, the extent of the fire area, and the fire floor itself if possible. All this information should be relayed via radio to the initial IC. Once a fire is confirmed, the IC should immediately call for a second or third alarm. Even with a small fire in a high-rise building, the complexity of the life hazard requires numerous personnel to assist in SAR and control the evacuation in order to avoid panic.

Never silence the building's audible fire alarm horn or siren until the activated alarm is verified. Although it is loud and obnoxious, and it can interfere with fire department communications, it is alerting the occupants of a fire. Once the alarm is determined to be false, the alarm can be silenced. In an actual fire, once the evacuation is complete and the life hazard mitigated, the IC can order the alarm to be silenced to improve fireground communications.

Identifying the Firefighting and Evacuation Stairways

The same recon team should then determine how many stairways access the fire floor, which stairway is closest to the fire, which one has the standpipe, and if hallways and stairways are contaminated by smoke. The recon team should also determine if any portion of the floor plan will require more than 200 feet of hose from the standpipe connection. The stairway with the most advantages should be designated the firefighting stairway. Usually it is the stairway with the standpipe, but that isn't always the case. In unusual circumstances, a stairway without the standpipe may have to be designated as the firefighting stairway because it's closer to the seat of the fire **FIGURE 15-12**. All this information should be related to the IC via radio. Ideally, the stairway closest to the fire means sooner water application. In modern high-rise buildings, there's usually more than one standpipe. The faster water that is applied to the fire, the sooner it can be contained and extinguished, meaning less of a beating for the firefighters and a reduced, if not eliminated, threat to trapped or exposed occupants. Once the fire is out, effective ventilation can begin. Everything gets better when the fire goes out.

The next responsibility for the recon team should be to search and ensure that the firefighting stairway is clear of fleeing occupants above the fire floor. Some fire departments call this team a **rapid ascent team (RAT)**. Any occupants in the firefighting stairway above the fire floor should immediately be directed back into the hallway and toward the evacuation stairway by the RAT. The entire firefighting stairway needs to be cleared, but especially from the fire floor up to the roof.

Usually the stairs opposite the firefighting stairway are designated as the evacuation stairway. In older buildings with a smoke or **fire tower**, it should be the preferred choice for the evacuation stairway. A fire tower is an enclosed stairway connected at each story by an outside balcony or a fireproof vestibule vented

FIGURE 15-12 In unusual circumstances, the stairway without the standpipe may have to be designated as the firefighting stairway because it's closer to the seat of the fire.
Courtesy of Mike Handoga.

to the outside. If there is no fire tower, the evacuation stairway should be pressurized by the building's HVAC system and free of smoke. Once the evacuation stairway is identified, the IC should make it known by announcing it over the building's public address (PA) system to all building occupants and firefighters. Instructions should be specific, utilizing the established designators familiar to the building occupants. The stairways have signage designating them "A," "B," and so on. An example of such a PA announcement is: "All occupants should use the B stairway to exit the building." Avoid using terms like *the north stairwell, the south stairwell,* or *the Bravo stairwell.* Using adjacent street names has also proven effective, for example, "Use the stairwell on the Broadway side of the building."

Evacuation Control

In keeping with the first incident priority of life safety, the initial IC needs to confirm and control the evacuation that is in progress. The post–September 11 human factor needs to be considered to avoid a panic. Sheltering occupants in place or convincing them to stay in their rooms or offices requires clear communications via the building's PA system. Even then, there may be a complete disregard of direction, and occupants who are self-evacuating can be anywhere; stairways could be congested with hundreds of people. The primary emphasis should be to search and evacuate occupants on the fire floor and the floor above the fire. The IC should check with building security or the dispatcher if there are any reports of people trapped or if there are disabled occupants who require assistance in evacuation. People may follow building training protocols to leave the affected floors and reenter the building several floors below the fire floor, or they might go to the roof or down to the street. Managing an evacuation in progress may require the use of several companies and requires the following considerations:

- What is the current status of the evacuation?
- Are the stairwells pressurized?
- Is there smoke in the stairwells?
- Are floor wardens involved in directing people?
- Are the proper floors being evacuated?
- Are there floors that still need to be evacuated?
- Is the entire building being dumped?
- Where are the building occupants being evacuated to?
- Is there a safe gathering area?
- Are occupants exiting the front of the building where they may be subjected to falling glass or burning debris?
- Are there reentry floors in the stairway?
- Can building occupants be moved to a sky lobby?
- Can they utilize a skybridge to an adjacent building?
- Are personnel positioned to control every entrance and exit to the building?
- Have safe evacuation routes been established to move occupants away from the building?
- Are building occupants exiting to extreme or inclement weather?
- Can they be moved to an underground parking garage for shelter?
- Are buses needed to shelter building occupants in severe weather?

An evacuation is not a rescue. It is a fast, fluid event that moves the largest number of able-bodied people as possible in an organized manner to an initial place of refuge.

Gaining Control of the Elevators

Gaining control and accounting for every elevator can require multiple companies. For example, The 76-story Columbia Center in Seattle, Washington stands at 937 feet (286 m). It has 14 elevator zones and 49 elevator cars. With the help of the maintenance supervisor, the status of all elevators needs to be confirmed, including freight, garage, and service elevators. If the automatic fire alarm system did not recall all the elevators to the lobby, it should be done manually. Every elevator should be searched and accounted for, then manually shut down except those that will be used for firefighting operations. Elevators in high-rise buildings should not be used for fires below the fourth floor because not much is gained (remember, staging is two floors below the fire). Valuable time and personnel resources are wasted just to ride up one or two floors. Elevators should not be used at all until it is verified that there is no smoke or water entering the shaft. Water can cause electrical shorts and malfunctions. The HVAC system should be able to pressurize the elevator shafts along with stairways to ensure that smoke is kept out. Elevator shaft pressurization also needs to be verified.

The initial IC should perform a risk-benefit analysis to determine if elevators will be used for the operation. Every mechanical advantage should be used to carry firefighters and equipment to upper floors to ward off fatigue before the fire attack. Firefighters should follow their department standard operating guidelines (SOGs) regarding their use in Phase I and Phase II operations.

Control of the HVAC System

After the standpipes, the HVAC system is a critical high-rise fire system, especially for systems designed for smoke management. The initial IC must rely on the recommendation of the building engineer or refer to the recommended procedures in the prefire plan to shut the system down or leave it running as designed.

Emergency Power Generators

A significant fire can affect the electrical power to the building. It must be determined if the building has an emergency power generator and whether it starts automatically or needs to be started manually. At the One Meridian Plaza Fire in Philadelphia, Pennsylvania, all electrical power was lost, which rendered the stairway pressurization fans, elevators, and fire pumps inoperable. The battery-powered emergency lighting failed quickly, and all firefighting operations were hampered by darkness within the hallways and stairwells. A large supply of batteries for flashlights is needed for extended operations in darkness. The running time for generators may be limited due to fuel supply, which should be monitored to ensure that fuel levels are adequate or if additional fuel needs to be brought in. Portable generators and portable lighting should also be brought to the staging area.

Access Keys

The FCC has access master keys, key cards, or fobs to gain access to floors, offices, utility rooms, mechanical rooms, electrical rooms, and a variety of occupancies within the building. When available, they should be distributed to teams and utilized. This saves time by providing quick and easy access without causing damage from traditional forcible entry tools and techniques. In residential high-rise buildings, they may not have master keys to individual private units, in which case, forcible entry must be used to gain access.

Communications

Sound-powered handsets are often available for distribution to specific crews in order to establish in-house communications. These phones can be used in the elevator cars and on every stairway landing. The phone is plugged into the wall jack and connects directly to the FCC. This allows division or group supervisors to communicate directly with the IC as well as with other division supervisors floor to floor. The benefit of the phones is that they are not subject to radio interference or dead zones within a steel Type I building. Typically, however, only a few handsets can operate simultaneously, and the handset wiring can deteriorate over time or be subjected to damage during a fire. At the First Interstate Fire in Los Angeles, California, these handsets had a lot of static and kept cutting out. In the end, they proved ineffective, and runners had to be utilized to ensure that messages were received without confusion.

Radio checks or status reports should be given periodically to ensure that messages are transmitted and received. Portable radio signals through repeaters are often lost in Type I construction. It may be necessary for interior crews to use a simplex (radio-to-radio) channel. Throughout the incident, company officers, divisions, and group supervisors must be mindful about moving toward a window to get a better signal if they cannot transmit messages over their portable radios. At the First Interstate Fire, one of the battalion chiefs in charge of the fire attack had to break out a window in order to achieve line-of-sight communications with the command post below. Cell phones are another way for division supervisors to communicate with the command post in steel high-rise buildings.

The Fire Pump

If the building has a fire pump, an indicator light for it will be on the fire alarm panel. If illuminated, it should be running. The fire pumps are usually found in the basement, but there could be a midlevel pump house to supply the upper floors of super-high-rise buildings. They can be electrically hardwired or powered by a diesel motor. Some have both, and the diesel motor serves as the backup pump in case of a power outage. Two firefighters should be assigned to ensure that the pump is operating properly. There is a discharge pressure gauge on the pump, and the pressure should be reported to the IC to note the capability of the pump. If low-pressure problems occur, the pump may need to be shut off so that the standpipe and sprinklers can be supplied by fire department pumpers. If a maintenance person knowledgeable about the system is available, that person should accompany the firefighters. Just because a high-rise building has a fire pump doesn't mean it is regularly tested or has been properly maintained. For example, if firefighters arrive and see flames from numerous windows or even from an entire floor of a fully sprinklered high-rise building, it is an indication that the fire pump is not supplying the sprinkler system properly or that the water tank level is low and shouldn't be relied on during the fire attack. In this case, the standpipe must be supplied exclusively by fire department pumpers.

Fire pumps for wet standpipe and automatic sprinkler systems may be supplied by gravity tanks, public water mains, pressurized reservoir tanks, or private water supplies. In addition to the supply flow requirements, tanks are also required to maintain the flow for a given period of time; for example, a tank may need to supply a fire attack for 120 minutes. These time periods should be noted during prefire inspections and included in the prefire plan document. Fire pumps are typically rated at 750 gpm.

The reservoir tank at the First Interstate Fire held 85,000 gallons (321,760 L) of water and was replenished by the city water main. This particular tank did not have the intake valves to be replenished by fire department pumpers. It was estimated that 4,000 gpm (15,142 L/m) were supplied to the risers, and 2,500 gpm (9,464 L/m) were flowing through the 20 fire attack hose lines. The city water main could not keep up with the flow demand and drained the tank to less than one-third of its capacity. Had the tank emptied out, fire department pumpers would have had to supply the entire system.

During a fire, there can be only one source of water supplying a standpipe system: either the building's fire pump or the fire department engine. Whichever one has the highest pressure becomes the source. The fire pump in a wet system draws water from the tank reservoir until a fire department engine increases its pressure beyond that of the fire pump. When this occurs, the building system's check valves close to keep the water from flowing back through the fire pump, and the engine apparatus supplies the standpipe system. If the engine pressure is reduced or insufficient, then it is pumping against a check valve held closed by the higher pressure from the fire pump, which again reengages and supplies the standpipe. Super-high-rise buildings may have zones requiring up to 500 psi for the systems. These buildings require fire pumps to obtain operating flows at these elevations because regular fire hose and engine apparatus cannot sustain these high pressures. Tall buildings can have more than one fire pump set to operate in tandem series. Initially, the driver should pump to the signage pressure (Figure 15-5). Some major cities have specialized high-pressure fire engines equipped with 3-stage pumps and high-pressure supply hose. These pumpers can double the water pressure of standard engines to 1,000 psi, surpassing the building's fire pump systems and eliminating the need for tandem pumping operations.

Directing Crews and the Initial Action Plan

A fire that requires the use of a standpipe system for water supply is most likely on a floor higher than the fourth story and is therefore some distance from the fire apparatus. Before they can attack the fire, firefighters have to carry all their equipment into the building. Often, the front entrance is fine, but certain fires, where glass and burning debris may be falling all around, may require the use of underground parking lots, tunnels, or loading docks to access the lobby of the building safely. These alternate access routes should be identified on the prefire plan.

The initial IC needs to decide where to place the first hose line. He or she should brief and direct the officers of the incoming crews to the firefighting stairway or to the up elevator to transport crews to staging, two floors below the fire floor. Many high-rise fires are beyond the reach of effective ground and exterior operations, so the initial strategy must start with a direct frontal attack on the fire floor from the stairway. The primary efforts are to advance hose lines to get water on the fire and to conduct search-evacuate-rescue (SER). Forcible entry in high-rise offices and residential units shouldn't require specialized tools. If aerial ladders and ground ladders are not used, engine company personnel may find themselves assigned to tasks that are normally the responsibility of the ladder company, and they are certainly capable of completing them. If smoke and heat makes the fire floor and the floors above untenable, ladder company members may find themselves assisting and supporting hose operations until the fire is knocked down and heat temperatures are reduced before they can resume SER and ventilation.

That being said, the priorities for the engine companies are to:

1. Get the initial hose line onto the fire floor and in operation. This may take two companies.
2. Stretch a backup line of equal or greater diameter to support the initial attack. This may take two companies. The backup line supports the position of the initial attack line, helps contain the fire from spreading, and protects the SER on the fire floor. The backup line can serve as the initial RIT as long as two members (minimum) remain outside the hazard area in the stairway prepared to provide assistance or rescue in case of an emergency.
3. Get an exposure line to the floor above the fire. This may take two companies.
4. Ready a charged hose line close to the fire floor for the RIT, and assign an entire company dedicated to rapid intervention.

The priorities for the ladder companies are to:

- Oversee the evacuation and initiate the primary SAR on the fire floor. Remove occupants to the evacuation stairway.

- Send a second team to assist in the primary search, evacuation, and rescue of the fire floor. Remove occupants to the evacuation stairway.
- Initiate the primary search, evacuation, and rescue of the floor above the fire. Remove occupants to the evacuation stairway.
- Support the pressurization of the firefighting stairway with positive-pressure ventilation (PPV) fans.
- Ensure that the rooftop access door or hatch remains closed during the fire attack to maintain stairway pressurization.
- Determine the operation status and effectiveness of the HVAC system and shut it down if needed.
- Search the attack stairway and the upper floors above the fire, notably the top of stairwells, if smoke is rising. The evacuation and the firefighting stairways should be checked throughout the fire incident.

Every fire is different, and the above sequences may be changed, but they all need to be addressed.

Engine Apparatus/Pumper Positioning

First-In Engine

While approaching the location, the driver can gather information by what he or she can visually observe from the cab. Newer fire codes require a hydrant within 100 feet (30.5 m) of an FDC. Otherwise, a hydrant should be within 300 feet (91.4 m) of an FDC. The best arrangement for supplying a standpipe system is to have a pumper positioned within 100 feet of the hydrant and 100 feet of the standpipe FDC so that only one or two sections need to be pulled from the hose bed. Positioning the first-in engine in front of the building keeps the supply lines to the standpipe system close to the apparatus and reduces friction loss, and avoids pumping through several hundred feet of hose **FIGURE 15-13**. Because of the amount of time it will take to get to the fire floor with all their equipment, firefighters do not have to assist in laying a supply to the standpipe connection. There is plenty of time for the driver to handle this alone. For safety, on side-mount pump panels, the hoses should be connected to the opposite side of the pump. If flames are visible and it appears that autoexposure is imminent, connecting to an FDC on the side below the fire may prove to be extremely dangerous for the pump operator and the exposed FDC supply hoses, which could be subject to falling burning debris and large shards of glass from broken windows **FIGURE 15-14**. In this case, the first-in engine may have to lay reverse to the next hydrant down the street. If there is another FDC on the opposite side of the building and the standpipes are interconnected, spotting the apparatus at that position would be safer. If the first-in engine is already hooked up to the FDC, the second- or third-in engine

FIGURE 15-13 Positioning the first-in pumper within 100 feet of the hydrant and 100 feet of the standpipe FDC reduces friction loss.

Photo by John Odegard.

FIGURE 15-14 If flames are visible and it appears that autoexposure is imminent, connecting to an FDC on the side below the fire may prove to be extremely dangerous.

Courtesy of Rick McClure (LAFD ret.).

should connect to the opposite FDC to provide a second, separate supply to the standpipe with a system of redundancy.

Signs should be provided to indicate whether a connection serves a standpipe system, a sprinkler system, or both. The Siamese connection or the FDC and the signage should be visible and unobstructed. In addition, the signage should indicate the area being serviced by the connection, and many have the required pressure that the driver should pump into the system. Pump to the posted pressure if the apparatus pump will supply the standpipe. Lacking this signage, pump to the outlet pressure of the building's fire pump (labeled at the pump or on the prefire plan). Pump to rooftop pressure by adding 5 psi per floor for friction loss and adding 25 psi for the standpipe Siamese friction loss. If the building is equipped with a fire pump, let the pump do its job. Pump 50 to 100 psi below the signage pressure.

Here are two helpful tricks for the driver to know if the building's fire pump is operating. First, put an ear against the FDC or a riser; the driver will be able to hear and feel the vibration of the fire pump when it is operating. The other tip is to close the supply line to the FDC. If the building's fire pump is running, the supply from the apparatus to the FDC should be static, and closing the discharge ball valve from the apparatus should be effortless. If there is noise and resistance closing the ball valve, that means water is flowing from the apparatus into the standpipe system and the fire pump has failed or is not operating properly. Another indication that water is flowing into the standpipe system is that the driver can see a drop on the discharge gauge. This means the standpipe must be supplied by the fire apparatus pump; hydraulic calculations for pressure and friction loss need to be adjusted for the standpipe, the elevation, and the hose lays.

The Siamese intakes have female swivel couplings that should spin freely; however, they are often painted over or rusted shut. They should also have a plug or a frangible disc cover to prevent trash and debris from entering the intake. Before hooking it up to the standpipe system, you should check the FDC intake for trash, rocks, and other debris **FIGURE 15-15**. People sometimes use these open Siamese appliances as trash receptacles, and plastic debris and small rocks can be pushed into the system, flow all the way to the nozzle, and clog the water stream. This is one reason why many firefighters prefer smooth-bore nozzles: small pieces of debris flow right through.

Many fire departments, including LAFD, have special high-pressure hose bundles to supply high-rise standpipes. At least two high-pressure supply lines should be connected into the Siamese intake. If more inlets are available, each inlet should receive a high-pressure supply line. The number of inlets indicates the flow available from the standpipe. The first line should be hooked to the left intake and charged to get water into the system quickly. The second line should then be connected to the right intake. This sequence was originally established because connecting the right side first meant that the right hose line would get in the way, thus slowing down use of a spanner wrench to tighten the left female swivel.

FIGURE 15-15 Before hooking up to the standpipe system, you should check the FDC intake for trash, rocks, and other debris.

Courtesy of Raul Angulo.

When the fire is severe, two fire department Siamese appliances can be hooked up to the standpipe FDC. This allows four lines to supply the system **FIGURE 15-16**. Many fire departments mandate that a single 3- or 4-inch large-diameter hose (LDH) connection be used instead of multiple 2½-inch hose connections. The goal is to get as much flow into the standpipe as possible; however, the LDH needs to be in high-pressure sections to withstand higher pump pressures, and a single supply line to an FDC could result in catastrophic loss of pressure if that single line fails or is cut by falling glass.

Once the FDC is hooked up, the driver should make all the connections to the hydrant by using as many ports and pump intakes as possible to maximize the intake water flow. The driver should announce "supply established" over the radio. The supply hose to the FDC should be protected from falling glass and debris with the use of ground ladders, covered with tarps and backboards, or by any means necessary. Traffic cones and fire scene tape should be used to mark off a safety area for the supply hoses into the FDC in case higher pressures rupture the hoses. An additional area, anywhere from 50 to 100 feet, should be cordoned off to create a safety zone for possible falling glass and debris.

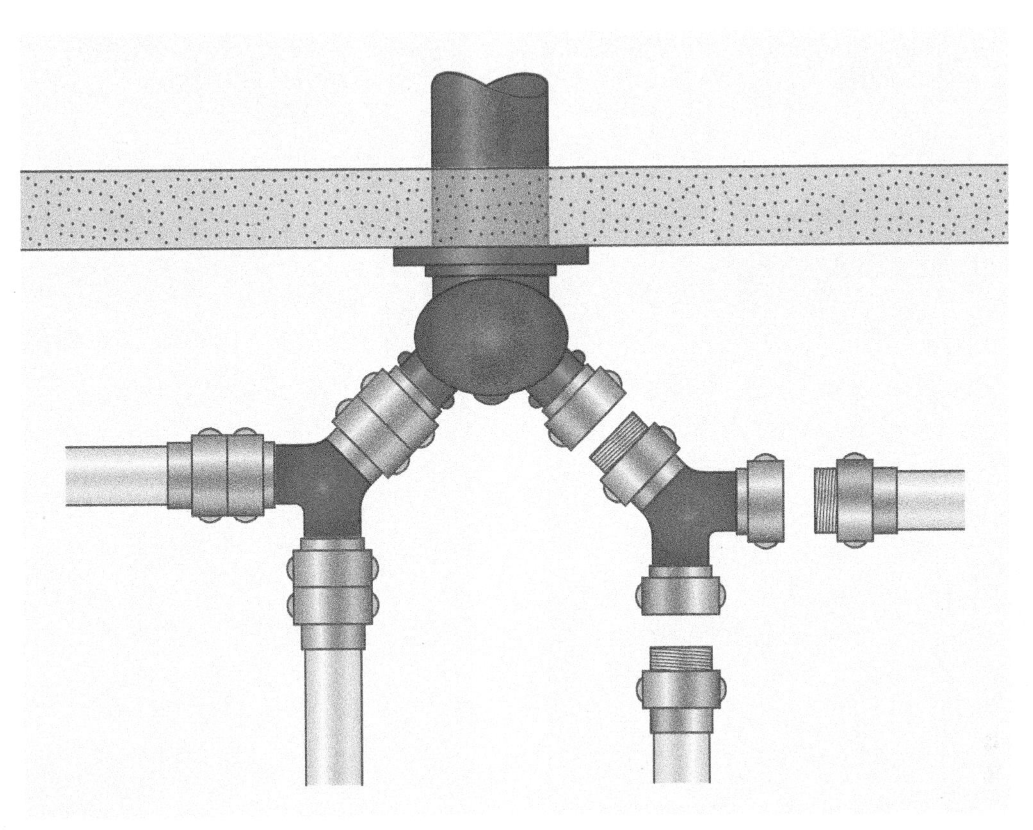

FIGURE 15-16 For a large fire, two fire department Siamese appliances can be hooked up to the standpipe FDC. This allows four high-pressure lines to supply the system.

At the First Interstate Fire, glass and debris covered a 100-foot area around the base of the building. LAFD tests have shown glass can "sail" roughly 200 feet out from a high-rise building.

Because the driver is on the exterior of the building for the duration, he or she is also in an advantageous position to see at least two sides of the building. Exterior fire and smoke conditions, persons in distress, or any unusual conditions should be reported immediately to the IC. The driver should consider personal safety and inform the IC when and where debris is falling. He or she can take refuge in the cab, or a fire department ambulance can be parked and staged next to the fire engine so the driver can take refuge inside while monitoring the apparatus pump.

Second- and Third-In Engines

Depending on the situation, the second-in engine can charge the standpipe on the opposite side of the building from a position outside the area where debris might fall. This assignment could also be given to the third-in engine. All the procedures listed above should be followed. An interconnected standpipe FDC should be supplied by a second pumper on a separate hydrant source and preferably a second water main. Whether there is a second FDC or just a single one, the forward first-in engine should also be supported by a tandem engine spotted at the next hydrant down the street **FIGURE 15-17**. A single or double (2) LDH supply should be laid to the first engine. Every port off the hydrant and intake into the second pumper should be utilized to maximize water flow. Setting up this tandem pumping configuration also provides a redundant water supply should any mechanical malfunction occur with the first-in engine. If falling glass or burning debris begins to rain down on the first-in engine, the pump can be left running and the driver can retreat to an area of refuge at the second engine. This in essence turns the forward engine into an unattended manifold. Any fire engines pumping into an FDC should be supported by an additional fire engine pumping in tandem series from another hydrant into that engine.

Water demands may be great throughout the incident, and a myriad of problems may develop. At some point, an engine officer should be designated as a water control officer and coordinate with a representative from the water department to ensure that municipal water main supplies are boosted.

Damaged or Blocked FDCs

On older buildings, FDCs may not be well maintained, and they could be damaged. In high-density

FIGURE 15-17 The forward first-in engine should be supported by a tandem series engine spotted at the next hydrant down the street.
Courtesy of Rick McClure (LAFD ret.).

downtown areas and alleys, the Siamese FDC can also be blocked by cars or delivery trucks. A solution to charge the standpipe when a supply line cannot be connected to the FDC intake is to lay a 2½-inch supply line into the building to the first floor stairway entrance, and connect the hose to the standpipe outlet valve using a 2½-inch double-female coupling on the first floor landing **FIGURE 15-18**. Whenever possible, a 2½-inch Siamese should be installed on the discharge valve using a double-female coupling so that two 2½-inch supply lines can be used to deliver water into the discharge valve outlet from the pumper. Before using LDH to charge the standpipe, ensure that it can withstand higher pumping pressures. For supplying a wet system, the supply line to the standpipe discharge valve should be hooked up and charged before the discharge valve on floor 1 is opened. This prevents water in the standpipe from running back into the supply line and possibly impeding the operation.

If a pressure reducing valve is attached to the outlet, this method will not be possible. PRVs are one-way

FIGURE 15-18 A solution to charging the standpipe when the FDC is blocked or damaged is to connect a 2½-inch supply line using a 2½-inch double-female coupling to the discharge outlet on the first floor.
A: Photo by Capt. Steve Baer; **B:** Photo by Capt. Steve Baer.

discharge valves; you cannot backflow water into a standpipe through a PRV. You will have to lay a supply line up the stairs—a labor-intensive operation consisting of two or three companies to establish a supply in the stairway to the lowest floor without a PRV or to the floor below the fire.

As a last resort, an aerial ladder or the waterway of a tower ladder can be used to create a temporary standpipe or to advance a hose line to upper floors or even the roof in low-rise buildings. However, this tactic should not become a preferred method or used in lieu of a perfectly functional standpipe. Aerial ladders and tower ladders are a limited specialized resource. Some departments have only one or two, and some departments don't have any. Using a million-dollar piece of apparatus to establish a temporary standpipe commits the ladder to supporting the attack and renders it useless for aerial rescues and elevated master streams—the designed purpose of this fire truck.

Stairwell Support Group

Fire department SOPs vary on the initial use of elevators for high-rise firefighting. If a department does not use elevators, a stairway support group needs to be set up immediately for the transfer of SCBA air cylinders and equipment to the staging floor. Next to an adequate water supply, the demand for air is paramount for firefighters. The need to stock the staging floor with fresh air cylinders as soon as possible cannot be overemphasized. The physical demands on firefighters limits them to approximately 20 minutes before they need a fresh bottle. Extra alarms need to be called for this task alone. The company officer assigned to stairway support must figure out the most effective relay system to move extra air cylinders from parked apparatus to staging. At the First Interstate Fire, over 600 air cylinders were hand-carried to the tenth floor.

In the initial stages, one firefighter can shuttle air cylinders and equipment two floors, but fatigue quickly becomes a factor if the stairway is hot and/or smoky; the summer stack effect can push smoke down to lower floors. Ideally, the relay should be one firefighter per floor. It also depends on what floor the staging is on. Having five companies or more assigned to stairwell support is not unreasonable. Stairwell support firefighters also need to be rotated to catch their breath. If the stairway is pressurized and clear of smoke, the group supervisor and the safety officer can make the decision to allow firefighters to remove SCBA, helmets, and coats during this operation. They can keep their PPE on their assigned floor landing at the ready. Ladder companies usually set up PPV fans at the base of the stairwell to assist in maintaining stair-shaft pressurization, which is also helpful in keeping the stairway support group cool.

In addition to shuttling air cylinders, the stairwell support group may be asked to establish a water supply or a secondary standpipe in the stairway using fire hose and shuttle equipment, and provide drinking water, flashlight batteries, medical supplies, and anything else that is needed in staging. At the DeWitt-Chestnut Fire in Chicago, Illinois, staging was on floor 34. At the One Meridian Plaza Fire in Philadelphia, Pennsylvania, staging was on floor 20. Due to the PRVs and complete failure of the fire pumps, the Philadelphia firefighters had to lay a 5-inch supply line up to the twenty-second floor, which took the stairwell support group about an hour to establish. Manually establishing a water supply in a stairwell is a physically taxing endeavor. To ensure enough hose is laid, it is best to carry rolled hose sections to the floor below the fire and unroll the hose down the stairwell. Using gravity to unroll hose is the fastest way.

Use of Elevators

The Empire State Building in New York City has 102 floors. The Willis Tower in Chicago, Illinois, has 110 floors. The round iconic US Bank skyscraper in Los Angeles, California, has 73 floors, and many high-rise buildings range between 40 to 60 stories. The use of elevators is essential in high-rise firefighting in order to get water to the fire as soon as possible. Even with high-speed elevators traveling at 1,200 feet/minute (366 meters/minute), getting water to a fire on the upper floors can take at least 15 to 20 minutes—if everything goes right. Consider how long it would take and how much physical energy firefighters would have left if they had to hike all the way up the stairs. We need to use the mechanical advantage of elevators to move firefighters and equipment to the staging floor whenever possible to prevent fatiguing firefighters before the fire attack even begins. Fire departments must have written SOGs that address the use of elevators during a high-rise incident. The main concerns are smoke in the hoistways or elevator shafts and water entering the hoistway; these can lead to electrical shorts and mechanical failures. These large vertical shafts are a primary route for heat and smoke spread. In newer buildings, pressurization fans are designed to pressurize the hoistways and thus keep them clear of smoke for firefighting operations, but not all HVAC systems pressurize the elevator shafts. This needs to be identified during the prefire review. In older buildings without hoistway pressurization, vertical ventilation of the hoistway at rooftop level can reduce the accumulation of smoke and heat on upper floors.

Elevators should not be used for fires below the fourth floor. Staging is set up two floors below the fire, and not much is gained by riding an elevator one or two floors. And the company doesn't lose a member to run the elevator car when all firefighters can use the stairs. The initial IC should consider the logic in resisting the use of elevators below the eighth floor but using elevators to the thirty-eighth floor. The benefits are obvious, and the accepted risk is the same, so why not use the mechanical advantage and save the strength of your firefighters for the battle? They're going to need it. It is interesting to note that in the NFPA high-rise study, the majority of high-rise fires in all the property classes started on floors no higher than the sixth floor, or below grade. Elevators should also not be used for fires on lower floors below the main entrance of a building for the same reason you shouldn't hook up to below-grade standpipe discharge connections. The fire is already below you. If smoke enters the elevator shaft, it can trap firefighters by filling the car with smoke.

There are two basic types of elevators: electric traction and hydraulic. Hydraulic elevators are usually found in low-rise or non-high-rise buildings. In this instance, the elevator mechanical room is in the basement. Electric traction elevators are used in actual high-rise buildings. These elevators use the giant hoist cables wrapped around drums along with cables attached to counterweights. The hoist motors and mechanical room are usually on the roof or the top floor of the building, although in modern high-rise buildings, elevator mechanical rooms can be located almost anywhere adjacent to the hoistway. There may be a low-rise bank elevator serving the lower floors; a mid-rise bank that serves only the middle floors; a high-rise bank, including express elevators to the uppermost floors, that serves only the upper floors; and freight elevators that typically serve nearly all floors of the skyscraper.

In high-rise fire situations, the elevators and stairways are managed by the company assigned to lobby control. With multiple elevators, there should be one designated as the up elevator and another as the down elevator. Between runs, the up elevator returns to the lobby, and the down elevator returns to the staging floor. The rest of the elevators should be shut down unless they are needed by the fire department. All members must understand and follow the elevator protocol for Phase I and Phase II operations. The IC must take special precautions and conduct a risk-benefit analysis before allowing firefighters to use a freight elevator in a fire situation because some freight elevators have manually operated doors and do not have Phase I or Phase II safety features. A firefighter would have to know how to operate a freight elevator manually. If they do have Phase II features, however, freight elevators can carry more weight than passenger elevators, and they can be beneficial in quickly transporting SCBA cylinders and equipment to the staging floor. Many freight elevators do not have an enclosed car design, and the elevating platform and guardrails are open to the hoistway, which is most likely not pressurized. The firefighter assigned to operating this type of elevator must check the shaft constantly for the presence of smoke.

Phase I elevator operation is tied to the building automatic fire alarm (AFA) system and occurs automatically if the alarm is set off by a smoke detector or pull station in elevator lobbies, shafts, elevator machinery rooms, or other system-designated areas. These elevator cars are automatically recalled to the main lobby. If they are not recalled, all elevators should be manually recalled, accounted for, searched, then shut off. The swift movement of the elevator car within the hoistway moves considerable air within the shaft, often bringing an odor of smoke when the car reaches the lobby. This may be your first physical indication that you actually have a fire. Phase I disables all floor buttons in the car and all call buttons on every floor so that the elevators cannot be used by any occupants. People who are in the elevator at the time of the alarm are carried down to the main lobby. If the alarm activation occurs in the main lobby, the elevators are recalled to an alternate designated floor, usually below the main lobby. This alternate floor recall is sometimes referred to as Phase III. The doors to the elevator car remain open on the ground floor until Phase II is initiated or the alarm is cleared, and the elevators are reset by the fire department to normal operations. Closing doors on unused elevators on the ground floor ensures optimization of shaft pressurization.

Phase II elevator operation occurs when the fire department actually takes control of the elevator car. The 3502 fire service key or barrel key should open the red elevator key box next to the floor 1 or main lobby elevator. Master access keys, keycards, or fobs may be kept in a Knox-Box next to the elevator or at the FCC of the building. The firefighter assigned to run the elevator should be equipped with full PPE, portable radio, fire phone to the FCC, and SCBA. In addition, this firefighter should keep a set of irons or at least a Halligan tool to force the car doors open in case of a power outage or emergency, a high-powered battle lantern to check the elevator shaft for smoke, a water-pressurized fire extinguisher, a spare SCBA cylinder, a hand-held multi-gas detector, and a baby ladder to reach the emergency hatch on the roof of the elevator car to escape if it stalls between floors.

Operating in Phase II

The firefighter assigned to run the elevator car in Phase II should check to make sure that all the command functions work with the appropriate buttons. In newer buildings, the elevator electronic eye or sensor should be disabled automatically so that unexpected smoke does not block or prevent the doors from closing. Opening and closing the car door works only with the manual buttons operated by the firefighter. A test run to the floor above the lobby should verify that Phase II is operating properly. The operator should note the maximum weight limitation of the car and avoid overcrowding the car with firefighters and equipment. Extra trips to carry firefighters and equipment to the staging floor is quicker than dealing with an elevator that stalls because the weight limit has been exceeded. Not only could you have an entire company stuck in an elevator, but valuable time and resources will have to be diverted from firefighting to deal with this embarrassing problem.

If time allows, the operator should stop the elevator a few floors below staging to point out the layout of the hallway or the floor, as well as the exits to additional stairways. This should take less than a minute and may be the only time crews get the chance to get their bearings. Some elevator operators may draw a floor plan on the interior of the elevator door as a reference to point out stairwells, fire locations, or other important features. Firefighters can expect to be ordered off the elevator two floors below the fire, then use the stairway from the floor below to hook up hose and gain entry to the fire floor.

It is extremely important that the elevator operator know which floor the fire alarm is on because he or she is the one manually operating the elevator car. Any confusion can have disastrous consequences because this firefighter could unintentionally take the car directly to the fire floor. An elevator should never be taken to the fire floor or any floor above the fire, and fire department personnel should not be allowed to break this rule. If the elevator is taken to the fire floor, the firefighters in the elevator could be exposed to flames, excessive heat, and/or toxic smoke when the door opens. This type of event occurred on February 8, 1978, at 04:36 hours at the Grosvenor House Fire in Seattle, Washington, when three firefighters took the elevator to the sixth floor of this 18-story residential high-rise. The fire began when an intoxicated occupant, who was heating some baked beans, fell asleep. When the alarm sounded, he made a feeble attempt to extinguish the fire in the kitchen, then left, leaving the front door to his apartment open. As the occupants evacuated, he met the fire department on the ground floor and stated to the battalion chief that the fire was in his unit but there was "no emergency . . . all is well." Dropping their guard, the three firefighters took the elevator to floor 6. When the doors opened, the thick smoke was down to the floor. In 1978, the existing fire code did not have language to deactivate the electronic eye while the building was in alarm. The thick smoke interrupted the sensor, keeping the doors from closing. All three firefighters, who were not wearing SCBA, quickly became disoriented and succumbed to the effects of carbon monoxide (CO). One firefighter dragged his partner and sought refuge in an unlocked apartment. The third firefighter, who was unconscious in the hallway needed to be dragged out by a rescue team. All three survived but suffered severe smoke inhalation.

The operator should check the shaft periodically for any smoke. The movement of the elevator car up and down the shaft can push around or draw in residual smoke. If smoke is detected or is noticeably increasing, stop the elevator at the nearest floor and get everyone out. Turn the elevator back to Phase I, which returns the elevator to the lobby. Note that the elevator shaft can allow CO or other nonvisible gases to drop down the hoistway. There have been cases when the elevator operator and lobby control personnel received significant doses of CO in or near the elevator shaft.

Firefighters should become completely familiar with the elevators in buildings to which they typically respond. New elevator security features that require fobs or security keycards, in addition to the elevator key, are being installed in high-rise buildings. The fire code is changing, making Phase I and Phase II elevator operations a little more complicated. In fact, in buildings with different elevator banks, not all cars return to the lobby in Phase I but may go to a different floor that is designated a safe refuge. Some buildings with more than one elevator in the same shaft create a challenge because the uppermost elevator must be recalled to the lobby and the lower elevator may not be able to access the uppermost floors. These features should be identified within the prefire plan. Company officers need to stay up to date on these building industry and fire code changes and understand that almost no new code requirements are retroactive to existing elevators.

Fire service access elevators (FSAEs) is a term for newly approved, "hardened," post–September 11 elevators designed for firefighter use, occupant evacuation, and high-rise firefighting and will become more common as new taller buildings are constructed. The cars and hoistways are built with reinforced construction that is not subject to the normal problems associated with regular elevator operations during a fire. The concrete shafts are pressurized, and the electrical operating components are insulated from smoke, heat, and water damage. These elevators have remote heat and smoke monitoring for all floor lobbies.

Many hospitals have specialized code blue elevators to move patients quickly between floors, much like a medical Phase II elevator operation. Whether these elevators are recalled or work independently of the automatic fire alarm system, or if the fire department has the ability to override a code blue elevator needs to be determined during prefire planning. An occupant evacuation operation (OEO) is a newer requirement in building codes that specifically provides ongoing elevator operations during an emergency for disabled occupants who cannot use stairs.

Expect the unexpected because sometimes you get a curve ball. An AFA came in for a new, partially occupied, residential high-rise building in downtown Seattle, Washington. The engine officer, who took lobby control, verified the alarm was on the rooftop (RT) level of this 40-story high-rise. However, none of the express elevators capable of traveling above floor 20 would return to the lobby; only the cars for floors 1 to 20 were operating. After troubleshooting all his options, he was getting ready to pull the trigger on the decision for his company to start climbing the 21 floors from 20 to 41 when the building manager showed up and explained that technicians were working on the RT display vent system. The manager had cell phone contact with the workers to verify that there was no fire and that the AFA system was probably set off by dust. The officer accepted the explanation but still needed a firefighter to verify it; however, the explanation provided by the building manager would not have solved his dilemma if there actually had been a fire on RT or a fire on another floor between 22 and 40 because the building was partially occupied. The building manager inadvertently offered that he turned the express elevators to off at night because he didn't want people going to the upper floors for security reasons. In this example, turning the cars to off disabled the Phase I and Phase II mode **FIGURE 15-19**. This situation had to be followed up with the involvement of the fire marshal. The point is that one would expect, when a new building goes into alarm, everything should work according to design, and the company officer should be able to carry out his or her responsibilities according to department policy. The takeaway is that not all civilian authorities are on the same page as the fire department. The actions of individuals can seriously interfere with or stall fire department operations.

FIGURE 15-19 This elevator serving the upper floors of a high-rise building was routinely shut off at night for security reasons, which disabled the Phase I and Phase II features in the occupied building.

Photo by Captain Steve Baer.

An elevator service technician from the elevator company should be summoned to respond to the command post during a major high-rise fire, but this person should also be consulted for regular hands-on training for accurate information and procedures on the proper operation of an elevator's system during a fire incident. Phase II operations can be utilized during medical calls. This legitimate use of Phase II develops confidence for use during a fire incident.

The Staging Floor

Typically, the staging floor, or staging, is set up two floors below the fire. If the fire is on floor 20, staging is on floor 18. All personnel coming up the stairway or from the elevator report first to staging for assignments. This is the forward gathering place for firefighters and equipment. All firefighting operations are launched from this area. Crews are also rotated and rehabbed on the staging floor. Experience has shown that the demands placed on the staging officer are great. A single engine company can be quickly overwhelmed, and it is recommended that two companies, if not three, be assigned to the staging floor—another reason to call quickly for additional alarms at a confirmed high-rise fire.

Hose, Nozzles, and Equipment

The first hose line into the building should be at least 200 feet long and of a diameter dictated by the size and intensity of the fire. The hose line should be able to reach any portion of the fire floor. Determine during pre-incident planning if longer hose lays are required. Only equipment that is absolutely necessary for the initial attack should be brought into the building by the first firefighters. There are various high-rise hose packs and hose carries for predetermined lengths and diameters of hose. Nozzle types and sizes vary. Hose can be shoulder-loaded (flaked), double-rolled, or donut-rolled. The most common are the pre-bundled, pre-tied, 50-foot sections with smooth-bore nozzles that are draped in a horseshoe bend over the SCBA cylinder of the firefighter.

Standpipe Bags

The following is a list of recommended equipment that should be placed in a standpipe bag:

- 2½-inch in-line pressure gauge
- 2½-inch 60° elbow (2)
- Spanner wrench (2)
- 18-inch pipe wrench
- Vice grips
- Wire brush
- Spare operating handwheel and handwheel wrench
- Reducers or increasers
- Webbing

The spare handwheel and wrench is used to open the standpipe valve if the handle is missing or if the valve cannot be opened by hand. Some departments use a 10- to 12-foot pigtail section of 2½- or 3-inch hose. FIGURE 15-20 shows an assortment of standpipe equipment and firefighters carrying the horseshoe hose bundles. Each department should select the tools that work best for its members. All hose, tools, and appliances should be lightweight.

Hose Diameters

Hose lines that are 1¾ inches are effective on most apartment high-rise fires with the typical household furnishings because the room size is compartmentalized by internal walls. However, a 2½-inch hose should also be available for backup in case the smaller hose lines are inadequate for controlling the fire or if the fire extends into the hallway. As in all firefighting operations, if the 1¾-inch hand line isn't knocking down and controlling the fire within 60 seconds, the 2½-inch backup hose should be used immediately. Two-and-half-inch hand lines should be anticipated for fighting fires in open-space office buildings and commercial skyscrapers, and in high-rise buildings that are not sprinklered. Many fire departments are opting for 2-inch attack hose. This is a happy medium, especially for departments with limited staffing. It flows more water than the 1¾-inch hose line but isn't as hard to handle as a 2½-inch hose. A 2-inch hose with a 15/16-inch smooth-bore tip operating at 75 psi yields

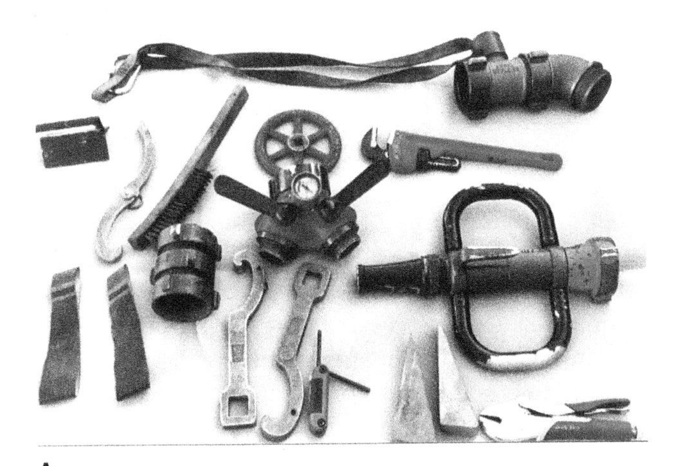
A

B

FIGURE 15-20 A. An assortment of standpipe equipment. B. Pre-bundled, 50-foot sections of hose with smooth-bore nozzles are draped in a horseshoe bend over the SCBA cylinders of firefighters.

A & B: Coutresy of Bernard J. Klaene.

approximately 230 gpm. A 2-inch hose with a smooth-bore 11⁄16-inch tip flows 212 gpm at 40 psi, 237 gpm at 50 psi, and 260 gpm at 60 psi.

Lightweight High-Rise Fire Hose

The North American Fire Hose Corporation has manufactured a new lightweight hose specially designed for high-rise firefighting. The Dura-Flow 800™ hose series has a thermoplastic, urethane inner core lining covered with an inner and outer woven Nylon 6-6 jacket. This advancement in fire hose technology reduces friction loss, which improves nozzle performance, and makes this hose lighter than regular structural firefighting hose. Reduced coil diameter makes the hose more flexible, so it can be packed in tighter folds. It also takes up less compartment space on the apparatus and is easier to carry. The woven Nylon 6-6 double jacket is abrasion-, cut-, and kink-resistant, and gives the hose its strength to maintain a 400-psi service test pressure. The nylon makes the charged hose slide more smoothly around bends, corners, and stair treads, so it is easier to advance **FIGURE 15-21**. Note the following comparisons:

- A 50-foot (15.2-m) length of 1¾-inch Dura-Flow 800™ hose weighs 14 pounds (6.4 kg) compared to its rubber-lined counterpart, which weighs 17 pounds (7.8 kg).
- A 50-foot section of 2-inch Dura-Flow 800™ hose weighs 17 pounds (7.8 kg) compared to its rubber-lined counterpart, which weighs 22 pounds (10 kg).
- A 50-foot length of 2½-inch Dura-Flow 800™ hose weighs 21 pounds (9.6 kg) compared to its rubber-lined counterpart, which weighs 26 pounds (11.8kg).

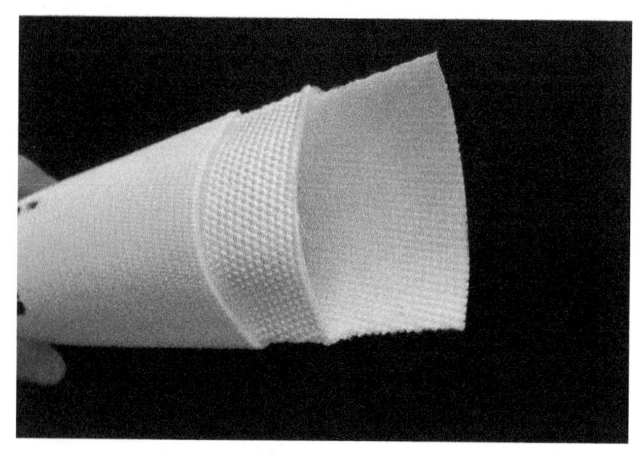

FIGURE 15-21 The Dura-Flow 800TM lightweight high-rise firefighting hose weighs significantly less than regular, rubber-lined, structural firefighting hose.

Photo by Mike Peterson, North American Fire Hose Corp.

If the fire attack team is carrying 200 feet of 2½-inch hose line, that's a weight-saving difference of 20 pounds!

Nozzles

The use of solid-stream, smooth-bore nozzles is often considered for high-rise fire attack because of their lower operating pressures. They can operate on low-pressure standpipe systems with a tip pressure of 50 psi. They have excellent penetration and good reach when they can hit the seat of the fire. Their disadvantage is they do not produce an efficient fog stream without gating down the shutoff, which reduces gpm flow. The closer the smooth-bore tip orifice gets to the diameter of the hose line, the less velocity the water stream has at the same flow rate. A trick of the trade is to remember that no nozzle should have a tip orifice opening greater than one-half (½) the diameter of its hose. So, a 2½-inch hose line should have a smooth-bore tip no greater than 1¼-inch, and a 1¾-inch hose line should have a smooth-bore tip no greater than 15⁄16-inch.

Standpipe outlet pressures vary by location and can range between 50 psi and 205 psi. If smooth-bore nozzles are used for fire attack, and standpipe pressures are adequate for low-pressure combination fog nozzles (75 psi), the combination fog nozzle should be used for the backup line and for the RIT line to protect firefighters. This anticipates conditions worsening. The entrained air from a fog pattern can move the flames away from the attack team as well as spray them down, keeping them cool as they back out. The fog spray can also act like a sprinkler head to cool overhead gases. Experience has shown that, when extreme fire conditions develop because of the wind, smooth-bore fire teams are not able to hold their ground because the solid stream doesn't absorb the radiant heat or push flames away from firefighters.

In many fire departments, there is resistance to using a combination fog nozzle in high-rise firefighting. A smooth-bore tip shoots a solid stream that doesn't easily break up. But consider that the solid stream is intentionally bounced off the walls or ceilings to extinguish the fire more effectively. Thus, one needs to ask, "Doesn't bouncing a solid stream off the walls, ceilings, or a solid object defeat the purpose of the stream they want maintain? Isn't breaking up a solid stream the function of a combination nozzle?"

Combination fog nozzles, including automatic and selectable gallonage nozzles, have an operating tip pressure between 75 and 100 psi. They have good penetration and reach with a straight stream, as well as an effective fog pattern when they're properly

A

B

FIGURE 15-22 TFT *Flip-Tip* nozzle: **A**. the combination fog nozzle and **B**. the smooth-bore tip.

Photos courtesy of Task Force Tips.

supplied, but they can have problems with PRVs and can be ineffective with low-pressure standpipes below 75 psi. Nozzle selection should be based in part on the building's minimum standpipe outlet pressures to deliver the desired gpm and reach. The selection is also based on how much flow and nozzle reaction can be controlled by the available firefighters and their ability to move the hose line (combat mobility).

Modern low-pressure, break-apart, high-rise nozzles are an excellent choice for both combination and smooth bore applications where the nozzle assembly includes a low-pressure (75 psi) combination spray nozzle with a rated discharge of 150 gpm coupled with a 15/16-inch solid-bore tip (50 psi) rated at 185 gpm. An example in new nozzle technology is the TFT *Flip-Tip* nozzle that provides all the features and benefits of the smooth-bore and spray tips by combining them into one. This nozzle assembly allows firefighters a choice of a spray nozzle or a solid smooth-bore nozzle for combatting a fire. The combination selection provides a single nozzle that offers fixed, selectable, or automatic gallonage into the tip that can be adjusted from straight stream to a wide fog pattern. It performs all the functions and features of any other combination nozzle because it is a genuine combination nozzle. If the fire situation warrants a harder-hitting, deeper-penetrating straight stream from a smooth-bore tip, the locking ring has a single-twist, push-down release. The combination tip is flipped down with a push of the palm, and the pivot-lock secures it in the down position. The nozzle is hinged in the middle, exposing the smooth-bore tip. There are nozzles for 1¾-inch and 2½-inch hose lines. Smooth-bore tips range from ⅞, 15/16, 1⅛, 1¼, to 1⅜ inches. This allows for flows of 50 to 300 gpm **FIGURE 15-22**.

In large open-floor areas, firefighters must be prepared to use master stream appliances for interior attack. A single- or two-inlet lightweight deluge set, made of modern aluminum alloys and equipped with a solid-bore tip, has proved most effective for such operations. Another new innovation in nozzle technology is the TFT *Blitzfire*. The *Blitzfire* is a 2½-inch lightweight, single-operator portable master stream that is ideal for initial interior attack. The master stream nozzle can flow up to 500 gpm (2,000 L/min). The monitor is compact and lightweight. The base model is designed for low-angle interior attack. The 10° to 46° elevation is perfect for directing a fire stream into any door or hallway of a high-rise during an initial attack **FIGURE 15-23**. The high-elevation (HE) version provides an even higher 86° elevation angle for tactical advantages, such as the ability to place a high-volume stream directly into an overhead or to shoot deep into the open floor space of a high-rise. Both models can be equipped with a unique water-driven turbine that drives the oscillating unit in a selectable 20°, 30°, or 40° sweeping motion. Both models can accommodate a 2½-inch combination spray nozzle tip or stacked smooth-bore tips. The *Blitzfire* portable

FIGURE 15-23 The lightweight, single-operator TFT *Blitzfire* portable monitor.

Photo courtesy of Task Force Tips.

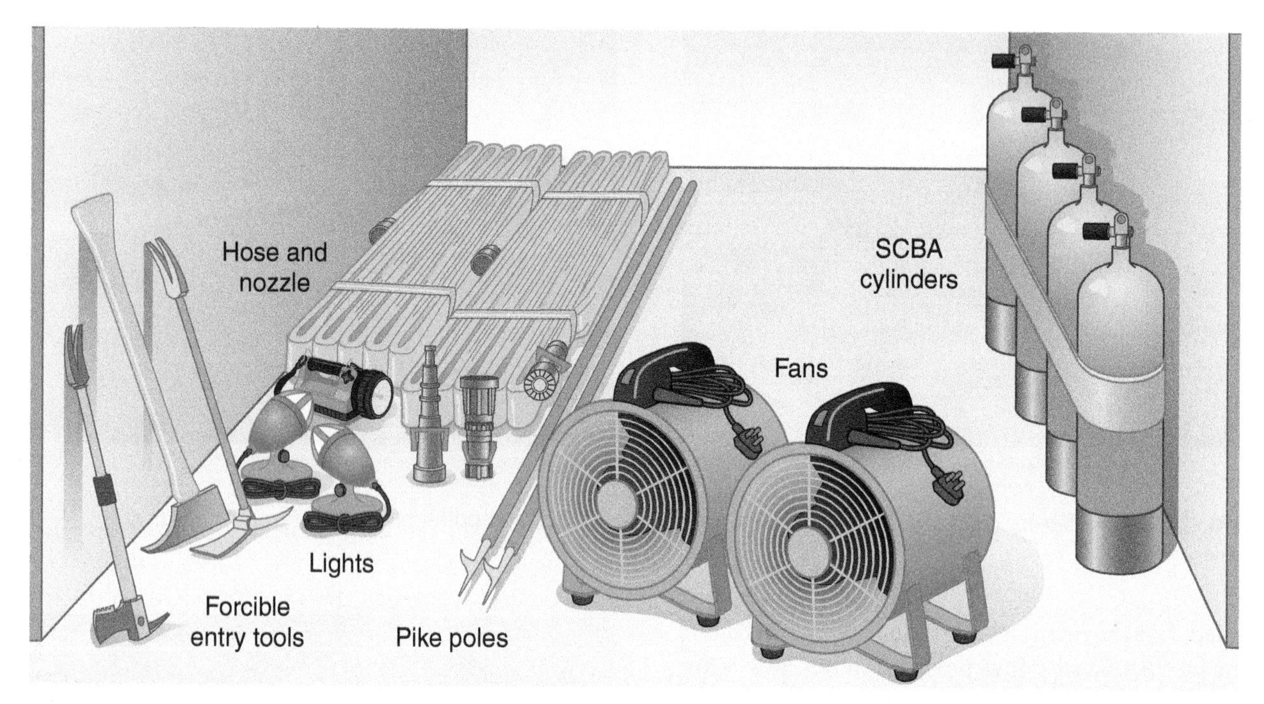

FIGURE 15-24 The equipment cache is set up in staging.

monitor has carbide-tip ground spikes along with two wide-leg stabilizers. The single 2½-inch female intake is low at the base, which keeps the supply hose low to the ground. These three components securely anchor the monitor, which absorbs the 250-pound nozzle reaction while flowing 500 gpm. There is little, if any, nozzle reaction for the firefighter to wrestle with or brace against, making this portable monitor excellent for offensive high-rise interior attacks. Once set, the monitor can be left unattended.

Equipment Cache

As firefighters report to staging, they must carry in hoses, extra SCBA cylinders, nozzles, search ropes, pickhead axes, TICs, and forcible entry tools. Conducting a primary search, attacking the fire, checking for fire extension, and performing other tasks may require firefighters to force their way into locked apartments, offices, or other secured areas of the building.

Other tools and equipment are eventually needed inside the building to support and assist in firefighting operations and should be placed in the equipment cache area of staging **FIGURE 15-24**. Pickhead axes are the most effective tool for taking out high-rise windows. Pike poles and drywall hooks are needed for checking the overhead, the plenum spaces, and overhaul. Gas and electric PPV fans and smoke ejectors are needed in setting up horizontal and vertical ventilation within the building after the fire is extinguished. Extra flashlight batteries and portable lighting may be necessary to illuminate smoke-filled or windowless areas of the building. Portable generators may be needed in case of electrical power failure. As additional companies arrive, a large supply of full SCBA cylinders will be needed for air exchange more than any other piece of equipment. Every member should bring at least one if not two extra SCBA cylinders when she or he reports to staging. Looping webbing around the cylinder stems and throwing it over a shoulder is an easy way to carry two bottles.

A medical area must also be set up in staging, or possibly three floors below the fire for a cleaner environment. In addition to a gurney and all the medical equipment needed by paramedics, a water canteen, jug, or cases of water bottles should be provided—the first thing firefighters need after being rotated out is clean drinking water. The physical condition of firefighters must be maintained during the incident, and firefighters engaged in firefighting activities need to be rotated, rehabbed, and medically monitored in a timely manner. Members must be rehabbed after working through two SCBA air cylinders. NFPA 1584, *Standard on the Rehabilitation Process for Members During Emergency Operations and Training Exercises*, defines the process for firefighter rehab. Additional alarms should be called early to ensure an adequate number of firefighters are onsite for rotating crews. There should be one crew in standby for every two crews assigned to a high-rise task.

Connecting to Standpipe Discharge Outlets

Standpipes provide a significant tactical advantage for quick and efficient direct attack evolutions and getting fast water to an elevated incipient and growth stage fire, as well as some fully developed fires, using 1¾-inch and 2½-inch hand lines. Other fully developed fires that will need more than 300 gpm to knock down and extinguish require the use of multiple hand lines or a lightweight portable master stream monitor or *Blitzfire* nozzle, both of which need to be supplied from the standpipe.

The discharge valve (outlet) is where firefighters hook up the hose to the standpipe, either directly onto the outlet or by first attaching a 2½-inch or 1½-inch wye. It is recommended that an inline pressure gauge be connected before the hose is attached **FIGURE 15-25**. Starting with the first floor, there is a 2½-inch discharge valve coming off the standpipe on every floor landing in the stairwell. Sometimes they are located on the half landing between floors. Every standpipe has a rooftop discharge manifold valve on the roof equal to the required flow to attack roof fires, roof deck and patio fires, and fires in rooftop machinery rooms, and for flowing water down the exterior sides of high-rise buildings to prevent autoextension and lapping. If not on the roof, the discharge valve will be located just inside the top stair landing at roof level. If the stairway continues down into a basement or lower multilevel parking garages, there will be additional below-grade discharge valves.

FIGURE 15-25 It is recommended that an inline pressure gauge be connected to the standpipe outlet before the hose is attached.

Photo by Captain Steve Baer.

Pressure-Reducing Valves (PRVs)

Pressure-reducing devices prevent dangerously high discharge pressures from the standpipe hose outlets. Two types of pressure-reducing devices can be installed on standpipes: flow-restricting devices and PRVs. A flow-restricting device controls the discharge pressure by restricting the flow through a reduced orifice plate. The specific pressure and discharge rate are determined when the standpipe is initially installed and flow-tested for certification. Flow-restricting devices do not reduce the static pressure (pressure with no water flowing). When water is flowing at the higher pressure, the flow rate is reduced because the water is passing through a smaller-diameter orifice plate within the valve.

PRVs limit the pressure on the downstream side at all flow rates. The PRV is set to deliver a specific pressure that will not be exceeded under any flow condition. Static and residual pressures remain set. These valves are often installed on the connections between the standpipe risers and the automatic sprinklers' sectional valves on the individual floors as well as on discharge hose outlets.

In super-high-rise buildings, the head pressure on lower floors can be tremendous. NFPA 13, *Standard for the Installation of Sprinkler Systems* deals with PRVs for sprinkler sectional valves. For standpipes and hose connections, NFPA 14, 7.2.3.2 states, "[W]here the static pressure at a 2½-inch (65-mm) hose connection exceeds 175 psi (12.1 bar), a listed pressure reducing device shall be provided to limit the static and residual pressure at the hose connection to no more than 175 psi." If flow pressures exceed 100 psi, then NFPA 14 requires that an approved device be installed at the discharge outlet to reduce pressures to a maximum of 100 psi (unless approved by the fire department, which may require a minimum pressure of 125 psi or more and a maximum pressure of up to 200 or more psi). Lower floor outlets have a higher static pressure, so expect PRVs to be installed with the higher setting to restrict the flow. Pressure settings decrease as the standpipe ascends to upper floors, which may have regular hose outlets without PRVs because of reduced inlet pressure due to the elevation. The determination about which pressure adjustments are required per floor is made during the system acceptance tests after the standpipe system is initially installed.

The 2½-inch outlets have the flow-restriction device or the PRV within the outlet housing or on the outside of the outlet housing. Most are set at the factory and cannot be adjusted in the field. Most of the time, PRVs operate as designed and do not present a problem in firefighting operations, but after the One Meridian Plaza Fire in Philadelphia, Pennsylvania, a failure to even recognize their existence became a national training concern. At the One Meridian Plaza Fire, human error resulted in deadly disaster. The PRVs were adjusted too high at the time of installation, and the discharge pressure was restricted to less than 60 psi, a pressure too low to provide an effective or functional stream for 1¾-inch or 2½-inch hand lines. The miscalibrated PRVs made it impossible for Philadelphia firefighters to launch an effective attack or for fire department pumpers to override the pressure restriction in the valves.

Adjusting a PRV is not a typical task for which firefighters train, and it is not part of the Firefighter I and II curriculum. Although some textbooks suggest adjusting or removing PRVs when necessary, this textbook does not. The methods for adjusting PRVs require specialized training and shouldn't be attempted by untrained personnel during an emergency incident. There are too many variables within a flowing standpipe system, and dismantling these valves may cause other problems with overpressurization. At the First Interstate Fire in Los Angeles, California, some of the PRVs failed and allowed pressures estimated at 500 psi to reach hand lines. Some hose lines ruptured, and others were nearly impossible to manage safely. Other alternative solutions should be implemented using fire department supply hose as a temporary standpipe.

The best way to find out if PRVs will be a problem during the fire attack is for the hose crew to flow the nozzle onto the floor below the fire after hooking up to the standpipe. In a relatively smoke-free environment, the company officer can check the volume, pressure, quality, and reach of the stream before entering the immediately dangerous to life or health (IDLH) area of the fire floor. If there is a problem with the PRVs, it will become apparent right here, before anyone gets hurt or killed. Fire department members, or at least company officers stationed in the downtown districts of major cities, should have some knowledge and experience on this subject, but part of any major high-rise incident management plan is to have a building engineer on site at the command post to help resolve such problems. If PRVs need to be adjusted, the adjustment should be coordinated with the building engineers.

Connecting to Below-Grade Standpipe Discharge Outlets

With fires in lower-level floors below the main entrance, hose lines should be laid down the stairs from the apparatus, even though there may be additional below-grade standpipe discharge connections. The reason for this is that the fire is already below you; heat and smoke may be pushing up the stairway as you're working your way down. If conditions change for the worse, you need the ability to back up and out of the stairway with a charged hose line to protect the attack team and to protect the stairway if rescues or evacuations are in progress. Head pressure is working in your favor when you are advancing below grade, so a combination fog nozzle is a better choice to protect firefighters and occupants. Additional firefighters on the main floor can assist in pulling the hose line back up the stairs, and this will not be possible if the hose line is connected to a below-grade discharge outlet. According to the NFPA *High-Rise Building Fires* report, 10% of all office and hotel high-rise fires start below grade—the common location for service areas.

Sometimes a below-grade standpipe connection may be the best choice. For example, some underground parking garages can go six levels below grade. A recon size-up and risk-benefit analysis needs to be conducted for vehicle fires occurring on these lower levels. Due to the amount of hose that may be required, these types of incidents warrant connecting to below-grade standpipe connections in the stairway.

House Lines and Fire Cabinets

With Class II standpipes, engine company personnel should not rely on house lines intended for use by the building occupants for fire attack. They should always use fire department hose. House lines should be tested at least once every 5 years but are often poorly maintained and not always actually tested. As a result, hose can be found partially charged, snarled, or in disarray in the cabinet or hose rack. Hose or nozzles could also be missing from the standpipe location. Unlined hose may be rotted. Valves that are never used or tested can be rusted and difficult to open by hand, or the handwheel may be missing. The only exceptions to this rule are for firefighters assigned to recon and searching for the seat of the fire, or those assigned to search, evacuation, and rescue above the fire floor who may be without the support of a fire department hand line. If for some reason they discover the fire, come across fire extension, or need the protection of a hose line, grabbing the house line makes sense and is better than doing nothing, but such an action should be followed up with a radio call for a hose line and crew. Experience has shown that ladder companies assigned to recon have used these house lines to provide fast water in extinguishing room fires before the engine company was able to get a hose line into operation. Such a case occurred at the Gateway Apartments and Townhouse Fire in San Francisco, California.

Dangers in the Firefighting Stairway

Many officers face the dilemma of whether to open the roof access door at the top of the stairway for vertical ventilation or leave it closed. Common sense says that it should be opened to let the pressurized heat and smoke out of the building. In a nonpressurized stairway, that may be the best tactic. In buildings that have pressurization fans for the stairway, it is not. In order for stairway pressurization to be maintained, the roof access door and the doors to each floor within the stairway need to remain closed. Modern pressurization systems can maintain a positive pressure with two and even three doors open in a stairwell; after that, system pressurization may not be effective in keeping smoke out of the stairwell. Occupants and residents propping hallway doors open on upper floors for convenience or air circulation, especially in summer months, is a common problem.

When firefighters make entry into the hallway from the stairway, the officer needs to watch what the smoke does. If pressure is sufficient within the stairwell, it may hold back the smoke, but conditions can change rapidly with any window failure or too many doors being opened within the firefighting stairwell. Thus, door control at the fire floor must be practiced to avoid allowing the air pressure in the stairway to decrease. PPV fans can be set up on lower levels of the stairway to help maintain stairway pressurization **FIGURE 15-26**. As soon as the offensive attack is initiated on the fire floor, you must understand that conditions in the stairway change quickly if stairway pressurization is not sufficient or maintained and if fire and smoke have entered the hallway. Flames, heat, and smoke immediately travel out the door and rise up the stairway to top floors. The stairway becomes a giant chimney with a lid **FIGURE 15-27**. Without door control, the stairway becomes an efficient high-low flow path when the roof access door is opened. Rapid movement of heat and smoke begins to develop **FIGURE 15-28**. If the fire has self-vented through a window, a wind-driven fire can push the flames, heat, and smoke up the stairwell, making it an extremely efficient high-low flow path with blowtorch force **FIGURE 15-29**. Having the rooftop door

FIGURE 15-26 In pressurized stairways, the roof access and hallway doors within the stairway must remain closed to maintain pressurization until the attack stairway is cleared of occupants. PPV fans at floor level help maintain stairway pressurization.

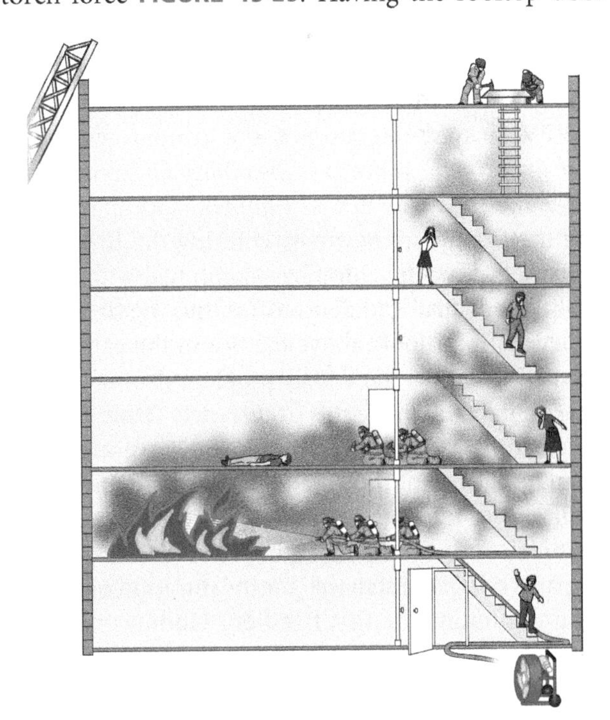

FIGURE 15-27 In a nonpressurized stairwell, or if pressurization is lost due to numerous open doors to the hallways, conditions in the stairway change quickly when fire attack is initiated. Heat and smoke immediately travels out the door and rises up the stairway to top floors.

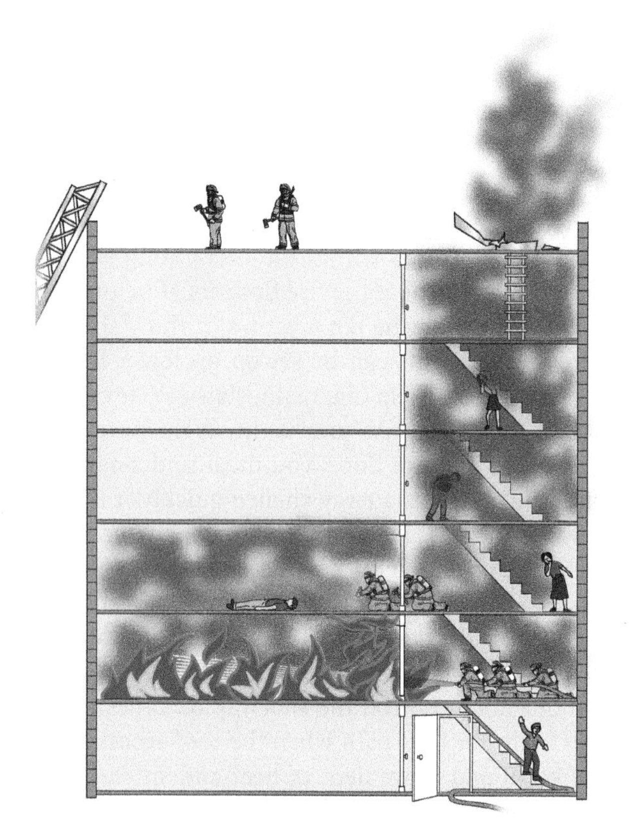

FIGURE 15-28 When the roof access door is opened, the stairwell leading to it becomes a rapid and efficient high-low flow path for heat and smoke.

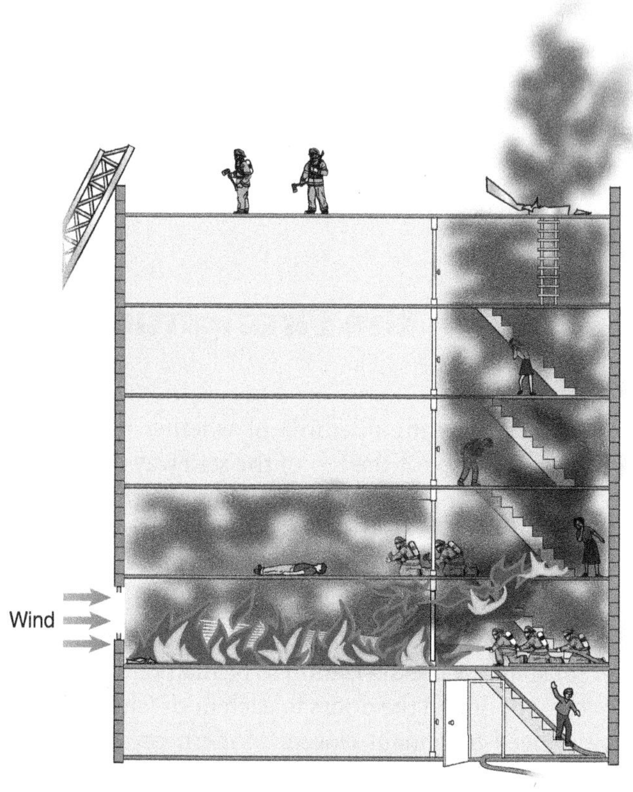

FIGURE 15-29 If the fire self-vents through a window, a wind-driven fire can push the flames, heat, and smoke up the stairwell, making it an extremely efficient high-low flow path with blowtorch force.

open increases the speed of the flow path, especially if PPV fans are operating at the ground-level entry. *But the primary concern is the ability for occupants to enter the "stairwell-turned-chimney."* Occupants fleeing the fire by unknowingly entering the firefighting stairway—now the chimney—combined with our inability to control and deny access into the firefighting stairway from floors above the fire in the early stages of the incident is the deadly problem. Company officers should anticipate this occurrence. Other than extinguishing the fire, no other actions save more lives than initiating an immediate search of the firefighting stairway for trapped occupants before the fire attack is commenced and implementing strategies to keep occupants out of the stairwell for the duration of the fire. At a minimum, the first five floor landings in the attack stairwell above the fire floor must be checked for occupants by a crew member before the hallway door is opened by the engine company to attack the fire.

Unless the entire fire attack stairwell can remain pressurized, connecting hose lines on the fire floor and above the fire floor landing should be avoided. If the attack line is taken off the fire floor and the stairway landing becomes untenable, control of the standpipe discharge valve is lost. Once the fire is sufficiently knocked down and thermal energy is taken away from the fire, additional lines can be taken off the fire floor standpipe for a backup line or for overhaul, but it should never be used for the initial fire attack **FIGURE 15-30**. Some departments are now using portable smoke curtains that can be quickly installed in doors between stairway and fire floors, significantly limiting smoke and heat from entering the stairwell.

Heavy smoke and flames in the hallway usually means the door to the fire unit is open, allowing the fire to spread. This can make the stairway above the fire floor untenable and deadly. New York City Firefighter John King of Engine Company 23 was killed in such a situation on December 27, 1961. As the crew was ready to make entry, he was flaking the excess hose on the stairs above, a practice we all use today to make it easier to advance the hose line onto the fire floor **FIGURE 15-31**. When the hallway door was opened, smoke and flames trapped Firefighter King on the stairs above the fire floor, killing him. On December 24, 1998, in New York City, four civilians died in a superheated, smoke-filled stairway. The fire started on floor 19, and the victims were found in the firefighting

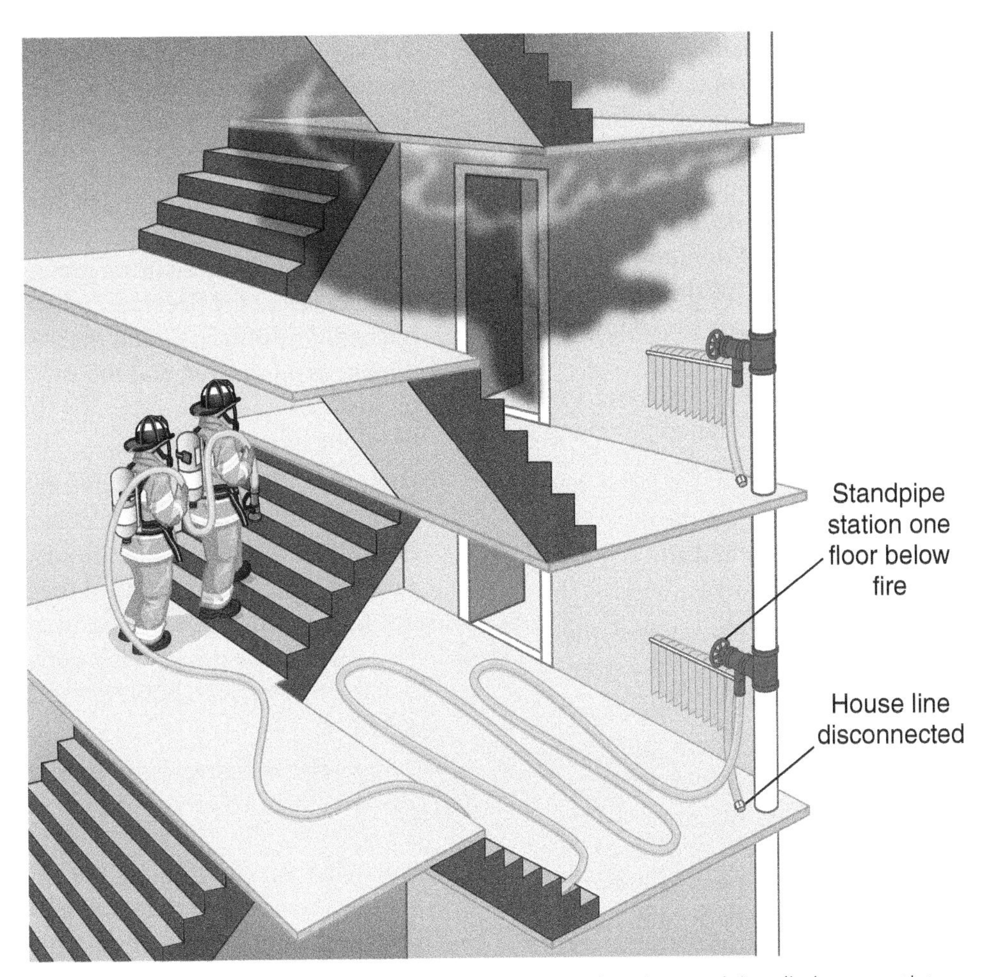

FIGURE 15-30 The first attack line should be connected to the standpipe discharge outlet one floor below the fire.

FIGURE 15-31 If the excess hose of the attack line is coiled or stretched to the upper stairs or landing, the officer should make sure that the entire crew is back on the hose line before opening the door to the fire floor.

stairwell on floors 27 and 29. On January 6, 2014, again in New York City, one man was killed, and another critically injured when they tried to escape a fire in their high-rise apartment. The fire started on floor 20. The two men unknowingly entered the firefighting stairwell on floor 38. As the firefighters opened the hallway door to advance to the fire floor, heat and smoke filled the upper stairwell: the men only made it down to floor 31.

The most tragic example was the Cook County Administration Building Fire on October 17, 2003, in Chicago, Illinois. A relatively small fire started in a supply room on floor 12 of this 35-story nonsprinklered high-rise. The supply room was directly adjacent to one of the two stairways. The fire started at 17:00. Within 3 minutes, smoke was visible from the exterior of the building. By 17:06, the windows had failed, and flames were visible from the street before the Chicago Fire Department arrived on scene. The 54-minute-long fire attack started at 17:16, and as the attack crew made its way onto floor 12, smoke and heat from the fire floor charged the attack stairway, trapping 13 civilians who were trying to escape. The stairway was not pressurized, nor did it have any smart locks, which would have allowed reentry into any of the floors. Fortunately, someone blocked open the door to floor 27, which allowed seven occupants to find refuge but not without smoke inhalation injuries requiring resuscitation. The other six died in the stairway above the fire floor. Unfortunately, it took over 90 minutes, when it was decided that the entire length of the fire attack stairway should be checked, before the first body was discovered.

These examples illustrate and reveal several important actions to consider during standpipe operations before initiating the offensive attack:

- Building occupants may not know the difference between a firefighting stairway and an evacuation stairway.
- If forcible entry to the fire floor is required, the door must be controlled until the upper floors of the firefighting stairway are cleared.
- Before the fire floor door is opened, a rapid ascent team (RAT) should search and check the upper levels in the firefighting stairwell for evacuees and quickly move them below the fire floor or move them inside an upper hallway and direct them to the evacuation stairway, but they must be moved out of the firefighting stairway before the attack begins. This understandably poses a dilemma when there are several floors above the fire. At minimum, the first five floors above the fire floor should be immediately checked for evacuees.
- The PA system located at the fire alarm panel should be used to clear the stairway and direct occupants to the evacuation stairway or direct them to remain in their rooms to shelter in place.
- Protect and shelter in place is the primary strategy for search teams. Rather than risk bringing occupants through an IDLH atmosphere, search teams should direct occupants back into their rooms.
- If a large number of occupants are flooding down the firefighting stairway, they have effectively stalled the fire attack. The company officer cannot initiate the attack by opening the door to the fire floor hallway without endangering the evacuees. They must wait for the stairwell above the fire to clear or look for another way to attack the fire.
- If an exposure line is needed on the floor above the fire, the firefighters involved in the fire attack might have to wait with the fire floor hallway door closed until the exposure line is stretched up the stairway into the hallway above in order to prevent firefighters from getting caught in the superheated stairway/chimney.
- If the excess hose of the attack line is coiled or stretched to the upper stairs or landing to assist in advancing the hose using gravity, this should be done *before* the fire floor hallway door is opened. The officer should make sure that the upper stairs are clear, and the entire crew is back on the hose line before opening the door to the fire floor (Figure 15-31).
- High-rise buildings often have more than one stairway with a standpipe. Consider taking the exposure lines to the upper floors and using a different standpipe to avoid the heat and smoke from the upper firefighting stairway. This may not be possible if it compromises the evacuation stairway, but it is another option to consider.
- In high-rise buildings with more than one stairway, the interior exposure line can be connected to the standpipe in the firefighting stairwell, then laid through the hallway below the fire floor, and then up to the floor above the fire, thus using a different stairwell and staying clear of the smoke and heat rising above the fire floor in the firefighting stairway. Although this tactic requires more hose, the lay should be fast because there is no smoke or heat to deal with. In this scenario, the hose can remain uncharged until it's time to make entry.
- Another option is to lay another 2½-inch hose line connected three or four floors below the fire floor and stretched up to the floor below the fire with a 2½-inch wye attached to it. Once charged, it provides two other standpipe outlets to connect to.

Initial Hose Operations in Residential High-Rise Buildings

This section of the chapter covers a variety of strategies and tactics to consider when executing the initial fire attack in residential high-rise buildings.

Beginning the Direct Attack

Hose lines that are 1¾ inches are effective on most apartment high-rise fires with the typical household furnishings; however, a 2½-inch hose should also be

FIGURE 15-32 As firefighters ascend the stairwell, they should ensure that all standpipe discharge valves on the lower floors are closed.

available for backup in case the smaller hose lines are inadequate for controlling the fire or if the fire extends into the hallway. As in all firefighting operations, if the 1¾-inch hand line isn't knocking down the fire in 1 minute, the 2½-inch backup hose should be used immediately. Backup lines should be equal to or greater than the hose diameter of the attack line utilizing a combination nozzle.

Ascending the Stairs

If firefighters ascend up the stairwell to the floor below the fire instead of using the elevators, it is a good (proactive) practice as they pass each landing to check that all standpipe discharge valves on the lower floors are closed **FIGURE 15-32**. If water is cascading down the stairway after the standpipe is charged, it is an indication that there are open discharge valves somewhere in the stairway, thus robbing water from the fire area on upper floors. If there is unexpected low pressure before the fire attack begins, it is another indication that lower floor valves are left open. One firefighter may have to run down the entire length of the stairway to make this check to close them, which wastes valuable time. After this check is made, the firefighter should take the elevator back up to staging, which is two floors below the fire floor, and rejoin the crew.

If elevators were used to take firefighters to the staging area two floors below the fire, the firefighters should advance by stairway to the fire floor. The first hose line must be connected to a standpipe discharge outlet in the enclosed stairwell, one floor below the fire. The hose should be coiled or spread out and up the stairs before it is charged. When charged, coiled hose is easy to advance and means less congestion in the stairway of the fire floor as the attack gets under way **FIGURE 15-33**. If 1¾-inch hose lines are to be used for initial attack, a wye can be placed on the outlet or attached to a short length of 2½-inch hose, and two attack lines can be hooked to it. If the hose connection is needed in the hallway with a Class II standpipe for firefighting, the house line and reducer coupling should be removed. Only in extreme rescue circumstances during a high-rise fire attack should the engine crew use the tenant hose found on the outlet instead of fire department hose. *Use your own hose.* Keep in mind that the flow and pressure in the house line outlet is typically much lower than that in firefighter standpipe outlets; therefore, attempting to operate fire hose from a house line outlet can result in inadequate flows and pressure.

FIGURE 15-33 When charged, coiled hose is easy to advance and means less congestion in the stairway of the fire floor.

Photo by Lt. Kevin O'Donnell.

Tactical Objectives

The tactical objectives on the fire floor of a high-rise building remain the same as in a one-story building:

- Forcible entry: access
- Fire attack line (first hose line)
- Backup line (second hose line also acts as initial RIT)

- Exposure line to floor above (third hose line)
- Rapid intervention line (fourth hose line)
- SER and evacuation
- Ventilation
- Salvage and overhaul

At a minimum, the four hose lines mentioned above must be deployed. Additional attack lines, backup lines, and interior exposure lines may be needed on the fire floor, and additional exposure lines may be needed on the floor or floors above the fire. Speed and maneuverability with the available personnel are extremely important for controlling fires in high-rise buildings. The size of the hand lines and nozzles still depend on the application of one of the two fire flow formulas, Royer/Nelson Iowa State University or National Fire Academy (NFA), but more than likely, 2½-inch hand lines should be anticipated for backup lines in fighting fires in high-rise buildings.

In most cases, not all hose lines operate simultaneously, but if 2½-inch lines are used and the standpipe is 4 inches in diameter, recognize that the combination of an attack line, backup line, exposure line, and RIT operating simultaneously might exceed the flow capacity of the standpipe. Most important, if three lines are flowing and the fourth line is opened, the pressure on all lines could drop, thus endangering the attack line crews and crews assigned to SER.

Rapid Intervention Teams (RITs)

There needs to be some clarification about when the RIT (or rapid intervention crew [RIC]) is assigned on the fire floor. Many feel that making RIT the second assignment after the initial attack team is a waste of firefighter resources early in the incident, when every firefighter should be utilized in getting sufficient water to the fire. If the backup line is the second assignment, and the exposure line is the third assignment, properly supporting the attack strategy should prevent a firefighter emergency that would require a RIT deployment. On the surface, this reasoning may sound logical, but there is an assumption that the RIT is standing around simply waiting for an emergency to happen while the attack crew is struggling on its own to get a line in position to attack the fire. Chapter 8 of NFPA 1500, *Standard on Fire Department Occupational Safety, Health, and Wellness Program* has a section, specifically section 8.8, called Rapid Intervention for Rescue of Members. Every fire officer should carefully study this section; it is the law. After an occupational fatality or line-of-duty death (LODD), our actions will be scrutinized for compliance to this standard and, without a doubt, we will be held accountable for willful noncompliance.

- Section 8.8.2 states: "In the initial stages of an incident where only one crew is operating in the hazardous area at a working structural fire, a minimum of four individuals shall be required, consisting of two members working as a crew in the hazard area and two standby members present outside the hazard area available for assistance or rescue at emergency operations where entry into the danger area is required." (This is the two in/two out rule that also applies to high-rise firefighting.)
- Section 8.8.2.9 states: "Once a second crew is assigned or operating in the hazard area, the incident shall no longer be considered in the 'initial stage' and at least one rapid intervention crew shall be deployed that complies with the requirements of 8.8.2."
- Section 8.8.2.5.1 states: "No one shall be permitted to serve as a standby member of the firefighting crew when other activities in which the firefighter is engaged inhibit the firefighter's ability to assist in or perform rescue, if necessary, or are of such importance that they cannot be abandoned without placing other firefighters in danger."
- Section 8.8.4 states: "A rapid intervention crew shall consist of at least two members and shall be available for immediate rescue of a member or a crew."
- Section 8.8.6 states: "The composition and structure of a rapid intervention crew shall be permitted to be flexible based on the type of incident and the size and complexity of the incident."

What does this all mean? In actuality, the NFPA standard gives you options with a tremendous amount of flexibility. The RIT needs to be viewed as supporting the attack. The standard never states that the RIT has to be a crew of four. It never states that they have to have a hose line. In fact, the only required equipment, in addition to full PPE with SCBA, is a universal air connection (UAC) to a supplied air source. RIT standby members (two minimum) can include the officer assigned to the floor division. The standard never states that they can't help. They can help lay the hose in the stairway, connect to the standpipe, bleed the air in the standpipe and the hose line, remove kinks, pull hose up the stairs, and help advance the charged line into the hallway from the stairway. They merely have to remain outside the hazard area but in close proximity, where they can drop whatever they're doing and immediately assist or rescue members who are operating inside the hazard area.

In a NIST experiment, a video was shot capturing the speed of fire development of commercial office furnishings provided in a typical office space. Within 20 seconds, thick black smoke was banking down one-quarter of the way from the ceiling. In 5 minutes, most of the furnishings were on fire, and thick black smoke was banked halfway down. In 7 minutes, flashover had occurred, and the room was completely involved in fire, with a zero survivability profile. Experience and case studies have already revealed that it takes between 15 and 20 minutes to get the first hose line in operation on upper floors after the arrival of the fire department. At the First Interstate Fire in Los Angeles, California; the One Meridian Plaza Fire in Philadelphia, Pennsylvania; and the Cook County Administration Building Fire in Chicago, Illinois, smoke and flames were already blowing out the windows when their respective fire departments arrived on scene. From arrival, it took Los Angeles firefighters 29 minutes to get the first line into operation on the twelfth floor. In other words, the first attack team may be entering a hazard area that is heavily involved in fire and they may require the immediate assistance of a rescue team in the first few minutes. Why would you hesitate to establish a RIT immediately after their entry and wait for the time it will take for the fourth company in the sequence to arrive on the fire floor before establishing a designated rescue team?

Consider the following evolutions for the first company assigned to fire attack. It can only be accomplished with 1¾-inch or 2-inch hose. Two firefighters cannot realistically advance and operate a 2½-inch hose line down an interior hallway and into a compartment in smoke, heat, and fire conditions. The fire attack crew must have a crew of four to enter the hazard area. If they connect on the floor below and bring a single attack line of 200 feet, the nozzle firefighter and his or her partner can make entry into the hallway or fire floor while the officer and the fourth firefighter remain on the landing and thus become the initial RIT. The officer becomes the fire floor division supervisor and the two in/two out rule is met.

If this crew of four can manage to bring two hose lines of 200 feet each, the attack line can be connected and charged from the floor below the fire while the second line, which is designated the RIT line, can be connected on the fire floor, coiled on the landing above the fire floor while all hallway doors remain closed, and charged, and the nozzle section can be brought down to the stairs just above the fire floor landing in a ready position. Then two firefighters enter the hazard area on the attack line while the officer and the fourth firefighter become the initial RIT; this time, however, they have a hose line for protection and for cooling members who may be rapidly exiting the fire floor. The two in/two out rule is met.

When the next company of four arrives, they can be given the backup line assignment to the attack crew. With more personnel, the 2½-inch line can enter the evolution. With two officers available, one remains division supervisor while the other enters the hazard area with the backup team. The arrival of the second company ends the initial stage of the incident, so a designated RIT company of two members minimum needs to be assigned. With six members now on the landing, the teams can be split two different ways depending on the hose:

1. One officer remains the division supervisor, with two firefighters on the backup line and three firefighters on the RIT line, *or*
2. One officer remains the division supervisor, with three firefighters on the backup line and two firefighters on the RIT line.

As more crews arrive, the RIT size can increase or be substituted for a ladder company or rescue company. A designated RIT hose line is not required by the NFPA 1500 standard, but if you're going into a hostile thermal environment to assist or rescue firefighters, wouldn't you want to protect yourself? It is doubtful that you would be able to even make entry without the protection of a hose line. The same rationale applies to ladder company members assigned to SER. There is no rule preventing firefighters assigned to SER from taking a hose line with them for protection and the protection of occupants. If there are no civilians on the fire floor, the only remaining unprotected occupants are the firefighters themselves. We must shift our mindset to operate with the greatest margin of safety and with the protection of charged hose lines rather than risk thermal assault without hose protection because that's how it was done in the past.

Hooking Up Below the Fire Floor

The firefighter hooking up to the standpipe hose outlet should open the valve before connecting the hose. This tactic quickly bleeds out the air in the standpipe and flushes any debris between the FDC and the outlet. Once water is flowing, the firefighter can close the valve and connect the in-line pressure gauge, then the hose. While the crew is hooking up to the standpipe and laying out the hose up to the fire floor landing, the company officer should look into the floor below the fire to get as much information as possible about the layout of the office/commercial open-floor space or the layout of the residential hallway. The officer should check for:

- Locations of other exits and stairways
- Location of the elevator lobby
- The number of rooms that are on the right
- The number of rooms that are on the left
- The length and height of the hallway
- Floor construction (industrial carpet, tile, or concrete)
- The existence of a plenum
- Door construction: wooden doors are easier to force than metal doors
- Direction in which doors swing (right or left, in or out)
- Wall construction, for example, gypsum board (Sheetrock) or concrete (wall breaching and reaching through for the door handle might be the easiest way to access a hallway or room)
- Open or closed floor plan
- The existence of a service room or utility closet (a common source of fires)
- The existence of cubicles
- The existence of wall partitions that will deflect hose streams or fall over from hose streams
- The existence of glass partitions and windows

A quick look inside a unit (apartment or office) would be even better because the layout in the fire unit will most likely be identical and aid in keeping you and your crew oriented. This survey can be quick and completed in less than a minute, or at least in the same time it takes the crew to connect to the standpipe and lay out the hose. It's worth the time because the floor plan of the fire floor may be the same. A quick scan of the hallway helps you remain oriented on the fire floor as well as plan for a secondary means of escape. If time allows, the entire crew should scan the floor as well. This opportunity can give some peace of mind as well as develop self-confidence before the firefighters enter a smoky hot environment with little or no visibility. The fire code requires buildings to have an evacuation floor map posted in elevator lobbies, but sometimes they are also posted in the stairway landings by the door. If so, they include a diagram of the floor plan **FIGURE 15-34**. This may mean taking the extra time to check out the floor plan map in the elevator lobby on the staging floor or the floor below the fire, but the time is worth it before proceeding to the fire floor.

Finally, before the nozzle section is stretched to the fire floor, the officer should have the attack team flow the nozzle and check the stream on the floor below the fire. Do not worry about water damage, which is the least of your worries. This tactic allows firefighters to bleed the line of air; flush the hose; and check the reach, flow, shape, and quality of the stream in a smoke-cleared hallway. Most important, if there is any problem with a PRV, it will present itself right here instead of on the fire floor, where crews may be caught in a hostile thermal environment with an inadequate stream. This is a very important safety check before starting the attack. When the crew is ready to make entry, the officer should announce on the radio, "Commencing fire attack." Not only is this a time mark for the IC, it is a warning that conditions in the attack stairwell are about to change.

1½-inch and 2½-inch Gated Wyes

The attack line and backup line can be connected to the standpipe on the floor below the fire using a 1½-inch or a 2½-inch gated wye. This gives the company officer options. Depending on the occupancy, whether the floors are compartmentalized, the fire load, the size of the fire, and if the building is fully sprinklered, a 1¾-inch attack line with a 1¾-inch backup line will sometimes be sufficient. For example, consider the fire load and square footage of a fire in a unit in a high-rise hotel that is sprinklered. There's one room with one or two beds, a nightstand, lamps, a TV, an easy chair, a desk or small table with a chair, chest of drawers and curtains—that's about it. This fire does not need a 2½-inch line for the initial attack. Two engine companies can manage a 1¾-inch attack line and backup line coming off a 1½-inch gated wye and handle this fire fairly quickly.

Using a 2½-inch gated wye can yield a number of combinations:

- Two 2½-inch attack lines
- One 2½-inch attack line and one 2½-inch backup line
- One 2½-inch line and two 1¾-inch lines coming off a 1½-inch gated wye
- Four 1¾-inch lines coming off two 1½-inch gated wyes for exposure protection to floors above
- Two 2½-inch lines to supply a portable monitor or *Blitzfire* master stream nozzle

These wye configurations add a tremendous amount of cantilevered weight to the standpipe discharge outlet and should be supported with webbing

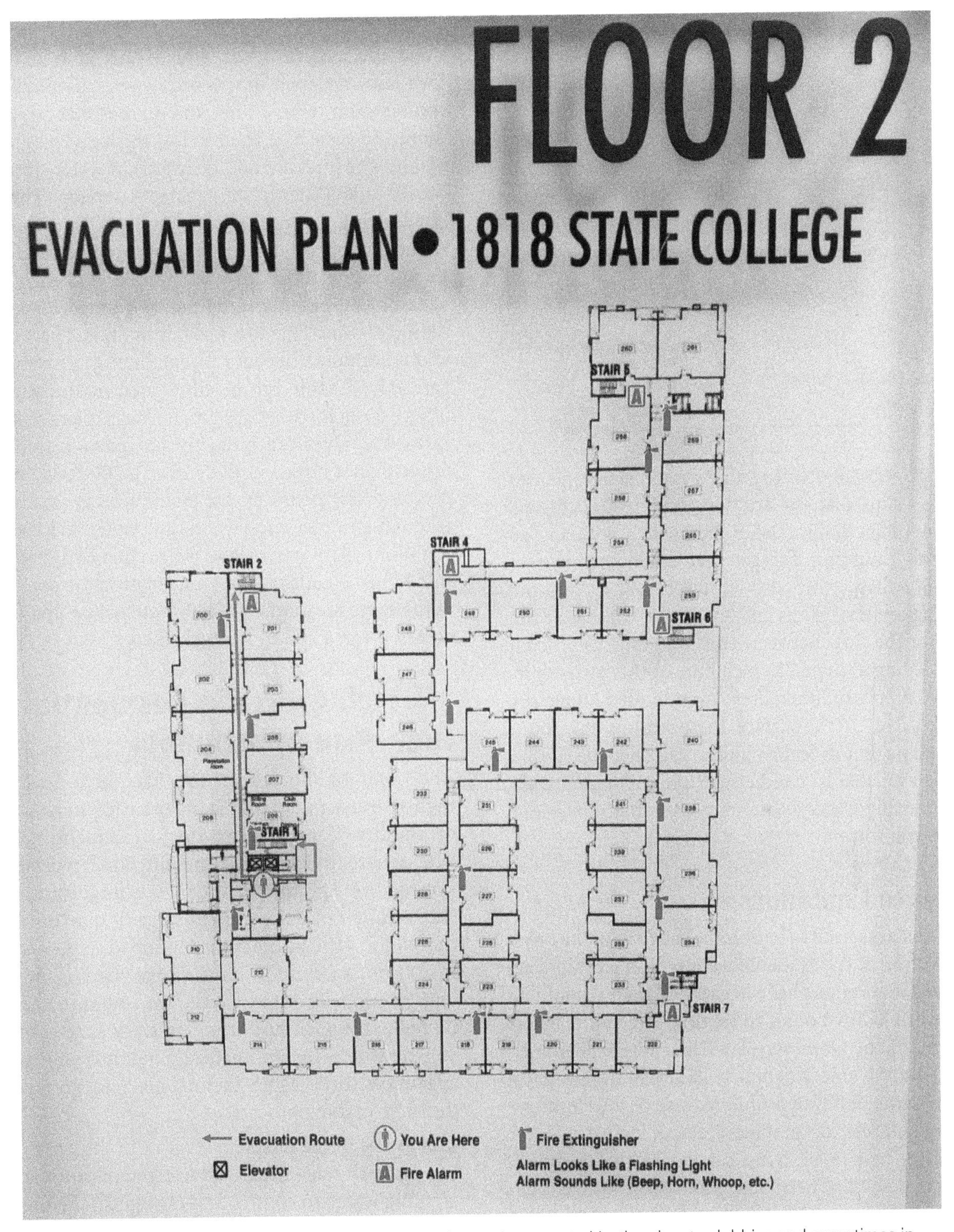

FIGURE 15-34 Evacuation maps include the floor plan and are posted in the elevator lobbies and sometimes in stairway landings.

Courtesy of Raul Angulo.

tied to the standpipe bracket. A better solution that many fire departments have implemented is to have what is called a pigtail: a 10- to 12-foot section of 2½-inch hose line that connects to the standpipe discharge valve, and then wyes, the in-line pressure gauge, or attack hose lines are connected to the male end of the pigtail. This relieves practically all the weight on the discharge valve and transfers it to the floor **FIGURE 15-35**. The pigtail also allows the hose connections to be moved to a corner of the floor landing, so all the connections aren't blocking the door and don't become an obstacle for firefighters.

FIGURE 15-35 A pigtail.
Courtesy of Raul Angulo.

Opponents of utilizing any wyes in standpipe evolutions claim that the friction loss created through these appliances diminishes the needed flow, and accidentally bumping the gate valve handle can inadvertently charge the hose bundle before it's laid out, creating a pile of kinks and spaghetti. Others say that splitting the flow between two lines isn't as efficient as supplying a single 2½-inch line, or that a standpipe cannot supply multiple lines coming off a single discharge valve. On the surface, these reasons may sound convincing, but in reality, unless you have the staffing readily available to handle multiple 2½-inch lines, it will probably create more maneuverability problems than it will solve.

Resource Limitations

Essential tasks need to be accomplished with the physical abilities of the available resources. If elevators are not used, firefighters have to carry all their equipment with extra SCBA bottles to staging, two floors below the fire. At the One Meridian Plaza Fire in Philadelphia, Pennsylvania, a complete failure of the electrical power to the building prohibited use of the elevators. All the suppression equipment, including extra SCBA cylinders, had to be hand-carried to the twentieth floor for staging. These firefighters were fatigued before they even started the fire attack. It is unreasonable to expect that all firefighters have the physical stamina to manage the weight of a charged 2½-inch hand line with the additional assault of smoke and heat after such a hike. You have to be flexible and allow the use of 1¾-inch lines. In Philadelphia, 1¾-inch hand lines were the initial attack lines. The inability to control the fire was due to the miscalibrated PRVs, not because firefighters selected 1¾-inch hand lines. They were flowing less than 60 psi, which is insufficient for some 2½-inch hose lays and nozzles as well.

The flow rate from a 4-inch properly supplied standpipe can easily be 600 to 750 gpm or more. Two-and-half-inch or 1½-inch wyes can allow for a combination of hose lines and nozzles that are within those 750 gpm. The friction loss through wyes is negligible. The lines are not likely to flow water simultaneously unless significant problems develop. The hose line flowing the most water should be the attack line. A backup line or an interior exposure line should not be flowing water unless it is needed, and then those streams will be intermittent and operated for a short period of time. The wyes provide options. The concern about accidentally charging the line by bumping the gate valve handle can be eliminated by the purchase of wye models with valve handle locks to prevent such incidents. Wyes can bleed the air out of a standpipe faster than a nozzle at the end of a 200-foot hose lay can, and any debris in the standpipe can easily pass through an open-end wye to flow water and clear the standpipe. The wyes allow for prebundled, predetermined hose loads of the same length to be connected at the same location, so a 200-foot backup line will be able to cover a 200-foot attack line.

1¾-inch Attack in Residential High-Rise Apartments

The common household fire load in a residential high-rise apartment unit is the same for an apartment on the first floor of a two-story apartment building. The square footage and the fire load typically do not require 300 gpm of water to extinguish a fire in the incipient or early growth stage. It may not be required for a fire in the fully developed stage if the fire load is not substantial and the apartment is in Type I construction. In other words, we wouldn't necessarily pull a 2½-inch line on a one-story apartment fire, so why would we automatically pull it on a high-rise apartment fire? More factors need to go into the decision-making process:

- A fire in the incipient stage or in the early growth stage needs fast water application, not big water, to control and extinguish the fire. This is the theory behind sprinklers, that is, small volumes of water applied quickly.
- Although the rapid deployment of a 1¾-inch hand line can get fast water on the fire, it should be backed up with a 2½-inch line.
- Residential high-rise units are compartmentalized and can often contain the fire to the room of origin.
- Getting quick water into a residential unit without opening the front door to accelerate

the fire can be accomplished by making a hole in the front door or the exterior Sheetrock wall within the hallway just large enough to insert and operate a nozzle.

- Residential high-rise fires behave differently than a fire in a wide area, open-space floor plan.
- Residential units often do not have a plenum.
- Modern high-rise insulated windows are very strong, with some being triple-paned. Unless they are exposed to direct flame contact or a wind-driven fire, they can retain the heat generated by household fire loads for a greater period of time than regular residential windows can.
- Everything hinges on wind conditions. Whether the winds are strong, light, or nonexistent must be determined. In high-rise fires, wind is the game-changer.
- Determine the direction of the wind. Is the fire on the windward side or leeward side of the building? Will the wind work in your favor or against you?
- In a decay fire, the tactic of making a hole through the front door or through the Sheetrock exterior wall of the hallway just large enough to insert and operate a nozzle can safely reduce the chance of a smoke explosion or the sudden rush of fresh air to reignite a smoldering fire.

While acknowledging the dangerous potential for fire spread in a residential high-rise, we must also acknowledge the tactical advantage gained with the speed and mobility afforded firefighters when using 1¾-inch lines. These hose lines provide the required gpm to extinguish the fire. Planning for the worst-case scenario doesn't mean we have to execute the tactics for one; you can have overkill even in a high-rise fire. These points are best illustrated in a high-rise apartment fire that occurred in San Francisco, California, and that is described in the next section.

Gateway Apartments and Townhomes Fire in San Francisco, California

On October 22, 2018, at 5:15 p.m., a three-alarm high-rise apartment fire occurred at the Gateway Apartments and Townhomes in the Financial District of San Francisco. The fire started in a corner unit on floor 12 and burned every corner unit above it, up to the sixteenth floor, primarily from autoexposure **FIGURE 15-36**. The Gateway is a 25-story, nonsprinklered, high-rise building that was built in 1965 and contains 196 residential units, with eight apartments per floor. At the time of the fire, the building was approximately 25% occupied because many of the residents had not come home from work yet. There was approximately a 10–15 mph prevailing wind.

FIGURE 15-36 The Gateway fire started in a corner unit on floor 12 and burned every corner unit above it to the sixteenth floor.

Photo courtesy of Jonathan Baxter, San Francisco Fire Department.

As mentioned, the fire started in a corner unit on floor 12. The fire alarm sounded in the building, and many occupants evacuated, although some disabled and elderly tenants in wheelchairs were unable to. Others decided to shelter in place and stay in their units. The corner units had balconies with a glass sliding door in addition to adjacent wall windows to maximize the view. Once the flames broke out the glass on floor 12, the fire started lapping to the floor above. The flames were so high, they made direct contact with the windows above, and these windows failed quickly. It didn't appear that the wind was working against the firefighters, and although conditions could change, it was not a wind-driven situation where flames would be pushed back onto the attack team. Instead, it was blowing lightly, sideways, counterclockwise around the exterior perimeter of the building **FIGURE 15-37**. This breeze pushed the lapping flames under the overhang of the balcony above. It was speculated that the balcony overhangs were enough of a horizontal

FIGURE 15-37 Although exterior windows failed in the Gateway Apartments and Townhomes Fire, the prevailing winds were light. A perfect scenario for a wind-driven fire was averted.

Photo courtesy of Jonathan Baxter, San Francisco Fire Department.

barrier for the flames that radiated heat and convected currents were allowed to accumulate at the top of the door, contributing to early failure of the glass sliders to the balcony. This theory makes sense. And so it went, unit-by-unit, floor-by floor, up to floor 16.

When the fire department arrived, firefighters charged the Class II standpipe (rated at 750 gpm), then climbed with all their equipment to the eleventh floor. The first attack team connected a 1¾-inch hand line with a smooth-bore tip to the standpipe on floor 11 and proceeded to floor 12. The building PA system was used to identify and direct occupants to the evacuation stairway, but the attack was delayed because occupants used the firefighting stairwell anyway. The company officer knew that opening the fire floor hallway could result in thick smoke and heat rushing up the stairway, endangering or injuring occupants, and thus did not open the door. After the upper stairway was cleared of fleeing residents, the firefighters made entry into the hallway for a direct attack on the fire. A second 1¾-inch hand line was taken off the wye and laid up to floor 13 for an exposure line above the fire. It was confirmed to all units that this building did have a thirteenth floor.

An SAR team was assigned to start on floor 12 where occupants were in the most danger and work its way up. When the team made it to the corner unit on floor 14 above the fire, the flames from autoexposure had reached the balcony. Without a hose line, they made a quick decision to go back into the hallway and grab the house line from the hose cabinet. The house line is a 100-foot single-jacket section of 1½-inch hose with an attached spray nozzle that is part of the Class II standpipe system. It is intended for occupant use (called first-aid use), and many texts state that it should never be used by fire department personnel for firefighting, but firefighters should also use common sense. Engine companies should not use these lines in lieu of fire department hose, but if ladder companies find themselves in a position of needing fast water for fire attack or protection, then by all means use the house line. That's what it's for! As it turned out, due to the actions of these quick-thinking firefighters and the use of the house line, the corner unit on floor 14, except for the balcony, was the only unit not damaged by the flames. However, the fire continued to lap up and extend to floor 15.

Because the wind wasn't working against the firefighters, smoke and heat conditions in the firefighting stairwell were tenable, and firefighters were able to quickly stretch 1¾-inch hand lines to floors 14, 15, and 16. With direct fire attacks happening simultaneously in the hallways and inside the fire units, the fires on all floors were extinguished in 45 minutes. The second-alarm companies were assigned to stretch 2½-inch backup lines to the five floors, but none of the 2½-inch lines were used in fighting the fire. The attacks were successful using 1¾-inch hand lines. Again, because the wind was working in their favor, horizontal cross-ventilation using the prevailing winds was effective and sufficient.

A rare but not unusual fireground hazard was discovered after the fires were knocked down: some of the glass windows that failed during the fire were curtain-wall windows that ran from floor to ceiling. The failure of these windows created an unobstructed opening and a severe and dangerous fall hazard to any unsuspecting firefighter performing an interior perimeter wall search or being unaware of their surroundings during overhaul. This illustrates the ever-present dangers on the elevated fireground. After extinguishment and before overhaul begins, a scene-safety survey, and a risk-benefit analysis needs to be conducted of the immediate fire area, and hazards like unobstructed openings should be cordoned off. Wooden studs, or fire-damaged furniture, like tables or sofas, can be used to fence off an open wall or windows with low sills. Fire scene tape should be used to flag the area, and all firefighters must be made aware of these severe fall hazards.

The Gateway Apartments and Townhomes Fire confirms some strategic and tactical points:

- Based on the common household fire loads, even in high-rise apartments, and the square footage and compartmentation of residential units, 1¾-inch hand lines are effective for combating residential high-rise fires when these lines are properly supplied.
- The speed and mobility of moving charged 1¾-inch hand lines made it possible for firefighters to stop autoexposure quickly and attack fires on five different floors. These actions would not have been so easy if 2½-inch hose lines were used because they would have required more flow from the standpipe and a lot more personnel and resources.
- The standpipe supplied the water flows for five separate 1¾-inch hand lines operating simultaneously.
- Two-and-a-half-inch hand lines were laid to back up the 1¾-inch lines; although they were never used, this is a sound tactic.
- Even though the PA system was used to direct residents to the evacuation stairway, occupants still evacuated using the firefighting stairway, a situation we must expect in any evacuation situation.
- The firefighters who saved the fourteenth-floor unit by using the house line should dispel the notion that fire personnel should never use a house line for fire attack. Use common sense.
- First-in companies hiked up the stairway with all their equipment to the eleventh floor, which was the floor below the fire. Elevators were not used until it was confirmed that there was no smoke in the hoistways and that the elevators operated properly. Then Phase II operations shuttled equipment to the staging floor.

Miraculously, there were no injuries, and only 30 units in the Gateway Apartments were unable to be reoccupied. This was a textbook example in residential high-rise firefighting where everything went right. The fact that the fires on five different floors were extinguished in 45 minutes with 1¾-inch hand lines is incredible. The extra time and personnel that would have been required to perform this same evolution with 2½-inch hand lines exclusively could have meant that the situation would have turned out much differently had the fire occurred in the middle of the night when the building was fully occupied and residents were asleep.

Single-Hose Method (2½-inch)

The method for taking a single 2½-inch hose line per floor can be an effective fire attack if sufficient firefighters are on hand to help advance the charged hose line. If not, this method can complicate the attack or cause it to fail. Consider the hose lines that need to be laid up the stairs and that you must allow 50 feet of hose per floor; the 200-foot attack line is connected on the floor below the fire. If the exposure line is needed on the floor above, it may need to be connected before the backup line to get that crew out of the firefighting stairway and into the hallway. That hose must connect two floors below the fire and needs to lay up to the floor above the fire; that is 150 feet of hose just for the stairs. The backup line must connect to the standpipe three floors below the fire and also requires 150 feet of hose for the stairs, plus an additional 150 feet to be able to reach the attack nozzle; that's 300 feet of charged 2½-inch hose—and moving it is not an easy task. The RIT line must be connected four floors below the fire, or perhaps from another standpipe. You can see how this situation can get complicated. Plus, you would need a sufficient number of companies to manage this much hose and weight **FIGURE 15-38**.

Some buildings, especially older ones, have a wide open well between floors within the stairway. Hose lines can be stretched vertically between landings in lieu of laying hose over the stairs. This method is essentially creating a vertical standpipe. It is quick and uses less hose but needs to be tied off at intervals to a

FIGURE 15-38 Moving charged 2½-inch hose lines up a stairway is extremely difficult without sufficient staffing. It's even harder with heat- and smoke-filled stairways and if the stairwell already has charged lines.

Courtesy of Michael Gala.

banister or railing to support the weight of the charged hose and couplings. Some wells are narrow, which may work with 1¾-inch hose lines, but might be too narrow for 2½-inch lines. There is a risk that the narrow well may act as a hose clamp when the hose is charged, restricting the water flow and preventing this line from advancing.

Another effective solution for the single-hose method to limit the amount of hose used in the stairway is connecting a 2½-inch hose line to the standpipe outlet four floors below the fire, laying that line up to the floor below the fire, and connecting a 2½-inch wye to the male end. Essentially, the officer creates a second standpipe in the firefighting stairway with two additional 2½-inch discharge ports, which gives the engine officer more options and helps avert problems encountered with hose layering within the stairway **FIGURE 15-39**. The RIT line is often the fourth hose laid. Connecting the RIT line to this wye is a good use for this hose lay; it brings the RIT line connection up to the floor below the fire and allows firefighters to use the same amount of hose as the attack line because it isn't being used up in the stairway.

Hose Layering

Hose layering also needs to be considered. We have all seen hose lays forcefully expand when the line is charged. Three charged 2½-inch lines in a single stairwell are unmanageable **FIGURE 15-40**. The first line laid, the hose on the bottom, is the attack line. A 50-foot section of charged 2½-inch hose weighs about 105 pounds (48 kg) and 200 feet of charged 2½-inch hose weighs about 420 pounds (181 kg). When the 2½-inch backup line and the 2½-inch exposure lines are charged, the weight of these two hoses and the friction against the jackets make it impossible for the attack line to advance **FIGURE 15-41**, which can be disastrous. If this fire cannot be extinguished with the firefighters available to position and maneuver two 2½-inch hoses effectively, a direct attack will most likely fail. You need the options and flexibility, as well as the speed and maneuverability, to use 1¾-inch hose lines in addition to the 2½-inch lines. Therefore, you need the wyes.

FIGURE 15-39 To limit the lengths of hose connected in the firefighting stairwell, a 2½-inch hose line can be connected to the standpipe outlet four floors below the fire. This creates a second standpipe in the firefighting stairway.
Courtesy of Raul Angulo.

FIGURE 15-40 Three charged 2½-inch lines in a single stairwell are unmanageable.
Courtesy of John Odegard.

FIGURE 15-41 The hose on the right is the attack line. This hose is on the bottom, indicating that it was the first line laid, and it is buried by the weight and friction of two additional 2½-inch lines. It will be extremely difficult if not impossible to advance without an army of firefighters.
Courtesy of John Odegard.

At the First Interstate Fire in Los Angeles, California, four stairwells were in the center core, each with a standpipe and one 2½-inch discharge valve per floor. All four standpipes were utilized to attack the fires on floors 12 through 16. In total, there were 20 hand lines in operation by 32 companies on five involved floors. Attack-line diameters were a combination of 1¾-inch, 2-inch, and 2½-inch hose, with a combined estimated flow of approximately 2,400 gpm. You can't perform this type of operation without using wyes.

New Standpipes

In newer high-rise buildings, architects and fire protection engineers have recognized the need to take multiple lines off the standpipe on a single floor. These new standpipes now have two separate 2½-inch discharge connections with handwheel gated valves permanently affixed to the standpipe. In essence, they have created a 2½-inch gated wye **FIGURE 15-42**. This

FIGURE 15-42 New standpipes have two separate 2½-inch discharge connections on each floor permanently affixed to the standpipe, thus creating a 2½-inch gated wye.
Photo by Capt. Steve Baer.

is now part of the City of Seattle Fire Code; other fire codes may follow this trend, thus ending any debate about the use of wyes. As for the proponents of using wyes, these new standpipes provide additional options for the use of multiple hand lines (not all hose lines flow simultaneously).

Hallway Corridor Standpipe Connections

In some high-rise buildings, the standpipe discharge valves were designed and placed in the hallway corridors on each floor. They are located just inside the entry from the stairway landing. There may be an additional hose outlet in the middle of long hallways supplied by an interior mid-riser. An additional standpipe may or may not be in the stairwell. The theory behind this design was that if all the firefighting operations were confined inside the hallway, it would prevent smoke from entering the stairwell because firefighters would not have to open the door from the stairs when making entry to the fire floor, thus keeping the stairs smoke-free for safe evacuation. Their theory is correct; unfortunately, no one included discussions with the fire department because we would have given them all the reasons why we do not connect attack lines on the fire floor. This design would actually be deadly in a wind-driven fire because firefighters would have no way to retreat without abandoning their hose line, and the hallway, along with the control of the water source for the attack hose connection, would become untenable. In one Los Angeles high-rise fire, the extreme heat in the hallway caused these hallway standpipe outlets to fail, in turn causing water problems throughout the building, including a cascade of water onto the staging floor.

If smoke or fire is encountered on the fire floor, hose lines should be stretched from a standpipe connection in the stairway and not from a standpipe connection in the hallway. If hallway standpipe outlets are all that are available, it should be taken from the floor below. This means that the door to the floor below the fire has to remain partially open for the hose, and the door to the fire floor is open so the attack team can advance, which totally defeats the original purpose of this design. Connecting additional lines to other floors will most likely contribute to the loss of adequate pressurization in the stairwell. Minimizing the draft of air in and the smoke exhausted out into the stairwell is the reason that door control is emphasized. Note that the hallway outlets use 2½-inch supply pipe, so there is additional friction loss in this connection, unlike that from a standpipe outlet connected directly from a 4- or 6-inch standpipe.

On the flip side, hallway corridor standpipe connections are excellent for connecting exposure lines to the floors above the fire. Although a building's architects were probably not designing this feature to protect exposure floors, it does allow for the safe connection of these hose lines without the exposure teams being inside the firefighting stairwell above the fire. Crews can bring their hose and equipment in from another stairway and safely set up the exposure operations and protect the search teams without being affected adversely by the direct attack in the firefighting stairway.

Although the control of the water source is still on the same floor, the margin of safety is acceptable because crews should be ahead of the fire. If the situation deteriorates and the floor is lost, it is because the fire below is gaining headway. All crews on the floor above the original fire must retreat to the evacuation stairwell, not the firefighting stairwell. Hose lines may have to be abandoned. Companies will have to regroup and go up to the next floor with new hose lines.

Stretching Hose from the Standpipe

The initial attack hose line must be connected to the standpipe discharge outlet in the stairwell one floor below the fire floor. The excess hose could be coiled at the mid-landing above the fire floor, or it could be pulled up the stairway past the fire floor toward the next floor landing, then reversed, making a large horseshoe bend, bringing the nozzle back toward the fire floor entrance before it is charged. The charged hose will come down the stairs more easily as the advance is made. Remember to accomplish these tasks *before* the hallway door to the fire floor is opened. If the hose is thrown down a stairway, it must be worked, or pulled up, after it is charged, which is not an easy task if this is 2½-inch hose. Use gravity to your advantage, especially with 2½-inch hose.

If the hallway is clear of smoke and the door to the fire unit is closed, the uncharged 2½-inch hose line should be stretched quickly down the corridor past the fire unit. Then make a large horseshoe bend and bring it back to the front door of the fire unit before the attack line is charged. This makes the advance into the fire unit easier. A V-split stretch also works. Care must be taken to ensure the uncharged hose line doesn't get pulled underneath the door between the threshold and the bottom rail. Again, when the hose line is charged, the threshold and the bottom of the door will act as a hose clamp restricting the flow or stopping it completely. Even worse, if the door closes over an uncharged hose line, once it is charged, it can effectively wedge the door in the closed position, trapping firefighters on the opposite side without water.

Making Entry and Attacking the Fire

If part of the hallway corridor is involved in fire, the hose line should be charged before it is advanced from the stairwell into the corridor. The engine officer should announce over the radio that the fire attack is commencing as a warning because atmospheric conditions are about to change. The nozzle and hose line should be worked to utilize the reach of the straight stream while the officer scans the area with the thermal-imaging camera (TIC) to get thermal readings. Maintaining the thermal balance aids in searching in poor visibility for any victims or locating the seat of the fire; however, if the intensity of the flames is growing and temperatures are climbing, the ceiling area must be cooled by deflecting the stream off the ceiling to prevent flashover.

Ideally, the seat of the fire should be hit with a straight stream, but if the room is involved, the Z-, T-, or clockwise-O-spray patterns should be used to cover all areas of the room, including the floor. Avoid rapid movements or whipping the nozzle around. Such movement entrains air towards the fire, increasing intensity, and the higher pressure created could send hot gases back on the attack team. Like any interior fire, if the fire isn't knocked down after 60 seconds of water application with a 1¾-inch hand line, bring up the 2½-inch backup line, and bring in an additional 2½-inch backup line or portable master stream.

Firefighters should always be prepared to fight their way into the corridor from the stairwell. When a long corridor is completely involved in fire, the hose line should operate from the hallway threshold and use a straight stream to reach as far into the hallway as possible. This cannot be done during wind-driven fires. The officer using the TIC can help the nozzle firefighter aim the stream precisely at the seat of the fire. The company officer must know if this condition is caused by a wind-driven fire. If extreme heat isn't driving the company out of the hallway, the IC or pump operator can be asked via portable radio to confirm visually if this fire has self-vented through the window. Flames will be visible from the exterior of the building. If they are visible, the crew must be pulled back to the safety of the stairwell. History and experience have demonstrated that even two 2½-inch hose lines with straight streams are ineffective against wind-driven fires. ***A note of warning:*** The fire must be sufficiently knocked down, if not extinguished, before windows are opened with a key or taken out in order to prevent a wind-driven fire from developing. The fire attack occurs without ventilation on fire floors, and ventilation doesn't commence until the fire is knocked down or extinguished. Water first, then vent.

Once the fire is knocked down in high-rise apartments or hotels, a straight stream at midlevel between the floor and ceiling might blow out any windows that were subjected to high heat, aiding in horizontal and hydraulic ventilation. But high-rise glass is strong; it may not work but it's worth a try because horizontal ventilation is challenging. It is best to ventilate to the outside from the room of origin. Once this exhaust portal is created, PPV fans can be used to bring in fresh air and accelerate the ventilation of smoke and fire gases. As conditions improve, all shards of glass should be removed to the inside of the fire room so that the window opening is totally clear. After knockdown, shut the line down and allow the smoke and steam to vent. Perform a primary search of the area and then follow with a secondary search by a different crew. The officer can reevaluate where hot spots are remaining. After final extinguishment, keep one crew to check the fire. Rotate crews; try to determine the cause of the fire; and perform a new size-up, safety survey, and risk-benefit analysis before salvage and overhaul begins.

Connect the RIT Line to the Fire Floor

The only hose outlet on the standpipe that isn't being utilized during the initial fire attack is the one on the fire floor landing. The attack line and the backup line are all coming from the standpipe outlets below. As mentioned earlier, any standpipe outlets used for hose lines above the fire floor during the fire attack depend on tenable smoke conditions within the firefighting stairwell, so exposure lines to the floor above the fire may come off a second standpipe. A RIT line should be the second hose line laid when the RIT is established, but if it isn't, the RIT line is the fourth line laid in the sequence. If you're using a single 2½-inch line per outlet method, that puts the RIT line connection four floors below the fire. A better solution is to connect the RIT line on the fire floor and coil the hose on the first stairway landing above the fire. This would depend on tenable conditions in the firefighting stairway with the hallway door closed as tightly as possible until the hose can be coiled and charged. The firefighter needs to stay low during this fast evolution. The charged nozzle could then be brought down to the first or second stair tread above the fire floor landing and left in a ready position unattended **FIGURE 15-43**. The RIT would stage in the hallway below the fire attack to stay out of the way. If a Mayday is called and the RIT is deployed, the team members can grab the charged hose line and enter the fire floor. Since this line is being advanced over (on top of) the existing hose lines, the RIT line will not be hampered by hose layering.

The RIT hose line doesn't necessarily have to enter the fire area to protect or save firefighters from death or injury. The first RIT can enter the hazard area and utilize the backup line to wet down and cool firefighters during a rescue. The line on the stairway can become the new backup line to support the rescue with a newly established second RIT. Consider the following scenario: If conditions suddenly change on the fire floor or extreme fire behavior starts to develop and the firefighters must bail out of the fire floor hallway without their hose line, it's because they're burning. They're trained to follow their hose line out, so these firefighters will be exiting to the fire floor stairway landing. This RIT line is in a ready position to protect their egress and flow water to cool them down—two good

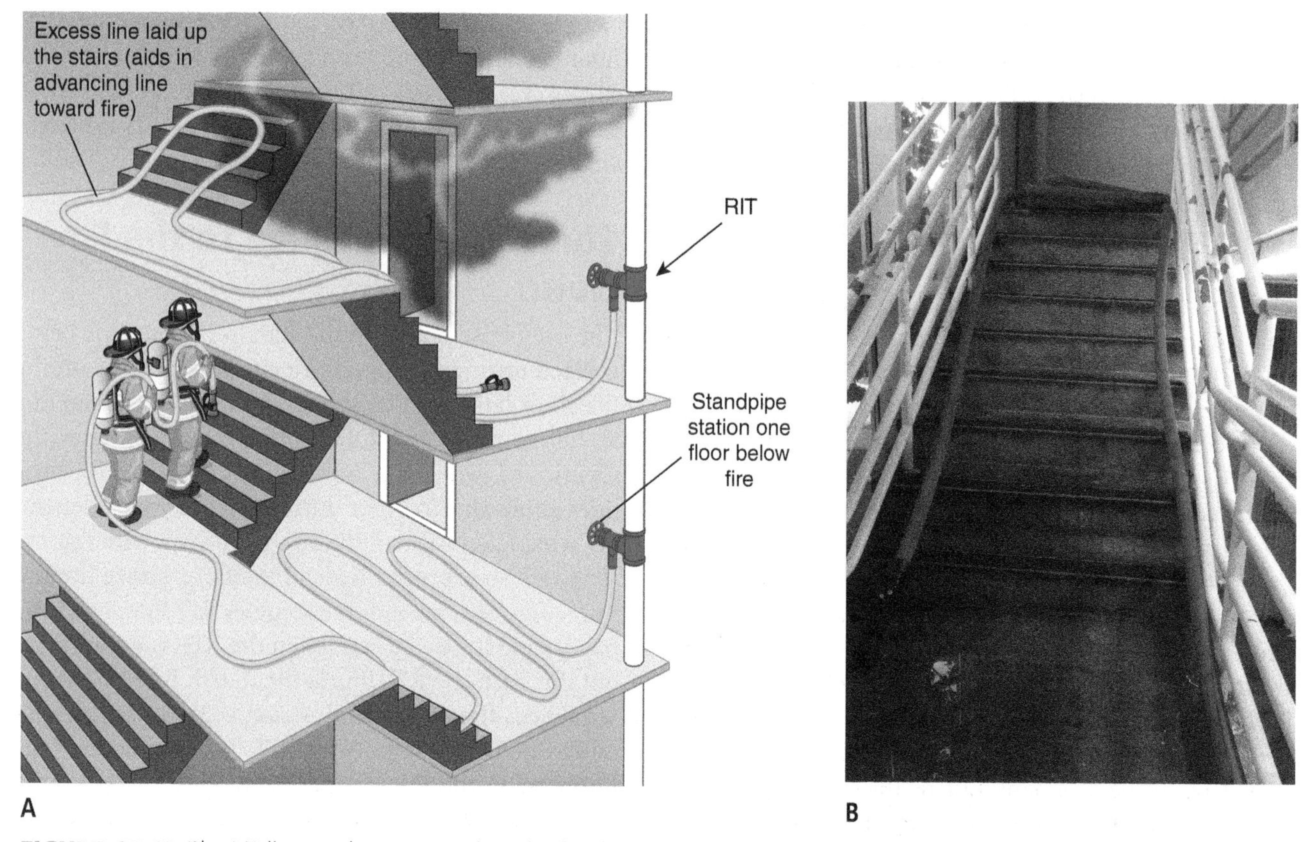

FIGURE 15-43 The RIT line can be connected on the fire floor hose outlet and coiled on the stairway landing above the fire floor. The combination nozzle can be left unattended in a state of readiness. The door to the fire floor should be closed as tightly as possible.

A: © Jones & Bartlett Learning; B: Photo by Lt. Kevin O'Donnell.

reasons to have a combination nozzle on the RIT line. A RIT firefighter simply needs to grab this hose line and assist the firefighters in clearing the threshold to safety **FIGURE 15-44**. Once the fire is extinguished and firefighters are no longer in danger of getting trapped, there is no reason that this hose can't be used for other overhaul activities.

FIGURE 15-44 The charged hose in the ready position for the RIT firefighter to protect a rapid egress of the attack crew.

Photo by Lt. Kevin O'Donnell.

Wind-Driven Fires in Residential High-Rises

Of the five property classifications in high-rise buildings, the ones that are least sprinklered are older residential high-rises. A wind-driven fire can be deadly in a residential high-rise. This scenario is created when the fire self-vents by breaking out a window on the windward side, or when a window is left open at the time of the fire. The wind, along with the fire, pressurizes the unit. As long as the front door of the unit remains closed, the flames, smoke, and gases pulsate while trying to escape against the wind through the open window. As soon as firefighters attempt to make a direct attack through the front door, a high-pressure to low-pressure flow path is created. Remember that high pressure always seeks the path of least resistance and moves toward the low pressure in an attempt to equalize. But in this scenario, the wind drives the flow path currents with extreme speed and temperatures well over 1,200°F (538°C). This creates a situation known as extreme fire behavior caused by a wind-driven fire. Firefighters cannot survive this event if they are caught in the flow path. This is the scenario that killed three New York City (FDNY) firefighters at the Vandalia Avenue Fire on December 18, 1998, in Brooklyn.

On December 10, 2009, in Chicago, Illinois, a fire started on floor 36 of the 43-story DeWitt-Chestnut Apartments. The female occupant of the fire unit was killed, and 12 others were injured. The occupant, who was trying to escape, collapsed at her front door. She blocked it open, which allowed the smoke and flames to enter the hallway. While Engine 98 was hooking up to the standpipe, the officer ran up to the fire floor to check fire conditions. As he entered the hallway, the door behind him shut. The ceiling temperatures were so hot that they started to warp the door. Conditions were rapidly changing for the worse, and the officer had to retreat quickly, but he could not open the hallway door through which he had entered. He ran as fast as he could to the opposite stairway and miraculously escaped without injury. The officer rejoined his crew on the floor below to initiate the fire attack. By this time, the hallway door was so warped that the firefighters needed forcible entry tools to pry it open. Engine 98 attempted to make entry to the fire floor but the firefighters were pushed back numerous times by high heat. Opening the door created a flow path that allowed a wind-driven fire to develop. They never got more than 10 feet (3 meters) into the hallway before bailing out. The convected heat currents flowing up the firefighting stairwell were estimated at well over 2,000°F. The fire on floor 36 completely melted the stairway identification placard on the floor 37 landing (Figure 15-7). These high-temperature convected currents were released every time the firefighters opened the hallway to attempt entry. Any firefighter or civilian in the stairway would have been killed. **FIGURE 15-45** shows the extent of the heat damage on floor 36. The floor-level temperatures were so hot that the shutoff bail on the nozzle of the abandoned hose line melted completely.

Before making entry into the hallway from the firefighting stairwell, the officer must determine smoke conditions in the hallway. This is easy if there is a wired glass window in the hallway fire door. A simple look can determine if the hallway is relatively clear of smoke or charged with smoke. The officer should check the edges around the door while it is closed for pressurized smoke pushing through the door jamb. Otherwise, the officer can open the door slowly to check hallway conditions, but she or he must be prepared to close the door quickly. Thick smoke in the hallway may be your only sign that the entry door into the fire unit has been left open. High heat and thick smoke under pressure are the first signs that the window in the fire room has self-vented.

FIGURE 15-45 Floor 36 of the DeWitt-Chestnut Fire in Chicago, Illinois.

Photo By Battalion Chief Michael Wielgat, Chicago Fire Department.

On commercial buildings where windows cannot be opened, crews on the exterior of the building can confirm if the window has self-vented. In residential buildings, a crew above the fire floor can open a window and visually check fire conditions below from a safe location. A driver on the outside perimeter, a spotter from an adjacent high-rise, or the IC at an exterior command post can also confirm, and relay via radio, if the fire has already self-vented through the window. Each of these personnel may be able to note a change in fire behavior and flame direction when the hallway door is opened. This may be the only way to determine if a wind-driven fire will be created when the hose team initiates a direct attack. If the spotter confirms that the fire has self-vented and the hallway is filled with smoke, the engine officer should not attempt to make entry into the hallway. The fire will have to be attacked from below, or a fire curtain will have to be deployed from the floor above the fire. You also cannot rule out arson, in which leaving the front door open to the unit is a deliberate attempt to spread the fire.

Fire Curtains

A fire curtain is basically a big, heavy, fire-resistant tarp. There are small 6-foot × 8-foot curtains and larger 10-foot × 12-foot curtains called fire blankets. The 6-foot × 8-foot fire curtain was designed for windows in residential occupancies. The 10-foot × 12-foot fire blanket can work on exterior glass curtain walls of commercial and office high-rise buildings, but the evolution is much more complicated and rarely attempted, even in New York City.

The purpose of this tool is to cover the exterior window of a fire room that has self-vented or was left open on the windward side of the building. The fire curtain covers the inlet portal of the flow path and prevents or stops the extreme fire behavior caused by a wind-driven fire. The window, in essence is closed, stopping the unlimited source of fresh air. This will lower the intensity of the flames and reduce the energy output, reducing the temperature. If the front door to the fire unit is closed, the attack crew should be able to enter the unit without the danger of a wind-driven fire developing. If the front door to the fire unit was left open, the curtain should stop the blowtorch flow path that exists in the hallway. Fire crews can advance into the fire unit from the hallway.

The fire curtain is heavy, so this is not a fast operation. A ladder crew has to gain access to deploy the curtain from the floor above or from the roof. It also requires another crew to stage in the unit below the fire. The ladder crew from above needs to drop two guide ropes attached to the fire curtain to the crew below. Then the tarp is dropped into place, suspended by two guide ropes from the top. The two crews work together to place the curtain in front of the window by pulling the respective ropes. Then the ropes are tensioned and tied off.

In a drill, this can be accomplished fairly quickly. In real-world conditions, strong winds can blow the ropes any which way, and the curtain can become like a sail in the wind. In high winds, if the ropes are not tensioned properly, the strong wind tends to push the curtain into the window. The crew working in the window below the fire may be struck by falling glass while the curtain is being shimmied into place. The crew working in the window above the fire or on the roof may end up in a thermal column. Nevertheless, a fire curtain is an effective solution to consider in combating a wind-driven fire in a residential high-rise where the fire has self-vented through a window. Once the fire is extinguished, the curtain can be removed to aid in horizontal ventilation by establishing an exhaust portal somewhere else on the fire floor—or into, up, and out of the stairwell.

The curtain doesn't work with balconies. The balcony railing and any patio furniture obstructs and prevents the curtain from creating a seal against a glass sliding door and the balcony ceiling overhangs prevent the curtain from being lowered or secured.

Residential High-Rise Nozzles

The high-rise nozzle, also called a floor-below nozzle, and the HydroVent nozzle are other methods for attacking a residential high-rise fire from the unit below when conditions for a wind-driven fire exist. The high-rise nozzle is an 8-foot (2.4 m) long, 1½-inch (3.8 cm) diameter aluminum pipe whose tip is bent back on itself approximately 68°. This bent-pipe J-configuration has a 1⅛-inch smooth-bore tip connected to the end and can flow approximately 250 gpm. The pipe is supplied by a single 2½-inch hose line taken off the standpipe. It has a ball-valve shutoff, so the hose can be moved to adjacent windows as needed. The nozzle is pushed out the window, and the straight stream is aimed into the fire window above. Rolling the nozzle back and forth covers the window area and shoots the straight stream across the ceiling. Deflecting the stream off the window header or ceiling knocks the fire down, and a spotter from the outside can confirm the proper positioning and effectiveness. Then the nozzle is shut down, and crews can make entry into the fire unit from the hallway for final extinguishment.

The HydroVent is a dual-attack suppression and ventilation nozzle that sprays water into the ceiling with one nozzle and hydraulically ventilates the fire simultaneously through a second nozzle, both attached to a single pipe. The nozzle is meant to be swung up and hung on the windowsill of the fire room by a single firefighter working at the window from inside the room below the fire. Once set and charged, the HydroVent nozzle can be left unattended. The suppression part of the nozzle is a round nozzle with four smooth-bore discharge holes that flow approximately 95 gpm toward the top of the heated compartment. This creates a sprinkler-type system in which the water converts to steam 1,700 times its volume. The water application knocks down the fire quickly and takes away the thermal energy, reducing interior temperatures. Operating simultaneously is the ventilation 95-gpm combination tip. By design, this part of the nozzle is already positioned on the pipe at the windowsill, aimed toward the outside for hydraulic ventilation and capable of pulling thousands of cubic feet per minute (cfm) of smoke, steam, and other gases of combustion out of the room, thus reducing interior temperatures even further. If the wind is too strong for ventilation to occur, it will still push all the spray and moisture along the flow path, cooling and extinguishing fire as it goes. An exterior spotter can confirm when the fire is knocked down, and crews can enter the fire unit from the hallway for final extinguishment.

The two nozzles are safer and faster than the fire curtain evolution because water application significantly cools the interior temperature of the fire room before the attack crew makes entry for final extinguishment.

Indirect Flanking Attack

If it is confirmed by observers on the exterior of the high-rise apartment that the fire room has self-vented and flames are exiting the window, and the engine officer has determined that the fire floor hallway is relatively clear of smoke with little if any heat, the officer can conclude that the front door to the fire unit is closed. However, a direct attack into the unit could create a wind-driven fire as soon as the firefighters open the door and the flow path reverses.

The safest method to attack this type of fire successfully is with an indirect attack that includes cutting a small hole just large enough to insert and operate the nozzle through the front door or the Sheetrock exterior wall of the unit from the hallway. If the decision is made to preserve the integrity of the door, the indirect attack can be made from the adjoining residential units. The first point is to ensure that the front door to the fire unit remains closed. Attack lines should be laid into both adjoining exposure units—to the right and to the left of the fire unit. A TIC can help determine which wall has the highest heat readings, and the indirect attack should be concentrated on that particular wall. Then, using piercing nozzles, an indirect attack can commence into the fire unit. Another indirect attack method is to breach the walls into the fire unit with holes just large enough to accommodate combination nozzles set to a medium fog, or distributor nozzles. This indirect attack is like introducing a giant sprinkler system into the unit. Backup lines should be laid to the exposure units to support the attack team because the nozzles will be through the wall, leaving the firefighters vulnerable should something happen.

The two factors that make this type of fire difficult to fight is the wind and the elevation. The fire load and the residential square footage require water flow rates similar to a one-story apartment or condominium. Therefore, multiple handlines, or even portable monitors operating from the adjoining units, should be able to flow the required gpm to extinguish this fire. Firefighters serving as observers from adjacent buildings or rooftops can let the attack crews know when the indirect attack has knocked down the fire. Then attack lines and backup lines can be repositioned for a direct attack and final extinguishment. Multiple PPV fans can also be set up in the hallway and at the entry of the fire unit to keep the wind at the back of the fire crew while the crew members overhaul the fire.

Indirect Attack from Above

If the ceiling of the fire room and the floor decking to the unit above the fire room are not concrete (for example, they are constructed of wood trusses), the fire can also be attacked indirectly from the unit above the fire room using piercing nozzles or distributor nozzles—tactics similar to those used on a basement fire. The ceiling and floor construction can be confirmed by pre-incident plans or by opening the ceiling in the hallway with a pike pole. A chainsaw can quickly make access holes from the floor above the fire, then the nozzles can be dropped in.

Commercial High-Rise Fire Attack

This section of the chapter contains firefighting strategies and tactics to be considered in Type I construction office and commercial spaces with large floor or open floor plans. High-rise buildings with a center- or side-core design have cores that contain the elevators, the elevator lobby, at least two stairwells, bathrooms, kitchen, breakroom, and utility closets. A labyrinth of office cubicles and offices surround the center core. The fire load of paper products, computers, office accessories, office furniture, and so on, are made of plastic and hydrocarbon-based synthetics, which provide a greater modern fuel load capacity with tremendous British thermal unit (BTU) heat release energy potential than other property classes. Other than the supporting columns, the floor space is wide open. Floor space can easily range from 20,000 ft^2 (1,858 m^2) to over 200,000 $feet^2$ (18,581 m^2). Unless controlled by automatic sprinklers, this open space provides ample oxygen for the fire to grow to a significant size before it even starts to become ventilation-limited—if it ever will. The fuel will probably be consumed (fuel-limited) before these fires run out of oxygen. Serious fires, causing death and injury to firefighters and occupants, have occurred in such buildings in the United States and elsewhere.

High-rise fires should be attacked as quickly as possible but, by their nature, getting the attack line to the seat of the fire can be difficult and time consuming, especially with fires in upper stories. Firefighters would be lucky to get a hose line in operation within 15 minutes of arrival, and 20 minutes is more realistic. Type I, fire-resistive construction is designed to contain a fire for an extended period of time. The maximum interior operational period is in Type I buildings.

The most unpredictable aspect in high-rise firefighting is the wind. Smoke movement within the building is determined by many factors, such as HVAC systems, summer and winter stack effect, prevailing winds, and others. It is a complicated subject, and few experts can predict with accuracy the causes and effects of firefighting actions.

Exterior Recon

A major problem in fighting high-rise fires is that firefighters cannot see the exterior upper perimeter when they are in a hallway corridor on the inside. They need eyes on the outside.

Perhaps it is time to introduce a new tactic in high-rise firefighting: sending a high-rise recon team equipped with binoculars into an adjoining high-rise building to the same floor so they have the closest line-of-sight horizontal perspective of the fire conditions across from them. It is similar to the important tactic of having the IC observe the C-side rear of a single-story structure fire. If visibility allows, the high-rise recon team can observe and immediately report via radio the fire and smoke conditions, wind and flame intensity, interior exposures, and fire load; confirm sprinkler operation; and, most important, coordinate with the fire attack team the cause and effect of fire behavior with their current actions. The high-rise recon team would also have the best vantage point to report the effectiveness of the attack as well as notify exposure teams on the floor above if the fire is being contained to the floor of origin or if and where it is spreading to the floor above. The high-rise recon team may also be able to locate visible trapped occupants in windows on the fire floor and floors above the fire. This new tactic makes sense and should be considered as a regular assignment at high-rise fires.

The photo in **FIGURE 15-46** was taken from floor 11 of an adjacent high-rise building at the First Interstate Fire. It clearly shows the fire conditions on floor 12. If you look carefully at the three windows to the right on floor 11 below the fire, it appears that a person is in the center window shrouded by smoke. The high temperature on May 4, 1988, in Los Angeles was 73°F (23°C). It is possible that a summer stack effect pushed smoke down below the fire floor. This picture was shot with 35-mm film and was not developed until the incident was over. The photographer did not realize what he had captured.

Direct Attack

If the fire department can set up and be ready to advance to the fire floor while the fire is in the growth stage, there is a good chance that the fire can be extinguished with one or two 2½-inch lines. A 2½-inch line equipped with a 1⅛-inch smooth-bore tip can flow 250 gpm, and the stream has about a 50-foot reach. This line should be able to extinguish about 2,500 square feet of fire.

Before making the attack on the fire floor, the company officer must again confirm the status of the firefighting stairwell above the fire. Occupants coming down the stairs can effectively stall the attack until they get below the fire. Do not open the stairwell door with civilians in the stairwell above the fire. Once the stairwell is clear, initiate the attack. The officer should check the door for pressurized smoke and be prepared to close it. Smoke pulsating around the edge of the door indicates a window has failed on the fire floor **FIGURE 15-47**. If the door is opened to the fire floor, the firefighting stairwell will be charged with smoke and heat. Pay attention to the smoke velocity and the density. Fast smoke movement reveals the direction of the flow path and which way the fire wants to go. Smoke and fresh air heading into the fire floor when the door is open indicates a fire in the growth stage and that the area isn't pressurized yet from the fire. The pressure in the stairway is greater than in the fire area. Following the path of the smoke and fresh air

FIGURE 15-46 The fire conditions on floor 12 of the First Interstate building are clearly visible. Light smoke is visible on floor 11, with what appears to be a trapped occupant shrouded by smoke.

Photo by Rick McClure, LAFD.

FIGURE 15-47 The officer should check the door for pressurized smoke and be prepared to close it. Smoke pulsating around the edge of the door indicates that a window has failed on the fire floor.

should take you right to the seat of the fire. If smoke is coming out into the stairwell, operate the 2½-inch hose line with a straight stream into the ceiling area. Use the reach of the stream to penetrate deep into the hallway **FIGURE 15-48**.

If the fire is substantial, the smoke and heat will make this stairwell untenable for firefighters trying to stretch an exposure line to the floor above. Taking an exposure line to the floor above via the firefighting stairwell may have to wait until the heat release rate on the fire floor subsides, which occurs after knocking down the fire.

Remember that smoke is fuel. Use a TIC to get heat readings. You have only one chance to knock down the fire with a direct attack. The direct attack can be augmented by operating the second 2½-inch line. Two 2½-inch attack lines are all that is realistic. Moving three or four 2½-inch lines through the hallway door is impossible to manage and advance. If two 2½-inch lines can't do the job, the fire is growing beyond the ability to overwhelm it with hand lines, and portable master streams will have to be utilized.

Pincer or Flank Attack

Another alternative attack is to use the center core of the floor to flank the fire. The center-core walls can shield crews from radiant heat, so one crew can go to the left and one crew can go to the right. They can utilize the reach of the streams to perform a pincer attack, hitting the seat of the fire from two sides without having opposing lines.

The fire may also surround the center core completely, and attacking from opposite stairwells may be necessary and effective without having opposing lines. In the First Interstate Fire, there were four stairways, each equipped with standpipes, and all four standpipes were used to supply the attack on the fire. Because each crew was operating from the center core toward the outer walls in their respective directions, there were no opposing lines.

FIGURE 15-48 If fire is coming into the stairwell, operate the 2½-inch hose line with a straight stream into the ceiling area.

Defensive Interior Attack

Portable Master Streams

Fire showing from numerous windows on a floor indicates that the building is not sprinklered and that the fire has gained substantial headway in large open areas. Time and energy shouldn't be wasted on attempting an attack with 2½-inch hand lines. Two 2½-inch lines require the efforts of six to eight firefighters and can flow only 500 gpm to 600 gpm at best, depending on the nozzle. They can be set up later as backup lines.

Master stream appliances should be put in service right away for the initial attack. The decision to use master streams is your last resort for a full direct attack. Such an attack is actually a defensive interior attack and should be viewed as such. The goal is to flow as many gpm of water as possible. The energy of this fire has to be knocked down to stop the intensity and spread of the fire; otherwise, the radiant heat will be unbearable. Pressurized smoke created on the fire floor seeks every possible vertical space to extend to. Autoexposure, or lapping, is the vertical extension of flames out the windows of a high-rise to the floor above. In this case, master streams have to be supplied from one or more standpipe outlets as required, with 2½-inch lines or LDH **FIGURE 15-49**.

The 500-gpm TFT *Blitzfire* monitor nozzle is an excellent tool for a defensive interior attack. The low-profile nozzle can be carried and placed in service by a single firefighter and, once set, it can be left unattended. It has a hydraulically powered, oscillating feature that gives it the ability to sweep the floor or aim the water stream into the overhead automatically.

If the fire cannot be knocked down with portable master streams, that's it. You're done. The fire has gone beyond your ability to supply sufficient gallonage to overwhelm the fire. It's time to back out and take a defensive posture in the stairway until the fire consumes the available fuel load. Firefighting efforts should now shift to attacking from the floor below and protecting exposures on the floors above. The hose line should cool the fire door to the fire floor entry to prevent warping. The fire rating for fire doors used on a 2-hour wall is 90 minutes. That time can be extended if the door and frame are constantly cooled and protected with a hose line.

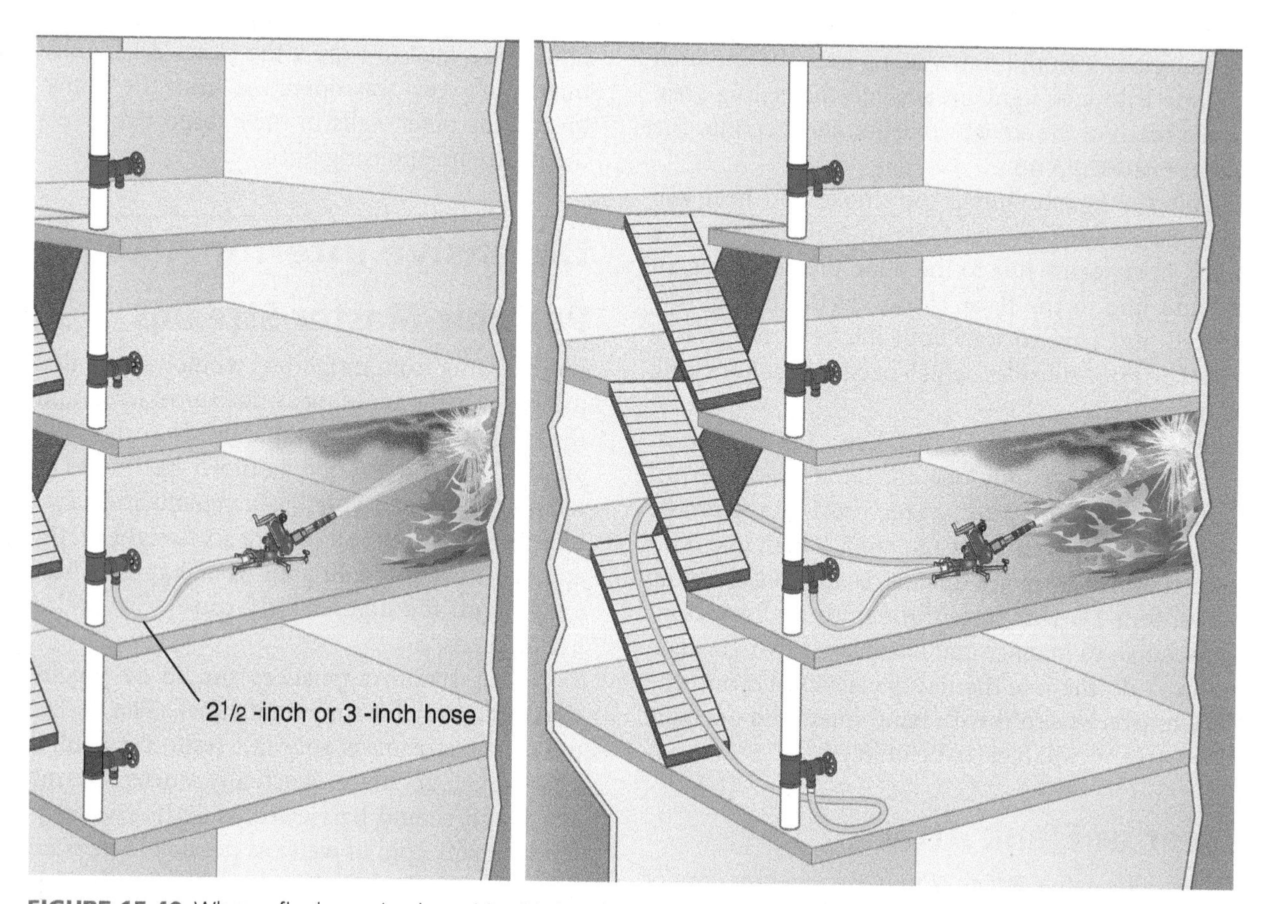

FIGURE 15-49 When a fire has gained considerable headway, a master stream should be used for a defensive interior attack. The monitor must be supplied from one or more standpipe outlets.

Ultimately, the decision to make a direct attack, a pincer or flank attack, or a defensive interior attack with master streams will be very clear. The decision is based on whether the interior heat conditions are tenable or not. If conditions are untenable, your decision should be no attack and let the fire burn itself out. The civilian survivability profile is zero, and there's no reason to risk and punish engine crews who will get burned or injured in an unwinnable attack when the only life hazards are themselves.

Relying on Type I Construction

Type I fire-resistive construction is the strongest construction firefighters encounter. Type I high-rise buildings and the vast array of fire load stuffed into them have to withstand and endure the strongest of natural elements and forces: wind, rain, snow, earthquakes, and fire. These steel structures do not contribute to the fire load and are coated to protect them from the effects of heat, unlike unprotected steel in Type II noncombustible construction. Unprotected steel can soften, elongate, and start to fail when temperatures rise above 1,000°F (538°C).

In Type I construction, the columns have a 4-hour fire rating; the beams are rated for 3 hours; and the floors, ceilings, and shafts have a fire rating of 2 hours. The hallway corridor fire doors are rated for 90 minutes. The fire load in Type I high-rise buildings can be great, as can the potential BTU heat energy production and heat release rate. This potential is designed into the building construction; thus, Type I high-rise office buildings are giant incinerators that can far outlast the burn time of any combustible material occupying the interior space. Let the fire on that floor burn itself out. Think of it as a controlled burn. Efforts should be concentrated on removing occupants above the fire and stopping fire extension.

At the First Interstate building, the fire extended at a rate estimated at 45 minutes per floor and burned intensely for approximately 90 minutes on each level. This resulted in two floors being heavily involved at any point during the fire. After 3 hours and 39 minutes of firefighting, five floors were destroyed by the fire; however, in the following months, structural engineers determined that the Type I fire-resistant building suffered no structural damage from the fire. At the One Meridian Plaza Fire in Philadelphia, Pennsylvania, nine floors were destroyed by fire. After 11 hours of uninterrupted fire in the building, the IC withdrew all crews from the building fearing a structural collapse, but it never happened. One Meridian Plaza was never reoccupied and was eventually demolished 8 years later.

No-Attack Strategy

A new strategy to consider when fire has gained control of a floor in a commercial office space high-rise with limited personnel and resources is the no-attack strategy. Time and effort can be spent on a fire attack only to realize that putting out the fire wasn't possible from the beginning. In a failed direct attack on the fire floor where interior conditions are untenable for firefighters, the civilian survivability profile is zero. Consider the benefits of a no-attack strategy when necessary, often after a failed direct attack. Keep the following points in mind for a no-attack strategy:

- Do not open the hallway corridor door to the fire floor so that the firefighting stairway remains clear of smoke, fire gases, and CO.
- Any occupants in the firefighting stairway are safe from heat and smoke.
- SER teams can move from floor to floor without the danger of being trapped by heat and smoke entering the firefighting stairwell.
- Evacuating occupants from upper floors can be expedited if stairwells stay clear of smoke and heat.
- There is no urgent need to vent the bulkhead at the top of the stairs.
- Any conditions that could create a wind-driven fire will not develop.
- Exposure teams can connect multiple hose lines from the firefighting stairway standpipe to protect the floor above the fire, without smoke or heat from the fire floor.
- PPV fans can assist in keeping the stairwell pressurized.

Once multiple exposure lines are in place on the floor above the fire, or whatever floor is necessary to get ahead of the fire, firefighters are in a position to contain the fire. A hose line needs to be in place to protect the fire doors at the entrances of the fire floor. Then the fire should be allowed to burn down and consume itself, removing the fuel leg of the fire tetrahedron. If there is another way to take the energy out of this fire to prevent extension and autoexposure by attacking from above or below the fire, those strategies and tactics should be implemented.

Attacking from Below the Fire Floor

As mentioned, stack effect can produce unpredictable smoke movement within the building. Although rare, sometimes smoke and fire can move down, so it is helpful to keep this in mind to avoid being surprised.

Let's use the fire behavior example from the First Interstate Fire. If the strategy for the fire floor is no attack, and the windows have self-ventilated, the estimated time for autoextension is 45 minutes per floor, and the burn-down time of the fuel in the fire is 90 minutes; that's a lot of time to try different options for fire attack from below. Attacking from below also allows firefighters to be in the safest atmospheric conditions while getting as close to the seat of the fire as possible.

Falling glass from the floors on fire was a dangerous, ongoing problem at the First Interstate and One Meridian Plaza fires. The film layer that coated and held together many of these large glass panes caught fire, and burning shards came sailing down. If the high-rise nozzle is deployed from below, windows have to be opened or broken. Because conditions should be almost normal below the fire floor, using the key to unlock the windows designed to be opened should prevent additional property damage and decrease the danger of more falling glass. If the glass needs to be broken, however, break it. A trick of the trade is to use a carbon dioxide (CO_2) extinguisher on the glass to freeze it, then shatter it with the point of a pickhead ax. Notify the IC and announce over the radio that selected windows will be broken for fire attack so personnel below can take cover. Include in the radio announcement the side of the building (A, B, C, or D). Wait for an "All clear" from the IC, then break out the windows that you need to break.

The High-Rise Emergency Response Offensive Nozzle (HEROPipe)

The HEROPipe is an 8- to 15-foot (2.4- to 4.6-m) telescopic, waterway high-rise nozzle that gives firefighters another opportunity for an exterior attack when the fire is well beyond the reach of aerial ladders and elevating platforms. The substantial portable monitor weighs about 80 pounds (36 kg) and is stored within a 7¾-foot (2.4-m) wheel-and-track-base unit. It can be wheeled inside any high-rise elevator. This system is designed to be assembled and operated from the safety of the floor below the fire. The unit secures to any Type I fire-resistive construction, including windows, balconies, and or glass curtain walls that extend from floor to ceiling. Many major cities have high-rise companies or squad units that carry special high-rise equipment. The setup time for the HEROPipe is negligible compared to the time spent on traditional high-rise tactics. Because firefighters are not battling heat and smoke, visibility is clear, and they may forgo using SCBA. A well-trained team can assemble and put the HEROPipe in service in less than 5 minutes.

The HEROPipe can be supplied by a single 2½-inch hose connected to the standpipe or by two parallel 2½-inch lines connected to a Siamese for additional gpm. The main waterway is a 3½-inch (8.9 cm) internal diameter pipe, and the telescoping extension is a 3-inch (7.6 cm) internal diameter pipe. When water is charged to the control valve, firefighters slide the wheeled waterway along the tracks and the counterweight of the hoses shoots the waterway up into position to the fire floor above.

The tip is an Elkhart Sidewinder EXM water cannon. This monitor tip is a giant combination nozzle that has varying degrees of fog patterns, including straight stream. The monitor is rated for 700 gpm at 80 psi. If a 1½-inch or 1⅞-inch smooth-bore tip is used, the HEROPipe can flow up to 900 gpm, with a penetrating stream reach of 145 feet (44.2 m) **FIGURE 15-50**. Six-inch standpipes with fire pumps, when properly supplied, can deliver anywhere from 500 to 1,000 gpm of water (600 to 750 gpm are more realistic attainable flows for high-rise firefighting). The remote-controlled monitor has a vertical sweep range of 60°, 20° below 0 level, and up vertically to 40°. It has a horizontal sweep range of 40°, 20° to the right, and 20° to the left. When the unit is set to automatic mode, it can be left unattended. The monitor tip oscillates up and down, and right to left simultaneously, within the respective degree ranges. A 2-hour and 4-hour battery pack runs the oscillating motor. The tip assembly can also accommodate a TIC with wireless image transmission to the operator. After the fire is knocked down, the TIC can detect remaining hot spots, and the monitor can be aimed directly to these spots. If the monitor is put into manual mode, the vertical angle can shoot straight up, 90°. The ability to shoot up 90° can be a significant autoexposure control measure for the lapping of flames to upper stories. After a section of the fire floor has been knocked down, the HEROPipe can be repositioned (because the unit is on wheels) to another window to attack the fire from a different position. With sufficient personnel, it can work the entire perimeter of the building.

FIGURE 15-50 With a 1½-inch or 1⅞-inch smooth-bore tip, the HEROPipe can flow up to 900 gpm.
Photo by Battalion Chief Michael Wielgat, Chicago Fire Department.

In a wind-driven fire, a unidirectional intake flow path is created through the fire floor, out through the hallway, and up the stairway, with temperatures reaching close to 2,000°F (1,093°C), and it can kill anything in its path. The HEROPipe can take advantage of the flow path. It is the only nozzle that can place an elevated, 900-gpm, heavy stream beyond aerial reach and into the window intake of the flow path. Wind currents carry and push all the steam and moisture through the flow path, extinguishing fire as it goes **FIGURE 15-51**.

In a wind-driven fire or in a substantial high-rise fire, this may be the quickest and safest tactic for getting water directly on the seat of the fire. Once water hits the fire, everything else gets better. In the DeWitt-Chestnut Fire, Chicago, Illinois, the fire was uncontrolled for over 3.5 hours. Due to the intense heat, firefighters had to wait until the fuel load burned down before they could extinguish it—6 hours after the fire started.

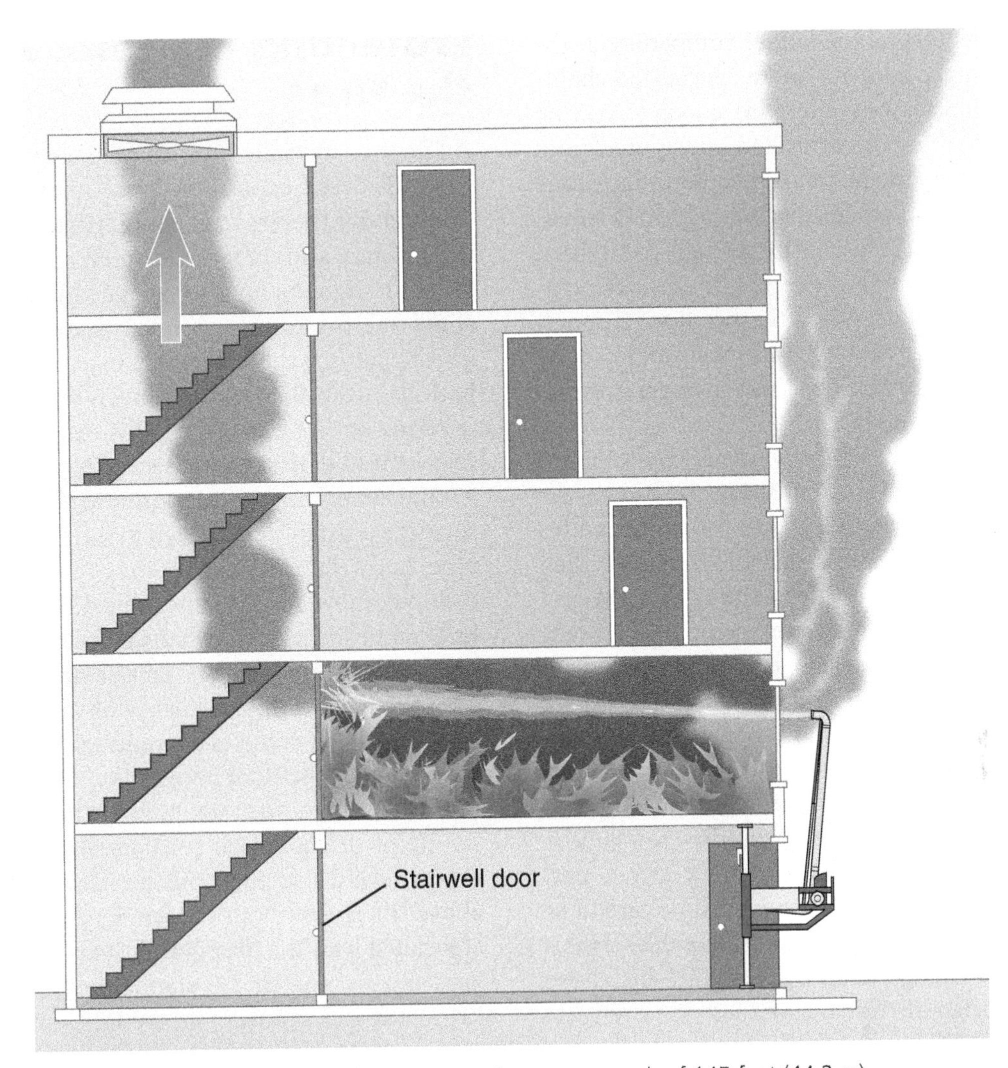

FIGURE 15-51 The HEROPipe has a penetrating stream reach of 145 feet (44.2 m).

The HEROPipe is a major investment for a fire department. Fire chiefs should investigate the possibility of obtaining grants or propose that insurance companies and building owners partner with them and invest in these portable monitors that will help protect their property. The HEROPipe and other high-rise nozzles could be an extension of the required fire protection systems that go into high-rise design for use by the fire department. According to the 2016 NFPA research report on *High-Rise Building Fires,* the property loss due to high-rise fires is, on average, $154 million annually. The property loss at First Interstate was $200 million, and One Meridian Plaza was a total loss. The direct property loss was $100 million, yet the building still had to be demolished, and litigation resulting from the fire amounted to an estimated $4 billion in civil damage claims. We must not settle for incremental half-measures but look to implement advanced nozzle technology that works now. All interested parties must try to adopt advanced nozzle technology that helps solve the problems associated with high-rise firefighting. Fire chiefs must try to adopt advanced nozzle technology that keeps firefighters safe. With a 900 gpm flow capacity, the HEROPipe is the right tool for this job, and it is the only nozzle that currently has the ability to deliver this flow.

Attacking from the Floor Above the Fire

Many previous firefighting textbooks showed a giant concrete-core drilling machine that was used by the FDNY back in the 1970s. This massive piece of equipment looked like a giant cannon and probably should have been set on tracks. It looked very heavy and difficult to put into operation. Although the idea of gaining access through a concrete floor had merit, the impracticability of the drill caused it to be eliminated from the regular inventory of the ladder companies. Technology has changed all that. Today, specialized chainsaws and circular saws come with concrete-cutting blades, and concrete-core drills and boring machines are powerful, compact, battery-operated units. Some are handheld units; others are set up on a small bracket. All can be carried in by a single firefighter. The feasibility of using a concrete-core drill reintroduces the method of accessing the fire floor through the concrete floor above—much like you would attack a basement fire from the floor above using a Bresnan distributor nozzle.

Engine and ladder company crews working on the exposure floor above can position themselves directly over the fire. Using long pike poles, the ladder crew can open the plenum space above so that they can see the exposed underside of the Q-decking and concrete floor, along with the supporting steel I-beam floor joists to the next floor above. The floor joist grid above is typically identical to the one that the ladder crew is standing on. Positioning themselves between the steel grid I-beams marks out the spot they need to drill to avoid the steel members. By checking the grid above them, members of the ladder crew can be sure they're simply boring through the concrete and the Q-decking **FIGURE 15-52**. They must be careful to avoid cutting through post-tension steel cables. Holes have to be large enough to drop down a 2½-inch hose line with a Bresnan distributor nozzle so the firefighters can create a 290-gpm sprinkler head onto the fire floor. This process can be repeated to create as many access holes as needed to drop additional distributor nozzles or cellar pipes. A hose line is required to cool the blade and concrete for all cutting operations, regardless of the cutting tool used.

FIGURE 15-52 By exposing the steel I-beam grid above them, firefighters can place themselves in the right spot to use a concrete-cutting tool on the Q-decking from the floor above the fire, then drop distributor nozzles into the fire area.

Courtesy of Raul Angulo.

Protecting the Floor Above the Fire

All floors, especially those immediately above the fire, are considered exposures. Hose lines need to get to floors above the fire to support SER teams conducting the primary search, forcible entry, and evacuation efforts. Firefighters must be assigned to check the floor above the fire for extension with hose lines and engage in battling the fire if it has extended. Whether the backup line or the exposure line is the second line laid depends on fire conditions. It is extremely important to get ahead of the vertical spread of the fire to contain it; otherwise, crews will continue to chase it. The exposure line is usually connected to the firefighting standpipe. But with heavy fire on the fire floor, the stairs to the floor above may be untenable. Unless the exposure line can be placed in the hallway before the fire attack begins, this hose line may first be needed to augment the fire attack until the energy of the fire is knocked down and temperatures are reduced; then it can be repositioned to the floor above.

If there is another stairwell in the building with a standpipe, it should be considered for supplying the exposure lines so they can be used to access floors above the fire without the smoke and heat problems associated with the firefighting stairwell. Once the life hazard is eliminated, any standpipe should be used for fire attack and exposure protection from any stairwell as needed. At the First Interstate Fire, all four stairways and all four standpipes were used to stop the autoexposure of flames lapping from floor to floor. If interior hallway standpipe discharge outlets are available, exposure hose lines can be connected on the exposure floor instead of the firefighting stairwell. Connecting to hallway standpipe outlets also leaves the evacuation stairway intact.

The First Interstate Fire occurred outside of standard business hours and there were approximately 50 occupants in the building. The need to maintain an evacuation stairwell is the highest priority if the fire occurs during regular business hours when the building is fully occupied. Every effort should be made to protect the evacuation stairway from smoke and heat and to prevent the smoke and heat from contaminating a second stairwell; thus, don't connect to the evacuation stairway standpipe in the early stages of the fire. At some point, the evacuation should be complete or sheltering in place is established, and getting lines above the fire will outweigh other concerns if the fire is to be stopped. At the First Interstate Fire, the concept of maintaining at least one stairway free of smoke so that it could be used for evacuation proved to be ineffective. This concept may be valid for less severe fires, but when the fire reaches the magnitude of the First Interstate Fire, all vertical shafts become potential chimneys. The ventilated vestibule design failed to keep heat and smoke out of the pressurized smoke towers and all the stairwells were charged with smoke.

Once exposure lines are in place and charged, the crews should prepare for the fire to extend vertically from autoexposure, or lapping from window to window **FIGURE 15-53**. Windows should not be broken out to prevent lapping. If they fail due to the fire, however, combination nozzles with a fog pattern must be used to cover the open space, cool the thermal column, and prevent flames from entering the exposure floor. An excellent tactic for controlling autoexposure is to get hose lines to the roof, or on several floors above the fire, and flow water down the exterior sides of the building. A cascading sheet of water down the exterior glass curtain walls keeps the glass cool and wet and effectively stops lapping.

Final Strategy at the First Interstate Fire

The First Interstate and One Meridian Plaza fires were two of the most spectacular towering inferno-type high-rise fires in US history, ranking after those of September 11, 2001. Each fire had its own unique problems. The One Meridian Plaza Fire started on floor 22. It burned for over 19 hours and consumed eight floors. Due to the concern for possible structural collapse, all companies were forced to abandon the building. When the fire reached the thirtieth floor, which had a functioning sprinkler system, 10 sprinkler heads finally controlled and extinguished the fire. The First Interstate Fire was a full-on fire attack. The fire started on floor 12 and destroyed the twelfth to the fifteenth floors in 3 hours and 45 minutes. According to the *USFA Technical Report Series on the First Interstate Fire*, the autoexposure flames were 30 feet high, and it took about 45 minutes for the fire to take hold of the next exposure floor. It was estimated that a total flow of 4,000 gpm was delivered by the standpipe risers, and the total effective fire flow provided by hose lines attacking the fire was approximately 2,400 gpm. With all the efforts of 32 companies operating 20 attack lines from all four stairways, it still wasn't enough to control the fire and put it out. It took about 90 minutes for the fire load to be consumed on each floor (which illustrates the concept behind Type I fire-resistant construction).

FIGURE 15-53 Firefighters should be assigned to check above the fire floor for fire spreading from window to window.

Once the water source was readied and positioned ahead of the fire in both the First Interstate and the One Meridian Plaza fires, it was controlled with the available water on that floor. Even a 30- to 35-gpm (per head) flow through 10 sprinkler heads in the One Meridian Plaza Fire was sufficient to extinguish the massive fire because extinguishment was ahead of the spread of the fire. The final strategy employed by the LAFD is textbook high-rise firefighting and worth reviewing in detail because it worked; for future fires, it should be implemented sooner rather than later. When it became evident from the exterior that they were going to lose the sixteenth floor, Deputy Chief Don Anthony, the IC, wanted to concentrate a full-out effort to attack the fire on floors 14 and 15 simultaneously to knock down the intense thermal energy and slow the speed of the fire while crews got ahead of the fire and established a defensive position on floor 16. At times, active suppression efforts were underway simultaneously on four levels as crews attempted to push the fire back from the central core to the perimeter of each floor. As more doors were opened, conditions in the stairways deteriorated, with heat and smoke going up and water cascading down. It required extreme effort by crews operating hand lines on heavily involved floors, with as many as four floors burning below them, but crews made it up onto floor 16, operated the attack lines, prepped the room, and flowed the floor. Then they waited for the fire to attack the sixteenth floor. Conceptually, this strategy goes back to the basic fire science covered in Chapter 2 of this text. They had been fighting a massive fire on the fully developed side of the fire time-temperature curve. Chief Anthony needed his firefighters and hose lines to be at the start of the fire time-temperature curve during the incipient and early growth stages on an uninvolved exposure floor. This strategy proved to be successful.

Preparing the Exposure Floor

All combustible material should be removed from around the windows and walls. This may include the entire perimeter of the exposure floor. Move items to the center of the room and away from shafts, ducts, and vertical channels. Combustible items include:

- Curtains, shades, and blinds
- Perimeter ceiling tiles
- Framed photos, artwork, and wall hangings
- Lamps and plants
- Office furniture, including desk chairs, tables, and filing cabinets
- Computers, printers, telephones
- Cubical dividers
- Carpet

Pull the ceiling around the perimeter to expose the plenum and the return-air channels. The perimeter walls should be bare. Open up any knee walls or spandrel panels below the glass, and expose all perimeter walls down to the studs. If possible, floors should be down to the concrete. Buildings with exterior glass curtain walls have a spandrel gap of several inches where the glass curtain walls are anchored to the concrete decking on each floor. These gaps are often left unsealed and are a vertical path for extension at the floor and ceiling level, and contribute to autoexposure. Any curtain wall spacing must be exposed, especially in the area around the floor slabs. Check or expose the window mullions (the vertical rods that join the large windows). At any point where smoke or flames start to show through the floor, flow water. Property conservation is not the concern on the floor above the fire. There are no half-measures here. We are trying to stop the fire, not preserve the floor.

Other vertical avenues for extension of fire include areas in the center core, which can include stairways, elevator shafts, utility shafts, mail chutes, garbage chutes, poke-throughs for communication and electrical lines, and plumbing pipe chases. The pipe chases are in the kitchen area, breakroom, and the men's and women's restrooms. These spaces need to be opened and wetted down. Walls inside utility closets and behind cabinets need to be open and wetted down. Do not rely merely on TIC readings to save a wall. If the fire is stopped, the whole floor will be remodeled, but if the fire extends, half-measures will amount to nothing. Soak everything down thoroughly.

Tools for Protecting the Exposure Floor

In addition to the hose lines, and combination nozzles the following tools are needed:

- TICs
- Pike poles and roof hooks
- Plaster hooks
- Pickhead axes
- Chainsaws and rescue saws
- Irons or forcible entry tools
- Baby ladders or attic ladders
- Spare SCBA bottles
- Portable generator and lights

This demolition work will need to be done on air so bringing spare SCBA cylinders up from staging saves time. Another primary avenue of smoke and fire spread are the HVAC ducts. Vent grates have to be removed. It may be necessary to pull down ductwork. In any case, firefighters need to be able to reach ceiling spaces, so baby ladders or attic ladders will be needed. There may be utility ladders available, but they may not be strong enough to hold the weight of a fully equipped firefighter. Once opened, all areas need to be checked with a TIC, with a hose line at the ready, or the fire may extend. Wet everything down.

Flowing the Floor

While firefighters are performing the above tasks, those assigned to the nozzle should **flow the floor**. Flowing the floor is wetting down the entire floor. High-rise floor systems are a combination of steel rods, corrugated steel, and concrete slabs. Heavy fire below can produce enough heat energy that radiates through by conduction to material on the floor above, primarily the floor adhesives and mastic used to secure commercial flooring and industrial grade carpet. Flowing the floor provides a water barrier to prevent extension. Streams should be aimed at the corners of the floor and perimeters where the floors meet the walls. You're not trying to create a lake, but a pond of standing water should be sufficient without concern for adding an excessive live load to the floor **FIGURE 15-54**. Any excess water will simply cross the door threshold and flow down the stairway. The nozzle should also flow the walls. Flowing the open wall spaces between the studs cools the temperatures of the perimeter as well as the temperatures of the fire gases seeping up. Cooling the glass panels with water should also be done, but the cool water against the hot glass may cause them to fail. Sufficient hose lines with combination nozzles must be in place to fight autoexposure. Broken-out

FIGURE 15-54 Flowing the floor is wetting down the entire floor. Standing water is a barrier preventing fire extension.

glass panels may cause additional flow path problems but should also help dissipate the heat and will help later with floor ventilation. A flow path with a wind-driven fire is not a concern yet because there should be no fuel burning on this floor for the wind to fan the flames. The fog patterns are necessary to keep lapping flames out of this area. The only fire load left on the floor is the office furnishings, and other combustible material that was piled in the center of the room. Hose it down. If it's wet, it won't burn. If you look at the One Meridian Plaza Fire case study, you can see that the fire was finally extinguished on the thirtieth floor by 10 sprinkler heads. Once the sprinklers fused, the office furnishings and combustible material were wet, and there was standing water on the floor. In a building without sprinklers, firefighters need to get ahead of the fire and create the conditions described here, which are the only preventative actions available. There are no other manual firefighting methods for protecting the exposure floor.

Checking for Extension Below the Fire Floor

Floors below the fire must also be checked for fire extension. Stack effect produces unpredictable convected air currents, and smoke and heat can actually travel down. If fire is observed below the fire floor, hose lines need to be stretched to these areas for fire control. In addition to fire extending to floors below the fire, property conservation should be considered early in the incident. Depending on their occupancies, high-rise buildings may contain valuable contents. Protecting these valuables, especially from water damage, reduces loss considerably. Property conservation on floors below the fire is usually labor-intensive, and it may be overlooked in the initial stages of the fire because of other tasks being conducted on the fireground. Extension below the fire may force the staging area to be relocated one or two more floors below.

Other Uses for Standpipe Systems

Although designed primarily for interior fire attack, standpipe systems can be used to attack fires in adjoining buildings and for exposure coverage.

Fire Attack in Adjoining Buildings

When fire has gained considerable headway in a building, firefighters may be kept from the fire floor by the intense heat. Under these conditions, if adjoining buildings are close enough to the fire building, hose streams developed from standpipe systems in nearby buildings could be used on the fire FIGURE 15-55. The hose streams can be directed across a court, across narrow streets and alleys, or from the roof of the adjoining building into higher floors of the fire building. Either 2½-inch hand lines or master stream appliances may be used for such operations. This tactic was used at the One Meridian Plaza Fire: Hose streams were set up from adjoining high-rise buildings within reach. It could not be used at the First Interstate building.

Exposure Protection

Fire on the roof and in upper floors of closely set buildings presents serious exposure problems. In some cases, aerial fire apparatus can be used for fire attack and exposure protection; however, when the fire is above the reach of aerial devices, almost all operations must depend on the standpipe system of the fire building and adjoining structures for water. Pumpers must be set up to pump into the standpipe systems of both the fire building and exposure buildings. Those of the exposed buildings must be charged to ensure proper water supplies if and when they are needed. This is especially important when only public waterworks supply the systems because the public waterworks pressure could be reduced to the point where attack lines supplied by these standpipes will be

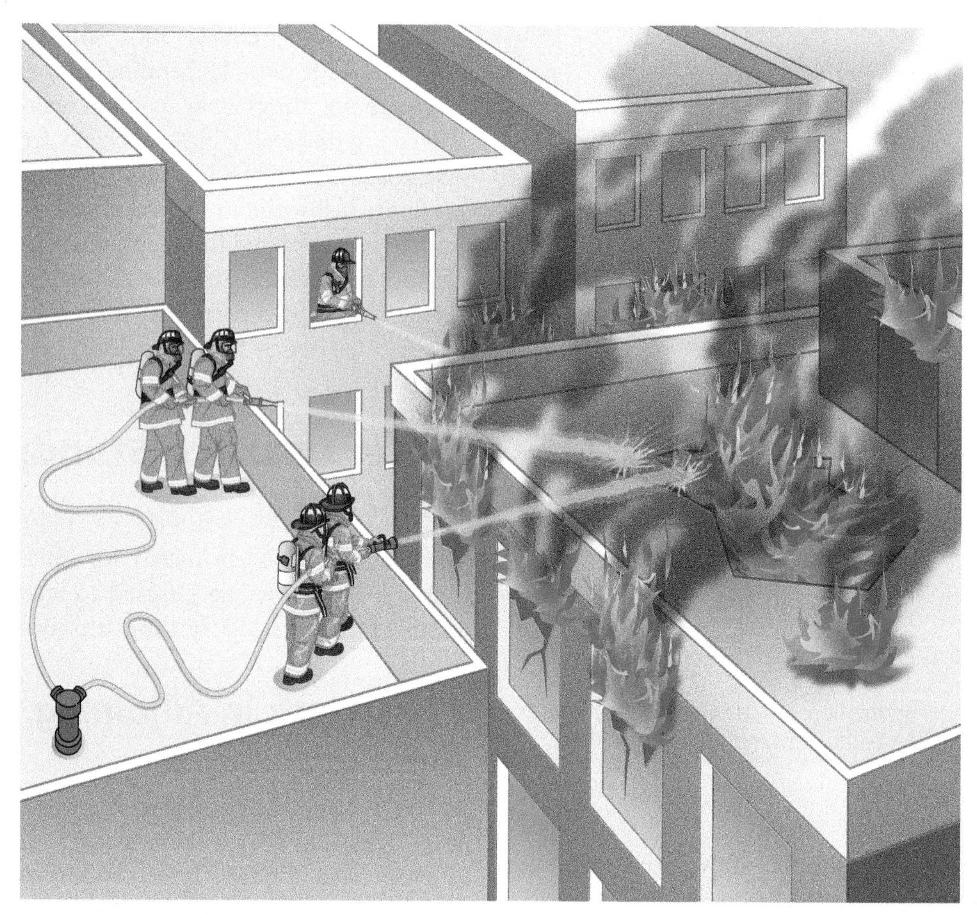

FIGURE 15-55 Standpipe systems may be used for fire attack and exposure protection from adjoining buildings.

ineffective. A simple tactic to protect exposure buildings is for crews to flow hose lines operating from the roofs and flow a cascading sheet of water down the exterior walls of the exposure building to keep them wet. This tactic can be accomplished with low water flows.

Use of Water from Uninvolved Buildings

Water from a gravity tank atop an uninvolved building can be drawn off and used to supply pumpers **FIGURE 15-56**. This is usually a measure of last resort; it is used, for example, when a very large fire requires many pumpers to operate simultaneously and the water main pressure may become dangerously low. To obtain the water, the pumper is parked as close as possible to the ground floor standpipe outlet. The 2½-inch hose or larger supply line is hooked up between the outlet and the pumper intake. The outlet is opened, and the pumper takes water from the gravity tank. Depending on the standpipe system, it could be more efficient to use LDH to support the operation.

Search-Evacuate-Rescue (SER)

Although SER is covered here, toward the end of this chapter, it is by no means a tactic that should wait for all the previously discussed tactics to be accomplished. On the contrary, it is implied that the search starts along with the evacuation, and it is implied that it may be initiated simultaneously with the initial fire attack. It might even be necessary for the engine company to forgo fire attack to concentrate on SER on the fire floor and with evacuation of large open-area floor spaces. Stairways may be congested with fleeing occupants, and the evacuation may be chaotic. There may be numerous disabled occupants who are in danger and who require immediate assistance. These actions may result in a delayed fire attack and an increased spread of the fire, along with additional property damage. It may also require a lot more time and resources to control the fire, but it is the right decision to make. A fast fire attack is important, but you also have the fire-resistive qualities of Type I construction working in your favor.

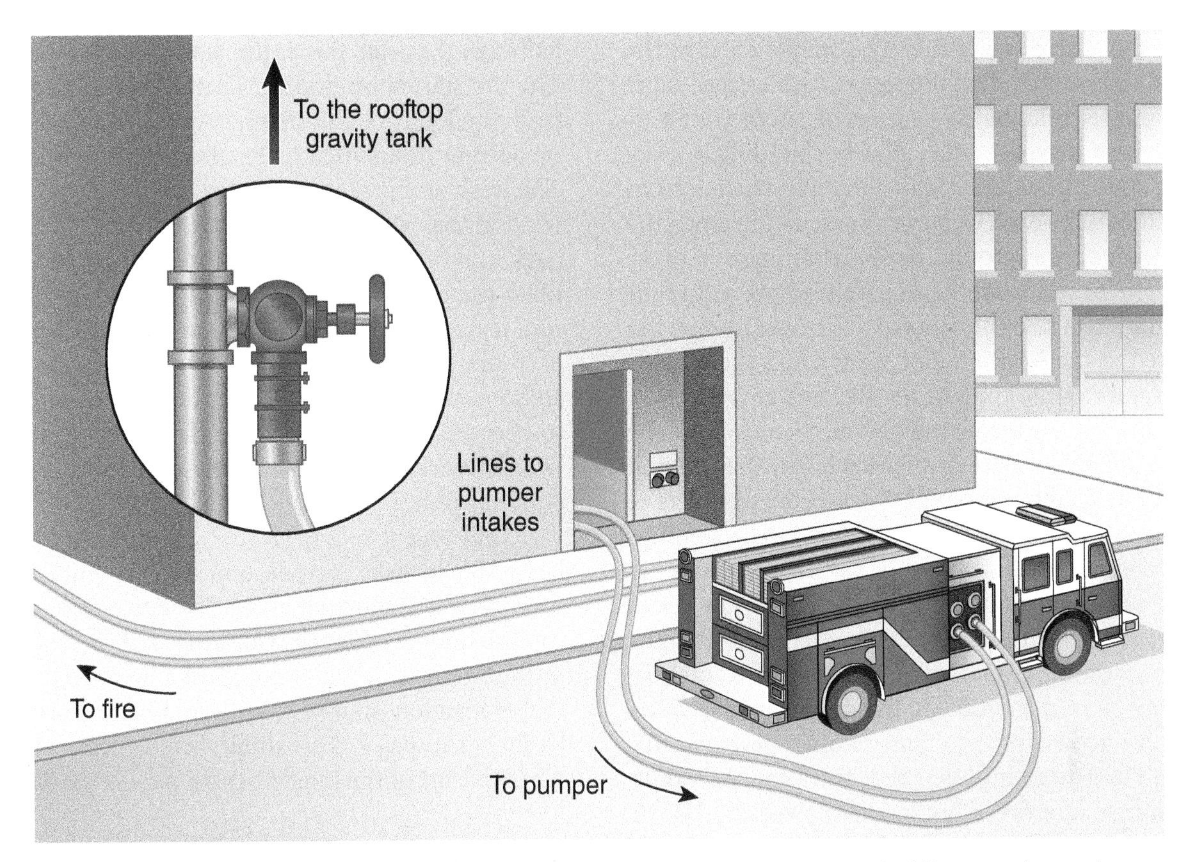

FIGURE 15-56 In water supply emergencies, standpipe gravity tanks from adjacent buildings may be used as a last resort to supply the pumpers.

Search and rescue (SAR) efforts in high-rise fires are different than those in regular structure fires. In a single-family residence, firefighters can expect to rescue one or two people and remove them from the burning structure. In a high-rise, that number can be 10, 20, 30, or perhaps hundreds of people may need assistance. This is why the word *evacuate* (and thus the letter *E*) is added to the common assigned term of SAR to form SER. By no means does this imply that rescues won't be made; they will, but the majority of the occupants need to be evacuated—that's how we rescue them.

Type I fire-resistive construction is the classification that can resist the fire for the longest time. It is designed so the fire load (fuel) is consumed before the building components start to suffer the effects of the fire. Therefore, the strategy for life safety in residential apartments, condominiums, and hotels is to shelter in place. This means that the occupants are safer staying in their units than trying to evacuate the building. The hallways and stairs may get charged with smoke. CO, which is colorless, odorless, and tasteless, can overcome an occupant in a hallway or stairwell that may appear to be free of smoke. This was the case at the MGM Grand Hotel Fire on November 21,1980, in Las Vegas, Nevada, where 85 people were killed.

The MGM Grand Hotel was a 26-story hotel with 2,076 rooms; at the time of the fire, 07:07, it was estimated that 5,000 people were in the hotel. The electrical fire started in a restaurant on the ground floor–casino level. Because the casino never closes, and staff members and customers are on the floor 24 hours a day, it was granted a fire code variance to omit sprinklers in the casino. The rationale was that with round-the-clock activity, any fire would be quickly noticed by security and reported to the fire department while trained staff put out the fire with portable fire extinguishers. However, The Deli restaurant was not open yet and there was no one in the immediate vicinity of the restaurant who took notice of the growing fire. Modern fuels, including furniture, decorations, wallpaper, foam padded chairs and booths, PVC pipe, slot machines, gambling equipment, and the flammable glues and adhesives to hold the plastic mirror ceiling tiles above the entire casino, contributed to rapid fire spread throughout the main floor. When flashover occurred, the flames were traveling 15 to 19 feet (4.7 to 5.8 m) per second. Eighteen people were killed in the casino level.

There were no smoke detectors, and the manual alarm pull stations failed or were not pulled. Residents were never notified of the alarm but awoke to the smell of smoke, sirens, commotion, and occupants yelling that there was a fire. The smoke entered the vertical shafts, including the elevator hoistways, stairwells, and seismic joints—a series of 1-foot wide vertical air plenum spaces that allow the building to sway during an earthquake and that ran unobstructed from the casino to the roof. Occupants evacuating down the stairs wedged the ground-floor doors open because they were locked from the stairwells to prevent reentry. Stack effect also contributed to fire spread. The fire occurred at the beginning of winter, and the outside temperature at the time of the fire was 38°F (3.3°C). When fire occurs below the neutral plane, and if the outside air is cooler than the inside air, the internal pressure created by stack effect draws the products of combustion toward any shafts or stairwell openings and pushes them vertically to upper levels.

The blistering summer temperatures of Las Vegas often triggered the smoke dampers to activate, shutting down the air-conditioning system to the hotel. To solve this nuisance problem, building engineers used lag screws to secure the dampers open permanently, probably not realizing the effect their actions would have in case of a fire. This action allowed the HVAC system to spread smoke all the way to the top floors of the hotel and throughout the building. The fire enshrouded the entire building in a massive smoke column that could be seen for miles.

In the end, 85 people died and 650 were injured. Sixty-seven of the deaths occurred in stairwells, stair landings, hallways, and elevator lobbies from floors 16 through 26. Of the 67, 25 occupants died in their rooms from smoke and CO being pushed through the HVAC system, although some showed no signs of smoke inhalation. Ten people died in elevators, and one person jumped and died. The fire never spread past the second level of the casino.

Shelter and Defend in Place

The MGM Grand Hotel Fire is a worse-case scenario, an example of what needs to be done to ensure life safety. Since the September 11 attacks on the World Trade Center and the Pentagon, and the ability to view graphic video footage of any high-rise fire on the Internet, like the Grenfell Tower Fire in London, England, it may be impossible to convince occupants that the safest thing to do is stay in their rooms or work areas. Their natural tendency is to flee. But convincing them is what you need to do to prevent them from inadvertently entering an IDLH atmosphere.

The Polo Club Condominium Fire occurred on October 31, 1991, in Denver, Colorado. The nonsprinklered, 20-story, residential high-rise had an atrium center core with interior balconies and hallways that ran the entire height of the building. The fire started on floor 7. There was nothing visible from the lobby or the atrium. As the fire attack commenced in room 702, the exterior window broke out. The fresh air currents exploded the fire and created a wind tunnel with blowtorch force, driving firefighters back into the atrium hallway. Intense heat and smoke filled the entire atrium. Louvered smoke dampers at roof level were open but could not handle the volume of smoke produced from the fire. The SAR strategy for adjacent units and on all the floors above the fire was to keep residents sheltered in place. Any evacuation would have exposed them to a massive building-wide chimney of heat and smoke.

Rapid ascent teams (RATs) are gaining in popularity. Their tasks include quickly searching the entire firefighter stairwell above the fire floor, ensuring rooftop doors and hatches are closed, directing people in the firefighting stairwell into hallways and toward the evacuation stairwell, and convincing occupants to shelter in place. This differs from reconnaissance, which is part of the interior size-up to locate the fire.

Search-and-Rescue Exterior Recon Team (SERT)

Search-and-rescue exterior recon teams (SERTs) and high-rise exterior recon search teams (HERSTs) are new concepts that should be incorporated into high-rise firefighting SOPs so that one or the other is assigned consistently. The most dangerous floors for trapped occupants are the fire floor and the floor above the fire. As was the case of the First Interstate Fire, stairways above the fire floor were untenable for search teams to climb. They had to wait until the fire was contained and knocked down. When there are exposure buildings of equal or greater height compared to the fire floor, a search recon team can enter these buildings with binoculars to get a line-of-sight view of the fire and smoke conditions, as well as check for trapped occupants visible on the fire floor and floors above. The exposure buildings do not have an IDLH atmosphere, so firefighters do not need a partner or SCBA. In theory, one company can split up and enter four separate exposure buildings to see the four sides of the fire building. The area where flames are visible should be checked first. Cell phone camera zoom capability is extremely detailed, and the IC can see human images that are sent to the command post in real time. Even the division supervisor on the fire floor or other designated positions can have these images available. When you consider how much time it would take for an interior search team to cover two floors, 30,000 square feet each, the time spent for this type of exterior recon is well worth it.

FIGURE 15-57 Person trapped on floor 37 at the DeWitt-Chestnut Fire in Chicago, Illinois.
Courtesy of Michael Wielgat.

FIGURE 15-57 clearly shows a person trapped on floor 37 at the DeWitt-Chestnut Fire in Chicago, Illinois. (Note the Christmas tree.) This person was rescued by Chicago firefighters. Occupants in the apartments below floor 36 sheltered in place. Smoke and fire conditions are also clearly visible in the figure. This photograph was shot with regular film and was not available for viewing at the time of the fire.

Aerial Drones

Drone technology is rapidly coming into play in the fire service. Using a drone to gain a close-up aerial perspective for size-up and exterior recon will become a common tactical tool in the near future. Helicopters can't get as close to a building as a drone can; however, there are still many variables preventing full implementation of drones. For example, how will they operate in inclement weather, high winds, thermal columns, and radiant heat from a fire? The aerial advantage of using helicopters for high-rise fires is obvious, but they also cause problems. At the First Interstate Fire, the constant loud noise of the rotors added to the noise levels within the building, making radio communications extremely difficult to hear. The helicopters also fanned the flames and created turbulent smoke movement around fire crews working in areas where the windows had already broken out.

Fire Dispatchers

The first point of contact with someone trapped in a high-rise is a 911 fire dispatcher. The caller may be able to tell the dispatcher the exact location of the fire as well as what's burning. The dispatcher should calm the caller and gather pertinent information, including the floor number, the room number, and the location of the caller inside the unit (i.e., bedroom, bathroom, closet) and the surrounding fire conditions (i.e., odor only, smoke, fire), then convince the caller that the safest thing to do is stay in her or his room. If the caller is in immediate danger of the fire inside the unit, or he or she lives in the unit above the fire and smoke conditions are getting worse, the caller should try to escape. But if the fire is anywhere else in the building, the caller should stay in the room. The IC must be given the floor and room number of the trapped occupant along with fire and smoke conditions. At the DeWitt-Chestnut Fire in Chicago, it was confirmed that three occupants were trapped on the roof because they called the dispatcher from that location. The dispatcher should reassure the caller that the fire department knows where they are and will come. Based on the information from the dispatcher, the IC should relay the floor and room number, along with the reported fire conditions to the SER team, including if the occupants are young, elderly, or disabled and require special assistance. Although the dispatcher is listening to all the radio communications on the fireground, the IC should make sure to clarify what information and instructions should be relayed to 911 callers by the dispatcher, for example, whether to shelter in place, the location of the evacuation stairway, avoiding the roof, keeping doors closed, opening windows. If the IC wants tenants on the exposure floor to evacuate, the dispatcher should know this information in case those occupants call 911 because then they can be instructed to evacuate to the proper stairway.

Public Address (PA) System

The initial company officer should utilize the building PA system to notify all the occupants to stay in their rooms unless their floor is in alarm. The officer should include the nature of the incident and what the fire department is doing to resolve it. Include any specific instructions for evacuation or to shelter in place. The more reassurance that can be given to occupants, the more they will cooperate. Don't lie or minimize the danger. If you choose to dump the whole building by evacuating all the tenants, then instruct them to the proper evacuation stairwells. A full evacuation may cause more problems than necessary. First, it is time consuming. The evacuation at the MGM Grand Hotel took 4 hours to complete. Next, where will evacuees gather? An outdoor location may be okay on a warm sunny day, but if it is 02:00, 10° F (-12° C), and snowing, you can't just send them out to the parking lot, especially if you're dealing with elderly occupants. Firefighters need to be trained in effective methods for sheltering and defending in place.

The use of handheld, battery-powered megaphones is important for the crews to manage a large evacuation. They should be carried on every fire apparatus, brought up to the equipment cache at staging, and used when directing those exiting the building to a shelter area away from the building. Hallway corridors can be extremely long. This device is an effective way to call occupants to the evacuation stairwell, call to check if anyone is in the hallway, call up and down the stairwells to see if anyone is there, and give instructions to occupants to stay in their rooms. This can be done by a firefighter announcing instructions while walking the hallway. The megaphone-amplified voice should be heard through closed hotel and apartment doors. Occupants can also be directed to turn on the television for complete updates from the PIO working with the local news station. There were 2,076 rooms in the MGM Grand Hotel, and all doors to any high-rise hotel units are locked. It is impractical to search every room with the initial units. And the doors should not be forced. If the message to occupants is that the safest place to remain is in their rooms, why would we force entry to search them? The occupants are most often safe in their rooms or apartments.

Traditional and Nontraditional Search Tools

Multiple master door keys should be obtained for SER units. These keys prevent needless property damage, and they are the fastest way to gain access to locked areas for the primary search and to remove people who cannot self-evacuate from exposure floors. These keys should be in the FCC. Forcible entry tools need to be carried by the SER team. In the MGM Grand Fire, as well as the One Meridian Plaza Fire, the doors in the stairwells were locked to prevent re-entry to the floors. Irons needed to be used to gain access to the floors above the fire. This adds considerable time to a search. One of the fastest and easiest ways to get through a door to a hallway or office is to breach the gypsum board wall, reach through, and open the door. This method uses less energy and causes less property damage than forcing through a heavy-duty fire-rated metal door. Obviously, this method doesn't work when the walls are concrete.

The MGM Grand Hotel Fire sheds light on additional tools that the SER team should consider using. Typically, SER teams are taught to perform a right-handed search and sweep the area with a tool. The scope was different here: although firefighters still needed a tool, the victims in the MGM Grand Hotel weren't hidden. They were found in the common entry egress paths. What these victims immediately needed was fresh air. The RAK is typically used for firefighter rapid intervention incidents, but they can be a valuable rescue tool for civilians.

It's time we start carrying RAKs in high-rise SER operations for civilian use. Bringing spare SCBA units for civilian use accomplishes the same thing. SCBA units would also serve as a backup if one or more SER team members run out of air.

Another tool that SER teams should bring is the handheld multi-gas detector. The smoke column can stratify when particulates cool in the stairwell and vertical shafts, but some fire gases can keep rising. CO is colorless, odorless, and tasteless. Some of the MGM Grand Hotel victims died in their rooms, but they showed no signs of smoke inhalation. They probably died from CO. Seventy-nine of the victims had CO levels that ranged from 25% to 66% saturation. Stairwells and hallways that appear clear can be charged with CO. Firefighters might be tempted to save their air and remove their facepiece, and they might suffer the effects of CO. Semiconscious occupants may appear to be injured, tired, or in shock when they're really suffering from the effects of CO. There is no way to confirm if a clear hallway or stairwell is really clear unless a gas monitor is carried in by the SER team. Again, what these victims of CO poisoning need immediately is fresh air. Lobby and staging areas should also be monitored for CO as both have been found to have high levels of CO due to stack effects.

SER Areas of Emphasis

The primary tactical objectives and areas of emphasis for SER teams are to:

- Check the firefighting stairwell above the fire floor for trapped occupants.
- Check the fire floor, especially isolated rooms where occupants may have sheltered in place.
- Check the rooms on both sides, adjacent to the fire room.
- Check elevator lobbies, starting with the fire floor. Although people shouldn't do so, some may still attempt to use the elevator to escape. Phase I would have recalled the elevators to the main lobby, so occupants may end up unconscious (or dead) waiting for an elevator on the fire floor.
- Check the floor above the fire, starting with the room directly above the fire.
- Check the evacuation stairwell.
- Check the top floors of the building.
- Direct wandering occupants to the evacuation stairwell.

- Direct occupants to stay in their rooms and shelter in place.
- Move wandering occupants into a residential unit for the duration.
- Check all stairways, hallways, corridors, and elevator lobbies on remaining floors above the fire.
- Check bathrooms and other rooms in the central core.
- Check hallway linen closets, utility rooms, and concession areas.
- Direct evacuees to a safe floor of refuge well below the fire floor and firefighting operations.
- Check all elevator cars.
- Check the freight elevators.
- Check the roof.
- Consider referring to a tenant list.

Once the fire is under control, a door-to-door, floor-by-floor inquiry can be made to check on the well-being of occupants. The multiple tasks that are required to perform a proper high-rise search, including carrying all the equipment mentioned above, is more than the typical two-person search team can handle. The IC needs to create a SER group with two or more companies assigned to perform the search in time to make a difference. Call for additional alarms.

SER in Open-Space High-Rise Office Buildings

Of all the property classes, high-rise office and commercial buildings are the most likely to be sprinklered. Their occupancy load is high during the day and low at night. In an open floor plan that is not compartmentalized, trained fire wardens, if they are available, move occupants and employees to a safe refuge area that can be five to ten floors below the fire, clear of fire department operations. Obviously open spaces do not allow you to shelter and defend in place if the floor is close to the fire. But once occupants are moved to safe refuge areas, they can remain there for the duration of the incident if necessary. It is not necessary to evacuate the entire building.

If the civilian fire wardens have cleared the floor, their work should still be verified by fire department personnel. If the floor is clear of smoke, there may be no need to perform a wall search of each office cubicle. Walking the main aisles and side aisles while calling out, perhaps with a handheld megaphone, should be sufficient. Individual offices, supply rooms, photocopy rooms, file rooms, utility closets, breakrooms, kitchens, and bathrooms should all be checked because people may be using earphones and be unable to hear announcements or alarms. Elevator lobbies should also be checked. ***Safety note:*** If an elevator car is found on an upper floor with the doors open, it is a sign of a malfunction. Check the car for occupants but do not enter the car.

Elevator cars are also searched for occupants at lobby level when they are recalled. If smoke is present, individual cubicles need to be searched. Special attention should be paid to making sure that the firefighting stairwell above the fire floor is clear of civilians.

Door Control

In buildings with pressurized stairways, door control during SER is important for maintaining pressurization. Leaving hallway doors open intentionally or unintentionally can reduce the pressure, which can allow smoke to enter the stairways. Some buildings have smart locks that automatically unlock all hallway doors in the stairway when the building goes into alarm, allowing for reentry onto any floor from the stairs. Other systems allow reentry every five floors. Some buildings do not have any reentry doors, and occupants have to exit the building on the ground level. Powerful pressurization fans in modern buildings can maintain pressure even when three doors to the stairway are left open, but that's about it.

It is understandable that firefighters want the ability to enter and reenter the floors from the stairwell, especially if the doors lock. The best tool for the job is the Jeromeo clamp: a simple spring-loaded clamp that can be found at any hardware store. The clamp is strong and inexpensive to replace. The best feature is that, unlike a door wedge, the Jeromeo clamp won't slip. You simply squeeze the clamp onto the edge of the door and let go **FIGURE 15-58**. You can place it at any height, and it won't get kicked loose by other firefighters. Placing the clamp close to the lock side of the door allows the door to close as tightly as possible while still having access. Placing the clamp on the hinge side of the door wedges the door open in case you need to pressurize the stairwell.

Firefighter Air Management

Keep in mind that fresh SCBA bottles are kept in the staging area, two floors below the fire. In the case studies of the MGM Grand Hotel, One Meridian Plaza, and the First Interstate fires, all the stairwells, even the evacuation stairway, became charged with heat and thick smoke. There were no smoke-free stairways. Many of the floors above the fire were also smoke filled.

FIGURE 15-58 The Jeromeo spring-loaded clamp.
Courtesy of Raul Angulo.

When more than one floor is on fire, and fire attack is happening on multiple floors (as was the case at the First Interstate Fire), heat conditions can become so high that all the stairwells are untenable. SER teams cannot access upper floors until the main fire is knocked down and temperatures are reduced. If crews are dropped off on the roof by helicopter, they may not be able to access the stairway because they are at the top of the chimney/thermal column in the stairwells.

SER crews given assignments well above the staging floor will most likely run out of air. The officer needs to monitor the SCBA regulators of all the members. In some conditions, it may be possible for firefighters to remain uncovered until the assigned floor is reached, thus saving their SCBA air supply. In smoke-filled stairways with tenable heat, however, crews may have to hike up several floors and use forcible entry to gain access to the hallways—all while being on air. The physical exertion uses up their air supply and narrows the window to perform the actual search. Some may be tempted to save their air and remove their facepieces. These firefighters run the risk of suffering from the effects of smoke inhalation and CO. The potential for getting too far in a hallway or getting too far up a stairwell, away from staging where the fresh SCBA bottles are stored, can get the SER team in trouble. Air management principles need to be practiced, which includes bringing extra air bottles if the assigned task requires considerable time and takes the crew far away from the staging area. Assign a stairway support company dedicated to the SER group just to bring extra air cylinders up to higher floors. The introduction to this chapter mentioned that, between 1977 and 1996, 16 firefighters died of traumatic injuries in high-rise fires. Nine of those 16 firefighters became disoriented and ran out of air while trying to find their way, ultimately dying of asphyxiation.

Remember, the three firefighters of E11 who died at the One Meridian Plaza Fire were given the task of ascending to the top of the stairway bulkhead to vent the smoke. They had to hike from floor 22 to floor 38, but they never made it before becoming disoriented and running out of air. One of the SAR teams consisting of eight firefighters also became disoriented and all members ran out of air in the mechanical room on floor 38. They were all rescued by the roof team. Like a scuba diver under water, every firefighter has to monitor his or her air consumption. Even the most straightforward tasks can become exhausting for firefighters during or after climbing 10 to 20 stories in heat and smoke. Fatigued firefighters consume more air in a shorter period of time, and it is the responsibility of the team leader or company officer to calculate if the assignment can be accomplished on a single bottle of air. Ideal air management is reaching the objective, performing the task, and returning from the objective with air to spare. But high-rise fires are never ideal. Returning to breathable air may be good enough. If not, taking an extra SCBA bottle may be the most important search tool of all **FIGURE 15-59**.

Air Standpipe Systems

One of the newest technological innovations for high-rise buildings and high-rise firefighting is the air standpipe system. Air standpipes are starting to appear in new high-rise construction around the country. Similar in concept

FIGURE 15-59 SER teams searching upper floors away from staging will most likely run out of air. A dedicated support company should be assigned to the SER group to bring extra air cylinders up to higher floors.

Courtesy of Rescue Air/Mike Gagliano.

FIGURE 15-60 The firefighter air replenishment system (FARS), or the air standpipe, can be supplied by the building's cascade unit.

Courtesy of Rescue Air/Mike Gagliano.

to water standpipes, air standpipes will be supplied by an air cascade system within the building FIGURE 15-60 or by a fire department mobile air unit FIGURE 15-61. With compressed air supplied to the entire system, firefighters will be able to trans-fill their SCBA air cylinders from the air station on designated floors, if not from any floor, without removing the facepiece. This concept is already starting to change the way SER and fire attack is performed in high-rise buildings.

Firefighter Air Replenishment System (FARS)

The following detailed description of the system is from Captain Mike Gagliano (Retired), of the Seattle (Washington) Fire Department:

The physical and logistical challenges for firefighters battling high-rise fires are enormous. Some of the best-trained fire departments in the country have met

FIGURE 15-61 The FARS air standpipe can also be supplied by a mobile air unit connected to the air standpipe the way a pumper connects to the FDC.

Courtesy of Rescue Air/Mike Gagliano.

their match when the fire is burning high in the sky. The variables of the historic high-rise fires covered in this book are many, but what doesn't change in these buildings is the two essential elements necessary to attack a fire, and rescue those at risk: water and air.

In the first half of the twentieth century, it became clear that new methods needed to be developed for fighting fires with an adequate water supply in tall buildings. Water tanks, standpipes, fire pumps, and sprinklers were designed to get water as quickly as possible to elevated locations. These are now standard systems in high-rise building design.

Unfortunately, the second necessary component has taken some time to catch up. But make no mistake, air is just as critical to effective fire and rescue operations as water. Unless you're planning to fight the fire from the outside, you are going to need lots of air. Air is what allows firefighters to do the work of firefighting and saving lives in superheated, toxic, carcinogenic, and smoke-filled environments. All the water in the world will not translate into success if firefighters don't have enough air to get the water where it needs to be. The best, most progressive air management program available is useless if there is no air to manage. The simple reality for firefighters is this: "You cannot save anyone if you can't breathe."

The answer to this difficult challenge is simple. Using the same delivery model as water standpipes, FARS delivers fresh air directly where it's needed most with no significant delay in the fire attack. FARS is a permanently installed building system that accomplishes rapid air resupply because it includes the following:

1. Air supply is accomplished by a mobile air unit supplying a fire department connection outside the building, or by a cascade air system within the building, or both.

2. Clean air is supplied through a fixed system of half-inch stainless steel piping and pumped to air delivery stations in the stairwell or other designated areas on upper floors.
3. Firefighters get air from a quick-connect panel in the stairwell using trans-fill connections, or from a rupture containment device that is typically placed within utility rooms or closets in the hallways.
4. The air within the self-contained piping distribution system is monitored around the clock for pressure and quality, and the system goes into alarm if any sensor detects a reading below acceptable NFPA standards.

FARS is a game-changer for rapid and efficient fire attack in high-rise buildings. Imagine the difference in your ability to attack the fire aggressively if air is being delivered to upper floors in the same manner as water in the standpipe. Instead of trying to carry cumbersome air bottles up the stairs or pack them into elevators that sometimes fail, the air is waiting for you at the point of attack. Instead of having to retreat and exchange air cylinders at staging, which is likely not even set up in the early stages of the incident, firefighters can move into the stairwell and immediately refill. Refilling at the quick-connect panels can be accomplished while firefighters are covered, without removing SCBA. Refill time takes between 30 seconds to 2 minutes depending on how many refills are occurring, so the fire attack can continue without significant delay. In addition, critical time is not spent using a labor-intensive bottle brigade that must climb the stairs to bring fresh air cylinders and other essential equipment up to staging.

FARS is currently installed in over 500 buildings in 10 states, with amendments to fire codes throughout the United States happening every year. The necessary requirements and recommendations are already part of numerous regulatory codes including:

- Appendix L of the 2018 ICC International Fire Code
- Appendix F of the 2018 Uniform Plumbing Code
- The 2018 National Fire Protection Association, NFPA-1

At the First Interstate Fire, over 600 bottles of air were needed by 383 firefighters while they fought the fire. Old video footage shows scene after scene of exhausted firefighters using hand trucks, carts, and endless lines of personnel carrying bottles to try and provide fresh cylinders for the fire attack. Fireground commanders were unanimous in proclaiming that air supply was a major factor in the firefight, and delays in air resupply slowed their efforts. Sadly, although this fire occurred in 1988, most fire departments do the exact same operations today with the same delays and difficulties. And those departments that cannot readily bring 383 firefighters, like the LAFD can, are in even bigger trouble.

The reality of resupplying air the old way, with bottle brigades and reliance on elevators, results in a very predictable outcome: *Air resupply at large buildings, without FARS, is slow and difficult and occurs at a much later stage in the fire than occurs with the immediate resupply available via FARS.*

When the late Chief Alan Brunacini was presented with the benefits of what FARS could do for his firefighters in the Phoenix (Arizona) Fire Department, he had a very simple answer: *yes*. When the political leaders hesitated, Chief Brunacini asked them simply to hold their breath while they considered his arguments for FARS because that was essentially what they were asking their firefighters to do, hundreds of feet up in the sky, at the biggest fire of their lives, with the delays of manual air resupply currently used. He got FARS for his Phoenix firefighters.

Immediate air delivered right where it is needed most is the reality of FARS. The answer for every fire chief and fire marshal in the United States should simply be "yes."

Ventilation

Ladder companies are traditionally responsible for ventilation, but engine companies may still have to assist with ventilation assignments. Due to the lack of exterior access and elevation, ventilation becomes complicated in high-rise buildings because they cannot be ventilated in the normal manner. The most significant factor facing firefighters is the wind because it cannot be controlled. Venting during a fire attack can create unpredictable air movement that can spread the fire and smoke throughout the fire floor and even the building. Venting can intensify the fire so severely that it creates a wind-driven fire and thus stalls the attack and prevents the search from continuing on the fire floor. Therefore, it is typically not performed until after the fire is substantially knocked down or extinguished.

One of the first assignments for the ventilation company should be to determine the status of the HVAC system. If the system is spreading smoke by the distribution system or the fresh air return shafts are contributing to fire growth, the system should be shut down. However, it is more important to determine if the HVAC system is pressurizing the stairways, the elevator shafts, the fire floor, and the floors above and below the fire. If the system is designed to sandwich the fire with pressurized stairs and floors, the system should be left in operation so it functions as designed. This information can be confirmed by reviewing the prefire plan or consulting face to face with the building engineer or maintenance supervisor.

The next assignment should be to place PPV fans at the base of the stairwells to augment the pressurization of the stairway shaft. If more fans are needed, they can be placed inside the stairway on the landing 10 floors apart or as needed. Another tactic to assist in maintaining stairway pressurization is to place a fan inside a lower floor hallway about 8 to 10 feet from the stairway door with the fan blowing toward the stairway. The hallway door remains open. These two assignments can be accomplished fairly quickly and can improve or maintain favorable conditions for evacuation without causing any negative effects for the fire attack teams.

Ventilation in low-rise commercial and residential buildings, those less than 20 floors, can still be done successfully with PPV fans and using the prevailing winds for horizontal cross-ventilation, but in mid-rise and high-rise buildings, the atmospheric phenomenon known as stack effect comes into play. This is not an easy subject to understand, and very few firefighters do, because it affects the entire building in so many different areas and in so many different ways. The firefighter at the end of the nozzle in a hallway within a 60-story high-rise is not contemplating the stack effect but rather is focused on advancing the line to the seat of the fire.

Stack effect is the vertical air flow within high-rise buildings caused by the temperature-created density differences between the building interior and exterior, or between two interior spaces (NFPA 92). Vincent Dunn defines stack effect as "the natural movement of air in a high-rise caused by the differences in temperature and atmospheric pressure inside and outside a sealed high-rise. Stack effect is responsible for the most forceful movement of smoke in a high-rise, and it is most pronounced in stairways and elevator shafts. Stack effect, when windows are broken out or opened, can create unusual smoke movement." If you've ever passed through the doors to a high-rise lobby or have passed through the doors of an observation deck of a high-rise, you may have experienced very strong wind gusts as you've entered or exited. Often, the wind exerts significant resistance against the door, requiring extra effort to push it open. This gives you a sense of the pressure differentials that exist in high-rise buildings.

Inside the building is an atmospheric column with a neutral plane within the center of the high-rise. This neutral plane is not constant and can fluctuate. When the outside temperature is cooler than the building interior temperatures, the internal pressure created from fires occurring below the neutral plane draws convected air currents, along with the heat and smoke, toward vertical shafts and stairwells. In fires occurring above the neutral plane, the pressure created from the fire moves convected air currents, along with the heat and smoke, away from the vertical shafts and toward the exterior walls of the building. The effects of smoke movement in fires occurring in the neutral plane is minimal, but the neutral plane can move. These effects are reversed depending on the time of year. In winter, the outside air is cold, while the inside of the building is heated; this is normal stack effect. Smoke from fires on lower floors rises **FIGURE 15-62**. In summer, outside temperatures are hot, while air conditioning keeps the interior climate of the building cool; this is reverse stack effect. Smoke from fires on upper floors can travel downward **FIGURE 15-63**. These temperature variables all affect convected air currents and the movement of smoke.

Now that your understanding of stack effect is crystal clear, here is the takeaway for the engine company. Convected air currents, or smoke movement within a high-rise, are unpredictable; there are many variables including:

- The intensity and heat energy of the fire
- The pressure created from the fire to move the smoke
- The condition and setting of the HVAC system, and whether it is shut down or contributing to the movement of smoke
- Outside temperature/inside temperature
- The wind
- Stack effect

An intense fire can develop sufficient heat and pressure that the heated smoke and gases rise several floors within the stairwell and inside vertical shafts. At some point the fire gases cool and the smoke stratifies. It could linger at a certain level or be pushed up or down by stack effect.

In older residential high-rise buildings, windows can be opened or closed. In commercial office buildings, many windows do not open, and those that do require a key. In any event, wind introduced from a high-rise window can intensify the fire, causing unpredictable and extreme fire behavior.

For engine companies, no air movement is better than unpredictable air movement. The strategy is for crews to attack the fire without ventilation. This will most likely be a brutal punishing endeavor. Once the

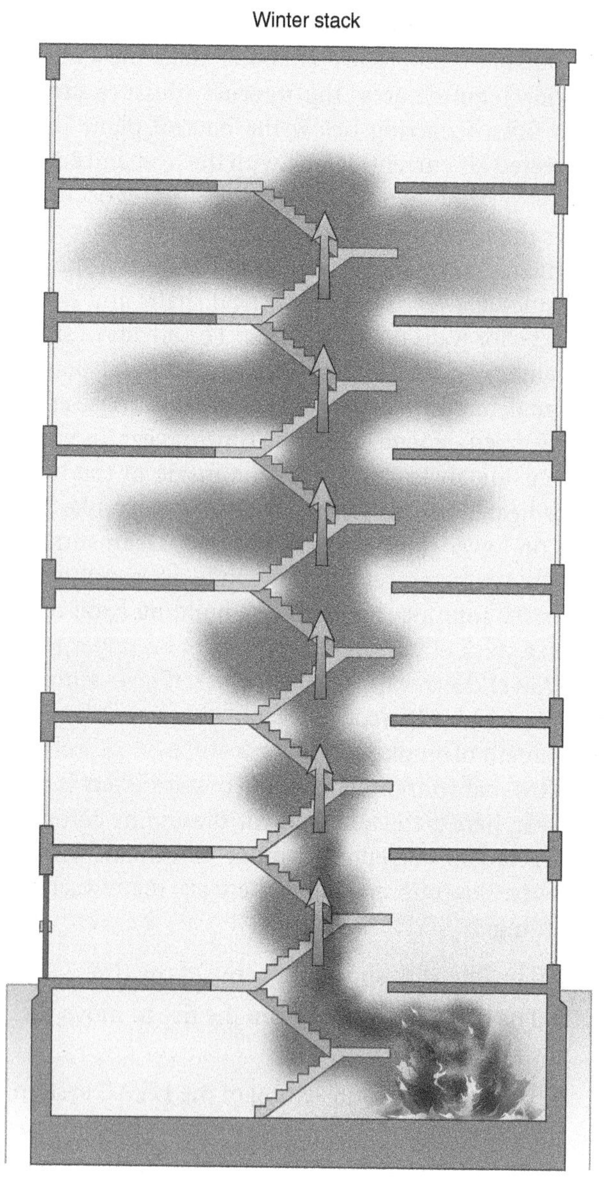

FIGURE 15-62 A winter stack effect occurs when the outside air is much cooler than the interior temperature. Smoke rises in this case in a normal stack effect.

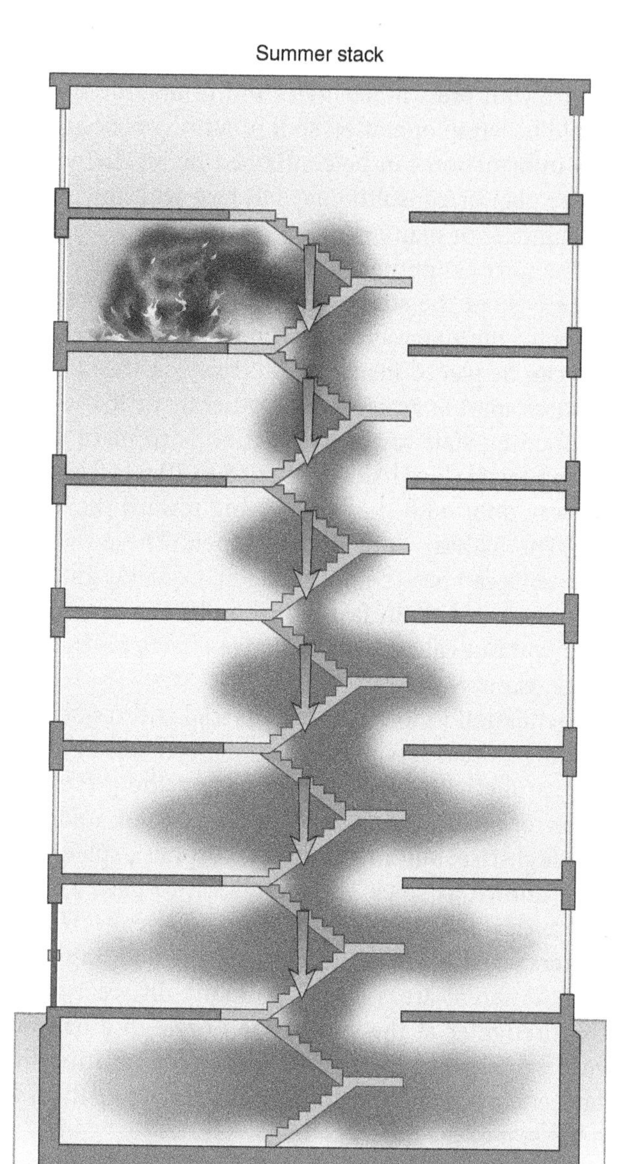

FIGURE 15-63 A summer stack effect occurs when the interior temperature is much cooler than the outside air. Smoke from fires on upper floors can push down in what is called reverse stack effect.

fire is sufficiently knocked down and extinguished, the production of heat and smoke ceases, and ventilation can begin. A combination of ventilation methods can be implemented to vent the existing smoke as quickly as possible. The concern for unpredictable smoke movement isn't as urgent because smoke production has ceased. Residual smoke can be ventilated without concern about additional smoke being generated. The wind is the enemy along with the fire when the fire is uncontrolled. Once the fire is extinguished, the threat is eliminated, and the wind becomes an ally in venting residual smoke **FIGURE 15-64**.

Horizontal ventilation needs to take place as soon as the fire is out to dissipate the heat, by opening or taking out windows. Taking out windows must first be

FIGURE 15-64 Once the fire is knocked down and extinguished, the production of heat and smoke will cease, and ventilation can begin.

coordinated with the IC. When the IC verbally confirms an "all clear below" over the radio, then crews can take out windows. This is an extremely dangerous job because, although ventilation is accomplished, a severe fall hazard has now been created.

Transitional ventilation is a series of tactics that starts with horizontal ventilation supported by PPV fans on the fire floor once the fire is extinguished. Consider where the wind is coming from and use it so it works in your favor. The pressurization in the evacuation stairway is maintained while the pressurization fans in the firefighting stairway are manually and mechanically shut down. The roof access door or the roof hatch in the firefighting stairway is opened. Firefighters must be positioned at the doors of the floor landings in both stairways to control the doors and to prevent unauthorized entry into the firefighting stairway. One firefighter opens the door onto the fire floor from the evacuation stairway, which increases the pressure on the fire floor. Then the door to the firefighting stairway on the fire floor is opened. The pressure differential along with the stack effect ventilate the smoke horizontally into the firefighting stairway, then it transitions and vertically ventilates the smoke out through the roof access portal at the top of the firefighting stairway. Once the smoke has cleared, the firefighting stairway door is closed, and the process is repeated up through the building.

Transitional ventilation should start with the fire floor and the first two floors above the fire floor. The company officer in charge of ventilation should evaluate the rest of the upper floors of the building for smoke conditions; it may be necessary to skip the rest of the floors and start transitional ventilation on the top floor, then work back down.

Once each floor is thoroughly vented, occupants who sheltered in place can be escorted safely by the SER team to the evacuation stairway and to a safe location outside the building. Although the building may be sound, maintenance or restoration issues may need to be handled before the building can be reoccupied.

Make sure to vent for CO. As smoke stratifies, fire gases that are lighter than air will continue to rise above the level of stratification. Firefighters should remain on air with SCBA until atmospheric readings are taken to ensure that CO levels are well below 35 ppm.

Post-Fire Operations

Once the fire is extinguished, the sprinkler system should be shut off; two firefighters should remain at the control valves to quickly turn the system back on if necessary. To avoid water damage, there is a tendency to prematurely shut down the sprinkler system. The emphasis should be on complete extinguishment before worrying about water damage. A premature shutdown of the sprinklers can allow the fire to regain intensity, causing further property damage and financial loss.

Overhaul should be performed in the usual manner. The monetary losses in high-rise fires are high, so try and protect the area of origin for fire investigation. Salvage and property conservation should be started during extinguishment efforts. In fires of such magnitude, it is unlikely there will be personnel to spare. The best strategy for salvage and property conservation is to put the fire out.

The Barrington Plaza fire on January 29, 2020 in Los Angeles, California demonstrates many of the points discussed in this chapter. The 26-story, non-sprinklered, residential building has no pressurized stairways. The fire had self-vented and was blowing out of a unit on floor 7 fueled by a 35-mph wind (see Figure 15-64). Many occupants did not hear a fire alarm but heard the sound of arriving fire engines or were notified by friends in the neighboring buildings, and one resident was notified by a family member who was watching the national news.

Firefighters climbing the firefighting stairway were met by fleeing occupants. The attack from the windward side was stalled until it was verified the occupants were out of the stairwell above the fire floor. The wind was blowing across the A side of the building from D to B and not directly into the unit; however, the wind drove heat and smoke into the stairways. It took many occupants more than 10 minutes to evacuate the stairway. Some collapsed on the stairs and needed to be rescued.

E37 initiated fire attack on floor 7 using a 2-inch hand line. The smoke and heat were down to the floor with zero-visibility and firefighters had to crawl on their bellies to reach the fire room. The attack was stalled again when they came across an unconscious male. The captain split the crew, designating one firefighter to effect rescue and leaving two firefighters to continue pushing forward, which they would not have been able to do if they were using 2½-inch hose. Exhausted from the heat and the smoke in a now untenable hallway, and almost out of air, E37 had to retreat back to the stairway landing.

The IC opted for a transitional attack, defensive to offensive. Interesting to note, it was made using one 1¾-inch handline operating off the tip of the aerial, spraying a straight stream into the unit using a combination nozzle. According to the fire chief, 335 firefighters had the fire out in 79 minutes. This fire also demonstrates why transitional attacks should be considered sooner rather than later.

After-Action REVIEW

IN SUMMARY

- A high-rise apartment or office building is a complex environment that presents several difficult challenges for firefighters to overcome when it is on fire. The primary goal is still to put the fire out quickly.
- High-rise fires are not quick incidents. They can last from hours to days, and they physically wear out crews.
- Without ground-level access and the ability to ventilate a high-rise fire quickly, intense sustained heat quickly fatigues firefighters during interior operations.
- Frequent rotation of crews helps prevent heat exhaustion.
- Elevation affects the physical stamina of firefighters and creates possible water pressure and supply problems.
- The secrets to success in fighting fires in high-rise buildings are prefire planning, consulting with the building or maintenance engineer, developing a prefire plan, and following the IC high-rise fire checklist.
- The first thing the company officer needs to determine when arriving at a high-rise fire is the location of the fire or the location of the fire alarm activation. Everything hinges on this information.
- Occupants who are self-evacuating can be anywhere, and stairways could be congested with hundreds of people.
- SER is a shift in the traditional term of (SAR) in high-rise firefighting because evacuation may be the next critical task related to the search that isn't a factor in other types of structural firefighting.
- The initial IC must gain control of the HVAC system, the elevators, and other building systems.
- The initial IC needs to decide where to place the first hose line. Due to the lack of ground operations and aerial access, the primary strategy for high-rise firefighting is a direct frontal attack with enough water to overwhelm the fire.
- One of the most dangerous situations created during the fire attack is the sudden change of conditions inside the firefighting stairway above the fire floor.
- The main strategy for SAR is defend in place, or shelter in place. This strategy for life safety in residential and commercial high-rise buildings means that the residents or occupants are safer staying in their units or work areas than they are trying to evacuate the building where the hallways and stairs can be charged with heat, smoke and CO.
- Wind is a major factor in firefighting operations. Wind can create unpredictable air currents and uncontrolled movement of smoke, leading to fire spread. Thus, ventilation usually takes place after the fire has been controlled or extinguished.
- A wind-driven fire can be a major problem when the exterior windows of a high-rise building fail. This can also lead to autoexposure: the lapping of flames to the floor above.
- The lack of ventilation creates the greatest physical stressor that firefighters must confront in high-rise firefighting: heat.
- The risk of injuries and death for firefighters is always greater in buildings that are not sprinklered.
- High-rise buildings without sprinkler systems require a massive amount of staffing and resources, and high-endurance efforts made by firefighters.
- After the fire has been knocked down and extinguished, efforts should be shifted immediately to ventilation, primary searches that weren't possible, and secondary searches.
- Unpredictable air currents are dangerous when there is a thermal threat of spreading the fire, but once the fire is out, crews can deal effectively with unpredictable air movement of residual smoke with relative safety.

KEY TERMS

approach assessment Part of the size-up, it includes a 360-degree assessment to view exterior conditions and as many sides of the building as possible while the apparatus is approaching the scene.

fire tower An enclosed stairway connected at each story by an outside balcony or a fireproof vestibule vented to the outside.

flow the floor The process of wetting down the entire exposure floor above the fire to prevent vertical fire spread.

plenum The space between the structural ceiling and the dropped ceiling of a floor/ceiling assembly; the underside of the floor above. It houses and conceals telecommunications cables for telephone and computer networks, and supports and conceals air ducts of the HVAC system.

rapid ascent team (RAT) A search team designated to quickly climb the entire firefighting stairway to ensure it is clear of any occupants fleeing from the fire floor landings.

shelter in place The strategy for life safety in residential and commercial high-rise buildings. Residents or occupants are safer staying in their units or work areas than trying to evacuate the building where the hallways and stairs can be charged with heat, smoke, and carbon monoxide.

stack effect The vertical air flow within high-rise buildings caused by the temperature-created density differences between the building interior and exterior, or between two interior spaces. (NFPA 92)

transitional ventilation From a pressurized stairway, smoke on a floor is directed horizontally to a nonpressurized stairway where the pressure differential causes the smoke to transition to vertical ventilation out through the rooftop portal.

REFERENCES

Ahrens, Mart. *High-Rise Building Fires*. NFPA Research November 2016. National Fire Protection Association, Quincy, MA.

Cook County Administration Building Fire, 69 West Washington, Chicago, Illinois. October 17, 2003, Heat Release Rate Experiments and FDS Simulations, NIST Special Publication SP-1021. Chicago, IL by National Institute of Standards and Technology, U.S. Department of Commerce, Gaithersburg, MD. Published July 2004.

Dunn, Vincent. *Command and Control of Fires and Emergencies*. Fire Engineering Books and Videos, 1999, by PennWell Corporation, Saddle Brook, NJ.

Dunn, Vincent. *Strategy of Firefighting*. Fire Engineering, 2007, by PennWell Corporation, Tulsa, OK.

Gustin, Bill. *What Every Firefighter Must Know About Fire Protection Systems, Part 1*. Fire Engineering, April Issue, 2018, PennWell Corporation, Fair Lawn, NJ.

Gustin, Bill. *What Every Firefighter Must Know About Fire Protection Systems, Part 2*. Fire Engineering, May Issue, 2018. PennWell Corporation, Fair Lawn, NJ.

Gustin, Bill. *Operating More Than One Hoseline from a Standpipe*. Fire Engineering, May Issue, 2015. PennWell Corporation, Fair Lawn, NJ.

Klaene, Bernard J. *Structural Firefighting Strategy and Tactics*. Jones & Bartlett Learning, Burlington, MA. 2016.

McGrail, David M. *Firefighting Operation in High-Rise and Standpipe-Equipped Buildings*. Fire Engineering, 2007. PennWell Corporation, Tulsa, OK.

Mensch, Amy E. George G. Cajaty Barbosa Braga, and Nelson P. Bryner. *Fire Exposures of Fire-Fighter Self-Contained Breathing Apparatus Facepiece Lenses*, NIST Technical Note (NIST TN) 1724. Published November 29, 2011 by National Institute of Standards and Technology, U.S. Department of Commerce, Gaithersburg, MD.

Mendes, Robert F. *Fighting High-Rise Building Fires*. Tactics and Logistics, National Fire Protection Association, 1975.

NFPA 14, *Standard for the Installation of Standpipe and Hose Systems*. National Fire Protection Association, Quincy, MA.

NFPA 1500, *Standard on Fire Department Occupational Safety, Health and Wellness Program*. National Fire Protection Association, Quincy, MA.

Norman, John. *Fire Officer's Handbook of Tactics*, 4th Edition. Fire Engineering, 2012. PennWell Corporation, Tulsa, OK.

O'Hagen, John T. *High Rise/Fire and Life Safety*. Dun Donnelley Publishing Corporation, New York, New York, 1977.

Reade Bush, J. Gordon Routley, U.S.FA-TR-082, April 1996 Technical Report Series. U.S. Fire Administration, FEMA, U.S. Department of Homeland Security, Emmitsburg, MD.

Routley, J. Gordon. *Interstate Bank Building Fire*. Los Angeles, California, USFA-TR-022, May 1988. U.S. Fire Administration, FEMA, U.S. Department of Homeland Security, Emmitsburg, MD.

Routley, J. Gordon, Charles Jennings, and Mark Chubb. *High-Rise Office Building Fire, One Meridian Plaza,* Philadelphia, Pennsylvania, USFA-TR-049, February 23, 1991. U.S. Fire Administration, FEMA, U.S. Department of Homeland Security, Emmitsburg, MD.

Seattle Fire Department Policy and Standard Operating Guidelines on High-Rise Firefighting, City of Seattle, Seattle, WA, July 25, 2018.

Special Report, Operational Considerations for High-Rise Firefighting.

Tracy, Jerry. *The High Rise Handbook. FDNY High-Rise Operations Symposium*, FDNY Training Academy, 2009, New York, NY.

CHAPTER 16

Salvage and Overhaul

LEARNING OBJECTIVES

- Describe the difference between salvage and overhaul.
- Explain the detrimental effects that a rekindle can have on the fire department and the community.
- Explain why the incident commander needs to perform a new size-up and risk-benefit analysis before overhaul begins.
- Describe the benefits of allowing the fire investigators to access the scene before overhaul begins.
- List the clues that fire investigators look for at the scene of the fire.
- Describe the actions that should be taken when a deceased fire victim is discovered.
- Describe the actions that should be taken when an incendiary device is discovered.
- List the criteria for performing a pre-overhaul safety inspection.
- Describe the health risks associated with smoke exposure during overhaul.
- Recognize the advantages of using foam and wet water during overhaul.
- Identify the basic principles of overhaul to ensure that the fire is completely extinguished.
- Describe the various ways to use salvage covers and tarps.
- Explain the company drive-by.

Introduction

Salvage often gets grouped with overhaul, but they are not the same thing. Salvage falls at the end of the RECEO/VS acronym (RECEO/VS stands for rescue, exposures, confinement, extinguishment, overhaul/ventilation and salvage), and salvage and overhaul tend to fall into the fourth incident priority of property conservation.

Salvage

Salvage comes from the Latin word *salvare,* meaning "to save." More specifically, salvage is saving something of value, or something useful, and extracting it from the rubble. It's saving property from further damage and destruction from fire, smoke, and water, and its roots go back to American colonial firefighting. The earliest salvage company in Manhattan, New York, was Hand in Hand, formed in 1780. The company had two main tools. One was a heavy-duty linen bag used to stuff the valuables and personal belongings of a client and remove them from the smoke and fire. The other tool was a bed key, which was used to disassemble beds—some of the most expensive pieces of furniture and personal property at that time. The most notable and well-organized salvage corps was the New York Fire Patrol, founded in 1835 by the New York Board of Fire Insurance Underwriters. Their mission was to discover fires and prevent losses to insured properties, primarily commercial properties. Many cities and insurance companies followed the New York model. Because patrols were funded by insurance premiums of policy holders who were paying for the services that also benefitted uninsured businesses and occupants, the National Board of Fire Underwriters voted to discontinue services and turn salvage responsibilities over to local fire departments. The New York Fire Patrol, one of the longest running salvage corps, finally disbanded in 2006.

Salvage starts with initial fireground actions. For example, checking a door to see if it's unlocked before using forcible entry to gain access is salvage. "Try before you pry" can prevent unnecessary damage to property. Taking out a window pane or breaking through a gypsum board wall instead of forcing or damaging a metal door to gain access is another example of salvage. It is much cheaper to replace a pane of glass or a panel of Sheetrock than a metal front door **FIGURE 16-1** and **FIGURE 16-2**. Depending on the situation, laddering a second-story open window instead of forcing a locked door to gain access causes zero damage. These are all salvage tactics.

The height of the emergency is during the fire. Once the fire is knocked down, the stabilization part of the incident priorities is in place and the emergency starts to wind down; therefore, firefighters should slow down and reevaluate the situation. This is typically when salvage and overhaul begin. If sufficient personnel are on scene, salvage operations can start at the same time as the firefight, but that's rarely the case. Usually there's a shortage of personnel, or there is just enough to cover offensive attack assignments. Once the fire is out and the urgency is over, crews can be reassigned. This is why salvage is often implemented during the overhaul phase of the fire. Rather than having a dedicated section on salvage, this chapter is peppered with tactical salvage recommendations and considerations. Anything that has to do with saving, protecting, preserving, removing, or covering property is a component of salvage.

FIGURE 16-1 Breaking through an ornate or metal door with an ax or irons can cause unnecessary damage if there are easier, less expensive ways to gain quick access.

FIGURE 16-2 Breaking a window adjacent to a locked door or breaching a gypsum wall is easy and may be less expensive to repair than a damaged metal door.

Overhaul

Overhaul is not an afterthought; in an analogy to the military, it is the turning point of the battle. If the enemy is not completely defeated, it retreats, regroups, and counterattacks.

After the main body of a fire is extinguished, the fire building still contains a significant volume of sparks, embers, and concealed areas of fire. **Overhaul** is essentially the careful and systematic examination of the areas that were on fire to determine whether there is any remaining fire, sparks, or embers that could cause the reignition or **rekindle** of burned material after the fire department has left the scene. The main purpose of overhaul is to make certain that no trace of fire remains to rekindle—in other words, overhaul is the assurance of complete extinguishment. Firefighters should follow the path of the fire and search the building for any signs of hidden or remaining fires, uncover them, and extinguish them. Even an area that *might have been* exposed to fire needs to be checked visually **FIGURE 16-3**. When all members of the fire department understand the behavior of fire, all can trace the possible paths of the fire **FIGURE 16-4**. Experience plays a big role here because firefighters need to tear apart floors, walls, ceilings, and attics, which is a form of justified property damage to ensure final extinguishment, while at the same time trying to salvage property **FIGURE 16-5**. These actions may have to be explained to the property owner who may not understand why firefighters are destroying more of their property even after the fire is out.

Rekindles

There's on old saying: "The lazy man works twice." Overhaul is no place for shortcuts; it must be systematic, thorough, and complete to prevent rekindles. A rekindle is not a less severe fire. Upon returning to the scene, the building, which is now familiar to firefighters, can cause them to become complacent to the dangers of structural weakness caused by the original fire. A rekindle may not be as spectacular as the original fire, but it has a major impact on the fire department and the community. Consider the following:

- A rekindle is a professional embarrassment to you, the incident commander (IC), and your fire department.
- It wastes the time, effort, and risk spent on the original fire and overhaul.
- It increases total out-of-service times of companies for the same incident.
- It takes away emergency resources from responding to other emergencies, and from protecting the rest of the city.
- It increases the risk of injury or death of firefighters for the same event.
- It increases the exposure of firefighters to the carcinogens found in smoke.
- It disrupts the business activities of the community (e.g., traffic, volunteer firefighters who have to leave their jobs and families for the second time).
- It can alarm community members needlessly; their first thought will be that an arsonist started the fire, not that it was caused by a lack of thoroughness from their local fire department.

A second purpose of overhaul is to leave the structure in as safe a condition as possible. Other individuals, such as fire investigators, utility personnel, property owners, and occupants, may need access to the building. Depending on the circumstances, the

FIGURE 16-3 During overhaul operations, firefighters use a thermal-imaging camera (TIC) and other tools to inspect walls, ceilings, attics, and other void spaces for any signs that the fire might not be completely extinguished.

FIGURE 16-4 When all members of the fire department understand fire behavior, all can trace the possible paths the fire may have taken.

FIGURE 16-5 Experience plays a big role in tearing apart floors, walls, ceilings, and attics to check for fire extention.

building could be partially occupied soon after the fire. Overhaul should be accomplished in a timely manner, but there is no need to rush or take unnecessary chances after the fire has been brought under control.

The IC must stop and perform another size-up and risk-benefit analysis, then develop an organized plan for conducting overhaul operations, identifying new problems but keeping the safety of firefighters paramount. Risk-benefit analysis must constantly be applied during overhaul because now the only life hazards on the fireground are the firefighters. It's not worth the injury to or death of any crew member for fire-damaged property. In the meantime, initial fire crews should be rotated out and replaced with fresh crews that are sufficient in number to hold the smoldering fire in check.

Rehabilitation

Rehabilitation, or **rehab**, is a cooling-down period for firefighters engaged in the initial fire attack. Setting up rehab can be simple or complex, depending on the size of the incident and current weather conditions. It should start at the engine in front of the structure where the attack lines came off the apparatus. The driver should have a small utility brush to brush off debris, ashes, chunks of drywall, and insulation that have fallen on a firefighter. Then the driver should use a small-diameter hose with a spray nozzle to hose off the firefighter, starting at the helmet, spraying down the shoulders and the self-contained breathing apparatus (SCBA) unit, and working down to the boots. This preliminary decontamination (decon) should be done while the firefighter is still on air and before the SCBA or any bunker gear is removed **FIGURE 16-6**.

FIGURE 16-6 The preliminary decontamination starts with the driver using a small-diameter hose with a spray nozzle to hose off the firefighter, head to toe, while the firefighter is still on air and before the SCBA or bunker gear is removed.

Courtesy of John Odegard.

After decon, the firefighters should drop their SCBA and remove helmet, hood, and coat to cool off. Medical personnel, support units, fire buffs, or auxiliary units often set up rehab stations with water, electrolyte replacement fluids, protein snacks, and the old standbys: coffee, hot chocolate, and doughnuts. The primary goal is to rehydrate every firefighter and replace lost fluids.

Medical Evaluation

The next stop is a quick medical evaluation to check vital signs, provide oxygen, and treat minor injuries **FIGURE 16-7**. The most important goal is to reduce the environmental stressors on the human body and reduce heat and body core temperature. The firefighters need to cool down and get their physical vital signs back into the normal ranges. According to the U.S. Fire Administration (USFA) website, approximately 58% of all firefighter line-of-duty deaths (LODDs) resulted from stress and overexertion. Having medical personnel check and record at least two sets of vital signs—blood pressure, heart rate, and respiration rate—of the on-scene firefighters can catch the first indications that something may be amiss. If after about 20 minutes, a firefighter's vital signs do not return to within normal limits, that individual may require further medical evaluation and/or transport to the emergency room. Once crews are rested, and empty SCBA bottles have been replaced by full ones, the crews can be reassigned for additional incident tasks.

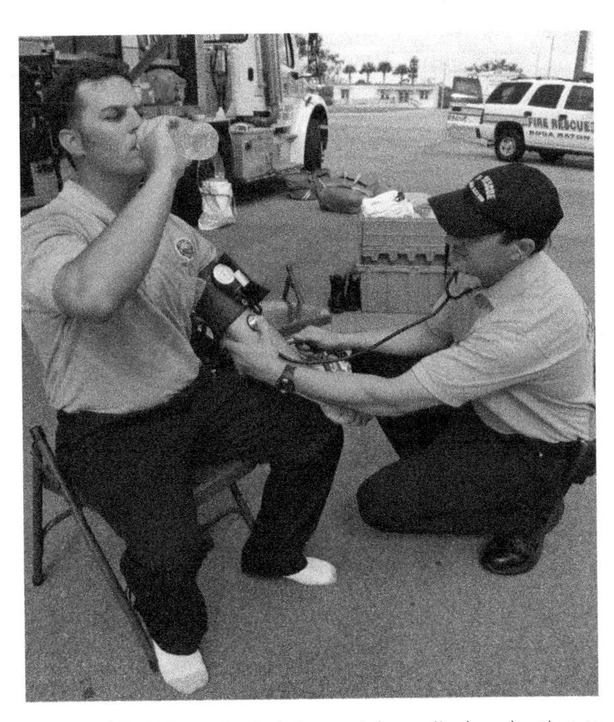

FIGURE 16-7 Part of rehab is a quick medical evaluation to check vital signs, provide oxygen, and treat minor injuries.

Inclement and Hot Weather

Every part of the country deals with extreme weather at some point during the year. During cold weather, efforts should be made to get crews out of the elements and into warm shelter. This can be as basic as requesting a public transportation bus that firefighters can use to get warm. During heat waves, fans, water misters, rehab cooling chairs, umbrellas and tents for shade, and trailers equipped with air-conditioning units are part of rehab to cool off overheated firefighters. The minimum criteria for developing and implementing the rehab process for firefighters is covered in NFPA 1584, *Standard on the Rehabilitation Process for Members During Emergency Operations and Training Exercises.*

Arson and Scene Preservation

Salvage operations can unintentionally remove evidence, and overaggressive overhaul often destroys evidence about the cause of a fire. Thus, the rehab period is a good time to get the arson investigators into the fire scene *before* overhaul starts. Many fire departments have a dedicated arson or fire investigator on duty or on call **FIGURE 16-8(A)**; others may have to rely on law enforcement or the state fire marshal **FIGURE 16-8(B)**. When there is a confirmed working fire, dispatchers should alert the arson investigator to respond; it's best to request one early in the incident to reduce the response time. The IC should call for the fire investigator as soon as it's obvious that the incident is significant. Suggested criteria for this determination are multiple alarm fires, fires with suspicious circumstances, incendiary patterns, a threat-to-burn occupancy, fire losses of $20,000 or greater, fires that result in a serious injury or death of a civilian or a firefighter, multiple set fires, fires in nursing homes and hospitals, fires involving incendiary devices or hazardous materials, and explosions.

Here are some points to consider for preserving the fire scene. Evidence takes many forms: some evidence is stationary and remains in place until disturbed; other forms are temporary, for example, flammable liquids that can evaporate or may be consumed quickly. The larger the fire and the longer it burns, the less chance there is that evidence will survive. Firefighters must understand and realize how their actions in forcible entry, search and rescue, fire attack, salvage, and overhaul can adversely affect the evidence at the fire scene. Consider how much material (possible evidence) is forcefully scattered with a 2½-inch smooth-bore nozzle aimed at the seat of the fire.

Some of the clues that arson investigators look for are burn patterns, multiple points of origin, trailers, flammable liquid fuel distribution, flammable liquid fuel containers, piles of newspapers, furniture piled together, the presence or absence of smoke detectors, and whether the detectors were hard-wired or battery operated. (Were there batteries in the smoke detectors?) Investigators check the position of the circuit breaker switches on the electrical panel; the position of shutoff valves on other utilities; the valves on the sprinkler system; and the position of knobs, switches, and controls on electrical and major appliances. And this is just a partial list; there are certainly much more criteria.

A

B

FIGURE 16-8 A. Many fire departments have a dedicated arson or fire investigator on duty or on call. **B.** A fire department may have to rely on law enforcement or the state fire marshal to investigate a fire.

A: Courtesy of John Odegard; B: © Glen E. Ellman.

Beware of the discovery of incendiary and explosive devices. Do not touch them. Evacuate the area and request the police bomb squad or a local military unit trained in the removal of such devices. Don't dismiss the possibility that a homegrown terrorist can turn your Smalltown, USA incident into a federal government investigation.

The most important scenario that can delay overhaul and require you to take exhaustive measures to preserve evidence at the fire scene is the discovery of a deceased fire victim. Limit the number of firefighters in the immediate area where the dead body or bodies were discovered and keep overhaul to a minimum to hold the smoldering fire in check. Make sure the firefighters do not move or remove the bodies. The incident may now be a crime scene. Fire investigators will let you know when the bodies can be removed. The only time removing deceased victims may be considered is when there is a danger of imminent structural collapse which would completely destroy the evidence. In such cases, the presence, location, and position of dead bodies must be well documented, including taking photographs if at all possible.

The dilemma of the competing interests between the IC, who wants to wrap up the incident and get units back in service, and the arson investigator, who prefers that no one disturbs the scene until his or her arrival and completion of the investigation, should be addressed ahead of time in department policy. There is a reasonable time frame for the IC to wait and there is an unreasonable time frame where the IC needs to begin overhaul and terminate the incident. The best compromise would be for an experienced chief officer to photograph the fire scene extensively, with all the relevant criteria, using a digital camera (not a cell phone); remove the memory card; and hand it over to a law enforcement officer to preserve the chain of evidence. (Additional online training on this subject is available free of charge from the International Association of Arson Investigators Inc. at www.cfitrainer.net.)

Pre-Overhaul Safety Inspection

During fire attack and related property conservation operations, firefighters work quickly and may need to take calculated risks to stabilize the incident and extinguish the fire. However, overhaul on the fire building takes place after the emergency is over, so there is no reason for rushing or taking unnecessary chances during overhaul. Overhaul should be completed systematically, and it starts with the IC, who must perform a new size-up and risk-benefit analysis, then develop an organized and safe plan for overhauling the fire building.

Experienced company officers and the safety chief should inspect the building thoroughly before firefighters begin the task of overhaul. The building has been damaged by fire and by the weight of the water used during the fire attack and extinguishment. There may be holes in the floors and the roof. Stairways may be damaged and may not be safe to use **FIGURE 16-9**. Portions of the building may be unsafe to enter under any condition. Structural integrity should determine whether firefighting activities should be conducted in all areas or only designated areas, or whether the entire building should be placed off limits. Areas that are hazardous and off limits must be identified. An example is a freestanding unstable wall; another is a precarious leaning chimney. Barrier tape, uncharged hose, and traffic cones should be used to create a collapse zone to identify and block access to these areas. Hazardous areas should also be announced on the radio so that all personnel can be aware of where they can and cannot work before overhaul begins. Precarious walls and leaning chimneys should be knocked over with master streams, if possible, to eliminate these hazards.

The extent of the pre-overhaul safety inspection, as well as the entire overhaul operation, depends on the extent of the fire. If the fire was a contents fire, confined to a room or two, an inspection of the fire area as well as the rooms above and below should be sufficient. If a large area of the building was involved in fire, then a more thorough inspection of the structure must be conducted. In many cases, the entire building needs to be examined, and safety hazards that might further threaten occupants, bystanders, and firefighters must be mitigated.

The reason for inspecting is simple: risk management. The building must be safe for firefighters to perform their tasks. Safety officers should ensure that unsafe areas are marked or taped off, and structurally unsafe areas of the building are placed off limits. If there is any doubt, don't hesitate to call for a structural engineer. Fire scene tape, traffic cones, and uncharged fire hose (the latter is abundant at any fire scene) are useful for marking and blocking off areas that are unsafe. Announce all hazards on the fireground over the radio. Firefighters cannot be safe unless the scene is safe.

Smoke Explosions

Firefighters working in overhaul areas must be well supervised. In spite of the unhurried atmosphere that should prevail during this phase of the operation,

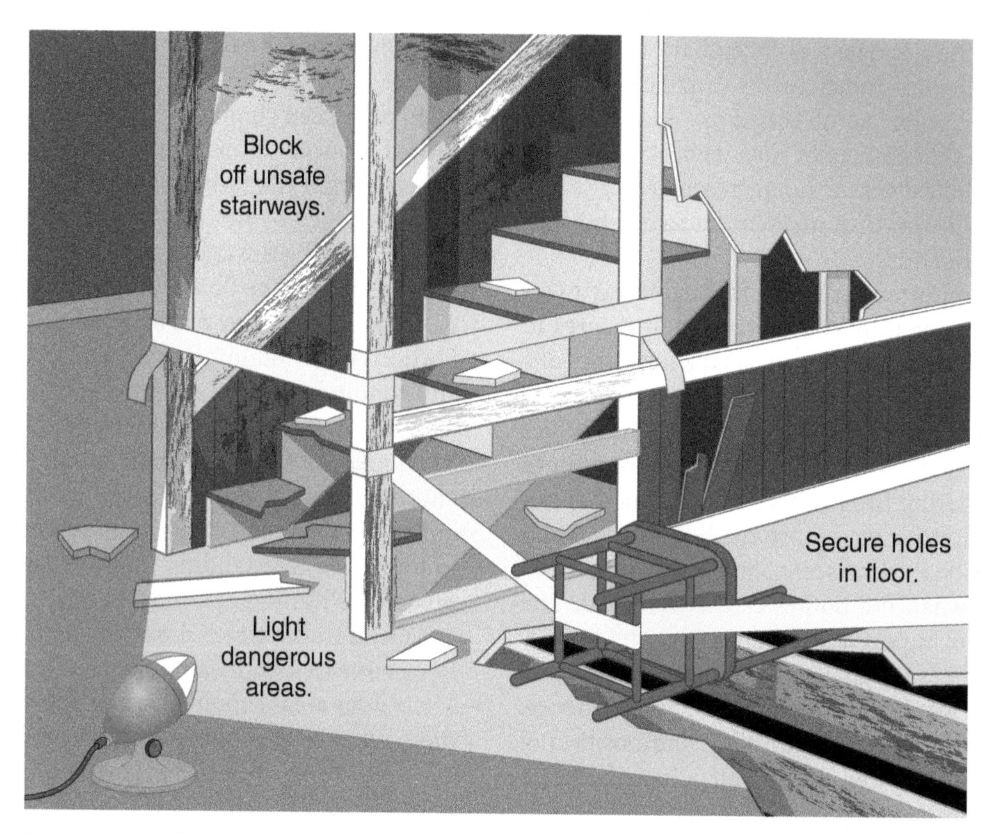

FIGURE 16-9 Before overhaul operations begin, the area must be inspected and necessary safeguards put in place. Portable lighting should be used to illuminate hazard areas.
Engine Company Fireground Operations, 3rd Edition, Harold Richman, Steve Persson, NFPA © 2008 Jones & Bartlett Publishers.

firefighters can become overly aggressive in overhauling so that they can finish the battle, and they can lose sight of the fact that there may be nothing left to save in the immediate fire area. Firefighters may also become complacent and drop their guard because they fail to recognize the dangers that exist after the fire has been knocked down. Many crew members presume that when the fire is in the smoldering decay stage and there are no more visible flames, the fire is no longer dangerous. Firefighter fatalities have occurred during the overhaul phase, and the injury rate for firefighters during this part of the operation is relatively high. Although more than 30 years old, the graphic 1986 case study of two Biloxi, Mississippi, firefighters illustrates the dangerous smoke conditions that exist when the fire is assumed to be contained and the emergency is basically over. Cold smoke explosions often occur during the overhaul phase. Although they are sometimes referred to as *cold smoke explosions*, the smoke isn't cold at all. It's cooling, but it is still within the flammable range.

A smoke explosion can happen at any time during a fire, but if it does occur, it typically occurs during the overhaul phase after the fire is out (**FIGURE 16-10**).

FIGURE 16-10 If a smoke explosion occurs, it typically happens during the overhaul phase of the fire.
Courtesy of John Odegard.

A smoke explosion is often confused with a backdraft. They are similar, but they are not the same. A backdraft has high-temperature fire gases that are too rich to burn and are confined within a pressurized space—all it needs for combustion is air. When air is introduced, often by an unwitting firefighter, the fire triangle is complete, and a backdraft occurs with explosive force. In a smoke explosion, smoke and fire gases have been pushed away from the fire, under pressure, along a path of least resistance throughout the building. Once the fire has been knocked down or extinguished, the thermal energy of the fire has been taken away; therefore, no more thermal pressure is being generated. The building is as pressurized as it is going to get from the fire and has pushed the smoke as far away from the seat of the fire as it can. However, the smoke and fire gases remain suspended within the structure. No more energy is being emitted, and temperatures begin to cool because firefighters have already made entry; there shouldn't be any more pressurized spaces. Although the smoke is cooling, it is well within the flammable range. A backdraft needs only oxygen to occur, while a smoke explosion needs only heat. All that is necessary is for the smoke to find an ignition source. That source is often found when firefighters dig out and expose hidden fires, which can flare up and cause a smoke explosion. The fire can reignite and burn the structure again, or the suspended smoke particles can rapidly burn with explosive force until the flammable vapors are consumed, at which point the fire goes out. Smoke explosions are prevented by hose lines wetting down burning material that is still emitting smoke, and with positive-pressure ventilation (PPV).

Biloxi, Mississippi, Fire Department Case Study

On October 20, 1986, a fire started in a large two-story rooming house sometime after 03:00 hours. A fire with very little showing upon arrival would end up taking the lives of Firefighter Carl Ohr and Firefighter Kurt Jacquet Sr. of the Biloxi Fire Department.

Built in 1926, the two-story wood-frame house had no fire stops between floors, and the open vertical channels between the studs ran from the cellar to the attic; in other words, it was classic balloon-frame construction. Based on interviews, firefighters did not recognize that they were dealing with balloon-frame construction at the time of the fire. (Always suspect balloon-frame construction in wood-frame houses built prior to 1950.) The house was converted illegally into seven separate apartments, each with small bedrooms, kitchens, and baths, creating a labyrinth of nooks and crannies. Besides the main electrical service drop to the original house, another illegal electrical source came into the house from the next-door neighbor's home. Two Romex® wires were buried through the yard between the two houses. They did not run through a junction box or circuit breaker and, over time, they acted like a heating element, causing a fire in the floor space between the first and second floor. Even though the utility company cut the main power to the house during the incident, the structure was still energized from these power lines.

Firefighters Ohr and Jacquet, along with Captain Brewer, on the first-in engine, were assigned search and rescue. After completing the primary search, they exited the house. The captain took the crew back in for a secondary search. They also brought in an attic ladder to help locate the source of the fire. After the secondary search, the entire crew exited the house. Captain Brewer went over to the command post to report an "all clear" for any civilian lives in the structure. There is video footage showing Firefighter Ohr and Firefighter Jacquet Sr. changing SCBA bottles at 03:45. In 1986, these would have been 30-minute bottles and, after making two interior searches, it can be assumed that they were in the structure at least 20 minutes.

For reasons unknown, Ohr and Jacquet reentered the structure a third time—without the knowledge of Captain Brewer and the IC. At this time, everyone believed the fire was contained. There was very little smoke inside the house on floor 1 and floor 2. However, the fire was raging in the attic. One can only deduce that the two firefighters didn't sense any danger; interior conditions didn't warrant taking in a charged hose line for protection because they believed that there wasn't any significant heat or smoke.

At some point during the third entry, Firefighter Ohr got in trouble and ran out of air. He was in a corner bedroom and tried to push out a window-mounted air conditioning unit but was unsuccessful and collapsed. Firefighter Jacquet came to his aid and attempted to drag Firefighter Ohr out of the structure but Ohr's SCBA straps became entangled on the bed and the chest of drawers. At this point, a news camera caught a significant amount of pressurized flames coming from a second-story bedroom window on the B-side of the house. Deemed a flashover, this sudden burst of flames created the thermal assault that killed the two firefighters. Their bodies were finally found at 05:05. Ohr had been overcome by the products of combustion and had a blood carbon monoxide (CO) level of 70%, which is lethal. Jacquet's CO level was only 15%, but he suffered burns to the esophagus and trachea from the superheated smoke.

Current practices and technological advances made in the fire service may have prevented the deaths of these two firefighters if the same incident occurred today. The development, implementation, and adherence to the incident command system (ICS) has all but eliminated freelancing. If it happens, it's a rare occurrence. Today, many departments use 45-minute SCBA bottles. The extra margin of air might have been all that Ohr and Jacquet would have needed to survive. Our structural firefighting gear now includes ear flaps, Nomex® hoods, and more sophisticated pants and coats; however, they're rated for 500° F (260° C) for a duration of about 5 minutes and any flashover is well above that. The use of thermal-imaging cameras (TICs) would have made the attic fire easier to detect and the downed firefighters easier to find. Finally, every firefighter is equipped with their own personal radio with an emergency button to transmit a Mayday, and all firefighters have had some kind of Mayday training. The two in/two out rule and the establishment of rapid intervention teams may well have led to a more favorable outcome. However, technological advancements and practices isn't the point. The point is that these exact fire conditions within the structure can play out the same way today as they did in 1986.

This case study from Biloxi offers two takeaways for the subject of overhaul. First, the Biloxi fire demonstrates the necessity of recognizing when you're dealing with balloon-frame construction. For a fire in this type of structure, you have to get a crew immediately into or over the attic space with a charged hose line to check for fire extension. Second, it seems that this was not a flashover but a cold smoke explosion. Remember, cold smoke does not mean the smoke is out of its flammable range. It simply needs an ignition source to bump it up a few degrees until it reaches ignition temperature. Smoke explosions can happen at any time, but they tend to happen during the overhaul phase. When the fire is building from the incipient stage through the fully developed stage, interior pressure is increasing. At the decay stage, the building is as pressurized as it can get, often pushing smoke a significant distance from the source of the fire.

Flashover is a high-temperature event, which is not what happened in Biloxi. Remember that Ohr and Jacquet didn't take a hose line in because it *wasn't* hot. Firefighter Jacquet's thermal burns to the esophagus and trachea occurred because he wasn't wearing his facepiece and he wasn't on air: he was trying to save Ohr while he was uncovered. The ignition source was obviously the flames in the attic, and the lingering smoke with high concentrations of CO on the second floor ignited with explosive force. Smoke is fuel, and it already exists in a vaporized state. When it ignites, it does so with lightning speed, and the effects at that point are very similar to a flashover. This is the condition company officers must be able to anticipate and recognize during overhaul so they can take the necessary steps to prevent it from occurring.

Air Monitoring

All firefighters should be in full structural protective clothing, including SCBA, and should be covered and on air while overhauling **FIGURE 16-11**. Firefighters without full personal protective equipment (PPE), including SCBA, should not be allowed inside the building. They should not be allowed to dress down during overhaul. CO levels have been shown to be higher during overhaul than in actual firefighting activities. The atmosphere in the areas where overhaul operations are performed should be monitored to determine CO levels, as well as any other toxic gases in the air. The handheld **multi-gas detector** and the TIC have been two of the greatest technological advancements for the fire service, and they are essential tools for overhaul and for ensuring firefighter safety. A multi-gas detector can measure oxygen-rich or oxygen-deficient atmospheres, CO levels, combustible gases, sulfur dioxide, nitrogen dioxide, and hydrogen cyanide levels. Many fire departments use the National Institute for Occupational Safety and Health (NIOSH)-recommended exposure limit for CO of 35 parts per million (ppm) for an 8-hour time-weighted average. If the CO readings are below 35 ppm, they allow firefighters to overhaul without SCBA. If the CO readings are above 35 ppm, they ensure that firefighters are wearing facepieces and are breathing air from the SCBA.

Smoke Particulates

Today's health risks associated with firefighter smoke exposure are better understood than they were even

FIGURE 16-11 All firefighters should be in full structural PPE clothing, including SCBA, and should be covered and on air while overhauling.

Courtesy of Lt. Mike Heaton.

FIGURE 16-12 Air-purifying respirators with full-face coverage are excellent for filtering particulates and dust in a non-IDLH environment, but they cannot be used in oxygen-deficient atmospheres.

Courtesy of Raul Angulo.

a decade ago. In addition to possible exposure to asbestos, smoke is filled with particulate matter (soot) and other carcinogenic chemicals that cannot be detected with a multi-gas detector—making even light smoke (depending on what was burning) dangerous to firefighters not wearing respiratory protection. Ninety-nine percent of smoke particulates are less than 1 micron (.001 mm) in diameter, and 97% of those particulates are too small to be seen with the naked eye. What may appear to be clean air to the eye isn't clean at all. Many fire investigators use full-face coverage, air-purifying respirators when determining the cause of a fire **FIGURE 16-12**. They are excellent for filtering particulates and dust in a non–immediately dangerous to life or health (IDLH) environment, but they have limitations and cannot be used in oxygen-deficient atmospheres. If you're looking to protect the health and longevity of your crews, there's no substitute for SCBA.

Below is an extensive list of carcinogenic chemical compounds and particulates found in smoke, many of which are flammable and add to the possibility of smoke explosions:

- CO
- Hydrogen sulfide
- Hydrogen chloride
- Carbon dioxide
- Chloroform
- Phosgene
- Acrolein
- Methyl methacrylate
- Formaldehyde and other aldehyde particulates
- Acid gases
- Sulfur dioxide
- Nitrogen oxide and dioxide
- Polycyclic aromatic hydrocarbons (PAHs)
- Polyvinyl chloride
- Benzene
- Toluene
- Styrene
- Dioxins
- Irritants
- Soot
- Certain heavy metals

In addition to ventilation, the use of PPV fans are extremely beneficial in introducing a flow of fresh air to aid in the removal of these carcinogens that are still off-gassing, even though there is no more visible smoke. The air can cause flare-ups of hidden embers and thus reveal their location for overhaul and final extinguishment.

Health Risks

The heightened awareness and understanding of the cancer risks associated with smoke have prompted many firefighters to make the personal decision to wear SCBA and remain on air throughout the entire overhaul. Progressive fire departments committed to the health and wellness of their members have also done away with the practice of dropping their SCBA when the CO reading is below 35 ppm, and have changed their department policy directing all firefighters to wear SCBA and remain on air until the overhaul is complete. In other words the practice of dropping the air mask is no longer optional.

The IC should also consider how many firefighters will be exposed to the overhaul health risks and for how long. Major university studies show that the carcinogenic compounds found in smoke permeate the bunker gear material and, combined with human perspiration, are absorbed into the skin. A decision needs to be made to "share the wealth," so to speak, and expose every member equally or limit the exposure to the absolute minimum number of firefighters needed to complete the overhaul.

Gloves

The fire room is filled with wood, metal, and plastic debris, much of it with sharp and jagged edges. The debris can include exposed nails, broken glass, hot surfaces, and burning embers. Hazards abound during overhaul. Firefighters must be in full PPE, SCBA, and *structural firefighting gloves*. There is a practice of firefighters exchanging their structural firefighting gloves for regular leather work gloves for better grip and dexterity after the fire is out. These leather gloves do not have the thermal rating or puncture resistance required by NFPA-compliant structural firefighting gloves and should not be allowed during overhaul. Company officers must help firefighters understand the constant hazards that surround them.

Portable Lighting

Once the fire is out and the emergency is over, there is no reason to work in the dark. Fire rooms and smoke-charred hallways can be pitch black after the electrical power has been cut, even if it is sunny outside. Portable lighting should be brought in to light up the interior scene. This assignment normally falls to the ladder company, but it needs to be done regardless. Generators and electrical cord reels should be brought in. String lights can be hung using nails or strung over furniture **FIGURE 16-13(A)**. Baby ladders and pike poles can be rigged like a stanchion to hang portable lights **FIGURE 16-13(B)**. Lights should be positioned as high as possible for maximum illumination; don't simply lay them on the floor.

The same goes for the exterior fire scene. Use apparatus floodlights to light up the perimeter. Care should be taken when firefighters are working on the roof or elevated heights at night. Powerful light beams aimed directly at the firefighters may temporarily blind them and actually make their situation worse.

Personnel

Personnel who have been fighting the fire for some time are tired and worn down by the physical activity, heat, smoke, and weight of their equipment. These individuals generally are not in the best shape to conduct a careful examination of the fire premises or immediately begin overhaul operations. They deserve a rest before reassignment. The IC's post-fire size-up and the pre-overhaul safety inspection is a good time for them to rest before extensive overhaul operations begin. These firefighters should be sent to a rehabilitation sector to rest, treat injuries, replace fluids, and have basic vital signs checked. Afterward, it should then be determined whether they are fit to return for reassignment. If these firefighters are needed later, they will be rested, alert, and better able to perform their duties. Fresh crews should be used to relieve firefighters who worked the initial extinguishment operations. If overhaul operations are extended, then additional or rehabbed firefighters should be assigned. Firefighters who were assigned to staging or to attending unused backup lines, or were on exposure protection assignments during the fire attack should be the first ones assigned to overhaul. Personnel arriving independently, such as off-duty career firefighters or additional arriving volunteers, are candidates for overhaul duties. If necessary, additional engine and/or ladder companies can be called to the fire scene for overhaul and the rotation of crews.

Personnel should be assigned in sufficient numbers and have the needed tools and equipment, including

A

B

FIGURE 16-13 **A**. String lights can be hung using nails. **B**. Baby ladders and pike poles can be rigged to hang portable lights. Lights should be positioned as high as possible for maximum illumination; don't simply lay them on the floor.

Courtesy of Raul Angulo.

the proper number of hose lines, to perform their overhaul assignments. Firefighters should be well supervised, and depending on the size and scope of the operation, safety officers should be assigned throughout the overhaul phase.

Control of Overhaul Personnel

Crew members who have not been inside the fire building during the firefighting operations may not know which areas have been damaged or where the structure may be weak. The outside appearance of the building might give them little information about the structural integrity of the interior. Thus, it is extremely important that all identified hazards be communicated to relief crews. Company officers must be diligent in supervising crews. There is no justifiable reason why a member should get hurt during overhaul. The accountability system should remain in use for the duration of the incident.

Firefighters performing overhaul should be assigned to a group or division supervisor depending on the layout of the fire building. Members should work in pairs (the buddy system) and maintain contact with each other and with the group supervisor of the overhaul operation. To ensure that the firefighter accountability system is maintained, no one should be allowed inside the fire building without first reporting to the division or group supervisor assigned to that area. Firefighters entering the building without reporting to the group supervisor are freelancing and could find themselves in serious trouble if no one is aware that they are inside.

This was a factor in the Biloxi case study, as already mentioned, and it also happened in Seattle, Washington, on July 12, 1987, at the abandoned and boarded-up Crest Apartments. A senior firefighter, who was the driver of one of the first listed engine companies, was separated from his crew when they were given an interior assignment. His apparatus was not needed for water supply, so he made the effort to rejoin his crew, who were already inside the fire building. A chief officer sent him into an uninvolved part of the structure to check some rooms. Not sensing any danger, he never took a partner. The firefighter entered a room that was boarded up from the inside with plywood. The fire never made it to this part of the building, but the pressurized smoke did. The company officer never knew his driver was inside the building, and the chief, who was running the fire, obviously distracted with other command responsibilities, wasn't keeping track of this individual firefighter, who, in his mind, was performing a nondangerous reconnaissance. The firefighter never made it out, however, because he ran out of air and died of smoke inhalation.

In this example, as well as in the Biloxi case study, incident management systems (IMS) and accountability systems weren't as developed as they are today. In the 1980s, only the officer had a portable radio. Neither the two Biloxi firefighters nor the Seattle firefighter had individual portable radios, so they had no communications to call for help. No one will ever know exactly what led to these firefighters getting into trouble they couldn't escape from. Today, every firefighter carries a personal radio. Each radio has an emergency button to call for help, and all firefighters have had some kind of Mayday training. Most likely, personal radios would have prevented these fatal events, and the outcomes would have been more favorable. Safety procedures, like the accountability system, were developed as a result of these types of incidents, so we need to adhere to them.

The other takeaway is that these two case studies happened after the fire was contained, during the overhaul phase. It doesn't matter how advanced our PPE is, the overhaul environment and atmosphere is the same today as it was in the 1980s when these two tragic examples occurred. And with the smoke by-products of modern fuels, the overhaul atmosphere is even more dangerous.

Disposable Buildings

The most important question an IC has to ask during offensive operations, and especially during the overhaul is, "At what point does this building no longer have any value?" The new construction market is not driven by firefighter safety; it's driven by capital cost savings (how can they construct a structurally sound building for the least amount of money). Well, most buildings are structurally sound—until they catch fire. Combine lightweight construction with the higher heat release rates of burning modern fuels, and the damage caused is cheaper to replace than to repair. In other words, a disposable building is one that is cheaper to tear down and replace with a new one than it is to repair.

Another question that the fire service has refused to consider on ethical, moral, or professional grounds, after the life safety incident priority has been mitigated, is, "If a building is insured, should we let it go?" In other words, should the fire service take a defensive posture and let it burn to the ground? This action would ensure that no firefighters are injured or killed fighting a fire inside a building that's going to be torn down anyway. Many in the fire service may answer, "That judgment call is not for us to make because the fire department's mission is to save lives and put out fires." Perhaps it is time for a national dialogue on this subject. If a building is proved to be insured and there are no lives at risk, future insurance policies may dictate defensive strategies only.

Here's an example to illustrate the point about risk management. On February 14, 2000, two Houston, Texas, firefighters died in the line of duty while fighting an early morning fire at a McDonald's restaurant when its roof collapsed. Within a couple of years of the Houston fire, my company, Engine 33, fought a similar early morning fire at the McDonald's in the Rainier Beach district of Seattle. With smoke and flames showing upon arrival, we made an aggressive interior attack and managed to extinguish the fire. We were proud of our efforts and, from our perspective, we felt we saved the building. We didn't suffer any firefighter injuries, and we didn't pay the price that the Houston Fire Department paid, but in both cases, the end result was the same: the damaged buildings were demolished by the McDonald's Corporation, and at least in Seattle, a new, more modern McDonald's restaurant was built on the same site.

When a major fire has been extinguished, there is a high probability that areas of the building are unsafe. Following standard operating guidelines (SOGs) for overhaul minimizes the chance of serious injury should an accident occur and ensures the accountability of members working inside the building. It will be easier for an IC or company officer to live with the decision of writing off a building than it will be to live with the aftermath of an LODD because the decision was made to send firefighters into a fire-damaged building that no longer had any value.

The Overhaul Operation and Work Assignments

Ordinarily, ladder companies are assigned tasks involving hand and power tools during overhaul. Engine companies advance and operate hose lines for extinguishing small fires and residual embers. If ladder company personnel are not on scene or are engaged elsewhere on the fireground, engine company personnel must perform both duties. In this case, the group supervisor should assign part of the overhaul group to perform ladder company tasks and the remainder of the group to perform as an engine company. Each firefighter then knows what task he or she has been assigned, and every firefighter, whether assigned to an engine or a ladder company, should know how to perform any overhaul task that is required.

If the group must be split up to perform its duties, there is an obvious way to make the split. Ladder companies are usually assigned to pull walls and ceilings in the fire room. They're also assigned to the floors above and rooms adjacent to the fire to check for extension as yet undetected. A crew should check the floor below the fire for extension and water damage. Engine company crews standing by with hose lines are called to these floors if hot spots are discovered. Jobs that are not specifically assigned as ladder company work or engine company work can be performed by either crew. If heavy material has to be removed from an area of the structure, everyone in the group is expected to lend a hand.

Securing Utilities

Controlling and securing utilities is pretty basic. The fire department has to deal with only four: water, gas, electric, and sometimes liquefied petroleum gas (LPG). Water lines can be damaged or severed at any point during the fire. When that happens, water runs freely, causing additional or unnecessary water damage to the occupancy. Faucets, refrigerators, washing machines, water heaters, and toilets all have a water shutoff valve. Mainline water valves are found inside, underneath, or outside the structure. It's just a matter of tracing the water line. The mainline can also be shut off at the water meter located between the sidewalk and the street. The water utility company typically handles the shutting off of a large water main. Shutting off the water is a responsible salvage tactic.

A natural gas leak at any point during a fire incident is extremely dangerous. If the gas has not been shut off as one of the initial tasks before a fire attack, it should be shut off before overhaul begins. The main supply to all gas meters, large and small, is usually fitted with a quarter-turn valve. Many fire department hand tools can accomplish this easy task. With fires involving multifamily occupancies, there is typically a large gas meter (to reduce the pressure coming in from the street) that feeds smaller gas meters to each individual unit. It may be necessary to shut off only the gas meters of the affected living units involved in the fire rather than shutting off the gas to the entire complex. Exercise safety and common sense. The majority of gas meters are located outside the building, usually on the sides or rear of the building. In many older Type III ordinary construction buildings built before 1950, however, the main gas meter can sometimes be found inside, usually in the basement. For obvious reasons, they present a more serious hazard. Remember that natural gas is lighter than air, so it rises and can accumulate in high spaces. These spaces need to be ventilated. Turning the gas back on to any occupancy should be handled by the gas utility company.

Electricity is another hazard to firefighters and, of all the utilities, is probably responsible for injuring or killing more firefighters through electrocution. Single-family dwellings, multifamily residential units, and light commercial occupancies all have a main circuit breaker switch that can be flipped at the main junction box. Depending on your local protocols, flipping the main circuit breaker would reasonably fall within the responsibility of the first-alarm assignments. Sending two firefighters to locate the main circuit breaker to cut the power is a fairly safe task. Flipping the main circuit breaker is no guarantee, however, that power has been cut to the structure.

Remember, in the Biloxi case study, the illegal wiring bypassed the junction box and circuit breakers is leaving the house electrically energized. The only way to be sure is to use an electrical **hot-stick voltage detector**. These handheld units are specially designed for first responders. Firefighters should never pull the meter. Although there are special covered meter pullers, this job is best left to the power utility company. The surest way to cut electrical power to a structure is to have the service drop, which runs to the weatherhead, disconnected by electric utility personnel **FIGURE 16-14**.

FIGURE 16-14 The surest way to cut electrical power to a structure is to have the utility company disconnect the service drop that runs to the weatherhead.

Courtesy of John Odegard.

Many homes use LPG or propane as their primary source for heating and cooking. These resources are no longer found exclusively in rural or suburban residential areas. Many neighborhoods now allow LPG tanks within the city limits of many urban and metropolitan areas. Residential LPG storage tanks come in five sizes: 120-, 250-, 325-, 500-, and 1,000-gallon tanks. The most popular sizes for single-family residences are the 250-gallon tanks. The larger tanks, from 325 to 500, to 1,000 gallons, are usually in the backyard, a safe distance from the house. In other words, they probably would not be the immediate exposure concern. But the 250-gallon tanks are more popular in denser neighborhoods because they don't take up much space. Expect these tanks to be right next to the exterior siding of the residence they serve.

The author had a personal experience with these tanks within the city limits of Seattle, Washington. Engine 33 was first-in to a fully involved house fire with exposure homes on side B and D, and two side-by-side, 250-gallon LPG tanks against the exterior D side of the burning structure. One firefighter took a 2½-inch exposure line to exposure B and another tailboard firefighter laid a 2½-inch exposure line for the LPG tanks. Once the driver was hooked up to the hydrant, he opened the deck gun to protect exposure D. Under the cover of a 2½-inch fog stream, the author approached the LPG tanks and shut the supply valves to the house.

All these tanks have a gate valve that needs to be twisted shut to secure the LPG supply into the structure. If it isn't done during fire attack, it should be done before overhaul. Remember, LPG is heavier than air, so if it's present, it will accumulate in low areas. Cellars and basements must be checked with a multi-gas detector, set to detect combustibles, before salvage and overhaul begins. Because CO may also be present in basements and cellars of buildings involved in fire, firefighters taking atmospheric readings must be protected with SCBA, be on air, and work with a partner in case they get in trouble.

Cleanup

Although good for public relations, cleanup is not a necessary part of overhaul. Debris that presents the possibility of a rekindle, such as burned structural members, furnishings, or equipment, should be removed and piled away from the building. Starting with the area exhibiting the most extensive fire damage, firefighters should check for possible clues to the ignition of the fire. Knowing the leading causes of residential fires can help firefighters deduce what led to the start of the fire. According to the USFA website, cooking, unintentional carelessness, open flames, smoking, and electrical malfunctions are the top five causes of residential fires. Don't overlook the fact that many fires are deliberately set. Indiscriminately tossing everything out the window and washing down everything in sight probably negates any chance of the fire investigators determining the cause of a suspicious fire. Building components or debris that fire investigators need to inspect should be left in place for examination if possible. A fire of incendiary origin can greatly alter overhaul operations or delay them until arson investigators can examine the fire scene. After they complete their work, the basic overhaul operation can continue.

If conditions warrant, and it is safe to do so, a building's maintenance team can perform cleanup with the IC's permission. As mentioned above, the exposure of firefighters to the cancer health risks of smoke may lead the IC to recommend that the owner hire a private contractor specializing in post-fire cleanup. The market competition of these private salvage and post-fire cleanup companies is growing. Many have radio scanners to monitor structure fires, and they often have a service contract representative on scene to solicit business from the owner even before the fire is out. From a risk management standpoint, this is called *transferring the risk.*

Removing Items from the Structure

Removing smoldering household items and furniture from a structure, especially mattresses and upholstery, can cause sudden flare-ups for unsuspecting firefighters. This is why firefighters should remain in full PPE with SCBA and structural firefighting gloves. Smoldering mattresses frequently burst into flames when they reach the outside fresh air. Thus, it is a good practice to wet down a salvage tarp and wrap it around the burned mattress. Wetting down the salvage tarp protects the canvas from burning embers and extends the life of the tarp. The wet salvage tarp excludes air from the mattress, so it doesn't burst into flames while a crew is carrying it through the structure. There should be a charged hose line at the debris pile to wet down all burned material thoroughly.

Using metal garbage cans with lids is also a good way to carry out burned debris. They protect the debris from getting a fresh air source so it doesn't flare up. Larger pieces of furniture may have to be broken apart into smaller manageable pieces inside the fire room before moving them outside. Avoid using the elevators to carry out burned mattresses and other large pieces of furniture unless it's absolutely necessary. A flare-up in an elevator car can be disastrous and send smoke up the shaft. If the elevator is to be used, utilize metal garbage cans with lids or wrap all burned debris in a wetted canvas tarp. One firefighter must be equipped with a water-pressurized fire extinguisher in case of flare-up.

FIGURE 16-15 Randomly throwing burned material out a window is extremely convenient, but use common sense.
Courtesy of Lt. Mike Heaton.

Avoid randomly throwing burned material out the window. Although this method is convenient, common sense and caution must be exercised. Firefighters walking around the perimeter have been injured after getting struck in the head because someone didn't check below before throwing debris out a window **FIGURE 16-15**. If the decision is made to throw material out a window, cordon off the area below with fire scene tape to warn firefighters of the hazard. Be aware that the wind can carry lighter debris and embers and that they can reignite and start a new fire downwind. As a firefighter, the author witnessed an officer learn this lesson the hard way. It involved a room fire on the top floor of a three-story apartment building. The large living room window frame was taken out, so it was wide enough to accommodate the burned smoldering sofa. Rather than overhauling the large sofa in the fire room, the officer ordered it to be thrown out the window. He emphasized to the crew to use enough force so it would clear the building. Well, it did, and as it fell toward the ground, it burst into flames, landed on its edge, and bounced back toward the apartment building, breaking the window of the first-floor apartment as it made entry. Can you imagine sitting in your living room, minding your own business, when suddenly a burning sofa crashes through your living room window? That lesson was never forgotten.

Overhaul Operations

Because less water is needed for overhaul than for firefighting operations, 2½-inch attack lines can be replaced with 1¾-inch attack lines for easier handling. Universal foam or **wet water** should be used during overhaul. Some apparatus are pre-plumbed with

FIGURE 16-16 Apparatus can be disconnected from the hydrant and legally parked to alleviate traffic congestion. Hose lines for overhaul can be taken off the hydrant.

Courtesy of Raul Angulo.

firefighting foam; if foam was used in the fire attack, it should also be used for overhaul. If it was not used, portable foam systems with built-in eductors should be used with foam or wet water agents. The foam breaks down the surface tension of water, allowing it to be absorbed quickly and penetrate deeper into the Class A fuels for complete extinguishment. With wet water, less water is used to extinguish the remaining fires, which cuts down on unnecessary water damage due to runoff. If companies are released from the incident, hose lines should be cleared from the streets and be moved to the curb or sidewalk so that returning fire apparatus do not run over them and the traffic flow can be restored in the area. If the apparatus is not needed for pumping, it can be disconnected from the hydrant and be legally parked to alleviate traffic congestion. The overhaul hose lines can be taken off the hydrant if there is sufficient water pressure **FIGURE 16-16**.

The basic purpose of the overhaul operation is to make sure that the fire is completely extinguished. This means that any area that could possibly have been in contact with fire or intense heat should be checked, whether the building is fire resistant or not. Many building materials resist fire, but vertical and horizontal channels, shafts, and voids make perfect avenues for fire spread from convection or conduction. During overhaul, which begins close to the area where firefighting operations ended, it is essential to look for signs of hidden fire **FIGURE 16-17**.

What to Look for and Where to Look

Overhaul should begin close to the area where firefighting operations ended. This is where a good knowledge of building construction really pays off. Firefighters should listen for the sound of fire, beware of high-heat areas, and look for flames and smoke in the area in which they are working. The TIC is one of the most important tools used in overhaul. The TIC enables firefighters to detect hot spots. During overhaul, it allows firefighters to identify hazards that may not otherwise be seen. Even the slightest amount of heat, on or behind surfaces, registers on the TIC. Crews can open spaces where heat has been detected and leave uninvolved areas intact. Concealed horizontal and vertical spaces, walls, and ceilings should be opened and inspected if a fire is suspected within. Areas that were opened during fire attack should also be rechecked. Portable lighting is useful when checking concealed spaces. If walls and ceilings have been in contact with fire and heat, they must be opened and checked for signs of fire. Any areas of concern should be wet down with foam or wet water **FIGURE 16-18**.

Salvage and property conservation should be practiced during overhaul operations. Before walls are opened and ceilings pulled, remove or collect simple furnishings and personal property. One trick of the trade is to open the chest of drawers and place as many small items as you can inside the drawers. Another method is to place salvageable objects on the bed and cover them using a tarp. Salvage covers should be used to cover and protect items that cannot be removed. Large salvage tarps and hallway runners should be used to cover the carpet before pulling Sheetrock, opening walls, and pulling ceilings. Every effort should be made to safeguard items in the building from further damage.

Ceiling spaces must be examined thoroughly. Any fire in the walls rises up the vertical wall spaces, then extends horizontally through the building, always taking the path of least resistance. False ceilings and ceiling tiles must also be checked and opened. Minimize damage when pulling ceilings; however, it is more important to extinguish the fire. The TIC is extremely useful in making the decision on whether to open a ceiling or wall space, based on the heat differential readings. But if you have any doubt, open the ceiling. Don't hold back because you're afraid to cause more damage—the building is already damaged. A rekindle causes significantly more damage and financial loss in the end.

A pike pole is an essential tool for the engine officer during fire attack. The officer needs to check the ceiling space above the attack crew to ensure that they're not being outflanked by a fire overhead. Shorter tools, like axes, are less cumbersome but don't provide the necessary reach to check the ceiling effectively. The

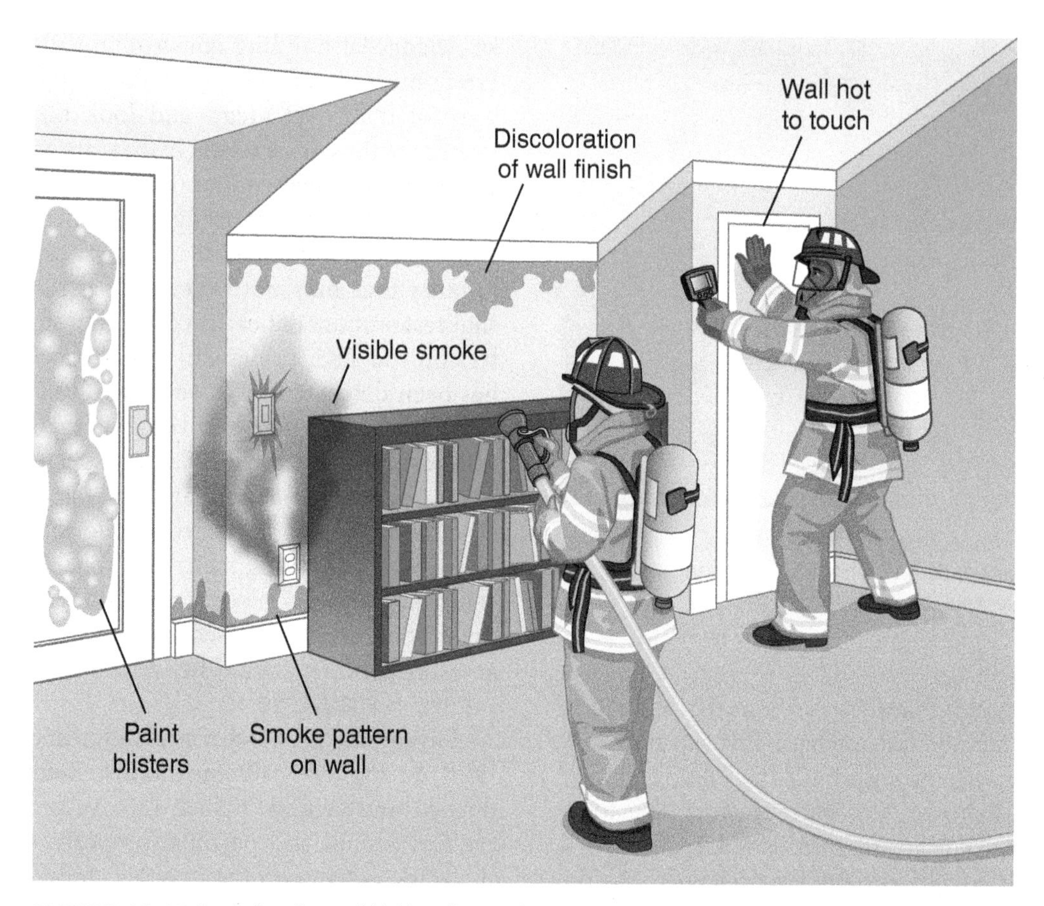

FIGURE 16-17 Look for signs of hidden fire, such as a wall that registers heat on the TIC or is hot to the touch, visible smoke, smoke patterns or discoloration on the wall, and blistering paint.

Engine Company Fireground Operations, 3rd Edition, Harold Richman, Steve Persson, NFPA © 2008 Jones & Bartlett Publishers.

FIGURE 16-18 When fire is suspected behind walls or ceilings, they should be opened until a clean area is found.

Engine Company Fireground Operations, 3rd Edition, Harold Richman, Steve Persson, NFPA © 2008 Jones & Bartlett Publishers.

same goes for overhaul. Use the longest pike pole you can get away with—it provides more control and allows you to stand clear of falling debris while you open a ceiling **FIGURE 16-19**. A firefighter using a short pike pole will be standing nearer, if not directly underneath, the ceiling being pulled. Extra caution must be taken when pulling ceilings in commercial structures. Heavy dead loads on the roof, like HVAC systems, may be barely supported by fire-damaged joists or rafters and may come crashing down on unsuspecting firefighters who are delivering impact loads while thrusting the pike pole, then pulling the ceiling down. A sudden release of this heavy machinery can be deadly for firefighters standing below it. The fire-damaged area below heavy roof loads should be cordoned off. It's best to keep applying water and foam instead of hitting weakened structural members with overhaul tools. Notify fireground personnel via radio that this hazard exists.

An inspection should also be made to determine whether sparks or embers have been carried into walls, partitions, and void spaces. Don't forget to check the base of the wall space. Burning embers and debris can fall down onto the bottom plate between the wall studs **FIGURE 16-20**. If a sign of smoke or a strong odor exists, a light water stream should be directed into the area to wet it down. Firefighters should look for steam if a hot spot has been encountered. An area with cobwebs probably has not seen the effects of heat and smoke; cobwebs tend to shrivel up in temperatures that are higher than normal. If a wall, ceiling, floor, or any other shaft shows signs of fire damage on examination, that area should be opened until the full extent of the fire damage is visible. Scrape off char until you get down to clean wood. Keep the hose line available to extinguish any remaining fire as the full area of the concealed space is exposed. All areas checked during overhaul must be checked again with the TIC before this task is considered complete. You want to leave everything cold and wet.

If the building has been insulated and fire has involved areas around the insulation, this material must be checked. The insulation must be removed in all areas of fire involvement until there is no indication of fire damage. Blown-in insulation is difficult to work with, and any areas covered with it present real problems. The smallest ember can travel to a remote part of an attic and smolder for hours in the insulation before it flares up and starts a rekindle. The dilemma for the company officer is determining how much insulation to remove. Firefighters may be tempted to leave well enough alone. This is rolling the dice; firefighters must ensure that there has been no fire extension. Utilize the TIC. Rolled-in fiberglass bats are easier to remove and check for hidden embers and, for the most part, are noncombustible.

Checking Above the Fire

Checking above the fire is carried out in a similar fashion as on the fire floor **FIGURE 16-21**. Use the TIC to locate heat differentials. In addition, baseboards may have to be removed to confirm if fire has traveled up through the walls and partitions. Any areas of concern should be opened up farther to expose fire extension, then wetted down with foam or wet water.

FIGURE 16-19 Use the longest pike pole you can to increase your area of safety from falling debris.

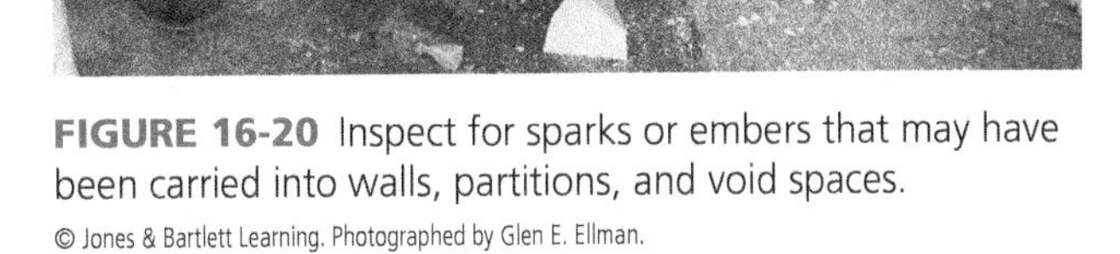

FIGURE 16-20 Inspect for sparks or embers that may have been carried into walls, partitions, and void spaces.

FIGURE 16-21 Checking above the fire is carried out in a similar fashion as on the fire floor.
Courtesy of Lt. Mike Heaton.

When fire has penetrated a ceiling space on the fire floor, it might have also extended into the floor assembly above. A full examination of this area may require the removal of flooring. This is especially important along walls and partitions and where shafts or other vertical spaces pass through the floor. If part of the floor must be removed, it should be taken up until a clean area shows the full extent of the fire. To hold damage to a minimum, the flooring should be removed as cleanly as possible. Cutting floors with power saws is usually faster and cleaner than using hand tools. In general, power saws are preferred over axes and other hand tools during cutting operations.

Vertical Shafts and Channels

Vertical fire spread is faster than horizontal spread. If it is apparent or suspected that fire has spread into a vertical shaft or channel, the space must be opened and checked. If openings into the space are available, they should be used for this purpose; otherwise, the space should be physically opened. Examine it for signs of smoke and fire. Use the TIC to determine heat differentials. A light water stream should be directed up or down into the area to wet possible hot spots.

Chimneys are vertical shafts and, although they are designed to hold heat and smoke, older chimneys can have loose bricks or broken mortar, leaving gaps. A chimney fire can spread into the attic space when the integrity of the chimney is compromised. Wooden structural members or any combustible material in contact with a chimney in the floor, ceiling, attic space, and roof assembly should be checked after a fire if the mortar isn't tightly sealed.

Lightweight parallel chord trusses used as flooring and ceiling joists create long, open horizontal channels. They are like small open cocklofts that can run the entire length of the room beneath the floor decking and above the ceiling drywall panels. It is important for the company officer to check these spaces so any fast-running fire doesn't outflank firefighters.

Firefighters should be assigned to check the bottom and the top of the shaft for fire and sparks. In balloon-frame construction, where there are no fire stops between floors, the vertical channels between the wall studs are open from the basement to the attic **FIGURE 16-22(A)**. It is often necessary (and easier, with less interior damage) to open up the exterior siding of the house to check the vertical void spaces **FIGURE 16-22(B)**. In balloon-frame construction, checking the attic should be the highest priority to get ahead of any fire spread rather than chase it.

Shafts that were opened or vented during fire control operations must be checked thoroughly during overhaul, even if the fire has been extinguished. The intense heat that was confined within the shaft might have rekindled the fire. For the same reason, any combustible material that came in direct contact with these shafts, such as wood floors, walls, attic spaces, or the roof, must be checked thoroughly **FIGURE 16-23**.

Cabinets and Compartments

When fire has definitely traveled inside walls and enclosed spaces; penetrated floor spaces, ceilings, and cabinets, including base and wall cabinets; or come in contact with these areas, they must be checked thoroughly. This is especially important when cabinets, such as those in residential kitchens, are involved. These units are usually constructed with a 3- to 5-inch enclosed space between the floor and the bottom shelf. In many cases, kitchens and bathrooms share the same enclosed shaft or wall space that carries electrical conduits, gas pipes, water pipes, plumbing, vents, and drains. Because the majority of vertical shafts and poke-throughs are in the kitchen area, the chance for weakened floor assemblies or localized floor collapse is greatest here **FIGURE 16-24**. Check the areas behind appliances. Check electric receptacle plugs and outlets connected to every major appliance. Many older homes still have laundry chutes from the bedroom hallway down to the basement, and dumbwaiters from an upper-level dining room down to the kitchen. These also need to be checked for fire extension.

Windows and Door Casings

When fire has involved window and door casings, they must be removed, and the concealed recesses checked for fire **FIGURE 16-25**. The wall needs to be opened and wet down if it appears that fire has

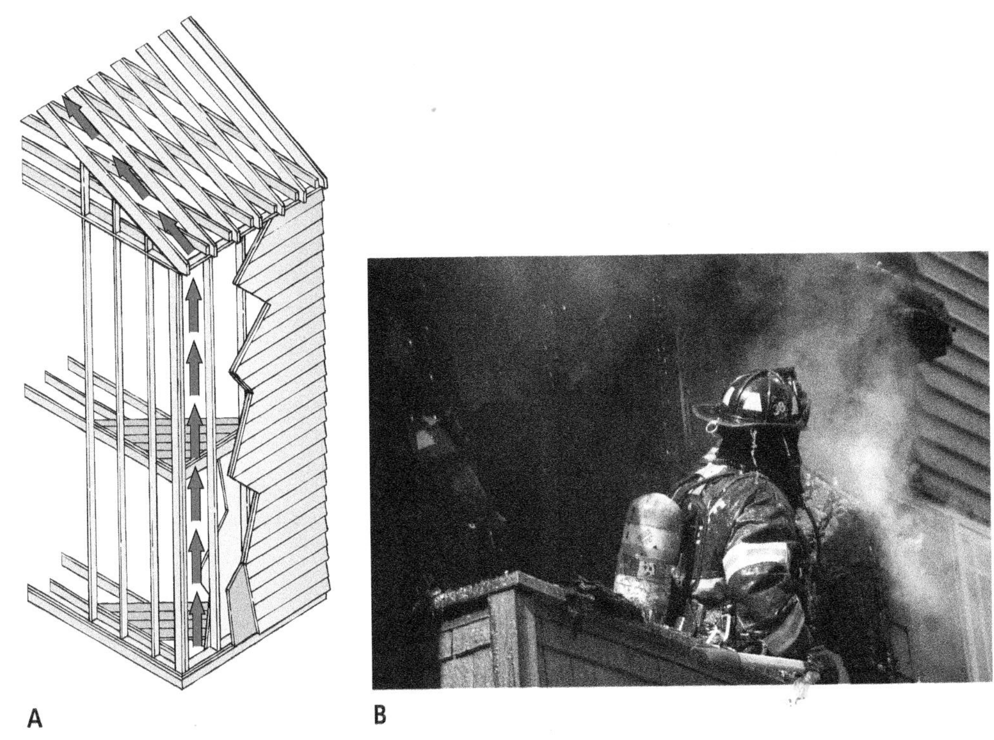

FIGURE 16-22 A. In balloon-frame construction, where there are no fire stops between floors, the vertical channels between the wall studs are open from the basement to the attic. B. It is often necessary and easier to open up the exterior siding of the house to check the vertical void spaces.

A: © Jones & Bartlett Learning; B: Courtesy of Lt. Mike Heaton.

traveled into this area. Continue opening the recess until you reach an area with no sign of fire extension. Wainscoting is handled similarly; it is removed until clean wood is found and the area is wet down as necessary. Use the TIC to confirm any heat differentials.

Debris Pile

Try to have only one debris pile located away from the structure. Have a hose line ready so the pile can be wetted down as firefighters bring out debris. Debris should be left cold and wet. Use a disposable tarp to cover the debris pile. The edges can be anchored with heavy material or dirt. Its purpose is to make a reasonable attempt to shield an attractive nuisance for children. Once the property is turned back over to the owners, the debris pile becomes their responsibility, but do your best to help them out. It's good customer service FIGURE 16-26.

Final Atmospheric Air Monitoring

Atmospheric readings should be taken one last time to check for CO, especially in multistory residences FIGURE 16-27. CO is tasteless, odorless, and colorless. If elevated levels register on a multi-gas detector, it is necessary to use PPV to clear it out. It is especially important to check the upper stories of low-rise and mid-rise apartment buildings. The author once had a significant fire in a restaurant on the first floor of a 10-story apartment building. The upper stories were unaffected by the smoke and fire. It was merely by chance that we took atmospheric readings on the upper floors because they appeared to be clear of smoke. We were surprised to find readings over 50 ppm of CO on floors 9 and 10. That could have been bad for residents reoccupying the building. Remember to vent for invisible CO, not just for smoke.

Water Removal

When taking up hose lines from the occupancy, it's easier to pull them completely out when the hoses are still charged. If couplings need to be broken, break them outside or out a window to limit water damage. Uncharged hose can be shoulder-loaded or lowered out a window. A quick and easy way for a single firefighter to get hose out of a two- or three-story house is to bring the male and female ends of the hose to the window and lower them to the ground together. The water will drain during this process. Slowly allow both couplings to rest on the ground to avoid damaging the male threads or the female swivel. Once the couplings

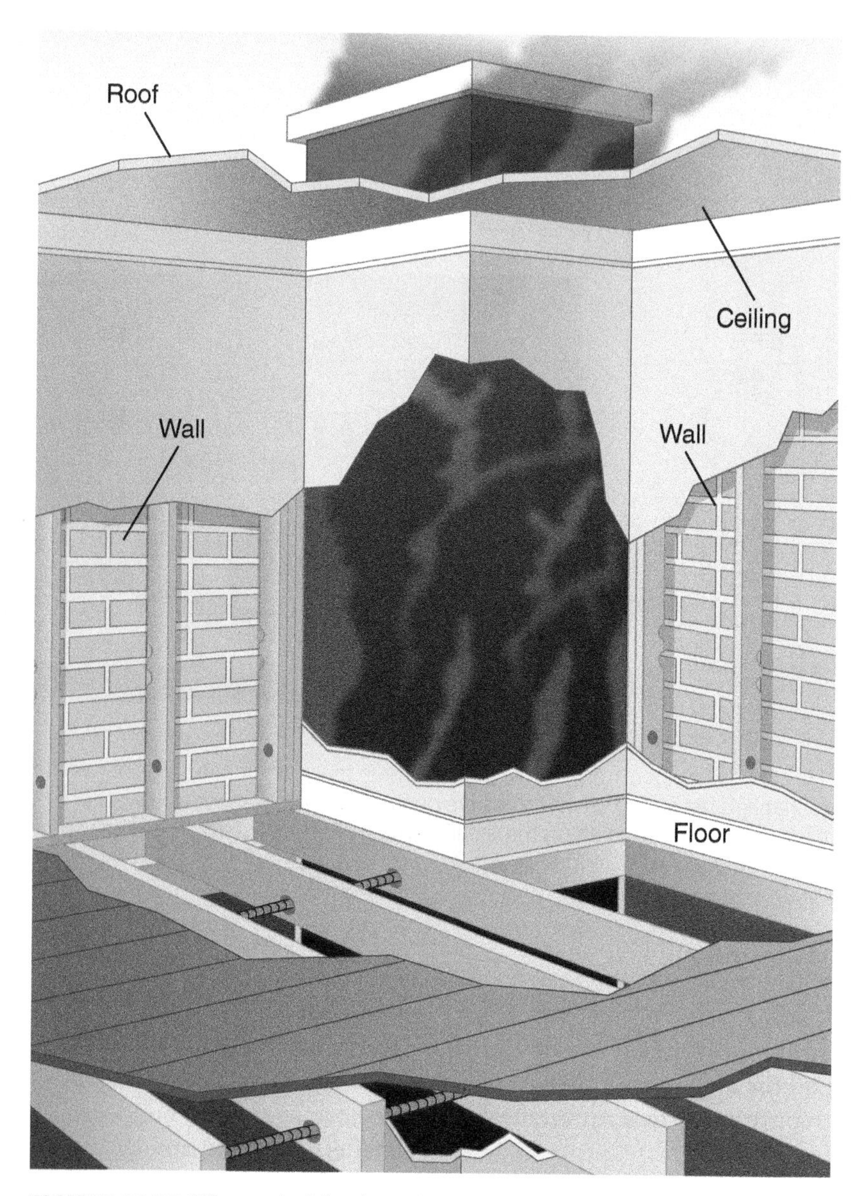

FIGURE 16-23 When a shaft has been the avenue of travel for a fire, the areas adjacent to or in direct contact with the shaft must be opened or checked with a TIC.

are resting on the ground without tension, simply throw the rest of the hose out the window.

During overhaul operations, excess water must be removed from the building. In some instances, this must be completed first to reduce the weight on roofs and floors so that they are safer to walk on. Consider the following: One gallon of water weighs 8.34 pounds (3.8 kg). One liter of water weighs 1 kilogram. One cubic foot of water weighs 62.4 pounds (28 kg). Every 1,000 gallons per minute (gpm) from a ladder pipe master stream adds 8,340 pounds (3,783 kg) to the structure every minute. One foot of standing water on a 20-foot × 20-foot roof or floor assembly adds 25,000 pounds (11,340 kg) of extra weight to the decking.

Water can be pushed and swept down vertical openings such as shafts, stairways, soil pipes of toilets opened at floor level, existing drains, and even back through hoses connected to Class I standpipes. When the building is heavily damaged and the water is deep, a hole can be made in outside walls, from windowsills to floor level, through which water can drain to the outside. Water follows gravity and the path of least resistance.

Portable pumps can also be used to remove water from low levels. If enough water has accumulated in a basement or other area to endanger firefighters if they enter, a safety officer should be assigned to that sector. Water accumulating in a large basement can be deep, and the possibility of drowning is a reality. The

FIGURE 16-24 Kitchen cabinets often conceal openings for utilities through which fire can travel.

Engine Company Fireground Operations, 3rd Edition, Harold Richman, Steve Persson, NFPA © 2008 Jones & Bartlett Publishers.

FIGURE 16-25 Window and door casings must be removed to check for hidden fire.

Engine Company Fireground Operations, 3rd Edition, Harold Richman, Steve Persson, NFPA © 2008 Jones & Bartlett Publishers.

FIGURE 16-26 Try to have only one debris pile located away from the structure. Have a hose line ready to wet down the pile as firefighters bring out debris.
Courtesy of Martin Grube/Fire Rescue TV.

FIGURE 16-27 Atmospheric readings should be taken one last time to check for CO, especially in multistory residences.
Courtesy of Rob Schnepp.

area should be marked, and all firefighters should be made aware of the danger. In removing water from a structure, as in all facets of overhaul, care should be taken to cause as little damage as possible. There is little point in doing a good job of firefighting and then unnecessarily damaging the building after the fire is out.

Not all water jobs are the result of firefighting. Water can enter a structure from broken water pipes and extreme weather. In commercial properties, sprinkler heads can be damaged or broken off completely by a forklift. The sprinkler system activates and sprays water until the valve is shut off or the sprinkler head is wedged **FIGURE 16-28**. The sprinkler head should always be wedged because it can take several minutes for the water to drain within the system. In the meantime, the sprinkler head continues to flow with substantial pressure, causing water damage. Sprinkler system piping can break after a major freeze when temperatures start to rise. The thawing process often cracks the water pipes, causing significant water damage to property. These types of service calls are also part of salvage. Firefighters need to intervene by placing salvage covers over vulnerable property; turning off the water; and using water vacuums, pumps, and squeegees to remove standing water from the premises **FIGURE 16-29**. Sprinkler heads should be replaced and the main valve turned back on to charge the system. Unless temperatures are below freezing, dry systems should be left charged and wet during the interim so the premises are still protected by a functioning wet system **FIGURE 16-30**. The sprinkler system must still be serviced by a fire protection systems technician. Remember, firefighters do not work in sewage-contaminated water. In such cases, the owner needs to call a plumber.

Securing the Property

For buildings that are still repairable, board up all doors and windows that cannot be locked. Cover holes in the roof from vertical ventilation with disposable tarps if doing so will prevent further damage from inclement weather. Otherwise, leave them open so off-gassing can occur and vent to the outside. Use common sense and risk management. Don't forgo safety. Do your best to prevent unauthorized entry from vagrants. Again, you're making a reasonable effort to secure the scene before it is turned back over to the property owner or before you simply leave the scene. Anyone who wants to gain access into a structure after the fire department leaves will do so regardless of our efforts. You're trying to limit the exposure to the fire department and provide the best customer service that is reasonable. The responsibility for properly securing the premises lies with the property owner.

Fire Watch

After a significant fire, the IC may decide to have a fire watch for a day or two. Make sure a fire watch log is maintained and signed by the on-duty company officer and subsequent relieving companies taking over the fire watch. This maintains the chain of scene control and security while the property is still under fire department control should this incident end up in a court of law.

A

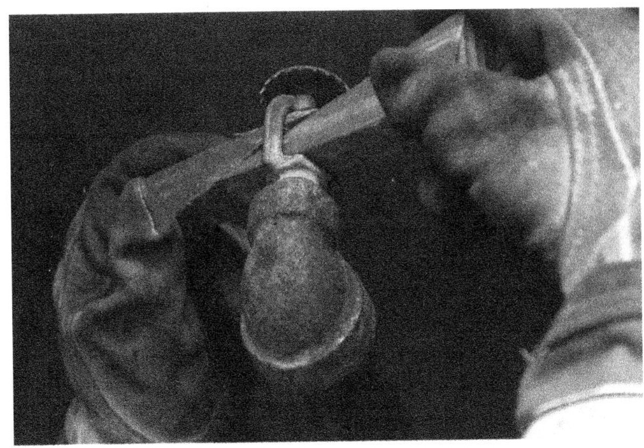

B

FIGURE 16-28 **A**. A sprinkler head sprays water until it is wedged or the valve is shut off. **B**. The sprinkler head should be wedged because it can take several minutes for the water to drain within the system.

FIGURE 16-29 Firefighters need to place salvage covers over vulnerable property, protecting them from fire, smoke, and water.

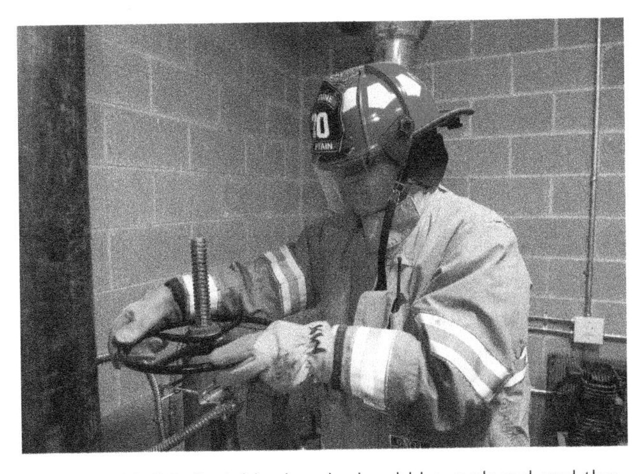

FIGURE 16-30 Sprinkler heads should be replaced and the main valve turned back on to charge the system.

Company Drive-By

If the fire occurred in your first-alarm district and the incident was terminated during the first half of your shift, make a point to drive by the scene every three to four hours. If you're returning to quarters after another alarm later in the day, go by way of the scene location to make sure the fire is still out. If you have any alarms in the middle of the night, drive by the fire scene and inspect the premises before returning to quarters. Trust me, you'll sleep better. The first thing the on-coming crew should do after apparatus and equipment checks is to take another ride out to the fire location for one last inspection.

Some Final Thoughts on Salvage

The professional philosophy that you need to teach your crews is this: Treat every house and piece of property as if it were their own. What may seem like junk to us may be valuable or sentimental to the occupants. Even though the fire may have been challenging and exciting to fight, there should be no high fives, laughing, or congratulating in front of the occupants and the public. A fire is a tragic and devastating event for a family and the neighborhood; enthusiastic displays of victory from firefighters will be misinterpreted as callous and can become a public relations nightmare for the fire department. Save outbursts of exhilaration for the firehouse.

Sometimes occupants escape only with the clothes on their backs. Here's a list of items they will be concerned about:

- Cell phone and chargers
- Wallets and purses
- Pets
- Prescription eyeglasses and medication
- Money, bank books, stocks, and bonds
- Precious metals like gold bullion or bars of silver
- Jewelry
- Keys
- Passports; certificates of birth, death, marriage, and divorce; wills
- Photographs
- Computers and hard drives
- Collectables
- Anything else the owners deem irreplaceable

Make every effort to retrieve these items.

Pets

Pets are an integral part of the family unit, but it must be understood by all that they do not carry the equal life value of a human being. A discussion within your department should take place on how much risk firefighters should take during the fire attack to search for pets, though it will never be officially required to do so because pets are often not cooperative, especially cats. Some pets will flee the scene if given the chance, but many tend to hide under beds or inside closets, so a search should be made for them during overhaul if the owner reports them missing. Pets will also suffer the effects of smoke inhalation and carbon monoxide, so if found, reasonable EMS efforts should be made to revive them. Often, they can be revived with mouth to snout forced-air ventilation or by using a nonrebreather mask with 100% oxygen. Deceased pets should be wrapped in a towel or sheet before being brought out. Handle them with dignity and respect. Bringing a deceased pet to the owner will be an extremely emotional moment requiring the utmost professional empathy.

Caged animals like guinea pigs and hedge hogs obviously cannot escape. If retrieved, the cages should be brought out to fresh air. Oxygen flowing through a canula can be aimed (blow-by) or inserted into the airway to help revive these animals. If possible, fish in aquariums should be removed and transferred to another container with clean water. If not, oxygen tubing can be inserted into the water of an aquarium flowing 100% oxygen to boost the O^2 level in the water.

Owner Walkthroughs

If the structure or home is deemed safe, and if the owners can emotionally handle it, walk them through the property to explain the work that was done and why it needed to be done. Many civilians don't understand the reason the fire department is destroying the house. Point out the dangerous areas to avoid, as well as the respiratory health hazards that will be present for a few days. This walkthrough should be quick to limit the owner's exposure. Let the owner know that power and utilities have been secured and that they should not attempt to turn them back on. Reconnecting utilities should be left to the insurance companies and the professional salvage companies to deal with.

The Federal Emergency Management Agency (FEMA) puts out a brochure, *After the Fire, Returning to Normal* for burned-out occupants. This brochure has a more extensive checklist on what to do next and how to handle the damage. These brochures should be stocked on every fire apparatus. Answer as many questions from the occupant as you can.

At some point, everyone you know will call 911 for a medical emergency—even other firefighters. But fewer people will need to call 911 for a structure fire within their home. For the citizens who unfortunately suffer a home structure fire, it is a devastating emotional event that could leave scars for life. Even though the fire department has authority and control of the scene, ICs can be very harsh with property owners and tenants because they were anxious to inspect or get back into their homes. We must understand that this is a reasonable concern on their part. Although the IC's concern is for the tenants' well-being and safety, at some point, the fire department is going to release the scene, put units back in service, and return to quarters. At that point, the owners and tenants will reenter their property and it will no longer be our concern. These people have to begin to rebuild their lives. Don't let their last interaction with the fire department be one of contempt and resentment. Go out of your way to show empathy and compassion and try to help in any way possible. Salvage and overhaul may be the only evidence property owners see of our service. It's very important to take pride in our work to make sure it leaves a professional impression.

After-Action REVIEW

IN SUMMARY

- Salvage starts with the arrival of the fire department. It can take place anytime during the fire incident. Salvage is saving something of value, or something useful, and extracting it from the rubble.
- Overhaul is essentially the careful and systematic examination of the areas that were on fire to determine whether there is any remaining fire, sparks, or embers that could cause the reignition, or rekindle, of burned materials. Firefighters should follow the path of the fire and search the building for any signs of hidden or remaining fires, uncover them, and extinguish them.
- Firefighters must realize how their actions in forcible entry, search and rescue, fire attack, salvage, and overhaul can adversely affect the evidence at the fire scene. If deceased fire victims are discovered, limit the number of firefighters in the immediate area, and keep the overhaul to a minimum.
- Overhaul on the fire building takes place after the emergency is over, so there is no reason for rushing or taking unnecessary chances. Overhaul should be completed systematically and starts with the IC, who must perform a new size-up and risk-benefit analysis, then develop an organized and safe plan for overhauling the fire building.
- A pre-overhaul safety inspection needs to be conducted throughout the entire fire area, and safety hazards that might further threaten firefighters need to be identified and mitigated. Announce all hazards on the fireground over the radio so all members are aware. Every effort should be made to determine that the structure is relatively safe for firefighters to proceed. All personnel must know where they can and cannot go. Areas that are off limits must be identified, and access to these areas must be marked or blocked off.
- If a smoke explosion occurs, it typically happens during the overhaul phase. It is often confused with a backdraft. Although they are similar, they are not the same. A backdraft needs only oxygen to occur, while a smoke explosion needs only heat.
- Once the fire is out and the emergency is over, there is no reason to work in the dark. Use portable lighting to light the interior scene.
- The most important question an IC must ask during offensive operations, and especially during the overhaul, is, "At what point does this building no longer have any value?" A disposable building is one that is cheaper to tear down and replace with a new one than it is to repair.
- Controlling and securing utilities is pretty basic. The fire department has to deal with only four: water, gas, electric, and sometimes LPG. If securing the utilities wasn't assigned during the initial fire attack, it should be done before overhaul begins.
- Universal foam or wet water should be used during overhaul. Foam breaks down the surface tension of water, allowing it to be absorbed and penetrate deeper into Class A fuels for complete extinguishment. With wet water, less water is used to extinguish the remaining fires, which cuts down on unnecessary water damage from runoff.
- The basic purpose of overhaul is to make sure that the fire is completely extinguished. Vertical and horizontal channels, shafts, and void spaces make perfect avenues for fire spread.
- A pike pole is an essential tool for the engine officer during overhaul. It provides the necessary reach to check the ceiling space above to ensure that the crew members are not being outflanked by a hidden overhead fire.
- Vertical fire spread is faster than horizontal fire spread. If it is apparent or suspected that fire has spread into a vertical shaft or channel, the space must be opened and checked. Firefighters must open and check any area that could possibly have had contact with fire or intense heat. Walls, ceilings, and floors may need to be opened.
- Lightweight parallel chord trusses used as flooring and ceiling joists create long, open horizontal channels. It is important for the company officer to check these spaces so that any fast-running fire doesn't outflank the firefighters. Use the TIC to confirm any heat differentials.

- In balloon-frame construction, checking the attic should be the highest priority to get ahead of any fire spread.
- All areas checked during overhaul must be checked one last time with the TIC before overhaul is considered complete.
- During salvage and overhaul operations, excess accumulation of standing water adds tremendous weight to an already damaged structure. One cubic foot of water weighs 62.4 pounds. One foot of standing water on a 20-foot × 20-foot room adds 25,000 pounds of weight to the floor system. This water must be drained to remove the stress load to the building.

KEY TERMS

hot-stick voltage detector A battery-operated handheld wand; if its tip is pointed to electrical wires and power outlets, it can detect electrical current as well as energized fixtures, appliances, and equipment.

multi-gas detector A handheld unit that measures the percentages of oxygen-rich or oxygen-deficient atmospheres and levels of CO, combustible gases, sulfur dioxide, nitrogen dioxide, and hydrogen cyanide.

overhaul The process of final extinguishment after the main body of a fire has been knocked down. All traces of fire must be extinguished at this time. (NFPA 402)

rehab *See* rehabilitation.

rehabilitation An intervention designed to mitigate the physical, physiological, and emotional stress of firefighting in order to sustain a fire department member's energy, improve performance, and decrease the likelihood of on-scene injury or death. (NFPA 1521) Also known as *rehab*.

rekindle A return to flaming combustion after apparent but incomplete extinguishment. (NFPA 921)

salvage A firefighting procedure for protecting savable property from further loss following fire. (NFPA 402)

wet water A surfactant wetting agent similar to Class A foam that, when mixed with plain water, reduces the water's surface tension so it can quickly absorb into dense, fibrous Class A fuels.

Glossary

A

alligatoring When wood pyrolyzes during a fire, creating black and both deep and shallow cracks along the surface of the wood that resembles the skin of an alligator.

approach assessment Part of the size-up, it includes a 360-degree assessment to view exterior conditions and as many sides of the building as possible while the apparatus is approaching the scene.

auto-exposure Vertical fire extension from floor to floor through the failure of exterior windows.

automatic sprinkler system A system of pipes filled with water that is under pressure; it discharges water immediately when a sprinkler head opens.

axial load A compression load imposed through the center of an object.

B

backdraft A deflagration (explosion) resulting from the sudden introduction of air into a confined space containing oxygen-deficient products of incomplete combustion. (NFPA 1403)

backup line An additional hose line used to reinforce and protect personnel in the event that the initial attack proves inadequate. (NFPA 1410)

balloon frame An older type of wood-frame construction in which the wall studs extend vertically from the basement of a structure to the roof without any fire stops.

ball valve Valve used on nozzles, gated wyes, and engine discharge gates. It consists of a ball with a hole in the middle of the ball.

bidirectional flow path A two-way or high-low air current of fresh air and smoke competing for the same opening. The lower air intake is fresh air from the outside being entrained or sucked in toward the seat of the fire. The high point or exhaust path is where the smoke and other heated gases are using the same opening (door or window) to escape, seeking equilibrium.

black fire A hot, high-volume, high-velocity, turbulent, ultra-dense black smoke that indicates an impending flashover or autoignition.

blind alley lay A scenario in which the attack engine forward lays a supply line from an intersection to the fire, and the supply engine reverse lays a supply line from the hose left by the attack engine to the water source. Also known as *split lay*.

blitz attack An aggressive fire attack that often utilizes a 2½-inch hand line or deck gun and occurs just prior to entry, search, and tactical ventilation. Also known as *defensive to offensive exterior attack*, *softening the target*, and *transitional attack*.

boiling liquid expanding vapor explosion (BLEVE) An explosion that occurs when pressurized liquefied materials (e.g., propane or butane) inside a closed container are exposed to a source of high heat. Eventually, the increasing pressure will exceed the container's ability to hold it.

bond When atoms and molecules have an affinity for one another, they form a chemical bond. The chemical bond is the "atomic glue" that holds these molecules together. Fire, or combustion, has its origin in the bonding of atoms and molecules.

bowstring truss A truss with a distinctive curved shape.

BRECEO/VS A memory aid to help company officers confirm or rule out the existence of a basement or below-grade occupancy; it stands for basement (or below-grade), rescue, exposures, confinement, extinguishment, overhaul/ventilation and salvage. The abbreviation places the letter B before the tactical objectives acronym, RECEO/VS.

C

chimney effect The area between a high-low pressure differential flow path. The fresh air intake portal is low, and the exhaust portal is high. As heat from the fire travels up through this flow path, the fire dynamics are the same as in a chimney. Temperatures can reach over 1,000°F (538°C). The chimney effect happens in stairwells, and under these conditions, the stairs can become a deadly trap for firefighters.

collapse zone The area around and away from the building where an exterior wall will land and scatter during a collapse. The minimum clearance from the building should be 1½ times the height of the building.

combination attack A type of fire attack employing both the direct and indirect attack methods.

combustion A chemical process of oxidation that occurs at a rate fast enough to produce heat and usually light in the form of either a glow or a flame. (NFPA 1)

compression force A force that causes material to be crushed or flattened axially through the material.

concentrated load A load applied on a small area.

conduction Heat transfer to another body or within a body by direct contact. (NFPA 921)

contemporary construction Buildings constructed since about 1970 that incorporate lightweight construction techniques and engineered wood components. These buildings exhibit less resistance to fire than older buildings.

convection Heat transfer by circulation within a medium such as a gas or a liquid. (NFPA 921)

coverage The assignment of companies to a particular side of the fire building or the fireground for size-up so that they accomplish any or all of the objectives of a firefighting operation.

crosslays Traverse hose beds. Also known as *bucket line*, *speed lay*, or *Mattydale lay*.

cross-lotting A fire department term referring to when an apparatus cannot lay supply hose and an LDH supply line needs to be shoulder-carried or hand-jacked several hundred feet to a water source. It requires four to six firefighters (or more) to be spread out equally along the length of the connected, but uncharged, supply line to reach the distant water supply. Also known as *hand jacking*.

D

dead load A load that consists of the weight of all materials of construction incorporated into the building, including, but not limited to, walls, floors, roofs, ceilings, stairways, built-in partitions, finishes, cladding, and other similarly incorporated architectural and structural items, and fixed service equipment, including the weight of cranes. (NFPA 5000)

defensive search A type of search conducted in the immediate vicinity of the door; used when heat conditions in a room indicate an imminent flashover and a hose line is not in place to cool the atmosphere.

defensive-to-offensive exterior attack A transitional attack that starts from the exterior of the structure to knock down and take the energy away from the fire. This is sometimes referred to as resetting the fire because the application of water cools the interior temperature, thus increasing the time period to flashover. Once the fire is knocked down, an interior offensive attack can begin. Also known as *blitz attack*, *softening the target*, *transitional attack*.

designed load A load in engineered structural designs has been planned for the intended use of the building.

diffusion When molecules circulate beyond the surface of the liquid and escape into the space within the container. Some molecules are reabsorbed back into the liquid, while others remain in a gaseous state within the container. If the gases are lighter than air and aren't confined inside the container, all the liquid eventually evaporates.

direct attack Firefighting operation involving the application of extinguishing agents directly onto the burning fuel.

disposable buildings Modern, lightweight construction that is cheaper to knock down and rebuild than to repair after it has been severely damaged by fire.

distributed load A load applied equally over a broad area.

divided hose bed A hose bed that is separated into two or more supply hose compartments running the length of the hose bed.

double-female coupling A hose adapter that is used to join two male hose couplings.

double-male coupling A hose adapter that is used to join two female hose couplings.

dry-barrel hydrant A type of hydrant used in areas subject to freezing weather. The valve that allows water to flow into the hydrant is located underground below the frost line, and the barrel of the hydrant is normally dry.

dry hydrant An arrangement of pipe permanently connected to a water source other than a piped, pressurized water supply system that provides a ready means of water for firefighting purposes and that utilizes the drafting (suction) capability of a fire department pump. (NFPA 1142)

E

eccentric load A compression load imposed off center to another object.

endothermic reaction A chemical reaction that absorbs heat or requires heat to be added.

equilibrium The chemical state of balance in various forms of matter. A state in which physical opposing forces or influences are in balance. For example, in a liquid, the point where there are as many molecules being liberated from a liquid as being reabsorbed by the liquid.

evaporation The state of conversion from a liquid to a gas under normal atmospheric conditions.

exothermic reaction A chemical reaction that results in the release of energy in the form of heat.

explosive fire growth A room-and-contents fire in a modern house often flashes over and then enters the decay stage due to decreased oxygen levels in the house, or it may use up the available oxygen before it reaches flashover. This produces a ventilation-limited fire where the fire goes into a dormant state or decay stage. In this decay stage, fire growth and fully developed stage is limited only by the lack of available oxygen. When air is introduced back into this environment, the fire can grow with explosive force.

explosive growth stage The second growth stage of a fire. In the initial growth stage, the room temperature is low. The explosive growth stage happens after the fire becomes ventilation-limited and the temperature is high and surrounding fuels are preheated. The introduction of air can lead to rapid fire growth with explosive force, often resulting in flashover.

exterior exposures Outside structures that can catch fire by radiant heat or direct flame contact from the original fire.

exposure hazard A condition that promotes the spread of fire if a fire starts in or reaches that area.

extreme fire behavior The sudden increase in oxygen combined with hot flammable gases from the fire that results in rapid fire growth and changes the direction of the flow path from bidirectional to unidirectional due to the pressure imposed by the wind. This is also called a *wind-driven fire*.

F

fast water Water delivered to the fire immediately through a preconnected hand line using tank water from the engine.

fiber-reinforced products (FiRPs) Plastic fibers mixed with wood to give the wood increased tensile strength.

fire dynamics The study of the characteristics of fire and the burning process, which includes how fires start, develop, and spread. It also includes the detailed study of fire science, chemistry, physics, heat transfer, fuels, materials, hydraulics, and building construction, and how these forces interact to affect fire behavior within a structure.

fire load The total energy content of combustible materials in a building, space, or area including furnishing and contents and combustible building elements. (NFPA 557)

fire patrol Other non-fire municipal employees equipped with a radio or cell phone assigned to driving neighborhood streets downwind from a large fire to look for falling embers and firebrands that could start new fires.

fire stopping The placement of horizontal wood members between the spaces of wall studs and flooring to prevent the vertical spread of fire from floor to floor.

fire tetrahedron A geometric shape used to depict the four components required for a fire to occur: fuel, oxygen, heat, and chemical chain reactions.

fire tower An enclosed stairway connected at each story by an outside balcony or a fireproof vestibule vented to the outside.

fire wall A wall separating buildings or subdividing a building to prevent the spread of fire and having a fire-resistance rating and structural stability. (NFPA 5000)

first-aid firefighting Part of the Class II standpipe system. The hose cabinet is equipped with a 2½-inch discharge valve, a reducer coupling to a 1½-inch male threads, 100 feet of 1½-inch hose line, and an attached spray nozzle designed for trained personnel.

flameover *See* rollover.

flame point The lowest temperature at which a liquid ignites and achieves sustained burning when exposed to a test flame in accordance with ASTM 92, Standard Test Method for Flash and Fire Points by Cleveland Open Cup Tester. (NFPA 1). Also known as *fire point.*

flashover A transition phase in the development of a compartment fire in which surfaces exposed to thermal radiation reach ignition temperature more or less simultaneously, and fire spreads rapidly throughout the space, resulting in full room involvement or total involvement of the compartment or enclosed space. (NFPA 921)

flash point The minimum temperature at which a liquid or a solid emits vapor sufficient to form an ignitable mixture with air near the surface of the liquid or the solid. (NFPA 115)

flow path The movement of heat and smoke from the higher pressure within the fire area toward the lower pressure areas accessible via doors, window openings, and roof structures. (NFPA 1410)

flow the floor The process of wetting down the entire exposure floor above the fire to prevent vertical fire spread.

forces Loads imposed on structural members that create stress and strain on the materials used to make the connections. Stress and strain are defined as forces applied through those materials.

fuel-limited fire A fire in which the heat release rate and fire growth are controlled by the characteristics of the fuel because there is adequate oxygen available for combustion. (NFPA 1410)

fully developed stage The stage of fire development where heat release rate has reached its peak within a compartment. (NFPA 1410)

G

gusset plate Connecting plate made of a thin sheet of steel used to connect the components of a truss.

gypsum board The generic name for a family of sheet products consisting of a noncombustible core, primarily gypsum with paper surfacing. (NFPA 5000)

H

heat flux The measure of the rate of heat transfer to a surface, typically expressed in kilowatts per meter squared (kW/m^2) or BTU/ft^2. (NFPA 268)

hose layering When two or more hand lines are laid over each other in the same hallway or stairway.

hot-stick voltage detector A battery-operated handheld wand; if its tip is pointed to electrical wires and power outlets, it can detect electrical current as well as energized fixtures, appliances, and equipment.

hydrant assist valve (HAV) A specialized type of valve that can be placed on a hydrant and allows another engine to increase the supply pressure without interrupting flow. Also known as *four-way valve.*

I

ice fog Frozen moisture in the air and in smoke.

ignition temperature Minimum temperature that a substance should attain in order to ignite under specific test conditions. (NFPA 402)

immediately dangerous to life and health (IDLH) Any condition that would pose an immediate or delayed threat to life, cause irreversible and/or adverse health effects, or interfere with an individual's ability to escape unaided from a hazardous environment. (NFPA 1670)

impact load A load that is in motion when it is applied.

indirect attack Firefighting operation involving the application of extinguishing agents to reduce the buildup of heat released from a fire without applying the agent directly onto the burning fuel. (NFPA 1145)

inorganic Being or composed of matter other than hydrocarbons and their derivatives, or matter that is not of plant or animal origin and does not contribute to the combustion process. (ASTM D1079: 2.1)

interior exposures Objects or rooms in the immediate vicinity of the original interior fire that are not yet burning. These rooms and objects exposed to radiant heat, conduction, convection, and direct flame contact cause the interior fire to spread to uninvolved areas.

L

laminar smoke flow Smooth or streamlined movement of smoke, which indicates that the pressure in the building is not excessively high.

legacy construction An older type of construction that used sawn dimensional lumber and was built before the 1970s.

liquid A fluid (such as water) that has no independent shape but takes the shape of its container. It has a definite volume, does not expand indefinitely, and is only slightly compressible.

live load The load produced by the use and occupancy of the building or other structure. It does not include construction or environmental loads such as wind load, snow load, rain load, earthquake load, flood load, or dead load. Live loads on a roof are those produced (1) during maintenance by workers, equipment, and materials and (2) during the life of the structure by movable objects such as planters and by people. (NFPA 5000)

M

master stream appliance Devices used to produce high-volume water streams for large fires. Most master stream appliances discharge between 350 gpm (1,325 L/min) and 1,500 gpm (5,678 L/min), although much larger capacities are available.

Also known as *deck gun, monitor, deluge*. Also referred to by a model or brand name, like the *Stinger* or the *Stang*.

matter Anything that occupies space; has mass, size, weight, or volume; and can be perceived by one or more senses. It cannot be destroyed and exists in three states: solids, liquids, and gases.

Mayday A verbal declaration indicating that a firefighter is lost, missing, or trapped and requires immediate assistance.

microbiologically influenced corrosion (MIC) The buildup of corrosion within the sprinkler pipes and heads that can limit or prevent water flow. MIC develops only in steel metal piping. Known in the firefighting industry as "MIC."

modern fuels The household furnishing, appliances, coverings, and decorations that are constructed from petroleum-based products, like plastics, foams, and synthetics.

movable exposures Exposures that can be protected by physically moving them away from the reach of the fire. Examples are trucks, tractors, automobiles, boats, yachts, and any other motorized equipment that can be driven or towed away from the fire.

multi-gas detector A handheld unit that measures the percentages of oxygen-rich or oxygen-deficient atmospheres and levels of CO, combustible gases, sulfur dioxide, nitrogen dioxide, and hydrogen cyanide.

mushrooming When convecting heat currents and smoke encounter a ceiling or other vertical barrier that prevents it from rising, it spreads out horizontally along the ceiling until it reaches a vertical wall; then it travels down toward the floor, being pushed by more heated pressurized air that is rising behind it.

N

neutral plane The interface at a vent, such as a doorway or a window opening, between the hot gas flowing out of a fire compartment and the cool air flowing into the compartment where the pressure difference between the interior and exterior is equal.

nonprotected steel Bare steel used in Type II non-combustible buildings for trusses, joists, and columns, without any fire-resistive material covering or coating the steel.

O

occupancy The purpose for which a building or other structure, or part thereof, is used or intended to be used. (NFPA 5000)

opposing lines The inadvertent positioning of two attack teams approaching each other on the same level from opposite sides of the structure.

organic Being or composed of hydrocarbons or their derivatives, or matter of plant or animal origin. (ASTM D1079: 2.1)

overhauling As opposed to post-fire operations, the term can also be used to manually attain a water supply by physically dragging or shoulder-loading LDH.

overhaul The process of final extinguishment after the main body of a fire has been knocked down. All traces of fire must be extinguished at this time. (NFPA 402)

oxidation Reaction with oxygen either in the form of the element or in the form of one of its compounds. (NFPA 53)

oxidizer Any material that readily yields oxygen or other oxidizing gas, or that readily reacts to promote or initiate combustion of combustible materials.

oxygen-limited fire A fire that depends on the amount of oxygen available for combustion.

P

parallel chord truss A truss in which the top and bottom chords are parallel.

pitched chord truss A type of truss typically used to support a sloping roof.

platform frame Construction technique for building the frame of the structure one floor at a time. Each floor has a top and bottom plate that acts as a firestop.

plenum The space between the structural ceiling and the dropped ceiling of a floor/ceiling assembly; the underside of the floor above. It houses and conceals telecommunications cables for telephone and computer networks, and supports and conceals air ducts of the HVAC system.

polar solvent A water-soluble flammable liquid such as alcohol, acetone, ester, and ketone.

pre-incident plan A document developed by gathering general and detailed data that is used by responding personnel in effectively managing emergencies for the protection of occupants, responding personnel, property, and the environment. (NFPA 1620)

pre-plumbed master stream appliance A master stream appliance that has a separate discharge pipe of adequate diameter that runs from the fire pump to the appliance.

primary search An immediate and quick search of the structures likely to contain survivors. (NFPA 1670)

pumper fire apparatus Fire apparatus with a permanently mounted fire pump of at least 750-gpm (3,000 L/min) capacity, a water tank, and hose body whose primary purpose is to combat structural and associated fires. (NFPA 1901)

pyrolysis A process in which material is decomposed, or broken down, into simpler molecular compounds by the effects of heat alone; pyrolysis often precedes combustion. (NFPA 921)

Q

quick incident action plan (QIAP) The initial action plan by the first-in company officer, who acts as the initial IC. This plan is the foundation of the incident action plan that will be expanded upon or modified after command is transferred to a chief officer.

R

radiation The combined process of emission, transmission, and absorption of energy traveling by electromagnetic wave propagation (e.g., infrared radiation) between a region of higher temperature and a region of lower temperature. (NFPA 550)

rapid ascent team (RAT) A search team designated to quickly climb the entire firefighting stairway to ensure it is clear of any occupants fleeing from the fire floor landings.

rapid intervention crew (RIC) A minimum of two fully equipped personnel on site, ready for immediate rescue of disoriented, injured, lost, or trapped rescue personnel. (NFPA 1006)

rebar Cold-drawn steel rods embedded to reinforce the tensile strength of a concrete slab when it will be used under tension.

rehabilitation An intervention designed to mitigate the physical, physiological, and emotional stress of firefighting in order to sustain a fire department member's energy, improve performance, and decrease the likelihood of on-scene injury or death. (NFPA 1521) Also known as *rehab*.

rehab *See* rehabilitation.

rekindle A return to flaming combustion after apparent but incomplete extinguishment. (NFPA 921)

rescue air kit (RAK) A prepackaged SCBA unit with a facepiece and 60-minute air bottle. The backpack assembly is omitted. It is a reliable, independent air source used in rapid intervention rescue and primary search.

resetting the fire Applying fast water on a fire that results in knockdown. Knocking down the fire takes away the heat and thermal energy of the fire and slows down rapid fire growth. The goal is to reset the time to flashover.

residual pressure The pressure that exists in the distribution system, measured at the residual hydrant at the time the flow readings are taken at the flow hydrants. (NFPA 24)

risk A combination of the probability and the degree of possible injury or damage to health in a hazardous situation. (NFPA 79)

risk-benefit analysis An assessment of the risk to rescuers versus the benefits that can be derived from their intended actions. (NFPA 1006)

rollover The condition in which unburned fuel from the fire has accumulated at the ceiling level to sufficient concentration, at or above the lower flammable limit, that it ignites and burns. This can occur without ignition of, or prior to the ignition of, other fuels separate from the origin. Also known as *flameover*. (NFPA 921)

room and contents fire A fire in the incipient stage and early growth stage. The only fuels burning are the furnishings in the room. Although the entire room is absorbing the heat of the fire, structural members and assemblies are still intact.

S

salvage A firefighting procedure for protecting savable property from further loss following fire. (NFPA 402)

secondary search A detailed, systematic search of an area conducted after the fire has been suppressed. (NFPA 1670)

second-tier problem A problem that is less urgent than a first-tier problem but still requires attention, for example, salvage, securing utilities, overhaul, discovering household hazardous materials, and water removal.

shear force The force that causes a material to be torn in opposite directions, perpendicular or diagonal to the material.

shelter in place The strategy for life safety in residential and commercial high-rise buildings. Residents or occupants are safer staying in their units or work areas than trying to evacuate the building where the hallways and stairs can be charged with heat, smoke, and carbon monoxide.

six-sided size-up *See* size-up. *Six-sided* refers to the four adjacent sides of a burning structure or compartment, plus the top level and bottom level.

size-up The process of gathering and analyzing information to help fire officers make decisions regarding the deployment of resources and the implementation of tactics. (NFPA 1410)

skid load One hundred fifty to 300 feet of prebundled 2½-inch hose carried in a hose slot. This hose is not preconnected, but it can be deployed quickly by connecting to any 2½-inch discharge port. The skid load can be used as a 2½-inch attack line, a backup line, an exposure line, or a quick supply to a master stream appliance, standpipe, or sprinkler connection.

smoke explosion A violent release of confined energy that occurs when a mixture of flammable gases and oxygen is present, usually in a void or other area separate from the fire compartment, and comes in contact with a source of ignition. In this situation, there is no change to the ventilation profile, such as an open door or window; rather, it occurs from the travel of smoke within the structure to an ignition source.

softening the target An aggressive offensive exterior fire attack that occurs just prior to entry, search, and tactical ventilation. Another expression for this term is "hitting it hard from the yard." Also known as *blitz attack, defensive-to-offensive exterior attack, transitional attack.*

spalling Chipping or pitting of concrete or masonry surfaces. (NFPA 921)

split lay A scenario in which the attack engine forward lays a supply line from an intersection to the fire, and the supply engine reverse lays a supply line from the hose left by the attack engine to the water source. Also known as *blind alley lay*.

stack effect The vertical air flow within high-rise buildings caused by the temperature-created density differences between the building interior and exterior, or between two interior spaces. (NFPA 92)

standpipe system An arrangement of piping, valves, and hose connections installed in a structure to deliver water for fire hose.

static pressure The pressure that exists at a given point under normal distribution system conditions measured at the residual hydrant with no hydrants flowing. (NFPA 24)

structure fire Any fire inside, on, under, or touching a structure. (NFPA 901)

suction hose A short section of supply hose used to supply water to the suction side of the fire pump.

supply Any water leaving the pump under pressure through a hose connected to a discharge port.

supply hose Hose designed for the purpose of moving water between a pressurized water source and a pump that is supplying attack lines. (NFPA 1961)

survivability profile An assessment that weighs the risks likely to be taken versus the benefits of those risks for the viability and survivability of potential fire victims under the current conditions in the structure.

T

tension force A force that causes a material to be stretched or pulled apart in line with the material.

thermal column A cylindrical area above a fire in which heated air and gases rise and travel upward. Also known as *thermal plume*.

thermal-imaging camera (TIC) A type of thermographic camera used in firefighting and search and rescue. By rendering infrared radiation as visible light, the camera allows firefighters to see areas of heat through smoke, darkness, or heat-permeable surfaces.

thermal layering The stratification (heat layers) that occurs in a room as a result of a fire. Also known as *heat stratification*.

thermal plume *See* thermal column.

thermoplastic materials Plastic material capable of being repeatedly softened by heating and hardened by cooling and that, in the softened state, can be repeatedly shaped by molding or forming. (NFPA 5000)

top-tier problem A problem on the fireground that requires immediate intervention. It may fall into one of the following seven categories: possible occupants, visible occupants, access problems, smoke, fire, possible presence of hazardous materials, and exposures.

torsional load A load imposed in a manner that causes another object to twist.

transitional attack An offensive fire attack initiated by an exterior, indirect hand-line operation into the fire compartment to initiate cooling while transitioning to interior direct fire attack in coordination with ventilation operations. Also known as *blitz attack, defensive-to-offensive exterior attack.*

transitional ventilation From a pressurized stairway, smoke on a floor is directed horizontally to a nonpressurized stairway where the pressure differential causes the smoke to transition to vertical ventilation out through the rooftop portal.

turbulent smoke flow Agitated, boiling, angry-looking smoke, which indicates great heat in the burning building. It is a precursor to flashover.

U

undesigned load The weight that a building supports for which the building was not designed or an added weight that was not anticipated. Buildings that are altered or refurbished for an occupancy other than the original intent can create an undesigned load.

unidirectional flow path A fresh-air intake opening or an exhaust exit port. It simply means the convected air currents are flowing in the same direction through the entire opening and exit port.

universal extinguishing foam (UEF) A firefighting extinguishing foam that works on all classes of fires.

V

value-time-size (VTS) The three risk-benefit assessment considerations made during size-up to justify the selection of a strategy.

vapor density The weight of an airborne concentration (vapor or gas) compared to an equal volume of dry air.

vaporization The transformation of a liquid to its vapor or gaseous state.

vapor pressure The pressure, measured in pounds per square inch absolute (psia), exerted by a liquid, as determined by ASTM D323, Standard Test Method for Vapor Pressure of Petroleum Products (Reid Method).

vent-enter-isolate-search (VEIS) A method of searching for fire victims that consists of selecting and opening a window to a bedroom or other living space; entering the room; closing the door to isolate the room from the fire, thus preventing a flow path from developing and allowing the room to ventilate naturally; quickly searching for any possible victims; and rescuing them out through the window and down a ladder.

ventilation-controlled fire growth model A six-stage model that includes the following stages: incipient stage, initial growth stage, ventilation-limited stage, explosive growth stage, fully developed stage, and decay stage.

ventilation-limited fire A fire in which the heat release rate and fire growth is controlled by the amount of air available to the fire. (NFPA 1403)

W

water shuttle apparatus corridor A circular, drive-through roadway that allows the tender to bring in and dump its water without backing up the apparatus or performing a three-point turn.

wet-barrel hydrant A type of hydrant that is intended for use where there is no danger of freezing weather and where each outlet is provided with a valve and an outlet. (NFPA 24)

wet water A surfactant wetting agent similar to Class A foam that, when mixed with plain water, reduces the water's surface tension so it can quickly absorb into dense, fibrous Class A fuels.

wood truss An assembly of small pieces of wood or wood and metal. A truss is a fake beam.

Index

A

acceptable losses, 15
access keys, 483
accountability, 11–12, 13*f*, 15, 476–477, 477*f*
  status board, 264, 265*f*, 477, 477*f*
  system, functional, 263–268, 265–268*f*
  worksheet, 264, 265*f*
acrolein, 116, 117*f*
aerial drones, 533
aerial ladders, 353, 353*f*, 426, 428*f*, 434, 435*f*, 440*f*
aerial master streams, 434, 435*f*
AESs. *See* automatic extinguishing systems (AESs)
AHJ. *See* authority having jurisdiction (AHJ)
air monitoring, 554, 554*f*, 565, 568*f*
air-purifying respirators, 555, 555*f*
air standpipe systems, 536–537, 537*f*
alarm, 275
  before, 275
  control panel, 480, 480*f*
  receipt of, 275, 276*f*
alligatoring, 42, 43*f*, 88, 88*f*
alpha side front, 196
apparatus, 194, 561*f*
  basic coverage responsibility. *See* coverage responsibility
  Bravo and Delta sides, 202–203, 203*f*
  central corridor construction, 208, 209*f*
  Charlie side rear. *See* Charlie side rear
  garden apartments, 206–208, 207*f*, 208*f*
  general front, rear, and side operations, 203, 204*f*
  high-rise buildings, 208
  mercantile areas, 206
  move-ups, 246
  pre-incident planning, 194
    building *vs.* structure, 194–195
  setback buildings, 204, 205*f*
  shopping malls, 204–205
  standard shopping centers, 205–206, 206*f*
  tracking the movement of, 246
  water tanks, 218
approach assessment, 479
assignment abbreviations, 262, 262*f*
atomic glue, 21
attached buildings, 198–199, 198*f*, 201–202, 201–202*f*
attack lines, sizes of, 324–325
  combination spray nozzles, 329
  and nozzles, 325–326
  solid *vs.* straight streams, 326–329, 327–329*f*
attic fires, 319
authority having jurisdiction (AHJ), 150
auto-exposure, 351
autoignition temperatures, 116
automatic dry standpipe, 445
automatic extinguishing systems (AESs), 453
automatic fire alarm control panel, 480, 480*f*
automatic smoke control panel, 480, 480*f*
automatic sprinkler systems, 443–444, 452–453, 453*f*, 471
  components of, 453, 454*f*
  deluge system, 458, 459*f*
  dry pipe system, 457, 458*f*
  exposure protection, 461
  fusible link and glass bulb sprinkler heads, 455–456, 455–456*f*
  microbiologically influenced corrosion (MIC), 456, 457*f*
  nozzle water flow and pressure problems in, 471–473
  out-of-service sprinkler systems, 461
  placing, 461–462, 462*f*
  pre-action system, 457–458
  shutting down, 461
  sprinkler heads, 453–455, 455–456*f*
  sprinkler response temperatures, 456, 456*f*, 457*t*
  sprinkler system failure, 461
  theory, 453, 454*f*
  water supply for sprinkler systems, 458–461, 459*f*, 460*f*
  wet pipe system, 456–457, 457*f*
automatic wet standpipe, 445
axial load, 82, 83*f*

B

backdraft, 51–53, 52*f*, 54*t*, 118*f*, 369–370, 369*f*
  for preventing, 370–371, 370*f*
backup lines, 337, 337*f*
  master streams for, 384, 384*f*
  personnel, 385–386, 385*f*
  positioning of, 379
    hose layering, 380, 381*f*
    hose order, 379–380, 379*f*, 380*f*
  purposes of, 337–338, 338*f*, 339*f*, 377*f*
  sizes of, 337–338, 338*f*, 339*f*, 377*f*, 381, 381*f*
    2½-inch hose lines, 383–384
    1¾-inch lines, 381–383, 382*f*, 383*f*
  use of, 384–385, 385*f*
balloon frame, 101, 104*f*
ball valve, 163, 164, 164*f*
basement fires, 15, 139–143*f*, 139–144, 312*f*, 337–338, 411, 412*f*
  Bricelyn Street Fire, 338
  carbon monoxide, 345
  cellar pipes and distributor nozzles, 343–344, 344*f*

basement fires *(Continued)*
controlling, 346–350, 346–350*f*
exterior window indirect/transitional attack, 343, 343*f*
first-floor collapse, 342–343
growth stage and fully developed stage basement fires, 411–414, 413*f*, 414*f*, 415*f*
identifying presence, 338–339
incipient stage basement fires, 411
making entry down stairs, 340–342, 341*f*, 342*f*
in Mercantile and Commercial Stores, 414–416, 415*f*
smoke behavior when making entry, 339–340, 339*f*, 340*f*
steam burns, 344–345
tactical considerations for, 144
basic hose loads, 153*f*
benzene, 116
bidirectional flow path, 39, 39*f*, 125
big-box stores, 5
Biloxi, Mississippi, Fire Department case study, 553–554
blackened windows, 120, 120*f*
black fire, 62
black smoke, 137, 137*f*
BLEVE. *See* boiling liquid expanding vapor explosion (BLEVE)
blind alley lay, 231
blitz attack, 133, 421, 422*f*
Blitzfire portable monitor, 168–169, 168*f*
blown-in insulation, 563
boiling liquid expanding vapor explosion (BLEVE), 25–26, 26*f*, 397
strategy for preventing, 26
boiling point, 27
bolt cutters, 166, 166*f*
bond, 21
bowstring truss, 87, 87*f*
boxes, series of, 5, 6*f*
breakaway nozzle, 4
BRECEO/VS, 10–11
Bresnan distributor nozzle, 142, 143*f*
Bricelyn Street Fire, 338
brick veneer wall, 100, 101*f*
British thermal units (BTUs), 7, 48, 119, 132, 181, 228
BTUs. *See* British thermal units (BTUs)
building collapse, anatomy of, 72, 72*f*
building construction, 72, 72*f*, 92
ceiling assemblies, 85
collapse times and operational periods, 94
concern for collapse determines strategy, 94
contents and fire loads, 73
fire-resistive floors, 84, 84*f*
floors and ceilings, 83
forces, 83, 83*f*
imposition of loads, 82–83, 83*f*
materials, 73–74
concrete, 74–75, 75*f*
engineered wood products, 80
fire-retardant wood, 80
glass, 77–79, 77–79*f*
gypsum board, 79
masonry, 74, 74*f*
plastics, 80–81
steel, 75–77, 75–77*f*
wood, 79–80
occupancy, 72–73
sprinkler regulations, 73
parapet walls and cornices, 91–92*f*
preparing for collapse, 92–94, 93–94*f*
roofs, 85, 85*f*
signs of collapse, 92
stairs, 90–91, 91*f*
strategic considerations, 95–97, 95–97*f*
terms and mechanics
concentrated load, 81
designed and undesigned loads, 82
distributed load, 81–82, 82*f*
fire load, 82
impact load, 82
types of loads, 81, 81*f*
time (T) in VTS
fire resistive, 98–99
heavy timber, 101–102, 102*f*, 103*f*
lightweight wood construction, 104–108, 106–108*f*
noncombustible, 99–100, 100*f*
ordinary, 100–101, 100–101*f*
plank and beam, 103–104, 106*f*
platform frame, 103, 105*f*
prefires and new construction inspections, 108
wood frame, 101–103, 104*f*
trusses, 86–87, 86–87*f*
steel, 89–90, 89–90*f*
wood, 87–89, 88*f*
types of, 97–98, 98*t*, 99*t*
void spaces, 90, 90*f*
wood-supported floors, 84–85, 84–85*f*
building material, under stress, 99*t*
burning process, 27–28, 28*f*
butterfly valve, 164, 164*f*

C

CAFSs. *See* compressed air foam systems (CAFSs)
calcination process, 105–106
campfire, 116, 116*f*
carbon black, 59
carbon monoxide (CO), 8, 8*f*, 27, 253, 345
ceiling assemblies, 85
center-core design, 469–470, 469*f*
Charlie side rear, 199
attached buildings, 201–201*f*, 201–202
detached buildings, 199–200*f*, 199–201
chemical chain reaction, 48
chemical cocoons, 5
chemical energy, 22, 22*f*
chimney effect, 140, 356
Chimney Snuffer, 169, 169*f*
class A foam, 177–178, 179–180, 180*f*, 181*f*
class B foam, 177–178
CO. *See* carbon monoxide (CO)
cockloft fires, 319
coiling, 337, 337*f*
cold smoke explosions, 552, 552*f*
collapse, 423*f*
determines strategy, 94
operational periods, 94
preparing for, 92–94, 93–94*f*
signs of, 92, 439
times, 12, 94
times for various floor assemblies, 98*t*
zone, 93, 93*f*, 94*f*, 421
color of smoke, 62

combination attack, 320
combination fog nozzles, advantage of, 158–160, 159*f*
combination nozzle fog patterns, 327
combination spray nozzle, 432
combined standpipe system, 445, 445*f*
combustion, 19, 22, 46, 46*f*
commercial fires, 15
communications, 483
composite wood, 80
compound gauge, 216, 216*f*
compressed air foam systems (CAFSs), 4, 177, 182, 182*f*
compression force, 83, 83*f*
concentrated loads, 81
concrete, 74–75, 75*f*
conduction, 34–36, 34–36*f*
construction inspections, 108
container size, 30
contemporary construction, 80
convected air currents, 29, 29*f*
convection, 36–38, 37–38*f*, 392, 393*f*
  exposures, 391–392
conventional wood flooring, 84
conventional wood-frame construction, 106*f*
coordinated fire attack, 130–136, 132–134*f*
cornices, 91–92*f*
correct hose-line position, 285, 285*f*
coverage assignments, 195–196, 196*f*
coverage problems, 194
coverage responsibility, 195
  alpha side front, 196
  attached buildings, 198–199, 198*f*
  coverage assignments, 195–196, 196*f*
  detached buildings, 196–197, 197*f*
  wide-frontage buildings, 197–198, 198*f*
covered malls, 205, 206*f*
cracked glass or crazing, 119, 119*f*, 137, 252, 300
cross-contamination, 7
crosslays, 153, 154*f*
cross-lotting, 216, 216*f*
cyanide, 253

D

dead load, 81, 81*f*
debris pile, 565, 568*f*
decision-making model, 11
decision-making process, 72
deck gun, 427*f*
decomposition, 22
defensive attack, 323–324
defensive fire, 422*f*
defensive interior attack, portable master streams, 521–522, 522*f*
defensive operations, 437, 438*f*
defensive search, 278
defensive-to-offensive exterior attack, 133
deluge sprinkler head, 459*f*
deluge system, 453, 458, 459*f*
density, 59, 62
designed loads, 82
detached buildings, 196–197, 197*f*, 199–200*f*, 199–201
DeWitt-Chestnut Fire in Chicago, 475, 475*f*, 517, 517*f*
diffusion, 24
direct attack, 316–317, 317*f*
direct-to-engine, 186–189, 187–191*f*
dispatchers, 395, 395*f*
disposable buildings, 254, 557–558
distributed load, 81–82, 82*f*
divided hose bed, 152, 152*f*
DNR orders. *See* do not resuscitate (DNR) orders
do not resuscitate (DNR) orders, 122
door control, 293, 535, 536*f*
double-female couplings, 165, 165*f*
double-male couplings, 165, 165*f*
drafting, 224*f*
dry-barrel hydrant, 214, 215*f*
dry hydrant, 214, 215*f*
dry pipe system, 457, 458*f*
dry pipe valve, 457, 458*f*
dry standpipe systems, 446*f*
dual-attack nozzle, 4
dump valve, 219, 220*f*

E

earthquake, 217, 217*f*
eccentric load, 83, 83*f*
EL. *See* explosive limits (EL)
electrical energy, 22–23
electricity, 389, 389*f*
electric power utility, 389, 389*f*
electrons, 21, 21*f*, 23
elevated master stream appliances, 426, 428*f*, 429*f*
elevated master streams, 437–438, 438*f*
elevated platforms, 429*f*, 434, 435*f*, 440*f*
elevators, 354, 482
  access keys, 483
  communications, 483
  control of HVAC system, 483
  emergency power generators, 483
  fire pump, 483–484
  operating in phase II, 491–492, 492*f*
  use of, 489–491
emergency medical services (EMS), 122
  annual required training, 3
  equipment, 202
  for medical emergencies, 2, 2*f*
  primary emergency response service, 3, 3*f*
emergency power generators, 483
EMS. *See* emergency medical services (EMS)
emulsifiers, 178–179, 178–179*f*
enclosed structure fire, 37
endothermic reaction, 21, 21*f*
engine apparatus/pumper positioning, 2, 223*f*
  first-in engine, 485–487, 485–487*f*
  second- and third-in engines, 487–489, 488*f*
engine companies, 213
  equipment, 150–151
  fire apparatus and equipment, 14
  members, 360*f*
  operations and responsibilities, 14–15
  personnel, 15
  primary responsibility of, 2, 2*f*
  standard equipment requirements for, 150
engine crew controls, 352, 352*f*
engineered wood
  products, 80
  trusses, 87, 88*f*
engine pressures, 429, 430*f*

engine pump speed, 225*f*
engines and pumpers
  discharges, 223–224
  intakes, 221–223, 223*f*, 224*f*, 225*f*
  pump speed and capacity, 224, 225*f*
  rated capacity, 221–223
  residual pressure, 224–225
entrained air, 132, 133*f*
equilibrium, 24
equipment
  advantages to class A foam, 179–180, 180*f*
  Blitzfire portable monitor, 168–169, 168*f*
  bolt cutters, 166, 166*f*
  CAFS, 182
  Chimney Snuffer, 169, 169*f*
  class A and class B foams, 177–178
  combination fog nozzles, 158–160
  combination spray, 158, 159*f*
  double-female couplings, 165, 165*f*
  double-male couplings, 165, 165*f*
  emulsifiers, 178–179, 178–179*f*
  engine company equipment, 150–151
  firefighter safety and foam, 180–181, 181*f*
  firefighting foam, 176–177
  fire hose, 153–157, 154–157*f*
  flamefighter and transformer piercing nozzles, 173–174, 174*f*
  ground ladders, 165–166, 166*f*
  history, science, and development of foam, 177
  hose alert hose restraint safety system, 174–176, 175*f*
  hose storage, 152–153, 152–153*f*
  hydrant assist valve (HAV), 162–163, 163*f*
  HydroVent™, 169–170, 169*f*
  master stream appliance, 160, 160*f*
  modern advancements in nozzle technology and equipment, 166–167, 167*f*
  nozzles, 157, 157*f*
  overhaul and rekindles, 182
  pre-plumbed master stream appliance, 160–161, 161*f*
  pump intake connections, 161–162, 161*f*, 162*f*
  soft-suction hookups, 161, 161*f*
  solid-stream smooth-bore tip, 157–158, 157–158*f*
  specialty high-rise nozzles, 170–173, 172*f*
  supply hose, 157
  TFT Flip-Tip, 167, 168*f*
  TFT vortex, 167–168, 168*f*
  thermal-imaging camera (TIC), 174, 174*f*
  2½-inch pump intake connections, 162, 163*f*
  2½-lnch fire hose, 155–156, 156*f*
  valves and hydrant gates, 163–164, 164–165*f*
evacuation control, 482
evaporation, 24, 27
evidence-based practices, 146
excessively long lays, 231, 231*f*
exothermic reaction, 21, 21*f*
explosive fire growth, 120
explosive growth phase, 54*t*
explosive growth stage, 31–32, 31–32*f*
explosive limits (EL), 27
exposure coverage, 388
exposure hazards, 388
exposure protection, 388*f*, 416, 461
  basement fires, 411, 412*f*
    growth stage and fully developed stage basement fires, 411–414, 413*f*, 414*f*, 415*f*
    incipient stage basement fires, 411
    in Mercantile and Commercial Stores, 414–416, 415*f*
  definition of, 388–389, 388*f*, 389*f*
  exterior exposures, 389–390, 390*f*
    coverage, 396–397, 397*f*, 398*f*
    fire patrols, 394–395, 395*f*
    initial response considerations, 390–391, 391*f*, 392*f*
    radiant heat exposures, 395–396, 395*f*, 396*f*
    Santana Row case study, 391–394, 393*f*
  horizontal spread, 417
  hose lines and nozzles
    positions, 400, 400*f*, 401*f*
    water streams, 397–400, 399*f*, 400*f*
  interior exposures, 400–403, 401*f*, 402*f*
    fire in concealed spaces, 403–404, 403*f*
    horizontal fire spread, 408–410, 409*f*
    open interior spread, 410–411, 410*f*
    property conservation *vs.* fire damage, 410
    vertical fire spread, 404–405, 404*f*, 405*f*, 406*f*
  vertical spread, 416–417, 416*f*
exterior exposures, 388, 389–390, 390*f*
  coverage, 396–397, 397*f*, 398*f*
  fire patrols, 394–395, 395*f*
  initial response considerations, 390–391, 391*f*, 392*f*
  radiant heat exposures, 395–396, 395*f*, 396*f*
  Santana Row case study, 391–394, 393*f*
extinguishing fires, UL/NIST recommendations for, 10
extinguishment, theory of, 9, 46, 47*f*
  eliminating available oxygen, 48
  interrupting chemical chain reaction, 48
  methods of, 47*f*
  reducing temperature, 46–48
  removing fuel, 48
extinguish structure fires, 2, 2*f*
extreme fire behavior, 125, 303, 304*f*

F

F-500, 4, 5
FACP. *See* fire alarm control panel (FACP)
fast water, 218, 310
FCC. *See* fire control center (FCC)
FDCs. *See* fire department connections (FDCs)
FDNY/NIST tests, 171–172
fiberboard, 80
fiber-reinforced products (FiRPs), 80
fire, 107*f*, 315*f*
  floor, 505–506, 507*f*
  in growth stage, 259, 259*f*
  hydrants, 448–449
  loads, 73, 82
  patrols, 394–395, 395*f*
  prevention inspections, 8
  pump, 483–484
  resistance ratings, 97, 98*t*
  resistive, 98–99
  room has, 137*f*
  in single-family residences, 203, 203*f*
  size and intensity of, 154*f*
  sprinklers, 472
  stopping, 103
  tower, 481
fire alarm control panel (FACP), 478
fire alarm systems, 8
fire apparatus, 2, 2*f*, 150
  and equipment, 14
  technology in manufacturing, 3

fire attack, 254, 255*f*, 468
fire behavioral sequence, 72
fire behavior events, 116–118, 116–118*f*, 119*f*
  backdraft, 51–53, 52*f*
  behavior of ventilation-limited fires, 53
  flashover, 50–51, 51*f*
  in modern structures, 56–57
  rapid fire growth, 53, 54*t*
  rollover, 50, 50*f*
  smoke explosions, 53–55, 55*f*
  thermal layering, 49–50, 49–50*f*
  wind-driven fire, 55–56, 56*f*
fire-behavior standpoint, 117
fire boat, 427*f*
fire code, 473, 473*f*
fire conditions, 195, 196*f*
fire control center (FCC), 479
fire curtains, 517–518
fire department connections (FDCs), 15, 141, 443, 473, 474*f*
fire departments, 9, 550, 550*f*
  rescue procedures used, 5
fire dispatchers, 533
fire dynamics, 19, 118–121, 119–121*f*, 124
  behavior events. *See* fire behavior events
  boiling liquid expanding vapor explosion (BLEVE), 25–26, 26*f*
  boiling point, 27
  burning process, 27–28, 28*f*
  classes of, 57–58, 57–58*f*
  decay stage, 33–34, 33*f*
  explosive growth stage, 31–32, 31–32*f*
  fire tetrahedron, 19, 19*f*
    fire chemistry, physics, and matter, 20–21*f*, 20–22
  fully developed stage, 32–33, 32*f*
  gases, 27, 27*f*, 45–46, 45*f*
  health and safety, 65–66
  incipient (ignition) stage, 28–29, 29*f*
  initial growth stage, 29–30, 29–30*f*
  liquids, 44–45, 45*f*
  modes of heat transfer
    conduction, 34–36, 34–36*f*
    convection, 36–38, 37–38*f*
  path of least resistance and flow paths, 39–40, 39*f*
  radiation, 40–41, 40–41*f*
  smoke. *See* smoke
  solids. *See* solids
  sources of heat energy. *See* heat energy
  theory of extinguishment. *See* extinguishment, theory of
  vapor density, 24, 25*f*, 25*t*
  vapor pressure, 24, 25*f*
  ventilation-limited stage, 31, 31*f*
firefighter/firefighting, 5, 7, 15, 24, 41, 120, 120*f*, 122, 123*f*, 301, 301*f*, 303, 385*f*, 449*f*, 467*f*
  accountability, 11–12, 13*f*, 15, 476–477, 477*f*
  air management, 535–536, 537*f*
  air replenishment system (FARS), 537–538
  deaths, 213, 213*f*, 213*t*
  experience, decades of, 9
  fatalities, 552
  foam, 176–177
  rescue drag, 306*f*
  safety and foam, 180–181, 181*f*
  safety considerations, 475–476, 475*f*, 476*f*
  stairway, 356–358, 357*f*, 358*f*, 481*f*
  technical advancements in, 5
fireground, incident priorities on, 10
fire growth, 123, 123*f*
  fuel-limited sequence of, 116
  in single-family residence, 368*f*
  stages of, 116, 116*f*
fire hose, 153–157, 154–157*f*, 226*f*
  friction loss characteristics of, 225
fireproof, 81
fire protection systems, 443–444, 443*f*, 462–463
  automatic sprinkler systems. *See* automatic sprinkler system
  standpipe systems. *See* standpipe systems
fire-resistive floors, 84, 84*f*
fire-retardant wood, 80
fire tetrahedron, 19, 19*f*
  fire chemistry, physics, and matter, 20–21*f*, 20–22
FiRPs. *See* fiber-reinforced products (FiRPs)
first-aid firefighting, 444, 445*f*, 449
first-alarm units, 15
first-in engine, 485–487, 485–487*f*
  initial role of, 11–12, 12*f*
fixed master stream appliances, 427*f*
FL. *See* flammable limits (FL)
flamefighter, 173–174, 174*f*
Flamefighter piercing nozzle, 173, 174*f*
flameover, 31, 54*t*
flame point, 44
flammable/explosive limits of gases, 27, 27*f*
flammable limits (FL), 27
flashlights, 115
flashover, 50–51, 51*f*, 54*t*, 117, 118*f*, 312–314, 554
  safety and survival, 314
flash point, 44
floors, and ceilings, 83
flow path, 39, 124–125
foam, 178
  class A, 177–178, 179–180, 180*f*, 181*f*
  class B, 177–178
  evolution and development of, 177
  hi-expansion, 177
  High-Expansion® (HI-EX®), 349–350
  history, science, and development of, 177
  low-expansion, 177
  manufacturers, 177
  medium-expansion, 177
  solution, premixed, 179
  universal extinguishing foam (UEF), 4, 4*f*
fog master stream, 437, 438*f*
fog nozzles, 4, 158*f*, 327*f*
  combination, 158–160
forces, 83
formaldehyde, 116
forward lay-dry, 184–186, 185*f*
forward lays, 229–231, 230*f*
forward lay-wet, 183–184, 184*f*
front intake on cab-forward apparatus, 162, 162*f*
front/rear intakes, 162, 162*f*
fuel-limited fire, 116
fuels, 19
  continuity of, 43–44
  household furnishings, 181
  orientation of, 43, 44*f*
  removing, 48
  supply, 30
fully developed stage, 32–33, 32*f*
functional accountability system, 263–268, 265–268*f*
fusible link, 455–456, 455–456*f*

G

garden apartments, 206–208, 207*f*, 208*f*
  interior design of, 206, 207*f*
gas detectors, 27, 27*f*
gases, 45–46, 45*f*, 59
  flammable/explosive limits of, 27, 27*f*
gate valves, 164, 164*f*, 451
glass, 77–79, 77–79*f*
  bulb sprinkler heads, 455–456, 455–456*f*
gloves, 555–556
gravity-fed tank system, 449
gravity tank, 449
Green Berets (Special Forces), 146
ground ladders, 165–166, 166*f*
gusset plate, 87
gypsum board, 79, 105, 107*f*

H

hallway corridor standpipe connections, 513–514
hard-suction hose, 161, 161*f*
HAV. *See* hydrant assist valve (HAV)
Hazard Control Technologies, Inc., 4
health risks, 555
  gloves, 555–556
  portable lighting, 556, 556*f*
heat buildup in fires, 7, 7*f*
heat energy, 23
  chemical, 22, 22*f*
  electrical, 22–23
  mechanical, 22
  nuclear energy, 23–24
  solar energy, 23
  sources of, 22
  static electricity, 23
heat flux, 34
heating, ventilation, and air-conditioning (HVAC)
  systems, 63, 468, 471, 483
  unit, 81
heat transfer, modes of
  conduction, 34–36, 34–36*f*
  convection, 36–38, 37–38*f*
heavy timber, 101–102, 102*f*, 103*f*
HEROPipe, 171, 172*f*, 173
  advantage of, 173
hidden fire, signs of, 561, 562*f*
hi-expansion foam, 177
High-Expansion® (HI-EX®) foam, 349–350
high-rise buildings, 208, 449*f*, 450, 469, 475
high-rise firefighting
  air standpipe systems, 536–537, 537*f*
  ascending stairs, 503, 503*f*
  attacking from floor above fire, 525–526, 526*f*
  automatic sprinkler systems, 471
    nozzle water flow and pressure problems in, 471–473
  beginning direct attack, 502–503
  building construction, 469
    center-core and side-core design, 469–470, 469*f*
    floor area, 470
    heating, ventilation, and air conditioning (HVAC) systems, 470–471
  checking for extension below fire floor, 529
  command priorities, 477–478
  commercial high-rise fire attack, 519
    direct attack, 520–521, 520*f*, 521*f*
    exterior recon, 520, 520*f*
    pincer/flank attack, 521
  conservative approach for smaller fire departments, 476
  dangers in firefighting stairway, 499–501*f*, 499–502
  defensive interior attack, portable master streams, 521–522, 522*f*
  door control, 535, 536*f*
  engine apparatus/pumper positioning
    first-in engine, 485–487, 485–487*f*
    second- and third-in engines, 487–489, 488*f*
  equipment cache, 496
  final strategy at first interstate fire, 527–528
    flowing the floor, 528–529, 529*f*
    preparing exposure floor, 528
    tools for protecting exposure floor, 528
  firefighter. *See* firefighter
  fire floor, 505–506, 507*f*
  gateway apartments and townhomes fire in San Francisco, California, 509–511, 510*f*, 511*f*
  hallway corridor standpipe connections, 513–514
  hose, nozzles, and equipment, 493–496, 493–496*f*
  initial size-up, 478–479
  lightweight high-rise fire hose, 494, 494*f*
  lobby control. *See* lobby control
  making entry and attacking fire, 514–515
  NFPA, 466–467, 466*f*, 473–474, 473*f*, 474*f*
  nozzles, 494–496, 495*f*
  1½- and 2½-inch gated wyes, 506–508, 509*f*
  1¾-inch attack in residential high-rise apartments, 508–509
  protecting floor above fire, 526–527, 527*f*
  rapid intervention teams (RITs), 503–505
  relying on type I construction, 522–523
    attacking from below fire floor, 523
    high-rise emergency response offensive nozzle (HEROPipe), 523–525, 524*f*, 525*f*
    no-attack strategy, 523
  RIT line to fire floor, 515–516, 516*f*
  search-evacuate-rescue (SER). *See* search-evacuate-rescue (SER)
  single-hose method (2½-inch), 511–512, 511*f*, 512*f*
    hose layering, 512–513, 512*f*, 513*f*
    new standpipes, 513, 513*f*
  specialized problems and hazards, 467–468, 467*f*
  staging floor, 492–493
  stairwell support group, 489
  standpipe bags, 493, 493*f*
  standpipe discharge outlets, connecting to, 197*f*, 497
    below-grade standpipe discharge outlets, 498–499
    house lines and fire cabinets, 498–499
    pressure-reducing valves (PRVs), 497–498
  stretching hose from standpipe, 514
  successful, 468–469
  tactical clarifications, 474–475
  tactical objectives, 503–504
  use of elevators, 489–491
    operating in phase II, 491–492, 492*f*
  uses for standpipe systems
    exposure protection, 529–530
    fire attack in adjoining buildings, 529, 530*f*
    water from uninvolved buildings, 530, 531*f*
  utilization and preservation of firefighters, 476
  variables in, 468
  ventilation, 538–541, 540*f*
    post-fire operations, 541

wind-driven fires in residential high-rises, 516–517, 517*f*
fire curtains, 517–518
indirect attack from above, 519
indirect flanking attack, 519
residential high-rise nozzles, 518
high-rise fires, 15
high-rise nozzles, 173
horizontal positive-pressure ventilation, 203
horizontal spread, 417
horizontal ventilation, 326, 327, 327*f*, 364–367, 365–367*f*
civilian life safety, 365
for fire control, 364
firefighter safety, 364
hose, 226, 226*f*
alert hose restraint safety system, 174–176, 175*f*
diameters, 493–494
forward and reverse lays, 229–231, 230*f*
large diameter hose (LDH), 228–229, 228*f*
layering, 355–356, 356*f*, 380, 381*f*, 512–513, 512*f*, 513*f*
nozzles, and equipment, 493–496, 493–496*f*
order, 379–380, 379*f*, 380*f*
storage, 152–153, 152–153*f*
supply line procedures, 229
3-inch and 3½-inch hose lines, 228
2½-inch hose line, 226–227, 227*f*
2½-inch hose lines, 226–228, 227*f*
two 2½-inch hose lines, 227–228
hose lines, 351, 352*f*
positions, 400, 400*f*, 401*f*
sizes and number of, 331–332, 332*f*, 333*f*
water streams, 397–400, 399*f*, 400*f*
hose stream placement, 283–285, 285*f*
Villa Plaza fire, 285–286
water supply and, 283, 284*f*
"hot smoldering" backdraft, 31
hot-stick voltage detector, 559
hot weather, 550
household furnishings, 5
human life thresholds, 60
HVAC. *See* heating, ventilation, and air-conditioning (HVAC)
hydrant assist valve (HAV), 162–163, 163*f*
hydrants, 214–216, 214–216*f*, 283, 284*f*
gate, 164, 165*f*
gates, 163–164, 164–165*f*
types of, 214, 214*f*
hydraulic ventilation, 132, 366, 366*f*
hydrocarbons, 20, 472
fuel molecules, 5
HydroVent™, 4, 4*f*, 169–170, 169*f*

I

IC. *See* incident commander (IC)
ice fog, 63
ICF. *See* insulated concrete formed (ICF)
ICS. *See* incident command system (ICS)
IDLH. *See* immediately dangerous to life and health (IDLH)
ignition temperature, 19
immediately dangerous to life and health (IDLH), 11, 32, 170, 278
atmosphere, 128, 128*f*
impact load, 82
imposition of loads, 82–83, 83*f*
IMS. *See* incident management system (IMS)
inadequate size-up, consequences of, 250
in-building pumps, 450
incident commander (IC), 8–9, 11, 12, 72, 195, 216, 238, 277, 311, 468
incident command system (ICS), 272
textbook, 268
versions of, 9
incident management system (IMS), 268
incident priority, 9–10, 10
incident stabilization, 9–10
incipient (ignition) stage, 28–29, 29*f*
inclement, 550
indirect attack, 142, 317–320, 318*f*, 319*f*, 320*f*
for attic and cockloft fires, 319
in residential high-rise fires, 319–320, 320*f*
ineffective master streams, 394
initial fire attack, 310
advancing hose line, 336–337, 337*f*
basement fires. *See* basement fires
basement sprinkler system using indirect attack, 345–346, 345*f*
choosing attack lines, 324
defensive attack, 323–324
direct attack, 316–317, 317*f*
effective stream operations during, 329–331
sizes and number of hose lines, 331–332, 332*f*, 333*f*
flashover, 312–314
safety and survival, 314, 314*f*
indirect attack, 317–320, 318*f*, 319*f*, 320*f*
initial attack lines, 315–316, 316*f*
lines, 334–336, 335*f*
locating fire, 316
making entry into fire room for, 358–359, 359*f*
master streams for, 332–334, 334*f*, 335*f*
protecting exposures, 350–351
rapid fire growth, 314–315, 315*f*
sizes of lines, 324–325
combination spray nozzles, 329
and nozzles, 325–326
solid *vs.* straight streams, 326–329, 327–329*f*
standpipe operations, 353–358, 354–358*f*
strategic and tactical areas for, 310–312, 311*f*, 312*f*, 313*f*, 314*f*
transitional attack, 320–323, 321*f*, 322*f*, 323*f*
upper floors, 351–353, 352*f*, 353*f*
ventilation. *See* ventilation
initial fireground assignments, 14
initial hose operations, safety first, 182–183
initial radio report, 259–260, 259*f*, 479
initial size-up, 247–249, 478–479
initial standby team (IST), 11
inorganic, 20
insulated concrete formed (ICF), 108
insulation, 30
interior direct attack, 316
interior exposures, 388, 400–403, 401*f*, 402*f*
fire in concealed spaces, 403–404, 403*f*
horizontal fire spread, 408–410, 409*f*
open interior spread, 410–411, 410*f*
property conservation *vs.* fire damage, 410
vertical fire spread, 404–405, 404*f*, 405*f*, 406*f*
interior stairways, 283, 284*f*
interior temperatures, 120, 120*f*
International Safety Instruments (ISI), 174, 174*f*
IST. *See* initial standby team (IST)

J

Jeromeo clamp, 299, 300*f*

K

Keokuk fire, 278
KO Fire Curtain, 127, 127*f*

L

ladder, 301, 301*f*, 302*f*
 roof, 306*f*
laminar smoke flow, 61, 61*f*
laminated glass, 77, 78*f*
laminated wood, 80
large diameter hose (LDH), 3, 152, 152*f*, 153*f*, 204, 216, 228–229, 228*f*, 425, 431
large-diameter supply hose, 223, 225*f*
LDH. *See* large diameter hose (LDH)
legacy construction, 80, 118
LEL. *See* lower explosive limit (LEL)
Lexan™, 5
life safety, 252
lighter materials, 36
lightweight gang-nail trusses, 108
lightweight high-rise fire hose, 494, 494*f*
lightweight hybrids, 108
lightweight truss construction, 106*f*
lightweight wood construction, 104–108, 106–108*f*
line-of-duty deaths (LODDs), 11, 72, 113, 182, 212, 238, 239, 250, 263, 318, 549
liquefied petroleum gas (LPG), 131
liquids, 24, 25, 44–45, 45*f*
little firefighters, 443
live fire experiments, 114, 114*f*
loads, types of, 81, 81*f*
lobby control
 directing crews and initial action plan, 484–485
 evacuation control, 482
 gaining control of elevators, 482–484
 identifying firefighting and evacuation stairways, 481–482, 481*f*
 verifying fire location, 479–481, 480*f*
LODDs. *See* line-of-duty deaths (LODDs)
long-lay supply lines, 217, 218*f*
Los Angeles Fire Department (LAFD) firefighters, 471
lower explosive limit (LEL), 27
low-expansion foam, 177
low-rise buildings, 354
LPG. *See* liquefied petroleum gas (LPG)

M

main fire, 392–393
manmade wood, 80
mansard roof assemblies, 206, 207*f*
manual dry standpipe, 445
manual wet standpipe, 445–447
manufactured buildings, 108
manufactured wood, 80
masonry, 74, 74*f*
massive flames, 131
master pump gauges, 216, 216*f*
master stream appliance, 160, 160*f*, 334, 335*f*, 421, 423, 424*f*, 428, 433*f*, 434, 434*f*, 436*f*
 adequate number of supply lines, 430–431, 432*f*
 defensive operations, 421, 422*f*, 423*f*, 424*f*
 directing, 434–435, 436*f*
 elevated master stream appliances, 426, 428*f*, 429*f*
 elevated master streams, 437–438, 438*f*
 fixed master stream appliances, 426, 427*f*
 portable master stream appliance, 424–426, 425*f*, 426*f*
 positioning, 433–434, 434*f*, 435*f*
 pumper-to-pumper operation, 430, 431*f*
 scenario 1, 429
 scenario 2, 429
 scenario 3, 430
 shutdown, 435, 437, 437*f*
 solid-stream *vs.* spray nozzles, 432–433, 432*f*, 433*f*
 structural collapse, 439, 439*f*
  collapse zones, 439–440, 440*f*
  signs of, 439
 transitional attack, 421, 421*f*, 422*f*
 use of, 432
 water supply for, 429, 430*f*
matter, 20–21*f*, 20–22
Mayday situation, 280
mechanical sources of heat energy, 22
mechanical ventilation, 365–366, 365*f*
mechanism of injury (MOI), 122
medium-expansion foam, 177
methane, 25
MIC. *See* microbiologically influenced corrosion (MIC)
microbiologically influenced corrosion (MIC), 456, 457*f*
minuteman hose load, 154, 155*f*
mobile ventilation unit (MVU), 171–172
mobile water supply apparatus (tankers and tenders), 218–220, 218*f*, 219–220*f*
modern construction, 6, 7*f*
modern fuels, 7, 7*f*, 43, 116
 homes furnished with, 8, 8*f*
modern lightweight construction, 84
modern residential construction, 78, 79*f*
MOI. *See* mechanism of injury (MOI)
movable exposures, 388, 388*f*
movement of apparatus, 246
multi-gas detector, 554
multiple pumpers, 222*f*
multiple wet systems, 450–451, 450*f*
mushrooming, 37, 38*f*
MVU. *See* mobile ventilation unit (MVU)

N

National Fire Protection Association (NFPA), 5, 97, 140, 150, 174–175, 175*f*, 194, 212, 280, 458, 473–474, 473*f*, 474*f*
 Command Sequence model, 244*f*
 equipment list for pumper fire apparatus, 151
 professional qualifications, 14–15
 Research Report, 466–467, 466*f*
National Fire Service Incident Management Consortium, 9
National Incident Management System/incident command system (NIMS/ICS), 8–9
National Institute for Occupational Safety and Health (NIOSH), 7
 Alert, 89
National Institute of Standards and Technology (NIST), 19, 467
 illustration, 140, 140*f*
natural ventilation, 367, 367*f*
neutral plane, 37
neutrons, 21, 21*f*
newer wood-frame buildings, 107*f*
NFPA. *See* National Fire Protection Association (NFPA)
NIOSH. *See* National Institute for Occupational Safety and Health (NIOSH)
NIST. *See* National Institute of Standards and Technology (NIST)

no-line-laid approach, 186–189, 187–191f
noncombustible construction, 99
noncombustible material, 97
noncombustible open-web steel joist, 89, 89f
nonprotected steel, 378
nozzles, 115, 154f, 157, 157f, 333, 494–496, 495f
combination spray, 158, 159f
firefighter, 130f
fog, 4, 158–160
and hose line, 359f
monitors with, 429f
positions, 400, 400f, 401f
pressure, 426
solid-stream smooth-bore tip, 157–158, 157–158f
specialty high-rise, 170–173, 172f
and streams, 326f
water flow and pressure problems in, 471–473
water streams, 397–400, 399f, 400f
nozzle technology, 4, 166–167, 167f
Blitzfire portable monitor, 168–169, 168f
Chimney Snuffer, 169, 169f
Task Force Tips® (TFT) Flip-Tip, 167, 168f
Task Force Tips® (TFT) vortex, 167–168, 168f
nuclear energy, 23–24

O

occupancy, 72–73
sprinkler regulations, 73
type of, 276f
offensive attack, 422f
offensive exterior attack, 134f, 143f
office search operation, 296, 297f
1¾-inch attack, 334, 335f
in residential high-rise apartments, 508–509
1½-inch gated wyes, 506–508, 509f
1¾-inch hose, 154f
1¾-inch lines, 381–383, 382f, 383f
One Meridian Plaza Fire, 453, 467
ongoing size-up, 268
operational risk assessment flowchart, 242, 243f
opposing lines, 348
organic compounds, 20
oriented search, 292
oriented strand board (OSB), 43, 80
OSB. *See* oriented strand board (OSB)
OS&Y. *See* outside stem-and-yoke (OS&Y)
out-of-service sprinkler systems, 461
outside stem-and-yoke (OS&Y), 451, 452f
overhaul operations, 182, 216, 216f, 547, 548f, 560–561, 561f
arson and scene preservation, 550–551, 550f
begin, 551, 552f
checking above the fire, 563–564, 564f
cabinets and compartments, 564, 565f
company drive-by, 569
debris pile, 565, 568f
final atmospheric air monitoring, 565, 568f
final thoughts on salvage, 569–570
owner walkthroughs, 570
pets, 570
vertical shafts and channels, 564, 565f, 566f
water removal, 565–568, 569f
windows and door casings, 564–565, 567f
rehabilitation, 549, 549f
inclement and hot weather, 550
medical evaluation, 549, 549f
rekindles, 547–549, 548f
what to look for, 561–563, 562f, 563f
where to look, 561–563, 562f, 563f
and work assignments, 558
cleanup, 559–560
removing items from structure, 560, 560f
securing utilities, 558–559, 559f
oxidation, 22
oxidizer, 21–22
oxygen, eliminating, 48
oxygen-limited fire, 117
oxygen supply, 30

P

paper manufacturing, 177
parallel chord truss, 87, 87f
parapet walls, 91–92f
Parker Doctrine, 144–146
particle board, 80
Passport firefighter accountability status board, 265, 266f
PA system. *See* public address (PA) system
personal alert safety system (PASS) devices, 115
personal protective equipment (PPE), 3, 49, 114, 146, 172–173, 314, 385, 475
personnel, 556–557
control of overhaul, 557
disposable buildings, 557–558
physician's orders for life-sustaining treatment (POLST), 122
piercing nozzle, 319f
pitched chord truss, 87, 87f
PIVs. *See* post-indicator valves (PIVs)
plank-and-beam construction, 103–104, 106f
plastics, 20, 62
combustibility of, 80
platform frame, 103, 105f
plenum, 470
plywood, 80
POGs. *See* policy and operating guidelines (POGs)
point of no return, 279
polar solvents, 48
policy and operating guidelines (POGs), 195
POLST. *See* physician's orders for life-sustaining treatment (POLST)
portable lighting, 556, 556f
portable master streams, 521–522, 522f
appliances, 424–426, 425f, 426f
portable radios, 115
portable water tank, 220f
POs. *See* possible occupants (POs)
positive-pressure ventilation (PPV), 125, 125f, 287, 294, 328
possible occupants (POs), 252, 253
post-indicator valves (PIVs), 451, 452f
PPE. *See* personal protective equipment (PPE)
PPV. *See* positive-pressure ventilation (PPV)
PRD. *See* pressure-restricting device (PRD)
pre-action system, 457–458
preconnected attack hose, 154f
prefabricated lumber, 5
prefires, 108
pre-incident coaching, 476
pre-incident planning, 194, 221, 244–245f, 388
building *vs.* structure, 194–195

pre-incident size-up, 242–246, 244–245*f*
  apparatus move-ups, 246
  tracking movement of apparatus, 246
  UL/NzIST factors, 246–247, 247*f*
premixed foam solution, 179
pre-overhaul safety inspection, 551, 552*f*
  air monitoring, 554, 554*f*
  Biloxi, Mississippi, Fire Department case study, 553–554
  health risks, 555
    gloves, 555–556
    portable lighting, 556, 556*f*
  smoke explosions, 551–553, 552*f*
  smoke particulates, 554–555, 555*f*
pre-plumbed deck guns, 425, 426*f*
pre-plumbed master stream appliance, 160–161, 161*f*
pre-plumbed monitors, 426, 426*f*
pressure levels, 216–217, 216*f*
pressure readings, 217, 217*f*
pressure-reducing valves (PRVs), 468
pressure-restricting device (PRD), 474
primary room search, 293
primary search, 287
problems, identifying, 13–14
professional safety attitude, 212, 212*f*
propane, 25, 25*f*
proper hose sizes, 289, 291*f*
property conservation
  vs. fire damage, 410
  operations, 15
protons, 21, 21*f*
PRVs. *See* pressure-reducing valves (PRVs)
public address (PA) system, 482, 533–534
public assemblies, 304, 304*f*
public waterworks, 448–449
pumper relays (tandem pumping), 231, 231*f*
  increasing water flow, 233–234, 234*f*
  relay pumping, 232–233, 233*f*
  setting up, 231–232, 232*f*
pumper-tankers, 219
pumper-to-pumper operation, 430, 431*f*
pump intake connections, 161–162, 161*f*, 162*f*
pump operators, 224
pyrolysis, 22, 42, 42*f*, 80

Q

QIAP. *See* quick initial action plan (QIAP)
quick initial action plan (QIAP), 9, 11, 12*f*, 197, 238, 246, 257, 259*f*, 262, 263*f*
  problems and developing, 260, 260*f*
  VTS/PSTR tactical worksheet, 258*f*

R

radiant heat exposures, 395–396, 395*f*, 396*f*
radiation, 40–41, 40–41*f*
radio, 10, 10*f*
RAK. *See* rescue air kit (RAK)
rapid ascent team (RAT), 481
rapid fire growth, 53, 54*t*
rapid intervention crew (RIC), 280–283
rapid intervention teams (RITs), 11, 272, 503–505
rapid ventilation, 294, 294*f*
RAT. *See* rapid ascent team (RAT)
rated capacity, 221–223
rebar, 75
rehabilitation, 549, 549*f*
  inclement and hot weather, 550
  medical evaluation, 549, 549*f*
  stations for firefighters, 3
rekindles, 182, 547–549, 548*f*
relay line, 232*f*
relay pumping, 232–233, 233*f*
relay supply line, 232*f*
reliable water supply, 127, 127*f*, 234, 234*f*
remote control modern fireboat monitors, 427*f*
rescue
  fire attack for, 287
  fire-resistant construction, 290–292
  hospitals, schools, institutions, 289–290, 291*f*
  multiple-family residences, 289, 289*f*, 290*f*
  profile considerations, 286–287
  single-family dwellings, 288–289, 288–289*f*
rescue air kit (RAK), 282, 293
rescue, exposures, confinement, extinguishment, overhaul/ventilation and salvage (RECEO/VS), 10–11, 286
resetting the fire, 273
residential basement fires, 310–311
residential basement windows, 280, 281*f*
residential fires, 15, 272
  behavior, 124
residual pressure, 224–225
resources, 11
reverse lays, 229–231, 230*f*
RIC. *See* rapid intervention crew (RIC)
risk, 239
  avoidance, 241
  priority, 241*f*
  transfer, 241
risk-benefit analysis, 9, 15, 72, 238, 260, 260*f*
risk management, 15, 239, 272
  control measures, 241–242
  evaluation, 239–241
  identification, 239
  monitoring, 242, 243*f*
  National Fire Academy (NFA) Command Sequence model, 242
  prioritization, 241–242, 241*f*
RITs. *See* rapid intervention teams (RITs)
rollover, 50, 50*f*, 278, 278*f*
roofs, 85, 85*f*
  components of, 86*f*
  ladder, 306*f*
room and contents fire, 95, 95*f*

S

salvage, 546, 546*f*, 547*f*
  final thoughts on, 569–570
Santana Row case study, 391–394, 393*f*
SCBA. *See* self-contained breathing apparatus (SCBA)
search
  indicating that a room has been searched, 298–299, 299*f*
  oriented, 292
  required tools, 292–293
  standard patterns, 295–297, 296*f*, 297*f*
  typical, 293–294, 294*f*
  value of standard procedure, 294–295
  what to check, 297–298, 298*f*
search and rescue, 272
  chronology of
    before alarm, 275
    at fire scene, 275–278, 276–277*f*
    Keokuk fire, 278

receipt of alarm, 275, 276*f*
defensive search tactics, 278–279, 279*f*
forcible entry effects on, 274–275, 274–275*f*
hose stream placement, 283–285, 285*f*
Villa Plaza fire, 285–286
primary search, 287
rapid intervention crew (RIC), 280–283
rescue profile considerations, 286–287
risk management, 272
safety considerations, 136–137, 136*f*
secondary search, 287
survivability profile, 272–273, 273*f*
smoke and room temperatures, 273–274
smoke toxicity, 274
two in/two out rule, 279–280, 281*f*, 282*f*
ventilation, 287, 287*f*
water supply and hose stream placement, 283, 284*f*
search-and-rescue exterior recon team (SERT), 532–533, 533*f*
search-evacuate-rescue (SER), 467, 530–532
aerial drones, 533
areas of emphasis, 534–535
fire dispatchers, 533
in open-space high-rise office buildings, 535
public address (PA) system, 533–534
shelter and defend in place, 532
traditional and nontraditional search tools, 534
secondary exposure fires, 393–394
secondary search, 287
second-tier problems, 13–14
self-contained breathing apparatus (SCBA), 3, 27, 114–115, 141, 293, 304, 305*f*, 467, 554, 554*f*
adoption of, 131
facepiece, 476, 476*f*
introduction of, 142, 142*f*, 143*f*
unit, 549
use of, 136
semiautomatic dry standpipe, 445
SER. *See* search-evacuate-rescue (SER)
SERT. *See* search-and-rescue exterior recon team (SERT)
setback buildings, 204, 205*f*
shear force, 83, 83*f*
shopping malls, 204–205
Siamese device, 431, 432*f*
side-core design, 469–470, 469*f*
significant cooling, 132
single engine hose lays, 183
direct-to-engine, 186–189, 187–191*f*
forward lay-dry, 184–186, 185*f*
forward lay-wet, 183–184, 184*f*
no-line-laid approach, 186–189, 187–191*f*
split lay, 186, 186*f*
single-family dwellings, 288–289, 288–289*f*
single-hose method, 355, 511–512, 511*f*, 512*f*
hose layering, 512–513, 512*f*, 513*f*
new standpipes, 513, 513*f*
single-intake FDC, 141, 141*f*
six-sided size-up, 238
size of fire, 13
size-up, 11, 238–239, 240*f*
declare your strategy, 256
functional accountability system, 263–268, 265–268*f*
identify your problems, 256–258, 258*f*
initial radio report, 259–260, 259*f*
making tactical assignments, 262–263, 262–263*f*
problems and developing, QIAP, 260, 260*f*
problems, identify, 258–259
selecting strategy and tactics, 260–262, 261–262*f*, 261*t*
setting up fireground, 252
value, 252–256, 253*f*, 255*f*
VTS and risk-benefit analysis, 260, 260*f*
size-up/preplan checklist, 239, 240*f*
Sling-Link, 304, 305*f*
small nozzles, 453
smoke, 7, 46, 46*f*, 58–59, 315*f*
components of, 59
curtains, 128, 128*f*
detector, 8, 8*f*
determine key elements, 60–63, 61–62*f*
explosions, 53–55, 54*t*, 55*f*, 181, 371, 371*f*, 551–553, 552*f*, 555
exposure, 7
fuel, 59–60
occupant survivability profile, 60
particulates, 554–555, 555*f*
predicting event, 63
rate of change, 63
reading through door, 63–65, 64–65*f*, 64*t*
and room temperatures, 273–274
shortcuts for reading, 64*t*
at structural fires, 58–59
fuel, 59–60
occupant survivability profile, 60
temperature, 122–123, 123*f*
toxicity, 122, 274
smoldering, 368
smooth-bore nozzles, 132, 132*f*, 328
smooth-bore solid stream, 432*f*
softening the target, 133
soft-suction hookups, 161, 161*f*
soft-suctions on side intakes, 161, 161*f*
SOGs. *See* standard operating guidelines (SOGs)
solar energy, 23
solid-fuel fires, 42–43
solids
characteristics of, 42–43, 43*f*
continuity of fuel, 43–44
orientation of fuel, 43, 44*f*
pyrolysis, 42, 42*f*
vs. straight streams, 326–329, 327–329*f*
surface-to-mass ratio, 43
solid-stream *vs.* spray nozzles, 432–433, 432*f*, 433*f*
spalling, 75, 75*f*
spare sprinkler heads, 462, 462*f*
specialty high-rise nozzles, 170–173, 172*f*
split lay, 186, 186*f*, 231
spray nozzles, 428
sprinkler heads, 453–455, 455–456*f*
components, 455, 455*f*
types of, 453, 454*f*
sprinkler regulations occupancy, 73
sprinkler response temperatures, 456, 456*f*, 457*t*
sprinkler systems, 115, 443, 451*f*
failure, 461
water supply for, 458–461, 459*f*, 460*f*
stairs, 90–91, 91*f*
Standard on Training for Initial Emergency Scene Operations, 5
standard operating guidelines (SOGs), 150, 151, 194, 212, 272, 294, 431–432, 443
standard operating procedures (SOPs), 11, 15, 195, 246, 467
standard search procedure, 294–295
standard shopping centers, 205–206, 206*f*

standard water supply evolution, 222*f*
standpipe discharge outlets, connecting to, 197*f*, 497
below-grade standpipe discharge outlets, 498–499
house lines and fire cabinets, 498–499
pressure-reducing valves (PRVs), 497–498
standpipe systems, 443, 444, 513, 513*f*
class I, 443*f*, 444
class II, 444, 445*f*
class III, 444–445
dry system features and uses, 446, 446*f*, 447*f*
evolutions, 354*f*
multiple wet systems, 450–451, 450*f*
public waterworks and fire hydrants, 448–449
sources, 449–450
stretching hose from, 514
types of, 445–446
valves, 451, 451*f*, 452*f*
water supply to wet standpipe systems, 447–448, 449*f*
wet system features and uses, 446–447, 448*f*
static electricity, 23
static water sources, 15, 217–218, 217–218*f*
steady smoke filling, 139, 139*f*
steam and white smoke, 436–437, 437*f*
steam burns, 318
steel, 75–77, 75–77*f*
gusset plates, 88, 88*f*
trusses, 89–90, 89–90*f*
strategy, 11, 260–262, 261–262*f*, 261*t*
streams, solid *vs.* straight, 326–329, 327–329*f*
stretching water usage, 181
structural collapse, 439, 439*f*
collapse zones, 439–440, 440*f*
signs of, 439
structural firefighting gloves, 555–556
structural firefighting nozzles, 4
structural integrity of building, 423, 424*f*
structure fire, 95, 116, 116*f*
suction hose, 216
supply, 216
hose, 157
line procedures, 229
surface-to-mass ratio, 43
survivability profile, 59, 121–122, 253, 272–273, 273*f*
smoke and room temperatures, 273–274
smoke temperature, 122–123, 123*f*
smoke toxicity, 122, 274

T

tactical worksheets, 11, 262, 263*f*
boards, 11–12
tactics, 11, 260–262, 261–262*f*, 261*t*
Task Force Tips® (TFT)
Blitzfire nozzle, 333, 334*f*, 353
Flip-Tip, 4, 4*f*, 167, 168*f*, 328, 328*f*
vortex, 167–168, 168*f*
Task Force Tips® (TFT)
Blitzfire nozzle, 333, 334*f*, 353
tasks, 11
technology
advancements in, 3–4*f*, 3–5
in manufacturing fire apparatus, 3
temperature, 22
and pressure, 24
reducing, 46–48
tempered glass, 77, 78*f*
tension force, 83, 83*f*
Texas Forest Service, 177
TFT Vortex, 328, 329*f*
thermal-imaging cameras (TICs), 3, 3*f*, 158, 174, 174*f*, 250–252, 251*f*, 278, 292, 303, 554
advantage of, 174
thermal imaging technology, 3
thermal layering, 49–50, 49–50*f*
thermal plume, 37, 37*f*
thermoplastic materials, 80
3-inch hose lines, 228
3½-inch hose lines, 228
360-degree walk-around size-up, 250
basements, 252
thermal-imaging cameras (TICs), 250–252, 251*f*
TICs. *See* thermal-imaging cameras (TICs)
time of day, 394
top-tier problems, 13
torsional load, 83, 83*f*
tower ladders, 434, 435*f*, 440*f*
traditional oriented search, 293
traditional time-temperature fire growth curve, 28*f*
transformer piercing nozzles, 173–174, 174*f*
transitional attack, 133, 310, 311*f*, 320–323, 321*f*, 322*f*, 323*f*, 421, 421*f*, 422*f*
defensive to offensive, 320–322, 321–322*f*
offensive to defensive, 322–323, 323*f*
transitional defensive, 422*f*
transitional exterior attack, 133
transitional strategy, 142
trusses, 86–87, 86–87*f*
roof and floor systems, 87
steel, 89–90, 89–90*f*
types of, 86, 86*f*
wood, 87–89, 88*f*
turbulent smoke flow, 61, 61*f*
2½-inch attack, 334, 335*f*
2½-inch backup line, 378, 379*f*
2½-inch fire hose, 155–156, 156*f*
2½-inch gated wyes, 506–508, 509*f*
2½-inch hose lines, 226–227, 227*f*, 383, 383*f*
two 2½-inch hose lines, 227–228
2½-inch hose lines, 383–384
2½-inch supply line, 227
typical search, 293–294, 294*f*

U

UL/NIST experiments, 8, 8*f*, 113–116, 114*f*, 145–146, 246–247, 247*f*
basement fires, 139–143*f*, 139–144
tactical considerations for, 144
fire behavior, 116–118, 116–118*f*, 119*f*
fire dynamics, 118–121, 119–121*f*, 124
Parker Doctrine, 144–146
reading windows, 137–138*f*, 137–139
recommendations for extinguishing fires, 10
residential fire behavior, 124
search and rescue safety considerations, 136–137, 136*f*
survivability profiles, 121–122
smoke temperature, 122–123, 123*f*
smoke toxicity, 122
ventilation, fire flow paths, and fires spread, 124–125, 125*f*, 126*f*
water application and coordinated fire attack, 130–136, 132–134*f*
wind-driven fires, 125–130, 127–130*f*

Underwriters Laboratories (UL), 5, 19
undesigned loads, 82
unidirectional flow path, 39, 39f, 125
universal extinguishing foam (UEF), 4, 4f
upper explosive limit (UEL), 27
upper floors, 351–353, 352f, 353f
  standpipe operations, 353–358, 354–358f
US Fire Administration (USFA) website, 549

V

values, 12
value, time, and size (VTS), 12–13, 72, 252, 260
  tactical worksheet, 260f
  time consideration category of, 72
valves, 163–164, 164–165f
  types of, 164, 164f
vapor density, 24, 25f, 25t
vaporization, 44
  of liquid fuels, 44–45
vapor pressure, 24, 25f
VEIS. *See* vent-enter-isolate-search (VEIS)
velocity (speed) of smoke, 61
vent-enter-isolate-search (VEIS), 136–137, 138f, 272, 301, 302f
  commercial structures and large-area search, 303
  emergency medical services (EMS), 307
  evolution, 138
  public assemblies, 303, 304f
  purpose of, 138
  rationale of, 303
  reading windows, 300–303, 301–302f
  rescue drags and carries, 303–305, 304–306f
  search and rescue safety considerations, 299–300, 300f
vent-enter-search (VES), rescue technique of, 136
ventilation, 287, 287f, 294, 313f, 318, 359–360, 360f, 361f, 538–541, 540f
  backdraft, 369–371, 369f, 370f
  defined, 123
  fire flow paths, and fires spread, 124–125, 125f, 126f
  horizontal, 364–367, 365–367f
  initiating coordinated attack, 373
  post-fire operations, 541
  smoke explosions, 371, 371f
  smoldering, and decay-stage fires, 367–368, 368f
  of smoldering fire, 371–372, 372f
  vertical, 360–363, 360–363f
ventilation-controlled fire growth model, 28, 28f
ventilation-limited fire, 5, 28, 28f, 79, 117, 117f, 123
  behavior of, 53
  sequence, 126f
  smoldering, 369, 369f
  stages of, 120–121, 121f
ventilation-limited stage, 31, 31f
verifiable occupants (VOs), 252, 257
vertical fire extension, 351, 352f
vertical fire spread, 404–405, 404f, 405f, 406f, 416–417, 416f
  control of, 406–408, 407f, 408f
  signs of, 405, 405f, 406f
vertical ventilation, 52
  cutting roof, 362–363, 363f
  hole, 108f
  louver cut, 363, 363f
  roof package, 361–362, 362f
victims
  attempting to jump from upper floors, 279, 279f
  on floor 6, 289, 290f
  primary search for, 316f
  at windows, 276, 277f
void spaces, 90, 90f
volume of smoke, 61
VTS. *See* value, time, and size (VTS)

W

water
  application, 130–136, 132–134f
  curtain, 395–396, 396f
  delivery of, 48
  gallons of, 181
  holds, surface tension of, 178, 178f
  main systems, 214–216, 214–216f
  motor gong, 462, 462f
  natural surface tension of, 179f
  shuttle apparatus corridor, 220–221, 221f
  streams, 397–400, 399f, 400f
water supply, 212
  engines and pumpers. *See* engines and pumpers
  evolution, 221, 222f
  hose. *See* hose
  and hose stream placement, 283, 284f
  inherent risks, 212–213, 213t
  mobile water supply apparatus (tankers and tenders), 218–220, 219–220f
  pumper relays (tandem pumping). *See* pumper relays (tandem pumping)
  source, 213
    apparatus water tanks, 218
    estimating attack lines, 217, 217f
    hydrants and water main systems, 214–216, 214–216f
    reading pressure levels, 216–217, 216f
    static water sources, 217–218, 217–218f
  typical water supply evolution, 221, 222f
  water shuttle apparatus corridor, 220–221, 221f
water supply for master stream appliances, 429, 430f
  adequate number of supply lines, 430–431, 432f
  pumper-to-pumper operation, 430, 431f
  scenario 1, 429
  scenario 2, 429
  scenario 3, 430
water vapor, 30, 30f, 119
wet-barrel hydrants, 214, 215f
wet pipe system, 456–457, 457f
wet standpipe, 451f
  systems, 445–447, 447f
wet water, 560
wide-frontage buildings, 197–198, 198f
wind, 127f
wind-driven fires, 55–56, 56f, 125–130, 127–130f, 516–517, 517f
  fire curtains, 517–518
  indirect attack from above, 519
  indirect flanking attack, 519
  residential high-rise nozzles, 518
wired glass, 78, 78f
wood, 22, 22f, 79–80
  studs, 103
  supported floors, 84–85, 84–85f
  trusses, 87–89, 88f
wood frame, 101–103, 104f
  construction, 5
  house, 20, 379, 379f